Advances in Plant Taxonomy and Systematics

Advances in Plant Taxonomy and Systematics

Editor

Lorenzo Peruzzi

MDPI • Basel • Beijing • Wuhan • Barcelona • Belgrade • Manchester • Tokyo • Cluj • Tianjin

Editor
Lorenzo Peruzzi
University of Pisa
Italy

Editorial Office
MDPI
St. Alban-Anlage 66
4052 Basel, Switzerland

This is a reprint of articles from the Special Issue published online in the open access journal *Biology* (ISSN 2079-7737) (available at: https://www.mdpi.com/journal/biology/special_issues/Plant_Taxonomy_Systematics).

For citation purposes, cite each article independently as indicated on the article page online and as indicated below:

LastName, A.A.; LastName, B.B.; LastName, C.C. Article Title. *Journal Name* **Year**, *Volume Number*, Page Range.

ISBN 978-3-0365-7558-2 (Hbk)
ISBN 978-3-0365-7559-9 (PDF)

Cover image courtesy of Lorenzo Peruzzi.

Contents

About the Editor

Lorenzo Peruzzi

Lorenzo Peruzzi is full professor of Systematic Botany at the Department of Biology of the University of Pisa, where he coordinates the activities of the PLANTSEED Lab and directs the Botanic Garden and Museum. He is interested in nomenclature, taxonomy, and systematics of vascular plants; Mediterranean flora; integrative taxonomy; evolution; biogeography; cytotaxonomy. He was Secretary of the Italian Botanical Society in 2014-2020 and is currently President of the Italian Society of Biogeography.

biology MDPI

Editorial

Advances in Plant Taxonomy and Systematics

Lorenzo Peruzzi

PLANTSEED Lab, Department of Biology, University of Pisa, Via Derna 1, 56126 Pisa, Italy; lorenzo.peruzzi@unipi.it

Systematics and taxonomy are basic sciences and are crucial for all applications dealing with living organisms [1]. Taxonomic classification schemes, sought by early scholars to reflect "natural systems" [2], are nowadays universally accepted to reflect actual systematic relationships among organisms.

Phylogenetic reconstructions based on molecular systematics have provided a stable classification system at class, order, and family levels for many plant groups (see, e.g., [3–5]). However, at the genus level, due to a lack of knowledge, many classifications are still unstable and a lot of taxonomic changes have been published [6], with species that are often recombined under different genera or synonymized with others. Taxonomy users, either in the scientific community or in wider society, perceive this as a relevant (and often not fully understood) problem [7,8]. However, these changes are the obvious consequence of an increase in systematic knowledge. In this respect, proposals and ideas to abandon Linnean taxonomy [9,10] have not been accepted so far. Fortunately, nomenclatural and taxonomic databases are becoming increasingly widespread and authoritative (see, e.g., [11]), meaning that this problem could be easily superseded.

At a microevolutionary level, an integrated taxonomic approach [12] using a number of independent lines of evidence [13] is needed to disentangle the complex systematic relationships among units of diversity [14].

Accordingly, on one side, there is the need to build sound taxonomic hypotheses using multiple lines of evidence (see, e.g., [15–18]); on the other hand, given the ongoing mass extinctions and the decline of taxonomists in academies [19,20], there is the need to speed up the recognition and description of biodiversity on earth. In this respect, citizen science could also be helpful [21], for instance in observing and capturing plant diversity with a coverage and frequency much higher than by just relying on academic scholars.

In the Special Issue "Advances in Plant Taxonomy and Systematics", all the topics previously mentioned were addressed in 15 high-quality and original studies, involving plant groups and researchers from all continents. In particular, the phylogeny and biogeography of Mammilloid cacti from Mexico (Cactaceae, eudicots) [22], Euphorbiaceae subfam. Acalyphoideae (Malpighiales, eudicots) in the Americas [23], *Astragalus* sect. *Stereothrix* (Fabaceae, Fabales, eudicots) [24] and *Veronica* subg. *Pentasepalae* (Plantaginaceae, Lamiales, eudicots) [25] from Eurasia were addressed. Whole plastome comparison revealed phylogenetic relationships in *Crassula* (Crassulaceae, Saxifragales, eudicots) [26] and in the family Magnoliaceae (Magnoliales, early branching angiosperms) [27]. The systematics of polyploid and/or apomictic species complexes was studied in European groups such as the *Ranunculus auricomus* complex (Ranunculaceae, Ranunculales, eudicots) [28], the *Sorbus austriaca* complex (Rosaceae, Rosales, eudicots) [29], *Crocus* ser. *Verni* (Iridaceae, Asparagales, monocots) [30], and *Leucanthemum* (Asteraceae, Asterales, eudicots) [31]. Integrated taxonomic approaches were followed for the characterization of the Asian palm genus *Bentinckia* (Arecaceae, Arecales, monocots) [32], for addressing infraspecific variability in the European *Armeria arenaria* (Plumbaginaceae, Caryophyllales, eudicots) [33], and for describing a new species endemic to Italy in *Adonis* sect. *Adonanthe* (Rancunculaceae) [34]. A thorough morphometric study dealt with the taxonomically

Citation: Peruzzi, L. Advances in Plant Taxonomy and Systematics. *Biology* **2023**, *12*, 570. https://doi.org/10.3390/biology12040570

Received: 4 April 2023
Accepted: 7 April 2023
Published: 9 April 2023

debated Mediterranean genus *Ophrys* (Orchidaceae, Asparagales, monocots), in which between 9 and over 400 species are recognized depending on the authors opinions [35], highlighting that "a serious challenge awaits writers of field guides to the European flora, as they struggle to summarise innumerable indistinguishable 'species' carved out of morphological continua". Finally, images shared by citizen scientists to the iNaturalist platform and on Facebook were particularly helpful, as they aided the identification of four out of the nine Australian species of the carnivorous genus *Drosera* (Droseraceae, Caryophyllales, eudicots) [36].

Conflicts of Interest: The author declares no conflict of interest.

References

1. Wheeler, Q. A taxonomic renaissance in three acts. *Megataxa* **2020**, *1*, 4–8. [CrossRef]
2. Lorch, J. The Natural System in Biology. *Philos. Sci.* **1961**, *28*, 282–295. [CrossRef]
3. Chase, M.W.; Reveal, J.L. A phylogenetic classification of the land plants to accompany APG III. *Bot. J. Linn. Soc.* **2009**, *161*, 122–127. [CrossRef]
4. The Angiosperm Phylogeny Group; Chase, M.W.; Christenhusz, M.J.M.; Fay, M.F.; Byng, J.W.; Judd, W.S.; Soltis, D.E.; Mabberley, D.J.; Sennikov, A.N.; Soltis, P.S.; et al. An update of the Angiosperm Phylogeny Group classification for the orders and families of flowering plants: APG IV. *Bot. J. Linn. Soc.* **2016**, *181*, 1–20. [CrossRef]
5. PPG, I. A community-derived classification for extant lycophytes and ferns. *J. Syst. Evol.* **2016**, *54*, 563–603. [CrossRef]
6. Nimis, P.L. A tale from Bioutopia. *Nature* **2001**, *413*, 21. [CrossRef]
7. Garnett, S.T.; Christidis, L. Taxonomy anarchy hampers conservation. *Nature* **2017**, *546*, 25–27. [CrossRef]
8. Morrison, W.R., III; Lohr, J.L.; Duchen, P.; Wilches, R.; Trujillo, D.; Mair, M.; Renner, S.S. The impact of taxonomic change on conservation: Does it kill, can it save, or is it just irrelevant? *Biol. Cons.* **2009**, *142*, 3201–3206. [CrossRef]
9. Casiraghi, M.; Galimberti, A.; Sandionigi, A.; Bruno, A.; Labra, M. Life With or Without Names. *Evol. Biol.* **2016**, *43*, 582–595. [CrossRef]
10. Brower, A.V.Z. Dead on arrival: A postmortem assessment of "phylogenetic nomenclature", 20+ years on. *Cladistics* **2020**, *36*, 627–637. [CrossRef]
11. Borsch, T.; Berendsohn, W.; Dalcin, E.; Delmas, M.; Demissew, S.; Elliott, A.; Fritsch, P.; Fuchs, A.; Geltman, D.; Guner, A.; et al. World Flora Online: Placing taxonomists at the heart of a definitive and comprehensive global resource on the world's plants. *Taxon* **2020**, *69*, 1311–1341. [CrossRef]
12. Dayrat, B. Towards integrative taxonomy. *Biol. J. Linn. Soc.* **2005**, *85*, 407–415. [CrossRef]
13. Turrill, W.B. The expansion of taxonomy with special reference to spermatophyta. *Biol. Rev. Biol. Proc. Camb. Phil. Soc.* **1938**, *13*, 342–373. [CrossRef]
14. Oberprieler, C. The Wettstein tesseract: A tool for conceptualising species-rank decisions and illustrating speciation trajectories. *Taxon* **2023**, *72*, 1–7. [CrossRef]
15. De Giorgi, P.; Giacò, A.; Astuti, G.; Minuto, L.; Varaldo, L.; De Luca, D.; De Rosa, A.; Bacchetta, G.; Sarigu, M.; Peruzzi, L. An integrated taxonomic approach points towards a single-species hypothesis for *Santolina* (Asteraceae) in Corsica and Sardinia. *Biology* **2022**, *11*, 356. [CrossRef] [PubMed]
16. Liu, L.; Astuti, G.; Coppi, A.; Peruzzi, L. Different chromosome numbers but slight morphological differentiation and genetic admixture among populations of the *Pulmonaria hirta* complex (Boraginaceae). *Taxon* **2022**, *71*, 1025–1043. [CrossRef]
17. Mátis, A.; Malkócs, T.; Kuhn, T.; Laczkó, L.; Moysiyenko, I.; Szabó, A.; Bădărău, A.S.; Sramkó, G. Hiding in plain sight: Integrative analyses uncover a cryptic *Salvia* species in Europe. *Taxon* **2023**, *72*, 78–97. [CrossRef]
18. Giacò, A.; Varaldo, L.; Casazza, G.; De Luca, D.; Caputo, P.; Sarigu, M.; Bacchetta, G.; Sáez, L.; Peruzzi, L. An integrative taxonomic study of *Santolina* (Asteraceae) from southern France and north-eastern Spain reveals new endemic taxa. *J. Syst. Evol.* 2022, *in press*. [CrossRef]
19. Britz, R.; Hundsdörfer, A.; Fritz, U. Funding, training, permits—The three big challenges of taxonomy. *Megataxa* **2020**, *1*, 49–52. [CrossRef]
20. Wheeler, Q. Are reports of the death of taxonomy an exaggeration? *New Phytol.* **2014**, *201*, 370–371. [CrossRef]
21. Fontaine, C.; Fontaine, B.; Prévot, A.-C. Do amateurs and citizen science fill the gaps left by scientists? *Curr. Opin. Insect.* **2021**, *46*, 83–87. [CrossRef]
22. Delil, A.; Chincoya, S.A.; Felipe Vaca-Paniagua, P.D.; Solórzano, S. Phylogenomics and biogeography of the Mammilloid clade revealed an intricate evolutionary history arose in the Mexican Plateau. *Biology* **2023**, *12*, 512. [CrossRef]
23. Külkamp, J.; Riina, R.; Ramírez-Amezcua, Y.; Iganci, J.R.V.; Cordeiro, I.; González-Páramo, R.; Lara-Cabrera, S.I.; Baumgratz, J.S.A. Systematics of Ditaxinae and related lineages within the subfamily Acalyphoideae (Euphorbiaceae) based on molecular phylogenetics. *Biology* **2023**, *12*, 173. [CrossRef] [PubMed]
24. Bagheri, A.; Maassoumi, A.A.; Brassac, J.; Blattner, F.R. Dated phylogeny of *Astragalus* section *Stereothrix* (Fabaceae) and allied taxa in the *Hypoglottis* clade. *Biology* **2023**, *12*, 138. [CrossRef] [PubMed]

25. Doostmohammadi, M.; Bordbar, F.; Albach, D.C.; Mirtadzadini, M. Phylogeny and historical biogeography of *Veronica* subgenus *Pentasepalae* (Plantaginaceae): Evidence for its origin and subsequent dispersal. *Biology* **2022**, *11*, 639. [CrossRef]
26. Ding, H.; Han, S.; Ye, Y.; Bi, D.; Zhang, S.; Yi, R.; Gao, J.; Yang, J.; Wu, L.; Kan, X. Ten plastomes of *Crassula* (Crassulaceae) and phylogenetic implications. *Biology* **2022**, *11*, 1779. [CrossRef]
27. Yang, L.; Tian, J.; Xu, L.; Zhao, X.; Song, Y.; Wang, D. Comparative chloroplast genomes of six Magnoliaceae species provide new insights into intergeneric relationships and phylogeny. *Biology* **2022**, *11*, 1279. [CrossRef]
28. Hodač, L.; Karbstein, K.; Tomasello, S.; Wäldchen, J.; Bradican, J.P.; Hörandl, E. Geometric morphometric versus genomic patterns in a large polyploid plant species complex. *Biology* **2023**, *12*, 418. [CrossRef]
29. Hajrudinović-Bogunić, A.; Frajman, B.; Schönswetter, P.; Siljak-Yakovlev, S.; Bogunić, F. Apomictic mountain whitebeam (*Sorbus austriaca*, Rosaceae) comprises several genetically and morphologically divergent lineages. *Biology* **2023**, *12*, 380. [CrossRef]
30. Raca, I.; Blattner, F.R.; Waminal, N.E.; Kerndorff, H.; Ranđelović, V.; Harpke, D. Disentangling *Crocus* series *Verni* and its polyploids. *Biology* **2023**, *12*, 303. [CrossRef]
31. Oberprieler, C.; Ott, T.; Vogt, R. Picks in the fabric of a polyploidy complex: Integrative species delimitation in the tetraploid *Leucanthemum* Mill. (Compositae, Anthemideae) representatives. *Biology* **2023**, *12*, 288. [CrossRef]
32. Kadam, S.K.; Mane, R.N.; Tamboli, A.S.; Gavade, S.K.; Deshmukh, P.V.; Lekhak, M.M.; Choo, Y.-S.; Pak, J.H. Cytogenetics, typification, molecular phylogeny and biogeography of *Bentinckia* (Arecoideae, Arecaceae), an unplaced Indian endemic palm from Areceae. *Biology* **2023**, *12*, 233. [CrossRef] [PubMed]
33. Tiburtini, M.; Astuti, G.; Bartolucci, F.; Casazza, G.; Varaldo, L.; De Luca, D.; Bottigliero, M.V.; Bacchetta, G.; Porceddu, M.; Domina, G.; et al. Integrative taxonomy of *Armeria arenaria* (Plumbaginaceae), with a special focus on the putative subspecies endemic to the Apennines. *Biology* **2022**, *11*, 1060. [CrossRef] [PubMed]
34. Conti, F.; Oberprieler, C.; Dorfner, M.; Schabel, E.; Nicoară, R.; Bartolucci, F. *Adonis fucensis* (*A.* sect. *Adonanthe*, Ranunculaceae), a new species from the Central Apennines (Italy). *Biology* **2023**, *12*, 118. [CrossRef] [PubMed]
35. Bateman, R.M.; Rudall, P.J. Morphological continua make poor species: Genus-wide morphometric survey of the European bee orchids (*Ophrys*, L.). *Biology* **2023**, *12*, 136. [CrossRef] [PubMed]
36. Krueger, T.; Robinson, A.; Bourke, G.; Fleischmann, A. Small leaves, big diversity: Citizen science and taxonomic revision triples species number in the carnivorous *Drosera microphylla* complex (*D.* section *Ergaleium*, Droseraceae). *Biology* **2023**, *12*, 141. [CrossRef] [PubMed]

biology

MDPI

Article

Phylogenomics and Biogeography of the Mammilloid Clade Revealed an Intricate Evolutionary History Arose in the Mexican Plateau

Delil A. Chincoya [1,2,*], Salvador Arias [3], Felipe Vaca-Paniagua [4,5,6], Patricia Dávila [7] and Sofía Solórzano [1,*]

1 Laboratorio de Ecología Molecular y Evolución, UBIPRO, FES Iztacala, Universidad Nacional Autónoma de México, Avenida de los Barrios 1, Los Reyes Iztacala, Tlalnepantla de Baz 54090, Estado de México, Mexico
2 Posgrado en Ciencias Biológicas, Unidad de Posgrado, Edificio D, 1° Piso, Circuito de Posgrados, Ciudad Universitaria, Coyoacán 04510, Ciudad de México, Mexico
3 Jardín Botánico, Instituto de Biología, Universidad Nacional Autónoma de México, Tercer Circuito Exterior, Ciudad Universitaria, Coyoacán 04510, Ciudad de México, Mexico
4 Laboratorio Nacional en Salud, Diagnóstico Molecular y Efecto Ambiental en Enfermedades Crónico-Degenerativas, FES Iztacala, Universidad Nacional Autónoma de México, Los Reyes Iztacala, Tlalnepantla de Baz 54090, Estado de México, Mexico
5 Subdirección de Investigación Básica, Instituto Nacional de Cancerología, Tlalpan 14080, Ciudad de México, Mexico
6 Unidad de Investigación en Biomedicina, FES Iztacala, Universidad Nacional Autónoma de México, Los Reyes Iztacala, Tlalnepantla de Baz 54090, Estado de México, Mexico
7 Laboratorio de Recursos Naturales, UBIPRO, FES Iztacala, Universidad Nacional Autónoma de México, Avenida de los Barrios 1, Los Reyes Iztacala, Tlalnepantla de Baz 54090, Estado de México, Mexico
* Correspondence: dela@comunidad.unam.mx (D.A.C.); solorzanols@unam.mx (S.S.)

Simple Summary: Cacti account for nearly 1440 species, most of them native to the American continent. These succulent plants are the most ubiquitous elements of the arid ecosystems. Mexico harbors the highest number of cacti species in the world (45%). Unfortunately, many of them are threatened by human activities. Although having this biodiversity relevance, presently the evolutionary processes of cacti have been poorly studied. Because the biological and conservation unit is the species, evolutionary studies provide relevant information. In this study, we analyzed how and when past events shaped the evolutionary relationships of 103 species. Our results showed that from 4.5 million years ago the arid regions of Mexico were the locations for abundant cacti speciation. From these lands, cacti have colonized most of the Mexican territories, the southern regions of the United States, as well as the Caribbean. The evolution of these plants was probably promoted by past temperatures that were comparable to the present ones. We identified different speciation and dispersal events in these fascinating plants. This study identified the Mexican Plateau as the place where the early stages of the evolutionary history of cacti occurred.

Abstract: Mexico harbors ~45% of world's cacti species richness. Their biogeography and phylogenomics were integrated to elucidate the evolutionary history of the genera *Coryphantha*, *Escobaria*, *Mammillaria*, *Mammilloydia*, *Neolloydia*, *Ortegocactus*, and *Pelecyphora* (Mammilloid Clade). We analyzed 52 orthologous loci from 142 complete genomes of chloroplast (103 taxa) to generate a cladogram and a chronogram; in the latter, the ancestral distribution was reconstructed with the Dispersal-Extinction-Cladogenesis model. The ancestor of these genera arose ~7 Mya on the Mexican Plateau, from which nine evolutionary lineages evolved. This region was the site of 52% of all the biogeographical processes. The lineages 2, 3 and 6 were responsible for the colonization of the arid southern territories. In the last 4 Mya, the Baja California Peninsula has been a region of prolific evolution, particularly for lineages 8 and 9. Dispersal was the most frequent process and vicariance had relevance in the isolation of cacti distributed in the south of Mexico. The 70 taxa sampled as *Mammillaria* were distributed in six distinct lineages; one of these presumably corresponded to this genus, which likely had its center of origin in the southern part of the Mexican Plateau. We recommend detailed studies to further determine the taxonomic circumscription of the seven genera.

Citation: Chincoya, D.A.; Arias, S.; Vaca-Paniagua, F.; Dávila, P.; Solórzano, S. Phylogenomics and Biogeography of the Mammilloid Clade Revealed an Intricate Evolutionary History Arose in the Mexican Plateau. *Biology* **2023**, *12*, 512. https://doi.org/10.3390/biology12040512

Academic Editor: Lorenzo Peruzzi

Received: 25 February 2023
Revised: 21 March 2023
Accepted: 24 March 2023
Published: 29 March 2023

Keywords: arid lands; biogeography; Cactaceae; colonization; *Mammillaria*; Mexican Plateau; Miocene; phylogenomics; Pleistocene; recent diversification

1. Introduction

The integration of the analytical frameworks of phylogenetics and biogeography allows analysis of the influence of biogeography on the evolutionary history of extant taxa, as well as to identify all those biogeographical events that promoted speciation [1]. The studies that incorporate these frameworks have inferred past biogeographical scenarios that have shaped the current geographical ranges of the species (e.g., [2,3]) and even large flora assemblies (e.g., [4–6]). Furthermore, they have enabled evaluation of the relative role of vicariance and dispersal in shaping the current geographical distribution of species [1,7,8]. Gradually, with the advances of high throughput sequencing technologies, more of these studies are using denser molecular sampling, which has made it possible to obtain confident phylogenetic trees that may serve to resolve close phylogenetic relationships [9]. Poor molecular sampling usually produces non-monophyletic trees, or discordance with phylogenies based on morphology [10]. In addition, the biogeographical data may support the establishment of taxonomic limits between species [10], and actually they have been identified as providing better auxiliary information than morphology to elucidate phylogenetic relationships [11].

In addition, it is recognized that global paleoclimatic changes have shaped the large current distribution patterns of the biota and caused extinctions at different geographical scales [12]. Furthermore, they influenced the expansion and contraction of the geographical distribution of the current extant species (e.g., [13,14]). Consequently, paleoclimate changes have been recognized as one of the most influential factors in shaping the world biodiversity patterns at large scales, but also for understanding the current local flora assemblies (e.g., [15]). On the other hand, the topography and the intricate local orography have also influenced the ecological, biogeographical, and evolutionary processes of the local biota [16]. All these events and processes that occurred in the past might have modified gene flow patterns, which gradually may cause population genetic divergence and eventually promoted speciation processes [17].

In the contemporary arid lands of the American continent, many complex assemblages of native local floras are found in which cacti taxa are the most ubiquitous elements. The nearly 1440 taxa grouped in Cactaceae [18] are recognized as a monophyletic group [19]. Today the evolutionary history of Cactaceae, particularly its origin and mode of speciation, are still considered enigmatic [20]. Due to the lack of fossil records of Cactaceae representatives, there is no direct evidence to date its origin. However, estimations based on molecular clock hypothesis have dated the origin of Cactaceae to nearly 28.8 million years ago (Mya) [21], or 32.11 Mya [22], and 35 Mya [23]. Accordingly, these estimations place the origin of Cactaceae in the Cenozoic Era, in the Paleogene period from the Late Eocene (~35 Mya) to the Middle Oligocene (~28 Mya). In addition, Arakaki et al. [23] concluded that unequal and inconstant speciation rates for 123 cacti sampled were explained by the environmental changes that occurred in the Miocene, based on the phylogenetic tree obtained with two loci, one from the nuclear genome (PHYC) and the other one from the chloroplast (trnK/matK). Accordingly, these authors suggested that there have been at least six main peaks of speciation in the evolutionary history of Cactaceae. These authors dated the earliest two speciation peaks to 25 Mya and 15 Mya, whereas the other four occurred in the last 8 million years. Furthermore, they showed that those last four peaks were contemporaneous to the decreases in atmospheric CO_2 that promoted global aridification, giving new ecological opportunities to cacti [23]. On the other hand, specialized paleoclimatic studies (e.g., [24,25]) have dated the decreases of atmospheric CO_2 from the Middle Miocene (14 Mya) to the Middle Pleistocene (0.8 Mya). These relatively low levels of CO_2 eventually caused a cooler and drier global climate, a phenomenon recognized as an aridification process [26].

In Cactaceae, the genus *Mammillaria* Haw is notable for its diversity [27], conservation concerns [28] and unresolved phylogenetic and taxonomic issues [18]. The taxonomy of *Mammillaria* has been controversial from its original description. In 1753, Charles Linnaeus described the type specimen as *Cactus mammillaris* L. and later it was renamed as *Mammillaria* in 1812 [29]. During its history, the genus *Mammillaria* has received 14 different names [27], reflecting the difficulty in achieving a clear taxonomic circumscription based almost entirely in external morphological traits. Throughout the last two centuries, numerous attempts have been made to organize the wide infrageneric morphological variation among taxa classified as *Mammillaria* (e.g., [30–32]. The most recent infrageneric classification was proposed by Hunt [18], who recognized eight subgenera and 15 series. The subgenus *Mammillaria* contains the highest number of species (117), followed by *Chilita* Orcutt (18) and *Krainzia* Backeb. (12); whereas *Cochemiea* Brandegee, *Dolichothele* (K.Schum.) Britton & Rose, *Mammillopsis* Morren, *Oehmea* Buxb., and *Phellosperma* Britton & Rose together add 16 species. For the purposes of the present study, we follow this last infrageneric classification system; nevertheless, phylogenetic support for these infrageneric classifications has not been tested.

Today, the global geographical distribution of *Mammillaria* ranges from the southern arid lands of the United States to the north of South America. Mexico has the highest documented diversity of *Mammillaria*. Nearly 20% of the species of *Mammillaria* are distributed in the Mexican arid lands of the southern part of the Chihuahuan Desert [33]. Only two of the 163 species currently recognized [18], *M. mammillaris* (L.) H. Karst and *M. nivosa* Link ex Pfeiff. are not documented in this country [33]. The genus *Mammillaria* is a rare taxon across Central America, as only four species are recorded in Guatemala (*M. albilanata* Backeb., *M. columbiana* Salm-Dyck, *M. ericantha* Link & Otto ex Pfeiff. and *M. voburnensis* Scheer), and two of them are also distributed across Nicaragua and Honduras (*M. columbiana* and *M. voburnensis*). Two more, *M. columbiana* and *M. mammillaris*, are documented in some small localities in the north of Venezuela and Colombia. In addition, four species (*M. columbiana*, *M. mammillaris*, *M. nivosa*, and *M. prolifera* (Mill.) Haw.) are recorded in the Caribbean islands [33].

The early phylogenetic studies carried out with *Mammillaria* reignited the unsolved discussion regarding its unclear taxonomic circumscription and its limits with taxa of another six genera (*Coryphantha* (Engelm.) Lem., *Escobaria* Britton & Rose, *Mammilloydia* Buxb., *Neolloydia* Britton & Rose, *Ortegocactus* Alexander, and *Pelecyphora* Ehrenb.). These six genera and *Mammillaria* compose the Mammilloid Clade [34]. Butterworth and Wallace [35] analyzed the phylogenetic relationships of 123 species of Mammilloid Clade (113 of them grouped in *Mammillaria)* based on two plastid loci (*rpl16* intron and the intergenic spacer *psbA-trnH*). Their phylogenetic tree showed abundant polytomies and low support bootstrap values. In addition, the sampled taxa of these six genera were grouped together with those of *Mammillaria*. Hence the authors concluded that this genus has a polyphyletic origin. Later, Crozier [36] used 10 plastid loci to analyze 157 cacti taxa; only 29 of them were *Mammillaria* taxa and 10 belonged to the six genera. The results of this study did not resolve the phylogenetic relationships of the sampled taxa; it also concluded non-monophyly for *Mammillaria*. In addition, Crozier [36] concluded that the monophyly of *Mammillaria* could only be obtained if: (1) the *Mammillaria* genus was expanded to include all the species currently grouped in the six genera; or (2) the genus *Mammillaria* includes only those species of the subgenus *Mammillaria* sensu Hunt [37]. Breslin et al. [38] recently sampled 93,808 bp of the large single copy (LSC) of the chloroplast genome from 78 cacti taxa, 52 of which were *Mammillaria* and 17 from five genera (*Coryphantha*, *Escobaria*, *Neolloydia*, *Ortegocactus*, and *Pelecyphora*). These authors concluded monophyly for *Mammillaria* by excluding all those species that were grouped in a distinct clade, which was composed of taxa in *Mammillaria*, *Neolloydia*, and *Ortegocactus*. In addition, it was proposed that all the species of this clade to be placed inside the genus *Cochemiea*.

In this study, we integrated phylogenomics and historical biogeography to elucidate the controversial evolutionary history of the group of seven genera of cacti (*Coryphantha, Escobaria, Mammillaria, Mammilloydia, Neolloydia, Ortegocactus,* and *Pelecyphora*) sensu Hunt [18]. We hypothesized that these taxa have a monophyletic origin, whose unique ancestor arose recently and rapidly evolved in response to past decreases in global temperature. The objectives were to evaluate the phylogenetic relationships of the studied species, to estimate their divergence times, and to identify the probable ancestral geographical distribution of the taxa studied in these seven genera; to discuss the possible effects of past global temperature and orographic events in the colonization and expansion of these cacti across the arid lands of Mexico; and finally we use our results to identify the taxonomic limits of the genera studied with emphasis on the taxa sampled in the genus *Mammillaria.*

2. Materials and Methods

2.1. Taxon Sampling

A total of 142 complete chloroplast genomes (cpDNA) of 103 taxa were analyzed (Table S1), of which 141 cpDNA belong to the tribe Cacteae (Cactoideae). The non Cactoideae taxon *Blossfeldia liliputana* Werderm. was included because it was identified as the sister species for the rest of the subfamily Cactoideae [36]. We compiled these cpDNA from the following sources: seven complete cpDNA of *Mammillaria* previously published [39], as well as the raw data of 86 genomes that were downloaded from NCBI site, which were linked to BioProject PRJNA671701 [38]. In addition, the whole complete chloroplast genomes of 49 taxa were de novo sequenced in this study. The tissue samples for 47 of these taxa were provided by the collection of the Botanical Garden of the Universidad Nacional Autónoma de México, whilst the tissues of *M. napina* J.A.Purpus and *M. huitzilopochtli* D.R.Hunt were obtained from completed research projects (SS). Among these 142 genomes, 132 represented seven of the genera (i.e., Mammilloid Clade): *Coryphantha, Escobaria, Mammillaria, Mammilloydia, Neolloydia, Ortegocactus,* and *Pelecyphora* (Table 1).

Table 1. Taxon diversity sampled for the seven genera. Taxonomic names and the total number of recognized taxa for the levels of genus and subgenus following Hunt [18]. Total number of the taxa and number of genomes analyzed (in silico plus those de novo sequenced); and the number of genomes de novo sequenced. NA indicates that the subgenus level is not recognized.

Genus	Subgenus	Number of Recognized Taxa	Total Number of Analyzed Taxa	Number of Genomes Analyzed	Number of Genomes de novo Sequenced
1. Coryphantha		42	10	11	3
	1.1 Coryphantha	26	8	9	2
	1.2 Neocoryphantha	15	2	2	1
2. Escobaria	NA	19	8	9	2
3. Mammillaria		163	70	105	37
	3.1. Chilita *	18	16	33	3
	3.2. Cochemiea *	3	3	10	2
	3.3 Dolichothele	6	2	2	2
	3.4. Krainzia	12	5	5	2
	3.5. Mammillaria	117	37	46	26
	3.6. Mammillopsis	1	1	1	0
	3.7. Oehmea	1	1	1	1
	3.8. Phellosperma	5	5	7	1

Table 1. *Cont.*

Genus	Subgenus	Number of Recognized Taxa	Total Number of Analyzed Taxa	Number of Genomes Analyzed	Number of Genomes de novo Sequenced
4. Mammilloydia	NA	1	1	1	1
5. Neolloydia *	NA	2	2	3	2
6. Ortegocactus *	NA	1	1	2	1
7. Pelecyphora	NA	2	1	1	1

* Taxa included in *Cochemiea* according to Breslin et al. [38].

In addition, the taxonomic sampling covered the whole geographical range of five of these genera (*Coryphantha*, *Mammilloydia*, *Neolloydia*, *Ortegocactus*, and *Pelecyphora*). In contrast, the geographical range of *Mammillaria* was not sampled in South America; and for *Escobaria* was not sampled the Caribbean. Of them, 105 specimens (70 taxa) corresponded to *Mammillaria*, which are currently distributed in continental and peninsular Mexican territories, as well as the southern parts of the USA and the Caribbean. We documented the geographical distribution in the Global Biodiversity Information Facility (GBIF) ((https://www.gbif.org/ (accessed on 20 March 2022)) (Figure 1). In order to reduce record density, we used the spThin package [40] to discard all those records with <1 km in separation distance. The geographical data of those remained records were hand-curated following to Hernández and Gómez-Hinostrosa [33]. In addition, we sampled as external groups 10 specimens of eight genera: *Acharagma* (N.P. Taylor) Glass, *Ariocarpus* Scheidw., *Blossfeldia* Werderm., *Cumarinia* (Knuth) Buxb., *Lophophora* J.M. Coult., *Stenocactus* (K. Schum.) A. Berger, *Strombocactus* Britton & Rose, and *Turbinicarpus* Backeb.

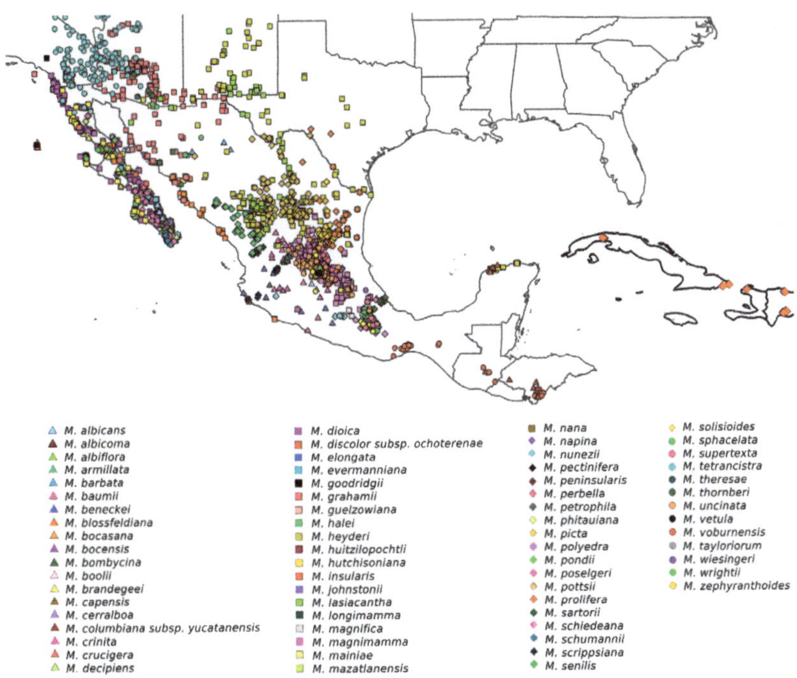

Figure 1. World geographical distribution of the 70 taxa of *Mammillaria*. Geographical distribution per taxon is showed in detail in Supplementary Materials (File S1).

2.2. DNA Extraction, cpDNA Enrichment, and High-Throughput Sequencing

For each of the 49 species de novo sequenced, 30–100 mg of frozen tissue was obtained to isolate 1 ug of gDNA with A260/280 ratio $\geq$ 1.7. The tissue samples were individually processed with the DNeasy plant mini kit (Qiagen, Hilden, Germany), following the manufacturer's instructions. To obtain an enriched proportion of chloroplast genome, these gDNAs were processed with the NEBNext Microbiome DNA Enrichment Kit (New England BioLabs, Ipswich, MA, USA) according to the kit's instructions. These enriched DNAs were used to prepare pair-end (PE) genomic libraries with the Nextera XT kit, with mean insert size of 400 bp, and were sequenced in MiSeq 2 $\times$ 300 cycles.

2.3. De Novo Assembly of Chloroplast Genomes

We assembled de novo the raw data of the 86 genomes attached to Breslin et al. [38] (Table S1); as well as the raw data of the 49 taxa de novo sequenced. These 135 genomes were filtered, trimmed and adapters were removed with TrimGalore version 0.4.3 [41]. The recovered reads with PHRED quality score $\geq$ 15 and length $\geq$ 80 bp were assembled with Get Organelle version v1.7.1 [42], using as a seed the cpDNA of *M. supertexta* (GenBank accession: MN508963.1) previously published [39].

2.4. Phylogenetic Relationships and Divergence Times

The 142 genomes (Table 1) were analyzed with BLAST version 2.5.0 [43] to identify common loci based on sequence similarity. All these common loci were aligned with MAFFT version v7.310 [44]. Because the genomes analyzed showed different structural arrangements, some sequences were not recovered, giving alignments with a high proportion of missing data; and other alignments showed low molecular variation. Thus, these two types of alignments were discarded before further phylogenetic analysis. Accordingly, only those alignments with sequences present for $\geq$70% of the studied taxa, $\geq$15% of proportion of informative sites (PIS) and a length of $\geq$200 bp were obtained. Accordingly, a total of 52 orthologous loci (Table S2) were identified and concatenated in a matrix of 48,869 bp used for the phylogenomic analysis and estimation of times of divergence. The matrix partitions and substitution models were estimated with ModelFinder [45], implemented in IQ-TREE2 version 2.1.4-beta [46]. The phylogenetic tree was generated with IQ-TREE2 using *B. liliputana* as the outgroup and running 10,000 ultra-fast bootstrap (UFBoot) replicates. Then, we estimated the evolutionary times of divergences using two secondary calibrations from previous estimations for the Cactaceae family [22]. In our first calibration, we used the crown age of 12.67–24.46 Mya estimated for the whole Cactoideae subfamily, and for the second calibration, we used the crown age of 4.86–10.63 Mya for the clade composed of the seven focus genera (*Coryphantha, Escobaria, Neolloydia, Mammillaria, Mammilloydia, Ortegocactus*, and *Pelecyphora*). We estimated the divergence times with BEAST version v2.6.6 [47], whose specific input file was constructed with BEAUti. In this input file was specified the GTR + I + Γ substitution model, which was estimated with Modeltest [48] according to AICc, a lognormal relaxed molecular clock, calibration points as uniform distributions, a Yule process tree prior, and 200,000,000 generations with a sampling frequency of each 2000 generations. In addition, convergence of parameter estimation was corroborated with Tracer version v1.7.2 [49], and the trees were summarized in a maximum clade credibility tree with TreeAnnotator version v2.6.3 [50]; 10% of the trees were discarded based on this final analysis.

2.5. Biogeographical Analysis

For the biogeographical analysis we documented the current geographical distribution for each of the 141 specimens (102 taxa) native to the arid lands of North America. As *B. liliputana* is endemic to South America it was discarded from the biogeographical analysis. The geographical data were compiled from GBIF ((https://www.gbif.org/ (accessed on 20 March 2022)). These data were verified by checking the geographical distribution of taxa reported from different sources [33,51,52]. The geographical distribution range of the

141 specimens was classified into the respective Mexican Floristic Provinces proposed by Rzedowski [53]. We estimated the ancestral geographical ranges based on the dated tree using the R package BioGeoBEARS version 1.1.1 [54] implemented in RASP4 v4.0 [55]. We evaluated four distinct models of the geographical range evolution for the 141 specimens: both the model of Dispersal-Extinction-Cladogenesis (DEC); and the likelihood version of Dispersal-Vicariance Analysis (DIVALIKE) were tested under two conditions, with and without the assumption of Founder-Event Speciation (+J) parameter. Finally, we plotted the changes estimations in the global surface air temperature (ΔT) in relation to the current values, previously published in the supplementary material (S4) of Herbert et al. [26].

3. Results

3.1. Evolutionary History of Cacti: Recent Divergence and Intricate Biogeography

The topologies of the phylogenetic tree (ML tree) (Figure 2) and the chronogram (BI tree) (Figure 3) were highly concordant; the only difference was found in the relationships of the small clade composed of *Ariocarpus*, *Strombocactus*, and *Turbinicarpus*. In the ML tree, the clade *Turbinicarpus-Strombocactus* was sister to *Ariocarpus*, whereas in the BI tree, the *Turbinicarpus-Ariocarpus* clade was sister to *Strombocactus*. In the biogeographical analysis, the DEC model (without +J parameter) was selected according to the value of AICcWt (File S2), however, this value was slightly higher than that obtained for the DEC+J model. In addition, these two models provided very similar estimations of the ancestral geographical distribution (Figures S1 and S2). The biogeographical analysis estimated a total of 135 dispersal events and 13 vicariant events (Figure 4).

The phylogenetic results (Figure 2) clearly identified for the 102 taxa (141 specimens) of the Cacteae tribe a common ancestor, which arose in the Mexican Plateau in the Late Miocene ~12.08 Mya (95% HPD: 7.73–16.82) (Figure 3). According to the temperatures taken from the bibliography [26], between 15 and 9 Mya there was a drastic decrease in global temperature of ΔT~8 °C. In this period, our results showed two key phylogenetic splits in Cacteae: the first one that separated *Stenocactus* from the remaining 101 taxa; followed by the second one that separated *Ariocarpus*, *Strombocactus* and *Turbinicarpus* from the remaining 96 taxa (Figure 4). Later, during a short period of nearly 2.7 million years (from 9 to 6.3 Mya), the temperature stayed stable (ΔT~0.1), and during this period two splits occurred. The first one is represented by the separation of *Acharagma* and *Lophophora*; the second one consists of the separation of the ancestor of *Cumarinia*. In this period (9 to 6.3 Mya), we identified the beginning of the complex evolutionary history of the 93 taxa belonging to the Mammilloid Clade (Figure 4). Moreover, these taxa continued their diversification processes during the next period of two million years (between 6.3 and 4.3 Mya), when the temperature again declined (ΔT~4.4 °C). These diversification processes continued and intensified during the last 4.3 Mya. In this last 4.3 million years, the temperature has not been steady; from 4.3 to 1 Mya (Late Pliocene to Pleistocene) a slight increase in temperature (ΔT~1 °C) was documented. However, in the last 1 million years, the temperature has decreased (ΔT~0.5 °C) (Figure 4).

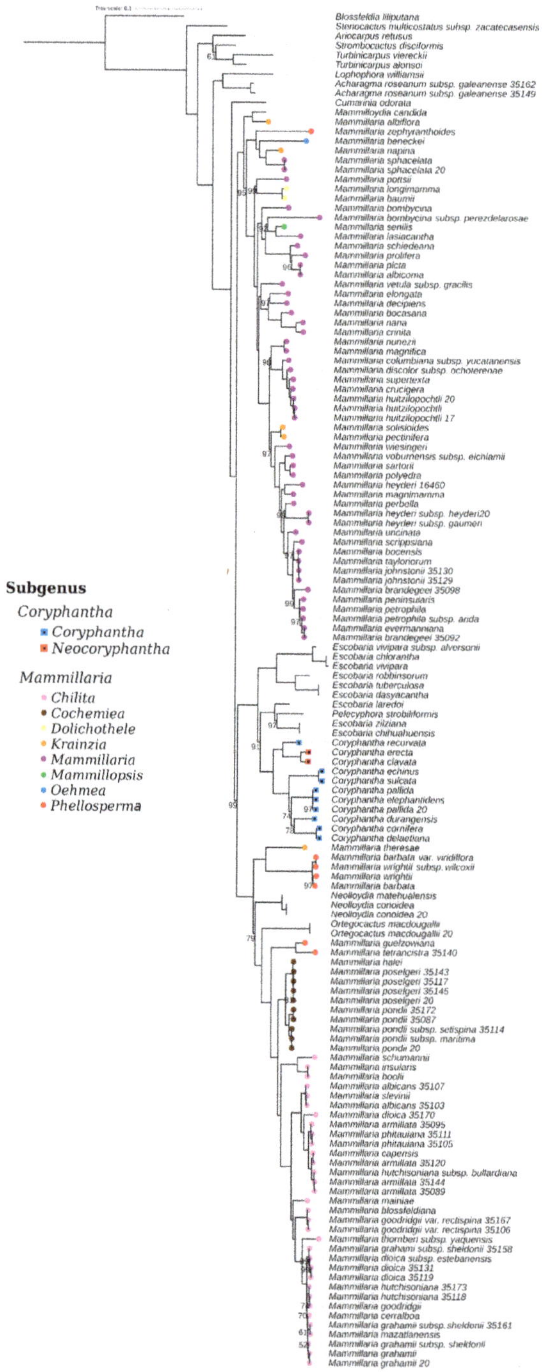

Figure 2. Phylogenetic tree estimated with IQ-TREE2 for the 103 taxa (142 specimens) using *B. liliputana* as the outgroup. Numbers below the nodes indicate UFBoot values < 100. Colored circles and squares indicate subgenera for *Mammillaria* and *Coryphantha*, respectively.

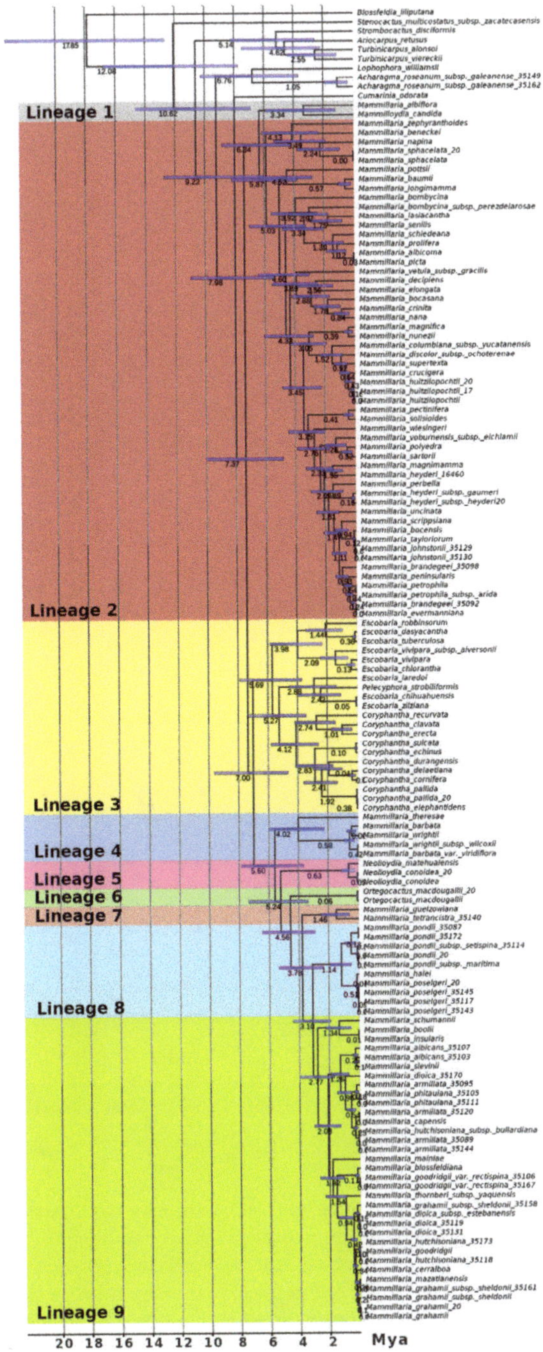

Figure 3. Chronogram estimated for the 103 taxa (142 specimens). The maximum clade credibility tree shows the divergence times estimated in BEAST. Blue bars represent 95% HPD intervals for the node ages. Shadow colors show the nine evolutionary lineages identified.

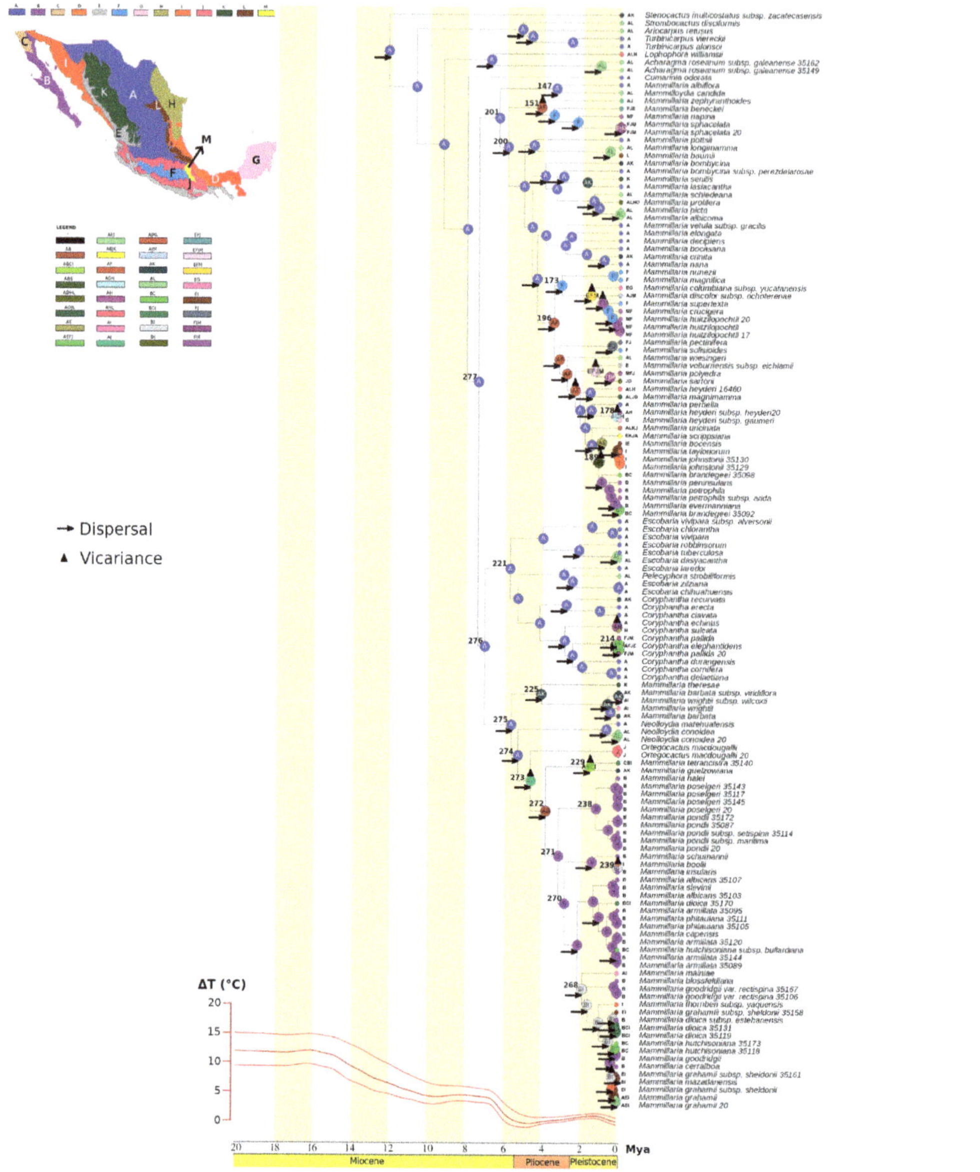

Figure 4. Ancestral geographical distribution of the North American taxa estimated on the chronogram. The estimation of the most likely ancestral distribution is represented in a colored circle for each node of the tree. The letters in the map and in the tree corresponded to Baja California (B), Balsas Basin (F), California (C), Gulf of Mexico Coast (D), Mexican Plateau (A), Meridional Serranias (J), Northeast Coastal Plain (H), Northwest Coastal Plain (I), Pacific Coast (E), Sierra Madre Occidental (K), Sierra Madre Oriental (L), Tehuacan Valley (M), and Yucatan Peninsula (G). The nodes of the main evolutionary events were numbered (see the text). The estimated events of dispersal and vicariance are indicated by arrows and triangles, respectively. The letters besides the tips indicate the current geographical distribution of the taxa. At the bottom of the figure were drawn the changes estimations in the global surface air temperature (ΔT).

The chronogram showed that the ancestor of the Mammilloid Clade originated nearly 7.37 Mya (95% HPD: 4.86–10.02 Mya). This early ancestor (node 277, Figure 4) arose at the end of the Miocene when the value ΔT of the temperature was low (ΔT~0.1 °C). This ancestor had as its probable ancestral geographical area the Mexican Plateau (Figure 4). Eventually, from this ancestor nine independent evolutionary lineages were derived (Figure 3); which profusely diversified in the last 4.3 Mya, when little increase-decrease of temperature occurred (Figure 4). This early common ancestor diverged into two new ancestors, one of which (node 276, Figure 4) was dated nearly 7 Mya (95% HPD: 4.67–9.64 Mya) (Figure 3). This ancestor had as its ancestral biogeographical scenario the Mexican Plateau (Figure 4), and from it evolved those taxa currently grouped in the six genera (*Coryphantha, Escobaria, Mammillaria, Neolloydia, Ortegocactus*, and *Pelecyphora*). The other ancestor (node 201, Figure 4) arose 6.34 Mya (95% HPD: 4.17–8.86 Mya) (Figure 3), probably also in the Mexican Plateau. From this lat ancestor evolved those taxa that were grouped in two genera: *Mammillaria* and *Mammilloydia*.

A conspicuous result obtained was that those 70 taxa (105 specimens) sampled as *Mammillaria* were distributed in two main independent clades (nodes 201 and 275, Figure 4). The immediate ancestors of these two clades originated in the past Mexican Plateau. However, those taxa derived from them clearly differ in their evolutionary history (Figure 4). Accordingly, the taxa sampled as genus Mammillaria, in fact, were distributed in six different and independent evolutionary lineages, each one with its own evolutionary history (Figures 3 and 4).

3.2. *Evolutionary History of the Nine Lineages*

3.2.1. Evolutionary Lineage 1

The short evolutionary lineage 1 was composed only of *Mammilloydia candida* (Scheidw.) Buxb. and *Mammillaria albiflora* Backeb, whose immediate ancestor was dated to nearly 3.34 Mya (95% HPD: 1.19–5.92 Mya). This ancestor (node 147, Figure 4) probably arose on the Mexican Plateau, during the Middle Pliocene, when the temperature underwent a slight increase (ΔT~1 °C) (Figure 4). At the present time, *M. candida* and *M. albiflora* are distributed in a small region on the southern region of the Mexican Plateau, and *M. candida* extends its geographical range to the northwest of this biogeographical area (Figure 4). Additionally, lineage 1 was identified as the phylogenetic sister to lineage 2 (Figure 2).

3.2.2. Evolutionary Lineage 2

The most probable ancestral geographical area for the immediate ancestor of lineage 2 was the Mexican Plateau (node 200, Figure 4). Lineage 2 arose nearly 5.87 Mya (95% HPD: 3.82–8.2), at the end of the Late Miocene, when the temperature decreased (ΔT~4.4 °C). However, most of the divergent processes in this lineage occurred in the last 4.3 million years, when a slight increase in the temperature (ΔT~1 °C) was followed by a slight decrease (ΔT~0.5 °C). This lineage grouped 45 of the sampled taxa, of which 37 taxa (82%) correspond to the subgenus *Mammillaria*, whereas the other eight taxa belong to five different subgenera (Figure 2). Three of these taxa (*M. napina, M. pectinifera* F.A.C. Weber, and *M. solisioides* Backeb.) corresponded to the subgenus *Krainzia*; two (*M. baumii* Boed and *M. longimamma* DC.) to the subgenus *Dolichothele*; one (*M. senilis* Lodd. ex Salm-Dyck) to *Mammilliopsis*; one (*M. beneckei* Ehrenb.) to the subgenus *Oehmea*; and finally, one species (*M. zephyranthoides* Scheidw.) to *Phellosperma*. Currently, 21 of these 45 taxa are endemic only to one of the 13 biogeographical areas (right-side letters beside the taxa in Figure 4); with the Mexican Plateau the area that has the highest number of endemics (eight taxa). In addition, most of the divergent events occurred in three biogeographical areas, which was the unique ancestral area or in conjunction with other ones: the Mexican Plateau (A) involved 66 % of the divergence events, the Balsas Basin (F) 35%, and the Tehuacan Valley (M) 13%.

Furthermore, biogeographical results suggested that such divergence processes were closely associated to the taxa dispersal towards new areas inside and outside of the Mexican

Plateau (Figure 4). Accordingly, in the lineage 2 the long-distance dispersal has been a common phenomenon during the last ~4 million years (Figure 4). During these long-distance dispersal events, it seems that the ancestors moved out of the Mexican Plateau and eventually displaced along different routes, either via continental arid lands or crossing the sea (Gulf of Mexico and Gulf of California). During the Pliocene, we identified two independent events of colonization (nodes 151 and 196; Figure 4) to the arid southern Mexican territories (F, J and M; Figure 4), where the colonizers eventually speciated in situ (nodes 151 and 173; Figure 4). The first long-dispersal event occurred 4.13 Mya, and the second one was dated 3.45 Mya. These two events occurred during a slight increase of temperature ($\Delta T \sim 1\ ^{\circ}C$) (Figure 4). These two colonization events took place from the Mexican Plateau (A) to the Balsas Basin (F), and from there to the adjacent areas of Tehuacan Valley (M) and Meridional Serranias (J) (Figure 4). In addition, we identified that during the Pleistocene, another two independent colonization events occurred towards the northern Mexican territories. The results indicate (Figure 4) that from the Mexican Plateau there was another dispersal route that took place along the foothills of Sierra Madre Occidental (A, K), crossing it, and reaching the Pacific slope of this Sierra. In addition, we identified a recent dispersal event dated nearly 1.11 Mya (node 189, Figure 4), in which an ancestor undertook a vicariant event, which separated two lineages, one of which diversified in the Baja California Peninsula (B) and the other in the northwest of continental Mexico (A, E) (Figure 4). On the other hand, on the eastern side of Mexico, another independent long-distance dispersal event was identified (node 178, Figure 4), crossing the Gulf of Mexico and reaching the Yucatán Peninsula nearly 0.16 Mya.

3.2.3. Evolutionary Lineage 3

We estimated the origin of the ancestor of this lineage (node 221, Figure 4) was in the Mexican Plateau (A) at 5.69 Mya (95% HPD: 3.72–7.89 Mya) (Figure 3). This lineage grouped 19 of the sampled taxa belonging to the genera *Coryphantha* (10), *Escobaria* (8), and *Pelecyphora* (1). Currently, 17 of these 19 taxa are distributed in the northern region of the Mexican Plateau (Figure 4), and only two taxa of *Coryphantha* are distributed in the southern arid lands (F, J and M, Figure 4), suggesting that a long-distance dispersal event allowed *Coryphantha* to reach the southern arid lands of Mexico (Figure 4). Therefore, in this lineage most of the past divergent processes were identified as on the Mexican Plateau, and eventually moving to northern and southern Mexico (Figure 4). The majority of these processes were dated to the Late Pliocene, when there was a slight temperature increase ($\Delta T \sim 1\ ^{\circ}C$) (Figure 4). Clearly, the phylogenetic relationships of this clade were fully resolved. These results recovered *Coryphantha* as monophyletic, whereas *Escobaria* is paraphyletic with respect to *Coryphantha* and *P. strobiliformis* (Figure 2).

These findings showed that the lineage 3 was the phylogenetic sister of a clade comprising six lineages (lineages 4–9) (Figure 3) that evolved from a common ancestor (node 275, Figure 4). This ancestor was dated to nearly 5.60 Mya (95% HPD: 3.62–7.84; Figure 3) and arose in the ancestral arid lands of the Mexican Plateau (Figure 4).

3.2.4. Evolutionary Lineage 4

The lineage 4 was derived from an ancestor (node 225, Figure 4) dated nearly 4.02 Mya (95% HPD: 2.27–6.01 Mya), which had its ancestral geographical area on the Mexican Plateau (A) and the foothills of the Sierra Madre Occidental (K). We identified an early split that separated subgenus *Krainzia* (*M. theresae* Cutak, Figure 2) from *Phellosperma* (*M. barbata* Engelm. and *M. wrightii* Engelm.). In this last subgenus, a recent divergence (0.58 Mya, Figure 3) was identified; presently the taxa of this lineage are distributed in the northwestern territories of the Mexican Plateau, the Sierra Madre Occidental and the Northwest Coastal Plain (A, K and I, Figure 4).

3.2.5. Evolutionary Lineage 5

Our results revealed a common ancestor (node 274, Figure 4) that originated lineage 5 and the other four independent lineages identified as lineages 6–9 (Figure 3). This ancestor was dated to 5.24 Mya (95% HPD: 3.38–7.38 Mya) and probably arose in the ancestral lands of the Mexican Plateau (Figure 4). Although the phylogenetic split of the ancestor (node 274, Figure 4) was dated to nearly 5 Mya, the origin of lineage 5 is very recent, as it was dated to 0.63 Mya (95% HPD: 0.23–1.13 Mya) (Figure 3), when a slight temperature decrease (ΔT~0.5 °C) occurred (Figure 4). The ancestral geographical area of this lineage was also the Mexican Plateau (A, Figure 4). In addition, this lineage currently groups the two taxa recognized in the genus *Neolloydia*, which are distributed in the Mexican Plateau (Figure 4). However, *N. matehualensis* Backeb. is endemic to the center of the Mexican Plateau, and *N. conoidea* (DC) Britton & Rose ranges from the southern to the northern range of the Mexican Plateau and reaches the southern arid lands of the USA.

3.2.6. Evolutionary Lineage 6

Lineage 6 was composed of the unique species recognized in the genus *Ortegocactus*. This lineage was derived from an old ancestor (node 273, Figure 4) dated nearly 4.56 Mya (95% HPD: 2.9–6.45 Mya) in the Early Pliocene, when the cooling period ended (Figure 4). This ancestor had as its probable ancestral geographical areas the Mexican Plateau and Meridional Serranias (A and J, Figure 4), and presently this lineage is endemic to the Meridional Serranias (J, Figure 4). Lastly, the results showed that the two sampled specimens of *O. macdougallii* Alexander recently diverged about 51,000 years ago (Figure 3).

3.2.7. Evolutionary Lineage 7

The ancestor of lineage 7 (node 229, Figure 4) was dated to 1.46 Mya (95% HPD: 0.6–2.48 Mya), during a time when the temperature increased slightly (ΔT~0.5 °C), and for this were estimated four probable ancestral areas (A, B, C and I, Figure 4). This lineage consists of only two northern native taxa; *M. guelzowiana* Werderm., which is endemic to the northwestern part of the continental Mexican territories; and *M. tetrancistra* Engelm. that is distributed in Baja California (B) and California (C), northwestern continental Mexican territories (I), and reaches the southern USA.

3.2.8. Evolutionary Lineage 8

Lineages 8 and 9 had a common ancestor (node 271, Figure 4) that arose in Baja California 3.1 Mya (95% HPD: 1.9–4.43 Mya) (B, Figure 4). In particular, the immediate ancestor of lineage 8 (node 238, Figure 4) was dated to 1.14 Mya (0.52–1.91 Mya) during the Pleistocene, concurrently with a small increase of temperature (ΔT~1 °C) (Figure 4). Lineage 8 grouped three taxa (*M. halei* Brandegee, *M. pondii* Greene, and *M. poselgeri* Hildm.) belonging to the *Cochemiea* subgenus.

3.2.9. Evolutionary Lineage 9

Lineage 9 grouped 16 taxa, all pertaining to the subgenus *Chilita* (Figure 2). Its immediate ancestor was dated to 2.77 Mya (95% HPD: 1.67–3.97 Mya) and its most probable ancestral area was Baja California (B, Figure 4). This lineage developed in the Late Pliocene when there was a slight increase in temperature (ΔT~1 °C). It diversified abundantly in Baja California. In this lineage, we identified two independent dispersal events from peninsular territories to the continental Northwest Coastal Plain (I, Figure 4). One of them occurred 1.82 Mya (node 268, Figure 4) and the other was dated to 0.01 Mya (node 239, Figure 4).

4. Discussion

4.1. Origin and Diversification of the Mammilloid Clade

The findings of this study revealed that the evolutionary history of the Mammilloid Clade (*Coryphantha, Escobaria, Mammillaria, Mammilloydia, Neolloydia, Ortegocactus*, and *Pelecyphora*) started ~7.5 Mya in the Miocene. During this epoch, there was a cooling trend, although the global temperature was still approximately 4–15 °C warmer than it is today [26]. The early and scarce divergence events that occurred in the Miocene were geographically restricted to the Mexican Plateau. However, during the last 4.5 million years, the cacti profusely diversified and expanded their distribution range to new areas when the global temperature was more similar to the present. Particularly, the main past colonization to new geographical areas (e.g., California, Northwest coastal plain, Pacific coast, Tehuacán Valley, and Yucatan peninsula) were dated to the last ~2.5 million years, in the Pleistocene. During this epoch various oscillations in temperature occurred [56] and have been associated with an aridity increase (e.g., [57,58]). Consequently, these cacti are modern taxa, with most of their evolutionary history occurring during the Plio-Pleistocene. In fact, the climatic oscillations in the Pleistocene were recognized as diversification driving forces for other land plants (e.g., [59–62]). Particularly for cacti, glacial [63,64] and interglacial [65,66] periods have been proposed as drivers of population processes, causing geographic contraction, isolation, and population divergence. Therefore, probably these climatic oscillations also promoted the diversification of the cacti studied here.

The Mexican Plateau has been considered geologically and climatically stable since ~15 Mya (Middle Miocene) [67]. Hence, we consider that such stability promoted the prolific speciation and colonization of cacti. However, there is a gradient of aridity along the Mexican Plateau, with the northern portion being drier than the central–southern one [68]. As cacti do not prosper in hyper-arid conditions [69], the relative "higher-humidity" at the southern end of the Mexican Plateau likely foster the ecological conditions for their abundance and speciation, which eventually led to geographic expansion. Accordingly, we postulated that the center of origin for the lineages 1 and 2 was the southern region of the Mexican Plateau, which previously was named as the Queretano-Hidalguense arid zone [70]. We based this hypothesis on the early phylogenetic split identified in these two lineages, and on their current geographical distribution. Consistent with this assumption, nearly 20% of the richness of the genus *Mammillaria* (sensu Hunt [18]) inhabit the arid lands of the Queretano-Hidalguense arid zone (Hidalgo, Guanajuato, and Querétaro) [33].

Based on similar reasoning, we inferred that the possible center of origin of lineage 3 might be the north of the Mexican Plateau. However, we recognized that more extensive taxonomic sampling is necessary to elucidate this issue. On the other hand, our results revealed that the taxa grouped in lineages 4, 7, 8 and 9 had an ecological and biogeographical affinity to northwestern Mexico. Considering that the Baja California area was the probable ancestral geographical area of lineages 8 and 9, these results suggest that this area was probably the center of origin and diversification for these lineages. Lastly, we do not discard the possibility that the small and enigmatic lineages 5 and 6 represent relicts of some phylogenetic lines that are mostly extinct. Population approaches may serve to elucidate their closest phylogenetic frontiers and recent hybridization (e.g., [17,71]). Therefore, we recommend application of this perspective to lineages 5 and 6, as well as for the sister lineages 1 and 2.

Our results also revealed that dispersal, not vicariance, was the most important past biogeographical process in these cacti. The abundant dispersal events may be related to the capacity of cacti to colonize and tolerate hostile environments (e.g., [72]), or successful seed dispersal strategies [73]. However, the data indicate that also vicariance had a relevant role in the taxa that currently are distributed in southern Mexico. Because the central portion of the Trans-Mexican Volcanic Belt (TMVB) was formed during the last three million years (plate 1 [74]), we address the TMVB as a biogeographical barrier for cacti. In fact, most of the events of colonization to southern Mexican territories were identified prior to 3 million years ago, thus the TMVB interrupted the connectivity between the arid lands of the Mexican Plateau and those of the Balsas Basin, Tehuacán Valley, and southern Meridional Serranias. In addition, the floristic affinities between the arid lands of the north and south of Mexico have been documented [53], suggesting that prior to the TMVB, the Mexican arid lands were connected from north to the south. Finally, our results showed that the arrival of cacti to the Baja California peninsula was due to dispersal and not by vicariance, as the colonization occurred later than the opening of the Gulf of California, which occurred nearly 12–6 Mya [75].

4.2. Taxonomic Contributions of the Phylogenetic Results

The findings of this study have explained the phylogenetic relationships of the 103 taxa, particularly the 70 taxa sampled from the genus *Mammillaria* sensu Hunt [18] were polyphyletic, as was identified previously (e.g., [35,36]). However, based on our results, the monophyly of this genus can be identified within a subset of the 70 taxa sampled as *Mammillaria*. We consider that the putative genus *Mammillaria* is represented by lineage 2, in which 85% of the taxa were from subgenus *Mammillaria*. Accordingly, monophyletic *Mammillaria* is not restricted exclusively to the *Mammillaria* subgenus, as Crozier [36] proposed, but also includes taxa of another five subgenera: *Dolichothele*, *Krainzia*, *Mammillopsis*, *Oehmea*, and *Phellosperma*. Recently, Breslin et al. [38] proposed a monophyletic circumscription of the genus *Mammillaria*, based on massive sequencing of the chloroplast genome and 52 taxa assumed to be from members of the genus. Because these 52 taxa exhibited polyphyletic relationships, these authors decided to exclude a substantial number of them in order to reach a monophyletic group.

In addition, Breslin et al. [38] proposed that the 36 taxa of the genus *Mammillaria* that were placed out of the monophyletic group, should be placed in the genus *Cochemiea* together with *N. conoidea* and *O. macdougallii*, although the species of these genera exhibit strong morphological variation (Table S1) [21]. Our results showed that lineages 4 to 9 were grouped in a distinct clade, independent of the clade that grouped lineages 1 and 2. These six lineages composed a monophyletic group (lineages 4–9). Although, these six lineages were grouped similarly to the clade named as *Cochemiea* by Breslin et al. [38] we do not agree to put together the taxa of these six lineages as our results showed strong disparities in the biogeographical history and ecologic affinities. Additionally, their strong morphological variations do not accomplish the unambiguous practical delimitation (i.e., taxonomic predictability) and stability that are required at the genus level [76]. We consider that based on a purely phylogenetic perspective, the proposal of Breslin et al. [38] to include in *Cochemiea* other taxa recognized as *Mammillaria*, *Neolloydia* and *Ortegocactus* is feasible. However, our results identified six lineages in the clade *Cochemiea* sensu Breslin et al. [38], and for us these may represent more than one genus: *Cochemiea* (lineages 7, 8 and 9), *Neolloydia* (5), *Ortegocactus* (6), and *Phellosperma* (4). These two contrasting stances exhibit the degree of subjectivity to establish the supraspecific taxonomic delimitation as has been discussed [76]. We consider that future phylogenetic studies are still necessary, and they must include specimens of the type *M. mammillaris*, and have a higher taxonomic sampling, especially of those taxa that are currently distributed in the west side of Mexican territories along the Pacific Coast. Consequently, we considered that the taxonomic circumscription of *Mammillaria* still remains unresolved. Lastly, our phylogenetic results partially supported the infrageneric classification of *Mammillaria* proposed by Hunt [18]. Accordingly, the taxa of the subgenera *Cochemiea* and *Chilita* were monophyletic. Although all the taxa of the subgenus *Mammillaria* were grouped, the monotypic and small subgenera (*Dolichothele*, *Mammillopsis* and *Oehmea*) as well as some taxa of *Phellosperma* and *Krainzia* were inserted among the species of the *Mammillaria* subgenus. In addition, because we included the raw data attached to Breslin et al. [38], we observed that in our phylogenetic tree, some of their specimens belonging to the same species were placed in discordant positions (*M. grahamii* subsp. *sheldonii* (Britton & Rose) D.R. Hunt (35158, 35161), *M. goodridgii* (35106, 35115, 35167), *M. albicans* Dawson (35107, 35103), *M. armillata* K.Brandegee (35093, 35144, 35089), *M. dioica* K.Brandegee (35170, 35131, 35119), and *M. heyderi* Muehlenpf. (16460)); this indicates the probable wrong taxonomic identification of their specimens, thus we overlooked these for the taxonomic discussion.

Recently, the study of Sánchez et al. [77] obtained a phylogenetic tree based on five chloroplast loci and eight morphologic characters. It showed that *Coryphantha* was a monophyletic genus, when excluding *C. macromeris* (Engelm.) Lem. Moreover, this last taxon and the taxa of *Escobaria* and *Pelecyphora* were grouped in the same clade, sister to *Coryphantha*, and were reclassified as a single genus (*Pelecyphora*). Our study also showed that *Coryphantha* is a monophyletic taxon, whereas *Escobaria* is paraphyletic with respect to *Coryphantha* and *Pelecyphora*. These discordances may be the result of distinct taxonomic and molecular sampling between the two studies. Nonetheless, it may be necessary to analyze morphological, ecological, and anatomical characters in order to solve these taxonomic issues.

5. Conclusions

We identified that the biogeographical processes, past climate conditions from the Miocene, and the recent emergence of the central portion of the TMVB strongly shaped the evolutionary history of the Mammilloid Clade (*Coryphantha, Escobaria, Mammillaria, Mammilloydia, Neolloydia, Ortegocactus*, and *Pelecyphora*). The past Mexican arid lands were key to providing ecological suitability for prolific cacti diversification. In these regions, they became abundant and ubiquitous elements of the arid flora. The large Mexican Plateau has been the primary evolutionary scenario for cacti, and this area is key to understand the diversity of cacti in Mexico, southern USA, Caribbean, and South America. Lastly, the Mexican territories harbor most of the world's richness of cacti, and it is urgent to protect these arid lands, particularly the region included in the northern part of Guanajuato, Hidalgo, and Querétaro, and southern of San Luis Potosí. Our findings indicate that the genus *Mammillaria* sensu Hunt [18] is taxonomically composed of distinct evolutionary lineages, whose phylogenetic relationships require more detailed studies to reach a precise taxonomic circumscription. In this light, we consider that it is premature to undertake nomenclatural changes in *Mammillaria, Mammilloydia, Neolloydia*, and *Ortegocactus* [38,78], and such changes will bring more confusion. Therefore, we recommend maintaining the conventional taxonomic classifications (e.g., [18]) until more robust studies are undertaken. In summary, we conclude that the taxonomic circumscription of the genus *Mammillaria* still needs more work, based on phylogenetic analyses encompassed with robust and detailed ecological studies of the current geographical distribution, past niche modeling, reproductive barriers, and a clear set of diagnostic morphological characters.

Supplementary Materials: The following supporting information can be downloaded at https://www.mdpi.com/article/10.3390/biology12040512/s1. Table S1: List and origin of the studied taxa; File S1: Geographical distribution of the 70 sampled taxa of *Mammillaria*; Table S2: List of 52 loci used for phylogenomic analysis; File S2: Output of the evaluation of six biogeographical models with BioGeoBEARS implemented in RASP4; Figure S1: Full results of ancestral geographical distribution estimated under the DEC model. Figure S2: Results of ancestral geographical distribution estimated under the DEC+J model.

Author Contributions: This article was conceptualized by S.S. as part of the Research Project IN228619 supported by PAPIIT-DGAPA; moreover, she supervised and administered the resources of this research, and got the funding resources. D.A.C. and S.S. planned, discussed, and agreed on the analytical approaches used in this research; D.A.C. carried out in full the statistical formal analysis. A partial support in massive sequencing was provided by F.V.-P.; S.A. contributed with tissue samples for most of de novo sequenced samples. D.A.C., S.S., F.V.-P. and P.D. wrote and prepared the original draft. All authors reviewed, commented, edited the last version of this manuscript. All authors have read and agreed to the published version of the manuscript.

Funding: This research was supported by UNAM-PAPIIT-IN228619.

Institutional Review Board Statement: Not applicable.

Informed Consent Statement: Not applicable.

Data Availability Statement: The raw data of the 49 samples de novo sequenced are available in the NCBI database under BioProject PRJNA934337. The SRA accessions for each sample are shown in Table S1.

Acknowledgments: This study was fully granted by UNAM-PAPIIT-IN228619. This article is part of the requirements for obtaining the doctoral degree of D.A.C. at the Posgrado en Ciencias Biológicas (Evolutionary Biology), Universidad Nacional Autónoma de México; she was funded by a grant from the Consejo Nacional de Ciencia y Tecnología CONACyT (855061). The sampling permissions were authorized to S.S. by SEMARNAT SGPA/DGVS/06880/16 for *Mammillaria huitzilopochtli* and SGP/DGVS/01833/08 for *M. napina*; whereas the rest of the studied species were provided from the cacti collection of the Botanical Garden of Instituto de Biología, UNAM. The massive sequencing was carried out by Clara E. Díaz-Velázquez and Aldo Hugo De La Cruz Montoya technicians of the Laboratorio Nacional en Salud: Diagnóstico Molecular y Efecto Ambiental en Enfermedades Crónico-Degenerativas, FES Iztacala, UNAM.

Conflicts of Interest: The authors declare no conflict of interest. The funders had no role in the design of the study; in the collection, analyses, or interpretation of data; in the writing of the manuscript; or in the decision to publish the results.

References

1. Ronquist, F.; Sanmartín, I. Phylogenetic methods in biogeography. *Annu. Rev. Ecol. Evol. Syst.* **2011**, *42*, 441–464. [CrossRef]
2. Xia, M.; Liu, Y.; Liu, J.; Chen, D.; Shi, Y.; Chen, Z.; Chen, D.; Jin, R.; Chen, H.; Peter, C.H.; et al. Out of the Himalaya-Hengduan Mountains: Phylogenomics, biogeography and diversification of *Polygonatum* Mill. (Asparagaceae) in the Northern Hemisphere. *Mol. Phylogenet. Evol.* **2022**, *169*, 107431. [CrossRef] [PubMed]
3. Rose, J.P.; Xiang, C.L.; Sytsma, K.J.; Drew, B.T. A timeframe for mint evolution: Towards a better understanding of trait evolution and historical biogeography in Lamiaceae. *Bot. J. Linn. Soc.* **2022**, *200*, 15–38. [CrossRef]
4. Liu, S.Y.; Zhu, H.; Yang, J. A phylogenetic perspective on biogeographical divergence of the flora in Yunnan, Southwestern China. *Sci. Rep.* **2017**, *7*, 43032. [CrossRef]
5. Qian, H.; Deng, T.; Ricklefs, R.E. Evolutionary assembly of the Arctic flora. *Glob. Ecol. Biogeogr.* **2022**, *31*, 396–404. [CrossRef]
6. Barros-Souza, Y.; Borges, L.M. Spatial-and lineage-dependent processes underpin floristic assembly in the megadiverse Eastern South American mountains. *J. Biogeogr.* **2023**, *50*, 302–315. [CrossRef]
7. Chen, L.Y.; Zhao, S.Y.; Mao, K.S.; Les, D.H.; Wang, Q.F.; Moody, M.L. Historical biogeography of Haloragaceae: An out-of-Australia hypothesis with multiple intercontinental dispersals. *Mol. Phylogenet. Evol.* **2014**, *78*, 87–95. [CrossRef]
8. Nge, F.J.; Kellermann, J.; Biffin, E.; Waycott, M.; Thiele, K.R. Historical biogeography of *Pomaderris* (Rhamnaceae): Continental vicariance in Australia and repeated independent dispersals to New Zealand. *Mol. Phylogenet. Evol.* **2021**, *158*, 107085. [CrossRef]
9. Liu, X.Q.; Xia, X.M.; Chen, L.; Wang, X.Q. Phylogeny and evolution of Cupressaceae: Updates on intergeneric relationships and new insights on ancient intergeneric hybridization. *Mol. Phylogenet. Evol.* **2022**, *177*, 107606. [CrossRef]
10. Heinrichs, J.; Hentschel, J.; Feldberg, K.; Bombosch, A.; Schneider, H. Phylogenetic biogeography and taxonomy of disjunctly distributed bryophytes. *J. Syst. Evol.* **2009**, *47*, 497–508. [CrossRef]
11. Oyston, J.W.; Wilkinson, M.; Ruta, M.; Wills, M.A. Molecular phylogenies map to biogeography better than morphological ones. *Commun. Biol.* **2022**, *5*, 521. [CrossRef] [PubMed]
12. Svenning, J.-C.; Eiserhardt, W.L.; Normand, S.; Ordonez, A.; Sandel, B. The influence of paleoclimate on present-day patterns in biodiversity and ecosystems. *Annu. Rev. Ecol. Evol. Syst.* **2015**, *46*, 551–572. [CrossRef]
13. Li, Z.Z.; Lehtonen, S.; Martins, K.; Gichira, A.W.; Wu, S.; Li, W.; Hu, G.W.; Liu, Y.; Zou, C.Y.; Wang, Q.F.; et al. Phylogenomics of the aquatic plant genus *Ottelia* (Hydrocharitaceae): Implications for historical biogeography. *Mol. Phylogenet. Evol.* **2020**, *152*, 106939. [CrossRef] [PubMed]
14. Lai, Y.J.; Li, S.J.; Wang, W.M. Evolutionary trends in leaf morphology and biogeography of Altingiaceae based on fossil evidence. *Palaeoworld* **2018**, *27*, 415–422. [CrossRef]
15. Ding, W.N.; Ree, R.H.; Spicer, R.A.; Xing, Y.W. Ancient orogenic and monsoon-driven assembly of the world's richest temperate alpine flora. *Science* **2020**, *369*, 578–581. [CrossRef]
16. Antonelli, A.; Nylander, J.A.A.; Persson, C.; Sanmartín, I. Tracing the impact of the Andean uplift on Neotropical plant evolution. *Proc. Natl. Acad. Sci. USA* **2009**, *106*, 9749–9754. [CrossRef] [PubMed]
17. Hewitt, G.M. Some genetic consequences of ice ages, and their role in divergence and speciation. *Biol. J. Linn. Soc.* **1996**, *58*, 247–276. [CrossRef]
18. Hunt, D.R. *The New Cactus Lexicon*; DH Books: Milborne Port, UK, 2006.
19. Nyffeler, R.; Eggli, U. Disintegrating Portulacaceae: A new familial classification of the suborder Portulacineae (Caryophyllales) based on molecular and morphological data. *Taxon* **2010**, *59*, 227–240. [CrossRef]
20. Walker, J.F.; Yang, Y.; Feng, T.; Timoneda, A.; Mikenas, J.; Hutchison, V.; Edwards, C.; Wang, N.; Ahluwalia, S.; Olivieri, J.; et al. From cacti to carnivores: Improved phylotranscriptomic sampling and hierarchical homology inference provide further insight into the evolution of Caryophyllales. *Am. J. Bot.* **2018**, *105*, 446–462.
21. Magallon, S.; Gomez-Acevedo, S.; Sanchez-Reyes, L.L.; Hernandez-Hernandez, T. A metacalibrated time-tree documents the early rise of flowering plant phylogenetic diversity. *New Phytol.* **2015**, *207*, 437–453. [CrossRef]

22. Hernandez-Hernandez, T.; Brown, J.W.; Schlumpberger, B.O.; Eguiarte, L.E.; Magallon, S. Beyond aridification: Multiple explanations for the elevated diversification of cacti in the New World Succulent Biome. *New Phytol.* **2014**, *202*, 1382–1397. [CrossRef] [PubMed]
23. Arakaki, M.; Christin, P.A.; Nyffeler, R.; Lendel, A.; Eggli, U.; Ogburn, M.; Spriggs, E.; Moore, M.J.; Edwards, E.J. Contemporaneous and recent radiations of the world's major succulent plant lineages. *Proc. Natl. Acad. Sci. USA* **2011**, *108*, 8379–8384. [CrossRef]
24. Pearson, P.N.; Palmer, M.R. Atmospheric carbon dioxide concentrations over the past 60 million years. *Nature* **2000**, *406*, 695–699. [CrossRef] [PubMed]
25. Tripati, A.K.; Roberts, C.D.; Eagle, R.A. Coupling of CO_2 and Ice Sheet Stability Over Major Climate Transitions of the Last 20 Million Years. *Science* **2009**, *326*, 1394–1397. [CrossRef]
26. Herbert, T.D.; Lawrence, K.T.; Tzanova, A.; Peterson, L.C.; Caballero-Gill, R.; Kelly, C.S. Late Miocene global cooling and the rise of modern ecosystems. *Nat. Geosci.* **2016**, *9*, 843. [CrossRef]
27. Anderson, E.F. *The Cactus Family*; Timber: Portland, OR, USA, 2001.
28. The IUCN Red List of Threatened Species. *Version 2022-1*. Available online: https://www.iucnredlist.org/search?taxonomies=111456&searchType=species (accessed on 1 November 2022).
29. Haworth, A.H. *Synopsis Plantarum Succulentarum*; Richard Taylor: London, UK, 1812.
30. Schumann, K. *Gesamtbeschreibung der Kakteen (Monographia Cactearum)*; Verlag von J. Neumann: Neudamm, Germany, 1898; pp. 472–601.
31. Hunt, D.R. Schumann and Buxbaum reconciled. *Cact. Succ. J. Gr. Brit.* **1971**, *33*, 53–72.
32. Lüthy, J.M. Taxonomische Untersuchung der Gattung *Mammillaria* Haw. Ph.D. Thesis, Universität Bern, Bern, Switzerland, 1995.
33. Hernández, H.M.; Gómez-Hinostrosa, C. *Mapping the Cacti of Mexico Part II: Mammillaria*; DH Books: Milborne Port, UK, 2015.
34. Butterworth, C.A.; Cota-Sanchez, J.H.; Wallace, R.S. Molecular systematics of tribe Cacteae (Cactaceae: Cactoideae): A phylogeny based on rpl16 intron sequence variation. *Syst. Bot.* **2002**, *27*, 257–270.
35. Butterworth, C.A.; Wallace, R.S. Phylogenetic studies of *Mammillaria* (Cactaceae)—Insights from chloroplast sequence variation and hypothesis testing using the parametric bootstrap. *Am. J. Bot.* **2004**, *91*, 1086–1098. [CrossRef] [PubMed]
36. Crozier, B.S. Systematics of Cactaceae Juss: Phylogeny, cpDNA Evolution, and Classification, with Emphasis on the Genus *Mammillaria* Haw. Ph.D. Thesis, University of Texas, Austin, TX, USA, 2005.
37. Hunt, D.R. Revised classified list of the genus *Mammillaria*. *Cact. Succ. J. Gr. Brit.* **1981**, *43*, 41–48.
38. Breslin, P.B.; Wojciechowski, M.F.; Majure, L.C. Molecular phylogeny of the Mammilloid clade (Cactaceae) resolves the monophyly of *Mammillaria*. *Taxon* **2021**, *70*, 308–323. [CrossRef]
39. Solórzano, S.; Chincoya, D.A.; Sanchez-Flores, A.; Estrada, K.; Díaz-Velásquez, C.E.; González-Rodríguez, A.; Vaca-Paniagua, F.; Dávila, P.; Arias, S. De novo assembly discovered novel structures in genome of plastids and revealed divergent inverted repeats in *Mammillaria* (Cactaceae, Caryophyllales). *Plants* **2019**, *8*, 392. [CrossRef] [PubMed]
40. Aiello-Lammens, M.E.; Boria, R.A.; Radosavljevic, A.; Vilela, B.; Anderson, R.P.; Bjornson, R.; Weston, S. spThin: Functions for Spatial Thinning of Species Occurrence Records for Use in Ecological Models. 2019. Available online: https://cran.r-project.org/web/packages/spThin/index.html (accessed on 15 March 2023).
41. Babraham Bioinformatics. Trim Galore. Available online: http://www.bioinformatics.babraham.ac.uk/projects/trim_galore/ (accessed on 21 April 2022).
42. Jin, J.J.; Yu, W.B.; Yang, J.B.; Song, Y.; DePamphilis, C.W.; Yi, T.S.; Li, D.Z. GetOrganelle: A fast and versatile toolkit for accurate de novo assembly of organelle genomes. *Genome Biol.* **2020**, *21*, 241. [CrossRef] [PubMed]
43. Zhang, Z.; Schwartz, S.; Wagner, L.; Miller, W. A greedy algorithm for aligning DNA sequences. *J. Comput. Biol.* **2000**, *7*, 203–214. [CrossRef]
44. Katoh, K.; Misawa, K.; Kuma, K.; Miyata, T. MAFFT: A novel method for rapid multiple sequence alignment based on fast Fourier transform. *Nucleic Acids Res.* **2002**, *30*, 3059–3066. [CrossRef]
45. Kalyaanamoorthy, S.; Minh, B.Q.; Wong, T.K.; Von Haeseler, A.; Jermiin, L.S. Model Finder: Fast model selection for accurate phylogenetic estimates. *Nat. Methods* **2017**, *14*, 587–589. [CrossRef] [PubMed]
46. Minh, B.Q.; Schmidt, H.A.; Chernomor, O.; Schrempf, D.; Woodhams, M.D.; Von Haeseler, A.; Lanfear, R. IQ-TREE 2: New models and efficient methods for phylogenetic inference in the genomic era. *Mol. Biol. Evol.* **2020**, *7*, 1530–1534. [CrossRef]
47. Bouckaert, R.; Vaughan, T.G.; Barido-Sottani, J.; Duchêne, S.; Fourment, M.; Gavryushkina, A.; Heled, J.; Jones, G.; Kühnert, D.; De Maio, N. BEAST 2.5: An advanced software platform for Bayesian evolutionary analysis. *PLoS Comput. Biol.* **2019**, *15*, e1006650. [CrossRef]
48. Posada, D.; Crandall, K.A. MODELTEST: Testing the model of DNA substitution. *Bioinformatics* **1998**, *14*, 817–818. [CrossRef]
49. Rambaut, A.; Drummond, A.J.; Xie, D.; Baele, G.; Suchard, M.A. Posterior summarisation in Bayesian phylogenetics using Tracer 1.7. *Syst. Biol.* **2018**, *67*, 901–904. [CrossRef]
50. Drummond, A.J.; Rambaut, A. BEAST: Bayesian evolutionary analysis by sampling trees. *BMC Evol. Biol.* **2007**, *7*, 214. [CrossRef]
51. Guzmán-Cruz, L.U.; Arias-Montes, S.; Dávila, P. *Catálogo de Cactáceas Mexicanas*; Universidad Nacional Autónoma de México, Comisión Nacional para el Conocimiento y Uso de la Biodiversidad: Ciudad de México, Mexico, 2007.
52. Dicht, R.; Lüthy, A. *Coryphantha: Cacti of Mexico and Southern USA*; Springer: Berlin, Germany, 2005.
53. Rzedowski, J. *Vegetación de México*; Limusa: Ciudad de México, Mexico, 1978.

54. Matzke, N. Probabilistic historical biogeography: New models for founder-event speciation, imperfect detection, and fossils allow improved accuracy and model-testing. *Front. Biogeogr.* **2015**, *5*, 242–248.
55. Yu, Y.; Blair, C.; He, X. RASP 4: Ancestral state reconstruction tool for multiple genes and characters. *Mol. Biol. Evol.* **2020**, *37*, 604–606. [CrossRef] [PubMed]
56. Lawrence, K.T.; Sosdian, S.; White, H.E.; Rosenthal, Y. North Atlantic climate evolution through the Plio-Pleistocene climate transitions. *Earth Planet. Sci. Lett.* **2010**, *300*, 329–342. [CrossRef]
57. Vuilleumier, B.S. Pleistocene changes in the fauna and flora of South America. *Science* **1971**, *173*, 771–780. [CrossRef] [PubMed]
58. Hoag, C.; Svenning, J.C. African environmental change from the Pleistocene to the Anthropocene. *Annu. Rev. Environ. Resour.* **2017**, *42*, 27–54. [CrossRef]
59. Janssens, S.B.; Knox, E.B.; Huysmans, S.; Smets, E.F.; Merckx, V.S. Rapid radiation of *Impatiens* (Balsaminaceae) during pliocene and pleistocene: Result of a global climate change. *Mol. Phylogenet. Evol.* **2009**, *52*, 806–824. [CrossRef]
60. DeChaine, E.G.; Wendling, B.M.; Forester, B.R. Integrating environmental, molecular, and morphological data to unravel an ice-age radiation of arctic-alpine Campanula in western North America. *Ecol. Evol.* **2014**, *4*, 3940–3959. [CrossRef]
61. Dorsey, B.L.; Gregory, T.J.; Sass, C.; Specht, C.D. Pleistocene diversification in an ancient lineage: A role for glacial cycles in the evolutionary history of *Dioon* Lindl. (Zamiaceae). *Am. J. Bot.* **2018**, *105*, 1512–1530. [CrossRef]
62. Nevado, B.; Contreras-Ortiz, N.; Hughes, C.; Filatov, D.A. Pleistocene glacial cycles drive isolation, gene flow and speciation in the high-elevation Andes. *New Phytol.* **2018**, *219*, 779–793. [CrossRef]
63. Cornejo-Romero, A.; Vargas-Mendoza, C.F.; Aguilar-Martínez, G.F.; Medina-Sánchez, J.; Rendón-Aguilar, B.; Valverde, P.L.; Zavala-Hurtado, J.A.; Serrato, A.; Rivas-Arancibia, S.; Pérez-Hernández, M.A.; et al. Alternative glacial-interglacial refugia demographic hypotheses tested on *Cephalocereus columna-trajani* (Cactaceae) in the intertropical Mexican drylands. *PLoS ONE* **2017**, *12*, e0175905. [CrossRef]
64. Ossa, C.G.; Montenegro, P.; Larridon, I.; Pérez, F. Response of xerophytic plants to glacial cycles in southern South America. *Ann. Bot.* **2019**, *124*, 15–26. [CrossRef] [PubMed]
65. Bonatelli, I.A.; Perez, M.F.; Peterson, A.T.; Taylor, N.P.; Zappi, D.C.; Machado, M.C.; Koch, I.; Pires, A.H.C.; Moraes, E.M. Interglacial microrefugia and diversification of a cactus species complex: Phylogeography and palaeodistributional reconstructions for *Pilosocereus aurisetus* and allies. *Mol. Ecol.* **2014**, *23*, 3044–3063. [CrossRef] [PubMed]
66. Ornelas, J.F.; Rodríguez-Gómez, F. Influence of Pleistocene glacial/interglacial cycles on the genetic structure of the Mistletoe Cactus *Rhipsalis baccifera* (Cactaceae) in Mesoamerica. *J. Hered.* **2015**, *106*, 196–210. [CrossRef]
67. Morafka, D.J. A historical biogeography of the Chihuahuan herpetofauna. In *A Biogeographical Analysis of the Chihuahuan Desert through Its Herpetofauna*; Dr. W. Junk: The Hague, The Netherlands, 1977; pp. 157–215.
68. UNESCO. *ATLAS of Arid Zones in Latin America and the Caribbean*; PHI UNESCO: Paris, France, 2010.
69. Guerrero, P.C.; Durán, A.P.; Walter, H.E. Latitudinal and altitudinal patterns of the endemic cacti from the Atacama Desert to Mediterranean Chile. *J. Environ.* **2011**, *75*, 991–997. [CrossRef]
70. Medellín-Leal, F. The Chihuahuan Desert. In *Reference Handbook on the Deserts of North America*; Bender, G.L., Ed.; Greenwood Press: Westfield, CT, USA, 1982; pp. 321–381.
71. Jürgens, N.; Oncken, I.; Oldeland, J.; Gunter, F.; Rudolph, B. *Welwitschia*: Phylogeography of a living fossil, diversified within a desert refuge. *Sci. Rep.* **2021**, *11*, 114. [CrossRef]
72. Puente, M.E.; Li, C.Y.; Bashan, Y. Rock-degrading endophytic bacteria in cacti. *Environ. Exp. Bot.* **2009**, *66*, 389–401. [CrossRef]
73. Bregman, R. Forms of seed dispersal in Cactaceae. *Acta Bot. Neerl.* **1988**, *37*, 395–402. [CrossRef]
74. Gómez-Tuena, A.; Orozco-Esquivel, M.T.; Ferrari, L. Igneous petrogenesis of the Trans-Mexican volcanic belt. *Geo. Soc. Amer. Spec.* **2007**, *422*, 129–181.
75. Ledesma-Vázquez, J.; Johnson, M.E.; Gonzalez-Yajimovich, O.; Santamaría-del-Angel, E. Gulf of California geography, geological origins, oceanography, and sedimentation patterns. In *Atlas of Coastal Ecosystems in the Western Gulf of California*; Johnson, M.E., Ledesma-Vázquez, J., Eds.; University Arizona Press: Tucson, Arizona, 2009; pp. 1–10.
76. Humphreys, A.M.; Linder, H.P. Concept versus data in delimitation of plant genera. *Taxon* **2009**, *58*, 1054–1074. [CrossRef]
77. Sánchez, D.; Vázquez-Benítez, B.; Vázquez-Sánchez, M.; Aquino, D.; Arias, S. Phylogenetic relationships in *Coryphantha* and implications on *Pelecyphora* and *Escobaria* (Cacteae, Cactoideae, Cactaceae). *PhytoKeys* **2022**, *188*, 115–165. [CrossRef]
78. Korotkova, N.; Aquino, D.; Arias, S.; Eggli, U.; Franck, A.; Gómez-Hinostrosa, C.; Guerrero, P.C.; Hernández, H.M.; Kohlbecker, A.; Köhler, M.; et al. Cactaceae at Caryophyllales.Org- A Dynamic Online Species-Level Taxonomic Backbone for the Family. *Willdenowia* **2021**, *51*, 251–270. [CrossRef]

biology

MDPI

Article

Systematics of Ditaxinae and Related Lineages within the Subfamily Acalyphoideae (Euphorbiaceae) Based on Molecular Phylogenetics

Josimar Külkamp [1,2,*], Ricarda Riina [2,*], Yocupitzia Ramírez-Amezcua [3], João R. V. Iganci [4,5], Inês Cordeiro [6], Raquel González-Páramo [2], Sabina Irene Lara-Cabrera [3] and José Fernando A. Baumgratz [1]

1 Programa de Pós-graduação em Botânica, Escola Nacional de Botânica Tropical, Instituto de Pesquisas Jardim Botânico do Rio de Janeiro. Rua Pacheco Leão 2040, Rio de Janeiro 22460-030, RJ, Brazil
2 Real Jardín Botánico (RJB), CSIC, Plaza de Murillo 2, 28014 Madrid, Spain
3 Laboratorio de Sistemática Molecular, Administración Santa María, Facultad de Biología, Universidad Michoacana de San Nicolás de Hidalgo, Morelia Michoacán 58091, Mexico
4 Programa de Pós-Graduação em Botânica, Universidade Federal do Rio Grande do Sul, Av. Bento Gonçalves 9500, Porto Alegre 91501-970, RS, Brazil
5 Programa de Pós-Graduação em Fisiologia Vegetal, Departamento de Botânica, Campus Capão do Leão, Universidade Federal de Pelotas, Caixa Postal 354, Pelotas 96010-900, RS, Brazil
6 Núcleo de Conservação da Biodiversidade, Instituto de Pesquisas Ambientais, Av. Miguel Stefano 3687, São Paulo 04301-012, SP, Brazil
* Correspondence: josimarbotanica@gmail.com (J.K.); rriina@rjb.csic.es (R.R.)

Simple Summary: This study represents the most comprehensive phylogenetic reconstruction of the plant subtribe Ditaxinae and related taxa within Acalyphoideae (Euphorbiaceae). The taxonomy of this group, mainly based in morphology, has long been controversial. Here, we present a new taxonomic classification at the genus and tribe ranks using a solid phylogenetic framework. We also provide key morphological synapomorphies supporting the main recovered clades.

Abstract: The subtribe Ditaxinae in the plant family Euphorbiaceae is composed of five genera (*Argythamnia*, *Caperonia*, *Chiropetalum*, *Ditaxis* and *Philyra*) and approximately 120 species of perennial herbs (rarely annual) to treelets. The subtribe is distributed throughout the Americas, with the exception of *Caperonia*, which also occurs in tropical Africa and Madagascar. Under the current classification, Ditaxinae includes genera with a questionable morphology-based taxonomy, especially *Argythamnia*, *Chiropetalum* and *Ditaxis*. Moreover, phylogenetic relationships among genera are largely unexplored, with previous works sampling <10% of taxa, showing Ditaxinae as paraphyletic. In this study, we inferred the phylogenetic relationships within Ditaxinae and related taxa using a dataset of nuclear (ETS, ITS) and plastid (*pet*D, *trn*LF, *trn*TL) DNA sequences and a wide taxon sampling (60%). We confirmed the paraphyly of Ditaxinae and *Ditaxis*, both with high support. Following our phylogenetic results, we combined *Ditaxis* in *Argythamnia* and upgraded Ditaxinae to the tribe level (Ditaxeae). We also established and described the tribe Caperonieae based on *Caperonia*, and transferred *Philyra* to the tribe Adelieae, along with *Adelia*, *Garciadelia*, *Lasiocroton* and *Leucocroton*. Finally, we discuss the main morphological synapomorphies for the genera and tribes and provide a taxonomic treatment, including all species recognized under each genus.

Keywords: Adelieae; *Argythamnia*; *Caperonia*; Caperonieae; *Chiropetalum*; Ditaxeae; *Ditaxis*; phylogenetics; *Philyra*

Citation: Külkamp, J.; Riina, R.; Ramírez-Amezcua, Y.; Iganci, J.R.V.; Cordeiro, I.; González-Páramo, R.; Lara-Cabrera, S.I.; Baumgratz, J.F.A. Systematics of Ditaxinae and Related Lineages within the Subfamily Acalyphoideae (Euphorbiaceae) Based on Molecular Phylogenetics. *Biology* **2023**, *12*, 173. https://doi.org/10.3390/biology12020173

Academic Editor: Lorenzo Peruzzi

Received: 27 December 2022
Revised: 17 January 2023
Accepted: 18 January 2023
Published: 21 January 2023

1. Introduction

The systematics of Euphorbiaceae Juss. have undergone substantial changes in the last two decades stemming from studies in molecular systematics. The family is currently classified into four subfamilies (Acalyphoideae Beilschmied, Cheilosoideae K.Wurdack & Petra

Hoffm., Crotonoideae Beilschmied and Euphorbioideae) [1–5]. Phylogenetic studies have led to updates in the systematics of Euphorbiaceae, where two biovulate subfamilies were segregated and elevated to the family level, Phyllanthaceae Martinov, Picrodendraceae Small and Putranjivaceae Endl. [2,6]. Acalyphoideae was recognized as a subfamily in 1975 [7] and currently comprises 14 tribes, 23 subtribes, 99 genera and approximately 1860 species [5,8]. The subfamily is distributed worldwide, except in polar regions, with greater diversity in tropical and subtropical areas [5,9–11].

In the classification proposed by Webster [5], the tribe Chrozophoreae (Müll.Arg.) Pax & K.Hoffm was composed of the subtribes Ditaxinae Griseb., Speranskiinae G.L.Webster and Chrozophorinae (Müll.Arg.) Pax & K.Hoffm. Ditaxinae was proposed in 1859 [12], with the genera *Argythamnia* P.Browne (=*Chiropetalum* A.Juss.), *Caperonia* A.St.-Hil. and *Ditaxis* Vahl ex A.Juss. Later, Müller [13] presented a classification for the tribe Acalypheae that consisted of 11 subtribes, including Chrozophorinae Müll.Arg. (containing *Argythamnia, Chiropetalum, Ditaxis* and *Philyra* Klotzsch) and Caperoniinae Müll.Arg. (including only *Caperonia*). Müller differentiated these subtribes based on the staminate flowers having a rudimentary ovary present at the apex of the staminal column in *Caperonia* and absent in the other genera.

In 1912, a new classification system called "Chrozophorinarum" was put forward by Pax and Hoffmann, wherein Ditaxinae was treated as a synonym of Chrozophorinae-regularis, which had been circumscribed with the genera *Aonikena* Speg., *Argythamnia, Caperonia, Chiropetalum, Chrozophora* Neck. ex A.Juss., *Ditaxis* and *Philyra* [14]. Webster re-established Ditaxinae to include *Argythamnia, Caperonia, Chiropetalum* (= *Aonikena*), *Ditaxis* and *Philyra* [7].

In its current circumscription, Ditaxinae consists of five genera (*Argythamnia, Caperonia, Chiropetalum, Ditaxis* and *Philyra*) and around 120 species of herbs, subshrubs, shrubs, and small trees, widely distributed in the New World (all genera) [5,15–21] and continental Africa and Madagascar (only *Caperonia*) [5,17,21].

Caperonia has approximately 35 herbaceous and subshrub species, of which 29 are distributed in the New World and seven in Africa/Madagascar. In the New World, it occurs from Mexico to central Argentina, with one species introduced in the southern United States. *Caperonia* has its greatest diversity in tropical and subtropical regions, and is the only genus of the tribe occurring in the Amazonian region, exclusively in marshy environments [5,14,17,21]. *Argythamnia* is composed of 19 species of perennial herbs, subshrubs to shrubs distributed in Central America (Caribbean and Mexico), where it is restricted to seasonally dry tropical forests and coastal vegetation [5,22]. *Chiropetalum* consists of 21 species of herbs and subshrubs and is disjunct between Mexico (two species) and South America (19 species), where it occurs from Peru to Patagonia, with its highest diversity in northern Argentina and southern Brazil [15,20,23]. *Chiropetalum* occurs in a variety of habitats, including a range of dry and humid forests, arid environments, grasslands and coastal vegetation. *Ditaxis* is the most species-rich genus of the subtribe with approximately 45 species, ranging from herbs to shrubs, and is widely distributed from the southern United States to northern Patagonia, in Argentina [18,24–26]. *Ditaxis* occupies various habitats, such as deserts and grasslands, but most species occur in seasonally dry tropical forests [18,25,26]. Finally, the monotypic genus *Philyra* (*P. brasiliensis* Klotzsch) is a shrub or a small tree. The genus is restricted to central and eastern South America, growing exclusively in seasonally dry tropical forests [5,19].

Argythamnia, Chiropetalum and *Ditaxis* form a group of great morphological complexity that has undergone many taxonomic changes. However, few studies have approached the three genera all together to understand their phylogenetic relationships [13,14,22,23,27]. *Argythamnia, Chiropetalum* and *Ditaxis* have sometimes been treated as subgenera of *Argythamnia s.l.* [22–24,27]. Currently, these taxa are treated at the genus level, but there is still disagreement among taxonomists. Recent studies, using DNA sequence data, have attempted to resolve the relationship among *Argythamnia, Chiropetalum* and *Ditaxis* [15,28], but their phylogenetic analyses revealed topologies with low support in some clades, preventing any taxonomic changes or updates. Similarly, two other phylogenetic studies have included terminals of Ditaxinae, but these did not exceed 10% of taxon sampling and yielded low-resolution phylogenies [3,29].

Cervantes and collaborators reconstructed the biogeographic history of Acalyphoideae based on a molecular phylogenetic analysis using the *petD*, *trn*L-F and *mat*K/*trn*K genetic regions [30]. Ditaxinae, even though represented by ~10% of the species, emerged as paraphyletic. Their results recovered *Philyra* as a sister to the tribe Adelieae G.L.Webster, and this clade was a sister to the *Argythamnia* + *Chiropetalum* + *Ditaxis* clade, as shown by Jestrow [29,31]. *Caperonia* emerged as a sister to all above taxa, albeit with low support [30].

Given the need for a solid phylogenetic and systematic framework for the subtribe Ditaxinae, in this study we established the following aims: (1) test the monophyly of Ditaxinae and its currently recognized genera, *Argythamnia*, *Caperonia*, *Chiropetalum*, *Ditaxis* and *Philyra*, using a comprehensive taxonomic and geographical sampling, including multiple accessions per species when possible; (2) circumscribe the recovered clades and identify potential morphological synapomorphies; (3) establish a suprageneric classification in the subfamily Acalyphoideae based on the recovered phylogenetic pattern in this study.

2. Material and Methods

2.1. Taxon Sampling and Outgroup Selection

Our sampling covered all currently recognized genera of subtribe Ditaxinae: *Argythamnia* (11 spp., 61% of the total), *Caperonia* (10 spp., 30%), *Chiropetalum* (18 spp., 86%), *Ditaxis* (35 spp., 77%) and *Philyra* (1 sp., 100%). Thus, our dataset included a total of 75 species, representing 60% of Ditaxinae. We also included five representatives of tribe Adelieae, the latter based on Jestrow's circumscription [29]. We used *Acalypha lanceolata* Willd., *Enriquebeltrania crenatifolia* (Miranda) Rzed., *Bernardia dichotoma* (Willd.) Müll.Arg., *Plukenetia penninervia* Müll.Arg., *P. volubilis* L. and *Seidelia triandra* (E.Mey.) Pax as outgroups, based on previous phylogenetic analyses [15,28,30]. Overall, our study sampled 86 species of the subfamily Acalyphoideae (Supplementary File S1: Table S1). The choice of outgroups also aimed to reconstruct the clades close to Ditaxinae following the study of Cervantes and collaborators [30]. The type species of each genus in Ditaxinae was sampled in our dataset. For the taxonomic treatment, type specimens were also analyzed to infer morphological similarities, mainly of taxa not represented in the phylogenetic analyses, in order to assess the preliminary generic assignment based on morphological similarities with the taxa represented in the phylogeny, as has been done in other complex groups within Euphorbiaceae [32–34].

We included samples collected in Africa, the Caribbean region, Central America, North America and South America. Plant tissues were preserved in silica gel, and vouchers were deposited in the herbaria BAA, FLOR, HUEFS, ICN, MA, MEXU, RB, SP, SPF and US (acronyms follow Thiers, continuously updated) [35]. Other tissue samples were obtained from herbarium specimens at BA, BAA, CA, CORD, CPAP, F, HUEFS, IEB, K, LPD, MA, MEXU, MO, MOL, RB, RSA, SI, SP, US and XAL (Supplementary File S1: Table S1). We also used 61 sequences (representing 22 species) from the US National Center for Biotechnology Information (NCBI) GenBank repository (https://www.ncbi.nlm.nih.gov/genbank). Voucher information and GenBank accession data are provided in Supplementary File S1: Table S1.

2.2. DNA Extraction, Amplification and Sequencing

DNA was extracted from silica-dried leaf tissue and herbarium material using the CTAB method [36] with some modifications [36] (see Supplementary File S2). The extracted DNA was quantified using a Qubit™ dsDNA BR Standard (Invitrogen). Samples with high concentrations (>20 ng/µL) were diluted (1:20, 1:50) depending on the concentration.

Three plastid (*trn*L-F, *trn*T-L, *pet*D) and two nuclear (ITS, ETS) genetic regions were sequenced (see Supplementary File S2: Table S2). PCR amplifications were conducted with 25 µL reactions (for thermocycler temperature protocols, see Supplementary File S2:

Table S3). Each reaction tube included MyTaq Red Mix (Bioline), H_2O, primers and genomic DNA. For samples that were difficult to amplify, PuReTaq Ready-To-Go PCR Beads (GE Healthcare) were used. PCR products were purified with ExoSap PCR Purification and sent for sequencing at MACROGEN (Macrogen, Madrid, Spain), using the same amplification primers (Supplementary File S2: Table S3).

The pherograms were edited manually in UGENE [37] and automatically aligned with MUSCLE, using the default parameters. Manual adjustments were made to each alignment matrix in UGENE, employing the similarity criterion. A 120 bp region was excluded from the analysis of the *trn*T-L data matrix due to an uncertain homology assessment in the alignment.

2.3. Phylogenetic Analyses

Evolutionary models of nucleotide substitution were selected based on maximum likelihood (ML) using the Akaike (AIC) [38] information criterion implemented in jModelTest v.2.1.10 [39,40]. Each marker was analyzed individually, and the models were GTR + I + G for ETS and ITS, TVM + I + G for *trn*L-F, TPM1uf + G for *trn*T-L and GTR + G for *pet*D. MrBayes does not allow implementing all of these models, and thus, we used the nearest and slightly more complex model, which was GTR + I + G for the nuclear regions and GTR + G for the plastid markers [41]. Bayesian inference (BI) appears to be more robust with respect to over-parametrization and more sensitive to infra-parametrization than the ML optimization used in jModelTest [42]. Each genetic region was analyzed individually based on BI and ML. Concatenated matrices with nuclear (ITS + ETS) and plastid markers (*trn*L-F + *trn*T-L + *pet*D) were also analyzed separately to check for possible incongruences in the topology, and finally, a matrix with all markers was analyzed with BI and ML approaches. Topological incongruence between nuclear and plastid regions was defined as the presence of clades with a posterior probability (PP) $\geq$ 0.95 in IB and bootstrapping support (BS) $\geq$ 70% in ML [43]. In the combined analysis using only one terminal per species, we prioritized keeping the terminals with at least one nuclear and one plastid region. Bayesian analyses consisted of two independent Markov Chain Monte Carlo (MCMC) runs of 50 million generations in MrBayes v.3.1.2 [44], sampling every 1000th generation, with 20% (first 10 million trees) discarded as burn-in. Output files were summarized with TreeAnnotator v.1.6.1 [45], and the performance of each analysis (effective sample sizes, ESS > 200) was evaluated using Tracer v.1.6 [46]. Phylogenetic trees for individual and combined markers reconstructed with BI and ML are presented. Maximum-likelihood analyses were performed with RAxML [47] on the concatenated supermatrix, under a GTRGAMA model with 1000 bootstrap replicates. All analyses were hosted at CIPRES Science Gateway [48].

3. Results

The aligned DNA matrix combining the five regions (ETS, ITS, *pet*D, *trn*T-L, *trn*L-F) was 3985 bp long and included 86 species (75 of Ditaxinae *s.l.*) and 223 terminals (there were species represented by more than one specimen and unidentified/unnamed specimens labeled as "sp."). A summary of each data partition and combined matrices is provided in Table 1. The marker *pet*D proved informative for the group. However, it was the region with the lowest taxonomic representation, as only recent tissue samples dried in silica gel could be amplified (Table 1). The analyses of the individual markers showed few cases of topological incongruences between the plastid and nuclear genome. However, in most cases, these incongruences did not have high support, and thus, the matrices (nuclear plus plastid datasets) were combined for the final analysis. Figure 1 represents the phylogenetic tree reconstructed when combining the five markers and the inclusion of one terminal per species. The phylogenetic analyses using all terminals (including multiple accessions) and individual and combined datasets are presented in Supplementary File S2: Figures S1–S9. The ML analysis did not show significant differences in tree topology when compared to the BI (Supplementary File S2: Figure S9).

Table 1. Descriptive statistics of the separate and combined DNA datasets used in the phylogenetic analyses.

	ITS	ETS	*trn*L-F	*trn*T-L	*pet*D	cpDNA Combined	nDNA Combined	All Markers Combined
Number of terminals/species	170/86	90/61	174/85	65/53	41/30	178/80	177/81	223/85
Aligned sequence length	750	394	1053	593	1195	2841	1144	3985
Missing data (%)	8.3	4.5	4.5	3.3	16.3	49.8	24.8	54.5
Model	GTR+I+G	GTR+I+G	TVMIV+G	TPM+I+G	GTR+G	-	-	-

Despite minor incongruences between different reconstructions, the genera *Argythamnia, Caperonia, Chiropetalum, Philyra* and *Adelia* were confirmed as monophyletic, whereas *Ditaxis* was paraphyletic in all reconstructions (Figure 1 and Supplementary File S2: Figures S1–S8). In contrast, the phylogenetic trees obtained from the analyses of the combined and individual markers presented some incongruence regarding the positioning of *Philyra, Adelia* and *Caperonia*. In the analyses of the *trn*L-F and cpDNA-combined datasets, *Philyra* emerged as a sister (PP = 1) to the clade *Caperonia* + *Adelia* + *Chiropetalum* + *Argythamnia* + *Ditaxis* (Supplementary File S2: Figures S3 and S6), whereas in *trn*T-L, *Philyra* formed a polytomy with *Adelia* (Supplementary File S2: Figure S1). In *pet*D and ETS, *Adelia* + *Philyra* was a sister of *Caperonia* + *Chiropetalum* + *Argythamnia* + *Ditaxis* with maximum support (Supplementary File S2: Figures S2 and S5). In the reconstructions based on ITS, ITS + ETS and the matrix with all markers combined, *Philyra* + *Adelia* emerged as a sister to *Argythamnia* + *Chiropetalum* + *Ditaxis*, while *Caperonia* emerged as a sister to the clade formed by all the five genera above (Figure 1 and Supplementary File S2: Figures S4 and S7–S9). Based on ETS only, *Caperonia* emerged as a sister to *Argythamnia* + *Ditaxis*, while *Chiropetalum* was recovered as a sister to *Caperonia* + *Argythamnia* + *Ditaxis*, both with low support (Supplementary File S2: Figure S5). In all other analyses, *Chiropetalum* emerged as a sister to *Argythamnia* + *Ditaxis* with high support (Figure 1 and Supplementary File S2: Figures S1–S4 and S6–S9). In all reconstructions, *Ditaxis* species were grouped into two clades (*Ditaxis* 1 and *Ditaxis* 2) separated by *Argythamnia s.s.* (Figure 1), leaving *Ditaxis* paraphyletic in its current circumscription. In ETS and ITS + ETS reconstructions (Supplementary File S2: Figures S5 and S7), the Andean species *Ditaxis jablonszkyana* Pax & K.Hoffm. and *D. malpighipila* (Hicken) L.C.Wheeler emerged as sisters to all other *Ditaxis* + *Argythamnia* species (PP = 1), whereas in all plastid reconstructions, these two species were recovered as sisters to clade *Ditaxis* 2 (PP = 1, Figure 1; Supplementary File S2: Figures S2–S4, S6 and S8–S9).

Phylogenetic trees generated from nuclear and plastid datasets, based on both BI and ML, supported the paraphyly of Ditaxinae as currently circumscribed (Figure 1 and Supplementary File S2: Figures S1–S8) due to the position of the representatives of the Adelieae tribe between the terminals of Ditaxinae. The results also reinforce that the Chrozophoreae tribe is polyphyletic in the current circumscription (Figure 1).

All species of *Chiropetalum* formed a single clade, and the geographically disjunct Mexican species emerged together with South American species. The largest clade of Ditaxinae (*Argythamnia* + *Ditaxis*) was recovered as the sister of *Chiropetalum*. *Argythamnia* species, all from the central region of the Americas (Caribbean, Central America and southern Mexico), resulted as the sister clade of *Ditaxis* species (*Ditaxis* 1 clade) with North American distribution (Figure 1). The clade *Ditaxis* 2, the sister of *Ditaxis* 1 + *Argythamnia*, included North American and all Central and South American species. The five African species/specimens of *Caperonia* sampled in the phylogeny (identified with * in Figure 1) were placed in two different clades (Figure 1). Species of the tribe Adelieae, exclusive to Central and South America, emerged as the sister clade of the monospecific genus *Philyra* (Figure 1).

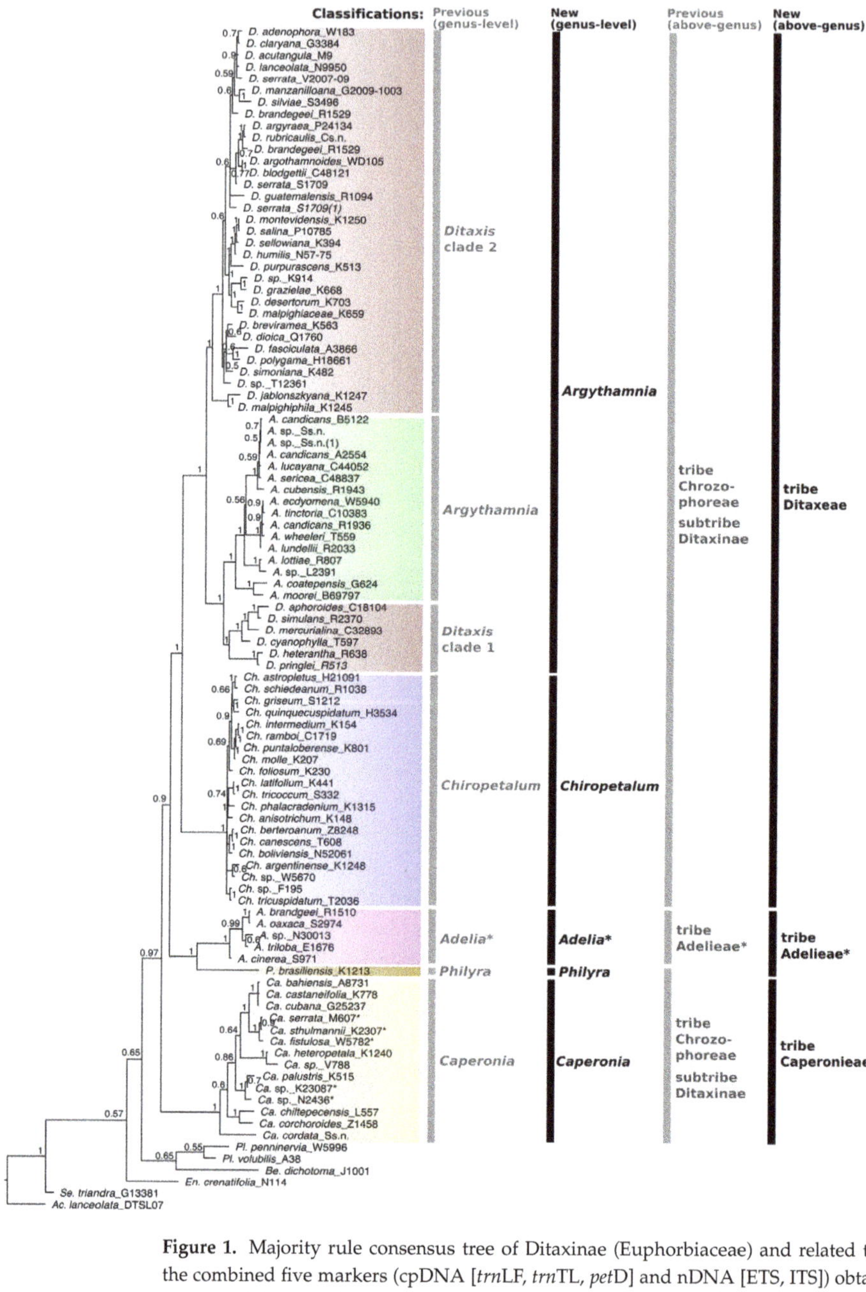

Figure 1. Majority rule consensus tree of Ditaxinae (Euphorbiaceae) and related taxa based on the combined five markers (cpDNA [*trn*LF, *trn*TL, *pet*D] and nDNA [ETS, ITS]) obtained through Bayesian inference. Bayesian posterior probabilities (PPs) are indicated on each branch. Vertical bars with labels on the right indicate the old (gray) [5] and the new (black) generic and suprageneric classifications. Asterisks in *Adelia* and Adelieae indicate that this clade is not fully represented here (several unsampled genera); we followed the classification proposed by Jestrow [29] (based on a complete generic sampling of Adelieae). To avoid confusion, some genera are abbreviated using two initial letters: Ac. = *Acalypha*, Be. = *Bernardia*, Ca. = *Caperonia*, Ch. = *Chiropetalum*, En. = *Enriquebeltrania*, Pl. = *Plukeneria*, Se. = *Seidelia*.

4. Discussion

This study presents the most comprehensive taxonomic and geographical sampling of Ditaxinae (ca. 60%) to date. In an attempt to solve the generic relationships among *Argythamnia*, *Chiropetalum* and *Ditaxis*, Ramírez-Amezcua [28] and Külkamp [15] sampled approximately 30% and 25% of the Ditaxinae species, respectively. Furthermore, the sampling of related groups (*Caperonia* and Adelieae) was less than 5%, precluding any suprageneric taxonomic decisions. As a result, our research provides a solid phylogenetic framework for new taxonomic delimitations at the genus and tribe levels.

4.1. Changes in Generic Delimitation

The relationship between *Argythamnia* and *Ditaxis* could not be resolved in previous studies, probably because of the relatively low (30%) taxon sampling and lack of phylogenetic support for some clades [15,28], while the genus *Chiropetalum*, albeit with low support, emerged as a separated clade in both studies. A recent phylogenetic reconstruction using a large representation of subfamily Acalyphoideae [30] also recovered a monophyletic *Chiropetalum*, in this case with maximum support. Here, in all reconstructions, *Chiropetalum* emerged as monophyletic and a sister to the clade containing *Ditaxis* and *Argythamnia*, with maximum support (PP = 1) (Figure 1). The high taxon sampling of *Chiropetalum* (90%) in our phylogenetic analyses gives us confidence in circumscribing the genus as a distinct taxon. However, further phylogenetic studies should sample *Chiropetalum patagonicum* (Speg.) O'Donell & Lourteig, since the species presents a remarkable divergent morphology (prostrate habit, absence of trichomes, petals of the staminate flower slightly lobed) from that of the rest of *Chiropetalum*. Ingram treated this species in the genus *Aonikena* Apeg. [23], whereas O'Donell & Lourteig classified it in *Chiropetalum* sect. *Aonikena* (Speg.) O'Donell & Lourt [49]. *Aonikena patagonica* would be well placed in *Chiropetalum* based on comparative morphology, but nevertheless, the inclusion of this species in further phylogenetic analyses is still required to definitively clarify its taxonomic placement. Based on our phylogenetic reconstruction and morphology studies, we identified several synapomorphies of *Chiropetalum*, including lobed petals in the staminate flowers (Figure 2C), stamens disposed in a whorl and fused at the base forming a column (Figure 2C) and the absence of petals in the pistillate flowers (Figure 2D), except for *C. tricuspidatum* (Lam.) A.Juss. and *C. argentinense* Skottsb., which have vestigial petals [23,26,49]. A few species of *Argythamnia s.s.* also have pistillate flowers without petals [22]. The presence of stellate trichomes is also a unique feature of *Chiropetalum*, but these trichomes are present in only 10 species (50%) [23,26].

Our results show that *Ditaxis* as currently recognized is paraphyletic, because the species of *Argythamnia s.s.* are nested within *Ditaxis* (Figure 1), a topology similar to the phylogenetic reconstruction in Ramirez-Amezcua [28]. The staminate flowers in *Argythamnia* have four (rarely five) petals and four (rarely five) free stamens, whereas in *Ditaxis* the staminate flowers present with five petals and 8–10 stamens united in a column. Thus, to avoid describing a new genus lacking morphological synapomorphies or a clear set of distinguishing characteristics, we expanded the circumscription of *Argythamnia s.s.* with the inclusion of the two clades of *Ditaxis* (clade 1 & 2; Figure 1) following, in part, Ingram's classification system [27]. Thus, *Argythamnia* in the circumscription proposed here is monophyletic and composed of three well-supported clades (Figure 1): (i) *Argythamnia s.s.*, (ii) *Ditaxis* clade 1, exclusive to North America, and (iii) *Ditaxis* clade 2, the most diverse clade of *Ditaxis s.s.*, with a distribution from North America to southern South America. In this new classification framework, *Argythamnia s.l.* is supported by the presence of petals in pistillate flowers (Figure 2A) (rarely absent) and entire petals (unlobed) in staminate flowers (Figure 2B). The presence of an apiculum on the seeds of *Argythamnia s.l.* should be studied further. Due to the lack of specimens with seeds for nine species of *Argythamnia s.l.*, that structure was little explored in this study. The seeds of the other genera of the tribes Ditaxeae, Adelieae and Caperonieae are globose rather than apiculate.

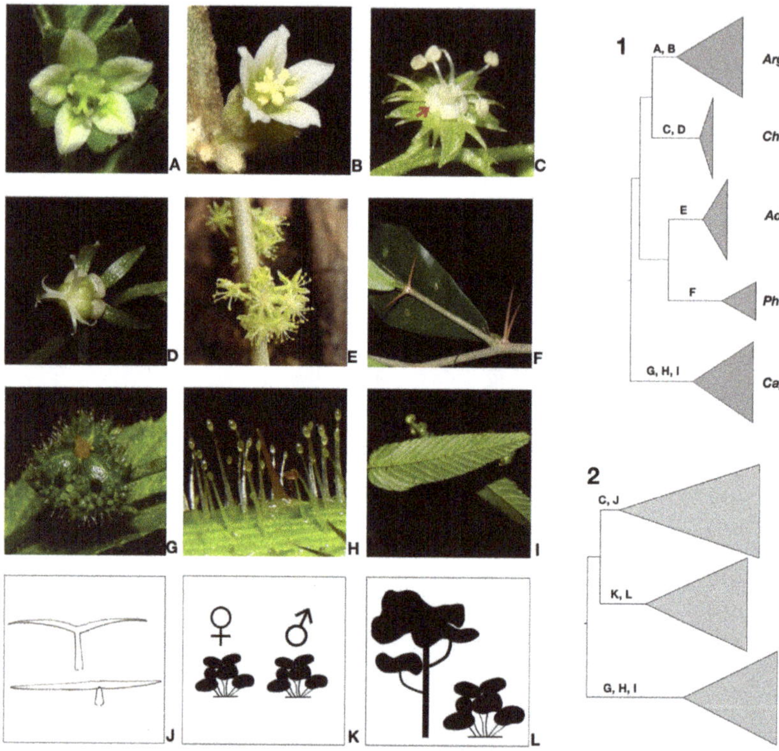

Figure 2. Schematic phylogenies of Ditaxinae and related taxa based on Bayesian inference using the combined five markers (cpDNA [*trn*LF, *trn*TL, *pet*D] and nDNA [ETS, ITS]). **1** & **2.** Generic (1) and suprageneric (2) classifications proposed here. Letters on branches indicate the morphological synapomorphies supporting each clade corresponding to the following illustrations. (**A**) Dichlamydeous pistillate flower of *Argythamnia desertorum*. (**B**) Staminate flower with entire petals of *Argythamnia desertorum*. (**C**) Staminate flower with lobed petals of *Chiropetalum phalacradenium* for cladogram 1 and floral nectaries for cladogram 2. (**D**) Monochlamydeous pistillate flower of *Chiropetalum phalacradenium*. (**E**) Monochlamydeous staminate flowers of *Adelia membranifolia*. (**F**) Pair of thorns below the leaves in *Philyra brasiliensis*. (**G**) Ovary with muricate surface in *Caperonia heteropetala*. (**H**) Glandular trichomes in *Caperonia heteropetala*. (**I**) Leaves with craspedodromous secondary veins in *Caperonia heteropetala*. (**J**) Malpighiaceous trichomes. (**K**) Dioecious sexual system. (**L**) Arboreal and shrubby habit.

Caperonia sensu Webster [5] is the only genus of Ditaxinae with an extra-New World distribution, with seven species occurring in tropical Africa and Madagascar. Here, we confirmed *Caperonia* as monophyletic, as suggested by Cervantes and collaborators [30], but with a broader taxon sampling. Pax & Hoffmann proposed two sections for *Caperonia*, *C.* sect. *Eucaperonia* ([nom. invalid.], autonym section = sect. *Caperonia*) and *C.* sect. *Aculeolatae* Pax & K.Hoffm. (taxa with prickles sampled in our phylogeny, *C. corchoroides* Müll.Arg., *C. cordata* A.St.-Hil., *C. heteropetala* Didr., *C. linearifolia* A.St.-Hil.) [14]. Based on morphology, we would have expected that these sections to be recovered in two clades in our phylogenetic analyses, but the presence of prickles appears to represent a plesiomorphic state, and some of the taxa studied have lost this state independently. However, we emphasize that *Caperonia* requires additional research with a larger taxonomic representation to clarify phylogenetic relationships, explore the need to establish an infrageneric classification and understand the origin and nature (multiple or single colonization events) of its amphi-Atlantic distribution pattern. When comparing *Caperonia* with phylogenetically closely related genera, its morphological divergence is marked by the presence of glandular trichomes (Figure 2H), a muricate ovary surface (Figure 2G) and parallel secondary veins (Figure 2I). These features are absent in all the other genera and are recognized here as synapomorphies for *Caperonia*. Another contrasting characteristic of *Caperonia* is its exclusive occurrence in marshy habitats [17,21], while all other related genera are found in desert or seasonally dry environments [15,16,19,22–25].

Philyra brasiliensis was originally the only species described in *Philyra*; however, the species was combined in *Ditaxis* by Baillon [50] and later transferred to *Argythamnia* by Müller [13]. Morphology does not support these classifications because *Philyra* lacks the synapomorphies recognized for *Argythamnia* + *Chiropetalum* (presence of floral nectaries and malpighiaceous trichomes). Moreover, *Philyra* is the only genus of the focal taxa having a pair of spines inserted on branches beneath the leaves (Figure 2F). Because of these unique characteristics, the species was treated again in *Philyra* [26]. The phylogenetic analyses of Jestrow and collaborators [31], Cervantes and collaborators [30] and our own results also support the circumscription of *Philyra* as a monospecific genus. The genus *Adelia* (sister to *Philyra*) includes some species with pointed branches, but it lacks the pair of spines below the leaves. *Adelia* is also distinguished from *Philyra* by its apetalous staminate flowers clustered in glomerules (Figure 2E), whereas in *Philyra*, the staminate flowers are dichlamydeous and grouped in racemes and the stamens (10–12) form a column with two whorls. Detailed phylogenetic information about *Adelia* can be found in previous studies focused on Adelieae that included a larger taxonomic representation [29,31,51,52].

4.2. Tribe Delimitation

Before our study, taxonomic affinities and phylogenetic relationships of subtribe Ditaxinae were uncertain mainly due to the poor taxon sampling in previous phylogenetic analyses [15,28,30,31,52]. Our results showed a robust topology (Figure 1), allowing us to propose a new classification. Ditaxinae has traditionally been assigned to the Chrozophoreae tribe [5,7,10]. Other phylogenetic analyses, however, revealed Chrozophoreae to be polyphyletic and Ditaxinae to be paraphyletic [2,30,31,51]. Here, we confirmed both results, with tribe Adelieae recovered as embedded among the terminals of Ditaxinae (Figure 1). *Argythamnia* (including *Ditaxis*) and *Chiropetalum* are part of Ditaxinae, which appear to be more closely related to each other than to *Caperonia* and *Philyra* (Figures 1 and 2).

Following our phylogenetic framework, we elevated Ditaxinae to the rank of tribe (Ditaxeae), including the genera *Argythamnia* (including *Ditaxis*) and *Chiropetalum* (Figures 1 and 2) and excluding *Caperonia* and *Philyra* (see the taxonomic treatment below). Tribe Ditaxeae is supported by two synapomorphies: the presence of floral nectaries (Figure 2C) and malpighiaceous trichomes (Figure 2J). Another important characteristic is the presence of a basal and suprabasal actinodromous venation pattern, which is very similar among taxa, but with small variations regarding the number of basal secondary veins (2–4) and the intensity of their impression on the leaf's surface. However, this character is not exclusive to Ditaxeae; some taxa in the tribe Adelieae also present a similar venation pattern. With the exclusion of subtribe Ditaxinae, the tribe Chrozophoreae is now circumscribed to include subtribes Speranskiinae and Chrozophorinae, which are exclusively paleotropical in their distribution.

We propose to circumscribe *Philyra* within tribe Adelieae (Figures 1 and 2), as suggested by Jestrow [31,51]. Traditionally, *Philyra* was circumscribed in Chrozophoreae and not in Adelieae, supported by the presence of petals in the pistillate and staminate flowers [5,10]. Now, tribe Adelieae comprises the genera *Adelia*, *Garciadelia* Jestrow & Jiménez Rodr., *Lasiocroton* Griseb., *Leucocroton* Griseb. and *Philyra*, which are united by two synapomorphies: the dioecious sexual system (rarely monoecious in *Leucocroton*) and the arborescent to shrubby habit (Figure 2K,L).

Systematists have always had difficulty placing *Caperonia*. Klotzsch [53] classified *Caperonia* in tribe Crotoneae Dumort., whereas Müller [13] placed it within tribe Acalypheae

Dumort., subtribe Caperoniinae Müll.Arg. Pax & Hoffmann [14], including the genus in subtribe Chrozophorinae, and Webster [7] classified *Caperonia* as part of tribe Chrozophoreae, subtribe Ditaxinae, where it remained until now. Here, we circumscribe *Caperonia* as a monogeneric tribe based on strong phylogenetic and morphological evidence. In the most recent phylogenetic reconstruction, based on plastid data only, *Caperonia* emerged as a sister to *Argythamnia* + *Chiropetalum* + *Ditaxis* + Adelieae [30]. Although we found that the position of *Caperonia* was incongruent (but with low support) among phylogenetic reconstructions based on individual plastid and nuclear markers (Supplementary File S2: Figures S1–S9), our combined analysis provides strong support for its position as a sister to Adelieae + Ditaxineae (as circumscribed here), justifying its treatment as a monogeneric tribe, Caperonieae.

The new tribe Caperonieae (see taxonomic treatment below) is supported by the presence of glandular trichomes (Figure 2H) and a muricate ovary surface (Figure 2G). We also highlight the presence of leaves with craspedodromous secondary veins (Figure 2I), heteromorphic petals in staminate flowers in most species and a thickened structure at the apex of the staminal column, identified by some authors as a rudimentary ovary (pistillode) [5]. However, ontogenetic studies are needed to understand the origin of this floral structure.

4.3. Taxonomic Treatment

The molecular phylogenetic results presented here support the establishment of a new classification for Ditaxinae, raising it from the subtribe to the tribe level (Ditaxeae), and including two well-supported clades composed of genera *Chiropetalum* and *Argythamnia*. We maintain tribe Adelieae, extending its circumscription to include the genus *Philyra*. We also elevate subtribe Caperoniinae to the tribe level, adding two new tribes to the subfamily Acalyphoideae. Furthermore, we expanded the circumscription of *Argythamnia* to include the two well-supported clades of *Ditaxis*, representatives that emerged as paraphyletic in our analyses. Future studies will be directed at refining this delimitation and possibly proposing infrageneric classification systems for *Argythamnia*, *Caperonia* and *Chiropetalum*. Here, we present the names and diagnosis of the tribes and genera recognized, as well as a list of all species recognized under each genus. The necessary infrageneric nomenclature combinations will be presented in future taxonomic studies. Species with phylogenetic data used in this study are marked with an asterisk (*) in the "species recognized" section of each genus below. In Supplementary File S3: Table S1, we present a summary of the new and previous classification of all taxa treated here.

1. CAPERONIEAE Külkamp & Riina, *stat. nov.*

Basionym: Caperoniinae Müll.Arg. (as 'Caperonieae'), Linnaea 34: 152. 1865.

Type, designated here: *Caperonia* A.St.-Hil.

Caperonia A.St.-Hil. Histoire des plantes les plus remarquables du Bresil et fu Paraguay 3/4: 244–247. 1825. *Ditaxis* sect. *Caperonia* (A.St.-Hil.) Baill. Adansonia, 4: 272. 1865.

Description: Monoecious, rarely dioecious; herbs, rarely subshrubs, annual or perennial; stems hollow; trichomes simple and glandular, sometimes prickly; stipules present; leaves alternate, petiolate or subsessile, penninerved, rarely palmatinerved, with craspedodromous secondary veins, margins serrate; inflorescences racemiform, bisexual or unisexual, bracteoles uniflorous, flowers dichlamydeous; staminate flowers with articulated pedicels; sepals 5, lanceolate, margin entire, pubescent or glabrous; petals 5, often unequal, glabrous, rarely pubescent, basally adnate to the staminal column; stamens 8–10 in two whorls, and pistillode on the column apex; floral nectaries absent; pistillate flowers proximal, dichlamydeous, sepals 5–6, equal or unequal, lanceolate to ovate, margin entire, pubescent, persistent in fruit; petals 5, usually equal, unequal or reduced; floral nectaries absent; ovary 3–locular, surface muricate, covered by glandular trichomes; style multifid; capsule verrucose, columella persistent; seeds one per locule, orbicular, foveolate, gray to black.

Distribution: *Caperonia* is distributed in the New World and Africa (continental Africa and Madagascar). The greatest diversity of *Caperonia* occurs in South America, mainly Brazil, with approximately 40% of the taxa (14). All *Caperonia* species occur in marshy environments [5,17,21].

Species recognized (35). **Africa/Madagascar** (7): *Caperonia fistulosa* Beille*, *C. latifolia* Pax, *C. palustris* (L.) A.St.-Hil.*, *C. rutenbergii* Müll.Arg., *C. serrata* (Turcz.) C.Presl.*, *C. stuhlmannii* Pax*, *C. subrotunda* Chiov. **America** (29): *Caperonia aculeolata* Müll.Arg., *C. altissima* Eskuche, *C. amarumayu* Külkamp & Cordeiro, *C. angustissima* Klotzsch, *C. bahiensis* Müll.Arg.*, *C. buettneriacea* Müll.Arg., *C. capiibariensis* Eskuche, *C. castaneifolia* (L.) A.St.-Hil.*, *C. castrobarrosiana* Paula & Hamburgo, *C. chiltepecensis* Croizat*, *C. corchoroides* Müll.Arg.*, *C. cordata* A.St.-Hil.*, *C. cubana* Pax & K.Hoffm.*, *C. gardneri* Müll.Arg., *C. glabrata* Pax & K.Hoffm., *C. heteropetala* Didr.*, *C. hystrix* Pax & K.Hoffm., *C. langsdorffii* Müll.Arg., *C. linearifolia* A.St.-Hil.*, *C. lutea* Pax & K.Hoffm., *C. maracaibensis* Külkamp & Cordeiro, *C. multicostata* Müll.Arg., *C. neglecta* G.L.Webster, *C. palustris* (L.) A.St.-Hil.*, *C. paraguayensis* Pax & K.Hoffm., *C. regnellii* Müll.Arg., *C. similis* Pax & K.Hoffm., *C. stenophylla* Müll.Arg. and *C. zaponzeta* Mansf.

2. DITAXEAE Külkamp & Riina, *stat. nov.*

Basionym: Ditaxinae Griseb., Fl. Brit. W. I., 43. 1859.

Type, designated here: *Ditaxis* Vahl ex A.Juss.

Description: Monoecious rarely dioecious; herbs, annual or perennial, and shrubs; branches erect, decumbent or prostrate; stipules present; leaves simple, alternate; venation actinodromous basal and suprabasal; margins serrate to entire; trichomes malpighiaceous, simple or stellate in both surfaces; racemes axillary, bisexual, rarely unisexual; pistillate flowers proximal and staminate distal; bracteoles uniflorous, lanceolate to ovate, pubescent, rarely glabrous; staminate flowers dichlamydeous; sepals (4–)5, linear to lanceolate, margin entire or serrated, pubescent or glabrous; petals (4–)5, entire, erose, laciniate or lobed, glabrous or pubescent, adnate to the staminal column; stamens 4–10, distinct or connate forming a column, stamens in one or two whorls; staminodes 0–5 at the top of the staminal column, pubescent or glabrous; floral nectaries 4–5, pubescent or glabrous; pistillate flowers dichlamydeous or monochlamydeous; sepals (4–)5(–6), linear, lanceolate, ovate or elliptic, pubescent rarely glabrous; petals 0–5, linear, lanceolate, oval, elliptic or rhomboid, pubescent or glabrous, margins entire, erose or laciniate; floral nectaries 5, adnate to the receptacle at the base of the ovary, glabrous, ciliate or pubescent; ovary pubescent, rarely glabrous; styles bifid or trifid, pubescent or glabrous; capsule 3–locular, smooth, pubescent rarely glabrous; seeds one per locule, orbicular to ovoid, apiculate or not, surface foveolate, smooth, undulate or reticulate, gray to black.

Distribution: Species of Ditaxeae are distributed throughout the New World, from the southern United States to Patagonia in the south of Argentina. There are two main centers of diversity for Ditaxeae: the first comprising southern North America, the Caribbean Islands and northern and central South America, and the second in northeastern Brazil [15,16,18,20,22–25,28].

Argythamnia P.Browne, Civ. Nat. Hist. Jamaica: 338. 1756.—Type: *Argythamnia candicans* Sw. = *Ditaxis* Vahl ex A.Juss., Euphorb. Gen. 27. 1824.—Type: *Ditaxis fasciculata* Vahl ex A.Juss.

Description: Monoecious, rarely dioecious herbs to shrubs, annual or perennial; trichomes malpighiaceous and simple; racemes bisexual, rarely unisexual; staminate flowers 2–15, dichlamydeous, sepals (4)5; petals (4)5, glabrous or pubescent, entire, erose or laciniate; stamens 4–10, distinct or connate, when connate arranged in two whorls; staminodes 0–5 at the top of the staminal column, pubescent or glabrous; floral nectaries 4 or 5, glabrous; pistillate flowers 1–4; dichlamydeous or monochlamydeous, sepals (4–)5(–6); petals 5, rarely 0, entire, erose or laciniate; floral nectaries 5, glabrous or ciliate; styles bifid or trifid; seeds orbicular to ovoid, apiculate, surface smooth, undulate or reticulate.

Distribution: *Argythamnia* is distributed throughout the New World, from southern United States to Patagonia. Greater diversity is found in southern North America, the

Caribbean Islands, northern and central South America and northeastern Brazil [16,18,22,24–26,28].

New combination

Argythamnia grazielae (Külkamp) Külkamp & Riina **comb. nov.**

≡ *Ditaxis grazielae* Külkamp, Phytotaxa 455(1): 154. 2020.

Type: BRAZIL. Bahia: Wanderley, 25 January 1996 (fl. fr.), *B.R. Chagas s.n.* (holotype: RB [RB00084882]!; isotypes: CEPEC [CEPEC131190]!, K [K001206888]!, MG!, NY [NY01183998]!, SPF [SPF196837]!).

Species recognized (68): *Argythamnia acaulis* (Herter ex Arechav.) J.W.Ingram, *A. acutangula* Croizat*, *A. adenophora* A.Gray*, *A. aphoroides* Müll.Arg.*, *A. argentea* Millsp., *A. argothamnoides* (Bertero ex Spreng.) J.W.Ingram*, *A. argyraea* Cory*, *A. arlynniana* J.W.Ingram, *A. brandegeei* Millsp.*, *A. breviramea* Müll.Arg.*, *A. calycina* Müll.Arg., *A. candicans* Swartz*, *A. claryana* Jeps.*, *A. coatepensis* (T.S.Brandegee) Croizat*, *A. cubensis* Britton & Wilson*, *A. cuneifolia* (Pax & K.Hoffm.) J.W.Ingram, *A. cyanophylla* (Wooton & Standl.) J.W.Ingram*, *A. depressa* (Greenm.) J.W.Ingram, *A. desertorum* Müll.Arg.*, *A. dioica* (Bonpland, Humboldt & Kunth) Müll.Arg.*, *A. ecdyomena* J.Ingram*, *A. erubescens* J.R.Johnst., *A. fasciculata* (Vahl ex A.Juss.) Müll.Arg.*, *A. fendleri* Müll.Arg., *A. grazielae* (Külkamp) Külkamp & Riina*, *A. guatemalensis* Müll.Arg.*, *A. haitiensis* (Urb.) J.W.Ingram, *A. haplostigma* Pax & K.Hoffm., *A. heterantha* (Zucc.) Müll.Arg.*, *A. heteropilosa* J.W.Ingram, *A. humilis* (Engelm. & A.Gray) Müll.Arg.*, *A. illimaniensis* (Baill.) Müll.Arg., *A. ingramii* Y.Ramírez-Amezcua & V.W.Steinm., *A. jablonszkyana* (Pax & K.Hoffm.) J.W.Ingram*, *A. katharinae* (Pax) Croizat, *A. lanceolata* (Benth.) Müll.Arg.*, *A. lottiae* J.W.Ingram*, *A. lucayana* Millsp.*, *A. lundellii* J.W.Ingram*, *A. macrantha* (Pax & K.Hoffm.) Croizat, *A. macrobotrys* (Pax & K.Hoffm.) J.W.Ingram, *A. malpighiacea* Ule*, *A. malpighiphila* (Hicken) J.W.Ingram*, *A. manzanilloana* Rose*, *A. mercurialina* (Nutt.) Müll.Arg.*, *A. microphylla* Pax, *A. montevidensis* (Didr.) Müll.Arg.*, *A. moorei* J.W.Ingram*, *A. oblongifolia* Urb., *A. pilosissima* (Benth.) Müll.Arg., *A. polygama* (Jacq.) Kuntze*, *A. pringlei* Greenm.*, *A. proctorii* J.W.Ingram, *A. purpurascens* S.Moore*, *A. rubricaulis* (Pax & K.Hoffm.) Croizat*, *A. salina* (Pax & K.Hoffm.) J.W.Ingram*, *A. sellowiana* (Pax & K.Hoffm.) J.W.Ingram*, *A. sericea* Griseb.*, *A. serrata* (Torr.) Müll.Arg.*, *A. silviae* Y.Ramírez-Amezcua & V.W.Steinm.*, *A. simoniana* (Casar.) Müll.Arg.*, *A. simulans* J.W.Ingram*, *A. sitiens* (T.S.Brandegee) J.W.Ingram, *A. stahlii* Urb., *A. tinctoria* Mill.* and *A. wheeleri* J.W.Ingram*.

Chiropetalum A.Juss., Ann. Sci. Nat. (Paris) 25: 21. 1832. —Type: *Chiropetalum lanceolatum* (Cav.) A.Juss.

Description: Monoecious herbs or subshrubs, perennial rarely annual; trichomes malpighiaceous, simple, stellate or rarely absent (*C. patagonicum*); racemes bisexual, rarely unisexual; staminate flowers 3–35, dichlamydeous, sepals 5; petals 5, glabrous, 3–7-lobed; floral nectaries 5, glabrous or pubescent; stamens 3–6, partially connate forming a column, anthers arranged in one whorl, staminodes absent; pistillate flowers 1–5, monochlamydeous, rarely dichlamydeous, sepals 5; petals usually absent, rarely 5; floral nectaries 5, glabrous or pubescent; styles bifid; capsule covered by simple, stellate and/or malpighiaceous trichomes; seeds orbicular, surface foveolate or rough.

Distribution: *Chiropetalum* is distributed in South America (19 species) and Mexico (2 species). Species richness is concentrated in the central region of South America, and the species presenting with the southernmost distribution is *C. patagonicum*, occurring in the Patagonia region of Argentina. Morphological and geographic details of each species can be found in studies of Ingram and Külkamp [15,20,23,26].

Species recognized (21): *Chiropetalum anisotrichum* (Müll.Arg.) Pax & K.Hoffm.*, *C. argentinense* Skottsb.*, *C. astroplethos* (J.W.Ingram) Radcl.-Sm. & Govaerts*, *C. berteroanum* Schltdl.*, *C. boliviense* (Müll.Arg.) Pax & K.Hoffm.*, *C. canescens* Phil.*, *C. cremnophilum* I.M.Johnst., *C. foliosum* Pax & K.Hoffm.*, *C. griseum* Griseb.*, *C. intermedium* Pax & K.Hoffm.*, *C. molle* (Klotzsch ex. Baill.) Pax & K.Hoffm.*, *C. patagonicum* (Speg.) O'Donell & Lourteig, *C. pavonianum* (Müll.Arg.) Pax, *C. phalacradenium* (J.W.Ingram) L.B.Sm. & Downs*, *C. puntaloberense* Alonso Paz & Bassagoda*, *C. quinquecuspidatum* (A.Juss.) Pax & K.Hoffm.*,

C. ramboi (Allem & Irgang) Radcl.-Sm. & Govaerts*, *C. ruizianum* (Müll.Arg.) Pax & K.Hoffm., *C. schiedeanum* (Müll.Arg.) Pax*, *C. tricoccum* (Vell.) Chodat & Hassl.* and *C. tricuspidatum* (Lam.) A.Juss*.

3. ADELIEAE G.L.Webster Taxon 24: 597. 1975

Type: *Adelia* L.

Description: Dioecious, rarely monoecious; trees to shrubs; pair of stipular spines absent, rarely present (*Philyra*); leaves alternate, simple; penninerved or actinodromous, basal and suprabasal, margins dentate to entire; trichomes simple or stellate; inflorescences axillary in racemes, glomerules or subpanicles, unisexual, rarely bisexual; staminate flowers monochlamydeous or dichlamydeous; sepals 4–5; petals absent, rarely 5; entire, pubescent; stamens 8–18(–30), filaments distinct or connate at base; staminode present or absent, floral nectaries absent; pistillate flowers monochlamydeous or dichlamydeous, sepals 5–6 lanceolate, ovate or elliptic, pubescent, rarely glabrous, persistent in fruit; petals 0–5, lanceolate, oval, elliptic or rhomboid, pubescent or glabrous; floral nectaries absent; ovary 3-locular, pubescent; ovules with inner integuments thick, outer thin to thick; styles bifid to multifid, pubescent or glabrous; capsule 3–locular; columella persistent; seeds one per locule, orbicular, surface foveolate, rough or smooth.

Distribution: Adelieae taxa are found from Mexico to Argentina, with three of the five genera endemic to the West Indies (*Garciadelia*, *Lasiocroton* and *Leucocroton*) [19,26,29,31,51,52,54,55].

Philyra Klotzsch, Archiv für Naturgeschichte 7(1): 199. 1841.—Type: *Philyra brasiliensis* Klotzsch

Description: Dioecious shrubs or treelets; paired stipular spines; leaves glabrous, venation pinnate, margin entire; bracts paleaceous, pubescent to glabrescent; staminate flowers dichlamydeous; sepals 5, petals 5; stamens 10–12, connate in a column with 2 whorls; staminodes 2 at the top of column, pubescent; pistillate flowers with pedicels larger than 12 mm long, dichlamydeous, petals 5, larger than sepals, styles multifid; capsule glabrous, columella persistent; seeds orbicular, surface smooth, gray to blackish.

Distribution: *Philyra* is distributed in northern Argentina, central and southern Paraguay and Brazil. In Brazil, this species occurs in the central–western region and the Atlantic coast, in the southeast and northeast of the country. For additional information, see Külkamp's studies [19,26].

Species recognized (1): *Philyra brasiliensis* Klotzsch*.

Adelia L., Syst. Nat. ed. 10, 2: 1298. 1759 (nom. cons.).—Type: *Adelia ricinella* L. (*typ. cons.*)

Distribution: *Adelia* has a continuous distribution from the southern United States to central South America. The greatest diversity of species is found in Mexico and Central America. Although a few species are widespread (e.g., *Adelia membranifolia* (Müll.Arg.) Chodat & Hassl.), most of them have a narrow distribution, and some are only known to be from a limited number of localities (e.g., *Adelia cinerea* (Wiggins & Rollins) A.Cerv., V.W.Steinm. & Flores-Olivera).

For a diagnosis, see De-Nova & Sosa and Jestrow's studies [31,51,55].

Species recognized (9): *Adelia barbinervis* Cham. & Schltdl., *A. brandegeei* V.W. Steinm.*, *A. cinerea* (Wiggins & Rollins) A.Cerv., V.W.Steinm. & Flores-Olvera*, *A. membranifolia* (Müll.Arg.) Chodat & Hassl., *A. oaxacana* (Müll.Arg.) Hemsl.*, *A. obovata* Wiggins & Rollins, *A. ricinella* L., *A. triloba* (Müll.Arg.) Hemsl.* and *A. vaseyi* (J.M.Coult.) Pax & K.Hoffm.

Garciadelia Jestrow & Jiménez Rodr., Taxon 59(6): 1809–1810. 2010.—Type: *Croton leprosus* Willd.

Distribution: *Garciadelia* has four species endemic to Hispaniola.

For a diagnosis, see Jestrow's studies [29,31].

Species included (4): *Garciadelia abbottii* Jestrow & Jiménez Rodr., *G. castilloae* Jestrow & Jiménez Rodr., *G. leprosa* (Willd.) Jestrow & Jiménez Rodr. and *G. mejiae* Jestrow & Jiménez Rodr.

Lasiocroton Griseb., Fl. Brit. W. I. 46. 1864.—Type: *Croton macrophyllus* Sw.

Distribution: *Lasiocroton* occurs in Cuba, Hispaniola, Jamaica and the Bahamas.

For a diagnosis, see Jestrow's studies [29,31].

Species recognized (7): *Lasiocroton bahamensis* Pax & K.Hoffm., *L. fawcettii* Urb., *L. gracilis* Britton & P.Wilson, *L. gutierrezii* Jestrow, *L. harrisii* Britton, *L. macrophyllus* (Sw.) Griseb. and *L. microphyllus* (A.Rich.) Jestrow.

Leucocroton Griseb. Abh. Königl. Ges. Wiss. Göttingen, 9: 20. 1861.—Type: *Leucocroton wrightii* Griseb.

Distribution: *Leucocroton* is restricted to serpentine soils of Cuba.

For a diagnosis, see studies by Borhidi and Jestrow [29,31,54].

Species recognized (26): *Leucocroton acunae* Borhidi, *L. anomalus* Borhidi, *L. bracteosus* Urb., *L. brittonii* Alain, *L. comosus* Urb., *L. cordifolius* (Britton & P.Wilson) Alain, *L. discolor* Urb., *L. ekmanii* Urb., *L. flavicans* Müll.Arg., *L. havanensis* Borhidi, *L. incrustatus* Borhidi, *L. linearifolius* Britton, *L. longibracteatus* Borhidi, *L. moaensis* Borhidi & O.Muñiz, *L. moncadae* Borhidi, *L. obovatus* Urb., *L. pachyphylloides* Borhidi, *L. pachyphyllus* Urb., *L. pallidus* Britton & P.Wilson, *L. revolutus* C.Wright, *L. sameki* Borhidi, *L. saxicola* Britton, *L. stenophyllus* Urb., *L. subpeltatus* (Urb.) Alain, *L. virens* Griseb. and *L. wrightii* Griseb.

Supplementary Materials: The following supporting information can be downloaded at: https://www.mdpi.com/article/10.3390/biology12020173/s1. Supplementary File S1: Supplementary table with taxa, voucher information and accessions used in this study. Supplementary File S2: Supplementary tables and figures including sequences of primers, protocols for PCR amplification and DNA extraction. Figures S1–S9 containing phylogenetic reconstructions of individual markers using Maximum Likelihood and Bayesian methods. Supplementary File S3: Supplementary table with summary of the new classification of Ditaxinae proposed here and the previous classification by Webster [5].

Author Contributions: J.K.: Conceptualization, formal analysis, investigation, discussion, writing (original draft). R.R.: Data resources, formal analysis, investigation, discussion, writing (original draft). Y.R.-A.: Data resources, investigation, discussion, writing (review and editing). J.R.V.I.: Investigation, discussion, writing (review and editing). I.C.: Investigation, discussion, writing (review and editing). R.G.-P.: Data resources, formal analysis. S.I.L.-C.: Data resources, writing (review and editing). J.F.A.B.: Investigation, discussion, writing (review and editing). All authors have read and agreed to the published version of the manuscript.

Funding: This research was funded in part by: Coordenação de Aperfeiçoamento de Pessoal de Nível Superior—Brasil (CAPES) Finance Code 001; Conselho Nacional de Desenvolvimento Científico e Tecnológico (CNPq) for financial support to I.C. (process 309917/2015-8), J.K. (process 141707/2020-8; 402157/2022-2), J.I. (process 311847/2021-8; 402157/2022-2); J.K. received support from SYNTHESYS+ (Synthesis of Systematic resources-process ES-TAF-8188); R.R. was supported by project grant PID2019-108109GB-I00 from MCIN/AEI/10.13039/501100011033/ and FEDER "A way to make Europe".

Institutional Review Board Statement: Not applicable.

Informed Consent Statement: Not applicable.

Data Availability Statement: The DNA sequence datasets generated for this study can be found in the NCBI GenBank website. Data on vouchers used in the phylogenetic analyses are included as Supplementary Material.

Acknowledgments: We thank the European SYNTHESYS+ for supporting this study and Real Jardín Botánico, CSIC (Madrid, Spain) for providing access to the herbarium (MA) and the laboratory of molecular systematics. R.G.P. thanks RJB-CSIC for allowing her to conduct an undergraduate student internship under the supervision of R.R. in 2021. We are thankful to all people who contributed with field sample collection, curators and technicians of the physical and virtual herbaria consulted (A, ACOR, B, BA, BAA, BC, BM, BKL, BR, C, CAS, CEPEQ, CGMS, CM, COL, COR, CORD, CPAP, CTES, E, ECT, F, FLOR, G, GB, GH, GOET, HAL, HAS, HBG, HBR, HUEFS, ICN, IEB, K, LIL, LP, M, MA, MBM, MEXU, MO, MOL, MPU, MVFA, MVJB, MVM, NY, P, PACA, PEL, R, RB, RSA, S, SGO, SI, SP, SPF, TUB, UC, UPCB, US and XAL). The authors also thank Victor Steinmann for his valuable contributions to earlier version of this manuscript.

Conflicts of Interest: The authors declare no conflict of interest.

References

1. Chase, M.W.; Zmarzty, S.; Lledo, M.D.; Wurdack, K.J.; Swensen, S.M.; Fay, M.F. When in doubt, put it in Flacourtiaceae: A molecular phylogenetic analysis based on plastid rbcL DNA sequences. *Kew Bull.* **2002**, *57*, 141–181. [CrossRef]
2. Wurdack, K.J.; Hoffman, P.; Chase, M.W. Molecular phylogenetic analysis of uniovulate Euphorbiaceae (Euphorbiaceae sensu stricto) using plastid rbcl and *trn*L-F DNA sequences. *Am. J. Bot.* **2005**, *92*, 1397–1420. [CrossRef]
3. Tokuoka, T. Molecular phylogenetic analysis of Euphorbiaceae *sensu stricto* based on plastid and nuclear DNA sequences and ovule and seed character evolution. *J. Plant Res.* **2007**, *120*, 511–522. [CrossRef]
4. Wurdack, K.J.; Davis, C.C. Malpighiales phylogenetics: Gaining ground on one of the most recalcitrant clades in the angiosperm tree of life. *Am. J. Bot.* **2009**, *96*, 1551–1570. [CrossRef]
5. Webster, G.L. Euphorbiaceae. In *The Families and Genera of Vascular Plants. Flowering Plants. Eudicots. Malpighiales*; Kubitzki, K., Ed.; Springer: Berlin/Heidelberg, Germany, 2014; Volume 11, pp. 51–216. ISBN 978-3-642-39417-1.
6. Wurdack, K.J.; Hoffmann, P.; Samuel, R.; Bruijn, A.; Bank, M.; Chase, M.W. Molecular phylogenetic analysis of Phyllanthaceae (Phyllanthoideae pro parte, Euphorbiaceae sensu lato) using plastid rbcL DNA sequences. *Am. J. Bot.* **2004**, *91*, 1882–1900. [CrossRef]
7. Webster, G.L. Conspectus of a new classification of the Euphorbiaceae. *Taxon* **1975**, *24*, 593–601. [CrossRef]
8. The Angiosperm Phylogeny Group; Chase, M.W.; Christenhusz, M.J.M.; Fay, M.F.; Byng, J.W.; Judd, W.S.; Soltis, D.E.; Mabberley, D.J.; Sennikov, A.N.; Soltis, P.S.; et al. An update of the Angiosperm Phylogeny Group classification for the orders and families of flowering plants: APG IV. *Bot. J. Linn. Soc.* **2016**, *181*, 12385. [CrossRef]
9. Govaerts, R.; Frodin, D.G.; Radcliffe-Smith, A. *World Checklist and Bibliography of Euphorbiaceae (and Pandaceae)*; Royal Botanical Gardens, Kew: London, UK, 2000; Volume 2, p. 1661, ISBN-10: 1900347849.
10. Radcliffe-Smith, A. *Genera Euphorbiacearum*; Royal Botanic Garden, Kew: London, UK, 2001; p. 464, ISBN-10: 1842460226.
11. Pscheidt, A.C.; Cordeiro, I. Synopsis of the tribe Hippomaneae (Euphorbiaceae) in São Paulo State. *Hoehnea* **2012**, *39*, 347–368. [CrossRef]
12. Grisebach, A.H.R. *Flora of the Britrish West Indian Island*; Rovell Reeve Co.: London, UK, 1859; Volume 2, pp. 193–315.
13. Müller, A. Euphorbiaceae: Vorläufige Mitteilungen aus dem für DeCandolle Prodromus bestimmten Manuscript über diese Familie. *Linnaea* **1863**, *34*, 1–225.
14. Pax, F.; Hoffmann, K. Euphorbiaceae—Acalypheae—Chrozophorinae. In *Das Pflanzenreich*, 4th ed.; Engler, A., Ed.; Engelmann: Leipzig, Germany, 1912; Volume 147, pp. 1–142.
15. Külkamp, J. *Chiropetalum* A. Juss. (Euphorbiaceae Juss.) no Brasil. Master's Thesis, Universidade Federal do Rio Grande do Sul, Porto Alegre, Brazil, 2018; p. 87.
16. Külkamp, J.; Iganci, J.R.V.; Cordeiro, I.; Baumgratz, J.F.A. *Ditaxis* (Euphorbiaceae) from the Brazilian Caatinga, including a new species. *Phytotaxa* **2020**, *455*, 152–160. [CrossRef]
17. Külkamp, J.; Caperonia in Flora e Funga do Brasil. Jardim Botânico do Rio de Janeiro. 2021. Available online: http://floradobrasil.jbrj.gov.br/reflora/floradobrasil/FB17471 (accessed on 5 August 2021).
18. Külkamp, J.; Ditaxis in Flora e Funga do Brasil. Jardim Botânico do Rio de Janeiro. 2021. Available online: http://floradobrasil.jbrj.gov.br/reflora/floradobrasil/FB25551 (accessed on 5 August 2021).
19. Külkamp, J.; Philyra in Flora e Funga do Brasil. Jardim Botânico do Rio de Janeiro. 2021. Available online: http://floradobrasil.jbrj.gov.br/reflora/floradobrasil/FB25551 (accessed on 5 August 2021).
20. Külkamp, J.; Iganci, J.R.V.; Cordeiro, I.; Chiropetalum in Flora e Funga do Brasil. Jardim Botânico do Rio de Janeiro 2021. Available online: http://floradobrasil.jbrj.gov.br/reflora/floradobrasil/FB29160 (accessed on 4 August 2021).
21. Külkamp, J.; Iganci, J.R.V.; Cordeiro, I.; Baumgratz, J.F. New species and occurrences of *Caperonia* (Euphorbiaceae) for South America. *Phytotaxa* **2021**, *529*, 86–92. [CrossRef]
22. Ingram, J.W. A revisional study of *Argythamnia* subgenus *Argythamnia* (Euphorbiaceae). *Gentes Herbarum* **1967**, *10*, 1–38.
23. Ingram, J.W. A revision of *Argythamnia* subgenus *Chiropetalum* (Euphorbiaceae). *Gentes Herbarum* **1980**, *11*, 437–468.
24. Ingram, J.W. A sectional revision of *Argythamnia* subgenus *Ditaxis* (Euphorbiaceae). Ph.D. Thesis, University of California, Berkeley, CA, USA, 1956; p. 115.
25. Ramírez-Amezcua, Y.; Steinmann, V.W. Revisión taxonómica de *Argythamnia* subgénero *Ditaxis* (Euphorbiaceae) en México. *Bot. Sci.* **2013**, *91*, 427–459. [CrossRef]
26. Külkamp, J. Sistemática da Subtribo Ditaxinae (Euphorbiaceae). Ph.D. Thesis, Escola Nacional de Botânica Tropical, Jardim Botânico do Rio de Janeiro, Rio de Janeiro, Brazil, 2022; p. 245.
27. Ingram, J.W. The generic limits of *Argythamnia* (Euphorbiaceae) defined. *Gentes Herbarum* **1980**, *11*, 427–436.
28. Ramírez-Amezcua, Y. Relaciones Filogenéticas en *Argythamnia* (Euphorbiaceae) Sensu Lato. Master's Thesis, Universidad Michoacana de San Nicolás de Hidalgo, Michoacán, Mexico, 2011; p. 64.
29. Jestrow, B.; Gutièrez, J.A.; Francisco-Ortega, J. Islands within islands: A molecular phylogenetic study of the *Leucocroton* alliance (Euphorbiaceae) across the Carribean Islands and within the serpentinite archipelago of Cuba. *J. Biogeogr.* **2012**, *39*, 452–464. [CrossRef]

30. Cervantes, A.; Fuentes, S.; Gutiérrez, J.; Magallón, S.; Borsch, T. Successive arrivals since the Miocene shaped the diversity of the Caribbean Acalyphoideae (Euphorbiaceae). *J. Biogeogr.* **2016**, *43*, 1773–1785. [CrossRef]
31. Jestrow, B.; Rodríguez, F.J.; Francisco-Ortega, J. Generic delimitation in the Antillean Adelieae (Euphorbiaceae) with description of the Hispaniolan endemic genus *Garciadelia*. *Taxon* **2010**, *59*, 1801–1814. [CrossRef]
32. Yang. Y.; Riina. R.; Morawetz, J.J.; Haevermans, T.; Aubriot, X.; Berry, P.E. Molecular phylogenetics and classification of *Euphorbia* subgenus *Chamaesyce* (Euphorbiaceae). *Taxon* **2012**, *61*, 764–789. [CrossRef]
33. Riina, R.; Peirson, J.A.; Geltman, D.V.; Molero, J.; Frajman, B.; Pahlevani, A.; Barres, L.; Morawetz, J.J.; Salmaki, Y.; Zarre, S.; et al. A worldwide molecular phylogeny and classification of the leafy spurges, *Euphorbia* subgenus *Esula* (Euphorbiaceae). *Taxon* **2013**, *62*, 316–342. [CrossRef]
34. Van Ee, B.W.; Riina, R.; Berry, P.E. A revised infrageneric classification and molecular phylogeny of New World *Croton* (Euphorbiaceae). *Taxon* **2011**, *60*, 791–823. [CrossRef]
35. Thiers, B. *Index Herbariorum: A Global Directory of Public Herbaria and Associated Staff*; New York Botanical Garden's Virtual Herbarium: New York, NY, USA, 2022; Available online: http://sweetgum.nybg.org/science/ih/ (accessed on 20 February 2022).
36. Doyle, J. DNA Protocols for Plants. In *Molecular Techniques in Taxonomy*; Hewitt, G.M., Johnston, A.W.B., Young, J.P.W., Eds.; Springer: Berlin/Heidelberg, Germany, 1991; pp. 283–293. ISBN 978-3-642-83962-7.
37. Okonechnikov, K.; Golosova, O.; Fursov, M. Unipro UGENE: A unified bioinformatics toolkit. *Bioinformatics* **2012**, *28*, 1166–1167. [CrossRef] [PubMed]
38. Akaike, H. A new look at the statistical model identification. *IEEE Trans. Autom. Contr.* **1974**, *19*, 716–723. [CrossRef]
39. Guindon, S.; Gascuel, O. A simple, fast and accurate method to estimate large phylogenies by maximum- likelihood. *Syst. Biol.* **2003**, *52*, 696–704. [CrossRef]
40. Darriba, D.; Taboada, G.L.; Doallo, R.; Posada, D. jModelTest 2: More models, new heuristics and parallel computing. *Nat. Methods* **2012**, *9*, 772. [CrossRef]
41. Masa-Iranzo, I.; Sanmartín, I.; Caruzo, M.B.R.; Riina, R. Skipping the Dry Diagonal: Spatio-temporal evolution of *Croton* section *Cleodora* (Euphorbiaceae) in the Neotropics. *Bot. J. Linn. Soc.* **2021**, *197*, 61–84. [CrossRef]
42. Ronquist, F.; Deans, A.R. Bayesian phylogenetics and its influence on insect systematics. *Annu. Rev. Entomol.* **2010**, *55*, 189–206. [CrossRef]
43. Alfaro, M.E.; Zoller, S.; Lutzoni, F. Bayes or bootstrap? A simulation study comparing the performance of Bayesian Markov chain Monte Carlo sampling and bootstrapping in assessing phylogenetic confidence. *Mol. Biol. Evol.* **2003**, *20*, 255–266. [CrossRef]
44. Ronquist, F.; Teslenko, M.; Van der Mark, P.; Ayres, D.; Darling, A.; Höhna, S.; Larget, B.; Liu, L.; Suchard, M.A.; Huelsenbeck, J.P. MrBayes 3.2: Efficient Bayesian phylogenetic inference and model choice across a large model space. *Syst. Biol.* **2012**, *61*, 539–542. [CrossRef]
45. Drummond, A.J.; Rambaut, A. BEAST: Bayesian evolutionary analysis by sampling trees. *BMC Evol. Biol.* **2007**, *7*, 214. [CrossRef]
46. Rambaut, A.; Drummond, A.J.; Xie, D.; Baele, G. Posterior Summarization in Bayesian Phylogenetics Using Tracer 1.7. *Syst. Biol.* **2018**, *67*, 901–904. [CrossRef] [PubMed]
47. Stamatakis, A.; Hoover, P.; Rougemont, J. A rapid bootstrap algorithm for the RAxML web servers. *Syst. Biol.* **2008**, *57*, 758–771. [CrossRef] [PubMed]
48. Miller, M.A.; Pfeiffer, W.; Schwartz, T. Creating the CIPRES Science Gateway for inference of large phylogenetic trees. In Proceedings of the Gateway Computing Environments Workshop (GCE), New Orleans, LA, USA, 14 November 2010; pp. 1–8. [CrossRef]
49. O'Donell, C.A.; Lourteig, A. Chrozophorae Argentinae. *Lilloa* **1942**, *8*, 37–81.
50. Baillon, H.E. D'Observations Botaniques. *Adansonia* **1864**, *4*, 289.
51. Jestrow, B.; Proctor, G.; Francisco-Ortega, J. *Lasiocroton trelawniensis* (Euphorbiaceae), a critically endangered species from the Cockpit Country of Jamaica, belongs to *Bernardia* (Euphorbiaceae). *Bot. Rev.* **2008**, *74*, 166–177. [CrossRef]
52. De-Nova, J.A.; Sosa, V. Phylogeny and generic delimitation of *Adelia* (Euphorbiaceae) inferred from molecular and morphological data. *Taxon* **2007**, *56*, 1027–1036. [CrossRef]
53. Bentham, G. Contribution towards a Flora of South America—Enumeration of Plants Collected by Mr. Schomburgk in British Guiana. *Lond. J. Bot.* **1843**, *2*, 42–52.
54. Borhidi, A. Taxonomic revision of genus *Leucocroton* (Euphorbiaceae). *Acta Bot. Hung.* **1991**, *36*, 13–40.
55. De-Nova, J.A.; Sosa, V.; Wurdack, K.J. Phylogenetic relationships and the description of a new species of *Enriquebeltrania* (Euphorbiaceae *s.s.*): An enigmatic genus endemic to Mexico. *Syst. Bot.* **2006**, *31*, 533–546. [CrossRef]

biology MDPI

Article

Dated Phylogeny of *Astragalus* Section *Stereothrix* (Fabaceae) and Allied Taxa in the Hypoglottis Clade

Ali Bagheri [1,*], Ali Asghar Maassoumi [2], Jonathan Brassac [3,4,5] and Frank R. Blattner [3,6,*]

1 Department of Plant and Animal Biology, Faculty of Biological Science and Technology, University of Isfahan, Hezarjarib Street, Isfahan P.O. Box 81746-73441, Iran
2 Botany Research Division, Research Institute of Forests and Rangelands, Agricultural Research, Education and Extension Organization (AREEO), Tehran P.O. Box 13185-116, Iran
3 Leibniz Institute of Plant Genetics and Crop Plant Research (IPK), D-06466 Gatersleben, Germany
4 Institute of Agricultural and Nutritional Sciences, Martin Luther University, D-06120 Halle, Germany
5 Julius Kühn Institute (JKI), D-06484 Quedlinburg, Germany
6 German Centre of Integrative Biodiversity Research (iDiv) Halle-Jena-Leipzig, D-04103 Leipzig, Germany
* Correspondence: a.bagheri@sci.ui.ac.ir (A.B.); blattner@ipk-gatersleben.de (F.R.B.)

Simple Summary: Biological taxonomic research deals with the grouping of organisms into entities that reflect their evolutionary history and relationships. In the species-rich plant genus *Astragalus*, the systematic grouping of many species changed several times during recent decades, which indicates problems in correctly recognizing relationships based on morphological characteristics. Here, we analyzed the relationships of *Astragalus* species from Iran and neighboring countries based on DNA sequences from three different loci. We found that species traditionally classified into two different sections of *Astragalus* occur intermingled in our phylogenetic trees instead of forming clear groups reflecting their taxonomic units. In addition, species thought to be only distantly related to the target species were found in this cluster. From this, we conclude that the currently used circumscription of taxonomic entities for these *Astragalus* species is false and should be abandoned. The reasons behind the the systematic classification problems of *Astragalus* include independent, parallel evolution or the loss of characteristics that were assumed to be unique and used to define certain systematic units. Thus, it is necessary to analyze the relationships of many *Astragalus* species to (i) identify traits useful for taxonomic classification and (ii) to understand the ecological and habitat differences driving their fast speciation.

Abstract: The *Astragalus* subgenus *Hypoglottis* Bunge, which consists of several sections, is one of the taxonomically most complicated groups in the genus. The *Astragalus* section *Stereothrix* Bunge belongs to this subgenus and is a significant element of the Irano-Turanian floristic region. A molecular phylogenetic analysis of this section and its closely related taxa using nuclear ribosomal DNA internal transcribed spacers (ITS) and external transcribed spacer (ETS) regions as well as plastid *mat*K sequences were conducted. Parsimony analyses and Bayesian phylogenetic inference revealed that the section is not monophyletic in its current form, as some taxa belonging to closely related sections such as *Hypoglottidei* DC. and the *Malacothrix* Bunge group within the sect. *Stereothrix* render it paraphyletic. Moreover, species groups belonging to sect. *Stereothrix* are placed in different clades within the phylogenetic tree of subgenus *Hypoglottis*, which indicates polyphyly, i.e., multiple independent origins of taxa placed in the sect. *Stereothrix*. Molecular dating of the group estimated an age of 3.62 (1.73–5.62) My for this assemblage with the major diversification events happening during the last 2 My. Many species groups separated only within the last 0.5 to 1 My. Based on morphological and molecular data, we discuss the phylogenetic relationships of the groups and synonymy of species. In addition, the included taxa of sect. *Hypoglottidei* are not monophyletic and include species belonging to sects. *Hololeuce*, *Koelziana*, *Malacothrix*, *Onobrychoideae*, and *Ornithodpodium* group within the sect. *Stereothrix* taxa. We conclude that only an analysis including all groups and nearly all species of the sections within the Hypoglottis clade can finally result in an new evolutionary-based system for these taxa.

Citation: Bagheri, A.; Maassoumi, A.A.; Brassac, J.; Blattner, F.R. Dated Phylogeny of *Astragalus* Section *Stereothrix* (Fabaceae) and Allied Taxa in the Hypoglottis Clade. *Biology* **2023**, *12*, 138. https://doi.org/10.3390/biology12010138

Academic Editor: Lorenzo Peruzzi

Received: 12 August 2022
Revised: 7 January 2023
Accepted: 9 January 2023
Published: 16 January 2023

Keywords: *Astragalus* subgenus *Hypoglottis*; Leguminosae; Iran; phylogeny; rapid diversification; section *Hypoglottidei*; section *Stereothrix*

1. Introduction

Astragalus L., harboring about 3000 species, is the largest genus among flowering plants [1,2]. The first comprehensive infrageneric classification of the genus *Astragalus* was done by Bunge, who described nine subgenus eras. Each subgenus, according to its habit and morphological characteristics, was subdivided into several sections [3,4]. Molecular investigations in the last two decades resolved a significant number of taxonomic problems at the sectional level [5–9]. However, the circumscription of some sections remains unresolved. In this study, we focus on taxa within the subgenus *Hypoglottis* Bunge, consisting of several sections. However, the taxonomic delimitation of sections and placement of many species are debated among researchers. The subgenus *Hypoglottis* contains woody and perennial herbaceous plants. The main taxa within this subgenus are sections *Hypoglottoidei* DC., *Malacothrix* Bunge, and *Stereothrix* Bunge, the latter being one of the most diverse sections within this subgenus. The major diagnostic characteristics of the section are the herbaceous caulescent growth form (rarely acaulescent), possession of basifixed hairs, imparipinnate leaves, stipules which are free from the petiole or shortly adnate to it, a non-inflated calyx, and rounded or emarginate wing blades [1]. According to the circumscription of Maassoumi [10–12], sect. *Stereothrix* is one of the medium-sized sections of *Astragalus*, with a total of 28 species.

Section *Stereothrix* has been taxonomically revised several times [1,10–16] but the taxonomic positions of some of the species suggested to belong to this section are unclear. In their account for Flora Iranica, Podlech et al. [13,14] treated sect. *Stereothrix* based on the diagnostic traits mentioned above. However, in a more recent revision of the genus, Podlech and Zarre [1] and Maassoumi [11,12,17] transferred a number of species from sect. *Stereothrix* to other closely related sections or vice versa (Tables 1 and 2).

Based on the recent comprehensive molecular studies on the entire genus by Azani et al. [6] and Su et al. [18] nine and ten clades have been inferred, respectively. One important clade in both studies is the Hypoglottis clade which includes annual and perennial species [6]. There is some evidence in these phylogenies that several sections of *Stereothrix* and *Hypoglottoidei* are non-monophyletic, as shown by Azani et al. [5,6], but more work is evidently needed. Here, rarely studied taxa belonging to different sections of the subgenus *Hypoglottis* (according to Bunge) and/or the Hypoglottis clade (according to Azani et al. [6] and Su et al. [18]), with a focus on representative species of sect. *Stereothrix*, were selected for molecular analysis to solve taxonomic problems and arrive at better insights into their systematic positions. For this, DNA sequences of the internal and external nuclear ribosomal DNA (nrDNA) spacers (i.e., the ITS and ETS regions) and the plastid *mat*K gene were used as molecular markers.

Table 1. Taxonomic treatment history of taxa belonging to sect. *Stereothrix* and its allies.

Species	Podlech et al., 2010 [13]	Podlech et al., 2012 [14]	Podlech and Zarre 2013 [1]	Maassoumi 2018 [11]	Maassoumi 2020 [12]	Current Study
A. altimontanus Podlech and Maassoumi	*Stereothrix*	*Stereothrix*	*Stereothrix*	*Stereothrix*	*Stereothrix*	*Hypoglottidei*
A. andabilensis Ranjbar and Mahmoudian	-	-	-	-	*Stereothrix*	*Stereothrix*
A. atricapillus Bornm.	*Hypoglottidei*	*Hypoglottidei*	*Hypoglottidei*	*Hypoglottidei*	*Hypoglottidei*	*Hypoglottidei*
A. badelehensis Maassoumi and Taheri	-	-	-	*Stereothrix*	*Stereothrix*	*Stereothrix*
A. barbatus Lam.	*Stereothrix*	-	*Stereothrix*	-	*Stereothrix*	*Stereothrix*
A. bavanatensis Zarre and Podlech	*Stereothrix*	*Stereothrix*	*Stereothrix*	*Stereothrix*	*Stereothrix*	*Stereothrix*
A. bojnurdensis Podlech	*Brachylobium*	*Brachylobium*	*Brachylobium*	*Hypoglottidei*	*Hypoglottidei*	*Hypoglottidei*
A. brachypetalus Trautv.	*Hypoglottidei*	*Hypoglottidei*	*Stereothrix*	*Hypoglottidei*	*Hypoglottidei*	*Hypoglottidei*
A. capito Boiss. and Hohen.	*Stereothrix*	*Stereothrix*	*Stereothrix*	*Stereothrix*	*Stereothrix*	*Stereothrix*
A. chamberlainianus H.Sümbül	-	-	*Stereothrix*	-	-	*Stereothrix*
A. daenensis Boiss.	*Brachylobium*	*Brachylobium*	*Brachylobium*	*Brachylobium*	*Brachylobium*	*Brachylobium*
A. damghanensis Podlech	*Stereothrix*	*Stereothrix*	*Stereothrix*	*Hypoglottidei*	*Stereothrix*	*Hypoglottidei*
A. doshman-ziariensis Maassoumi and Podlech	*Stereothrix*	*Stereothrix*	*Stereothrix*	*Stereothrix*	*Koelziana*	*Stereothrix*
A. hakkariensis Podlech	*Stereothrix*	-	-	*Stereothrix*	*Stereothrix*	*Stereothrix*
A. herbertii Maassoumi	*Hypoglottidei*	*Hypoglottidei*	*Hypoglottidei*	*Hypoglottidei*	*Hypoglottidei*	*Hypoglottidei*
A. inexpectatus Maassoumi and Podlech	*Malacothrix*	*Malacothrix*	*Malacothrix*	*Malacothrix/ Plagiophaca*	*Plagiophaca*	*Malacothrix*
A. koelzii Barneby	*Stereothrix*	*Stereothrix*	*Stereothrix*	*Koelziana*	*Koelziana*	*Koelziana*
A. ledinghamii Barneby	*Stereothrix*	*Stereothrix*	*Stereothrix*	*Stereothrix*	*Koelziana*	*Stereothrix*
A. leucothrix Freyn and Bornm.	-	-	*Stereothrix*	-	*Stereothrix*	*Stereothrix*
A. longirostratus Pau (=*A. perpexus* Maassoumi)	*Hhypoglottidei*	*Hemiphaca*	*Oroboidei*	*Hypoglottidei*	*Hypoglottidei*	*Hypoglottidei*
A. mahmutlarensis Podlech	-	-	*Stereothrix*	-	*Stereothrix*	*Stereothrix*
A. mahneshanensis Maassoumi and Moussavi	-	*Stereothrix*	*Stereothrix*	*Stereothrix*	*Stereothrix*	*Stereothrix*
A. montis-alamkuhi Maassoumi	-	-	-	*Stereothrix*	*Stereothrix*	*Stereothrix*
A. montismishoudaghi Sheikh Akbari Mehr, Ghorbani and Maassoumi	-	-	-	*Stereothrix*	*Stereothrix*	*Stereothrix*
A. montis-varvashti Podlech	*Stereothrix*	*Stereothrix*	*Stereothrix*	*Stereothrix*	*Stereothrix*	*Stereothrix*
A. nezva-montis Podlech and Zarre	*Hypoglottidei*	*Hypoglottidei*	*Hypoglottidei*	*Plagiophaca*	*Plagiophaca*	*Malacothrix*
A. nurensis Boiss. and Buhse (=*A. pish-chakensis* Maassoumi)	*Hypoglottidei*	*Hypoglottidei*	*Hypoglottidei*	*Hypoglottidei*	*Hypoglottidei*	*Hypoglottidei*
A. penetratus Maassoumi	*Brachylobium*	*Brachylobium*	*Brachylobium*	*Brachylobium*	*Brachylobium*	*Brachylobium*
A. plagiophacos Maassoumi and Podlech	*Plagiophaca*	*Plagiophaca*	*Plagiophaca*	*Plagiophaca*	*Plagiophaca*	*Plagiophaca*
A. podosphaerus Boiss. and Hausskn.	*Stereothrix*	*Stereothrix*	*Stereothrix*	*Stereothrix*	*Stereothrix*	*Stereothrix*
A. pseudocapito Podlech	*Stereothrix*	*Stereothrix*	*Stereothrix*	*Stereothrix*	*Stereothrix*	*Stereothrix*
A. rimarum Bornm.	*Hypoglottidei*	*Hypoglottidei*	*Hypoglottidei*	*Hypoglottidei*	*Hypoglottidei*	*Hypoglottidei*
A. saganlugensis Trautv.	*Hypoglottidei*	*Hypoglottidei*	*Hypoglottidei*	*Hypoglottidei*	*Hypoglottidei*	*Hypoglottidei*
A. setosulus Gontsch.	-	-	*Stereothrix*	-	*Stereothrix*	*Stereothrix*
A. sphaeranthus Boiss.	*Stereothrix*	*Stereothrix*	*Stereothrix*	*Stereothrix*	*Stereothrix*	*Stereothrix*
A. yazdii (Vassilcz.) Podlech and Maassoumi	*Brachylobium*	*Brachylobium*	*Brachylobium*	*Brachylobium*	*Brachylobium*	*Brachylobium*

Table 2. A summary of the establishment of the relevant sections associated with the *Stereothrix* clade over the years in chronological order.

Section	Established by	Year
Hypoglottidei	DC.	1825
Oroboidei	A.Gray	1864
Stereothri	Bunge	1868
Malacothrix	Bunge	1868
Hemiphaca	Bunge	1868
Brachylobium	Boiss.	1872
Koelziana	Širj. and Rech.f.	1953
Plagiophaca	Maassoumi and Podlech	1989

2. Materials and Methods

2.1. Taxon Sampling

Herbarium dried leaf materials for DNA extraction from the *Astragalus* sect. *Stereothrix* and closely related taxa comprising most of the type specimens (16 species means about 45% of the total species sequenced here) were obtained from the relevant collections of the herbaria MSB, TARI, and W (herbarium acronyms follow Thiers [19]). In total, we included 83 individuals representing 60 species comprising 22 sects. *Stereothrix* and 29 species of the other related sections, plus 6 species from taxonomically distant taxa, including *A. annularis* (sect. *Annulares*), *A. echinops* and *A. alopecias* (sect. *Alopecuroidei*), *A. hymenostegis* (sect. *Hymenostegis*), *A. glaucacanthus* (sect. *Poterion*), and *A. compactus* (sect. *Rhacophorus*). Species from the genus Oxytropis DC. (*O. aucheri* and *O. pilosa*) as well as *Colutea persica* were included as outgroups. In addition, we obtained 26 sequences from GenBank for completing our datasets. Voucher specimen information and GenBank sequence accession numbers for the examined taxa are listed in Table S1 (Supplemental online information).

2.2. DNA Extraction, PCR Amplification, and DNA Sequencing

Total genomic DNA was extracted from dried herbarium leaf tissue with a DNeasy Plant DNA Extraction Kit (Qiagen) according to the instructions of the manufacturer. Lysis time was doubled in comparison to Qiagen's protocol to account for the dry state of the herbarium-derived leaves. After DNA extraction, we checked DNA quality and concentration on 1.5% agarose gels. For the internal transcribed spacer (ITS) region, including ITS1, 5.8S rDNA, and ITS2, amplifications were done using primers ITS-A and ITS-B [20]. In addition, for old herbarium materials, the internal primers ITS-C and ITS-D, binding in the 5.8S rDNA [20], were used together with the before mentioned ITS primers for separate amplification of ITS1 and ITS2. For the 5′ external transcribed spacer (ETS) region upstream of the 18S rDNA, amplifications were done using primers ETS-cis2F and 18S-ETS [21]. Finally, for the chloroplast matK gene, amplifications were done using the primer pairs trnK685F/matK832R and matK4LaF/trnK2R* [22]. PCR amplification protocols for all markers by Bagheri et al. [7] were followed. Both nuclear regions and the plastid gene were directly Sanger sequenced on an ABI 3730 XL using the amplification primers.

2.3. Sequence Alignments

Forward and reverse sequences of ETS, ITS, and matK were assembled in CHROMAS v. 2.6.6 [23], manually corrected where necessary, and afterward aligned using MUSCLE version 3.8.425. Five datasets were generated, namely, each region separately, the concatenation of both nrDNA regions (ETS + ITS), and the three regions concatenated (ETS + ITS + matK).

2.4. Phylogenetic Inferences

Maximum parsimony (MP) analyses were conducted in PAUP* 4.0a169 [24] using a two-step heuristic search, as described by Blattner [25] with 1000 initial random addition sequences (RAS). To test clade support, bootstrap analyses were run on all datasets with resampling 1000 times with the same settings as before, except that we did not use the initial RAS step. PAUP* was also used to infer the best-fitting model of sequence evolution for the three marker regions (Table 3) using the Akaike information criterion (AICc).

Table 3. Characteristics of the analyzed datasets.

	ETS	ITS	ETS + ITS	*mat*K	ETS + ITS + *mat*K
Alignment lengths	282	618	900	1153	2053
Constant characteristics	252	459	711	1044	1755
Variable characteristics	30	159	189	109	298
Parsimony-informative characteristics	7	99	106	61	167
MP tree lengths	208	279	318	150	711
Consistency index (CI)	0.78	0.75	0.75	0.76	0.76
Retention index (RI)	0.82	0.86	0.86	0.85	0.86
Sequence evolution model (No. of categories)	F81(1)	SYM + Γ(6)	TrN + I + Γ(6)	TVM + I + Γ(6)	TVM + I + Γ(6)

Bayesian phylogenetic inference (BI) was conducted in MRBAYES 3.2.6 [26] on the partitioned dataset specifying the respective models of sequence evolution for each data partition. In BI, two times four chains were run for 5 million generations for all datasets specifying the respective model of sequence evolution. In all analyses, we sampled a tree every 500 generations. Converging log-likelihoods, potential scale reduction factors for each parameter, and inspection of tabulated model parameters in MRBAYES suggested that stationary had been reached in all analyses. The first 25% of trees of each run were discarded as burn-in.

2.5. Incongruent Length Difference Test

The congruence of the nrDNA and plastid datasets was evaluated using the partition homogeneity or incongruent length difference test (ILD) of Farris et al. [27] in PAUP*. The test was run using the heuristic search option including the simple addition sequence and TBR branch swapping with 1000 homogeneity replicates with the elimination of invariant characteristics [28].

2.6. Divergence Time Estimation

The clade ages and divergence times among the investigated taxa were estimated using the crown-age for *Astragalus* of 14.36 million years (My). This age was obtained by Azani et al. [6] from a dating analysis based on ITS and chloroplast sequences from 110 representatives of *Astragalus* and papilionoid legumes from the Hologalegina clade. Azani et al. [6] used two calibration points inferred from Lavin et al.'s dating analysis of Leguminosae using 12 legume-specific fossils [29]. We used BEAST 2.7.0 [30,31] to analyze the partitioned sequences. The site model and the phylogeny were co-estimated in a single Bayesian analysis as offered by the BEAST package BMODELTEST [32]. This package not only reduces the number of steps to perform the phylogenetic analysis by integrating the model-testing phase in the main analysis but also incorporates the site model uncertainty into the phylogenetic posterior distribution. We used the uncorrelated log-normal relaxed clock, as provided by the BEAST package optimized relaxed clock (ORC) [33,34], and the calibrated Yule prior [35,36]. Monophyly of the *Astragalus* clade was enforced and the node was defined with a normal-distributed prior (mean = 14.36, stdev = 2). Three independent analyses were run for 20 million generations each, sampling every 5000 generations. TRACER 1.7.2 was used to assess the convergence of the analyses and most parameters reached an effective sample size (ESS) of at least 200, indicating a good mixing of the Markov chain Monte Carlo (MCMC). The runs were combined with LOGCOMBINER (part of the BEAST software) discarding the initial 25% of each run as burn-in. A maximum clade credibility (MCC) tree was summarized with TREEANOTATOR (part of the BEAST software) using the option "Common Ancestor heights" for the nodes.

3. Results

3.1. Phylogenetic Analyses

The aligned nrDNA ETS, ITS, combined dataset of ETS + ITS, matK, and combined dataset of ETS + ITS + matK matrices comprised 282, 618, 900, 1153, and 2053 bp across 83 accessions, respectively. The ILD test ($p = 0.04$) suggested no significant length incongruence between the nuclear and plastid markers, therefore, we also analyzed them as a combined dataset. The independent MP and BI analyses of the ITS + ETS and matK datasets produced consistent results differing only regarding the phylogenetic resolution of the obtained trees, which is higher in the nuclear dataset compared to matK. Hence, because of similar results of the analyses, only the total evidence Bayesian tree of the three combined marker regions along with its posterior probabilities (PP) and bootstrap values >70% [37] from MP analysis is shown here (Figure 1). Differences between the BI and MP analyses occurred at three positions in the tree, where BI resolved relationships with low support values that in the MP strict consensus tree resulted in polytomies. Characteristics of the analyzed datasets are provided in Table 3.

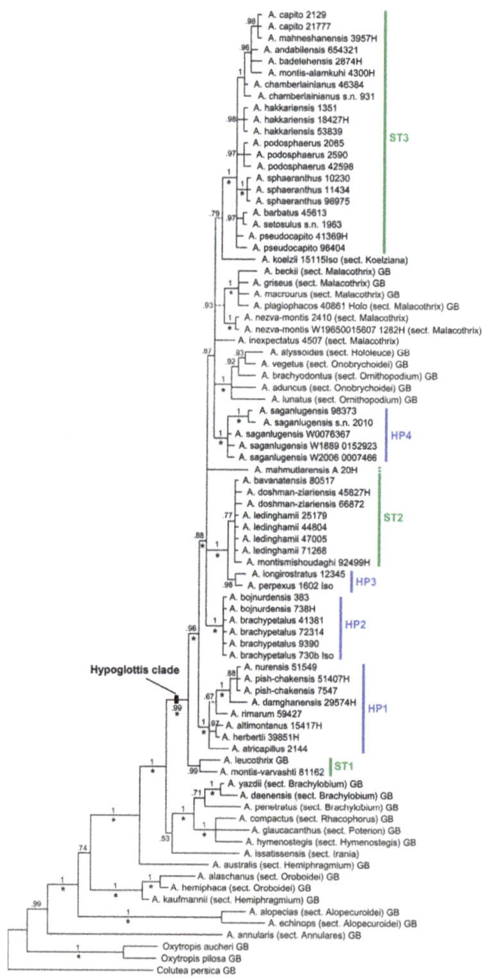

Figure 1. Bayesian phylogenetic tree based on the combined data matrix consisting of ETS, ITS, and *mat*K sequences. Numbers at branches indicate the Bayesian posterior probabilities. The tree topology

is identical to the strict consensus tree of maximum parsimony (MP) analysis except for three clades (indicated by dashed branches) that were not recovered in MP. MP bootstrap support >70% is indicated by asterisks (*) at the respective branches. Sectional affiliation of species outside sects. *Stereothrix* (ST1–ST3) and *Hypoglottidei* (HP1–HP4) are given in brackets after the species' names. GB indicates sequences that were obtained from the GenBank nucleotide database.

3.2. Phylogenetic Reconstructions and Age Estimates

Bayesian phylogenetic inference and Maximum Parsimony based on the combined dataset (ETS + ITS + *mat*K) resulted in a tree with several strongly to moderately supported subclades (Figure 1) within the large so-called Hypoglottis clade of Azani et al. [6] and Su et al. [18]. The two large sections within subgen. *Hypoglottis*, i.e., sects. *Hypoglottidei* (members marked as HP1–HP4 in Figure 1) and *Stereothrix* (ST1–ST3), are polyphyletic. Furthermore, taxa of sects. *Onobrychoidei*, *Ornithopodium*, and *Hololeuce* forming the subgen. *Cercidothrix* Bunge group inside the Hypoglottis clade. The former two sections are not monophyletic. Regarding the subgen. *Hypoglottis* sections, where we included multiple species, relationships of taxa are not completely resolved.

Non-monophyly was also detected outside the Hypoglottis clade where *A. australis* and *A. kaufmannii*, the two analyzed members of the sect. *Hemiphragmium*, are nested within two different clades.

In our BI tree, not all species belonging to taxonomically distant sections are resolved as sister taxa of the Hypoglottis clade. Species from sects. *Ornithopodium*, *Onobrychoidei*, and *Hololeuce* of subgen. *Cercidothrix* were found nested within this clade so that the boundaries of subgen. *Hypoglottis* in the current circumscription is also not clear. The remainder of the taxa belonging to subgen. *Hypoglottis* form a large polytomy, with members of the sect. *Stereothrix* mostly placed in three groups categorized as ST1–ST3 (Figure 1) and intermingled with taxa of other sections, mostly from sections *Hypoglottidei* and *Malacothrix*.

Age estimations for the clades within our set of *Astragalus* taxa (Figure 2) arrived at a crown age of 3.62 (1.73–5.62) My for the Hypoglottis clade, with the major diversification events within sects. *Stereothrix* and *Hypoglottidei* occurring during the last 2 My and many species originating only during the last 500,000 years.

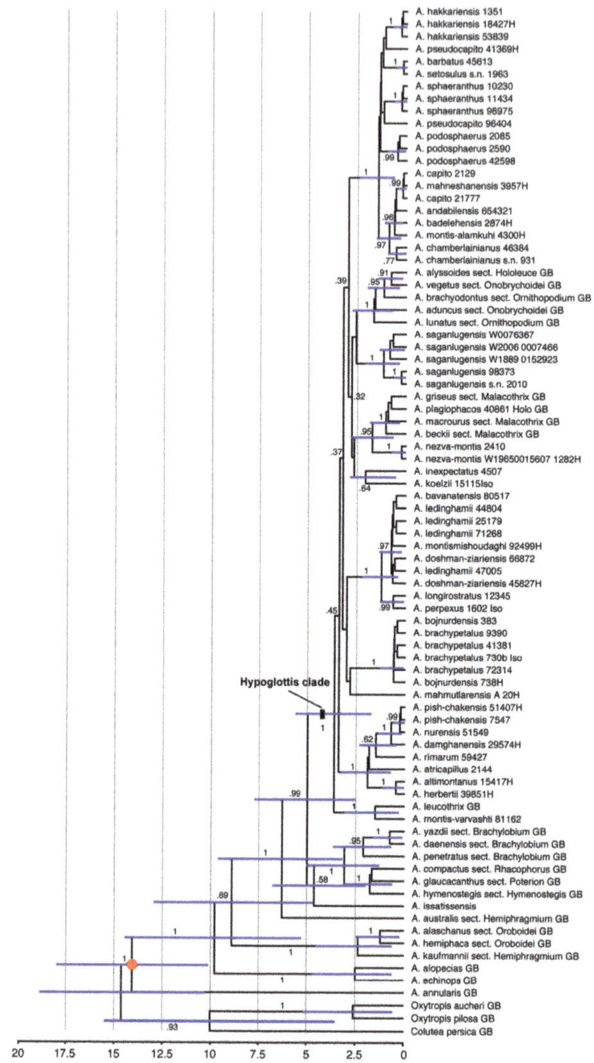

Figure 2. Dated phylogeny calculated from the combined ETS, ITS, and *mat*K sequences using the calibrated Yule prior and a secondary calibration on the crown group of *Astragalus* of 14.6 million years (red dot). Node bars indicate 95% highest probability density intervals (HPD) for the ages. The scale provides a timeline in million years before the present. Numbers along the branches give posterior probabilities.

4. Discussion

4.1. Non-Monophyly of Bunge's Traditional Subgenus Hypoglottis vs. Monophyly of the Hypoglottis Clade

Subgenus *Hypoglottis* was established by Bunge [3,4] as comprising sects. *Stereothrix, Hypoglottidei, Malacothrix,* and *Dasyphyllium*. Our results confirm that these sections are closely related and belong to the Hypoglottis clade, an informal unit that most closely resembles Bunge's subgen. *Hypoglottis*. Ranjbar and Karamian [38] transferred two other sections (*Hemiphaca*, also treated as sect. *Oroboidei*, and *Hemiphragmium*) to subgen. *Hypoglottis*. However, our results show that these taxa clearly fall outside the Hypoglottis clade and that sect. *Hemiphragmium* is even not monophyletic (Figure 1). In contrast, species of sects. *Hololeuce, Onobrychoidei,* and *Ornithopodium,* all belong to the subgen. *Cercidothrix,* grouped within the Hypoglottis clade, rendering subgen. *Hypoglottis* sensu Bunge paraphyletic. For these outgroup taxa, we included only a few species so we cannot draw further conclusions here.

Maassoumi et al. [39], in their infrageneric system of *Astragalus,* classified all the above taxa in Clade VII, including taxa belonging to subgen. *Hypoglottis* and *Cercidothrix* together with some annual species from sects. *Sesamei, Hispiduli, Dipelta, Mirae, Platyglottis, Heterodontus, Ankylotus, Pentaglottis, Ophiocarpus,* and New World aneuploids (Neo-*Astragalus*) [39]. This placing is also reflected in other studies [5,6,18]. There is considerable previous evidence in other papers that subgenus *Cercidothrix* is not monophyletic [18], and that the type species is actually in a completely different clade (Hamosa clade). By considering these approaches, Clade VII in Maassoumi et al. [39] and the Hypoglottis clade can be assumed to be monophyletic, particularly as the latter receives high support values (PP 1, MP bootstrap support 100%) in our analysis.

4.2. Divergence Times and Fast and Young Diversification

The Hypoglottis clade is characterized by large polytomies. The lack of phylogenetic resolution within this part of our data (Figure 1) is interpreted as evidence for the rapid and simultaneous evolutionary radiation of the involved taxa about 3 (1.9–4) My ago in the Upper Pliocene, resulting in a so-called hard polytomy in the phylogenetic trees. While morphological differentiation within this clade is partly pronounced, the morphological radiation was not accompanied by a similar variation in the molecular marker regions we used for our study. A Middle Pliocene (~4 My ago) diversification in the Irano-Turanian steppe regions was suggested for sect. *Hymenostegis* [7]. Azani et al. [6] also reported the divergence of the main clades of *Astragalus* from the Middle Miocene to the Pleistocene. In their study, the divergence time estimate for the crown age of the Hypoglottis clade is 8.36 (6.29–10.54) My. However, here, we arrived at a younger age estimate for the clade harboring sects. *Stereothrix* and *Hypoglottidei* with a crown age of 3.62 (1.73–5.62) My. The discrepancy can be explained by the taxa included in the analyses. Azani et al. [6] considered basal species originating from Eastern Asia as representatives of the Hypoglottis clade, while here the focus was on the Irano-Turanian floristic elements. The extant species of these groups are mostly estimated to have originated during the last 0.5 to 1 My when the climate fluctuated repeatedly between Pleistocene glacial and interglacial periods, resulting in changes between cold and dry and warmer and more humid conditions in western Asia. Plant populations in the Hypoglottis clade migrating to cope with changing conditions might have contributed to geographic isolation- and vicariance-driven speciation.

4.3. Non-Monophyly of Sections Stereothrix, Hypoglottidei, and Their Allied Taxa

Our analysis shows that neither sects. *Stereothrix* nor *Hypoglottidei* are monophyletic in their current circumscription. The taxa belonging to sect. *Stereothrix* are placed in three distinct subclades (ST1–ST3) in the main tree (Figure 1), while the examined taxa for the sect. *Hypoglottidei* fall into four subclades (HP1–HP4).

Section *Stereothrix* subclade ST1, consisting of *A. montis-varvashti* and *A. leucothrix,* is the sister group to all other taxa in the Hypoglottis clade studied here; the first species is an endemic taxon of northern parts of Turkey, which grows at a 500–800 m elevation, while the latter grows in the northern parts of Iran on Varvasht Mountain at a 4000 m elevation as an alpine species. Based on morphology, the taxa clearly belong to sect. *Stereothrix,* but in our phylogenetic analyses they are remote from the core of the sect. *Stereothrix* (ST3) taxa and their relations with other members of this section remain unclear.

The second subclade (ST2) includes four species (*A. bavanatensis, A. doshman-ziariensis, A. ledinghamii,* and *A. montismishoudaghi*) belonging to this section. All taxa occur at low elevations between 700 and 2500 m in the southwestern parts of Iran. In contrast, most species are at the core of the sect. *Stereothrix* (ST3), adapted to high-elevation habitats in the central and southwestern parts of Iran. In addition to the distinct geographical distribution (mostly central and southwestern Iran instead of northern Iran), this subclade shares some morphological characteristics, including the calyx covered with white hairs mixed with few black hairs, inflorescences that are long and cylindrical and only rarely globose, plus vegetative parts that are mostly covered with very asymmetrical hairs. Species in this group are also not always monophyletic. For example, *Astragalus montismishoudaghi* from northwestern Iran groups in a polytomy with, among others, four individuals of *A. ledinghamii* from the southwestern parts of Iran (Figure 1: ST2). In addition to their different distribution areas, their morphology is also slightly but consistently different (connected stipule to petiole vs. free; obovate standard vs. rhomboid; plus some additional differences in quantitative characteristics such as having a shorter stem height, stipule length, calyx length, and calyx teeth in *A. montismishoudaghi*). We interpret this as characteristics of very young species (Figure 2) which just started to differentiate from their close relatives in the northwest of Iran.

The third subclade (ST3) of sect. *Stereothrix* (PP 0.99, BS 84%) contains the type species of the section (*A. barbatus*). Most taxa occur at relatively high elevations of the Alborz Mountains and in the north and northwest of Iran plus southeastern Turkey. Only two species, namely *A. sphaeranthus* and *A. podosphaerus*, occur in the alpine zone of the Zagros Mountains in western Iran (in contrast to ST2 members). In addition to their shared geographical distribution, the core clade of sect. *Stereothrix* (ST3) is defined by their dense, multifloral, and globose inflorescences, and having leaflets with densely tomentose and spreading hairs.

In this subgroup, *A. badelehensis* was previously assumed to be synonymous with *A. capito* [1], but they differ from each other in some morphological characteristics, including wings blades rounded at the apex (non-obliquely emarginate), peduncle up to 3.5 cm, covered with subappressed hairs (vs. peduncle 0.5–2 cm, covered with spreading hairs), calyx covered with spreading white hairs (non-spreading white hairs mixed with few shorter black hairs). Our phylogenetic tree also shows that they group in different clades; therefore, we recognize *A. badelehensis* as a distinct species.

Section *Hypoglottidei* is, with more than 50 species [1,12,40], one of the medium-sized groups within *Astragalus*. Similar to sect. *Stereothrix*, the taxa belonging to sect. *Hypoglottidei* do not form a monophyletic group in our study, although only 14 species of this section were included. The first subclade (HP1) includes seven species and forms the core of the sect. *Hypoglottidei*. These species are distributed in northern Iran in the Alborz Mountains. Two taxa (*A. altimontanus* and *A. damghanensis*) belonging to the sect. *Stereothrix* group here, indicating non-monophyly of both taxa.

According to Maassoumi [10], the specimen number "Wendelbo & Assadi" 29574, (MSB and TARI) was considered to belong to *A. haematinus*, which is now a synonym of *A. nurensis* (sect. *Hypoglottidei*) [1,12]. Podlech [41] described the new species *A. damghanensis* based on the above-mentioned specimen and put it in sect. *Stereothrix*. Here, we analyzed this specimen and showed that the systematic position of *A. damghanensis* is incorrect in sect. *Stereothrix* as it groups with sect. *Hypoglottidei* taxa. In general, the species belonging to sect. *Stereothrix* grow in alpine areas, while taxa of sect. *Hypoglottidei* grow at lower elevations (*A. damghanensis* grows at around 450 m). Moreover, morphologically, *A. damghanensis*, by having a tubular calyx with subulate teeth and distinctly incised wing apices, shares sect. *Hypoglottidei* characteristics. Taking into account the evidence of the habitat, distribution, and morphology of this species, together with its position in our molecular analysis, we can state that it is much closer to the traditional sect. *Hypoglottidei* species than to sect. *Stereothrix*.

The second subclade (HP2) consists of *A. brachypetalus* (with four individuals) and *A. bojnurdensis* (with two individuals). The first species is widely distributed in northeastern Iran and Turkmenistan while the latter is restricted to only a small area in northeastern Iran. Both species have important common morphological features (long calyx teeth and dense to lax globose inflorescences) which separate them from sect. *Hypoglottidei*. According to Podlech and Zarre [1], these taxa should either be placed in sect. *Stereothrix* or sect. *Brachylobium* based on their morphological characteristics. We assume them to be closer to sect. *Hypoglottidei* taxa (as in Maassoumi [11,12]), however, more studies are needed to determine the exact taxonomic and phylogenetic position of these species.

The third subclade (HP3) including *A. longirostratus* and *A. perpexus* (the latter one is synonymous) from sect. *Hypoglottidei* is placed here with high support values as a sister group to ST2, the species that were transferred by Podlech et al. [14] and Podlech and Zarre [1] from sect. *Hypoglottidei* to sect. *Oroboidei* and sect. *Hemiphaca*, respectively (see Tables 1 and 2). This taxon grows in the Zagros Mountains in Lorestan, Chaharmahal and Bakhtiari, and Isfahan provinces of Iran. It is easily distinguishable from other members of sect. *Hypoglottidei* by having deeply incised and bicornuate wing petals. For us, the status of this species is not finally resolved and future studies on this taxon are needed.

The fourth subclade of sect. *Hypoglottidei* taxa (HP4) falls within a large polytomy together with clade ST3, plus species from diverse sections including sects. *Onobrychoidei*, *Hololeuce*, *Ornithopodium*, and *Malacothrix*. It is formed by *A. saganlugensis* with five individuals. This species occurs mostly in Turkey, Armenia, Azerbaijan, and a small area in northwestern Iran [42]. Foliaceous and green stipules are unique features of this species. Finally, the polytomy harboring HP4 and ST3 also contains species of sect. *Malacothrix* and sections belonging to subgen. *Cercidothrix* (sects. *Ornithopodium*, *Onobrychoidei*, and *Hololeuce*). Section *Malacothrix*, with more than 150 species [1,11], is one of the largest groups within *Astragalus*. In this study, we included just a few species in our dataset. Additionally, *A. nezva-montis* from sect. *Hypoglottidei* and *A. plagiophacos* from sect. *Plagiophaca* are nested in this subclade. More recently, Maassoumi [11] transferred two taxa (*A. nezva-montis* and *A. inexpectatus*) to section *Plagiophaca*, but here, we considered them as members of sect. *Malacothrix*. Our results support the notion that not only does the sectional division of *Astragalus* seem to be partly questionable but that some subgenera also might not reflect the evolutionary history of the taxa [6,18,21].

One remarkable species is *A. koelzii* that, in Figure 1, is sister to ST3 and, in Figure 2, is sister to *A. inexpectatus*, although in both cases with very low support. This species, by having unifoliolate leaves, is easily discernable from other members of sect. *Stereothrix*. It grows in an oak forest (*Quercus brantii*) in the Khuzestan province of Iran. Sirjaev and Rechinger [43] placed it in the monotypic sect. *Koelziana*, but Podlech et al. [14] and Podlech and Zarre [1] included this monotypic section as synonyms of sect. *Stereothrix*. Recently, Maassoumi [11] revived sect. *Koelziana* as a separate section within *Astragalus*. Here, in our molecular study, we included material taken from the type specimen of this taxon. Our efforts to find more individuals of this species in the vicinity of the type locality unfortunately failed. Maassoumi [12] transferred two other species (*A. doshman-ziariensis* and *A. ledinghamii*) from sect. *Stereothrix* to sect. *Koelziana*, a relationship that our results do not support. It is certain that *A. koelzii*, with its different morphological features, is closely related but distinct from other members of sect. *Stereothrix*, which supports a monotypic sect. *Koelziana*, but a definitive interpretation of the phylogenetic and taxonomic position of this taxon needs further study.

5. Conclusions

Our phylogenetic analysis focusing on rarely-studied species from the Irano-Turanian flora confirms that the infrageneric classification of *Astragalus* in sects. *Stereothrix* and *Hypoglottidei* is false. We also find clear evidence that non-monophyly is far-reaching regarding the sections and even subgenera within the Hypoglottic clade. This finding is in accord with earlier studies, resulting in similar groupings identified as non-monophyletic [5,6,18]. However, an increase in taxonomical sampling seems to have the highest priority to uncover the extent to which these groups are non-monophyletic and eventually define monophyletic units within this clade. Although we remark here on changes regarding the sectional affiliation of certain critical taxa, it is obvious that, due to repeated parallel evolution and/or loss of morphological traits and the young age of many species (mostly less than 1 My old), it is not possible to classify the examined taxa into the existing morphology-defined sections. What can be concluded is that the fast biological radiation resulting in high species numbers of *Astragalus* is ongoing in different geographical areas of western Asia, where diverse climatic conditions might contribute to speciation. However, this alone cannot be the main driver of diversification, as other plant groups co-occurring with the local *Astragalus* species do not show similar species richness in the study area. With regard to the intrageneric system for the analyzed taxa, we can only suggest abandoning the current system and merging all of the above-mentioned sections into a larger and monophyletic entity. To achieve the goal of a comprehensive circumscription not only in the Hypoglottis clade but probably also in many other *Astragalus* series from western to central Asia, the use of genome-wide DNA sequences seem necessary to increase the resolution within the phylogenetic trees and better discern hard polytomies from badly resolved tree parts due to a low number of available characteristics [6,44]. Only based on such a resolved dataset might we arrive at a better understanding of the reasons for the rapid speciation in *Astragalus* and the evolutionary trajectory of the morphological and ecological characteristics that might define infrageneric groups.

Supplementary Materials: The following supporting information can be downloaded at: https://www.mdpi.com/article/10.3390/biology12010138/s1, Table S1: Examined taxa information analyzed in this study including species and section names, locality, and vouchers plus GenBank accession numbers for the sequences. Dataset S1: The file "SI_Stereothrix_matK_ITS_ETS.nex" contains the aligned sequences of *mat*K, ITS, and ETS in Nexus format.

Author Contributions: Conceptualization, A.B. and A.A.M.; methodology, A.B. and F.R.B.; data analysis, J.B. and F.R.B.; resources, A.B. and A.A.M.; data curation, A.B.; writing—original draft preparation, A.B.; writing—review and editing, F.R.B.; visualization, F.R.B.; funding acquisition, A.B. All authors have read and agreed to the published version of the manuscript.

Funding: This research was funded by the Iran National Science Foundation (INSF), grant number 99004956 to A.B. Costs for open access publishing were funded by the Deutsche Forschungsgemeinschaft (DFG, German Research Foundation, grant 491250510) and the Leibniz Association's Open Access Fund.

Institutional Review Board Statement: Not applicable.

Informed Consent Statement: Not applicable.

Data Availability Statement: The analyzed DNA sequences are available through GenBank.

Acknowledgments: We would like to thank the Iranian National Science Foundation (INSF), the University of Isfahan, and the Leibniz Institute of Plant Genetics and Crop Plant Research (IPK) for supporting this study. We also thank the herbaria MSB, TARI, and W for permission to use leaf materials of specimens for the molecular studies.

Conflicts of Interest: The authors declare no conflict of interest.

References

1. Podlech, D.; Zarre, S. *A Taxonomic Revision of the Genus Astragalus L. (Leguminosae) in the Old World*, 1st ed.; Naturhistorisches Museum Wien: Vienna, Austria, 2013; Volume 1–3, p. 2439.
2. Maassoumi, A.A.; Ashouri, P. The hotspots and conservation gaps of the mega genus *Astragalus* (Fabaceae) in the Old-World. *Biodivers. Conserv.* **2022**, *31*, 2119–2139. [CrossRef]
3. Bunge, A. Generis Astragali species Gerontogeae. Pars prior, claves diagnosticae. *Mem. Pres. Acad. Imp. Sci. St. Petersb. Septième Série* **1868**, *11*, 1–140.
4. Bunge, A. Generis Astragali species Gerontogeae. Pars altera, Specierum enumeration. *Mem. Pres. Acad. Imp. Sci. St. Petersb. Septième Série* **1869**, *15*, 1–245.
5. Azani, N.; Bruneau, A.; Wojciechowski, M.F.; Zarre, S. Molecular phylogenetics of annual *Astragalus* (Fabaceae) and its systematic implications. *Bot. J. Linn. Soc.* **2017**, *184*, 347–365. [CrossRef]
6. Azani, N.; Bruneau, A.; Wojciechowski, M.F.; Zarre, S. Miocene climate change as a driving force for multiple origins of annual species in *Astragalus* (Fabaceae, Papilionoideae). *Mol. Phylogenet. Evol.* **2019**, *137*, 210–221. [CrossRef]

7. Bagheri, A.; Maassoumi, A.A.; Rahiminejad, M.R.; Brassac, J.; Blattner, F.R. Molecular phylogeny and divergence times of *Astragalus* section *Hymenostegis*: An analysis of a rapidly diversifying species group in Fabaceae. *Sci. Rep.* **2017**, *7*, 14033. [CrossRef]
8. Kazempour Osaloo, S.; Maassoumi, A.A.; Murakami, N. Molecular systematic of the genus *Astragalus* L. (Fabaceae): Phylogenetic analyses of nuclear ribosomal DNA internal transcribed spacers and chloroplast gene *ndh*F sequences. *Plant Syst. Evol.* **2003**, *242*, 1–32. [CrossRef]
9. Kazempour Osaloo, S.; Maassoumi, A.A.; Murakami, N. Molecular systematics of the Old World *Astragalus* (Fabaceae) as inferred from nrDNA ITS sequence data. *Brittonia* **2005**, *57*, 367–381. [CrossRef]
10. Maassoumi, A.A. *The genus Astragalus in Iran*; Research Institute of Forests and Rangelands: Tehran, Iran, 1989; Volume 2, p. 386.
11. Maassoumi, A.A. *Fabaceae: Astragalus III. Flora of Iran, No. 145.*; Research Institute of Forests and Rangeland: Tehran, Iran, 2018; p. 766.
12. Maassoumi, A.A. Astragalus of World. 2020. Available online: http://astragalusofworld.com (accessed on 20 July 2022).
13. Podlech, D.; Zarre, S.; Maassoumi, A.A.; Ekici, M.; Sytin, A. *Flora Iranica, Papilionaceae VI, Astragalus L. IV & Barnebyella Podlech*; No. 178; Akademische Druck-u. Verlagsanstalt: Graz, Austria, 2010; p. 430.
14. Podlech, D.; Maassoumi, A.A.; Zarre, S. *Flora Iranica, Papilionaceae VII, Astragalus L. V.*; No. 179; Akademische Druck-u. Verlagsanstalt: Graz, Austria, 2012; p. 312.
15. Mahmoodi, M.; Maassoumi, A.A.; Noroozi, J. A new alpine species and a new record of *Astragalus* sect. *Stereothrix* (Fabaceae) from Iran, with comments on the phytogeography of the section. *Willdenowia* **2013**, *43*, 263–270.
16. Ranjbar, M.; Mahmoudian, B.; Vitek, E. A synopsis of genus *Astragalus* L. sect. *Stereothrix* Bunge (Fabaceae). *Biol. Diversity Conserv.* **2013**, *6*, 114–133.
17. Maassoumi, A.A.; Mahmoodi, M.; Bagheri, A. *Astragalus* sect. *Irania* (Fabaceae): A new monotypic section from Iran. *Iran. J. Bot.* **2020**, *26*, 100–110.
18. Su, C.; Duan, L.; Liu, P.; Liu, J.; Chang, Z.; Wen, J. Chloroplast phylogenomics and character evolution of eastern Asian *Astragalus* (Leguminosae): Tackling the phylogenetic structure of the largest genus of flowering plants in Asia. *Mol. Phylogenet. Evol.* **2021**, *156*, 107025. [CrossRef] [PubMed]
19. Thiers, B. Index Herbariorum: A Global Directory of Public Herbaria and Associated Staff. 2022. Available online: http://sweetgum.nybg.org/science/ih/ (accessed on 1 July 2022).
20. Blattner, F.R. Direct amplification of the entire ITS region from poorly preserved plant material using recombinant PCR. *Biotechniques* **1999**, *29*, 1180–1186. [CrossRef] [PubMed]
21. Riahi, M.; Zarre, S.; Maassoumi, A.A.; Wojciechowski, M.F. Towards a phylogeny for *Astragalus* section *Caprini* (Fabaceae) and its allies based on nuclear and plastid DNA sequences. *Plant Syst. Evol.* **2011**, *293*, 119–133. [CrossRef]
22. Wojciechowski, M.F.; Lavin, M.; Sanderson, M.J. A phylogeny of legumes (Leguminosae) based on analysis of the plastid *mat*K gene resolves many well-supported subclades within the family. *Am. J. Bot.* **2004**, *91*, 1846–1862. [CrossRef]
23. Chromas. Available online: https://technelysium.com.au/wp/chromas/ (accessed on 10 July 2022).
24. Swofford, D.L. *PAUP*: Phylogenetic Analysis Using Parsimony (*and Other Methods)*; Sinauer Associates: Sunderland, MA, USA, 2002.
25. Blattner, F.R. Phylogeny of *Hordeum* (Poaceae) as inferred by nuclear rDNA ITS sequences. *Mol. Phylogenet. Evol.* **2004**, *33*, 289–299. [CrossRef]
26. Ronquist, F.; Teslenko, M.; Van der Mark, P.; Ayres, D.L.; Darling, A.; Höhna, S.; Huelsenbeck, J.P. MrBayes 3.2: Efficient Bayesian phylogenetic inference and model choice across a large model space. *Syst. Biol.* **2012**, *61*, 539–542. [CrossRef]
27. Farris, J.S.; Källersjö, M.; Kluge, A.G.; Bult, C. Testing significance of incongruence. *Cladistics* **1994**, *10*, 315–319. [CrossRef]
28. Cunningham, C.W. Can three incongruence tests predict when data should be combined? *Mol. Biol. Evol.* **1997**, *14*, 733–740. [CrossRef]
29. Lavin, M.; Herendeen, P.S.; Wojciechowski, M.F. Evolutionary rates analysis of Leguminosae implicates a rapid diversification of lineages during the Tertiary. *Syst. Biol.* **2005**, *54*, 575–594. [CrossRef]
30. Bouckaert, R.R.; Heled, J.; Kühnert, D.; Vaughan, T.; Wu, C.-H.; Xie, D.; Suchard, M.A.; Rambaut, A.; Drummond, A.J. BEAST 2: A software platform for Bayesian evolutionary analysis. *PLoS Comput. Biol.* **2014**, *10*, e1003537. [CrossRef] [PubMed]
31. Bouckaert, R.R.; Vaughan, T.G.; Barido-Sottani, J.; Duchêne, S.; Fourment, M.; Gavryushkina, A.; Heled, A.; Jones, G.; Kühnert, D.; De Maio, N.; et al. BEAST 2.5: An advanced software platform for Bayesian evolutionary analysis. *PLoS Comput. Biol.* **2019**, *15*, e1006650. [CrossRef] [PubMed]
32. Bouckaert, R.R.; Drummond, A.J. bModelTest: Bayesian phylogenetic site model averaging and model comparison. *BMC Evol. Biol.* **2017**, *17*, 1–11. [CrossRef] [PubMed]
33. Zhang, R.; Drummond, A. Improving the performance of Bayesian phylogenetic inference under relaxed clock models. *BMC Evol. Biol.* **2020**, *20*, 54. [CrossRef] [PubMed]
34. Douglas, J.; Zhang, R.; Bouckaert, R.R. Adaptive dating and fast proposals: Revisiting the phylogenetic relaxed clock model. *PLoS Comput. Biol.* **2021**, *17*, e1008322. [CrossRef]
35. Drummond, A.J.; Ho, S.Y.W.; Phillips, M.J.; Rambaut, A. Relaxed phylogenetics and dating with confidence. *PLoS Biol.* **2006**, *4*, e88. [CrossRef]
36. Heled, J.; Drummond, A.J. Calibrated tree priors for relaxed phylogenetics and divergence time estimation. *Syst. Biol.* **2012**, *61*, 138–149. [CrossRef]

37. Hillis, D.H.; Bull, J.J. An empirical test of bootstrapping as a method for assessing confidence in phylogenetic analysis. *Syst. Biol.* **1993**, *42*, 182–192. [CrossRef]
38. Ranjbar, M.; Karamian, R. *Astragalus* sect. *Astragalus* (Fabaceae) in Iran, complementary notes with a key to the species. *Nord. J. Bot.* **2002**, *22*, 177–182. [CrossRef]
39. Maassoumi, A.A.; Kazempour Osaloo, S.; Sotoodeh, A. A new infrageneric grouping of *Astragalus* (Fabaceae). *Iran. J. Bot.* **2016**, *22*, 88–100.
40. Maassoumi, A.A. *Old World Check-List of Astragalus*, 1st ed.; Research Institute of Forests and Rangeland: Tehran, Iran, 1998; p. 617.
41. Zarre, S.; Podlech, D.; Taeb, F. New species of the genus *Astragalus* L. (Fabaceae), mainly from the "Flora Iranica" area. *Feddes Repert.* **2005**, *116*, 54–79. [CrossRef]
42. Bagheri, A.; Maassoumi, A.A.; Vitek, E. Rediscovery of *Astragalus saganlugensis* (Fabaceae, Galegeae) in Iran after 184 years. *Phytotaxa* **2018**, *350*, 297–299. [CrossRef]
43. Sirjaev, G.; Rechinger, K.H. Astragali novi iranici I. (Rechingeri iter iranicum secundum No. 28.). *Österr. Akad. Wiss. Math.-Nat. Kl. Anz.* **1953**, *90*, 114–125.
44. Raca, I.; Blattner, F.R.; Waminal, N.E.; Kerndorff, H.; Ranđelović, V.; Harpke, D. Disentangling a polyploid complex within *Crocus* series *Verni*. *Biology* 2023, *in press*.

biology

Article

Phylogeny and Historical Biogeography of *Veronica* Subgenus *Pentasepalae* (Plantaginaceae): Evidence for Its Origin and Subsequent Dispersal

Moslem Doostmohammadi [1], Firouzeh Bordbar [1,*], Dirk C. Albach [2,*] and Mansour Mirtadzadini [1]

[1] Department of Biology, Faculty of Science, Shahid Bahonar University of Kerman, Kerman P.O. Box 76169-133, Iran; doost@sci.uk.ac.ir (M.D.); mirtadz@uk.ac.ir (M.M.)

[2] Institut für Biologie und Umweltwissenschaften, Carl von Ossietzky-Universität Oldenburg, 26111 Oldenburg, Germany

* Correspondence: bordbar@uk.ac.ir (F.B.); dirk.albach@uni-oldenburg.de (D.C.A.)

Simple Summary: The Irano-Turanian phytogeographical region is considered a biodiversity reservoir for adjacent regions. The present phylogeographic study suggests that *Veronica* subgenus *Pentasepalae* originated in the Iranian plateau and was dispersed via a North African route to the Mediterranean and the Euro-Siberian regions. These findings highlight the importance of the Iranian plateau as a center of origin for many temperate plant species. Our results also resolve several taxonomic and phylogenetic issues surrounding the Southwest Asian species of this subgenus.

Abstract: *Veronica* subgenus *Pentasepalae* is the largest subgenus of *Veronica* in the Northern Hemisphere with approximately 80 species mainly from Southwest Asia. In order to reconstruct the phylogenetic relationships among the members of *V.* subgenus *Pentasepalae* and to test the "out of the Iranian plateau" hypothesis, we applied thorough taxonomic sampling, employing nuclear DNA (ITS) sequence data complimented with morphological studies and chromosome number counts. Several high or moderately supported clades are reconstructed, but the backbone of the phylogenetic tree is generally unresolved, and many Southwest Asian species are scattered along a large polytomy. It is proposed that rapid diversification of the Irano-Turanian species in allopatric glacial refugia and a relatively high rate of extinction during interglacial periods resulted in such phylogenetic topology. The highly variable Asian *V. orientalis*–*V. multifida* complex formed a highly polyphyletic assemblage, emphasizing the idea of cryptic speciation within this group. The phylogenetic results allow the re-assignment of two species into this subgenus. In addition, *V. bombycina* subsp. *bolkardaghensis*, *V. macrostachya* subsp. *schizostegia* and *V. fuhsii* var. *linearis* are raised to species rank and the new name *V. parsana* is proposed for the latter. Molecular dating and ancestral area reconstructions indicate a divergence age of about 9 million years ago and a place of origin on the Iranian Plateau. Migration to the Western Mediterranean region has likely taken place through a North African route during early quaternary glacial times. This study supports the assumption of the Irano-Turanian region as a source of taxa for neighboring regions, particularly in the alpine flora.

Keywords: alpine species; chromosome number; Irano-Turanian region; biogeography; rapid radiation; *Veronica*

Citation: Doostmohammadi, M.; Bordbar, F.; Albach, D.C.; Mirtadzadini, M. Phylogeny and Historical Biogeography of *Veronica* Subgenus *Pentasepalae* (Plantaginaceae): Evidence for Its Origin and Subsequent Dispersal. *Biology* **2022**, *11*, 639. https://doi.org/10.3390/biology11050639

Academic Editor: Lorenzo Peruzzi

Received: 25 March 2022
Accepted: 19 April 2022
Published: 21 April 2022

1. Introduction

The Irano-Turanian phytogeographical region (IT region) is one of the richest floristic regions in the Holarctic kingdom. It is the center of origin and diversification of many xeromorphic taxa, particularly several large taxonomic groups including *Astragalus* L., *Cousinia* Cass., *Acantholimon* Boiss., *Silene* L. and *Euphorbia* L., with many species being endemic within its territory [1–5]. The complex configurations of topography and climate, which created isolated populations accompanied by a dampened impact of quaternary glaciations with a lower rate of extinction, have been hypothesized to be the major factors responsible for the rich diversity and high endemism of the IT region [6]. The IT region is also considered a major source of taxa for the neighboring regions. Many plant lineages have been suggested to start their radiations, at least partly, in the IT region and eventually colonized adjacent areas (examples are: *Aethionema* [7], Arabideae [8], *Atraphaxis* [9], *Calophaca* [10], *Centaurea* [11], *Ferula* [12], *Hesperis* [13], *Lagochilus* [14], PPAM clade of Poaceae [15], *Scrophularia* [16] and *Sisymbrium* [17]). Particularly, several Mediterranean (M) taxa (among others) trace their origins back to the IT region, according to some recent phylogeographical studies. For instance, Manafzadeh et al. [18] suggested that the genus *Haplophyllum* Juss. has its cradle in the Central Asian part of the IT region, and after in situ diversification, it started to invade the Eastern and Western Mediterranean region, where it gave rise to daughter species. Moreover, Malik et al. [19] hypothesized that the diversification process in *Artemisia* subgenus *Seriphidium* Besser ex Less. started in the Tian–Shan, Pamir and Hindu Kush mountain ranges and subsequently expanded into the Mediterranean. Finally, in the genus *Gagea* Salisb., the Mediterranean region has been shown to be colonized multiple times from the IT region [20]. Based on limited sampling, it was suggested that this east-to-west directional dispersal is also present in *Veronica* subgenus *Pentasepalae* (Benth.) M.M.Mart.Ort., Albach and M.A.Fisch. [21]. Alpine species in the north and west of Iran (in Alborz, Kopet–Dagh and Zagros mountains) are considered ancestral relict plants of the subgenus *Pentasepalae* and are important from a biogeographical point of view to understand morphological trends in the diversification of the group. With its wide Eurasian distribution, *V.* subgenus *Pentasepalae* constitutes an excellent biological model to investigate floristic relationships and the biogeographical history of IT and M regions, specifically for plants living at high mountain and alpine elevations. Here, we report a comprehensive phylogenetic analysis of *V.* subgenus *Pentasepalae* based on nuclear ribosomal internal transcribed spacer (ITS) sequence variation, to determine the major clades of the subgenus with the most comprehensive sampling to date and to infer the origin of this prominent Northern Hemisphere temperate plant species group.

Veronica L. is a species-rich genus of Plantaginaceae (sensu APG III [22] and IV [23]) comprising more than 250 species of annuals and perennial herbs in the Northern Hemisphere in addition to about 180 shrubby species of the *Hebe*–complex from the Southern Hemisphere [21,24]. According to the most recent circumscription, the genus *Veronica* is divided into 12 subgenera [25,26]. *Veronica* subgenus *Pentasepalae* is a well-defined monophyletic subgenus, according to the phylogenetic analysis of nuclear and plastid DNA sequence data [21,25], and is the largest subgenus of *Veronica* in the Northern Hemisphere, comprising about 80 species [27] with several recent additions from Europe [28,29] and Turkey [30]. The species of this subgenus are distributed from Morocco and the Iberian Peninsula to the Altai Mountains in Central Asia (Figure 1), with a center of diversity in Turkey and northern Iran and comprising most of the perennial species of *Veronica* in SW Asia (representative taxa are shown in Figure 2). Based on molecular, morphological and karyological data, species of this subgenus are categorized into four subsections (*V.* subsect. *Pentasepalae* Benth., *V.* subsect. *Armeno–Persicae* Stroh, *V.* subsect. *Orientales* (Wulff) Stroh and *V.* subsect. *Petraea* Benth.), while nine species (mostly from Iran) are treated with uncertain position [27]. The first subsection is a relatively well-studied group of about 23 taxa, generally distributed in Europe with some representatives in Asia (Turkey, Caucasus and Siberia) and North Africa [28,29,31–33]. It is a monophyletic lineage based on nuclear and plastid DNA sequences, but support for monophyly is lower when the Siberian *V. krylovii* Schischk. is included [31]. Parallel evolution of morphological characters and the existence of morphologically intermediate forms, hybridization and polyploidization makes this subsection one of the most taxonomically challenging groups within the genus *Veronica* [29,31,32]. The other three subsections together with those nine unclassified species are distributed in SW Asia, with some members reaching SE Europe. There are several relict and isolated species in the alpine and subalpine regions of the Alborz, Kopet–Dagh, Zagros, Caucasus, Taurus and East Anatolian mountain ranges that are well delimitated and usually represent specified morphological characters. These species are important constituents of the alpine vegetation in this region, some of them reaching subnival/nival zones including *V. aucheri* Boiss., which happens to be the highest vascular plant species in Iran, climbing up to about 4800 m in Damavand Mountain, central Alborz ([34]; personal observations). However, although many of these alpine species are morphologically well defined, their phylogenetic relationships are not yet resolved, and only scarce genetic data of limited accessions are available. On the other hand, the species at lower elevations are morphologically more polymorphic and belong to the *V. orientalis* Mill.–*V. multifida* L. complex. These two species and their allies are mainly distributed from Jordan through central Turkey to Armenia and Western Iran. Highly polymorphic morphological characters in both vegetative and reproductive organs make the species in this complex difficult to identify. Previous molecular studies on this complex in the western part of its range demonstrated that *V. orientalis* is not monophyletic and no clear biogeographic patterns were indicated [35,36]. The eastern populations in Northern and Western Iran have not yet been investigated.

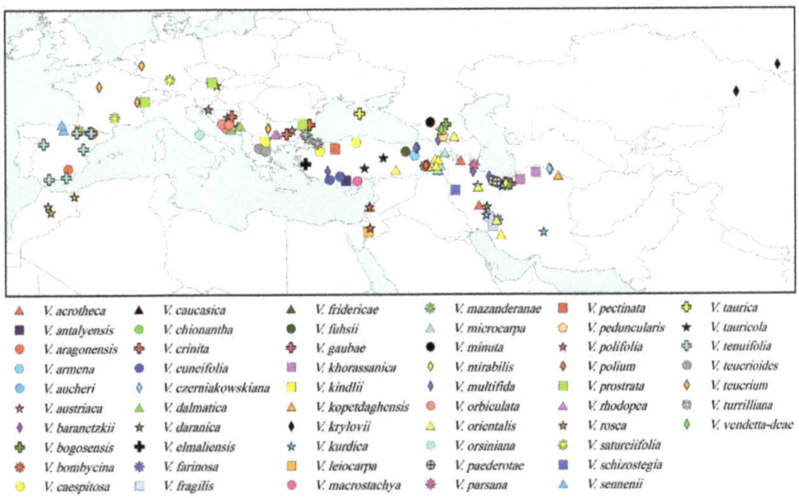

Figure 1. Map of the samples of *Veronica* subgenus *Pentasepalae* used for phylogenetic analyses in this study. Those specimens not present in the map are based on cultivated material (Table S1).

This study aims to investigate the phylogenetic relationships among species of *Veronica* subgenus *Pentasepalae* to study their biogeographical patterns and further delimit their taxonomic and geographic ranges. Our specific aims are: (1) to infer phylogenetic relationships and ascertain major clades within the subgenus while testing the accuracy of previous classifications, (2) to infer the number of origins of the *V. orientalis*–morphotype from within the subgenus, (3) to provide more information on ploidy level variation among the Iranian species and (4) to explore the spatiotemporal evolution of the subgenus, especially its place of origin, historical migration routes and diversification patterns.

Figure 2. *Cont.*

Figure 2. Morphological diversity of *Veronica* subgenus *Pentasepalae* in the Iranian plateau. (**A**) *V. chionantha*, (**B**) *V. farinosa*, (**C**) *V. orientalis*, (**D**) *V. rechingeri*, (**E**) *V. aucheri*, (**F**) *V. czerniakowskiana*, (**G**) *V. kurdica* subsp. *filicaulis*, (**H**) *V. paederotae*, (**I**) *V. gaubae*, (**J**) *V. fragilis*, (**K**) *V. mazanderanae*, (**L**) *V. schizostegia*, (**M**) *V. kurdica* subsp. *kurdica*, (**N**) *V. kopetdaghensis*, (**O**) *V. daranica*, (**P**) *V. khorassanica* and (**Q**) *V. mirabilis*. Photos (**A–K**,**Q**) by M. Doostmohammadi, (**L**,**M**,**O**,**P**) by M. Mirtadzadini and (**N**) by H. Moazzeni.

2. Materials and Methods

2.1. Plant Material

Sampling was both taxonomically (all type species of sections and subsections described within the subgenus) and geographically (Figure 1) comprehensive, including 63 out of about 80 accepted species (79%) covering the entire distribution range of the subgenus. Samples for the molecular studies included dried plant specimens collected during fieldworks mainly in Iran and tiny fragments taken from herbarium specimens deposited at E, FUMH, MIR, OLD and TUH. The newly generated sequences were complemented with previously published ITS sequences of the species mostly belonging to *V.* subsection *Pentasepalae* [31]. The morphologically variable *V. orientalis*–*V. multifida* complex was represented from different geographical sites (16 accessions). Plants were identified following the taxonomy of Borissova [37], Fischer [38], Fischer [39] and Saeidi-Mehrvarz [40], and the accepted names in the nomenclatural revision of Rojas-Andrés et al. [28] and Rojas-Andres and Martínez-Ortega [32] were applied for species of *V.* subsect. *Pentasepalae*. *Veronica chamaedrys* L. (*V.* subgenus *Chamaedrys* (W. D. J. Koch) Buchenau) and *V. polita* Fr. and *V. campylopoda* Boiss. (*V.* subgenus *Pocilla* (Dumort.) M. M. Mart. Ort., Albach and M. A. Fisch.) were used as outgroups for the phylogenetic analysis [31,41]. In total, 144 ITS sequences (of which 61 were newly generated here) were included in the analyses. Voucher information, the source of material and GenBank accession numbers are given in Table S1. Divergence time analysis was carried out on a subset of this matrix including only one accession per species of *V.* subgenus *Pentasepalae* (63 in total) together with several samples from other subgenera of *Veronica* and other sister genera as outgroup, according to Surina et al. [42] and Meudt et al. [43].

2.2. DNA Extraction, Amplification and Sequencing

New sequences were generated in two laboratories using different protocols. For some sequences, total genomic DNA was extracted following the modified CTAB protocol [44]. The quality of the extracted DNA was checked on 1.2% TBE–agarose gels, and the amount of DNA was estimated using a spectrophotometer at 260 nm. The ITS region (ITS1, 5.8S rDNA and ITS2) was amplified using the primer pair ITS–A and ITS–B [45]. PCR condition for ITS amplification were an initial denaturation (3 min at 94 °C) followed by 38 cycles of denaturation (30 s at 94 °C), annealing (40 s at 53 °C), extension (1 min at 68 °C) and a final extension step (10 min at 70 °C). Reactions were carried out in the volume of 30 µL, containing 12 µL deionized water, 15 µL Taq DNA polymerase master mix Red (Ampliqon; Tris–HCl pH 8.5, (NH4)2SO4, 4 mM MgCl2, 0.2% Tween 20, 0.4 mM each of dNTP, 0.2 units/µL Ampliqon Taq DNA polymerase, inert red dye and stabilizer), 0.75 µL each primer and 1.5 µL 1:60 diluted template DNA. Sequencing reactions were performed using the same PCR primers. For the rest of sequences, DNA was isolated from herbarium or silica-gel-dried leaves using the DNeasyTM Plant Mini kit (Qiagen GmbH, Hilden, Germany), following the manufacturer's instructions. The quality of the extracted DNA was checked on 0.8% TBE-agarose gels, and the concentration was measured spectrophotometrically with a GeneQuant RNA/DNA calculator (Pharmacia, Cambridge, U.K.). Following the protocol of Sonibare et al. [36], the ITS region was amplified using the ITS-A and ITS-4 primers [46,47]. The products were purified using QIA quick PCR purification and gel extraction kit (Qiagen GmbH, Hilden, Germany) following the manufacturer's protocols. The same primers used for PCR amplification were also used for the sequencing reactions by commercial sequencing companies.

2.3. Sequence Alignment, Phylogenetic Reconstruction and Dating

Sequences were initially aligned using MAFFT v. 6.0 [48] and edited manually using PhyDE v. 0.9971 [49]. Insertions and deletions (indels) were coded as binary characters using the simple indel coding approach, as implemented in SeqState v. 1.4.1 [50]. Phylogenetic analyses were conducted using Maximum Parsimony (MP), Maximum Likelihood (ML) and Bayesian Inference (BI). Maximum Parsimony (MP) analyses were performed using heuristic searches in PAUP* v. 4.0b10 [51] in combination with parsimony Ratchet [52] in PRAP [53]. Ratchet settings included 1000 iterations with 25% of the positions randomly unweighted (weight = 2) during each replicate and 100 random additional cycles. Tree lengths and homoplasy indices (consistency (CI), retention (RI) and rescaled consistency (RC) indices) were calculated in PAUP* [51]. Jackknife (JK) support was estimated in PAUP by conducting a single heuristic search within each 10,000 replicates using the Tree Bisection and Re-connection (TBR) branch-swapping algorithm and a deletion of 36.79% characters in each replicate. A strict-consensus tree was constructed from all saved trees. The best model of molecular evolution was found using jModelTest v.2.1.10 [54]. The GTR+Γ+I model was found to fit best with the ITS region according to the Akaike information criterion (AIC). Maximum likelihood (ML) tree inference and bootstrapping (BS) were conducted with RAxML v. 1.5b1 [55]. The model was set to GTRGAMMAI, and bootstrap analyses were carried out with 1000 replicates. Bayesian inference (BI) was conducted using MrBayes v.3.2.6 [56]. Two parallel runs of four MCMC chains including three heated and one cold chain were run simultaneously for 10 million generations, sampling every 500 generations. After removing 25% of the sampled trees as burn-in, a 50% majority-rule consensus tree was constructed.

Divergence times were estimated using BEAST v 1.10.4 [57]. The model GTR+Γ was used in the analysis. The BEAST.XML input file was generated using BEAUTi v 1.10.4 [57]. Rate evolution was modeled in an uncorrelated lognormal relaxed clock framework [58] to allow for rate variation among lineages. A Yule tree prior was used, as recommended for species-level phylogenies in the BEAST manual. Nodes were calibrated following Surina et al. [42] and Meudt et al. [43], which applies age estimates based on palaeobotanical, geomorphological and fossil data. According to these calibration points, (1) the

Plantago L./*Aragoa* Kunth stem was constrained to be monophyletic using an exponential distribution with a mean of 1 and an offset of 19.4 million years ago (Mya), and (2) for the crown age of Aragoa (which was also constrained monophyletic), a uniform age prior was set that spans from 3.3 to 0 Mya. Three separate MCMC analyses were run for 40 million generations each, sampling every 1500th generation. Convergences of the chains and estimated sample sizes (ESSs) were confirmed to be sufficiently high (>200) in Tracer v 1.7.1 [59]. Independent runs were combined by LogCombiner1.10.4, and the first 10% of the generations from each run were discarded as burn-in. TreeAnnotator v 1.10.4 was used to compute the maximum clade credibility tree (MCC tree) with node heights being the median of the age estimates.

2.4. Ancestral Area Reconstruction

To reconstruct the biogeographical history of *V.* subgenus *Pentasepalae*, eight major geographical areas were defined following the known distribution patterns of species: (A) Iranian plateau including Alborz, Kopet–Dagh and Zagros mountain chains together with highlands of Kerman; (B) Caucasus extended northward to Crimea; (C) Anatolia and Levant; (D) Altai and Tarbagatai Mountains of Central Asia; (E) Northwest Africa; (F) Iberian Peninsula; (G) Mediterranean Europe; and (H) Euro-Siberian region. Ancestral range estimation was inferred using the maximum clade credibility tree file generated in BEAST, representing 63 species of *V.* subgenus *Pentasepalae* included in this analysis with only one accession per species and excluding the outgroups using RASP (Reconstruct Ancestral State in Phylogenies) v 4.2 [60]. The analyses were run under a Statistical Dispersal–Vicariance (S–DIVA) approach [61] and the Bayesian Binary MCMC Method (BBM) [62]. The BBM analysis was run for 5,000,000 cycles sampling every 1000 cycles under the estimated F81+Γ model with a null root distribution. The maximum number of possible ancestral areas was set to three for both analyses.

2.5. Chromosome Counting

Mitotic chromosome counts were conducted using seeds sampled from herbarium specimens and germinated in petri dishes in the laboratory following [63]. Actively growing root tips were pretreated with α–monobromonaphthalene for five h at 4 °C, then rinsed in distilled water, fixed in Carnoy solution (3:1 absolute ethanol: glacial acetic acid) and stored at 4 °C until use. Hydrolysis was conducted with 1 N HCl at 60 °C for 1 min, stained in aceto-iron hematoxylin at 30 °C for 2 h and then squashed in a drop of 45% acetic acid. All mounted slides were screened under an Optika B–500 light microscope, chromosome numbers of at least five cells (for each individual) were determined, and well-spread metaphase plates were photographed using an OPTIKAM HDMI–4083.13 microscope photomicrograph system.

3. Results

3.1. Phylogenetic Analyses and Divergence Time Estimates

The aligned nrITS data matrix comprised 144 sequences (141 ingroups) and 608 characters including 52 coded indels, 156 potentially parsimony informative sites and 115 variable uninformative sites. Maximum parsimony analyses resulted in 625 most parsimonious trees with a length of 737, a consistency index of 0.478 and a retention index of 0.746. Topologies of the Maximum Likelihood and Maximum Parsimony analyses were largely congruent with that of the Bayesian inference, except for the position of some weakly supported terminal nodes. Therefore, only the results from the Bayesian analyses are shown here, which is better resolved and better supported, along with posterior probabilities as well as respective ML and MP bootstrap values. The phylogenetic tree (Figure 3) strongly supports the monophyly of *V.* subgenus *Pentasepalae* (node A, PP = 0.98). Within the subgenus, *V.* subsection *Pentasepalae* (excluding *V. krylovii*, node N, PP = 1) and *V.* subsection *Petraea* (excluding *V. vendetta–deae* Albach and *V. baranetzkii* Bordz., node M, PP = 0.99) were resolved as monophyletic. However, members of *V.* subsection *Orientales* and *V*. subsection *Armeno–Persicae* were scattered along the tree and formed several small assemblages. *Veronica czerniakowskiana* Monjuschko and *V. fragilis* Boiss. from the Iranian plateau branched at the base of tree, being sister to a polytomic clade (node B, PP = 0.71) containing the rest of the species of the subgenus. Within this polytomy, several clusters of species are detectable. A highly supported clade including *V. kurdica* Benth. and allies was reconstructed (node L, PP = 0.95), distinct from members of *V. orientalis*. Different accessions of *V. orientalis*–*V. multifida* and allies assembled in two moderately supported clades (node J, PP = 0.61 and node I, PP = 0.91), while one accession (V. orientalis–3) did not group to any species. The annual species *V. mazanderanae* Wendelbo and *V. gaubae* Bornm. cluster together in a highly supported group (node K, PP = 0.94), and the only representative of *V. mirabilis* Wendelbo shows a sister–group relationship to another Alborz endemic, *V. aucheri* (node C, PP = 1). The backbone of the subgenus did not resolve well, and several species from SW Asia did not form any supported clade, whereas other species grouped in some low-to-moderately supported clades.

The maximum clade credibility chronogram inferred with BEAST from the dataset with 63 taxa (Figure 4) with the topologies obtained from the 50% majority rule consensus cladogram is somewhat different from our Bayesian analysis. Along with the basally branching *V. czerniakowskiana* and *V. fragilis*, the rest of the species are clustered in four major clades. Based on these results, the estimated divergence (stem) and diversification (crown) ages of *Veronica* subgenus *Pentasepalae* are, respectively, ca. 9.07 (95% HPD: 11.92–6.56) and 7.01 (95% HPD: 9.57–4.64) Mya. These estimates seem older than the stem age of 7.06 Mya (million years ago) and crown age of 4.94 Mya reported by [43]. The crown age of the four clades (A, B, C and D) was dated to be at 4.57 Mya (95% HPD: 6.35–3.21), 3.34 Mya (95% HPD: 5.06–1.7), 3.63 Mya (95% HPD: 5.08–2.39) and 3.15 Mya (95% HPD: 4.52–2.005), respectively.

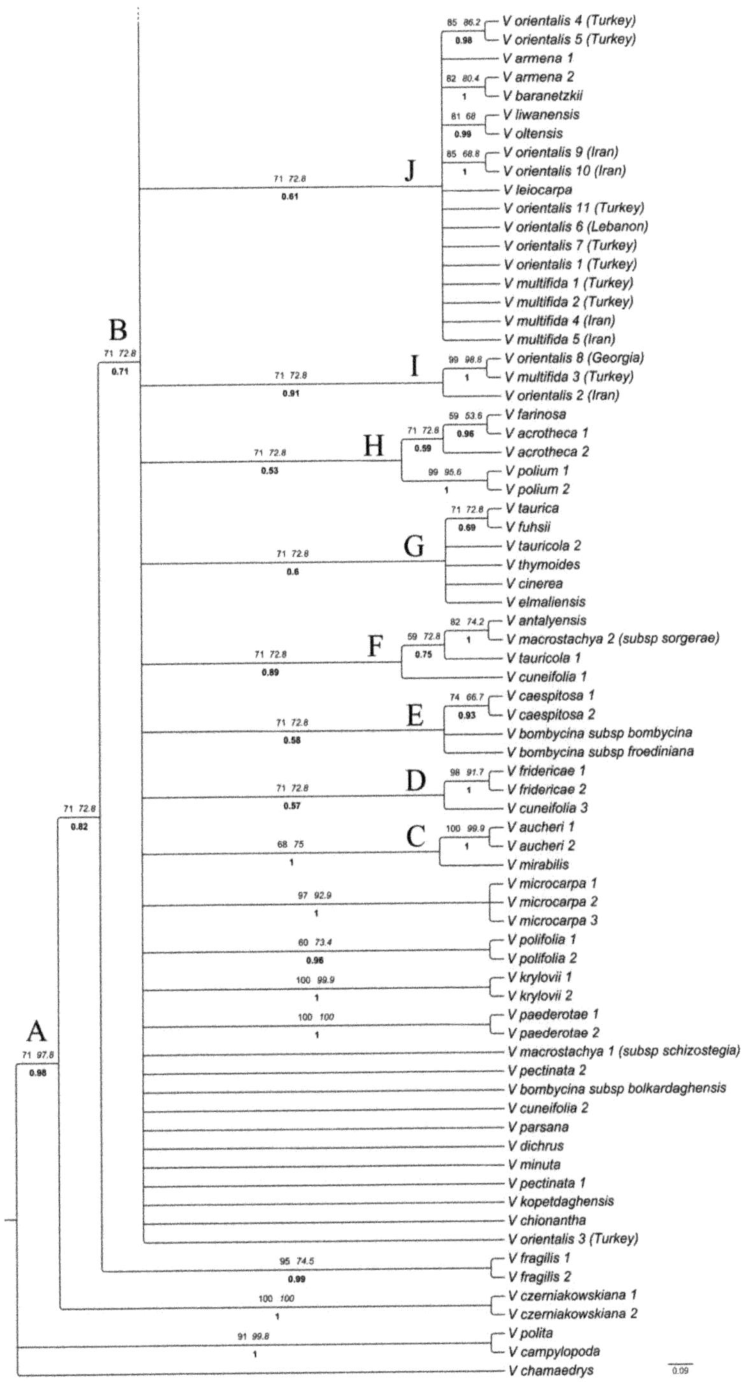

Figure 3. *Cont.*

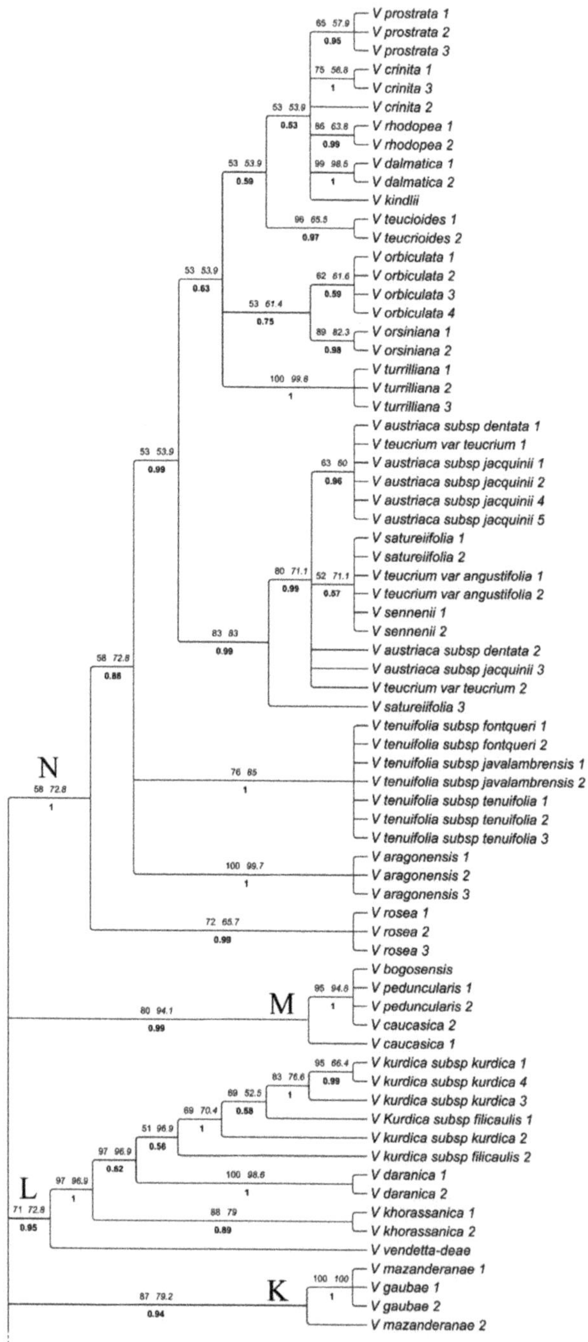

Figure 3. Fifty percent majority-rule consensus phylogenetic tree obtained from the Bayesian analysis of the ITS region for *Veronica* subgenus *Pentasepalae*. Posterior probabilities obtained from BI (boldface) are shown below branches, and bootstrap support values for the same nodes found in ML analysis (regular) and jackknife support values from MP analysis (italics) are indicated above branches.

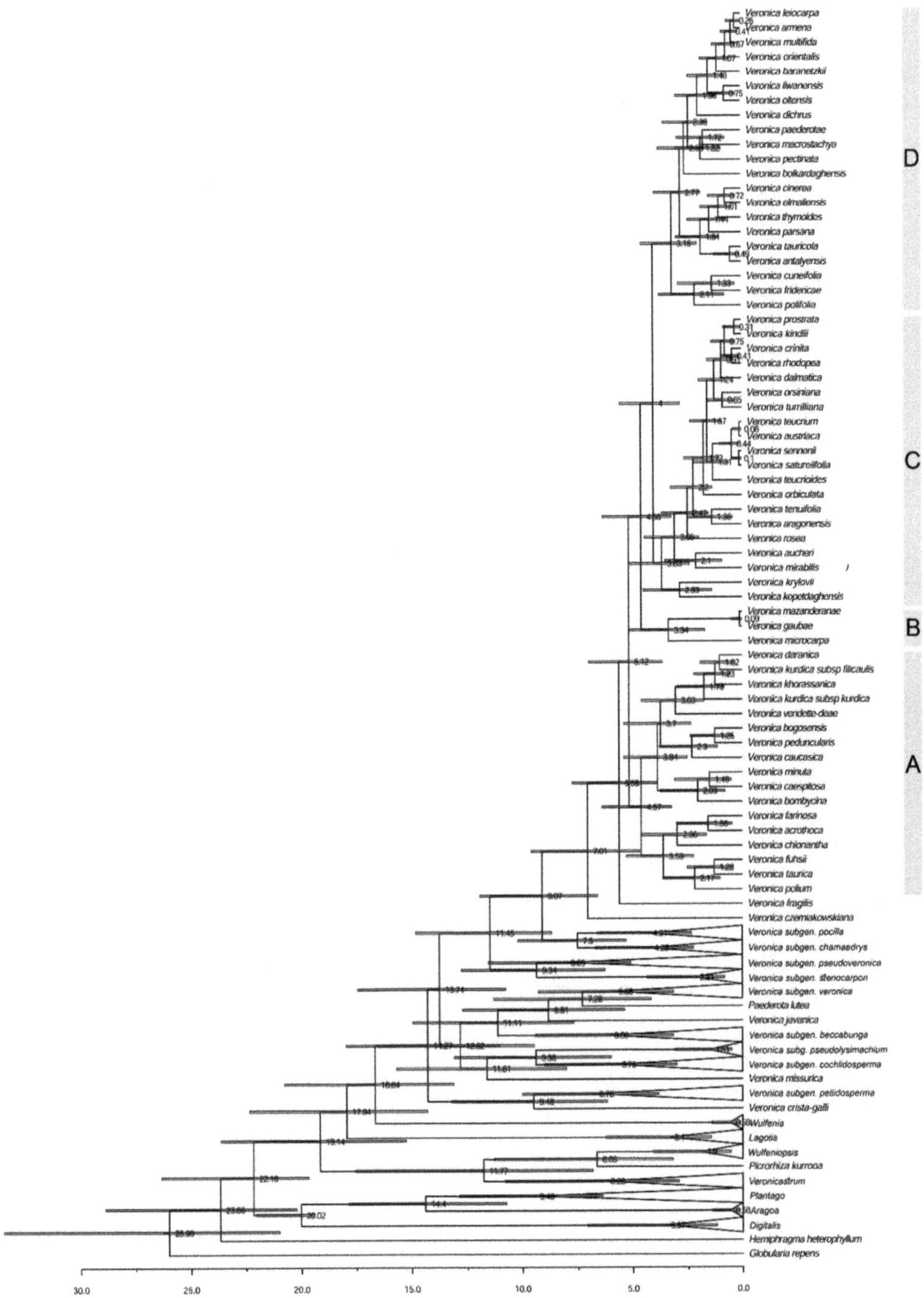

Figure 4. Dated phylogenetic tree of *Veronica* subgenus *Pentasepalae* retrieved from BEAST. Estimated divergence age values are represented for each node. Letters (A–D) on the right correspond to the four reconstructed clades.

3.2. Historical Biogeography Reconstructions

For ancestral area reconstruction, the results estimated from S–DIVA and BBM analyses are largely similar for major clades, with slight differences at a few nodes. Therefore, only the results of the BBM reconstruction are provided here (Figure 5), since it better explains the spatiotemporal radiation of the subgenus. Both BBM and S–DIVA analyses suggested the Iranian plateau (area A) as the most probable ancestral area of *V.* subgenus *Pentasepalae*, where diversification of many species took place and distributions in several other areas can be regarded as dispersal from this region. The Most Recent Common Ancestor (MRCA) areas of clades A, B and C were nested in the Iranian plateau (area A) with respective probabilities of 33%, 40% and 91%, whereas the MRCA area of clade D was in Turkey (area C, 92%). The results of the ancestral area reconstruction indicated that *V.* subgenus *Pentasepalae* required a total of 42 dispersals, 13 vicariance and 1 extinction event to reach its current distribution range.

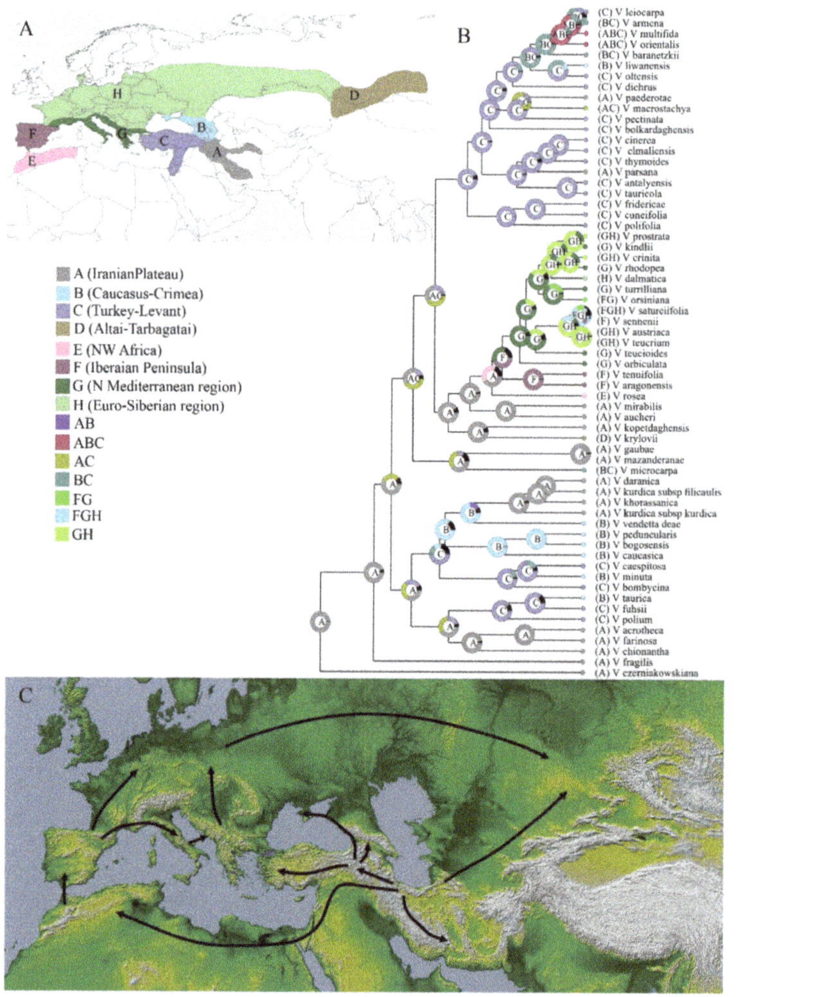

Figure 5. Biogeographic history of *Veronica* subgenus *Pentasepalae*. (**A**) Visual representation of the eight operational areas, as stated in the text. (**B**) Ancestral area reconstruction performed with BBM analysis. (**C**) Dispersal events and radiation from the Iranian Plateau on the basis of ancestral area reconstruction.

3.3. Chromosome Numbers

Somatic chromosome numbers of 10 populations from *V.* subgenus *Pentasepalae* were counted (Figure 6, Table 1). Chromosome numbers of six taxa are reported here for the first time (i.e., *V. acrotheca* Bornm., *V. khorassanica* Czerniak., *V. kurdica* subsp. *kurdica*, *V. kurdica* subsp. *filicaulis* (Freyn) Fischer, *V. schizostegia* (=*V. macrostachya* subsp. *schizostegia* (Bornm.) Fischer) and *V. rechingeri* Fischer), and new chromosome counts of *V. microcarpa* Boiss. (diploid) and *V. orientalis* (hexaploid) confirm previous ploidy level estimations based on flow cytometry [36].

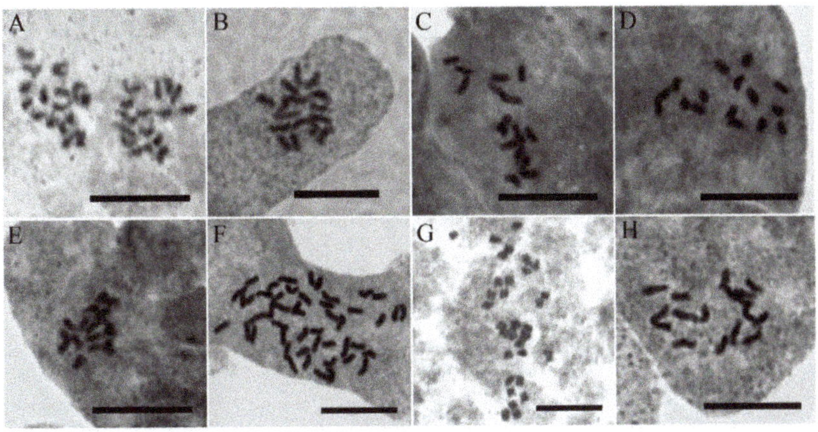

Figure 6. Metaphase plates of the representative accessions of *Veronica* subgenus *Pentasepalae*: (**A**) *V. acrotheca* (3975); (**B**) *V. khorassanica* (3976); (**C**) *V. kurdica* subsp. *filicaulis* (3563); (**D**) *V. kurdica* subsp. *kurdica* (3971); (**E**) *V. microcarpa* (3972); (**F**) *V. orientalis* (3593); (**G**) *V. rechingeri* (3634); and (**H**) *V. schizostegia* (3977). Scale bars = 10 μm.

Table 1. Voucher information and chromosome numbers of studied *Veronica* species.

NO.	Taxon Name	Chromosome Number/Ploidy Level	Locality	Herbarium (Voucher)
1	*V. acrotheca*	$2n = 16/2x$	Iran: Lorestan, Lake Gahar	M. Mirtadzadini and al. (3975 MIR)
2	*V. khorassanica*	$2n = 16/2x$	Iran: Semnan, Shahrud, Northwest of Mojen Waterfall	M. Mirtadzadini (3976 MIR)
3	*V. kurdica* subsp. *filicaulis*	$2n = 16/2x$	Iran: Bakhtiari, Tshoghakhor, Mt. Kallar	M. Mirtadzadini (3563 MIR)
4	*V. kurdica* subsp. *filicaulis*	$2n = 16/2x$	Iran: Kerman, Mt. Bahr-Aseman	F. Rezanejad (3511 MIR)
5	*V. kurdica* subsp. *kurdica*	$2n = 16/2x$	Iran: Damavand, East of Lake Tar	M. Mirtadzadini (3971 MIR)
6	*V. kurdica* subsp. *kurdica*	$2n = 16/2x$	Iran: Ghazvin, West of Abe-garm, Kise-jin to Dashtak	M. Mirtadzadini (3984 MIR)
7	*V. microcarpa*	$2n = 16/2x$	Iran: Azarbiajan, Jolfa, above St. Stephanus Church	M. Mirtadzadini (3972 MIR)
8	*V. orientalis*	$2n = 48/6x$	Iran: Azarbaijan, near Takab	M. Doostmohammadi (3593 MIR)
9	*V. rechingeri*	$2n = 32/4x$	Iran: Mazandaran, Kalardasht, Vandarbon to Tange-Galu	M. Doostmohammadi and A. Ghorbanalizadeh (3634 MIR)
10	*V. schizostegia*	$2n = 16/2x$	Iran: Kordestan, South of Marivan, Dezli to Nowsud	M. Mirtadzadini (3977 MIR)

4. Discussion

4.1. Relict Species of Veronica in the Irano-Turanian Region with No Immediate Relatives

Our phylogenetic analysis of *V.* subgen. *Pentasepalae* reveals the monophyly of the European species of *V.* subsect. *Pentasepalae* and its phylogenetic relationships as being resolved quite well. However, the backbone of the phylogenetic tree is not resolved, and many individuals of species from Southwest Asia are scattered along a large polytomy with the exception of the first-branching *V. czerniakowskiana* and *V. fragilis*. The pattern of a first-branching *V. czerniakowskiana* and the remaining species in a polytomy has already been found in one of the first phylogenetic analyses of *Veronica* [21]. Poorly resolved phylogenies are usually interpreted as the result of rapid diversification [64,65]. Rapid plant species radiations have been shown to occur in biodiversity hot spots and specifically in regions that have experienced radical climatic and geological changes [66–68]. Probably the most prominent area in Eurasia with rapid diversification is the Qinghai–Tibet Plateau (QTP), but a number of studies have also suggested that rapid radiation has occurred in arid steppes of Southwest Asia and around the Mediterranean basin [3,64,65,69]. The Irano-Turanian members of *V.* subgenus *Pentasepalae* fit well into this pattern. Meudt et al. [43] provided evidence that *V.* subgenus *Pentasepalae* (along with *V.* subgenus *Pseudolysimachium*) has the highest diversification rates among the subgenera of *Veronica* in the Northern Hemisphere, and in *V.* subgenus *Pentasepalae,* the Asian species clearly exhibit a higher diversification rate relative to the European species. Our dated phylogeny represents a diversification age of ca. 7 Mya for the crown of the subgenus. From this point forward, nearly 2 million years went by until a dramatic change in species radiation occurred (ca. 5.12 Mya, crown age of species other than *V. czerniakowskiana* and *V. fragilis*). This time estimate roughly corresponds with an active mountain building in the Iranian plateau (ca. 5 Mya, [70–72]). Topographic heterogeneity, as the result of the uplift of the Iranian plateau, potentially increased the degree of isolation of plant populations and thereby may have triggered the rapid allopatric speciation. High numbers of species of *V.* subgenus *Pentasepalae* on the Iranian plateau with quite similar ecological niches and their tendency to narrow endemism also implies that geographical factors (such as allopatry), rather than evolutionary adaptation to different ecological niches, have been the major force of speciation in Southwest Asia. Furthermore, polyploidy seems to be rare among these relictual species since 2/3 of the species for which ploidy is known are diploid, and all polyploids tend to be widespread, lowland taxa such as *V. orientalis* and *V. austriaca* rather than relictual species [27,31,36] (Table 1). Therefore, rapid radiation of the IT species due to uplift of the high mountains is likely the reason for the poorly resolved phylogenetic tree. In addition, a possible high rate of extinction among the IT species has left several relictual species isolated, outside other assemblages in the ITS tree. In a paleoecological study, Djamali et al. [6] demonstrated that *Cousinia*, a typical member of the IT region, was continuously well represented in pollen assemblages of glacial periods, suggesting that this genus not only survived but was even more abundant during glacial periods of the IT region. They argued that the dampened impact of quaternary glaciations compared to higher latitude European mountains resulted in lower rates of extinction in the IT region during cold periods. However, it should not be neglected that many cold–adapted species were prone to extinction during interglacial warm periods. It is an almost universal pattern that suggests during interglacial periods the cold-adapted species undergo an upward migration to deglaciated high-altitude interglacial refugia. However, it is not the only response of plant species to interglacial warming. Many species that cannot cope with this migration in such a short period of time, particularly the lineages on isolated lower mountain tops or with low seed dispersal capacity (such as *Veronica*), have to persist in situ or otherwise are condemned to extinction. We envision that this event has happened to several IT species of *Veronica*. Several morphologically well-separated, relictual species occur in mountains of Iran, Caucasus and Eastern Turkey and have no morphologically closely related taxa. The long branch length in the ITS tree of many of these species (not shown) suggest that they are also genetically distinct and quite isolated. Many of these species are chasmophytic, growing preferentially in rock crevices, such as *V. chionantha* Bornm., *V. fragilis*, *V. czerniakowskiana*, *V. aucheri*, *V. kopetdaghensis* Fedtsch. and *V. gaubae*. Establishing a new population on recently deglaciated rocks or in open grassland is presumably more difficult and stressful for alpine species, which likely worsened the situation for these species. In summary, we hypothesize that the immediate relatives of many extant species of the Irano-Turanian region have gone extinct due to warm–dry conditions of interglacial times, leaving several morphologically isolated species in sheltered, azonal habitats.

4.2. Phylogeny and Systematics

Although the European members of *V.* subgenus *Pentasepalae* have been the subject of several phylogenetic studies [29,31,33,73], for many Iranian and Turkish species (except for some members of *V. orientalis* complex [35,36]), no sequence has been available, and no extensive DNA-based study has been published for Southwest Asian species until now. The present study is the first comprehensive molecular phylogenetic study of *V.* subgenus *Pentasepalae* across SW Asia, supplemented with previous results on European species, representing the most complete overview of species relationships in the whole subgenus. Despite the fact that only 20% of the species were omitted, we may have missed a larger part of the genetic diversity of the subgenus since many species are polyphyletic and geographically more comprehensive; thus, intraspecific sampling may be necessary to cover the diversity of the clade. The relict species *V. czerniakowskiana* and *V. fragilis* from Kopet–Dagh and Zagros Mountains form an early-branching clade in the Bayesian and BEAST trees, sister to an unresolved clade comprising the remaining taxa of the subgenus. These two species are taxonomically too little-known, and no certain affinities can be recognized for them [39]. Our efforts for counting their somatic chromosome numbers were also not successful. Within the large polytomy of the Bayesian tree, several small to large clades are resolved that will be discussed in order below. One interesting group is the clustering of *V. mazanderanae* and *V. gaubae* with two accessions each (node K). These two species are the only annual species of the whole subgenus. Annual life form has evolved multiple times independently in the genus *Veronica*, usually associated with a dysploid reduction in base chromosome numbers [27,41]. *Veronica mazanderanae*/*V. gaubae* are another example of independent origin of annual life history in *Veronica*, albeit at least *V. mazanderanae*, by retaining their base chromosome number ($2n = 16$ in *V. mazanderanae*, [27,63]). In contrast to other annual species of *Veronica*, *V. mazanderanae* and *V. gaubae* are relatively long-lived annual species surviving for about 3–4 months in relatively stable conditions in sub-alpine elevations of Alborz mountain range. These species have a long phenological period, and one can see both ripe capsules and young flowers in one individual. We propose that moist, stable habitats offer a long period of reproduction to some durable annuals, which consequently face less selection for a reduction of their base chromosome number. *Veronica mazanderanae* was formerly assigned to *V.* subgenus *Pellidosperma*, but its morphological and karyological characters fit well to *V.* subgenus *Pentasepalae*. The present phylogenetic analysis confirms its placement in *V.* subgenus *Pentasepalae*.

Veronica aucheri, recovered as monophyletic with two accessions, is sister to *V. mirabilis* in a highly supported clade (node C). *Veronica aucheri* is acknowledged as the highest dwelling vascular plant species of Iran, reaching up to about 4800 m in Damavand Mountain, central Alborz [34]. The species has a varied taxonomic history, being classified in sect. *Pocilla* by Boissier [74], Bentham [75] and Römpp [76] based on suggested annuality, terminal inflorescence and foliaceous lower bracts. Bornmüller and Gauba [77] drew attention to the similarity to *V. gaubae*. Elenevskij [78] and Fischer [39] hypothesized a close relationship with *V. bogosensis* Tumadz. from the Northern Caucasus, a species here, and Albach et al. [21] related it to other Caucasian species around *V. peduncularis*. The second species, *V. mirabilis*, is another alpine species restricted to a few localities in Central Alborz. We could not find any reliable morphological synapomorphy that relates these two species to each other. Due to its long corolla tube (6–10 mm), *V. mirabilis* was traditionally classified in *V.* sect. *Paederotoides* Benth., together with *V. paederotae* Boiss. [39], though these two species are highly differentiated in their other characters (e.g., leaf shape, length of filament, length of style, length of capsule). *Veronica paederotae* is not closely related to *V. mirabilis* here but is found isolated in the polytomy (Figure 3). We suggest that some similar pollinators might underpin the convergent evolution of floral traits in *V. mirabilis* and *V. paederotae*, yet we need field observations on pollinators to prove it.

Three species of *V.* subsection *Petraea*, including *V. bogosensis*, *V. caucasica* M. Bieb. and *V. peduncularis* M. Bieb., form a highly supported clade (node M), whereas *V. vendetta–deae* is clustered with the *V. kurdica* species group. In previous studies [21,31], *V. vendetta–deae* represented a sister group to the three formerly mentioned species, albeit with moderate support (60% BS, 73% pp, respectively). These phylogenetic studies, however, lacked representatives of *V. kurdica* or its allied species. The alternative position of *V. vendetta–deae* suggests a probable phylogenetic relationship between the *V. kurdica* species group and species of *V.* subsection *Petraea*. This relationship is also reconstructed in our BEAST tree (Figure 4) and supported by these two regions being geographically adjacent. Members of *V.* subsection *Petraea* are generally distributed in the Caucasus highlands extending west to northeast of Turkey and south to north of Iran [39], whereas *V. kurdica* and relatives (*V. daranica* Saeidi, *V. khorassanica*) are distributed along the Alborz and Zagros Mountains of Iran reaching the southern parts of the Caucasus with its northernmost populations, although these populations have not been sampled here. Our single accession of *V. baranetzkii*, previously considered a member of *V.* subsect. *Petraea*, is grouped with the *V. orientalis–V. multifida* complex. Future phylogenetic studies along the distribution range of *V. baranetzkii* and other species of the subsection not sampled here (*V. petraea* Steven, *V. umbrosa* M. Bieb., *V. filifolia* Lipsky, *V. borisovae* Holub) are required to test the circumscription of the subsection.

Morphological variations among populations of *V. kurdica* were classified under two different subspecies that are well differentiated both taxonomically and geographically [39]. *Veronica kurdica* subsp. *kurdica* differs from *V. kurdica* subsp. *filicaulis* generally in length of pedicel (4–8 mm vs. 1.5–4 mm), length of style (3.5–4.5 mm vs. 1.5–3.5 mm) and corolla color (dark blue vs. pink). The geographical ranges of these two subspecies are also well separated with the type of subspecies distributed throughout the Alborz Mountains in Northern Iran (although some unverified samples propose its occurrence in Armenia as well), whereas *V. kurdica* subsp. *filicaulis* is restricted to Zagros Mountains and highlands of Kerman in Southeastern Iran. *Veronica kurdica* subsp. *filicaulis* was initially published as a distinct species (*V. filicaulis* Freyn) based on specimens from alpine habitats of Oshtoran–kuh Mountain (central Zagros) [79]. The southeastern-most populations in highlands of Kerman were also described as another species, *V. kermanica* Parsa [80]. Later, Fischer [39] reduced them to subspecific rank of *V. kurdica*. Our phylogenetic analyses corroborated the close relationship between the two taxa, but different accessions of the two subspecies (totally six accessions) are intermingled (node L). Another species morphologically similar and closely related to *V. kurdica* is *V. daranica* and Ghahreman. This recently published species was originally compared in its diagnostic characters with *V. davisii* Fischer [81] and was consequently placed in *Veronica* subgenus *Beccabunga* (Hill) M. M. Mart. Ort., Albach and M. A. Fisch in the recent circumscription of genus *Veronica* [27]. Our morphological studies on the type specimen, a new gathering at the locus classicus and a new population in Bakhtiari province revealed that *V. daranica* is neither related to *V. davisii* nor to any other species of *V.* subgenus *Beccabunga* but is in fact morphologically similar to *V. kurdica* subsp. *filicaulis* and is only slightly differentiated from *V. kurdica* subsp. *filicaulis* by its dense, compact growth form (which is probably due to its habitat, usually growing in crevices of rocks) and thinner petals (1–2 mm vs. 2–3 mm). Our phylogenetic studies confirm this relationship as *V. daranica* is nested within *V. kurdica* clade (Figure 3). Therefore, we assign *V. daranica* now to *V.* subgenus *Pentasepalae*. The third species of this well-supported (97% BS, 1 PP) clade, *V. khorassanica*, can be regarded as a vicariant of *V. kurdica* subsp. *kurdica* in the Eastern Alborz extending to Kopet–Dagh Mountains.

Differentiation of *V. kurdica* from *V. orientalis* is sometimes controversial, and their morphological similarities have been discussed previously [39]. The growth form proves to be a good differential character, as already mentioned by Fischer [39], in which *V. kurdica* is distinguishable from *V. orientalis* in its strongly-branched, low-lying, long, occasionally even rooted, thin stems, having no central stem, while *V. orientalis* forms compact branches rising from a strong central caudex. The results presented here (Figures 3 and 4) reveal that *V. kurdica* is not related to *V. orientalis* and is actually phylogenetically more closely related to species of *V.* subgenus *Petraea*. The superficial resemblance of *V. kurdica* to *V. orientalis,* particularly in specimens from high elevations, is therefore the result of convergent evolution of morphological traits. Probably, the *V. orientalis*–*V. multifida* complex is the taxonomically most challenging group in Southwest Asia. Species related to this complex are not precisely delimitated and are difficult to identify because of their high plasticity in morphological characters. Leaf shape and indumentum are important diagnostic characters for distinguishing members of this complex, but these traits have sometimes evolved convergently in different species, in response to habitat conditions. Besides, different ploidy levels and hybridization with other species resulted in many intermediate forms that complicate species boundaries. *Veronica orientalis* as currently circumscribed is distributed from Syria and Central Turkey through Georgia and Armenia to north of Iran and extends southward to Southern Zagros mountain and to Jordan. The other species, *V. multifida,* grows further west reaching Bulgaria, but in its eastern part has a roughly overlapping range of distribution and grows sympatrically in many habitats with *V. orientalis*. Previous phylogenetic efforts for taxonomic and geographic delimitation of this complex showed that *V. orientalis* is not monophyletic in Turkey [36]. The present phylogenetic tree extends that analysis demonstrating that *V. orientalis* has at least three different origins, confirming the polyphyletic nature of this species and therefore suggesting that other taxa can be split from the broadly circumscribed *V. orientalis*. Albach and Al-Gharaibeh [35] recognized *V. polifolia* Benth., *V. leiocarpa* Boiss and *V. orientalis* as distinct species in the Levant. In the present Bayesian tree, *V. polifolia* was recovered as monophyletic and distinct from the other two species, whereas *V. leiocarpa* was assembled together with Levantine and other representatives of *V. orientalis,* suggesting a polyploid origin of this octoploid from lower-ploid taxa in *V. orientalis*. In addition to *V. leiocarpa*, our analysis revealed several other species being closely related to *V. orientalis* and *V. multifida,* including: *V. armena* Boiss., *V. liwanensis* Koch, *V. oltensis* Woronow and *V. baranetzkii*. In its southern distribution range, *V. orientalis* has some relationships with dry adapted *V. leiocarpa* and *V. polifolia* from Jordan and Lebanon, and in the northern-most regions, it is related to cold–wet adapted Caucasian *V. oltensis, V. liwanensis, V. baranetzkii* and *V. armena*. These species are all poorly studied with unsettled taxonomic and geographic borders. Sonibare et al. [36] discussed a scenario regarding repeated expansion and contraction cycles in populations of *V. orientalis,* in response to climatic oscillations, which resulted in several ploidy levels in contact zones of expanding populations; some of them are now widely distributed (hexaploids). The same process of expansion–shrinkage of species ranges has also been proposed to be responsible for the exceptional richness in the genus *Astragalus* from the same area of Northwestern Iran–Eastern Turkey [69]. Closely related species of *V. orientalis* (mentioned above) might have been the result of these climatic cycles between dry and more humid conditions. These climatic shifts could drive diversification of species through allopatric speciation in fragmented, isolated subpopulations of a formerly widespread species, both at the southern and northern margins of distribution of *V. orientalis*. Similarly, *V. multifida* is also split into two lineages, confirming the hypothesis of convergent evolution of pinnatifid leaves in this species, as seen in other species of *Veronica* [21]. This strengthens the idea that the name *V. multifida* is in fact an umbrella covering at least two different species [36]. The fact that different ploidy levels have been found in the species [27], similar to *V. orientalis,* suggests that here, similarly, even more than two taxa are included. In any case, taxonomic delimitation of *V. orientalis* and allies is left for a future comprehensive study with better coverage of the genomes and populations, as well as in-depth analyses of ploidy levels and morphological variation.

Two other Iranian species, *V. acrotheca* and *V. farinosa* Hausskn., are morphologically similar and share a partly sympatric range of distribution in the Alborz and Zagros mountains. Both contain pinnatifid leaves and turgid capsules. In fact, they are differentiated mainly based on upward versus downward curved leaf hairs, shape of calyx (linear vs. lanceolate) and occurrence of a small mucro at the base of the style in *V. acrotheca* [39,82]. In our phylogenetic tree, two accessions of *V. acrotheca* are paraphyletic in relation to *V. farinosa*, suggesting a probable conspecificity of *V. acrotheca* with *V. farinosa*. However, since our samples from these two species are restricted to only three, taxonomic and geographic delimitation of *V. acrotheca* and *V. farinosa* still remain unresolved and, based on the morphological differences, should continue to be recognized at the species level. A weakly supported clade (node H) relates *V. polium* Davis from the southeast of Turkey to *V. acrotheca* and *V. farinosa*, despite being morphologically and geographically distant.

Morphological similarities and probable close relationships among the Turkish *V. fuhsii* Freyn and Sint., *V. thymoides* Davis, *V. elmaliensis* Fischer, *V. cinerea* Boiss. and *V. tauricola* Bornm. have been previously discussed by Fischer [38]. Their close relationship (although with weak support) is reconstructed in a clade together with *V. taurica* from the Crimean Peninsula (node G). In this way, *V. taurica* Willd. separates from the geographically close Caucasian species and rather shows a relationship to the Irano-Turanian species of central Turkey. Several Turkish species did not form monophyletic groups, and their different accessions are scattered in different clades. In a prominent example, one accession each of *V. cuneifolia* D. Don, *V. tauricola* and *V. macrostachya* Vahl together with the only representative of *V. antalyensis* Fischer clustered in a moderately supported clade (node F). The other accessions of these species are distributed in other subclades. The nonmonophyletic nature of several taxonomic entities highlights the need for a critical morphological review of Turkish species. In the Iberian and the Balkan Peninsula, it has been shown that cryptic taxa are present in *V.* subgenus *Pentasepalae* [29,33], and it is highly likely to find such cryptic taxa also in Turkey using highly variable molecular markers. In addition, the fact that there are several species in Turkey containing infra-specific taxa highlights the high morphological variation and complex taxonomy of *V.* subgenus *Pentasepalae* in this region. One of these polymorphic species is *V. macrostachya*, including four subspecies [38,39]. Three subspecies (i.e., *V. macrostachya* subsp. *macrostachya*, *V. macrostachya* subsp. *mardinensis* (Bornm.) Fischer and *V. macrostachya* subsp. *sorgerae* Fischer) are distributed mainly in Southern Turkey, Northern Syria and Lebanon and are not morphologically well separated with some intermediate populations [38]. However, the fourth subspecies (*V. macrostachya* subsp. *schizostegia*) is quite uniform throughout its range and is separated from other subspecies by several morphological characters. It is geographically restricted to mountainous areas along the Iran–Iraq borders and adjacent Kurdistan of Iraq. We included representatives of the latter subspecies and of *V. macrostachya* subsp. *sorgerae* in our phylogenetic analyses, which demonstrated that they are not closely related. On one hand, *V. macrostachya* subsp. *sorgerae* is nested in a clade of Turkish species and has a sister–group relationship with *V. antalyensis* from Southern Turkey, whereas *V. macrostachya* subsp. *schizostegia* did not group closely to any other specific species. Based on this, we reached the conclusion that *V. macrostachya* subsp. *schizostegia* is markedly different from the other three subspecies and deserves to be raised to species rank (see taxonomic treatment). A more in-depth morpho-molecular analysis will resolve the taxonomic situation of the other three subspecies. Another case is *V. bombycina* Boiss. having three subspecies. Two of them (*V. bombycina* subsp. *bombycina* and *V. bombycina* subsp. *froediniana* Rech.) assembled weakly with *V. caespitosa* Boiss. in the more easterly distributed clade D (Figures 3 and 4), whereas *V. bombycina* subsp. *bolkardaghensis* M.A.Fisch. is nested in an isolated position in the polytomy (Figure 3) or in the more western clade D (Figure 4). This supports our a priori suspicion that these are two species, which is formalized in the taxonomic treatment section. Other examples of non-monophyletic species are *V. cuneifolia* and *V. tauricola* with accessions scattered along the phylogenetic tree, although being morphologically similar. There are many endemic species of *V.* subgenus *Pentasepalae* in Turkey, and finding new cryptic taxa further highlights the importance of this region as a hotspot and center of diversification of the subgenus.

The clade corresponding to members of *V.* subsection *Pentasepalae* (node N) receives high support, and relationships among species of this clade are also largely resolved. In a previous phylogenetic analysis, Rojas-Andrés et al. [31] discussed that support for monophyly of this subsection was lower when *V. krylovii* was included. In agreement with these observations, *V. krylovii* did not cluster with *V.* subsection *Pentasepalae* in the present phylogenetic analysis. Being endemic to the Altai Mountains and South Siberia, *V. krylovii* is situated at the northern- and eastern-most margin of the subgenus. Despite being left ungrouped in our Bayesian tree in the BEAST analysis, *V. krylovii* forms a sister relationship with *V. kopetdaghensis* from the northeast of Iran. Reconstructed relationships among members of *V.* subsection *Pentasepalae* largely corresponds to the ITS phylogeny of Rojas-Andrés et al. [31], with only slight differences in a few shallow nodes and some node support. For a detailed discussion on phylogenetic relationships of *V.* subsection *Pentasepalae,* we refer to Rojas-Andrés et al. [31] and Padilla-García et al. [29].

Among the individual species that did not form any cluster, *V. fuhsii* var. *linearis* Parsa is worth mentioning. This variety was described from Talesh highlands in the Northwestern Alborz [80]. Our studies on type materials and a new gathering from the type locality revealed that this variety is not related to other Iranian species except for some superficial resemblance to *V. multifida*. An association to *V. fuhsii* from Northeastern Turkey was also not verified. Morphologically it has some similarities to the Turkish species *V. elmaliensis* Fischer, using the taxonomic key in the Flora of Turkey [38]. According to morphological characteristics of *V. fuhsii* var. *linearis*, we consider that this taxon merits the specific rank, being endemic to Northern Iran with probable relatives in Eastern Turkey (see taxonomic treatment).

4.3. *Origin of V. Subgenus Pentasepalae: Out of the Iranian Plateau*

Our analyses yielded a divergence time (origin age) of ca. 9 Mya for *V.* subgenus *Pentasepalae* and a place of origin on the Iranian plateau, more specifically in the mountain ranges of Alborz, Kopet–Dagh and Zagros mountains. This time estimate is consistent with the late Miocene global cooling and drying [83]. On a more local scale, the geological hypothesis [72] suggests that the major uplift of the Iranian plateau has taken place 15–12 Mya, which resulted in a more continental climate under the predominant global cooling climate. The increasing aridity and continentality has probably triggered the origin of *V.* subgenus *Pentasepalae*. The initial split in the subgenus took place in ca. 7 Mya, and diversification of major clades occurred between 5.6 to 4 Mya, which coincides with another uplift and active mountain building in the Iranian plateau in about 5 Mya [70–72], implying that these mountain uplifts have probably played a major role in allopatric speciation of the subgenus. Many temperate plants that are now widely distributed across the Northern Hemisphere have been hypothesized to have originated in the Qinghai–Tibet Plateau (QTP) and adjacent regions and then migrated to other regions of the Northern Hemisphere [84–88]. Likewise, the tribe Veroniceae with nine genera and about 500 species has most likely originated in the QTP, with four of its genera restricted to this region [42,88,89]. West of the QTP, mountains of the Irano-Turanian region have been shown to act as a secondary center of speciation and diversification [90], so that numerous species groups, particularly xerophytes, originated and started their radiations there [3,7–10,12–17,19,20,91]. This is also the case for *V.* subgenus *Pentasepalae,* which originated in western parts of the IT region in the Iranian plateau based on our analyses (Figure 5). High mountains of Iran are part of the Irano-Anatolian biodiversity hotspot [92], harboring a high concentration of endemic species [93,94] and likely the center of origin for several xeromorphic taxa. However, the biogeography, diversification and evolutionary history of these plants are still little known.

4.4. Dispersal and Vicariance

According to the biogeographical analysis, several major migration routes within *V.* subgenus *Pentasepalae* can be recognized: dispersal from the Iranian plateau to Anatolia and then to Caucasus and Crimea, and several back migrations from the Caucasus and Anatolia to Iranian highlands; dispersal to the Altai mountains of Central Asia and also to North Africa and the Western Mediterranean region, and from there to the Euro-Siberian area. Close relationships and multiple dispersals to Turkish and Caucasus Mountains from the Iranian highlands were expected, due to the geographical proximity of these three regions. Furthermore, our study revealed that floristic exchange among the Iranian highlands and Caucasus and Turkish mountains were not unidirectional, and several waves of migrations from Turkish or Caucasus Mountains into the Iranian highlands are detectable. After dispersal to Anatolia in the common ancestor of clade D, a diversification took place around *V. orientalis* and *V. multifida*. Likewise, a broadly defined *V.* subsection *Petraea* (including relatives of *V. kurdica*) originated in the Caucasus and dispersed backward to Alborz and Zagros by members of *V. kurdica* group reaching highlands of Kerman, the southern-most limit of the whole subgenus. The Crimean endemic *V. taurica* was shown to have originated from an ancestor in Northeastern Turkey.

There is a remarkable dispersal from the Iranian plateau to Altai Mountains of Central Asia in about 2.8 Mya. Current distribution of *V. krylovii* in the Tarbagatai Mountains of Kazakhstan and the Altai Mountains of Russia is a long disjunct (about 3000 km) from its sister species in our analysis, *V. kopetdaghensis* of Northeastern Iran (Figures 1 and 4). This disjunct pattern of distribution between high mountains of Central Asia with either mountains of Southeastern or Northern Iran has previously been addressed in several taxa [95,96]. One paramount example for such kind of distribution is the genus *Paraquilegia* Drumm. and Hutch. (Ranunculaceae) with about 11 species in Central Asian mountains north up to the Altai Mountains, and *P. caespitosa* (Boiss. and Hohen.) Drumm. and Hutch. being endemic to central Alborz, north of Iran [97]. This pattern of disjunction has been suggested to be due to vicariance and results from the postglacial warming, which has forced the alpine plants to higher elevations and fragmented their formerly continuous ranges [95,98]. However, this hypothesis has never been tested through phylogeographic analysis for any of these disjunctly distributed species. The estimated age for the common ancestor of *V. krylovii* and *V. kopetdaghensis* is about 2.8 Mya, which corresponds to the beginning of glacial–interglacial cycles that started at about 2.6 Mya in the early Quaternary [99]. Considering also the lack of evidence for frequent long-distance dispersal of *Veronica* seeds in general weakens the idea of long-distance dispersal in this case. It seems reasonable to suggest that the once widely distributed ancestor of these two species at lower elevations had to move upward to the Altai and Kopet–Dagh mountains during interglacial periods, and intermediate populations in the dry lowlands of Turkmenistan and Kazakhstan went extinct. Following this, we argue that the occurrence of *V. krylovii* in Central Asia is likely due to vicariance rather than long-distance dispersal, a scenario that might be correct for other species with the same distribution pattern in the mountains of Iran and highlands of Central Asia and Himalaya as well. Nevertheless, the absence of *V. krylovii* or *V. kopetdaghensis* or their relatives from Pamir and Tian–Shan Mountains needs to be explained by the extinction of intermediate populations in these drier mountains and the lack of refugia there.

As mentioned before, examples of biotic migration from the Irano-Turanian region to the Mediterranean and Euro-Siberian regions are numerous, but neither of these migrations occurred at the same window of time nor took place through the same migratory route. Colonization of the Western Mediterranean from Eastern Mediterranean–Western Asian species are traditionally attributed to a North Mediterranean pathway via Southern Europe [18,20,84], whereas other studies offer an alternative dispersal route from North Africa for some taxa [11,100,101]. Our analysis proposes a dispersal from the Iranian plateau (Central Alborz Mountain) to Northwest Africa and successively to the Iberian Peninsula, thereafter to the Mediterranean area of Southern Europe and then to Central

Europe. This pattern is compatible with a North African migration route for *V.* subgenus *Pentasepalae*. Migration from the Iranian plateau to Northwest Africa is relatively young (ca. 2.5 Mya) and likely happened during the early glacial cold climates, when North Africa had more favorable climatic conditions and the northern side of the Mediterranean Sea (Europe) was glaciated. Afterward, in the interglacial period, the cold-adapted species of *V.* subgenus *Pentasepalae* migrated to higher elevations and settled in disjunct sub-populations in the Atlas Mountains of Northwestern Africa and the highlands of the Iberian Peninsula [73], while populations of Northeastern Africa vanished in response to warm climate. The role of Northwestern Africa and the Iberian Peninsula as a Pleistocene refugia for both warm-adapted and cold-adapted species (including *Veronica*) has been highlighted in several studies [73,102,103], and the close floristic affinity of Northwestern Africa to the Irano-Turanian region has previously been mentioned in some classic floristic publications [104]. Zohary [105] in his review on the geobotanical foundations of the Middle East included the southern foothills of the Atlas Mountains in Northwestern Africa to the Irano-Turanian region as a distinct province, the Mauritanian steppes, characterized by several typical Irano-Turanian species such as: *Noaea mucronata* (Forssk.) Asch. and Schweinf., *Fraxinus xanthoxyloides* (G. Don) Wall. and ex. A.DC., *Achillea santolina* Sibth. and Sm., *Salvia balansae* Noe ex. Coss., *Centaurea carolipauana* Fern.Casas and Susanna, *Artemisisa herba–alba* Asso and *Ferula tingitana* L. Although this opinion was later rejected and this region was accepted as a transitional region between Mediterranean and Saharo–Sindian regions [106], the relatively high numbers of Irano-Turanian species or species with their affinities in the Irano-Turanian region emphasizes the close floristic connection between Northwestern Africa and the Irano-Turanian region. Although the time and dispersal route of the range expansion of some species groups such as *Centaurea* [11], *Haplophyllum* [18] and *Delphinium* [107] are different, most probably several other species have had a similar evolutionary history as *Veronica* subgenus *Pentasepalae* and took the same migration route from the Irano-Turanian region to Northwestern Africa.

Repeated expansion–retraction events in response to Pleistocene climatic oscillations probably resulted in various ploidy levels in contact zones among different cytotypes and acted as a biodiversity driver in the Northern Balkan Peninsula and areas further west in the Mediterranean region. This polyploidization event associated with genome downsizing through which polyploid species gain novel features that make them better able to tolerate the colder and wetter conditions of higher latitudes might have contributed to the colonization of new habitats in the Euro-Siberian cold conditions, while the diploid progenitors have been confined to refugial areas of the Mediterranean region [31,108].

4.5. Taxonomic Treatment

Veronica bolkardaghensis (M.A.Fisch.) Albach **comb. and stat. nov.**

≡ *Veronica bombycina* subsp. *bolkardaghensis* M.A.Fisch. in Pl. Syst. Evol. 128: 294 (1977).

Holotype: Turkey, Konya: districto Ermenek, in monte Yelibel Dag inter oppida Ermenek et Konya, in rupibus calcareis et glareosis usque ad cacumen, 2080–2350 m", A. Huber-Morath no. 8613, 10. Jun. 1948 (BASBG); isotypes: G! (343616), WU! (0070354)

Diagnosis: *Veronica bolkardaghensis* resembles *Veronica bombycina* in its habit of dense cushions with white, densely tomentose indumentum and growing among alpine scree. *Veronica bolkardaghensis* differs from *Veronica bombycina* in rather subtle morphological characters such as the clearly revolute leaves, the longer calyx (3–7 mm vs. 2.5–3.5 mm) and calyx shape (widely ovate vs. oblong). However, it is likely that closer inspection would reveal further subtle differences. Further evidence for separation is geography since *Veronica bolkardaghensis* is a more western element from Southern Turkey, whereas *Veronica bombycina* is a more eastern element from Southeastern Turkey (subsp. *froediniana*) south to Lebanon Mts. and Anti-Lebanon Mts. (subsp. *bombycina*).

Distribution: Turkey, Taurus Mts of southern Anatolia.

Veronica schizostegia (Bornm.) Doostm. and Bordbar **comb. and stat. nov.**

≡ *Veronica aleppica* var. *schizostegia* Bornm. in Feddes Repertorium 9: 113 (1910) ≡ *Veronica aleppica* subsp. *schizostegia* (Bornm.) Bornm. in Beih. Bot. Centralbl. 28: 480 (1911) ≡ *Veronica macrostachya* var. *schizostegia* (Bornm.) Riek in Feddes Rep., Beih. 79: 27 (1935) ≡ *Veronica macrostachya* subsp. *schizostegia* (Bornm.) M. A. Fischer in Flora Turkey and East Aegean Islands 6: 744 (1978)

Lectotype (designated by Fischer (1981), p. 136), second step lectotype (designated here): Iraq, Kurdistan, in monte Kuhi–Sefin supra pagum Schaklava (ditionis Erbil), 1000 m. 21. 5. 1893 J. Bornmüller 1628 (B! 0278579); Isolectotypes (designated here): B! (0278578), BP! (347767), BR* (BR0000005423033), JE! (152), WU! (0029659), W! (1895–1676), OXF!, P! (P03529531, P03529532).

Diagnosis: *Veronica schizostegia* differs from the morphologically similar *V. macrostachya* subsp. *mardinensis* mainly in loosely hairy leaves (vs. densely tomentose gray leaves), longer and denser inflorescences having only glandular hairs (vs. loose inflorescences with both glandular and eglandular hairs) and longer peduncles (2–4 cm vs. 1–2 cm).

Distribution: Western Iran, close to border with Iraq and adjacent highlands of Kurdistan of Iraq and Turkey.

Veronica parsana Doostm. and Bordbar **nom. nov.**

≡ *Veronica fuhsii var. linearis* Parsa in Flore de l'Iran 4: 437 (1949)

Lectotype (designated here): Iran: Gilan, Talesh area, Aspina, 1700 m. 27.07.1941, anonymous collector, TEH 779 (4986)! Isolectotype (designated here): TEH 4987!

Diagnosis: *Veronica parsana* is morphologically similar to *V. elmaliensis* Fischer but differs in loosely hairy stems (vs. densely whitish–hirsute hairs), longer styles (4–5 mm vs. 3.5–4 mm) and longer pedicles (5–8 mm vs. 2–6 mm).

Distribution: Endemic to Western Alborz mountain chain in Talesh highlands.

Etymology: *Veronica parsana* is named after Ahmad Parsa (1907–1997), one of the first Iranian botanists. He made a significant contribution to the plant taxonomy of Iran by writing the first Flora for Iran.

Note: Here we raised the taxonomic rank of *Veronica fuhsii var. linearis* to a specific level, but in order to avoid a homonymy with *Veronica linearis* (Bornm.) Rojas-Andrés and M.M.Mart.Ort [28], a new name was needed.

5. Conclusions

This study supports the monophyly of *V.* subgenus *Pentasepalae* and re-assigns *V. mazanderanae* and *V. daranica* to the subgenus. Several well-supported clades are reconstructed within a poorly resolved main clade, which is interpreted as the result of rapid radiation in the Irano-Turanian region as well as probable high rate of extinction during interglacial periods that left several relict and isolated species in Southwest Asia. Our phylogeographical analysis of *V.* subgenus *Pentasepalae* indicates that the subgenus originated in the Iranian Plateau (including Alborz, Zagros and Kopet-Dagh Mountains) approximately 9 Mya and then dispersed out of the Iranian Plateau to other parts of Eurasia. A North African route is proposed for the migration of derived species to the Western Mediterranean region during early Quaternary glacial times.

Supplementary Materials: The following supporting information can be downloaded at https://www.mdpi.com/article/10.3390/biology11050639/s1, Table S1. Details of specimens of *Veronica* included in this study, including locality, herbarium information and GenBank accession numbers.

Author Contributions: Conceptualization, M.D. and D.C.A.; supervision, F.B., M.M. and D.C.A.; field work, M.D. and M.M.; data curation, M.D., F.B. and D.C.A.; analysis, M.D. and F.B.; writing—original draft preparation, M.D.; writing—review and editing, F.B., M.M. and D.C.A. All authors have read and agreed to the published version of the manuscript.

Funding: This research was funded by the Iran National Science Foundation (INSF), grant number 99025662.

Institutional Review Board Statement: Not applicable.

Informed Consent Statement: Not applicable.

Data Availability Statement: All herbarium specimens used in this study are kept in the collections of different institutions (see Table S1 for details).

Acknowledgments: The Herbarium curators and staff at E, FUMH, IRAN, MIR, OLD, TARI, TUH, Gilan University Herbarium and Herbarium of Research Centre of Agriculture and Natural Resources of Kerman are highly appreciated for their help during visits and in some cases for providing leaf particles. We are particularly grateful to Amir Talebi for help in plant collection, Atefeh Ghorbanalizadeh for help during field work and preparing the maps, Elham Hatami for help in the phylogenetic analysis and Eike Mayland-Quellhorst for help in the molecular lab.

Conflicts of Interest: The authors declare no conflict of interest.

References

1. López-Vinyallonga, S.; Mehregan, I.; Garcia-Jacas, N.; Tscherneva, O.; Susanna, A.; Kadereit, J.W. Phylogeny and Evolution of the *Arctium-Cousinia* Complex (Compositae, Cardueae-Carduinae). *Taxon* **2009**, *58*, 153–171. [CrossRef]
2. Azani, N.; Bruneau, A.; Wojciechowski, M.F.; Zarre, S. Miocene climate change as a driving force for multiple origins of annual species in *Astragalus* (Fabaceae, Papilionoideae). *Mol. Phylogenet. Evol.* **2019**, *137*, 210–221. [CrossRef] [PubMed]
3. Moharrek, F.; Sanmartín, I.; Kazempour-Osaloo, S.; Nieto Feliner, G. Morphological Innovations and Vast Extensions of Mountain Habitats Triggered Rapid Diversification Within the Species-Rich Irano-Turanian Genus *Acantholimon* (Plumbaginaceae). *Front. Genet.* **2019**, *9*, 698. [CrossRef] [PubMed]
4. Jafari, F.; Zarre, S.; Gholipour, A.; Eggens, F.; Rabeler, R.K.; Oxelman, B. A new taxonomic backbone for the infrageneric classification of the species-rich genus *Silene* (Caryophyllaceae). *Taxon* **2020**, *69*, 337–368. [CrossRef]
5. Pahlevani, A.H.; Liede-Schumann, S.; Akhani, H. Diversity, distribution, endemism and conservation status of *Euphorbia* (Euphorbiaceae) in SW Asia and adjacent countries. *Plant Syst. Evol.* **2020**, *306*, 80. [CrossRef]
6. Djamali, M.; Baumel, A.; Brewer, S.; Jackson, S.T.; Kadereit, J.W.; López-Vinyallonga, S.; Mehregan, I.; Shabanian, E.; Simakova, A. Ecological implications of *Cousinia* Cass. (Asteraceae) persistence through the last two glacial–interglacial cycles in the continental Middle East for the Irano-Turanian flora. *Rev. Palaeobot. Palyno.* **2012**, *172*, 10–20. [CrossRef]
7. Mohammadin, S.; Peterse, K.; van de Kerke, S.J.; Chatrou, L.W.; Dönmez, A.A.; Mummenhoff, K.; Pires, J.C.; Edger, P.P.; Al-Shehbaz, I.A.; Schranz, M.E. Anatolian origins and diversification of *Aethionema*, the sister lineage of the core Brassicaceae. *Am. J. Bot.* **2017**, *104*, 1042–1054. [CrossRef]
8. Karl, R.; Koch, M.A. A world-wide perspective on crucifer speciation and evolution: Phylogenetics, biogeography and trait evolution in tribe Arabideae. *Ann. Bot.* **2013**, *112*, 983–1001. [CrossRef]
9. Zhang, M.L.; Sanderson, S.C.; Sun, Y.X.; Byalt, V.V.; Hao, X.L. Tertiary montane origin of the Central Asian flora, evidence inferred from cpDNA sequences of *Atraphaxis* (Polygonaceae). *J. Integr. Plant Boil.* **2014**, *56*, 1125–1135. [CrossRef]
10. Zhang, M.L.; Wen, Z.B.; Fritsch, P.W.; Sanderson, S.C. Spatiotemporal Evolution of *Calophaca* (Fabaceae) Reveals Multiple Dispersals in Central Asian Mountains. *PLoS ONE* **2015**, *10*, e0123228. [CrossRef]
11. Font, M.; Garcia-Jacas, N.; Vilatersana, R.; Roquet, C.; Susanna, A. Evolution and biogeography of *Centaurea* section *Acrocentron* inferred from nuclear and plastid DNA sequence analyses. *Ann. Bot.* **2009**, *103*, 985–997. [CrossRef]
12. Panahi, M. Biogeographic reconstruction of the genus *Ferula* inferred from analyses of nrDNA and cpDNA sequences. *Iran. J. Bot.* **2019**, *25*, 79–94. [CrossRef]
13. Eslami-Farouji, A.; Khodayari, H.; Assadi, M.; Çetin, Ö.; Mummenhoff, K.; Özüdoğru, B. Phylogeny and biogeography of the genus *Hesperis* (Brassicaceae, tribe Hesperideae) inferred from nuclear ribosomal DNA sequence data. *Plant Syst. Evol.* **2021**, *307*, 17. [CrossRef]
14. Zhang, M.L.; Zeng, X.Q.; Sanderson, S.C.; Byalt, V.V.; Sukhorukov, A.P. Insight into Central Asian flora from the Cenozoic Tianshan montane origin and radiation of *Lagochilus* (Lamiaceae). *PLoS ONE* **2017**, *12*, e0178389. [CrossRef]
15. Soreng, R.J.; Gillespie, L.J.; Boudko, E.A.; Cabi, E. Biogeography, Timing, and Life History Traits in the PPAM clade: Coleanthinae (syn. Puccinelliinae), Poinae, Alopecurinae superclade, Miliinae, and Avenulinae and Phleinae (Poaceae, Pooideae, Poeae). *J. Syst. Evol.* **2022**. Online ahead of print. [CrossRef]
16. Scheunert, A.; Heubl, G. Against all odds: Reconstructing the evolutionary history of *Scrophularia* (Scrophulariaceae) despite high levels of incongruence and reticulate evolution. *Org. Divers. Evol.* **2017**, *17*, 323–349. [CrossRef]
17. Žerdoner Čalasan, A.; German, D.A.; Hurka, H.; Neuffer, B. A story from the Miocene: Clock-dated phylogeny of *Sisymbrium* L. (Sisymbrieae, Brassicaceae). *Ecol. Evol.* **2021**, *11*, 2573–2595. [CrossRef]
18. Manafzadeh, S.; Salvo, G.; Conti, E.A. tale of migrations from east to west: The Irano-Turanian floristic region as a source of Mediterranean xerophytes. *J. Biogeogr.* **2014**, *41*, 366–379. [CrossRef]
19. Malik, S.; Vitales, D.; Qasim Hayat, M.; Korobkov, A.A.; Garnatje, T.; Vallès, J. Phylogeny and biogeography of *Artemisia* subgenus *Seriphidium* (Asteraceae: Anthemideae). *Taxon* **2017**, *66*, 934–952. [CrossRef]

20. Peterson, A.; Harpke, D.; Peterson, J.; Harpke, A.; Peruzzi, L. A pre-Miocene Irano-Turanian cradle: Origin and diversification of the species-rich monocot genus *Gagea* (Liliaceae). *Ecol. Evol.* **2019**, *9*, 5870–5890. [CrossRef]
21. Albach, D.C.; Martínez-Ortega, M.M.; Chase, M.W. *Veronica*: Parallel morphological evolution and phylogeography in the Mediterranean. *Plant Syst. Evol.* **2004**, *246*, 177–194. [CrossRef]
22. Angiosperm Phylogeny Group III. An update of the Angiosperm Phylogeny Group classification for the orders and families of flowering plants: APG III. *Bot. J. Linn. Soc.* **2009**, *161*, 105–121. [CrossRef]
23. Angiosperm phylogeny Group IV. An update of the Angiosperm Phylogeny Group classification for the orders and families of flowering plants: APG IV. *Bot. J. Linn. Soc.* **2016**, *181*, 1–20. [CrossRef]
24. Fischer, E. Scrophulariaceae. In *The Families and Genera of Vascular Plants*; Kadereit, J., Ed.; Springer: Berlin/Heidelberg, Germany, 2004; Volume 7, pp. 333–433.
25. Albach, D.C.; Martínez-Ortega, M.M.; Fischer, M.A.; Chase, M.W. A new classification of the tribe Veroniceae—problems and a possible solution. *Taxon* **2004**, *53*, 429–452. [CrossRef]
26. Garnock-Jones, P.; Albach, D.C.; Briggs, B.G. Botanical names in Southern Hemisphere *Veronica* (Plantaginaceae): Sect. *Detzneria, sect. Hebe*, and sect. *Labiatoides. Taxon* **2007**, *56*, 571–582. [CrossRef]
27. Albach, D.C.; Martínez-Ortega, M.M.; Delgado, L.; Weiss-Schneeweiss, H.; Özgökce, F.; Fischer, M.A. Chromosome Numbers in Veroniceae (Plantaginaceae): Review and Several New Counts. *Ann. Missouri Bot. Gard.* **2008**, *95*, 543–566. [CrossRef]
28. Rojas-Andrés, B.M.; Rico, E.; Martínez-Ortega, M.M. A nomenclatural treatment for *Veronica* subsect. *Pentasepalae (Plantaginaceae sensu APG III) and typification of several names. Taxon* **2016**, *65*, 617–627. [CrossRef]
29. Padilla-García, N.; Rojas-Andrés, B.M.; López-González, N.; Castro, M.; Castro, S.; Loureiro, J.; Albach, D.C.; Machon, N.; Martínez-Ortega, M.M. The challenge of species delimitation in the diploid-polyploid complex *Veronica* subsection *Pentasepalae. Mol. Phylogenet. Evol.* **2018**, *119*, 196–209. [CrossRef] [PubMed]
30. Yaylaci, Ö.K.; Sezer, O.; Özgisi, K.; Özturk, D.; Erkara, İ.P.; Koyuncu, O.; Ocak, A. A new *Veronica* (Plantaginaceae) species from central Anatolia, Turkey. *Phytotaxa* **2018**, *362*, 55–67. [CrossRef]
31. Rojas-Andrés, B.M.; Albach, D.C.; Martínez-Ortega, M.M. Exploring the intricate evolutionary history of the diploid–polyploid complex *Veronica* subsection *Pentasepalae* (Plantaginaceae). *Bot. J. Linn. Soc.* **2015**, *179*, 670–692. [CrossRef]
32. Rojas-Andres, B.M.; Martínez-Ortega, M.M. Taxonomic revision of *Veronica* subsection *Pentasepalae* (Veronica, Plantaginaceae sensu APG III). *Phytotaxa* **2016**, *285*, 1–100. [CrossRef]
33. Martínez-Ortega, M.M.; Delgado, L.; Albach, D.C.; Elena-Rosselló, J.A.; Rico, E. Species Boundaries and Phylogeographic Patterns in Cryptic Taxa Inferred from AFLP Markers: *Veronica* subgen. *Pentasepalae (Scrophulariaceae) in the Western Mediterranean. Syst. Bot.* **2004**, *29*, 965–986. [CrossRef]
34. Noroozi, J.; Körner, C. A bioclimatic characterization of high elevation habitats in the Alborz mountains of Iran. *Alp. Bot.* **2018**, *128*, 1–11. [CrossRef]
35. Albach, D.C.; Al-Gharaibeh, M. Systematics and biogeography of *Veronica* subg. *Pentasepalae from the Levant. Willdenowia* **2015**, *45*, 5–14. [CrossRef]
36. Sonibare, M.A.; Armagan, M.; Özgökce, F.; Yaprak, A.E.; Mayland-Quellhorst, E.; Albach, D.C. Analysis of taxonomic and geographic patterns of Turkish *Veronica orientalis* using nuclear and plastid DNA and morphological data. *Plant Syst. Evol.* **2014**, *300*, 645–664. [CrossRef]
37. Borissova, A.G. *Veronica In Flora of the U.S.S.R.*; Shishkin, B.K., Bobrov, E.G., Eds.; Akademii Nauk SSSR: Moscow, Russia, 1955; Volume 22, pp. 373–557.
38. Fischer, M.A. Scrophulariaceae. In *Flora of Turkey and the East Aegean Islands*; Davis, P.H., Edmondson, J.R., Mill, R.R., Parris, B.S., Eds.; Edinburgh University Press: Edinburgh, UK, 1978; Volume 6, pp. 458–784.
39. Fischer, M.A. Veronica. In *Flora Iranica 147 (Scrophulariaceae I)*; Rechinger, K.H., Ed.; Akadem-ische Druck- u, Verlagsanstalt: Graz, Austria, 1981; pp. 52–165.
40. Saeidi-Mehrvarz, S. Veronica. In *Flora of Iran 68*; Assadi, M., Maassoumi, A.A., Khatamsaz, M., Mozaffarian, V., Eds.; Research Institute of Forests and Rangelands Publications: Tehran, Iran, 2011. (In Persian)
41. Albach, D.C.; Martínez-Ortega, M.M.; Fischer, M.A.; Chase, M.W. Evolution of Veroniceae: A Phylogenetic Perspective. *Ann. Mssouri Bot. Gard.* **2004**, *91*, 275–302. Available online: https://www.jstor.org/stable/3298609 (accessed on 1 December 2021).
42. Surina, B.; Pfanzelt, S.; Einzmann, H.J.R.; Albach, D.C. Bridging the Alps and the Middle East: Evolution, phylogeny and systematics of the genus *Wulfenia* (Plantaginaceae). *Taxon* **2014**, *63*, 843–858. [CrossRef]
43. Meudt, H.M.; Rojas-Andrés, B.M.; Prebble, J.M.; Low, E.; Garnock-Jones, P.J.; Albach, D.C. Is genome downsizing associated with diversification in polyploid lineages of *Veronica*? *Bot. J. Linn. Soc.* **2015**, *178*, 243–266. [CrossRef]
44. Doyle, J.J.; Doyle, J.L. A rapid DNA isolation procedure for small quantities of fresh leaf tissue. *Phytochem. Bull.* **1987**, *19*, 11–15.
45. Blattner, F.R. Direct Amplification of the Entire ITS Region from Poorly Preserved Plant Material Using Recombinant PCR. *BioTechniques* **1999**, *27*, 1180–1186. [CrossRef]
46. Sang, T.; Crawford, D.J.; Stuessy, T.F. Documentation of reticulated evolution in peonies (*Paeonia*) using internal transcribed spacer sequences of nuclear ribosomal DNA: Implications for biogeography and concerted evolution. *Proc. Natl. Acad. Sci. USA* **1995**, *92*, 6813–6817. [CrossRef]

47. White, T.J.; Bruns, T.D.; Lee, S.; Taylor, J. Amplification and Direct Sequencing of Fungal Ribosomal RNA Genes for Phylogenetics. In *PCR Protocols: A Guide to Methods and Applications*; Innis, M., Gelfand, D., Sninsky, J., White, T., Eds.; Academic Press: San Diego, CA, USA, 1990; pp. 315–322.
48. Katoh, K.; Toh, H. Recent developments in the MAFFT multiple sequence alignment program. *Brief. Bioinform.* **2008**, *9*, 286–298. [CrossRef]
49. Muller, K.; Muller, J.; Quandt, D. PhyDE®: Phylogenetic Data Editor, Version 0.9971. 2010. Available online: http://www.phyde.de/index.html (accessed on 1 March 2021).
50. Müller, K. SeqState: Primer design and sequence statistics for phylogenetic DNA datasets. *Appl. Bioinform.* **2005**, *4*, 65–69. [CrossRef]
51. Swafford, D.L. *PAUP*: Phylogenetic Analysis Using Parsimony (*and Other Methods)*; Version 4.0.b10; Sinauer Associates: Sunderland, MA, USA, 2002.
52. Nixon, K.C. The Parsimony Ratchet, a New Method for Rapid Parsimony Analysis. *Cladistics* **1999**, *15*, 407–414. [CrossRef]
53. Müller, K. PRAP—computation of Bremer support for large data sets. *Mol. Phylogeneti. Evol.* **2004**, *31*, 780–782. [CrossRef]
54. Darriba, D.; Taboada, G.L.; Doallo, R.; Posada, D. jModelTest 2: More models, new heuristics and parallel computing. *Nat. Methods* **2012**, *9*, 772. [CrossRef]
55. Silvestro, D.; Michalak, I. raxmlGUI: A graphical front-end for RAxML. *Org. Divers. Evol.* **2012**, *12*, 335–337. [CrossRef]
56. Ronquist, F.; Teslenko, M.; van der Mark, P.; Ayres, D.L.; Darling, A.; Höhna, S.; Larget, B.; Liu, L.; Suchard, M.A.; Huelsenbeck, J.P. MrBayes 3.2: Efficient Bayesian Phylogenetic Inference and Model Choice across a Large Model Space. *Syst. Biol.* **2012**, *61*, 539–542. [CrossRef]
57. Suchard, M.A.; Lemey, P.; Baele, G.; Ayres, D.L.; Drummond, A.J.; Rambaut, A. Bayesian phylogenetic and phylodynamic data integration using BEAST 1.10. *Virus Evol.* **2018**, *4*, vey016. [CrossRef]
58. Drummond, A.J.; Ho, S.Y.W.; Phillips, M.J.; Rambaut, A. Relaxed Phylogenetics and Dating with Confidence. *PLoS Biol.* **2006**, *4*, e88. [CrossRef]
59. Rambaut, A.; Suchard, M.A.; Xie, D.; Drummond, A.J. Tracer v1.7. 2014. Available online: http://beast.bio.ed.ac.uk/Tracer (accessed on 20 January 2022).
60. Yu, Y.; Harris, A.J.; Blair, C.; He, X. RASP (Reconstruct Ancestral State in Phylogenies): A tool for historical biogeography. *Mol. Phylogenet. Evol.* **2015**, *87*, 46–49. [CrossRef] [PubMed]
61. Yu, Y.; Harris, A.J.; He, X. S-DIVA (Statistical Dispersal-Vicariance Analysis): A tool for inferring biogeographic histories. *Mol. Phylogenet. Evol.* **2010**, *56*, 848–850. [CrossRef] [PubMed]
62. Ronquist, F.; Huelsenbeck, J.P. MrBayes 3: Bayesian phylogenetic inference under mixed models. *Bioinformatics* **2003**, *19*, 1572–1574. [CrossRef] [PubMed]
63. Doostmohammadi, M.; Mirtadzadini, M.; Bordbar, F. Chromosome numbers of some annual *Veronica* (Plantaginaceae) with report of *V. tenuissima* from flora of Iran. *Feddes Repert.* **2021**, *132*, 158–166. [CrossRef]
64. Valente, L.M.; Savolainen, V.; Vargas, P. Unparalleled rates of species diversification in Europe. *Proc. Biol. Sci.* **2010**, *277*, 1489–1496. [CrossRef]
65. Bell, C.D.; Mavrodiev, E.V.; Soltis, P.S.; Calaminus, A.K.; Albach, D.C.; Cellinese, N.; Garcia-Jacas, N.; Soltis, D.E. Rapid diversification of *Tragopogon* and ecological associates in Eurasia. *J. Evolution. Biol.* **2012**, *25*, 2470–2480. [CrossRef]
66. Richardson, J.E.; Pennington, R.T.; Pennington, T.D.; Hollingsworth, P.M. Rapid diversification of a species-rich genus of neotropical rain forest trees. *Science* **2001**, *293*, 2242–2245. [CrossRef]
67. Hughes, C.; Eastwood, R. Island radiation on a continental scale: Exceptional rates of plant diversification after uplift of the Andes. *Proc. Natl. Acad. Sci. USA* **2006**, *103*, 10334–10339. [CrossRef]
68. Sauquet, H.; Weston, P.H.; Anderson, C.L.; Barker, N.P.; Cantrill, D.J.; Mast, A.R.; Savolainen, V. Contrasted patterns of hyperdiversification in Mediterranean hotspots. *Proc. Natl. Acad. Sci. USA* **2009**, *106*, 221–225. [CrossRef]
69. Bagheri, A.; Maassoumi, A.A.; Rahiminejad, M.R.; Brassac, J.; Blattner, F.R. Molecular phylogeny and divergence times of *Astragalus* section *Hymenostegis*: An analysis of a rapidly diversifying species group in Fabaceae. *Sci. Rep.* **2017**, *7*, 14033. [CrossRef]
70. Axen, G.J.; Lam, P.S.; Grove, M.; Stockli, D.F.; Hassanzadeh, J. Exhumation of the west-central Alborz Mountains, Iran, Caspian subsidence, and collision-related tectonics. *Geology* **2001**, *29*, 559–562. [CrossRef]
71. Lacombe, O.; Mouthereau, F.; Kargar, S.; Meyer, B. Late Cenozoic and modern stress fields in the western Fars (Iran): Implications for the tectonic and kinematic evolution of central Zagros. *Tectonics* **2006**, *25*, TC1003. [CrossRef]
72. Mouthereau, F.; Lacombe, O.; Vergés, J. Building the Zagros collisional orogen: Timing, strain distribution and the dynamics of Arabia/Eurasia plate convergence. *Tectonophysics* **2012**, *532–535*, 27–60. [CrossRef]
73. Padilla-García, N.; Machon, N.; Segarra-Moragues, J.G.; Martínez-Ortega, M.M. Surviving in southern refugia: The case of *Veronica aragonensis*, a rare endemic from the Iberian Peninsula. *Alp. Bot.* **2021**, *131*, 161–175. [CrossRef]
74. Boissier, E. *Diagnoses Plantarum Orientalum Novarum*; B. Hermann: Leipzig, Germany, 1844.
75. Bentham, G. Scrophulariaceae. In *Prodromus Systematis Naturalis Regni Vegetabilis*; de Candolle, A., Ed.; Victor Masson: Paris, France, 1846; Volume 10, pp. 448–491.
76. Römpp, H. Die Verwandtschaftsverhältnisse in der Gattung *Veronica*. *Repert. Spec. Nov. Regni Veg.* **1928**, *50*, 1–171.

77. Bornmüller, J.; Gauba, E. Florae Keredjensis fundamenta (Plantae Gaubaeanae Iranica). Supplementum. *Repert. Spec. Nov. Regni Veg.* **1942**, *51*, 209–235.
78. Elenevsky, A.G. *Sistematika i Geografiya Veronik SSSR i Prilezhashchikh Stran*; Nauka: Moscow, Russia, 1978.
79. Freyn, J. Ueber neue und bemerkenswerte orientalische Pflanzenarten (suite et fin). *Bull. Herb. Boissier* **1897**, *5*, 781–804.
80. Parsa, A. *Veronica L. in Flore de l'Iran*; Publications du Ministere de L'Education: Teheran, Iran, 1949; Volume 4, pp. 423–466.
81. Saeidi-Mehrvarz, S.; Ghahreman, A.; Assadi, M. A new species of *Veronica* (Scrophulariaceae) from Iran. *Nord. J. Bot.* **2001**, *21*, 581–584. [CrossRef]
82. Fischer, M.A. *Veronica fridericae*, a new species from Eastern Turkey. *Plant Syst. Evol.* **1984**, *144*, 67–71. [CrossRef]
83. Miao, Y.; Herrmann, M.; Wu, F.; Yan, X.; Yang, S. What controlled Mid–Late Miocene long-term aridification in Central Asia?—Global cooling or Tibetan Plateau uplift: A review. *Earth Sci. Rev.* **2012**, *112*, 155–172. [CrossRef]
84. Kadereit, J.W.; Licht, W.; Uhink, C.H. Asian relationships of the flora of the European Alps. *Plant Ecol. Divers.* **2008**, *1*, 171–179. [CrossRef]
85. Zhang, M.; Kang, Y.; Zhou, L.; Podlech, D. Phylogenetic origin of Phyllolobium with a further implication for diversification of *Astragalus* in China. *J. Integr. Plant Boil.* **2009**, *51*, 889–899. [CrossRef]
86. Xu, T.; Abbott, R.J.; Milne, R.I.; Mao, K.; Du, F.K.; Wu, G.; Ciren, Z.; Miehe, G.; Liu, J. Phylogeography and allopatric divergence of cypress species (*Cupressus* L.) in the Qinghai-Tibetan Plateau and adjacent regions. *BMC Evol. Biol.* **2010**, *10*, 194. [CrossRef]
87. Jia, D.R.; Abbott, R.J.; Liu, T.L.; Mao, K.S.; Bartish, I.V.; Liu, J.Q. Out of the Qinghai-Tibet Plateau: Evidence for the origin and dispersal of Eurasian temperate plants from a phylogeographic study of *Hippophaë rhamnoides* (Elaeagnaceae). *New Phytol.* **2012**, *194*, 1123–1133. [CrossRef]
88. Li, G.D.; Kim, C.; Zha, H.G.; Zhou, Z.; Nie, Z.L.; Sun, H. Molecular phylogeny and biogeography of the arctic-alpine genus *Lagotis* (Plantaginaceae). *Taxon* **2014**, *63*, 103–111. [CrossRef]
89. Buhk, N.; Zhao, L.; Li, H.; Albach, D.C. Molecular systematics and morphometrics in *Veronica* subsect. *Canae (Plantaginaceae). Plant Syst. Evol.* **2015**, *301*, 1967–1979. [CrossRef]
90. Huang, J.; Yang, L.Q.; Yu, Y.; Liu, Y.M.; Xie, D.F.; Li, J.; He, X.J.; Zhou, S.D. Molecular phylogenetics and historical biogeography of the tribe Lilieae (Liliaceae): Bi-directional dispersal between biodiversity hotspots in Eurasia. *Ann. Bot.* **2018**, *122*, 1245–1262. [CrossRef]
91. Sherafati, M.; Kazempour-Osaloo, S.; Khoshsokhan-Mozaffar, M.; Esmailbegi, S.; Staedler, Y.M.; Manafzadeh, S. Diversification of Cynoglossinae genera (Cynoglosseae-Boraginaceae) in the western Irano–Turanian bioregion. *Botany* **2021**, *99*, 541–553. [CrossRef]
92. Mittermeier, R.; Gil, P.; Hoffmann, M.; Pilgrim, J.; Brooks, T.; Mittermeier, C.; Lamoreux, J.; Da Fonseca, G. *Hotspots Revisited: Earths Biologically Richest and Most Endangered Terrestrial Ecoregions*; Cemex 315; Conservation International: Sierra Madre, CA, USA, 2005.
93. Noroozi, J.; Talebi, A.; Doostmohammadi, M.; Rumpf, S.B.; Linder, H.P.; Schneeweiss, G.M. Hotspots within a global biodiversity hotspot-areas of endemism are associated with high mountain ranges. *Sci. Rep.* **2018**, *8*, 10345. [CrossRef]
94. Noroozi, J.; Talebi, A.; Doostmohammadi, M.; Manafzadeh, S.; Asgarpour, Z.; Schneeweiss, G.M. Endemic diversity and distribution of the Iranian vascular flora across phytogeographical regions, biodiversity hotspots and areas of endemism. *Sci. Rep.* **2019**, *9*, 12991. [CrossRef]
95. Noroozi, J.; Akhani, H.; Breckle, S.W. Biodiversity and phytogeography of the alpine flora of Iran. *Biodivers. Conserv.* **2008**, *17*, 493–521. [CrossRef]
96. Doostmohammadi, M.; Talebi, A.; Noroozi, J. The Yazd–Kerman Massifs. In *Plant Biogeography and Vegetation of High Mountains of Central and South-West Asia*; Noroozi, J., Ed.; Springer International Publishing: Cham, Germany, 2020; pp. 151–183.
97. Noroozi, J.; Pauli, H.; Grabherr, G.; Breckle, S.W. The subnival–nival vascular plant species of Iran: A unique high-mountain flora and its threat from climate warming. *Biodivers. Conserv.* **2011**, *20*, 1319–1338. [CrossRef]
98. Doostmohammadi, M.; Kilian, N. *Lactuca pumila* (Asteraceae, Cichorieae) revisited-additional evidence for a phytogeographical link between SE Zagros and Hindu Kush. *Phytotaxa* **2017**, *307*, 133–140. [CrossRef]
99. Gibbard, P.L.; Head, M.J.; Walker, M.J.C. The Subcommission on Quaternary Stratigraphy. Formal ratification of the Quaternary System/Period and the Pleistocene Series/Epoch with a base at 2.58 Ma. *J. Quat. Sci.* **2010**, *25*, 96–102. [CrossRef]
100. Sanmartin, I. Dispersal vs. vicariance in the Mediterranean: Historical biogeography of the Palearctic Pachydeminae (Coleoptera, Scarabaeoidea). *J. Biogeogr.* **2003**, *30*, 1883–1897. [CrossRef]
101. Pérez-Collazos, E.; Sanchez-Gómez, P.; Jiménez, F.; Catalán, P. The phylogeographical history of the Iberian steppe plant *Ferula loscosii* (Apiaceae): A test of the abundant-centre hypothesis. *Mol. Ecol.* **2009**, *18*, 848–861. [CrossRef]
102. Garciá-Aloy, S.; Daniel, V.; Cristina, R.; Isabel, S.; Pablo, V.; Julián Molero, B.; Peris, K.; Juan José, A.; María Luisa, A. North-west Africa as a source and refuge area of plant biodiversity: A case study on *Campanula kremeri* and *Campanula occidentalis*. *J. Biogeogr.* **2017**, *44*, 2057–2068. [CrossRef]
103. Gil-López, M.J.; Segarra-Moragues, J.G.; Casimiro-Soriguer, R.; Ojeda, F. From the Strait of Gibraltar to northern Europe: Pleistocene refugia and biogeographic history of heather (*Calluna vulgaris*, Ericaceae). *Bot. J. Linn. Soc.* **2022**, *198*, 41–56. [CrossRef]
104. Davis, P.H.; Hedge, I.C. Floristic links between NW Africa and SW Asia. *Ann. Naturhist. Mus. Wien* **1971**, *73*, 43–57.
105. Zohary, M. *Geobotanical Foundations of the Middle East*; Gustav Fischer Verlag: Stuttgart, Germany, 1973; Volumes 1 and 2.

106. Léonard, J. *Contribution à Létude de la Flore et de la Végétation des Déserts d'Iran, Fascicule. 8 (Étude des Aires de Distribution les Phytochories, les Chorotypes) and Fascicule. 9 (Considerapays Phytogeographiques Sur les Phytochories Irano-Touranienne, Saharo-Sindienne et de la Somalie-paysMasai)*; Jardin Botanique National de Belgique: Meise, Belgium, 1988 and 1989.
107. Xiang, K.L.; Aytac, Z.; Liu, Y.; Espinosa, F.; Jabbour, F.; Byng, J.W.; Zhang, C.F.; Erst, A.S.; Wang, W. Recircumscription of *Delphinium* subgen *Delphinium* (Ranunculaceae) and implications for its biogeography. *Taxon* **2017**, *66*, 554–566. [CrossRef]
108. Rojas-Andrés, B.M.; Padilla-García, N.; de Pedro, M.; López-González, N.; Delgado, L.; Albach, D.C.; Castro, M.; Castro, S.; Loureiro, J.; Martínez-Ortega, M.M. Environmental differences are correlated with the distribution pattern of cytotypes in *Veronica* subsection *Pentasepalae* at a broad scale. *Ann. Bot.* **2020**, *125*, 471–484. [CrossRef]

biology

MDPI

Article

Ten Plastomes of *Crassula* (Crassulaceae) and Phylogenetic Implications

Hengwu Ding [1,†], Shiyun Han [1,†], Yuanxin Ye [1], De Bi [2], Sijia Zhang [1], Ran Yi [1], Jinming Gao [1], Jianke Yang [1], Longhua Wu [3] and Xianzhao Kan [1,4,*]

1 Anhui Provincial Key Laboratory of the Conservation and Exploitation of Biological Resources, College of Life Sciences, Anhui Normal University, Wuhu 241000, China
2 College of Landscape Engineering, Suzhou Polytechnic Institute of Agriculture, Suzhou 215000, China
3 CAS Key Laboratory of Soil Environment and Pollution Remediation, Institute of Soil Science, Chinese Academy of Sciences, Nanjing 210008, China
4 The Institute of Bioinformatics, College of Life Sciences, Anhui Normal University, Wuhu 241000, China
* Correspondence: xianzhao@ahnu.edu.cn; Tel.: +86-139-5537-2268
† These authors contributed equally to this work.

Simple Summary: Plastids are semi-autonomous plant organelles which play critical roles in photosynthesis, stress response, and storage. The plastid genomes (plastomes) in angiosperms are relatively conserved in quadripartite structure, but variable in size, gene content, and evolutionary rates of genes. The genus *Crassula* L. is the second-largest genus in the family Crassulaceae J.St.-Hil, that significantly contributes to the diversity of Crassulaceae. However, few studies have focused on the evolution of plastomes within *Crassula*. In the present study, we sequenced ten plastomes of *Crassula*: *C. alstonii* Marloth, *C. columella* Marloth & Schönland, *C. dejecta* Jacq., *C. deltoidei* Thunb., *C. expansa* subsp. *fragilis* (Baker) Toelken, *C. mesembrianthemopsis* Dinter, *C. mesembryanthoides* (Haw.) D.Dietr., *C. socialis* Schönland, *C. tecta* Thunb., and *C. volkensii* Engl. Through comparative studies, we found *Crassula* plastomes have unique codon usage and aversion patterns within Crassulaceae. In addition, genomic features, evolutionary rates, and phylogenetic implications were analyzed using plastome data. Our findings will not only reveal new insights into the plastome evolution of Crassulaceae, but also provide potential molecular markers for DNA barcoding.

Abstract: The genus *Crassula* is the second-largest genus in the family Crassulaceae, with about 200 species. As an acknowledged super-barcode, plastomes have been extensively utilized for plant evolutionary studies. Here, we first report 10 new plastomes of *Crassula*. We further focused on the structural characterizations, codon usage, aversion patterns, and evolutionary rates of plastomes. The IR junction patterns—IRb had 110 bp expansion to *rps19*—were conservative among *Crassula* species. Interestingly, we found the codon usage patterns of *matK* gene in *Crassula* species are unique among Crassulaceae species with elevated ENC values. Furthermore, subgenus *Crassula* species have specific GC-biases in the *matK* gene. In addition, the codon aversion motifs from *matK*, *pafI*, and *rpl22* contained phylogenetic implications within *Crassula*. The evolutionary rates analyses indicated all plastid genes of Crassulaceae were under the purifying selection. Among plastid genes, *ycf1* and *ycf2* were the most rapidly evolving genes, whereas *psaC* was the most conserved gene. Additionally, our phylogenetic analyses strongly supported that *Crassula* is sister to all other Crassulaceae species. Our findings will be useful for further evolutionary studies within the *Crassula* and Crassulaceae.

Keywords: *Crassula*; Crassulaceae; plastome; codon usage; codon aversion; DNA barcoding; evolutionary rates; phylogeny

Citation: Ding, H.; Han, S.; Ye, Y.; Bi, D.; Zhang, S.; Yi, R.; Gao, J.; Yang, J.; Wu, L.; Kan, X. Ten Plastomes of *Crassula* (Crassulaceae) and Phylogenetic Implications. *Biology* **2022**, *11*, 1779. https://doi.org/10.3390/biology11121779

Academic Editor: Lorenzo Peruzzi

Received: 3 November 2022
Accepted: 5 December 2022
Published: 7 December 2022

1. Introduction

The family Crassulaceae comprises approximately 1400 species in 34 genera and three subfamilies (Crassuloideae Burnett, Kalanchoideae A. Berger, and Sempervivoideae Arn.) [1–7].

These subfamilies can be further subdivided into seven major clades: Crassula (Crassuloideae), Kalanchoe (Kalanchoideae), and the other five clades (Sempervivum, Leucosedum, Aeonium, Acre, and Telephium), which form the largest subfamily Sempervivoideae [3–6,8]. The genus *Crassula*, with about 200 accepted species, is the only unique genus in the clade Crassula, the second-largest genus of Crassulaceae, and significantly contributes to the diversity of Crassulaceae [3,9,10]. Previous taxonomic revision of *Crassula* recognized two subgenera: *Crassula* L. and *Disporocarpa* Fischer & C.A. Mey. [7,11,12]. The monophyly of the subgenus *Crassula* was well supported in two recent molecular phylogenetic studies [9,10]. Nevertheless, the monophyly of subgenus *Disporocarpa* is still controversial [9,10]. Thus, more evidence and further investigations are required to clarify the phylogenetic relationships of *Crassula*.

Plastids are semi-autonomous plant organelles which have many vital functions, such as photosynthesis, stress response, and storage [13]. In angiosperms, the plastid genome (plastome) generally exhibits a conserved quadripartite circular structure with a size of 120–170 kb, comprising two single copy regions (larger and small regions, namely LSC and SSC, respectively) and two inverted repeat regions (IRs) [14–16]. Owing to the low level of recombination, uniparental inheritance, and without interference from paralogs, plastome has been extensively utilized as a super-barcode for plant species identification and evolutionary studies [17–24]. Due to the rapid development and widespread application of high-throughput sequencing technologies (such as Illumina, PacBio, and Nanopore sequencing technologies), an increasing number of complete Crassulaceae plastomes (more than 70 sequences) have been deposited in public databases. However, within the *Crassula*, only one plastome has been reported to date [6]. The lack of plastome data has limited the progress in investigating the evolutionary history of *Crassula*. Therefore, more plastome data from *Crassula* are needed to address this issue.

Codon usage bias (CUB), indicating the preferential utilization of synonymous codons in protein-coding genes (PCGs), has evolved via combined effects of genetic drift, mutation, and natural selection [17,25–28]. Owing to different species having diverse codon usage patterns, investigations of CUB can reveal phylogenetic relationships between species [17,25–28]. Codon aversion is defined as the codon which is not used in a certain gene [29–31]. The codon aversion motif is phylogenetically conserved in some lineages [29–31]. Interestingly, our recent reports in Macaronesian species (Crassulaceae) and *Bletilla* Rchb.f. species (Orchidaceae Juss.) have suggested that plastid CUB and codon aversion patterns might harbor phylogenetic signals [17,26]. Therefore, the analyses of plastid genes in codon-usage aspects might broaden our understanding of the phylogeny of both *Crassula* and Crassulaceae.

Evolutionary rate, calculated by the ratio (dN/dS) of nonsynonymous rate (dN) and synonymous rate (dS), can quantify the intensity of the selective force acting on a PCG [32–34]. The evolutionary rate can also reflect the pattern of natural selection (dN/dS value >1, =1, and <1 indicate positive, neutral, and purifying selection, respectively) [33–35]. The dN/dS values in different genes are variable, which might be influenced by many factors, such as protein function, population size, generation time, and DNA-repair efficiency [36,37]. The dN/dS values of plastid genes have been measured in many plant lineages, and most values were lower than 1, indicating plastid genes were mainly under the purifying selection [13,22,38–40]. Currently, the detailed rates and patterns of plastid genes were largely unknow in Crassulaceae. Knowledge of the evolutionary rates and patterns will shed light on how the diversifying selection affected the plastome evolution in Crassulaceae.

To address these issues, we newly sequenced and assembled the plastomes of ten *Crassula* species (*C. alstonii*, *C. columella*, *C. dejecta*, *C. deltoidea*, *C. expansa* subsp. fragilis, *C. mesembrianthemopsis*, *C. mesembryanthoides*, *C. socialis*, *C. tecta* and *C. volkensii*) using Illumina sequencing technology. Together with the public data, we performed comprehensive analyses to investigate (1) structural characterizations of *Crassula* plastomes, (2) unique CUB and codon-aversion patterns for Crassula plastomes, (3) evolutionary rates and patterns of plastid genes of Crassulaceae, and (4) phylogenetic relationships among Crassulaceae species. Our findings will not only shed new insights into the plastome evolution of Crassulaceae, but also provide potential molecular markers for DNA barcoding.

2. Materials and Methods

2.1. Sample Collection, DNA Extraction, and Sequencing

The fresh leaf samples of ten *Crassula* species were collected from greenhouses of Anhui Normal University, with the voucher codes KL01739, KL01709, KL01449, KL01646, KL02048, KL01731, KL01444, KL01653, KL01657, and KL01688 for *C. alstonii*, *C. columella*, *C. dejecta*, *C. deltoidea*, *C. expansa* subsp. *fragilis*, *C. mesembrianthemopsis*, *C. mesembryanthoides*, *C. socialis*, *C. tecta*, and *C. volkensii*, respectively. The Plant Genomic DNA kit (Tiangen, Beijing, China) was used for Genomic DNA extraction. Furtherly, a TruSeq DNA PCR-Free Library Prep Kit (Illumina, San Diego, CA, USA) was employed for library construction. Then, these libraries were sequenced using the Illumina Hiseq X Ten (Illumina, San Diego, CA, USA) platform.

2.2. Plastome Assembly, Genome Annotation, and Comparative Genomic Analysis

All resulting high-quality clean reads were assembled by using GetOrganelle 1.7.5 [41] with the plastome of *C. perforata* Thunb. (NC_053949) [6] as reference. The plastomes were initially annotated with the online program GeSeq [42] and then checked manually. Bowtie 2.4.1 [43] and Chloroplot [44] were utilized for the sequencing depth estimation and the drawing of a gene map, respectively. Genome comparisons were visualized using mVISTA [45] in Shuffle-LAGAN mode. In order to detect highly variable regions (HVRs) among plastomes, the sliding-window nucleotide diversity (π) values were measured in DnaSP v6.12 (window length = 600 bp, and step size = 200 bp) [46]. The contiguous sliding windows with higher π values ($\pi > \pi_{mean}$ + 2 standard deviation) were merged as a HVR [47,48]. The contraction and expansion of IR regions at the junctions of plastomes were subsequently plotted using R package IRscope V0.1.R (Viikki Plant Science Centre, University of Helsinki, Helsinki, Finland) [49].

2.3. Codon Usage and Aversion Indices Analyses

To investigate the codon usage indices, we used CodonW v.1.4.2 (Peden, University of Nottingham, Nottingham, UK) to calculate the values of relative synonymous codon usage (RSCU), and the effective number of codons (ENC) of plastid genes (length $\geq$300 bp) among 87 Crassulaceae species (10 of which are new in this study, Table S1). The RSCU value for a codon represents the observed frequency divided by the expected frequency (RSCU >1 implies a codon use higher than expected, and vice versa) [50]. The RSCU heatmap was rendered using TBtools 1.098 [51]. In addition, the ENC values, ranging from 20 (extreme bias) to 61 (no bias), quantify the level of CUB of synonymous codons [52]. Furtherly, the parity rule 2 (PR2) plot was performed according to the two formulas: GC-bias = [G3/(G3 + C3) | 4] and AT-bias = [A3/(A3 + T3) | 4] (" | 4" means 4-fold degenerate synonymous codons, and G3, C3, A3 and T3 denotes nucleotide composition at the 3rd codon sites, respectively) [53,54]. The points lying at the centre of plot (AT bias = 0.5 and GC bias = 0.5) indicate no bias, whereas the off-centred points reflect the direction and extent of bias [53,54]. Moreover, the codon aversion motifs harboring strong phylogenetic implications were identified by using CAM v.1.02 [31].

2.4. Nucleotide Substitution Rate Analyses

The 79 PCGs from 87 species of Crassulaceae were employed to evaluate the evolutionary rates (Table S1). The percentage of variable sites (PV) and average π values were measured with DnaSP v6.12 (Departament de Genètica, Universitat de Barcelona, Barcelona, Spain) [46]. The nucleotide substitution rates, including dN, dS, and dN/dS, were inferred with PAML v4.9 [55] under F3X4 and M0 model.

2.5. Phylogenetic Implications Analyses

Phylogenetic relationships among 87 Crassulaceae species were inferred by maximum-likelihood (ML) and Bayesian inference (BI) methods, based on 79 PCGs (Data S1). Recent studies of Lu et al. [9] and Bruyns et al. [10] revealed a sister relationship between Crassulaceae and Haloragaceae R.Br. Therefore, two species of Haloragaceae (*Myriophyllum aquaticum* (Vell.) Verdc., NC_048889 and *Myriophyllum spicatum* L., NC_037885) were selected as outgroups. Multiple sequence alignments were generated using MAFFT v7.505 in PhyloSuite v1.2.1 with codon model [56]. The best-fit nucleotide substitution models were evaluated with ModelTest-NG v0.1.7 [57]. Subsequently, we employed RAxML-NG 1.1 [58] and MrBayes v3.2.7a [59] for ML and BI analyses, respectively. For ML analyses, the reliabilities were assessed with 1000 bootstrap replicates, and the convergence was evaluated by using parameter "–bsconverge" in RAxML-NG package (Computational Molecular Evolution Group, Heidelberg Institute for Theoretical Studies, Heidelberg, Germany). For BI analyses, four independent Markov chains and two independent runs (running for 10,000,000 generations, and sampling every 1000 generations) were conducted, with Tracer 1.7.1 (Institute of Evolutionary Biology, University of Edinburgh, Edinburgh, UK) [60] for the convergence. After discarding the first 25% trees as burn-in, the remaining 75% trees were used to estimate the consensus tree and Bayesian posterior probabilities.

3. Results

3.1. Plastome Organizations and Structural Features

Based on bowtie2 mapping, totally 3,246,461, 1,740,232, 3,915,950, 2,632,260, 2,801,895, 504,972, 5,440,628, 3,398,113, 1,877,288 and 1,530,319 paired reads were mapped to the plastomes of *C. alstonii* (coverage: 3344.02×), *C. columella* (coverage: 1762.18×), *C. dejecta* (coverage: 4020.78×), *C. deltoidei* (coverage: 5284.33×), *C. expansa* subsp. *fragilis* (coverage: 5796.17×), *C. mesembrianthemopsis* (coverage: 1010.16×), *C. mesembryanthoides* (coverage: 5557.15×), *C. socialis* (coverage: 3486.98×), *C. tecta* (coverage: 1918.07×), and *C. volkensii* (coverage: 3161.69×), respectively. The new complete plastomes of ten species of *Crassula* (accession numbers: OP729482–OP729487 and OP882297–OP882300) were typical circular and quadripartite biomolecules (Figure 1), with sizes ranging from 144,855 bp to 146,060 bp. These plastomes contains LSC (78,303–79,707 bp), SSC (16,568–16,871 bp), and IR (24,810–24,878 bp). The overall GC contents of *Crassula* plastomes were between 37.73% and 38.32%. Notably, the GC contents of IR regions (42.93–43.15%) were found to be higher than those of in the LSC (35.75–36.51%) and SSC regions (31.67–32.40%). In addition, these plastomes contain 134 genes, including 85 PCGs, 37 tRNA genes, 8 rRNA genes and 4 pseudogenes. Among these genes, 6 PCGs, 7 tRNA genes, 4 rRNA genes, and one pseudogene (*ycf15*), were completely duplicated in the IR regions (Table 1).

Furthermore, based on the results obtained with mVISTA, in all plastomes investigated it was found that the IR and coding regions (exons, tRNAs, and rRNAs) are more conserved than SC and conserved non-coding regions (CNS), respectively (Figure 2). Additionally, the results also revealed that 3 plastomes (labelled 8–10) of subgenus *Disporocarpa* exhibited higher divergences than 7 plastomes (labelled 1–7) of subgenus *Crassula*, when compared with the reference.

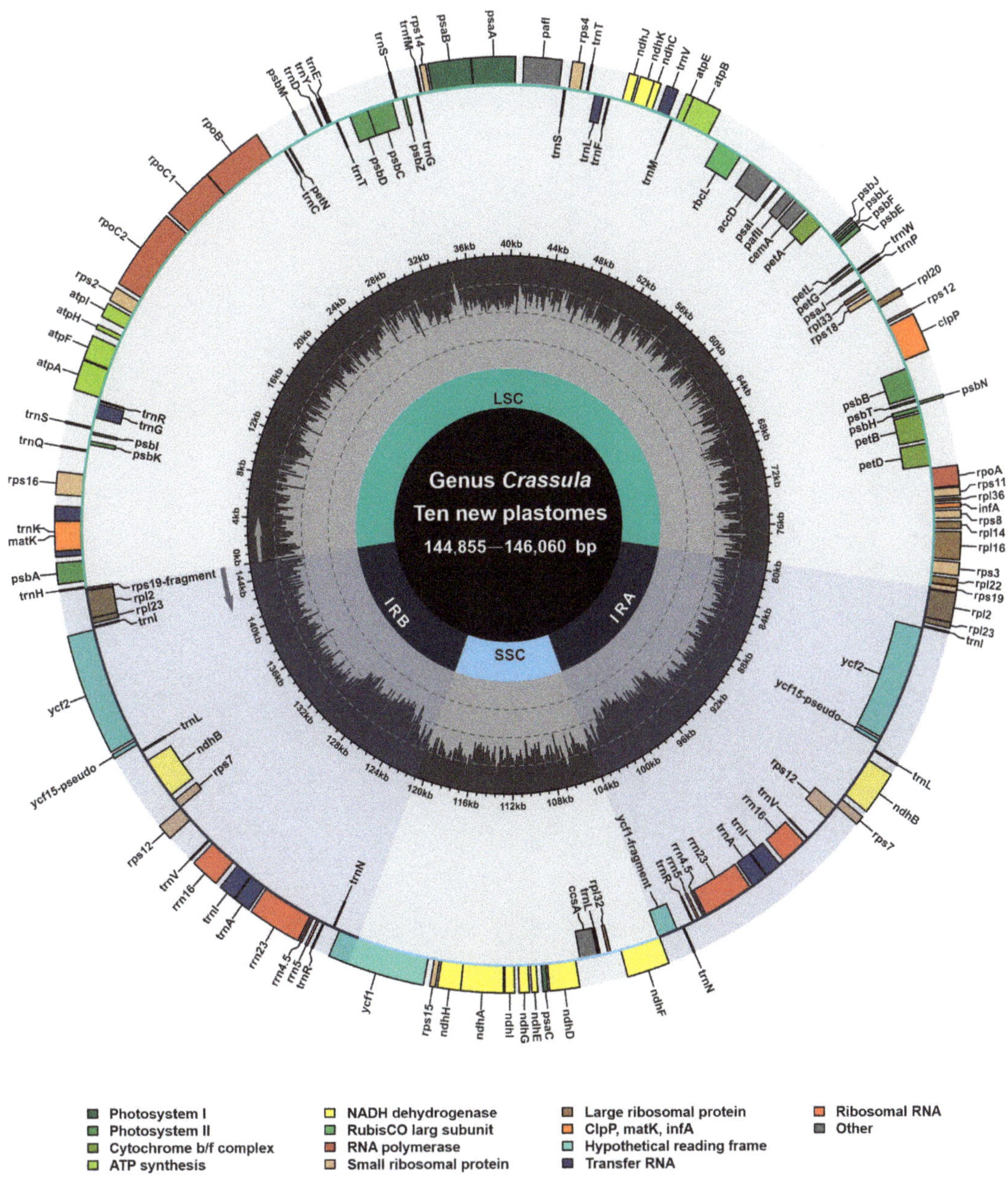

Figure 1. Annotation map of ten new plastomes from *Crassula* species. Directed with arrows, genes that are listed inside and outside of the circle are respectively transcribed clockwise and counterclockwise. Different colors represent different functional groups.

Table 1. Plastome characteristics of *Crassula* species.

Species	Accession Number	Size (bp)				GC Content (%)				Gene Number				
		Genome	LSC	IR	SSC	Genome	LSC	IR	SSC	Total	PCGs	tRNA	rRNA	Pseudo
C. alstonii †	OP729482 *	145,507	79,304	24,813	16,577	37.82	35.86	42.96	31.85	134 (19)	85 (6)	37 (7)	8 (4)	4 (2)
C. columella †	OP729483 *	145,554	79,332	24,827	16,568	37.83	35.86	42.95	31.86	134 (19)	85 (6)	37 (7)	8 (4)	4 (2)
C. dejecta †	OP729484 *	145,870	79,495	24,859	16,657	37.73	35.76	42.93	31.67	134 (19)	85 (6)	37 (7)	8 (4)	4 (2)
C. mesembrianthemopsis †	OP882297 *	146,046	79,664	24,854	16,674	37.78	35.78	42.97	31.89	134 (19)	85 (6)	37 (7)	8 (4)	4 (2)
C. mesembryanthoides †	OP729485 *	146,057	79,707	24,858	16,634	37.78	35.78	42.94	31.9	134 (19)	85 (6)	37 (7)	8 (4)	4 (2)
C. perforata †	NC_053949	145,737	79,465	24,810	16,652	37.75	35.75	42.97	31.77	134 (19)	85 (6)	37 (7)	8 (4)	4 (2)
C. socialis †	OP729486 *	145,842	79,549	24,828	16,637	37.79	35.8	42.95	31.86	134 (19)	85 (6)	37 (7)	8 (4)	4 (2)
C. tecta †	OP729487 *	146,060	79,662	24,854	16,690	37.78	35.78	42.97	31.88	134 (19)	85 (6)	37 (7)	8 (4)	4 (2)
C. deltoidea #	OP882298 *	145,175	78,580	24,862	16,871	38.13	36.25	43.08	32.32	134 (19)	85 (6)	37 (7)	8 (4)	4 (2)
C. expansa subsp. Fragilis #	OP882299 *	144,969	78,449	24,856	16,808	38.30	36.51	43.14	32.4	134 (19)	85 (6)	37 (7)	8 (4)	4 (2)
C. volkensii #	OP882300 *	144,855	78,303	24,878	16,796	38.32	36.51	43.15	32.4	134 (19)	85 (6)	37 (7)	8 (4)	4 (2)

† These species belong to subgenus *Crassula*. #, These speciesbelong to subgenus *Disporocarpa*. *, The new plastomes were generated in this study. (n), The number of genes located on IRs.

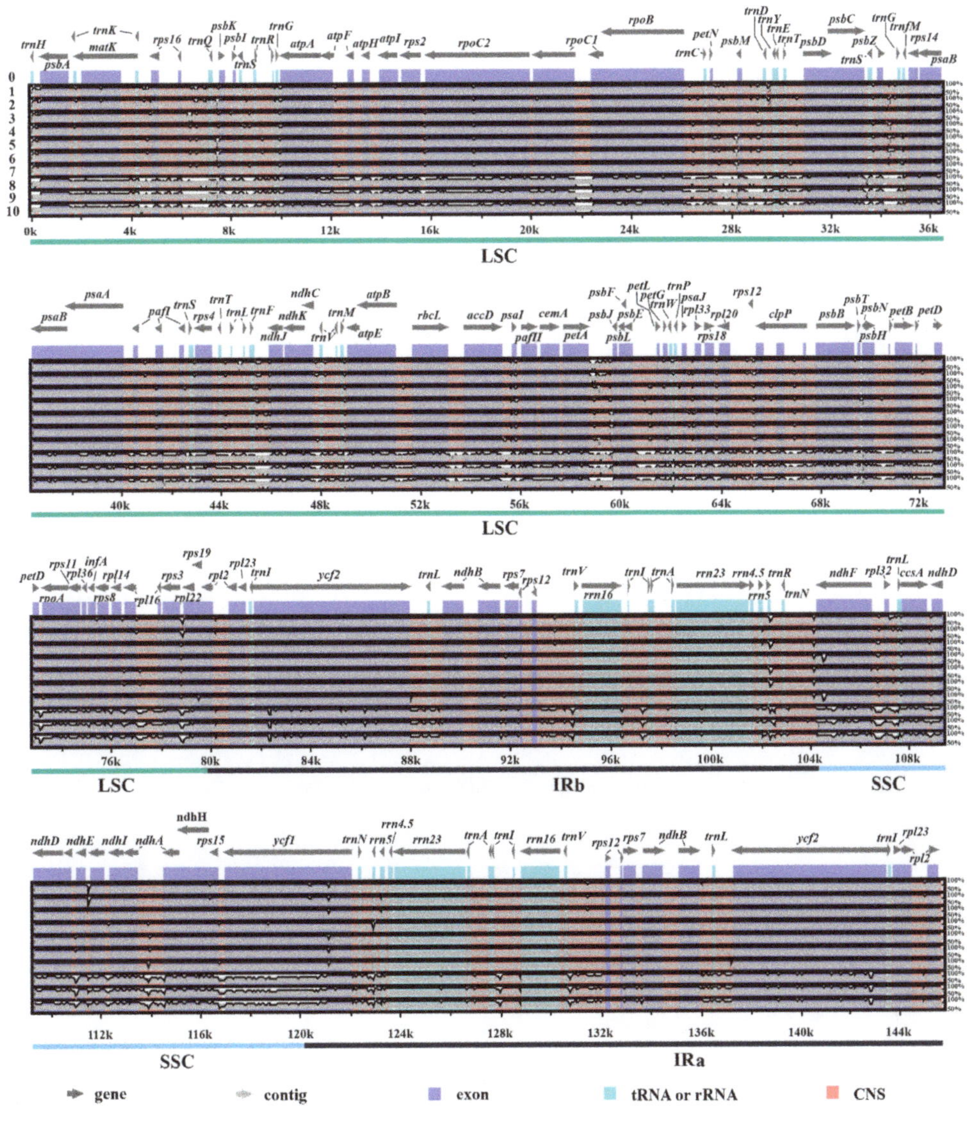

Figure 2. Structure comparisons of ten new *Crassula* plastomes using the mVISTA program. Y-scale represents the percent identity between 50% and 100%. The labels 0 to 10 indicate *C. perforata* (reference), *C. alstonii*, *C. columella*, *C. dejecta*, *C. mesembryanthoides*, *C. tecta*, *C. mesembrianthemopsis*, *C. socialis*, *C. volkensii*, *C. expansa* subsp. *fragilis*, and *C. deltoidei*, respectively.

The sliding-window-based π values estimated for 11 plastomes of *Crassula* ranged from 0.00073 to 0.10315 (Table S2 and Data S2). The mean π value and its standard deviation were 0.02978 and 0.01954, respectively. Thus, a total of 11 HVRs were identified with relatively high variability (π > 0.06886) (Figure 3). These HVRs containing high π values (0.06912–0.08653) and abundant variable sites (111–559) might be used as potential DNA barcodes for species identification within *Crassula* (Table 2).

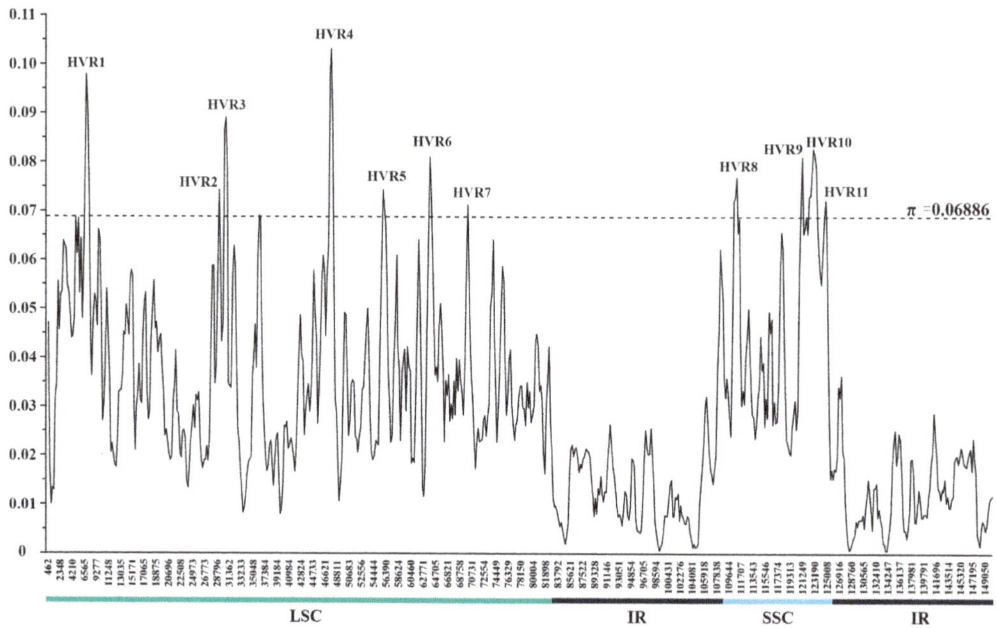

Figure 3. Sliding-window analysis of the plastomes of 11 *Crassula* species (window length: 600 bp; step size: 200 bp). x-axis: position of the midpoint of a window; y-axis: π value of each window. Regions with higher π values (π > 0.06886) were considered as HVRs.

Table 2. The HVRs identified in the plastomes of 11 *Crassula* species.

HVR	Coordinates	Region Size (bp)	π Value	Variable Sites	Annotations
HVR1	6259–8197	1939	0.07521	240	*rps16–trnQ-UUG*
HVR2	28,684–29,678	995	0.07427	119	*petN–psbM*
HVR3	29,942–31,262	1321	0.07064	182	*psbM–trnD-GUC–trnY-GUA*
HVR4	35,624–36,637	1014	0.06912	114	*psbZ–trnG-GCC*
HVR5	47,003–48,289	1287	0.08653	223	*trnL-UAA–trnF-GAA–ndhJ*
HVR6	55,345–56,290	946	0.07518	158	*rbcL–accD*
HVR7	63,273–64,156	884	0.07652	164	*psbE–petL*
HVR8	69,495–70,212	718	0.07133	111	*clpP–psbB*
HVR9	109,945–111,098	1154	0.06969	192	*ndhF–rpl32–trnL-UAG*
HVR10	120,441–123,490	3050	0.07259	559	*rps15–ycf1*
HVR11	124,303–124,908	606	0.07227	119	*ycf1*

In our current study, all 11 plastomes of *Crassula* displayed similar IR junction patterns (Figure 4). The SSC/IRa borders are located in the coding regions of *ycf1* gene, resulting in the fragmentations of *ycf1* (*ycf1*-fragment) in IRb regions. Moreover, *ndhF* genes were discovered to occur mainly in SSC, and partly in IRb, regions. Notably, *rps19* genes are located at the LSC/IRb junctions, with extension into the IRb regions for 110 bp. Similarly, *trnH* genes lie at the IRa/LSC junctions, with uniform 3 bp-sized expansions to the IRa regions.

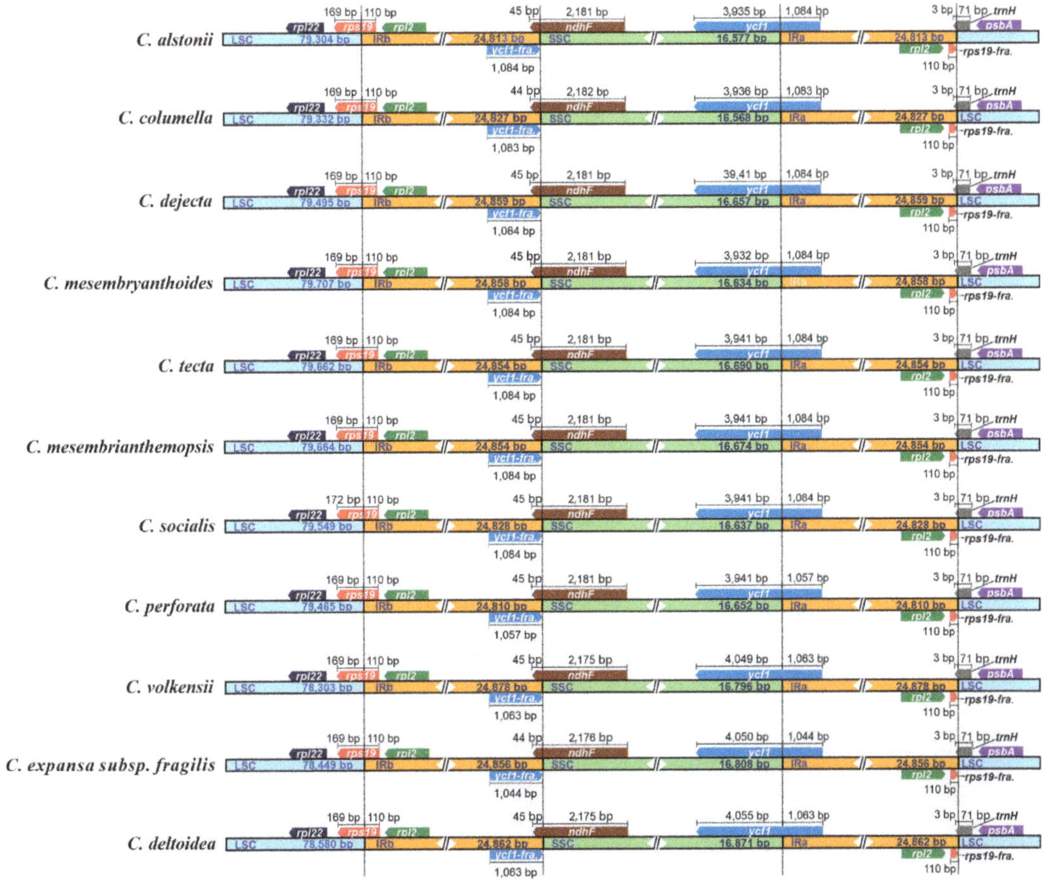

Figure 4. Comparisons of LSC, SSC, and IR region borders among plastomes of 11 *Crassula* species. Blue, orange and green blocks represent the LSC, IR and SSC regions, respectively. Gene boxes represented above the block were transcribed clockwise and those represented below the block were transcribed clockwise. "fra." is the abbreviation of "fragment".

3.2. Codon Usage and Aversion Patterns

To compare the patterns of codon usage and aversion between *Crassula* and other Crassulaceae species, four analyses (RSCU, ENC, PR2-plot, and codon aversion motif) of 53 plastid genes (length $\geq$300 bp) were performed.

The overall RSCU values ranged from 0.32 (CTC or AGC) to 2.07 (TTA) among Crassulaceae species (Table S3). Similar with other Crassulaceae species, seven taxa of *Crassula* exhibited significant preference for A/T-ending codons over G/C-ending codons in plastid genes (Figure 5). Importantly, the RSCU heatmap showed two subgenera within the *Crassula*: subgenus *Disporocarpa* included *C. expansa* subsp. *fragilis*, *C. deltoidea* and *C. volkensii*; subgenus *Crassula* consisted of the remaining eight taxa (Figure 5).

The ENC values ranged from 30.83 (*ndhC* in *Sedum sarmentosum* Bunge) to 57.74 (*ndhJ* in *C. volkensii* and *C. expansa* subsp. *fragilis*) among Crassulaceae species (Table S4). Generally, ENC values $\leq$35 indicate high codon preference [52,61,62]. The results show that most of the ENC values (99.48%) were higher than 35, indicating a weaker bias. Most surprisingly of all, we detected the ENC values of *matK*, from the Crassula clade, are significantly higher than those of all other clades (Table S4 and Figure 6). It might prove to be a unique feature for *Crassula* species. To further verify this finding, more sampling data and comprehensive analyses are need in future studies.

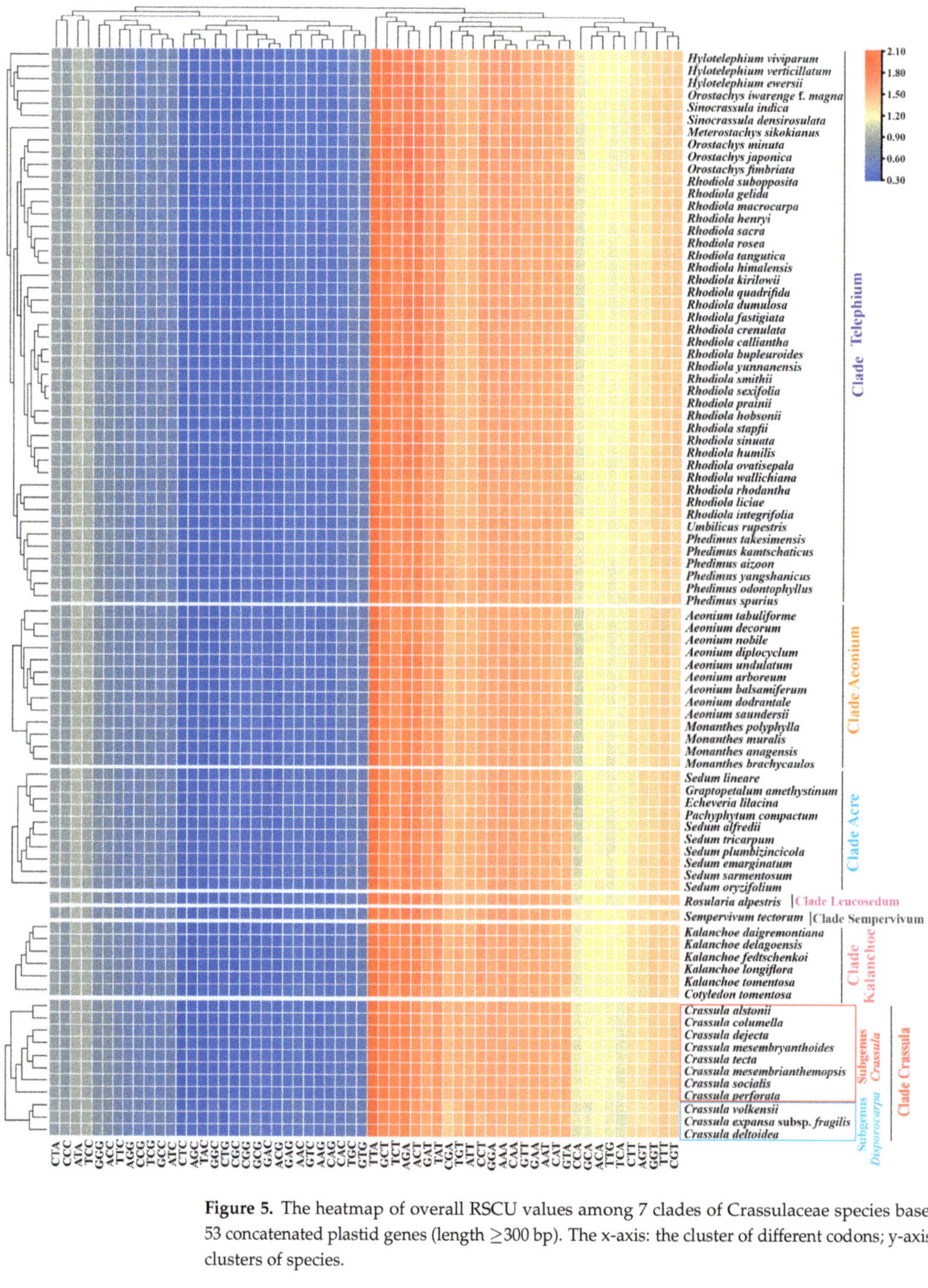

Figure 5. The heatmap of overall RSCU values among 7 clades of Crassulaceae species based on 53 concatenated plastid genes (length ≥300 bp). The x-axis: the cluster of different codons; y-axis: the clusters of species.

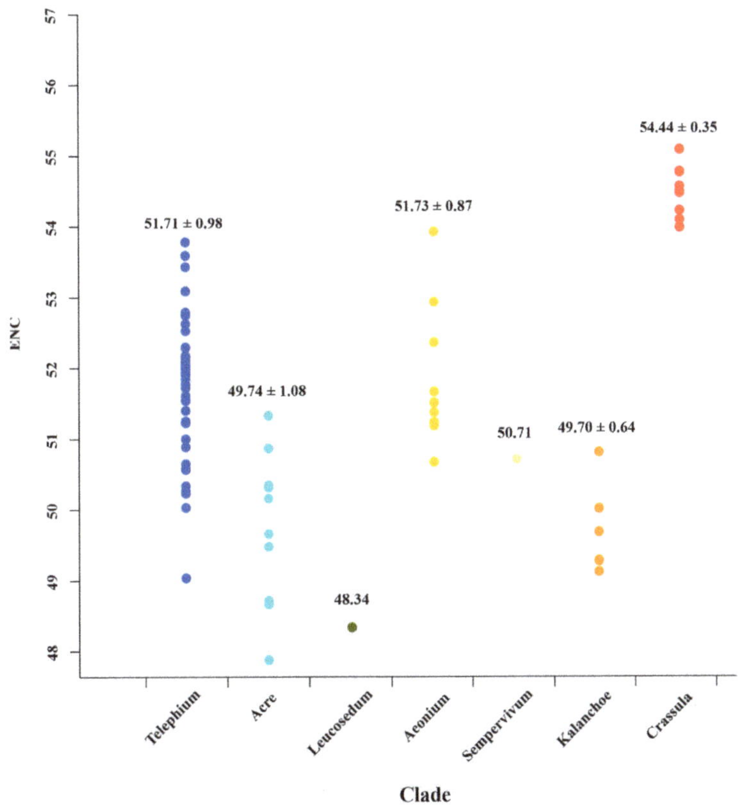

Figure 6. The ENC value distributions of *matK* for 7 clades of Crassulaceae. The mean values with standard deviations are labeled.

The PR2 plots of *matK* and 52 other PCGs are presented in Figures 7 and S1, respectively. These results indicated the nucleotide usage at the 3rd codon site of 4-fold degenerate codons is uneven in different genes. For example, *rps14*, *clpP*, *psbA*, and *pafII* prefer to use A/G, A/C, T/C, and T/G in 4-fold degenerate sites, respectively (Figure S1). In addition, these unbalanced utilizations were also found in different species (Figure S1). Obviously divergent GC-biases were observed in *matK* genes between species of subgenus *Crassula* and others. Specifically, all GC-biases of clades from Kalanchoideae and Sempervivoideae, plus subgenus *Disporocarpa*, were less than 0.5. On the contrary, all these values for the subgenus *Crassula* were higher than 0.5, which might be unique characteristic for subgenus *Crassula*. Moreover, species with close relationships had identical nucleotide biases. For example, *C. alstonii* and *C. columella* had identical AT-biases (0.4074) and GC-biases (0.5455). Similar phenomena could also be observed in *C. mesembryanthoides* and *C. tecta* (AT-biases = 0.4286, and GC-biases = 0.5455).

Owing to the codon aversion motifs containing phylogenetic implication, we analyzed codon aversion patterns of genes among Crassulaceae species. Except for *rpoB*, *rpoC2*, *ycf1* and *ycf2*, codon aversion motifs were found in the remaining 49 genes (Table S5). It is worth noting that 27 and 16 unique codon aversion motifs were detected for species of subgenus *Crassula* and subgenus *Disporocarpa*, respectively (Table 3), which might be used as potential biomarkers for species identification. Further to this, 8 consensus motifs might be considered as the feature of genus *Crassula* (Table 3). Moreover, the codon aversion motifs from 3 genes (*matK*, *pafI* and *rpl22*) could also divide 11 species into two subgenera (subgenus *Crassula* and subgenus *Disporocarpa*) (Figure 8), which is congruent with results from RSCU heatmap.

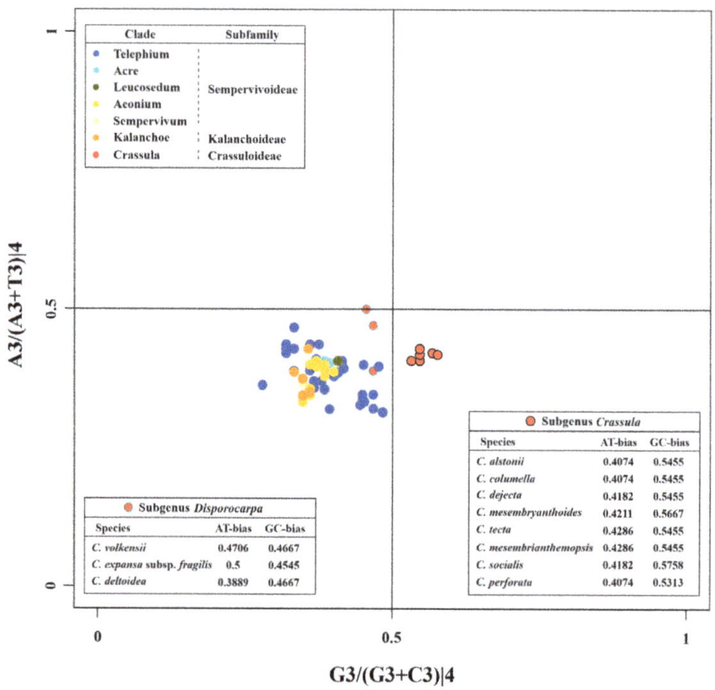

Figure 7. The PR2 plot of *matK* of Crassulaceae. Different colors represent different clades. Within the clade Crassula (or genus *Crassula*), red circles with black edges and cyan edges represent species of subgenus *Crassula* and *Disporocarpa*, respectively.

Table 3. The specific codon aversion motifs of *Crassula* within Crassulaceae.

Gene	Specific Codon Aversion Motifs		
	Subgenus *Crassula*	**Subgenus *Disporocarpa***	**Consensus for Genus *Crassula***
accD	TGC		
atpB	TGT	CGG	
atpI	TGT, CTG, CTA,		
cemA	GGG		
ccsA	GGC		
ndhA	CAT, CCG, CAG, CGG	CAT, CCG	CAT, CCG
ndhE	CTC		
ndhI	TCT	TCT	TCT
ndhJ	TGT, ACG		
ndhK	TGC, CAC, CTG	TGC, CAC, CTG	TGC, CAC, CTG
pafII	AAG	AAG	AAG
petB	AAC, AGA	CAG	
petD	GGC	GCC	
psbB		CGA	
psbD		ACG	
rbcL	GCG, CCC, AGG		
rpl16	GAT	GAT	GAT
rpl20		TGT	
rps3		GCG, GTC	
rps4	TCT	CGG	

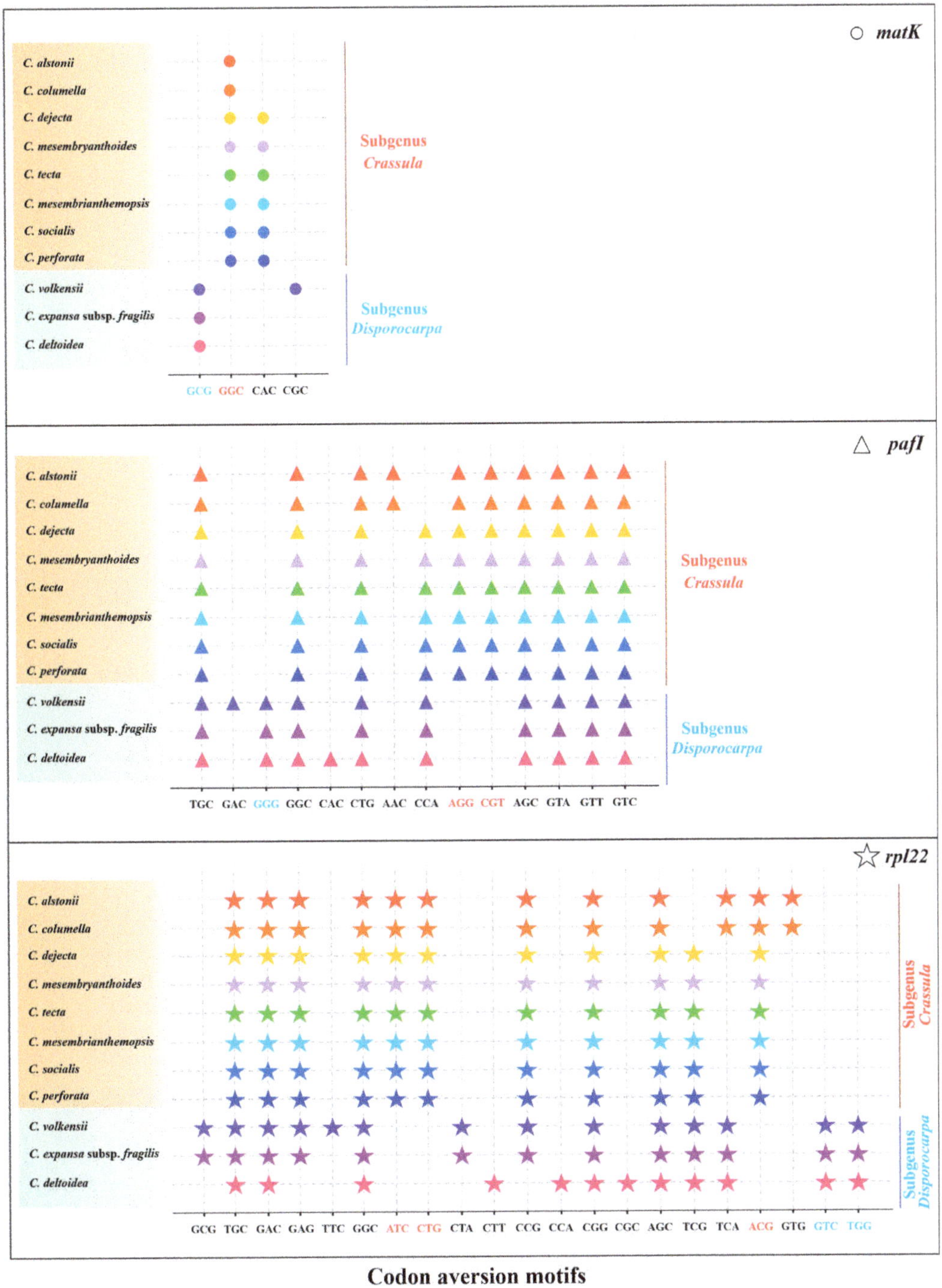

Figure 8. The specific codon aversion motifs of *matK*, *pafI* and *rpl22* gene for the 11 species of *Crassula*. The dots marked in different colors represent different species. Codons in red and green were specific for subgenus *Crassula* and subgenus *Disporocarpa*, respectively.

3.3. Evolutionary Rates and Patterns

The π (0.00447–0.0914) and PV (4.91–37.52%) values of 79 plastid PCGs of Crassulaceae species were plotted in Figure 9a. Two genes, referring to *ycf1* (π = 0.0914, PV = 35.78%) and *matK* (π = 0.08239, PV = 37.52%), had obviously higher π and PV values than those of the other 77 genes, indicating they might accumulate more mutations than other plastid genes. The detailed data are listed in Table S6.

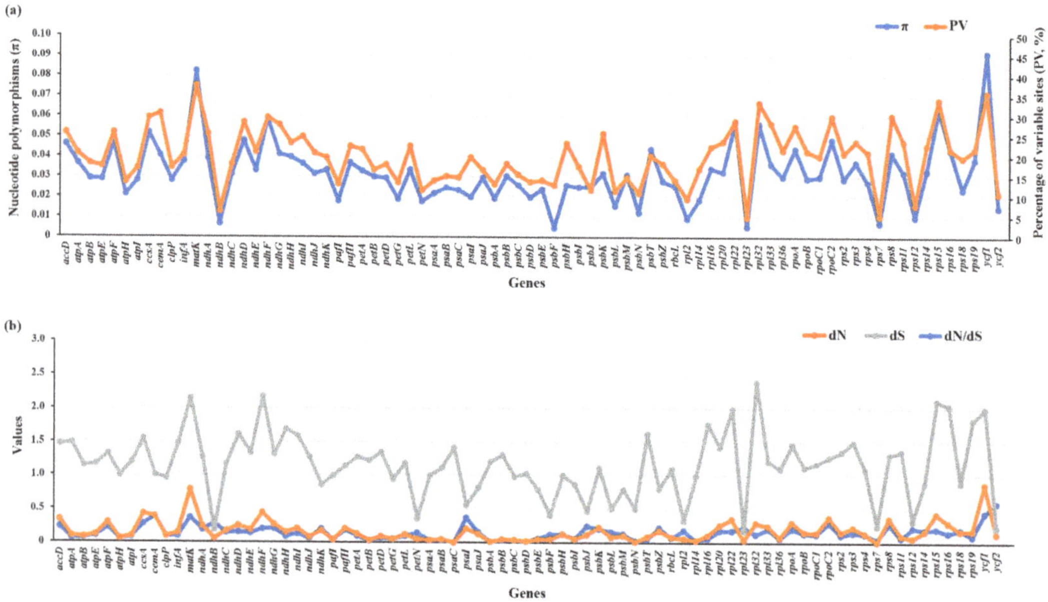

Figure 9. Sequence polymorphism among 79 PCGs of 87 Crassulaceae species. (**a**) Nucleotide diversity (π) and percentages of variable sites (PV). (**b**) Estimations of nonsynonymous (dN) and synonymous (dS) substitution rates and the dN/dS.

To further quantify the evolutionary rates of PCGs, the nucleotide substitution rates, including dN, dS and dN/dS, were calculated (Figure 9b, Table S6). The dN values ranged from 0 to 0.8671, with higher dN values for *ycf1* (dN = 0.8671) and *matK* (dN = 0.7804) than for others. Compared with dN values, the dS values had relatively wide ranges (0.177–2.3917), resulting in corresponding dN/dS ratios (0–0.5891) of less than 1. This finding indicates the plastid genes from Crassulaceae appear to be evolving under a purifying selective constraint. Among 79 plastid PCGs, *ycf2* is the most rapidly evolving gene, with the highest ratio (dN/dS = 0.5891), followed by *ycf1*, *cemA*, *psaI*, and *matK*. By contrast, *psaC* was the most conserved gene with the lowest ratio (dN/dS = 0).

3.4. Phylogenetic Implications

To investigate the evolutionary relationships among 87 species of Crassulaceae, phylogenetic analyses were performed. After a model test, GTR + G4 and GTR + I+G4 were inferred as the optimal substitution models for most genes (the detailed models can be seen in Table S7). As shown in Figure 10, the trees inferred from two methods displayed the same topology.

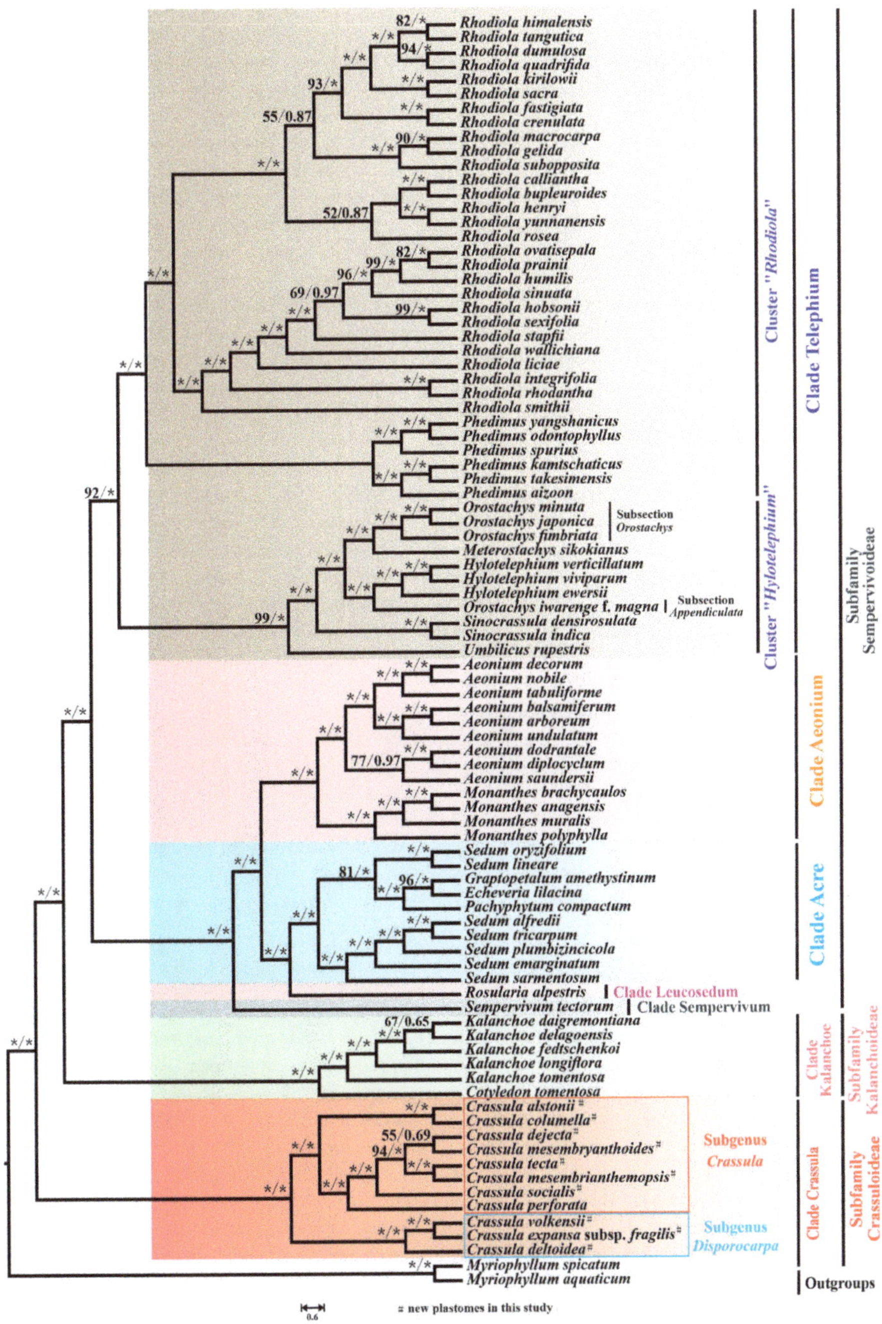

Figure 10. The phylogenetic tree of 87 Crassulaceae species based on ML and BI methods. The maximum likelihood bootstrap (BS) and Bayesian posterior probability (PP) values for each node are indicated; * indicates 100% bootstrap or 1.00 PP.

Ten species of *Crassula* that we sequenced, together with *C. perforate*, form the well-supported clade Crassula, which is sister to all other Crassulaceae species (maximum likelihood bootstrap [BS] = 100 and bayesian posterior probability [PP] = 1.00). In addition, our phylogenetic tree indicated that this monophyletic clade could be clustered into two subgenera: subgenus *Disporocarpa* harbored *C. volkensii*, *C. expansa* subsp. *fragilis* and *C. deltoidea* ([BS] = 100 and [PP] = 1.00). Subgenus *Crassula* included the remaining eight *Crassula* species (*C. alstonii*, *C. columella*, *C. dejecta*, *C. mesembryanthoides*, *C. tecta*, *C. mesembrianthemopsis*, *C. socialis*, and *C. perforata*) ([BS] = 100 and [PP] = 1.00). Within subgenus *Crassula*, two species (*C. alstonii* and *C. columella*) were sister to six other species (*C. dejecta*, *C. mesembryanthoides*, *C. tecta*, *C. mesembrianthemopsis*, *C. socialis*, and *C. perforata*) ([BS] = 100 and [PP] = 1.00). Further, *C. tecta* and *C. mesembrianthemopsis* formed the well-supported sister taxa ([BS] = 100 and [PP] = 1.00). Unfortunately, the sister group of *C. dejecta* and *C. mesembryanthoides* had relatively weak support ([BS] = 55 and [PP] =0.69). Due to the limited plastome data within *Crassula*, there are many unsolved phylogenetic problems in this clade. Therefore, more samples are needed to solve this issue.

As expected, the six species from genus *Kalanchoe* Adans. and genus *Cotyledon* L. formed the monophyletic clade Kalanchoe (or subfamily Kalanchoideae) ([BS] = 100 and [PP] = 1.00). The remaining 70 species, belonging to the subfamily Sempervivoideae, can be further grouped into 5 distinct clades: Acre, Aeonium, Leucosedum, Sempervivum, and Telephium. In detail, 7 *Sedum* L. species and 3 species from other genera respectively (*Graptopetalum* Rose, *Echeveria* DC., and *Pachyphytum* Link, Klotzsch & Otto) formed a well-supported clade Acre ([BS] = 100 and [PP] = 1.0). However, it is clear from our results that *Sedum* is not monophyletic, with some other taxa embedded within this genus.

In addition, 13 species from genus *Aeonium* Webb & Berthel. and genus *Monanthes* Haw. make up the clade Aeonium ([BS] = 100 and [PP] = 1.0). Furthermore, due to sampling in this study, only a single species form Leucosedum and Sempervivum clades, and full resolution of relationships within these clades requires sufficient molecular sequences. Notably, the clade Telephium, with 45 species, consists of clusters "*Rhodiola*" and "*Hylotelephium*" [63] ([BS] = 92 and [PP] = 1.0). Within cluster "*Hylotelephium*", non-monophyly of *Orostachys* Fisch. was observed. Three *Orostachys* species, (*O. japonica* (Maxim.) A.Berger, *O. minuta* (Kom.) A.Berger, and *O. fimbriata* (Turcz.) A.Berger) belonging to the Subsection *Orostachys* [63] ([BS] = 100 and [PP] = 1.0), were sister to *Meterostachys sikokianus* (Makino) Nakai, while *O. iwarenge* f. *magna* Y.N.Lee (Subsection *Appendiculata*) and three *Hylotelephium* H.Ohba species formed a group with strong support ([BS] = 100 and [PP] = 1.0).

4. Discussion

Ten new plastomes of *Crassula* were reported in the present study. Combined with available data from public database, we conducted comprehensive analyses, including plastome organizations, codon usage and aversion patterns, evolutionary rates, and phylogenetic implications.

The expansion and contraction of IR regions are common evolutionary events and have been considered as the main mechanism for the length variation of angiosperm plastomes [64–66]. In our study, we performed comparative analyses among *Crassula* plastomes, and found that the IRb regions had uniform length (110 bp) expansions to the *rps19* gene. This 110-bp expansion had been also observed in *Aeonium*, *Monanthes*, and most other taxa of Crassulaceae in our recent study [17]. This finding indicated that the conserved IR organization might act as a family-specific marker for Crassulaceae species.

Interestingly, it was reported that *rps19* genes were completely located in the LSC regions in *Forsythia suspensa* (Thunb.) Vahl, *Olea europaea* Hoffmanns. & Link L., and *Quercus litseoides* Dunn [67,68], and were fully encoded by the IR regions in *Polystachya adansoniae* Rchb.f., *Polystachya bennettiana* Rchb.f., and *Dracaena cinnabari* Balf.f. [69,70]. There are several mechanisms that might explain the IR expansion and contraction [71–73]. For instance, Goulding et al. [71] proposed that short IR expansions may occur by gene conversion events, whereas large IR expansions involved in double-strand DNA breaks. In order to better reveal the mechanisms of IR expansion and contraction, more extensive investigations in Crassulaceae and Saxifragales are required.

Investigations of codon usage patterns could reveal phylogenetic relationships between organisms [25,74]. In particular, 11 species of *Crassula* can be divided into two subgenera from the RSCU heatmap, which agreed with the results of phylogenetic analyses. This finding further demonstrates that RSCU values contain phylogenetic implications [75–80]. Additionally, we observed codon usage patterns are gene-specific and/or species-specific, reflected in diversified ENC values and various distribution patterns in PR2 plots. Interestingly, we found the codon usage patterns of *matK* gene in *Crassula* species are unique among Crassulaceae species with elevated ENC values. Furthermore, the GC-biases of *matK* gene with specific preference (>0.5) might be the particular feature for subgenus *Crassula*. Due to rapid evolutionary rate, high universality, and significant interspecific divergence, the *matK* gene has been broadly used in plant evolutionary studies as one of the core DNA barcodes [9,10,81–84].

Codon aversion, a novel concept proposed by Miller et al. [29–31], is an informative character in phylogenetics. Specifically, the codon aversion motifs in orthologous genes are generally conserved in specific lineages [29–31]. To date, these analyses have only been performed in a few plant plastomes [17,26]. For example, the specific codon aversion motifs of *rpoA* gene could distinguish not only the two genera (*Aeonium* and *Monanthes*), but also the three subclades of *Aeonium* in our recent report [17]. In this work, genus-specific and subgenus-specific codon aversion motifs were identified for 11 *Crassula* species. These findings suggest codon aversion pattern could be used as a promising tool for phylogenetic study.

Generally, the dN/dS ratios of genes could reflect the extent of selection pressures during evolution [22]. Here, the dN/dS values of plastid PCGs ranged from 0 to 0.5891 within Crassulaceae, indicating all plastid genes were under purifying selection. Among these values, elevated dN/dS ratios were found for *ycf1* (0.4349) and *ycf2* (0.5891). Similarly, high dN/dS ratios of these two genes were also observed in other families, such as Asteraceae Bercht. & J.Presl [38], Mazaceae Reveal [22], and Musaceae Juss. [13]. The *ycf1* gene was related to protein translocation [85]. The *ycf2* gene is necessary for cell viability, but the detail function is still unknown [86]. Why *ycf1* and *ycf2* evolve relatively fast is interesting. The possible reason put forwarded by Barnard-Kubow et al. [87] considered that relaxed purifying selection or positive selection on *ycf1*, *ycf2* and some other genes might result in the development of reproductive isolation and subsequent speciation in plants. Therefore, the results suggested that *ycf1* and *ycf2* might play important roles in the divergence of Crassulaceae.

Our phylogenetic tree divided 87 species into 3 subfamilies and 7 clades. The clade Crassula is sister to all other 6 clades, which agrees with the phylogeny reported by Gontcharova et al. [4], Chang et al. [6], and Han et al. [17]. Furtherly, 11 *Crassula* species could be furtherly divided into two subgenera, which generally accords with the morphological differences (floral shape) reported by Bruyns et al. [10] (Table S8). Nevertheless, there are still some unsolved phylogenetic problems within Crassulaceae. The first problem is that the plastid phylogeny of *Crassula* is not entirely clear due to the limited data. According to the classification proposed by Tölken [11,88], 11 and 9 sections were respectively identified in subgenus *Crassula* and subgenus *Disporocarpa*. However, Bruyns et al. [10] indicated that most sections were not monophyletic. Moreover, subgenus *Disporocarpa* recently has been regarded as a paraphyletic group [9,10]. The second is the genus *Sedum*, which is not monophyletic in our study, agreeing with the widely accepted viewpoint [3–5,89,90]. Finally, the genus *Orostachys* has been demonstrated to be non-monophyletic based on plastid data, which is consistent with previous analysis based on nuclear internal transcribed spacers (ITS) data [63]. In order to better understand the phylogeny of *Crassula* or Crassulaceae, more data are needed for the further detailed analyses.

5. Conclusions

In the present study, 10 new plastomes of *Crassula* species were reported. These plastomes exhibited identical gene content and order, and that they contained 134 genes (130 functional gene and 4 pseudogenes). The 11 identified HVRs with relatively high variability ($\pi > 0.06886$) might be used as potential DNA barcodes for species identification within *Crassula*. The unique expansion pattern, where the IRb regions had uniform length (110 bp) boundary expansions to *rps19*, might become a plesiomorphy of Crassulaceae. According to RSCU values, the A/T-ending codons were favored in plastid genes. Most importantly, we found the codon usage patterns of the *matK* gene in *Crassula* species are unique among Crassulaceae species with elevated ENC values. Furthermore, subgenus *Crassula* species have specific GC-biases in the *matK* gene. In addition, the codon aversion motifs from *matK*, *pafI* and *rpl22* contained phylogenetic implications within *Crassula*. Compared with other Crassulaceae species, 27 and 16 unique codon aversion motifs were detected for subgenus *Crassula* and subgenus *Disporocarpa*, respectively. Additionally, the evolutionary rates analyses indicated all plastid genes of Crassulaceae were under purifying selection. Among these genes, *ycf1* (dN/dS = 0.4349) and *ycf2* (dN/dS = 0.5891) were the most rapidly evolving genes, whereas *psaC* (dN/dS = 0) was the most conserved gene. Finally, our phylogenetic analyses strongly supported *Crassula* is sister to all other Crassulaceae species. Our results will be benefit for further evolutionary studies within the *Crassula* and Crassulaceae.

Supplementary Materials: The following supporting information can be downloaded at: https://www.mdpi.com/article/10.3390/biology11121779/s1, Figure S1: PR2 plots of 52 plastid genes; Table S1: Accession numbers of Crassulaceae plastomes sampled in this study; Table S2: The sliding window-based π values estimated by 11 plastomes of *Crassula*; Table S3: The RSCU values of codons among Crassulaceae species; Table S4: The ENC values of 53 plastid genes of Crassulaceae; Table S5: Plastid codon aversion motifs among Crassulaceae species; Table S6: Evolutionary rates of 79 plastid genes of Crassulaceae; Table S7: The best evolutionary models for 79 plastid genes; Table S8: The morphological characteristics of 11 *Crassula* species; Data S1: Phylogenetic datasets of 79 plastid genes; Data S2: Sequence matrix of 11 plastomes of *Crassula* for sliding-window analysis.

Author Contributions: Conceptualization, supervision, and project administration, X.K.; resources and validation, L.W. and D.B.; data curation, Y.Y. and J.G.; investigation and formal analysis, S.Z. and J.Y.; software, S.H. and R.Y.; methodology and writing—original draft, H.D.; funding acquisition, H.D., D.B. and X.K; writing—review & editing, X.K. All authors have read and agreed to the published version of the manuscript.

Funding: This study was supported by the Opening Foundation of National Engineering Laboratory of Soil Pollution Control and Remediation Technologies, and Key Laboratory of Heavy Metal Pollution Prevention & Control, Ministry of Agriculture and Rural Affairs (Grant no. NEL&MARA-003), the Basic Research Program (Natural Science Foundation) of Jiangsu Province (Grant no. BK20211078), and the Scientific Research Project Foundation of Postgraduate of the Anhui Higher Education Institutions of China (Grant no. YJS20210136).

Institutional Review Board Statement: Not applicable.

Informed Consent Statement: Not applicable.

Data Availability Statement: The sequence data generated in this study are available in GenBank of the National Center for Biotechnology Information (NCBI) under the access numbers: OP729482–OP729487 and OP882297–OP882300.

Acknowledgments: We kindly acknowledge three anonymous reviewers for the fruitful and critical comments.

Conflicts of Interest: The authors declare no conflict of interest.

References

1. Thiede, J.; Eggli, U. Crassulaceae. In *The Families and Genera of Vascular Plants*; Kubitzki, K., Ed.; Springer: Berlin/Heidelberg, Germany, 2007; pp. 83–118.
2. Smith, G.F.; Figueiredo, E.; Van Wyk, A.E. *Kalanchoe (Crassulaceae) in Southern Africa: Classification, Biology, and Cultivation*; Academic Press: London, UK, 2019.
3. Messerschmid, T.F.; Klein, J.T.; Kadereit, G.; Kadereit, J.W. Linnaeus's folly–phylogeny, evolution and classification of *Sedum* (Crassulaceae) and Crassulaceae subfamily Sempervivoideae. *Taxon* **2020**, *69*, 892–926. [CrossRef]
4. Gontcharova, S.; Gontcharov, A. Molecular phylogeny and systematics of flowering plants of the family Crassulaceae DC. *Mol. Biol.* **2009**, *43*, 794–803. [CrossRef]
5. Mort, M.; O'Leary, T.; Carrillo-Reyes, P.; Nowell, T.; Archibald, J.; Randle, C. Phylogeny and evolution of Crassulaceae: Past, present, and future. *Schumannia* **2010**, *6*, 69–86.
6. Chang, H.; Zhang, L.; Xie, H.; Liu, J.; Xi, Z.; Xu, X. The Conservation of Chloroplast Genome Structure and Improved Resolution of Infrafamilial Relationships of Crassulaceae. *Front. Plant Sci.* **2021**, *12*, 631884. [CrossRef] [PubMed]
7. Eggli, U. *Illustrated Handbook of Succulent Plants: Crassulaceae*; Eggli, U., Ed.; Springer: Berlin/Heidelberg, Germany, 2003.
8. Van Ham, R.C.; Hart, H.T. Phylogenetic relationships in the Crassulaceae inferred from chloroplast DNA restriction-site variation. *Am. J. Bot.* **1998**, *85*, 123–134. [CrossRef]
9. Lu, M.; Fradera-Soler, M.; Forest, F.; Barraclough, T.G.; Grace, O.M. Evidence linking life-form to a major shift in diversification rate in *Crassula*. *Am. J. Bot.* **2022**, *109*, 272–290. [CrossRef]
10. Bruyns, P.; Hanáček, P.; Klak, C. *Crassula*, insights into an old, arid-adapted group of southern African leaf-succulents. *Mol. Phylogenet. Evol.* **2019**, *131*, 35–47. [CrossRef]
11. Tölken, H.R. A revision of the genus *Crassula* in southern Africa. *Contrib. Bolus. Herb.* **1977**, *8*, 1–595.
12. Mort, M.E.; Randle, C.P.; Burgoyne, P.; Smith, G.; Jaarsveld, E.; Hopper, S.D. Analyses of cpDNA *matK* sequence data place *Tillaea* (Crassulaceae) within *Crassula*. *Plant Syst. Evol.* **2009**, *283*, 211–217. [CrossRef]
13. Pandey, A.; Chaudhary, S.; Bhat, B. The Potential Role of Plastome Copy Number as a Quality Biomarker for Plant Products using Real-time Quantitative Polymerase Chain Reaction. *Curr. Genom.* **2022**, *23*, 289–298. [CrossRef]
14. Wicke, S.; Schneeweiss, G.M.; Depamphilis, C.W.; Müller, K.F.; Quandt, D. The evolution of the plastid chromosome in land plants: Gene content, gene order, gene function. *Plant Mol. Biol.* **2011**, *76*, 273–297. [CrossRef] [PubMed]
15. Ren, W.; Guo, D.; Xing, G.; Yang, C.; Zhang, Y.; Yang, J.; Niu, L.; Zhong, X.; Zhao, Q.; Cui, Y. Complete chloroplast genome sequence and comparative and phylogenetic analyses of the cultivated *Cyperus esculentus*. *Diversity* **2021**, *13*, 405. [CrossRef]
16. Jian, H.-Y.; Zhang, Y.-H.; Yan, H.-J.; Qiu, X.-Q.; Wang, Q.-G.; Li, S.-B.; Zhang, S.-D. The complete chloroplast genome of a key ancestor of modern roses, *Rosa chinensis* var. *spontanea*, and a comparison with congeneric species. *Molecules* **2018**, *23*, 389.
17. Han, S.; Bi, D.; Yi, R.; Ding, H.; Wu, L.; Kan, X. Plastome evolution of *Aeonium* and *Monanthes* (Crassulaceae): Insights into the variation of plastomic tRNAs, and the patterns of codon usage and aversion. *Planta* **2022**, *256*, 35. [CrossRef] [PubMed]
18. Ravi, V.; Khurana, J.; Tyagi, A.; Khurana, P. An update on chloroplast genomes. *Plant Syst. Evol.* **2008**, *271*, 101–122. [CrossRef]
19. Yang, X.; Huang, Y.; Li, Z.; Chen, J. Complete plastid genome of *Primula calliantha* Franch.(Primulaceae): An alpine ornamental plant endemic to Hengduan Mountain, China. *Mitochondrial DNA Part B* **2021**, *6*, 2643–2645. [CrossRef] [PubMed]
20. Yin, X.; Qian, C.; Yan, X.; Fang, T.; Fan, X.; Zhou, S.; Ma, X.-F. Will the artificial populations be sustainable? A genetic assessment on *Caragana korshinskii* afforestation in the semiarid regions of North China. *Eur. J. For. Res.* **2022**, *141*, 105–116. [CrossRef]
21. Zhou, H.; He, W.; Liu, X.; Tembrock, L.R.; Wu, Z.; Xu, D.; Liao, X. Comparative Analyses of 35 Complete Chloroplast Genomes from the Genus *Dalbergia* (Fabaceae) and the Identification of DNA Barcodes for Tracking Illegal Logging and Counterfeit Rosewood. *Forests* **2022**, *13*, 626.
22. Zeng, S.; Li, J.; Yang, Q.; Wu, Y.; Yu, J.; Pei, X. Comparative plastid genomics of Mazaceae: Focusing on a new recognized genus, *Puchiumazus*. *Planta* **2021**, *254*, 99. [CrossRef]
23. Tian, X.; Guo, J.; Zhou, X.; Ma, K.; Ma, Y.; Shi, T.; Shi, Y. Comparative and Evolutionary Analyses on the Complete Plastomes of Five *Kalanchoe* Horticultural Plants. *Front. Plant Sci.* **2021**, *12*, 705874. [CrossRef]
24. Ding, H.; Zhu, R.; Dong, J.; Bi, D.; Jiang, L.; Zeng, J.; Huang, Q.; Liu, H.; Xu, W.; Wu, L. Next-Generation Genome Sequencing of *Sedum plumbizincicola* Sheds Light on the Structural Evolution of Plastid rRNA Operon and Phylogenetic Implications within Saxifragales. *Plants* **2019**, *8*, 386. [CrossRef] [PubMed]
25. Parvathy, S.T.; Udayasuriyan, V.; Bhadana, V. Codon usage bias. *Mol. Biol. Rep.* **2022**, 539–565. [CrossRef] [PubMed]
26. Han, S.; Wang, R.; Hong, X.; Wu, C.; Zhang, S.; Kan, X. Plastomes of *Bletilla* (Orchidaceae) and Phylogenetic Implications. *Int. J. Mol. Sci.* **2022**, *23*, 10151. [CrossRef] [PubMed]
27. Zhang, Y.; Shen, Z.; Meng, X.; Zhang, L.; Liu, Z.; Liu, M.; Zhang, F.; Zhao, J. Codon usage patterns across seven Rosales species. *BMC Plant Biol.* **2022**, *22*, 65. [CrossRef] [PubMed]
28. Yang, J.; Ding, H.; Kan, X. Codon usage patterns and evolution of HSP60 in birds. *Int. J. Biol. Macromol.* **2021**, *183*, 1002–1012. [CrossRef] [PubMed]
29. Miller, J.B.; Hippen, A.A.; Belyeu, J.R.; Whiting, M.F.; Ridge, P.G. Missing something? Codon aversion as a new character system in phylogenetics. *Cladistics* **2017**, *33*, 545–556. [CrossRef] [PubMed]
30. Miller, J.B.; McKinnon, L.M.; Whiting, M.F.; Ridge, P.G. Codon use and aversion is largely phylogenetically conserved across the tree of life. *Mol. Phylogenet. Evol.* **2020**, *144*, 106697. [CrossRef]

31. Miller, J.B.; McKinnon, L.M.; Whiting, M.F.; Ridge, P.G. CAM: An alignment-free method to recover phylogenies using codon aversion motifs. *PeerJ* **2019**, *7*, e6984. [CrossRef]
32. Alba, M.M.; Castresana, J. Inverse relationship between evolutionary rate and age of mammalian genes. *Mol. Biol. Evol.* **2005**, *22*, 598–606. [CrossRef]
33. Pajkos, M.; Dosztányi, Z. Functions of intrinsically disordered proteins through evolutionary lenses. *Prog. Mol. Biol. Transl. Sci.* **2021**, *183*, 45–74.
34. Jeffares, D.C.; Tomiczek, B.; Sojo, V.; Reis, M.D. A beginners guide to estimating the non-synonymous to synonymous rate ratio of all protein-coding genes in a genome. *Methods Mol. Biol.* **2015**, *1201*, 65–90. [PubMed]
35. Kryazhimskiy, S.; Plotkin, J.B. The population genetics of dN/dS. *PLoS Genet.* **2008**, *4*, e1000304. [CrossRef] [PubMed]
36. Li, X.; Li, Y.; Sylvester, S.P.; Zang, M.; El-Kassaby, Y.A.; Fang, Y. Evolutionary patterns of nucleotide substitution rates in plastid genomes of *Quercus*. *Ecol. Evol.* **2021**, *11*, 13401–13414. [CrossRef] [PubMed]
37. Bromham, L.; Hua, X.; Lanfear, R.; Cowman, P.F. Exploring the relationships between mutation rates, life history, genome size, environment, and species richness in flowering plants. *Am. Nat.* **2015**, *185*, 507–524. [CrossRef]
38. Su, Y.; Huang, L.; Wang, Z.; Wang, T. Comparative chloroplast genomics between the invasive weed *Mikania micrantha* and its indigenous congener *Mikania cordata*: Structure variation, identification of highly divergent regions, divergence time estimation, and phylogenetic analysis. *Mol. Phylogenet. Evol.* **2018**, *126*, 181–195. [CrossRef]
39. Tang, H.; Tang, L.; Shao, S.; Peng, Y.; Li, L.; Luo, Y. Chloroplast genomic diversity in *Bulbophyllum* section *Macrocaulia* (Orchidaceae, Epidendroideae, Malaxideae): Insights into species divergence and adaptive evolution. *Plant Divers.* **2021**, *43*, 350–361. [CrossRef]
40. Zhang, C.; Li, S.-Q.; Xie, H.-H.; Liu, J.-Q.; Gao, X.-F. Comparative plastid genome analyses of *Rosa*: Insights into the phylogeny and gene divergence. *Tree Genet. Genomes* **2022**, *18*, 20. [CrossRef]
41. Jin, J.-J.; Yu, W.-B.; Yang, J.-B.; Song, Y.; DePamphilis, C.W.; Yi, T.-S.; Li, D.-Z. GetOrganelle: A fast and versatile toolkit for accurate de novo assembly of organelle genomes. *Genome Biol.* **2020**, *21*, 241. [CrossRef]
42. Tillich, M.; Lehwark, P.; Pellizzer, T.; Ulbricht-Jones, E.S.; Fischer, A.; Bock, R.; Greiner, S. GeSeq–versatile and accurate annotation of organelle genomes. *Nucleic Acids Res.* **2017**, *45*, W6–W11. [CrossRef]
43. Langmead, B.; Salzberg, S.L. Fast gapped-read alignment with Bowtie 2. *Nat. Methods* **2012**, *9*, 357–359. [CrossRef]
44. Zheng, S.; Poczai, P.; Hyvönen, J.; Tang, J.; Amiryousefi, A. Chloroplot: An online program for the versatile plotting of organelle genomes. *Front. Genet.* **2020**, *11*, 576124. [CrossRef]
45. Frazer, K.A.; Pachter, L.; Poliakov, A.; Rubin, E.M.; Dubchak, I. VISTA: Computational tools for comparative genomics. *Nucleic Acids Res.* **2004**, *32*, W273–W279. [CrossRef] [PubMed]
46. Rozas, J.; Ferrer-Mata, A.; Sánchez-DelBarrio, J.C.; Guirao-Rico, S.; Librado, P.; Ramos-Onsins, S.E.; Sánchez-Gracia, A. DnaSP 6: DNA sequence polymorphism analysis of large data sets. *Mol. Biol. Evol.* **2017**, *34*, 3299–3302. [CrossRef] [PubMed]
47. Dong, W.; Liu, J.; Yu, J.; Wang, L.; Zhou, S. Highly variable chloroplast markers for evaluating plant phylogeny at low taxonomic levels and for DNA barcoding. *PLoS ONE* **2012**, *7*, e35071. [CrossRef] [PubMed]
48. Bi, Y.; Zhang, M.-F.; Xue, J.; Dong, R.; Du, Y.-P.; Zhang, X.-H. Chloroplast genomic resources for phylogeny and DNA barcoding: A case study on *Fritillaria*. *Sci. Rep.* **2018**, *8*, 1184. [CrossRef]
49. Amiryousefi, A.; Hyvönen, J.; Poczai, P. IRscope: An online program to visualize the junction sites of chloroplast genomes. *Bioinformatics* **2018**, *34*, 3030–3031. [CrossRef]
50. Sharp, P.M.; Li, W.-H. An evolutionary perspective on synonymous codon usage in unicellular organisms. *J. Mol. Evol.* **1986**, *24*, 28–38. [CrossRef]
51. Chen, C.; Chen, H.; Zhang, Y.; Thomas, H.R.; Frank, M.H.; He, Y.; Xia, R. TBtools: An integrative toolkit developed for interactive analyses of big biological data. *Mol. Plant* **2020**, *13*, 1194–1202. [CrossRef]
52. Wright, F. The 'effective number of codons' used in a gene. *Gene* **1990**, *87*, 23–29. [CrossRef]
53. Sueoka, N. Translation-coupled violation of Parity Rule 2 in human genes is not the cause of heterogeneity of the DNA G+ C content of third codon position. *Gene* **1999**, *238*, 53–58. [CrossRef]
54. Sueoka, N. Two aspects of DNA base composition: G + C content and translation-coupled deviation from intra-strand rule of A = T and G = C. *J. Mol. Evol.* **1999**, *49*, 49–62. [CrossRef] [PubMed]
55. Yang, Z. PAML 4: Phylogenetic analysis by maximum likelihood. *Mol. Biol. Evol.* **2007**, *24*, 1586–1591. [CrossRef] [PubMed]
56. Zhang, D.; Gao, F.; Jakovlić, I.; Zou, H.; Zhang, J.; Li, W.X.; Wang, G.T. PhyloSuite: An integrated and scalable desktop platform for streamlined molecular sequence data management and evolutionary phylogenetics studies. *Mol. Ecol. Resour.* **2020**, *20*, 348–355. [CrossRef] [PubMed]
57. Darriba, D.; Posada, D.; Kozlov, A.M.; Stamatakis, A.; Morel, B.; Flouri, T. ModelTest-NG: A new and scalable tool for the selection of DNA and protein evolutionary models. *Mol. Biol. Evol.* **2020**, *37*, 291–294. [CrossRef] [PubMed]
58. Kozlov, A.M.; Darriba, D.; Flouri, T.; Morel, B.; Stamatakis, A. RAxML-NG: A fast, scalable and user-friendly tool for maximum likelihood phylogenetic inference. *Bioinformatics* **2019**, *35*, 4453–4455. [CrossRef]
59. Ronquist, F.; Teslenko, M.; Van Der Mark, P.; Ayres, D.L.; Darling, A.; Höhna, S.; Larget, B.; Liu, L.; Suchard, M.A.; Huelsenbeck, J.P. MrBayes 3.2: Efficient Bayesian phylogenetic inference and model choice across a large model space. *Syst. Biol.* **2012**, *61*, 539–542. [CrossRef]
60. Rambaut, A.; Drummond, A.J.; Xie, D.; Baele, G.; Suchard, M.A. Posterior summarization in Bayesian phylogenetics using Tracer 1.7. *Syst. Biol.* **2018**, *67*, 901–904. [CrossRef]

61. Chen, Z.; Zhao, J.; Qiao, J.; Li, W.; Li, J.; Xu, R.; Wang, H.; Liu, Z.; Xing, B.; Wendel, J.F. Comparative analysis of codon usage between *Gossypium hirsutum* and *G. barbadense* mitochondrial genomes. *Mitochondrial DNA Part B* **2020**, *5*, 2500–2506. [CrossRef]
62. Yuan, M.L.; Zhang, L.J.; Zhang, Q.L.; Zhang, L.; Li, M.; Wang, X.T.; Feng, R.Q.; Tang, P.A. Mitogenome evolution in ladybirds: Potential association with dietary adaptation. *Ecol. Evol.* **2020**, *10*, 1042–1053. [CrossRef]
63. Gontcharova, S.; Artyukova, E.; Gontcharov, A. Phylogenetic relationships among members of the subfamily Sedoideae (Crassulaceae) inferred from the ITS region sequences of nuclear rDNA. *Russ. J. Genet.* **2006**, *42*, 654–661. [CrossRef]
64. Li, X.; Li, Y.; Zang, M.; Li, M.; Fang, Y. Complete chloroplast genome sequence and phylogenetic analysis of *Quercus acutissima*. *Int. J. Mol. Sci.* **2018**, *19*, 2443. [CrossRef] [PubMed]
65. Turmel, M.; Otis, C.; Lemieux, C. The chloroplast and mitochondrial genome sequences of the charophyte *Chaetosphaeridium globosum*: Insights into the timing of the events that restructured organelle DNAs within the green algal lineage that led to land plants. *Proc. Natl. Acad. Sci. USA* **2002**, *99*, 11275–11280. [CrossRef] [PubMed]
66. Jansen, R.K.; Ruhlman, T.A. Plastid Genomes of Seed Plants. In *Genomics of Chloroplasts and Mitochondria*; Bock, R., Knoop, V., Eds.; Springer: Dordrecht, The Netherlands, 2012; pp. 103–126.
67. Wang, W.; Yu, H.; Wang, J.; Lei, W.; Gao, J.; Qiu, X.; Wang, J. The complete chloroplast genome sequences of the medicinal plant *Forsythia suspensa* (Oleaceae). *Int. J. Mol. Sci.* **2017**, *18*, 2288. [CrossRef] [PubMed]
68. Li, Y.; Wang, T.-R.; Kozlowski, G.; Liu, M.-H.; Yi, L.-T.; Song, Y.-G. Complete Chloroplast Genome of an Endangered Species *Quercus litseoides*, and Its Comparative, Evolutionary, and Phylogenetic Study with Other *Quercus* Section *Cyclobalanopsis* Species. *Genes* **2022**, *13*, 1184. [CrossRef] [PubMed]
69. Jiang, H.; Tian, J.; Yang, J.; Dong, X.; Zhong, Z.; Mwachala, G.; Zhang, C.; Hu, G.; Wang, Q. Comparative and phylogenetic analyses of six Kenya *Polystachya* (Orchidaceae) species based on the complete chloroplast genome sequences. *BMC Plant Biol.* **2022**, *22*, 177. [CrossRef]
70. Celiński, K.; Sokołowska, J.; Fuchs, H.; Maděra, P.; Wiland-Szymańska, J. Characterization of the Complete Chloroplast Genome Sequence of the Socotra Dragons Blood Tree (*Dracaena cinnabari* Balf.). *Forests* **2022**, *13*, 932. [CrossRef]
71. Goulding, S.E.; Olmstead, R.G.; Morden, C.W.; Wolfe, K.H. Ebb and flow of the chloroplast inverted repeat. *Mol. Gen. Genet.* **1996**, *252*, 195–206. [CrossRef]
72. Wang, R.J.; Cheng, C.L.; Chang, C.C.; Wu, C.L.; Su, T.M.; Chaw, S.M. Dynamics and evolution of the inverted repeat-large single copy junctions in the chloroplast genomes of monocots. *BMC Evol. Biol.* **2008**, *8*, 36. [CrossRef]
73. Zhu, A.; Guo, W.; Gupta, S.; Fan, W.; Mower, J.P. Evolutionary dynamics of the plastid inverted repeat: The effects of expansion, contraction, and loss on substitution rates. *New Phytol.* **2016**, *209*, 1747–1756. [CrossRef]
74. Chi, X.; Zhang, F.; Dong, Q.; Chen, S. Insights into comparative genomics, codon usage bias, and phylogenetic relationship of species from Biebersteiniaceae and Nitrariaceae based on complete chloroplast genomes. *Plants* **2020**, *9*, 1605. [CrossRef]
75. Ma, Q.; Li, C.; Wang, J.; Wang, Y.; Ding, Z. Analysis of synonymous codon usage in *FAD7* genes from different plant species. *Genet. Mol. Res.* **2015**, *14*, 1414–1422. [CrossRef] [PubMed]
76. Tats, A.; Tenson, T.; Remm, M. Preferred and avoided codon pairs in three domains of life. *BMC Genom.* **2008**, *9*, 463. [CrossRef] [PubMed]
77. Liu, J.; Zhu, D.; Ma, G.; Liu, M.; Wang, M.; Jia, R.; Chen, S.; Sun, K.; Yang, Q.; Wu, Y. Genome-wide analysis of the synonymous codon usage patterns in *Riemerella anatipestifer*. *Int. J. Mol. Sci.* **2016**, *17*, 1304. [CrossRef] [PubMed]
78. Xu, Q.; Chen, H.; Sun, W.; Zhu, D.; Zhang, Y.; Chen, J.-L.; Chen, Y. Genome-wide analysis of the synonymous codon usage pattern of *Streptococcus suis*. *Microb. Pathog.* **2021**, *150*, 104732. [CrossRef] [PubMed]
79. Kattoor, J.J.; Malik, Y.S.; Sasidharan, A.; Rajan, V.M.; Dhama, K.; Ghosh, S.; Bányai, K.; Kobayashi, N.; Singh, R.K. Analysis of codon usage pattern evolution in avian rotaviruses and their preferred host. *Infect. Genet. Evol.* **2015**, *34*, 17–25. [CrossRef]
80. Brandão, P.E. The evolution of codon usage in structural and non-structural viral genes: The case of Avian coronavirus and its natural host *Gallus gallus*. *Virus Res.* **2013**, *178*, 264–271. [CrossRef]
81. Yu, J.; Xue, J.H.; Zhou, S.L. New universal *matK* primers for DNA barcoding angiosperms. *J. Syst. Evol.* **2011**, *49*, 176–181. [CrossRef]
82. Li, F.-W.; Kuo, L.-Y.; Rothfels, C.J.; Ebihara, A.; Chiou, W.-L.; Windham, M.D.; Pryer, K.M. *rbcL* and *matK* earn two thumbs up as the core DNA barcode for ferns. *PLoS ONE* **2011**, *6*, e26597. [CrossRef]
83. Surveswaran, S.; Gowda, V.; Sun, M. Using an integrated approach to identify cryptic species, divergence patterns and hybrid species in Asian ladies' tresses orchids (*Spiranthes*, Orchidaceae). *Mol. Phylogenet. Evol.* **2018**, *124*, 106–121. [CrossRef]
84. Ashokan, A.; Leong-Škorničková, J.; Suksathan, P.; Newman, M.; Kress, W.J.; Gowda, V. Floral evolution and pollinator diversification in *Hedychium*: Revisiting Darwin's predictions using an integrative taxonomic approach. *Am. J. Bot.* **2022**, 1410–1427. [CrossRef]
85. Kikuchi, S.; Bédard, J.; Hirano, M.; Hirabayashi, Y.; Oishi, M.; Imai, M.; Takase, M.; Ide, T.; Nakai, M. Uncovering the protein translocon at the chloroplast inner envelope membrane. *Science* **2013**, *339*, 571–574. [CrossRef] [PubMed]
86. Drescher, A.; Ruf, S.; Calsa Jr, T.; Carrer, H.; Bock, R. The two largest chloroplast genome-encoded open reading frames of higher plants are essential genes. *Plant J.* **2000**, *22*, 97–104. [CrossRef] [PubMed]
87. Barnard-Kubow, K.B.; Sloan, D.B.; Galloway, L.F. Correlation between sequence divergence and polymorphism reveals similar evolutionary mechanisms acting across multiple timescales in a rapidly evolving plastid genome. *BMC Evol. Biol.* **2014**, *14*, 268. [CrossRef] [PubMed]

88. Tölken, H.R. *Flora of Southern Africa*; Leistner, O.A., Ed.; National Botanical Institute: Pretoria, South Africa, 1985; Volume 14, pp. 1–244.
89. Nikulin, V.Y.; Gontcharova, S.B.; Stephenson, R.; Gontcharov, A.A. Phylogenetic relationships between *Sedum* L. and related genera (Crassulaceae) based on ITS rDNA sequence comparisons. *Flora* **2016**, *224*, 218–229. [CrossRef]
90. Carrillo-Reyes, P.; Sosa, V.; Mort, M.E. Molecular phylogeny of the *Acre* clade (Crassulaceae): Dealing with the lack of definitions for *Echeveria* and *Sedum*. *Mol. Phylogenet. Evol.* **2009**, *53*, 267–276. [CrossRef] [PubMed]

biology

MDPI

Article

Comparative Chloroplast Genomes of Six Magnoliaceae Species Provide New Insights into Intergeneric Relationships and Phylogeny

Lin Yang [1,2,†], Jinhong Tian [1,2,†], Liu Xu [1,2], Xueli Zhao [1,2], Yuyang Song [3,*] and Dawei Wang [1,2,*]

1 Key Laboratory for Forest Resources Conservation and Utilization in the Southwest Mountains of China Ministry of Education, Southwest Forestry University, Kunming 650224, China
2 Key Laboratory for Forest Genetics and Tree Improvement and Propagation in Universities of Yunnan Province, Southwest Forestry University, Kunming 650224, China
3 Department of Forestry, Agricultural College, Xinjiang Shihezi University, Shihezi 832003, China
* Correspondence: syy.agr.shzu@126.com (Y.S.); wangdawei@swfu.edu.cn (D.W.)
† These authors contributed equally to this work.

Simple Summary: Magnoliaceae is one of the most endangered families of angiosperms. The systematic classification of Magnoliaceae has been controversial for a long time due to minor differences in morphology. In the present study, six new chloroplast genomes of Magnoliaceae were sequenced, and the 37 published chloroplast genomes of the family were subjected to phylogenetic analyses. The results showed that all these chloroplast genomes possess the typical quadripartite structure with a conserved genome arrangement and gene structures, yet their lengths varied due to the expansion/contraction of the IR/SC boundaries. Phylogenetic relationships within Magnoliaceae were determined using complete cp genome sequences. These findings will provide a theoretical basis for adjusting the phylogenetic position of Magnoliaceae at the molecular level.

Abstract: Magnoliaceae plants are industrial tree species with high ornamental and medicinal value. We published six complete chloroplast genomes of Magnoliaceae by using Illumina sequencing. These showed a typical quadripartite structure of angiosperm and were 159,901–160,008 bp in size. A total of 324 microsatellite loci and six variable intergenic regions (Pi > 0.01) were identified in six genomes. Compared with five other genomes, the contraction and expansion of the IR regions were significantly different in *Manglietia grandis*. To gain a more thorough understanding of the intergeneric relationships in Magnoliaceae, we also included 31 published chloroplast genomes of close relative species for phylogenetic analyses. New insights into the intergeneric relationships of Magnoliaceae are provided based on our results and previous morphological, phytochemical and anatomical information. We suggest that the genus *Yulania* should be separated from the genus *Michelia* and its systematic position of should be restored; the genera *Paramichelia* and *Tsoongiodendron* should be merged into the genus *Michelia*; the genera *Pachylarnax* and *Parakmeria* should be combined into one genus. These findings will provide a theoretical basis for adjusting the phylogenetic position of Magnoliaceae at the molecular level.

Keywords: Magnoliaceae; chloroplast genome; phylogenomics; intergeneric relationship

Citation: Yang, L.; Tian, J.; Xu, L.; Zhao, X.; Song, Y.; Wang, D. Comparative Chloroplast Genomes of Six Magnoliaceae Species Provide New Insights into Intergeneric Relationships and Phylogeny. *Biology* **2022**, *11*, 1279. https://doi.org/10.3390/biology11091279

Academic Editor: Lorenzo Peruzzi

Received: 20 July 2022
Accepted: 25 August 2022
Published: 28 August 2022

1. Introduction

Chloroplasts are critical plant organelles that play a prominent role in photosynthesis [1]. Chloroplast genomes (cp genomes) are highly conserved because of the genetic replication mechanisms of uniparent inheritance and the relatively high level of genetic variation resulting from the low selective pressure, making them useful for revealing phylogenetic relationships [2]. With the development of Illumina and assembly technologies, the cp genomes of an increasing number of species have been published [3–5]. These cp genomes provide valuable information about species identification, trait improvement, genealogical geography and the conservation of endangered species [6–8].

Magnoliaceae is one of the most endangered families of angiosperms, and it was listed under Class II National Protection in China [9]. It is considered a key material indispensable for exploring the origin of angiosperms and also an important component of tropical to temperate evergreen broadleaf in deciduous broadleaf forests, which are ecologically important [10]. Magnoliaceae plants are industrial tree species with high medicinal value [11]. The leaves, flowers and bark of them are rich in monoterpenes and sesquiterpenes, which have good anti-tumor-promoting and anti-carcinogenesis activities and are used to treat inflammation and ulceration diseases [12].

The current methods for distinguishing the taxonomic position of Magnoliaceae mainly consider anatomical and morphological aspects [13]. The systematic classification of Magnoliaceae has been controversial for a long time [14]. A total of 12 genera were classified in the narrow concept of Magnoliaceae for the first time by Dandy in 1964 [15], and afterward, it was split into 16 and 18 genera according to the characters of stomatal pores on the leaf epidermis and polygamous flower, respectively [16,17]. A few years later, some taxonomists suggested that Magnoliaceae should be divided into two genera (*Magnolia* L. and *Michelia* L.) based on their main morphological traits, while the remaining 16 genera should be combined with both [18]. In summary, the main controversial differences in the classification of Magnoliaceae are the merging or separation of intergeneric relationships [14]. In our study, we reconstructed the phylogenetic relationship among the genera *Yulania* Spach, *Michelia* L., *Paramichelia* Hu, *Tsoongiodendron* Chun, *Pachylarnax* Dandy and *Parakmeria* Hu & W.C.Cheng by using 37 species of Magnoliaceae to carry out a sequence alignment and phylogenetic analysis of cp genomes. These results provide a molecular-level basis to determine the systematic taxonomic position of Magnoliaceae species.

2. Materials and Methods

2.1. Plant Materials and DNA Sequencing

The young green and disease-free leaves of 6 species for Magnoliaceae were collected from natural distribution areas (Table 1). The plant species was identified by *Assoc. Prof.* Jianhua Qi (College of Forestry, Southwest Forestry University), and the voucher specimens were stored at the Key Laboratory for Forest Resources Conservation and Utilization in the Southwest Mountains of China Ministry of Education (2020Y18), Southwest Forestry University, Kunming, China. DNA extraction and sequencing were performed according to a previous study by Wang et al. [19].

Table 1. The sampling area and information of six species of the Magnoliaceae family.

Genus	Species	Protection Grade	Sampling Area	Longitude/Latitude
Manglietia	*Manglietia crassipes* Y.L.Law	-	Guangxi, China	109°50′ E/23°40′ N
	Manglietia grandis Hu & W.C.Cheng	II	Yunnan, China	104°33′ E/22°48′ N
	Manglietia hookeri Cubitt & W.W.Sm.	-	Yunnan, China	99°55′ E/21°10′ N
	Manglietia ventii N.V.Tiep.	II	Yunnan, China	102°10′ E/24°23′ N
Yulania	*Yulania kobus* (DC.) Spach	-	Yunnan, China	102°10′ E/24°23′ N
	Yulania soulangeana (Soul.-Bod.) D.L.Fu	-	Yunnan, China	102°10′ E/24°23′ N

Notes: II: Endangered

2.2. Chloroplast Genome Assembly and Annotation

The cp genome sequences of *Manglietia dandyi* (MF990567) were used as a reference sequence to assemble the 6 cp genomes of Magnoliaceae using MEGA5.1(Mega Limited, Auckland, New Zealand) [20]. The annotation of the 6 cp genomes was performed via Genious 8.1.3 with sequences of other closely related species. The method of genome annotation was the same as Zheng et al. [21]. The sequences of 6 cp genomes were deposited in GenBank NCBI (MW415418, MW415419, MW415420, MW415421, MW415416 and MW415417). The cp genome map was drawn using OGDRAW37 [22].

2.3. Sequence Divergence, Genome Comparison and Single-Sequence Repeat Analysis

The 6 cp genomes of the Magnoliaceae were sequenced performed using the VISualization Tool in Shuffle-LAGAN mode for Alignments [23]. We used the DnaSPv. 5.0 software (J. Rozas et al., Barcelona, The Kingdom of Spain) to set the parameter to a window length size of 600 bp and the distance between each locus to 200 bp to measure nucleotide diversity (Pi) [24]. The 6 cp genome sequences were uploaded to the online IRscope software to visualize their IR/SC boundaries using the .gb format [25]. The simple sequence repeat (SSR) markers were searched by surveying six genomic sequences of the Magnoliaceae using MISAv program (http://genome.lbl.gov/vista/index.shtml, accessed on 15 March 2022) [26].

2.4. Phylogenetic Analysis

Sequence alignment was performed using the newly assembled 6 cp genomes and 25 closely related cp genomes, with 6 species of the genera *Illicium* L., *Kadsura* Kaempf. ex Juss. and *Schisandra* Michx. added for analysis, which were downloaded from the NCBI (Table S1). Phylogenetic analyses were performed according to a study of Wu et al. [27].

3. Results

3.1. Characteristics of the Six cp Genomes

The six cp genomes of Magnoliaceae are similar to other angiosperms (Table 2 and Figure 1). The complete cp genome is between 159,901 and 160,008 bp in length, exhibiting a classic four-partition structure with an SSC region (18,800–18,803 bp), LSC region (87,753–88,534 bp h) and two IR regions (26,207–26,602 bp). Six cp genomes contained 131 genes (86 protein-coding genes, 37 tRNA genes and 8 rRNA genes), which include 44 photosynthesis genes, 58 translation-related genes and 11 other genes (Table S2).

Table 2. Summary of chloroplast genome characteristics of six Magnoliaceae chloroplast genomes.

Species	*Manglietia crassipes*	*Manglietia grandis*	*Manglietia hookeri*	*Manglietia ventii*	*Yulania kobus*	*Yulania soulangeana*
Total length (bp)	159,901	160,008	159,905	159,950	159,778	159,778
LSC length (bp)	87,959	88,534	87,973	88,008	87,840	87,753
SSC length (bp)	18,800	18,803	18,776	18,800	18,734	18,734
IR length (bp)	26,571	26,207	26,578	26,571	26,602	26,602
Overall GC content (%)	39.3	39.3	39.3	39.3	39.3	39.3
Total gene number	131	131	131	131	131	131
GenBank accession	MW415418	MW415419	MW415420	MW415421	MW415416	MW415417

3.2. Comparative Genomic, IR Expansion and Contraction, and SSR Analysis

To investigate the levels of sequence polymorphism, the six cp genomes of Magnoliaceae species were compared (Figure 2). The results showed that the structures, orders and contents of these six cp genomes were all conserved. The Pi values of these six genomes ranged from 0 to 0.0153. Although aligned sequences showed relatively low divergences, some hotspot regions with high variation were also identified. The variable regions with Pi exceeding 0.01 in the six cp genomes were *ndhF-trnL-UAG, ndhD-ndhE, rpl32-trnL-UAG, petG-psaJ, psaC-ndhA, trnF-* and *GAA-ndhK* (Figure 3).

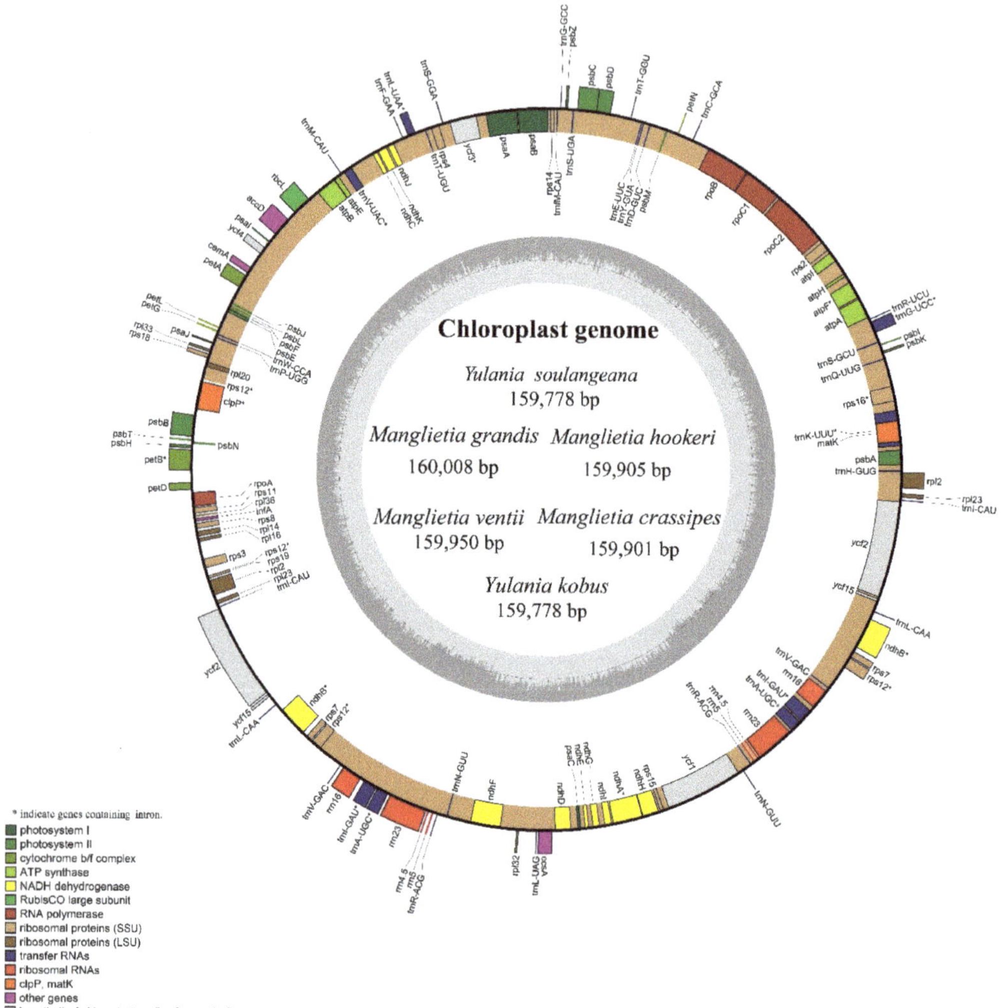

Figure 1. Structural map of the six chloroplast genomes of Magnoliaceae species.

The six CP genomes' IR/LSC and IR/SSC boundary structures were compared (Figure 4). The results showed that the IR boundaries of the cp genomes of the six Magnoliaceae species were comparatively conserved. Only the *rpl2* gene of *Manglietia grandis* expanded to the LSC region, with an expansion length of 308 bp, and the *rpl2* genes of the remaining five species were in the IRb region. Among them, the distributions of genes on the IRb/SSC and SSC/IRa boundaries were similar for the *ndhF* and *ycf1* genes, and the length of the *ycf1* gene on the SSC/IRa boundary ranged from 5558 to 5594 bp, all of which were pseudogenes. The characteristics of SSRs in six cp genomes were analyzed, a total of 324 repeats were certified in six genomes, and most SSRs included the A/T rather than the G/C motif (Figure 5b and Table S3). Mononucleotide repeats were the most abundant SSR in all the species; pentanucleotide repeats were the least abundant. The analysis of long repeats in six species revealed more forward and palindromic repeats than reverse and complementary repeats (Figure 5C,D).

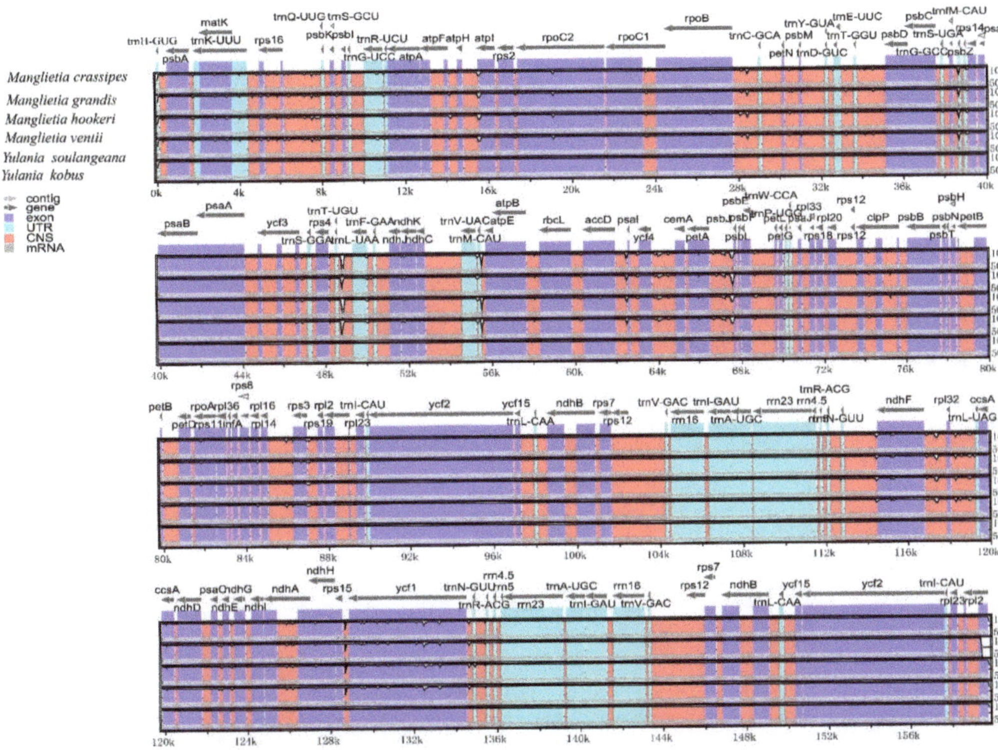

Figure 2. Alignment of whole chloroplast genome sequences from the six Magnoliaceae species.

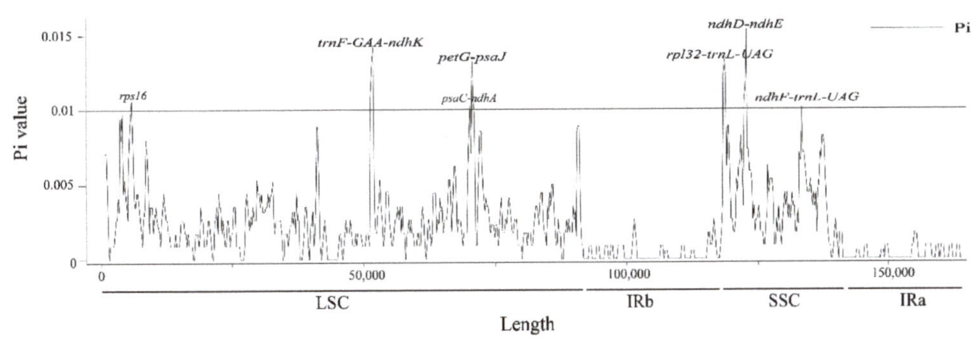

Figure 3. Sliding window analysis of the whole chloroplast genome nucleotide diversity (Pi) of the six Magnoliaceae species.

3.3. Phylogenetic Analysis

Phylogenetic analysis of six species was performed using the ML method (Figure 6); the results showed that most of the nodes had 100% bootstrap values. The phylogenetic tree showed that the 37 species of Magnoliaceae can be broadly divided into two clusters. Among them, the genera *Yulania*, *Paramichelia*, *Michelia*, *Tsoongiodendron*, *Alcimandra* Dandy, *Pachylarnax*, *Parakmeria*, *Woonyoungia* Y.W.Law, *Manglietia* Blume, *Talauma* Juss. and *Liriodendron* L. were clustered into one group, and the genera *Illicium*, *Kadsura* and *Schisandra* were also clustered into one group. In our phylogenetic tree, *Pachylarnax*, *Parakmeria* and *Michelia* were closely related to *Paramichelia* and *Tsoongiodendron*, but the genera *Illicium*, *Kadsura* and *Schisandra* are not clustered into a group with Magnoliaceae species. In addition, we also compared the Flora Reipubicae Popularis Sinicae (FRPS) and Flora of China (FOC) plant classifications in Magnoliaceae, finding that the taxonomic statuses of the genera *Paramichelia*, *Tsoongiodendron*, *Pachylarnax* and *Parakmeria* were different.

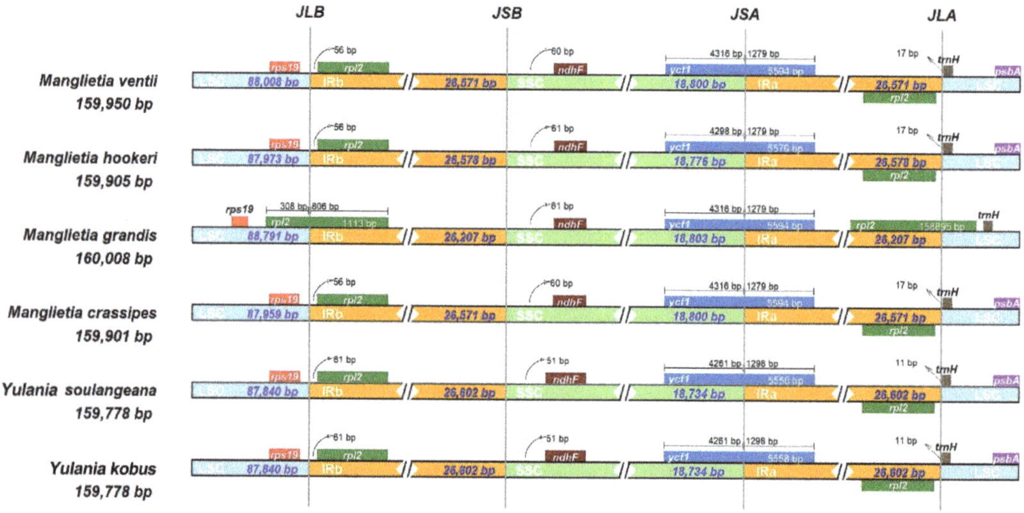

Figure 4. Comparison of the border regions of the six chloroplast genomes of Magnoliaceae. Note: Different genes are denoted by colored boxes. The gaps between the genes and the boundaries are indicated by the base lengths (bp).

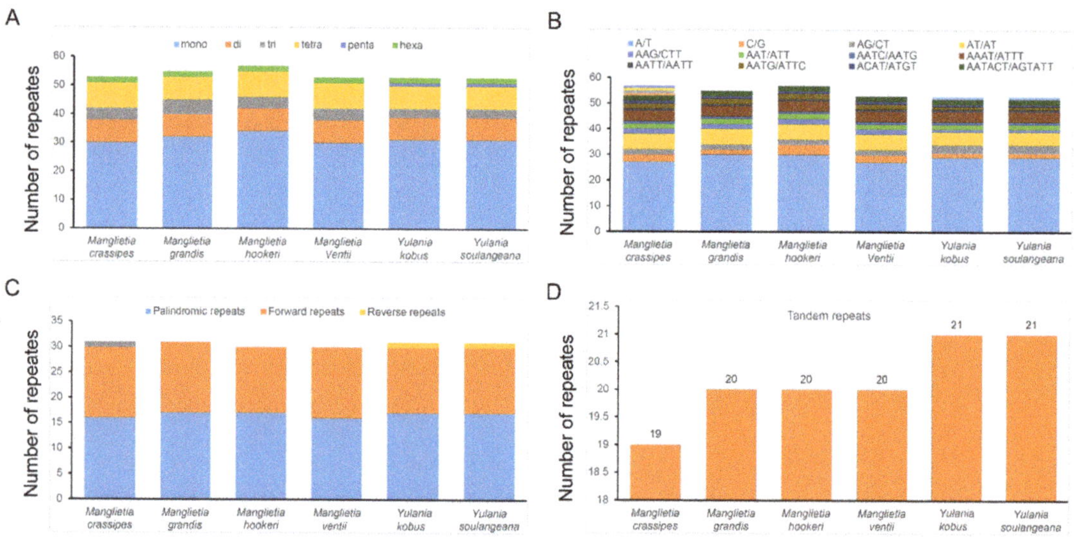

Figure 5. Comparison of repeats in six species of Magnoliaceae family. Note: (**A**) The type frequency of different SSR types. (**B**) The type frequency of SSR motifs in different repeat class types. (**C**) The type frequency of different repeat types. (**D**) The type frequency of dispersed repeat sequences.

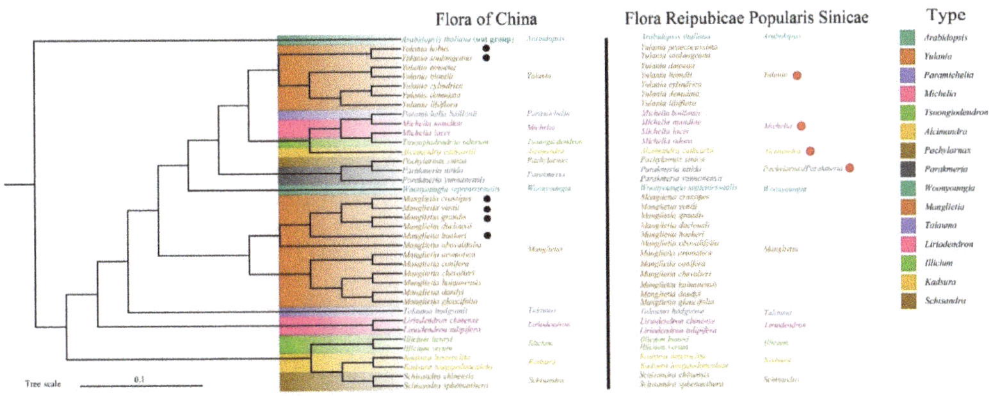

Figure 6. Maximum likelihood (ML) phylogenetic tree based on 37 complete chloroplast genomes of Magnoliaceae. Note: Black circles indicate the six species of Magnoliaceae in this study. The red circles indicate the genera whose phylogenetic positions were discussed in this study.

4. Discussion

Here, for the first time, we present cp genomes for six Magnoliaceae species, including four *Manglietia* species and two *Yulania* species. These cp genomes are consistent with the characteristics of most angiosperm species [28], and did not differ significantly from each other in terms of structure and length (159,901–160,008 bp). In addition, we found that the mean contents of AT and GC in these six cp genomes were 61.7% and 39.3%. In the genome, the higher the AT content, the lower the DNA density, and the sequences were more prone to denaturation and mutation [29]. Therefore, we speculated that the six cp sequences of Magnoliaceae were somewhat mutagenic and their chloroplast gene sequences might be more prone to variation than those of other species.

The results of the IR boundary analysis showed that the contraction of the IR region (26,207 bp) was the most pronounced in *Manglietia grandis*, with an expansion of the *rpl2* gene in its IR to LSC region of 308 bp, while the *rpl2* genes of the other five species were intact and located in the IR region. This indicated that the boundary change of LSC /IR is the dominant factor affecting the expansion and contraction of the cp genome IR region of *Manglietia grandis*. However, such an expansion is small, and no important expansions or contractions were observed in these cp sequences. This result is similar to the expansion of the chloroplast genomes of other Magnoliaceae species in the IR region [30], but different from the contraction of Zingiberaceae and Arecaceae [31,32]. This indicates that different species have evolved under the influence of different factors, resulting in different degrees of expansion and contraction of IR/SC boundaries, thus showing the diversity in genome length and boundaries [33].

The varied SSRs in cp genomes have a greater taxonomic distance between them than nuclear and mitochondrial genomes; they are widely used in studies of the genetic diversity and germplasm resources of plant populations [34]. We identified 324 SSRs in cp genomes of six Magnoliaceae species, most of which had mononucleotide repeats composed of A/T. These SSRs can be used to develop microsatellite markers for genetic diversity and evolution analyses [35]. We also screened a total of seven highly variable regions through nucleotide diversity analysis. Among them, four were located in the LSC region and three in the SSC region. This indicates that the LSC and SSC regions of these six Magnoliaceae species have high nucleotide variability, and these highly variable regions can be used as potential polymorphic molecular markers for evolutionary studies [36].

These six cp genome sequences were phylogenetically analyzed with their 31 relatives; the results showed that species of Magnoliaceae clustered in a group, and the genera *Illicium*, *Schisandra* and *Kadsura*, which do not belong to Magnoliaceae, were divided into a separate group. This result is consistent with the classification of Angiosperm Phylogeny Group (APG IV) system [37]. Meanwhile, the most of nodes had high bootstrap values in our phylogenetic tree, and the results of phylogenetic analysis for monophyletic genera are consistent with previous studies, indicating that the phylogenetic tree in this study is reliable [13,38]. The aim of this study was to determine the intergeneric relationships within Magnolioideae, as the systematic classification of Magnoliaceae has been controversial for a long time [14]. It has previously been demonstrated that the genus *Yulania* is included in the genus *Michelia* due to its pre-branching characteristics [39]. However, the contents of volatile oils obtained from flower and the pit vessel characteristics of wood of these two genera were significantly different in subsequent studies [40]. In particular, reproductive isolation was discovered due to the discontinuity of geographical distribution of the two genera; the genus *Yulania* was separated from the genus *Michelia* [41,42]. This result is consistent with the phylogenetic analysis conducted in our study. Furthermore, it was consistent with previous conclusions inferred from *Matk* and *ndhF* sequences [43]. Similar results were reported for the Bupleurum family, with new insights into its phylogenetic status provided through assessing the cp genomes and morphological characteristics of fruits and leaves [34,44]. We thus suggest that the genus *Yulania* should be separated from the genus *Michelia*, and its systematic position should be restored.

In the present study, the genera *Paramichelia*, *Tsoongiodendron* and *Michelia* were clustered into one clade, which is identical to the results of another phylogenetic analysis based on molecular markers [45]. This strongly supports the idea of a close relationship between these three genera. It has been argued that the genera *Paramichelia* and *Tsoongiodendron* should be separated from the genus *Michelia* according to the different characters of the ripe fruit carpels [17]. This tiny difference is considered by traditional taxonomists to be the result of parallel evolution [46]. In other words, these three genera come from the same ancestor and therefore show the same trend in evolution [47]. Based on all this evidence, we share the view that the genera *Paramichelia* and *Tsoongiodendron* should be merged into the genus *Michelia*. Similarly, Flora of China suggested adjusting the genera *Paramichelia* and *Tsoongiodendron* to genus-level status in the systematic position [48].

The genus *Pachylarnax* was established based on its polygamous flower [49], and it is considered to be more closely related to the genus *Manglietia* [50]. This argument was not consistent with the result of the phylogenetic analysis in our study; we suggested that, compared with *Manglietia*, the genus *Parakmeria* is more closely related to *Pachylarnax*. Meanwhile, this view is also consistent with the results of the phylogenetic analysis using the B-class MADS-box gene [51]. Additionally, the genera *Pachylarnax* and *Parakmeria* both have the high-taxonomic-value characteristic of curling young leaves [52]. We thus recommend that the genera *Pachylarnax* and *Parakmeria* should be combined into one genus. Furthermore, based on all the results related to phylogenetic relationships, we compared the two classifications and found that the FRPS can locate the species attribution more precisely than FOC in Magnoliaceae.

5. Conclusions

This study reports the complete cp genome sequence of six Magnoliaceae species: *M. crassipes*, *M. grandis*, *M. hookeri*, *M. ventii*, *Y. praecocissima* and *Y. soulangeana*. New insights into the intergeneric relationships in the Magnoliidae family are provided by combining our findings with previous studies. We recommend that the genus *Yulania* should be separated from the genus *Michelia*, and the systematic position of *Yulania* should be restored; the genera *Paramichelia* and *Tsoongiodendron* should be merged into the genera *Michelia*; and the genera *Pachylarnax* and *Parakmeria* should be combined into one genus. These results provide a theoretical foundation for the phylogenetic position of Magnoliaceae at the molecular level.

Supplementary Materials: The following supporting information can be downloaded at: https://www.mdpi.com/article/10.3390/biology11091279/s1, Table S1: Chloroplast genomes of 32 species for phylogenetic analysis; Table S2: Chloroplast genome-encoded gene types and functional classification for six species of the family. Table S3: Simple sequence repeats in the six chloroplast genomes of the family Magnoliaceae.

Author Contributions: All authors contributed to the study conception and design. L.Y.: Conceptualization, Methodology, Formal analysis, Writing—original draft; J.T.: Conceptualization, Methodology, Formal analysis, Writing—original draft; L.X.: Validation, Investigation, Software. X.Z.: Methodology, Investigation; Y.S.: Writing—review and editing, Formal analysis; D.W.: Writing—review and editing, Formal analysis, Funding acquisition. All authors have read and agreed to the published version of the manuscript.

Funding: This research was supported by the Fund of Ten-Thousand Talents Program of Yunnan Province, grant number YNWR–QNBJ–2020-230.

Institutional Review Board Statement: Not applicable.

Informed Consent Statement: Not applicable.

Data Availability Statement: The datasets generated during and/or analyzed during the current study are available in the NCBI repository, [https://www.ncbi.nlm.nih.gov/] (accessed on 5 March 2022). The datasets generated during and/or analyzed during the current study are available from the corresponding author on reasonable request. All data generated or analyzed during this study are included in this published article [and its supplementary information files].

Acknowledgments: The authors are grateful to Jianhua Qi (College of Forestry, Southwest Forestry University) for plant species identified.

Conflicts of Interest: The authors declare no conflict of interest.

References

1. Saldaña, C.L.; Rodriguez-Grados, P.; Chávez-Galarza, J.C.; Feijoo, S.; Guerrero-Abad, J.C.; Vásquez, H.V.; Maicelo, J.L.; Jhoncon, J.H.; Carlos, I.; Arbizu, C.I. Unlocking the complete chloroplast genome of a native tree species from the Amazon basin, capirona (*Calycophyllum spruceanum*, rubiaceae), and its comparative analysis with other ixoroideae species. *Genes* **2022**, *13*, 113. [CrossRef] [PubMed]
2. Shen, J.; Li, X.; Chen, X.; Huang, X.L.; Jin, S.H. The complete chloroplast genome of *Carya cathayensis* and phylogenetic analysis. *Genes* **2022**, *13*, 369. [CrossRef] [PubMed]
3. Antunes, A.M.; Soares, T.N.; Targueta, C.P.; Novaes, E.; Guedes-Coelho, A.S.; Campos-Telles, M.P. The chloroplast genome sequence of *Dipteryx alata* vog. (Fabaceae: *Papilionoideae*): Genomic features and comparative analysis with other legume genomes. *Braz. J. Bot.* **2020**, *43*, 271–278. [CrossRef]
4. Feng, S.; Zheng, K.; Jiao, K.; Cai, Y.C.; Chen, C.L.; Mao, Y.Y.; Wang, L.Y.; Zhan, X.R.; Ying, Q.C.; Wang, H.Z. Complete chloroplast genomes of four *Physalis species* (Solanaceae): Lights into genome structure, comparative analysis, and phylogenetic relationships. *BMC Plant Biol.* **2020**, *20*, 242. [CrossRef] [PubMed]
5. Zhou, T.; Zhu, H.; Wang, J.; Xu, Y.C.; Xu, F.S.; Wang, X.M. Complete chloroplast genome sequence determination of *Rheum* species and comparative chloroplast genomics for the members of rumiceae. *Plant Cell Rep.* **2020**, *39*, 811–824. [CrossRef]
6. Duan, Y.Z.; Zhang, K. Comparative analysis and phylogenetic evolution of the complete chloroplast genome of *Ammopiptanthus*. *Acta Bot. Boreali-Occident. Sin.* **2020**, *40*, 1323–1332.
7. Xie, B.B. Phylogeography Study of Siraitia Grosvenorii. Master's Thesis, Guangxi Normal University, Guilin, Guangxi, China, 2019.
8. Zhao, S.; Deng, L.H.; Chen, F. Codon usage bias of chloroplast genome in *Kandelia obovate*. *J. For. Environ.* **2020**, *40*, 534–541.
9. Ngoc, A.L.; Nurhayat, T.; Alden, S.E.; Nguyen, D.H.; Kendra, P.E. Insecticidal and attractant activities of *Magnolia citrata* leaf essential oil against two major pests from diptera: *Aedes aegypti* (culicidae) and *Ceratitis capitata* (tephritidae). *Molecules* **2021**, *26*, 2311. [CrossRef]
10. Timmis, J.N.; Ayliffe, M.A.; Huang, C.Y.; William, M. Endosymbiotic gene transfer: Organelle genomes forge eukaryotic chromosomes. *Nat. Rev. Genet.* **2004**, *5*, 123–135. [CrossRef]
11. Huang, Y.B.; Wang, J.M.; Li, R.X. Evaluation of cold resistance and physiological mechanism of seven *Manglietia* species. *J. Trop. Subtrop. Bot.* **2021**, *29*, 519–529. [CrossRef]
12. Sun, B.Y.; Liu, Y.; Xue, Y.Y.; Zhang, Y. The anti-tumor activity validation of *Magnolia sargentiana.* and *Manglietia hookeri.* on base of reverse molecular docking. *Comput. Appl. Chem.* **2019**, *36*, 337–344.
13. Zhu, B.; Qian, F.; Wang, X.S.; Liu, Y.L. The phylogeny of magnoliaceae based on chloroplast genome. *J. Biol.* **2022**, *39*, 53–58. [CrossRef]

14. Sima, Y.K.; Lu, S.G.; Han, M.Y.; Ma, H.F.; Xu, T.; Xu, L.; Hao, J.B.; Li, D.; Chen, S.Y. Advances in magnoliaceae systematic research. *J. West China For. Sci.* **2012**, *41*, 116–127. [CrossRef]
15. Hutchinson, J. *The Genera of Flowering Plants*; Clarendon Press: Oxford, UK, 1964; pp. 44–57. [CrossRef]
16. Baranova, M.A.; Jeffrey, C.; Wu, Q.G. *Leaf Epidermis Features of Magnoliaceae and Their Systematic Significance, with Special Reference to the Genera Parakmeria, Manglietiastrum and Woonyoungia*; Science Press: Beijing, China, 2000; pp. 143–152.
17. Liu, Y.H.; Fan, H.M.; Chen, Z.Y. *Proceeding of the International Symposium on the Family Magnoliaceae, Guangzhou, China, 18–22 May 1998*; Science Press: Beijing, China, 2000; pp. 3–13.
18. Chen, D.Q.; Xu, F.X. Observation of pollen morphology of two species from *Talauma*. *Guihaia* **2008**, *3*, 317–319. [CrossRef]
19. Wang, Y.C.; Zhou, H.Y.; Liu, X.Q. Chloroplast genomic variation in Euonymus *maackii* Rupr. and its differentiation time in euonymus. *Forests* **2022**, *13*, 265. [CrossRef]
20. Kumar, S.; Stecher, G.; Tamura, K. MEGA7: Molecular evolutionary genetics analysis version 7.0 for bigger datasets. *Mol. Biol. Evol.* **2016**, *33*, 1870–1874. [CrossRef]
21. Zheng, W.; Zhang, H.; Wang, Q.M.; Gao, Y.; Zhang, Z.H.; Sun, Y.X. Complete chloroplast genome sequence of *Clivia miniata* and its characteristics. *Acta Hortic. Sin.* **2020**, *47*, 2439–2450. [CrossRef]
22. Tillich, M.; Lehwark, P.; Pellizzer, T.; Ulbricht-Jones, E.S.; Fischer, A.; Bock, R.; Greiner, S. GeSeq-versatile and accurate annotation of organelle genomes. *Nucleic Acids Res.* **2017**, *45*, 6–11. [CrossRef]
23. Mu, Y.N. Phylogenic and Taxonomic Research of *Hypolepisin* China Based on the Complete Chloroplast Genome. Ph.D. Thesis, Harbin Normal University, Harbin, China, 2020. [CrossRef]
24. Librado, P.; Rozas, J. DnaSPv5: A software for comprehensive analysis of DNA polymorphism data. *Bioinformatics* **2009**, *25*, 1451–1452. [CrossRef]
25. Huang, X.; Tan, W.; Li, F.; Liao, R.; Guo, Z.; Shi, T.; Gao, Z. The chloroplast genome of *Prunus zhengheensis*: Genome comparative and phylogenetic relationships analysis. *Gene* **2021**, *793*, 145751. [CrossRef]
26. Thiel, T.; Michalek, W.; Varshney, R.K.; Graner, A. Exploiting EST databases for the development and characterization of gene-derived SSR-Markers in barley (*Hordeum vulgare* L.). *Theor. Appl. Genet.* **2003**, *106*, 411–422. [CrossRef] [PubMed]
27. Wu, L.W.; Cui, Y.X.; Wang, Q.; Xu, Z.C.; Wang, Y.; Lin, Y.L.; Song, J.Y.; Yao, H. Identifcation and phylogenetic analysis of fve *Crataegus* species (Rosaceae) based on complete chloroplast genomes. *Planta* **2021**, *254*, 14. [CrossRef] [PubMed]
28. Han, C.Y.; Ding, R.; Zong, X.Y.; Zhang, L.J.; Chen, X.H.; Qu, B. Structural characterization of *Platanthera ussuriensis* chloroplast genome and comparative analyses with other species of Orchidaceae. *BMC Genom.* **2022**, *23*, 84. [CrossRef]
29. Li, L.X.; Peng, J.Y.; Wang, D.W.; Duan, A.A. Analysis of chloroplast genome characteristics phylogeny and codon preference in *Docynia longiunguis*. *Chin. J. Biotechnol.* **2022**, *38*, 328–342. [CrossRef]
30. Guzmán-Díaz, S.; Núñez, F.A.A.; Veltjen, E.; Asselman, P.; Larridon, I.; Samain, M.-S. Comparison of magnoliaceae plastomes: Adding neotropical *Magnolia* to the discussion. *Plants* **2022**, *11*, 448. [CrossRef] [PubMed]
31. Gao, B.M.; Yuan, L.; Tang, T.L.; Hou, T.; Pan, K.; Wie, N. The complete chloroplast genome sequence of *Alpinia oxyphylla* Miq. and comparison analysis within the Zingiberaceae family. *PLoS ONE* **2019**, *14*, e0218817. [CrossRef] [PubMed]
32. Santana, A.; Gomes, T.; Nimz, T.; Vieira, L.N.; Guerra, M.P.; Nodari, R.O.; Souza, E.M.; Pedrosa, F.O.; Rogalski, M. The complete plastome of macaw palm [*Acrocomia aculeata* (Jacq.) lodd. ex mart.] and extensive molecular analyses of the evolution of plastid genes in arecaceae. *Planta* **2018**, *247*, 1011–1030. [CrossRef]
33. Dong, W.; Liu, J.; Yu, J.; Zhou, S. Highly variable chloroplast markers for evaluating plant phylogeny at low taxonomic levels and for DNA barcoding. *PLoS ONE* **2012**, *7*, e35071. [CrossRef]
34. Alzahrani, A.; Albokhari, J.; Yaradua, S.; Abba, A. Comparative analysis of chloroplast genomes of four medicinal capparaceae species: Genome structures, phylogenetic relationships and adaptive evolution. *Plants* **2021**, *10*, 1229. [CrossRef]
35. Saina, J.K.; Li, Z.Z.; Mekbib, Y.; Gichira, A.W.; Liao, Y.Y. Transcriptome sequencing and microsatellite marker discovery in *Ailanthus altissima* (Mill.) Swingle (Simaroubaceae). *Mol. Biol. Rep.* **2021**, *48*, 2007–2023. [CrossRef]
36. Menezes, A.; Resende-Moreira, L.C.; Buzatti, O.; Nazareno, A.G.; Carlsen, M.; Lobo, F.P.; Kalapothakis, E.; Lovato, M.B. Chloroplast genomes of *Byrsonima* species (malpighiaceae): Comparative analysis and screening of high divergence sequences. *Sci. Rep.* **2018**, *8*, 2210. [CrossRef] [PubMed]
37. Angiosperm Phylogeny Group; Chase, M.W.; Christenhusz, M.J.; Fay, M.F.; Byng, J.W.; Judd, W.S.; Soltis, D.E.; Mabberley, D.J.; Sennikov, A.N.; Soltis, P.S.; et al. An update of the angiosperm phylogeny group classification for the orders and families of flowering plants: APG IV. *Bot. J. Linn. Soc.* **2016**, *181*, 1–20.
38. Wang, Y.-B.; Liu, B.-B.; Nie, Z.-L.; Chen, H.-F.; Chen, F.-J.; Figlar, R.B.; Wen, J. Major clades and a revised classification of *Magnolia* and magnoliaceae based on whole plastid genome sequences via genome skimming. *J. Syst. Evol.* **2020**, *58*, 673–695. [CrossRef]
39. Figlar, R.B. Proleptic branch initiation in michelia and magnolia subgenus Yulania provides basis for combinations in subfamily magnolioideae. In Proceedings of the International Symposium on the Family Magnoliaceae, Guangzhou, China, 18–22 May 1998; Science Press: Beijing, China, 2000; pp. 14–25.
40. Fu, D.L.; Zhao, D.X.; Sun, J.H.; Xu, J.W.; Wang, J. Component of volatile oils in 10 chinese species of *Yulania* genus and its phylogenetic implicatior. *Sci. Silvae Sin.* **2005**, *41*, 68–74.
41. Gong, X.; Pang, Y.Z.; Yang, Z.Y. The cross-compatibility of magnoliaceae. *Plant Divers.* **2001**, *23*, 339–344.

42. Dong, S.-S.; Wang, Y.-L.; Xia, N.-H.; Liu, Y.; Liu, M.; Lian, L.; Li, N.; Li, L.-F.; Lang, X.-A.; Gong, Y.-Q.; et al. Plastid and nuclear phylogenomic incongruences and biogeographic implications of *Magnolia* s.l. (Magnoliaceae). *J. Syst. Evol.* **2022**, *60*, 1–15. [CrossRef]
43. Kim, S. Phylogenetic relationships in family magnoliaceae inferred from *ndhF* sequences. *Am. J. Bot.* **2001**, *88*, 717–728. [CrossRef]
44. Xie, X.N.; Huang, R.; Li, F.; Tian, E.; Li, C.; Chao, Z. Phylogenetic position of *Bupleurum sikangense* inferred from the complete chloroplast genome sequence. *Gene* **2021**, *798*, 145801. [CrossRef]
45. Wang, M.; Peng, J.Q.; Cao, J.W.; Cao, F.X.; Li, J.J.; Xue, C.; Wu, Y. Studies on genetic diversity of 48 wild species of magnoliaceae in Yunnan. *J. Trop. Subtrop. Bot.* **2020**, *28*, 277–284. [CrossRef]
46. Xu, F.X.; Wu, Q.G. Pollen morphology of *Paramichelia* and *Tsoongiodedron* from China. *Acta Bot. Boreali-Occident. Sin.* **1995**, *6*, 3.
47. Jin, H.; Shi, S.H.; Pan, H.C.; Huang, Y.L.; Zhang, H.D. Phylogenetic relationships between *Michelia* (Magnoliaceae) and its related genera based on the *matk* gene sequence. *Acta Sci. Nat. Univ. Sunyatseni* **1999**, *38*, 91–97. [CrossRef]
48. Wu, Z.Y. Sinicae agendae academiae sinicae edita. In *Flora Reipublicae Popularis Sinicae*; Science Press: Beijing, China, 2022.
49. Hu, X.X.; Zheng, W.J. *Parakmeria*, a new genus of Magnoliaceae of southwestern China. *J. Syst. Evol.* **1951**, *1*, 1. Available online: http://journal.ucas.ac.cn/CN/Y1951/V1/I1/1 (accessed on 18 March 2022).
50. Wu, Z.Y. The areal-types of Chinese genera of seed plants. *Plant Divers.* **1991**, *13*, 1. Available online: https://journal.kib.ac.cn/EN/Y1991/V13/IS4/1 (accessed on 27 March 2022).
51. Xiong, Q.; Zhang, M.; Yu, D.P. Cloning and phylogenetic analysis of b-class mads-box gene in *Parakmeria omeiensis*. *Mol. Plant Breed.* **2018**, *16*, 5540–5548. [CrossRef]
52. Sima, Y.K.; Wang, J.; Cao, L.M.; Wang, B.Y.; Wang, Y.H. Prefoliation features of the magnoliaceae and their systematic significance. *J. Yunnan Univ.* **2001**, *23*, 71–78.

biology

MDPI

Article

Geometric Morphometric Versus Genomic Patterns in a Large Polyploid Plant Species Complex

Ladislav Hodač [1,2,†], Kevin Karbstein [1,2,†], Salvatore Tomasello [1], Jana Wäldchen [2], John Paul Bradican [1] and Elvira Hörandl [1,*]

1 Department of Systematics, Biodiversity and Evolution of Plants (with Herbarium), Albrecht-von-Haller Institute for Plant Sciences, University of Göttingen, 37073 Göttingen, Germany; lhodac@uni-goettingen.de (L.H.); kevin.karbstein@uni-goettingen.de (K.K.); salvatore.tomasello@uni-goettingen.de (S.T.); johnpaul.bradican@uni-goettingen.de (J.P.B.)

2 Department of Biogeochemical Integration, Max Planck Institute for Biogeochemistry, 07745 Jena, Germany; jwald@bgc-jena.mpg.de

* Correspondence: elvira.hoerandl@biologie.uni-goettingen.de

† These authors contributed equally to this work.

Simple Summary: Plant species complexes with hybridization and asexual reproduction often exhibit complex morphological patterns, which is problematic for classifications. Here, we analyze geometric morphometric, genomic, and ecological data with comprehensive statistics to evaluate phenotypic variation in the Eurasian *Ranunculus auricomus* complex. Genomic clusters correspond largely to morphological groupings, but most described asexual hybrid taxa cannot be discriminated from each other. Phenotypic variation is more influenced by genomic composition than by climatic conditions, and the phenotypic variation of asexual hybrids resembles a mosaic of intermediate and transgressive phenotypes. Our results support a taxonomic revision of the complex.

Abstract: Plant species complexes represent a particularly interesting example of taxonomically complex groups (TCGs), linking hybridization, apomixis, and polyploidy with complex morphological patterns. In such TCGs, mosaic-like character combinations and conflicts of morphological data with molecular phylogenies present a major problem for species classification. Here, we used the large polyploid apomictic European *Ranunculus auricomus* complex to study relationships among five diploid sexual progenitor species and 75 polyploid apomictic derivate taxa, based on geometric morphometrics using 11,690 landmarked objects (basal and stem leaves, receptacles), genomic data (97,312 RAD-Seq loci, 48 phased target enrichment genes, 71 plastid regions) from 220 populations. We showed that (1) observed genomic clusters correspond to morphological groupings based on basal leaves and concatenated traits, and morphological groups were best resolved with RAD-Seq data; (2) described apomictic taxa usually overlap within trait morphospace except for those taxa at the space edges; (3) apomictic phenotypes are highly influenced by parental subgenome composition and to a lesser extent by climatic factors; and (4) allopolyploid apomictic taxa, compared to their sexual progenitor, resemble a mosaic of ecological and morphological intermediate to transgressive biotypes. The joint evaluation of phylogenomic, phenotypic, reproductive, and ecological data supports a revision of purely descriptive, subjective traditional morphological classifications.

Keywords: apomixis; genomics; geometric morphometrics; polyploidy; *Ranunculus auricomus*; taxonomically complex groups (TCGs)

Citation: Hodač, L.; Karbstein, K.; Tomasello, S.; Wäldchen, J.; Bradican, J.P.; Hörandl, E. Geometric Morphometric Versus Genomic Patterns in a Large Polyploid Plant Species Complex. *Biology* **2023**, *12*, 418. https://doi.org/10.3390/biology12030418

Academic Editor: Lorenzo Peruzzi

Received: 1 February 2023
Revised: 6 March 2023
Accepted: 7 March 2023
Published: 9 March 2023

1. Introduction

Polyploidy and hybridization are regarded as key factors for plant evolution [1–5]. Polyploidy, the presence of more than two chromosome sets within a cell, has several positive evolutionary consequences. Multiple gene copies allow for higher gene expression along with higher physiological (and thus phenotypic) flexibility in relation to abiotic and biotic environmental conditions [6,7]. Polyploids thus often perform better in past glaciated areas, under climatic change, or in the colonization of new ecosystems [8–10]. In addition, hybridization, the fusion of previously diverged subgenomes, leads to new genetic combinations, increased heterozygosity, hybrid vigor, buffering of deleterious mutations, and changes in secondary metabolites [11,12]. Nevertheless, newly formed polyploids may have reduced fertility due to meiotic errors [13–15], but they can escape hybrid sterility via asexual reproduction and/or selfing [5,9,13,16]. Apomixis, the asexual reproduction via seeds, occurs in c. 19% of families and c. 2% of genera in flowering plants [17,18]. Hybridization is probably the main trigger of apomixis [16,19,20]. Apomixis is heritable and genetically controlled but usually facultative because it represents a modification of the sexual pathway [20–22]. The extant positive side-effects of polyploidy and hybridization are 'fixed' over generations and can foster the establishment of apomictic lineages in new or stressful environments (e.g., in previously glaciated areas; [9,23,24]).

Taxonomy, i.e., documenting, classifying, naming, and understanding the diversity of life, represents a cornerstone of biological research [25–27]. More than two million eukaryotic species have been described thus far, but many species remain undiscovered or unnamed [28–30]. Species are the fundamental units of evolutionary and biodiversity research (e.g., ecology or nature conservation). Traditional plant taxonomy has a long historical background and was based until the 1970s almost exclusively on morphological distinctness (reviewed by [31]). The subjectivity of defining "distinctness" by descriptive methods, and the recognition of different evolutionary processes leading to distinct entities have led to many different species concepts and pluralistic views [31]. Phylogenetic lineage concepts can be further problematic in cases of reticulated evolution [1]. For hybridizing complexes with few intermediates, cluster species concepts based on phenetic or genetic similarity have been proposed [32]. To better recognize evolutionary processes, modern authors consider species as separate genetic ancestor-descendent lineages, a concept that applies to diploids, polyploids, sexuals, and asexuals [33–36]. Criteria from previous concepts should now be applied to analyze and describe the evolutionary role and circumscription of lineages, e.g., their persistence in time and space or phenotypic differentiation, which is still an obstacle [1,33,37–40]. The current era of genomics has enabled astonishing breakthroughs in high-throughput sequencing (HTS) of DNA, computation capabilities, and bioinformatics, resulting in a plethora of new evolutionary insights and subsequent taxonomic revisions and species descriptions in the plant kingdom [4,36,39,41]. Despite all this progress, awareness is increasing that not all lineages necessarily represent species. The currently most accurate model for species delimitation ("Multispecies Coalescent", or MSC) tends to oversplit groups into many species [35,39,42]. For example, information on geographical isolation can provide insight into whether observed lineages represent populations or species [39]. Additional criteria are therefore needed for the formal classification of lineages. Recognition of genetic and/or morphological clusters is another timely approach for species delimitation and can be applied to phenotypic and genetic data regardless of the mode of reproduction and the presence/absence of crossing barriers [32,43]. Consequently, an integrative taxon-omics approach that combines taxonomy with 21st-century '-omics' (HTS) and other data sources (e.g., morphology, reproduction, or ecology) excludes discipline-dependent failure rates and is thus considered to be the gold standard in species delimitation [44–47].

Taxonomically complex groups (TCGs) [48] offer a unique opportunity to study flowering plant evolution. TCGs are groups of related individuals that are characterized by various biological factors that complicate the delimitation of species [48,49]. Apomictic polyploid complexes fit the definition of TCGs; they link intricate microevolutionary processes such as polyploidization, hybridization, and asexuality with macroevolutionary patterns [3,45,50]. Sexually diploid parents usually generate hundreds of hybrid, polyploid hybrid, and/or apomictic derivatives multiple times throughout time and space [38,51–54]. Particularly, the combination of polyploidy and hybridization (allopolyploidy) frequently shows higher degrees of (epi)genomic and transcriptomic changes than polyploidy alone (autopolyploidy) [3,7,55–57] and is thus more likely to create biotypes with novel phenotypic features [58–60]. In nature, many distinct autopolyploid cytotypes remain unnamed and hence unrecognized due to only minor morphological differences compared to diploid progenitors [57,61,62]. Concerning apomictic polyploid complexes, [38] reviewed four alternative approaches for a case-by-case classification: (i) classify the obligate sexual progenitors as species; (ii) merge them and highly facultative apomictic lineages into a single species; (iii) treat the main hybrid clusters of the facultative apomicts as species. If the parentage of allopolyploid apomicts can be reconstructed, then designating apomicts as nothotaxa [63] can be a useful approach to formally separate from sexual species [51,64–66]; and (iv), classify obligate apomictic lineages as agamospecies. While options (i) and (ii) have been applied in several genera (reviewed by [38]), the challenge remains for (iii) polyploid complexes comprising hundreds and thousands of described taxa with uncertain taxonomic circumscriptions. Only a few case studies using integrative taxon-omics and a combination of ancestor-descendant lineage and cluster criteria have been published thus far (e.g., [39,54,66]). Classification is highly dependent on the degree of apomixis and the stability of lineages in polyploid complexes. For instance, [51] made substantial progress in untying highly reticulate relationships and genome evolution in facultative to obligate apomictic polyploid complexes, but recognizing distinct lineages and their morphotype was nearly impossible due to innumerous reticulations producing large network-like clusters. Another issue for phylogenomic, as well as phenotypic, reconstructions arises when a sexual progenitor is not sampled or presumed to be extinct, and consequently, its morphotype remains unknown [51,52,67]. For instance, sexual progenitors for some agamospecies are completely unknown (e.g., *Alchemilla*, [68]).

In the last years, many researchers working in the field of integrative taxon-omics focused on bringing their plant model systems into the era of genomics, utilizing either (sub)genomic datasets (e.g., restriction-site associated DNA sequencing (RAD-Seq) or target enrichment of nuclear genes (TEG)) and/or a combination of the different genomic, nuclear gene, and plastid regions (e.g., *Cardamine, Leucanthemum, Ranunculus auricomus, Rubus*, or *Salix*) [36,51,69–74]. Gathering information to infer lineage characteristics and subsequent species delimitation, e.g., shared or distinct morphotypes, is still regarded as an important criterion for species delimitation [31,36,37,51]. However, this is often done using traditional morphological descriptions, morphometrics, or character evolution approaches to modeling character state changes within a phylogeny [31,75]. In the past, purely descriptive traditional morphological classification led to subjective descriptions of hundreds to thousands of morphotypes as species due to minor morphological differences in TCGs (e.g., [76,77]), a practice that was particularly prevalent in apomictic polyploid complexes (e.g., [50,78,79]). Delimitation that only relies on single, partly author-dependent 'diagnostic' characters bears the danger of subjective, irreproducible taxonomic classifications. Therefore, analysis using multiple characters is preferred to more objectively characterize different phenotypes [31,39,80]. Another challenge for species delimitation is the exclusion of non-relevant variation of characters (e.g., allometry or asymmetrical development of organs), which is a relevant factor in plants due to large phenotypic plasticity in response to environmental factors [31,81–84].

Geometric morphometrics (GM, or GMM by recent publications; e.g., [85,86]) tackles the aforementioned issues through the exact, objective, and fine-scale evaluation of shapes and shape changes via landmarks (i.e., anatomical loci) [87–90]. This approach has been applied across many disciplines (e.g., botany, paleontology, medicine, or engineering), and uses a collection of multivariate statistical analysis to visualize Cartesian coordinate data [83,91]. In plant research, leaf shapes were frequently analyzed for species characterization and delimitation [88,92]. However, GM approaches can also be easily extended to other structures possessing shared biologically homologous regions in a specific study group, e.g., receptacle shape in [39], or 3D flower shape in [93,94]. In general, morphological shape changes are associated with (epi)genetic variation and environmentally related responses [88,95–98]. Nevertheless, to the best of our knowledge, no study so far has inferred interspecific, landmark-based, multi-trait shape changes across large geographic scales for species delimitation. Moreover, GM is also able to support final taxonomic decisions based on HTS data, particularly in TCGs where morphological differences are hard to assess with traditional morphological approaches [39,99,100]. Combinations of phylogenetic and -omic data with landmark-based GM approaches are effective in disentangling intricate plant species relationships [39,101]. However, in plant species complexes, the most challenging aspect is the aforementioned delimitation of young allopolyploid derivatives. In TCGs with porous genomes (i.e., some genomic regions are protected from interspecific gene flow, whereas others are not) and hybridization, mosaic-like character combinations and conflicts of morphological data with molecular phylogenies are a major problem for classification (reviewed, e.g., by [38,101,102]). Consequently, it remains to be tested whether these landmark-based GM approaches using multi-trait data are suitable for resolving highly reticulate TCGs.

Concerning species delimitation in TCGs, different phylogenomic approaches are available to efficiently resolve intricate relationships. RAD-Seq collects non-coding and coding regions across the entire genome and delivers thousands to hundreds of thousands of loci and SNPs [103,104]. This provides a particularly powerful method for tackling TCGs characterized by low genetic divergence, reticulations, and ILS [51,69,71]. However, RAD-Seq loci are usually short and insufficiently informative to allow for reliable allele phasing. Retrieving allelic information and discriminating homoeologous loci is particularly crucial for accurate inferences of reticulate polyploid relationships [67,105–107]. Single-copy nuclear genes assembled from TEG are often longer than RAD-Seq loci, enabling the segregation of alleles at a single locus (i.e., phasing), and thus MSC approaches [104,108,109]. Therefore, phylogenomic analyses conducted with TEG datasets can more clearly delimit the genetic structure of polyploid complexes, differentiate between allo- and autopolyploid evolution, and determine the parentage of a single polyploid [51,73,107]. In addition, data from plastid regions or entire plastomes (CP) can be easily gained from TEG off-target reads [110,111]. Together with RAD-Seq and TEG, these help to identify reticulations (nuclear-plastid discordances), homoploid speciation, extinction of sexual progenitors, allopolyploidization events, and/or maternal progenitors of polyploids [51,73,112,113]. Despite all this progress, detailed morphological characterizations of all those lineages/clusters found by modern phylogenomic approaches are often missing in these studies. Consequently, there is a need to inform these lineages/clusters and their evolutionary reconstructions by using detailed morphological characteristics obtained from comprehensive landmark-based, multi-trait GM datasets informed by subgenomic data. Additionally, knowledge is missing on which genomic dataset (RAD-Seq, TEG, or CP) best fits observed morphological differentiation.

The *Ranunculus auricomus* plant complex is a model system for apomixis research but also for studying the evolution of phylogenetically young TCGs [13,21,51,64,87,88,114,115]. Within the genus *Ranunculus*, the group falls into a large clade, with its closest relatives occurring in North America and Central Asia [116,117]. The distribution of taxa ranges from Greenland to Europe, Northern Asia, and Alaska; it spans arctic, boreal, temperate, and Mediterranean climate zones [118–120]. Taxa occupy various habitats—from stream- and riverside habitats, alluvial to humid deciduous forests, extensively used swampy to semi-dry meadows, and waysides [50,121–123]. The complex comprises more than 800 taxa [124,125] that were predominantly described by applying descriptive morphological species concepts (e.g., [122,126–129]). The existence of two remarkably different morphotypes already led Linnaeus in 1753 [130] to classify the complex into two different species: *R. auricomus* L. from Western Europe, characterized by dissected basal leaves, and *R. cassubicus* L. from North Poland or further east (Siberia), with large non-dissected basal leaves [130,131] (Figure 1A,B). In the 19th century, intermediate morphotypes between these two taxa occurring in Central Europe, Sweden, and Finland were described as *R. fallax* (Wimm. & Grabowski) Sloboda [132], and in 1922 some dwarf arctic-alpine morphotypes from Siberia were discriminated as *R. monophyllus* Ovcz. [133]. These four morphotypes established a widely used classification of four main species with several subspecies [122,134], which was used in many European floras. Subsequently, hundreds of different, partly only locally occurring, morphospecies have been described, connecting these four core groups by endless intermediates (e.g., [122,126,127,135–138]). However, these morphology-based species concepts failed either due to an inability to split the morphotype continuum or the presence of intricate evolutionary processes [122,134,139,140]. Consequently, the complex is often treated as an agglomerate in regional floras (e.g., [141]), neglecting its biodiversity.

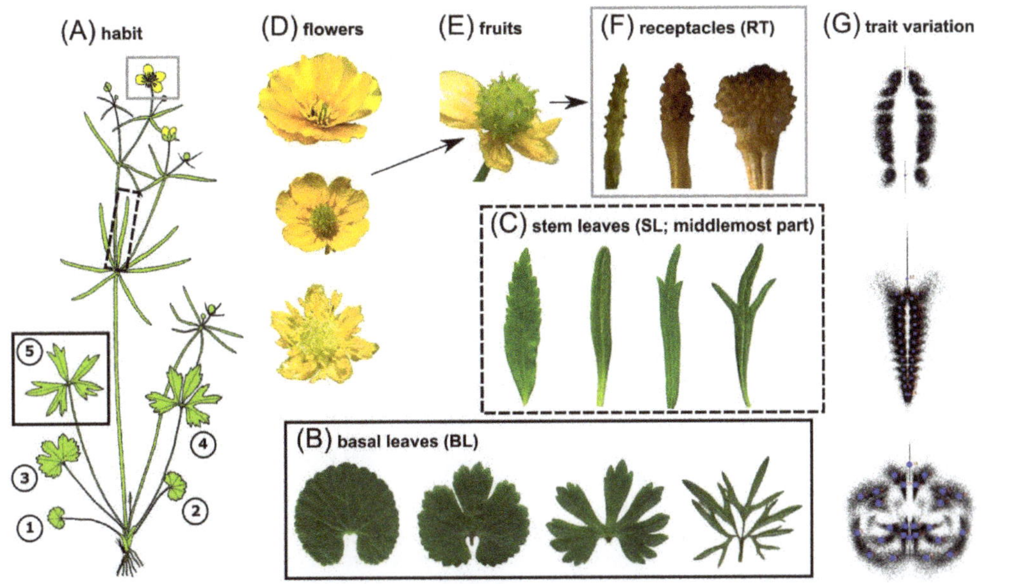

Figure 1. Morphological variation of taxonomically informative traits within the *R. auricomus* species complex (see also Figure S1A–C for leaf venation and landmark configurations per trait). (**A**) Illustration of a typical *R. auricomus* individual with morphological traits highlighted in boxes: basal leaf cycle (black box; 1–5, 1–4 = early spring leaves, 5 = most dissected leaf at anthesis) and stem leaves with the middlemost segment (black-dotted box) and reproductive structures (grey box; flower, fruit, and receptacle at the fruiting stage). Figure source: The figure was taken from [50], which is published under the Creative Commons Attribution License (**B**) Variability of basal leaves among taxa ranging from undivided (left) to broad three-lobed or dissected (center) to strongly dissected forms (right). (**C**) Variability of the middlemost segment of a stem leaf among taxa, ranging from broadly lanceolate (with teeth; left) to narrowly linear forms (with dissection; right). (**D**) Variability of petal variation, ranging from five to 15 (mostly sexuals; the uppermost flower) or reduced or absent forms (mostly polyploid apomicts; the lowermost flower). (**E**) Collective fruit in the ripening process: Concerning polyploid apomicts, a single collective fruit can contain achenes with either sexually or apomictically produced seeds. (**F**) Variability of receptacles among taxa, ranging from narrowly large (left) to smaller roundish receptacles (right). (**G**) The variation of the evaluated traits captured by GM is shown as clouds of 2D landmarks. Photographs of Figure 1B–F: © Kevin Karbstein.

The *R. auricomus* complex is composed of a few, mainly diploid sexual progenitors and hundreds (>800) of polyploid apomictic derivatives. Sexual species are characterized by complete flowers, whereas obligate facultative apomicts exhibit rather reduced flowers with fewer or no petals (the petaloid nectary scales of *Ranunculus* are here conveniently called 'petals') [24] (Figure 1D). Taxa have a heterophyllous basal leaf cycle, i.e., usually starting with a non-dissected to three-lobed spring leaf or a basal sheath. The subsequent leaves are more and more dissected and appear during anthesis, but such dissected leaves can also be missing under unfavorable environmental conditions; non-dissected to three-lobed leaves appear during the fruiting stage and persist over summer and autumn (Figure 1A,B; development and homology of three different types of leaf cycles are explained by [120,121,142]). Phylogenomic analyses based on subgenomic data (RAD-Seq, TEG) and GM revealed five geographically isolated, genetically distinct sexual progenitors [39] (Figure 2A,B). Speciation took place ca. 830,000–580,000 years ago and was triggered by vicariance processes during a time frame of severe climatic fluctuations [143]. Based on previous studies [88,121,144], [39] developed a landmarking scheme for the taxonomically most informative traits: (i) the most-dissected basal leaves in the leaf cycle during anthesis; (ii) the central part of the lowermost stem leaf; and (iii) receptacle at fruit stage (Figure 1A–F, Figure S1). The diploid and partly autotetraploid species *R. cassubicifolius* and the diploid, probably homoploid hybrid species *R. flabellifolius* are distributed in Central and Eastern Europe and are characterized by a leaf cycle without dissected basal leaves, but during anthesis, the non-dissected to three-lobed 'summer' leaves are already present; *R. cassubicifolius* has broad lanceolate stem leaf segments, whereas *R. flabellifolius* forms a fan-shaped stem leaf with connate segments [51,115,127,128,145,146] (Figure 1B,C; Figures 6 and S8 in [39]). The other sexual species are characterized by a heterophyllous leaf cycle with dissected basal leaves at anthesis. The diploid *R. envalirensis* and the only exclusively tetraploid (probably allotetraploid) sexual *R. marsicus* inhabit restricted ranges in the Southern European mountain systems. These dwarf species show basal leaves with three- to five-lobed or dissected segments and linear stem leaf segments (with sinuses in the case of *R. marsicus*; [39,51,128,147,148]). In contrast, *R. notabilis* is widely distributed in the Illyrian lowlands, is taller, and has rather narrowly lobed or dissected basal leaves and mostly linear stem leaf segments [39,128,137].

Further comprehensive phylogenetic and phylogenomic studies demonstrated that the evolutionary history of the *R. auricomus* complex is substantially shaped by hybridization among sexual progenitors combined with polyploidization [51,64,87,88,115]. Recently, an integrative approach based on subgenomic data (RAD-Seq, TEG, CP), ploidy, and reproductive data with appropriate polyploid bioinformatic tools revealed (i) that only five diploid sexual progenitor species (including an unknown progenitor) probably generated a large number of diverse polyploid apomicts; three to five allopolyploid genetic clusters including progenitor species were characterized by substantial post-origin genome evolution and subgenome dominance [51]. However, it is unclear whether these clusters can also be morphologically recognized. The study revealed further that almost all previously described morphospecies were polyphyletic and did not represent stable ancestor-descendant lineages. The question remains whether these hybrid biotypes (provisorily treated as nothotaxa) would exhibit specific phenotypic variation that would be more extreme or new, or the ability to settle new abiotic and biotic environments compared to their progenitor species. Such morphotypes could eventually result from transgressive segregation and hybrid speciation [149,150]. Transgressive segregation might have occurred in the initial, mostly sexually formed *R. auricomus* hybrid generations [16].

Consequently, we aim at addressing the following questions in this study: (1) Do the genetic clusters found by [51] correspond to morphological clusters? Which genomic dataset (genomic, nuclear, or plastid) is most congruent with the morphological clustering? (2) Is the GM approach of [39] able to delineate the polyploid apomicts from each other and the sexual species? Which are the most informative traits? Do any of the described nothotaxa form well-differentiated morphological clusters? (3) Are morphological shape changes associated with environmental factors or, rather, with genetic factors? (4) Are polyploid apomicts inside or outside the morphospace or ecological niche of the diploid sexual progenitors? We will focus here on the huge diversity of temperate to submeridional taxa that were genetically analyzed by [51], whereas arctic-alpine dwarf forms ('*R. monophyllus*') but also Mediterranean taxa of the complex will be the subject of upcoming studies.

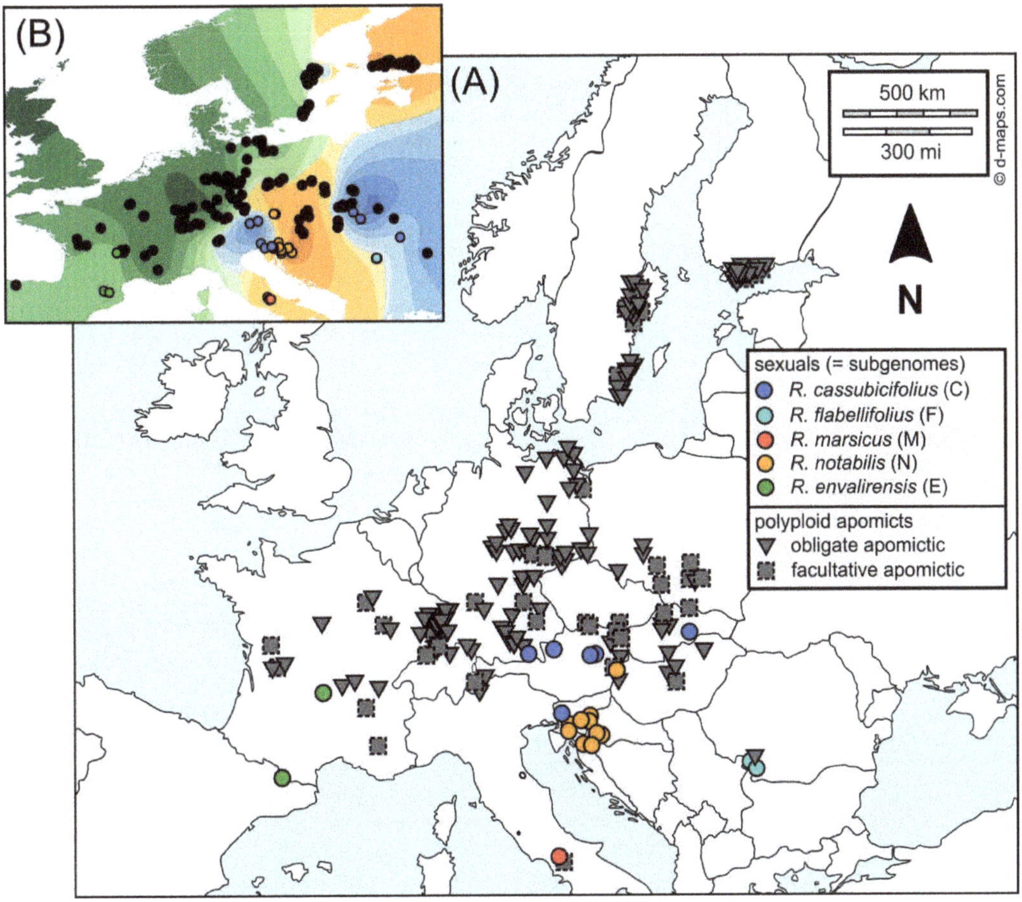

Figure 2. Sampling localities of studied *Ranunculus auricomus* populations across Europe (details in Table S1). (**A**) Symbols represent the reproduction modes of populations (colored circles = diploid to tetraploid sexual; dark gray solid triangles = polyploid obligate apomicts; dark gray dashed squares = polyploid facultative apomicts; [24]). The color scheme was also applied to Figures 3–7, and 9. The original map was downloaded from https://d-maps.com/ (accessed on 8 October 2020), created by [24,51] which are published under the Creative Commons Attribution License, and modified herein. (**B**) Geographic map illustrating the ancestry coefficients of the likeliest genomic resolution (K = 3 RAD-Seq clusters) across Europe according to sNMF analyses published in [51]. These genetic clusters are comparable to RADpainter clusters, which are used here in statistical analyses (Figures 3–7 and 9). Black circles indicate apomictic polyploids, whereas colored circles represent sexual species (similar population sampling as in [51]), and geographic regions (with polyploids) are colored according to the dominant genomic contribution from the respective sexual progenitor species. The figure was created by [51] (published under the Creative Commons Attribution License) and modified herein.

2. Materials and Methods

2.1. Study Locations and Material Sampling

In the present study, we included 28 populations of all four diploids and one tetraploid sexual species (see taxonomic treatment in [39]) and 192 populations of the ca. 75 most widespread tetra-, penta-, and hexaploid apomictic *R. auricomus* taxa (flow cytometric ploidy and reproduction mode measurements published in [24] and deposited in FigShare https://doi.org/10.6084/m9.figshare.13352429 (accessed on 7 March 2023). A sampling of garden plants took place from 2013 to 2018, totaling 220 populations across temperate and submeridional Europe (Figure 1, Table S1). Per population, we recorded altitude, GPS coordinates, and habitat, and collected herbarium specimens. Details about locations, ploidy, reproduction modes, samples per population, and further genomic and environmental characteristics are given in Table S1. Subpopulations from the same locality were treated as separate populations in subsequent statistical analyses because they are characterized by different taxa (morphotypes). Sampled living plants were kept in the Old Botanical Garden at the University of Göttingen under controlled environmental conditions (garden beds with similar solar radiation and water supply) for GM analyses. Individuals were cultivated in 1.5 l pots with Fruhstorfer Topferde LD 80. Voucher specimens were deposited in the herbarium of the University of Göttingen (GOET).

2.2. Genomic and Environmental Data Analysis

Wet lab work, data filtering, assembly, parameter optimization, and bioinformatic data evaluation concerning RAD-Seq, TEG, and CP data are described in detail in [24,39,51]. Demultiplexed RAD-Seq and TEG raw reads are deposited in the Sequence Read Archive (SRA) of NCBI (https://www.ncbi.nlm.nih.gov/bioproject/627796 (accessed on 7 March 2023); https://www.ncbi.nlm.nih.gov/bioproject/628081 (accessed on 7 March 2023)). To clarify which genomic dataset (genomic, nuclear, or plastid) best explains the morphological clustering, we used the (phylo)genomic results of [51] that comprise the same sexual and apomictic populations investigated herein. Consequently, we grouped the GM dataset (see the section below) according to the found clades/clusters in [51]. The following naming of clades/clusters corresponds to the respective sexual progenitor found in each clade/cluster.

The RAD-Seq datasets were applied to a genetic structure (sNMF, [151,152]; 1 SNP/locus (unlinked SNPs), 33,165 loci, 33,165 SNPs, and 55% missing data) and genetic similarity (RADpainter+fineRADstructure, [153]; 97,312 loci, 438,775 SNPs, and 74% missing data) analysis. The sNMF analysis is based on an unlinked single nucleotide polymorphism (SNP) alignment (33,165 loci, 194,083 SNPs, 55% missing data). Ancestry coefficients were calculated with method 'max', i.e., at each point, the cluster for which the ancestry coefficient was maximal. The sNMF analysis showed three clusters, i.e., a Western European cluster containing the sexual diploid progenitor *R. envalirensis* (E) and related polyploid apomicts, a Central-Eastern European cluster containing the sexual diploid progenitors *R. notabilis* and *R. flabellifolius* (and tetraploid *R. marsicus*, N + F+M), and related polyploid apomicts, and an Eastern European cluster containing the sexual diploid progenitor *R. cassubicifolius* (C) and related polyploid apomicts, as the likeliest genetic resolution. RADpainter also inferred the same three genetic clusters, although a few incongruences were observed (e.g., between clusters E and N + F+M). The TEG dataset was utilized in a STACEY species delimitation analysis [154], using the most informative, nonhomoplasious, and free-from-paralog-sequences 48 nuclear genes, including allele phasing across all ploidy levels as described in [51]. The STACEY analyses inferred five genetic clusters, i.e., clusters each containing *R. cassubicifolius* (C), *R. flabellifolius* (F), *R. marsicus* (M), *R. notabilis* (N), and *R. envalirensis* (E) with related polyploid apomicts. These results, in contrast to RAD-Seq, better delimit progenitor species and their related polyploid apomicts. The CP dataset was used for a maximum likelihood (ML) tree analysis (RAxML_NG, [155]) based on 71 plastid regions (representing ca. 50% of the expected plastome length), containing at least 50% of samples per region, as described in [51]. The ML tree of plastid data analysis exhibited four genetic clades/haplotype groups, i.e., a clade containing *R. cassubicifolius* and *R. flabellifolius* (C + F) and related polyploid apomicts; a clade only with *R. envalirensis*-related polyploid apomicts (including an unknown and probably extinct *R. envalirensis*-related Central European progenitor U); a clade containing *R. envalirensis* (E), and a clade containing *R. notabilis* and *R. marsicus* (N + M). These results thus substantially differ from RAD-Seq and TEG results and suggest a reticulate evolution of the diploid progenitor *R. flabellifolius* (F) from

R. cassubicifolius (C) as one putative parent and of the tetraploid *R. marsicus* from at least *R. notabilis* (N), respectively. The ML tree analysis exhibited four genetic clades/haplotype groups, i.e., a clade containing *R. cassubicifolius* and *R. flabellifolius* (C + F) and related polyploid apomicts; a clade only with *R. envalirensis*-related polyploid apomicts (including an unknown and probably extinct *R. envalirensis*-related Central European progenitor U); a clade containing *R. envalirensis* (E); and a clade containing *R. notabilis* and *R. marsicus* (N + M). These results thus substantially differ from RAD-Seq and TEG results and suggest a reticulate evolution of the diploid progenitor *R. flabellifolius* (F) from *R. cassubicifolius* (C) as one putative parent, and of tetraploid *R. marsicus* from at least *R. notabilis* (N), respectively.

The gathering of environmental data from both in situ records and WorldClim databases version 2 [156] including data standardization, is described in detail in [24]. All populations of the concatenated GM dataset are characterized by the following abiotic environmental factors: GPS coordinates (longitude and latitude), altitude (meters above sea level, m. a.s.l.), bioclimatic variables 1–19 in 2.5 min resolution (temperature, precipitation, and their respective seasonality variables), and solar radiation in 2.5 min resolution (kJ m^{-2} day^{-1}). We removed autocorrelated variables ($r > 0.8$, Figures S2–S6) from the modeling procedure [157], using the R-package 'corrplot' version 0.92 [158] and R version 4.2.0 [159].

2.3. Geometric Morphometric (GM) Data Analysis

2.3.1. Data Collection and Preparation

The GM dataset is composed of fresh material sampled from living garden plants and material from herbarium specimens of the same populations (Table S1). We added specimens from different herbaria to supplement the datasets with type material (Table S1, see also [39]). Following the approach of [39] (but see also [88,121,144]), we collected the taxonomically most informative traits of *R. auricomus* individuals, i.e., basal and stem leaves during anthesis and receptacles during the fruiting stage (Figure 1B,C,F). Collections were regularly checked against type specimens to ensure accurate selection. We only used individuals that are characterized by basal leaf, stem leaf, and receptacle traits. As a rule, and as far as possible, several basal leaves, stem leaves, and receptacles were recorded per plant individual, and eight plant individuals were recorded for each population on average. From April to May 2018 and 2019, we harvested, on average, three fresh basal and stem leaves per flowering plant. Leaves were scanned immediately after sampling in 400 dpi resolution using CanoScan LiDE 220 (Canon, Ota, Japan) and Epson Perfection V500 Photo (Seiko Epson, Suwa, Japan) scanners. To increase the statistical robustness but also the accuracy of GM analysis, we additionally digitized the taxonomically most informative leaf traits of selected herbarium specimens with the Herbscan Light Box (including a digital camera with 50.6 megapixels) of the GOET herbarium. Herbarized plant material might exhibit allometric shape changes during the drying process [83,160]. However, the number of analyzed herbarium scans was relatively small compared to the garden material in our dataset. Because of the careful selection of non-type and type herbarium material, which also captures in situ specific phenotypic plasticity, the inclusion of these data makes the statistical analyses more robust and accurate. Moreover, we collected three receptacles per individual on average at the fruiting stage from June to August 2018 and 2019. Receptacles were digitized with 10–15-fold magnification under a Leica M125 microscope (Leica Microsystems, Wetzlar, Germany).

In total, the concatenated dataset comprised 4070 basal leaves, 4148 stem leaves, and 3472 receptacles based on 1858, 1880, and 1587 individuals, respectively (images and landmark files are stored in Figshare https://doi.org/10.6084/m9.figshare.21393375) (accessed on 7 March 2023). Information for five sexual taxa, 64 apomictic polyploid taxa with taxonomic assignment, and another 17 apomictic polyploid taxa without taxonomic assignment ('cf', or 'indet') were recorded. The total dataset comprises 2048 individuals from 220 populations. The majority (73%) of digitized plant material was derived from garden cultures (University of Göttingen) and was supplemented by herbarium specimens (27%). In total, all five sexual taxa and 37 apomictic taxa were represented by at least two populations (Table S1). Concerning the 64 morphologically assignable apomictic taxa, 30 taxa were represented by three or more populations. In general, three different leaf cycles are recognized within the *R. auricomus* complex [120,121,142,161] that roughly fit the three observed genomic RAD-Seq clusters (see [51] and Figure 4 in [120]). Taxa of cluster 1 usually have no dissected leaves at anthesis, which appear in clusters two and three on separate shoots (such additional shoots with dissected leaves can also be missing in stressed, small individuals of clusters 2 and 3, see [120,142,161,162], but such individuals were not included here). To avoid missing data for basal leaves (BL) of cluster 1, we took the functionally equivalent most-dissected summer leaves, which appear during early anthesis (of the next shoot; see [161]) for joint analyses with clusters 2 and 3, as in [39] (see also trait selection for GM analyses below). We additionally evaluated populations of genetic cluster 1 and genetic clusters 2 and 3 in some analyses separately (e.g., Figures 5 and 6).

2.3.2. Digitalization of Traits and Extraction of Shape Variables

Image processing and the creation of TPS files followed the strategies described in [39,88]. The concatenated GM dataset of the sexual, di- to tetraploid populations was already published by [39] and added to the polyploid apomicts evaluated for the first time in this study. Herein, 2D landmark data of basal leaves (twenty-six landmarks), stem leaves (eight landmarks, twenty semilandmarks), and receptacles (nine landmarks, ten semilandmarks) were recorded using TpsDig version 1.4.0 [163]. The TPS-formatted raw datasets consisted of 4070 basal leaf configurations (BL), 4148 stem leaf configurations (SL), and 3472 receptacle configurations (RT). The three morphometric datasets were subjected to Procrustes superimpositions in TpsRelw version 1.70 [163] and MorphoJ version 1.07d [164] as described in [39] and only the symmetric component was further used to extract shape variables. Because most of the subsequent data analyses were based on population-level comparisons, the GM datasets were first averaged accordingly. Before the extraction of shape variables, landmark configurations were averaged across the same traits within each plant and across multiple plants within each population. Thus, for each population, we obtained symmetrized and averaged basal leaf, stem leaf, and receptacle configurations, each containing information from several plant individuals. Shape variables were calculated as scores of the symmetrized averaged landmark configurations (population means) on the shape principal components, also known as relative warps (RWs). In some analyses (e.g., in trait covariation analysis, PLS), the traits were analyzed separately, and in others, they were concatenated into a single morphometric dataset (e.g., in multi-group discriminant analyses).

2.3.3. Genomic Clusters and Morphological Groups

We performed a multi-group discriminant analysis (Canonical Variates Analysis, CVA) of single-trait and concatenated GM datasets to investigate which genomic dataset best reflected the morphological differentiation. In the CVAs, we compared the morphometric distances between population clusters whose composition was inferred from analyses of genome-wide RAD-Seq data (three-cluster scenario), nuclear TEG (five-cluster scenario), and plastomes (four-cluster scenario). For these comparisons of morphological groupings, morphometric data for 66 populations were used, for which all three NGS datasets were available. Wherever the three traits were analyzed separately in the software MorphoJ, the Procrustes and Mahalanobis distances of the group centroids were calculated, including permutation tests of significance. In concatenated trait analyses using the software PAST version 4.11 [165], differences between group centroids were captured by Euclidean metrics and approved by permutation tests (NP-MANOVA, two-group permutation test). Results showed (see below) that RAD-Seq (RADpainter) clusters best explained the observed morphological differentiation in single-trait and concatenated GM analyses (Table S2). Consequently, we used the three genetic clusters inferred from RADpainter analysis as the grouping for subsequent GM analyses.

2.3.4. Covariation of Traits, Taxonomic Resolution, and Shape Changes along Genomic Gradients within Clusters

The three traits (BL, SL, and RT) were examined for the independence of their shape variation. The basic question of whether, for example, the basal leaves vary completely independently of the stem leaves was analyzed. We tested whether there is a significant covariance structure between any two traits, employing a partial least squares (PLS) analysis in the software MorphoJ. Trait covariance analyses were performed separately for each of the three observed RADpainter clusters. The significance of the covariance was determined by permutation tests, and the corresponding morphological trends of the traits were visualized as wireframe graphs. To study the resolution of described *R. auricomus* taxa within observed morphometric clusters, we conducted a CVA of 24 agamospecies with three or more sampled populations. Within each RAD-Seq cluster, the three morphological traits were analyzed separately to investigate their ability to distinguish between agamospecies.

To study morphological shape changes along genomic gradients, we calculated regression models between observed morphotypes and genomic background based on the RAD-Seq similarity matrix concerning all 220 populations. For each polyploid population, the RADpainter method was first used to determine how the four diploid/sexual subgenomes (C, E, F, and N) were represented in its polyploid apomictic genome. The percentages of the four subgenomes were used as predictor variables in a regression analysis to model the associated shape change of basal leaves, stem leaves, and receptacles. In other words, the regression model predicted the appearance of the traits depending on their genetic background. The regression models in the software TpsRegr64 version 1.50 [163] can visualize changes in the traits along gradients of given variables. Goodall's F-test statistic was applied to the regression model, and its significance was determined by permutation tests with 10,000 rounds.

2.3.5. Shape-Environment and Shape-Genomics Association Models

To infer the sources of morphological shape variation (e.g., environment or genetics; [81,82,95,166]), we calculated distance matrix-based multiple regression models (MRM) using the R package 'ecodist' version 2.0.9 [167]. First, we ensured that environmental factors and shape principal component axes (relative warps) were non-autocorrelated among all traits and for single traits ($r > 0.8$, Figures S2–S6) using the R-package 'corrplot' version 0.92. We transformed shape principal ordination components among all traits and per single trait into distance matrices based on Euclidean distances. Second, we transformed non-autocorrelated environmental characteristics of populations among all factors and per single factor into distance matrices based on Euclidean distances. Third, we imported the raw RADpainter similarity matrix into R and transformed it into a distance matrix using Euclidean distances. The normal distribution of distance matrices was checked by applying the basic R functions 'qqnorm' and 'qqplot'. In all cases, we inferred non-normally distributed data. Finally, we used 211 populations with exact overlapping shapes, environments, and genomic RAD-Seq (RADpainter) information ($3 \times 22{,}155$ data entries). We calculated four linear MRMs based on scaled (unit variance) variables, 1000 permutations, and Spearman rank correlations due to non-normally distributed data. A general MRM using shape distances as response and environmental and genomic distances (and their interaction) as explanatory variables, and three more detailed MRMs using shape distances of BL, SL, and RT as response variables and all single environmental factors and genomic distances as explanatory variables.

Subsequently, the inferred significant environmental variables were used to model their effect on a basal leaf, stem leaf, and receptacle phenotypes. The regression models of the association between shapes and variable gradients were computed in the software TpsRegr64 version 1.50, the model fit was tested by permutation tests with 10,000 rounds.

2.3.6. Ancestral Shape Reconstruction

The approach of Section 2.3.4 reconstructed a three-lobed to -dissected ancestral BL type for each genomic cluster (see Results). To verify this finding and to model in detail the ancestral basal leaf shape at the root of the European *R. auricomus* complex, we performed the squared-change parsimony analysis [168,169] for reconstructing the ancestral BL shape based on a phylogenomic tree (inferred from RAD-Seq data; for the phylogenomic methodology see [51]), using MorphoJ. The ancestral shape reconstruction utilized only individuals with exactly overlapping GM and RAD-Seq data, i.e., six samples of *R. cassubicifolius* (non-dissected leaf morphotype), two samples of *R. flabellifolius* (non- and slightly dissected leaf morphotypes), two samples of *R. marsicus* (dissected leaf morphotype), two samples of *R. envalirensis* (dissected leaf morphotype), and twelve samples of *R. notabilis* (dissected leaf morphotype). After computing the shape changes across all nodes in the phylogenomic tree, we summarized them using the evolutionary principal components analysis (EPCA) in MorphoJ, to extract the most important shape-shifts in the BL morphological evolution.

2.3.7. Inferring Morphological and Genomic Differentiation in an Ecological Context, and Intermediary Versus Transgressive Hybrid Patterns

Due to their taxonomic importance and discriminative power, we explored the BL variation at three different levels (among clusters 1–3, among apomicts and sexuals within clusters, and hybrids and their genomic progenitors according to results in [51]) to infer intermediacy versus transgressive hybrid patterns. We employed a set of different analyses in MorphoJ: (1) principal components analysis (PCA) to explore the main shape trends in a common morphospace of different apomictic clusters or among apomicts and sexuals, (2) canonical variates analysis (CVA) and two-group discriminant analysis (DA) to test predefined groups for their morphological differentiation (and mean classification accuracy), and (3) partial least squares (PLS) analysis to put the phenotypic variation of the apomicts and/or their progenitors into the context of associated environmental factors. We selected four environmental covariates that exhibited the strongest association with BL variation, namely altitude, BIO3 (isothermality), BIO8 (mean temperature of the wettest quarter), and BIO18 (precipitation of the warmest quarter). The resulting PLS scatter plots showed the BL shape variation (PLS 1 ordination axis from shape data) against an ordination axis extracted from the environmental variables. The PLS analysis calculates the size of the covariation between the two linked datasets and provides a permutation p-value (10,000 rounds) for the significance of the covariance model. The PLS analysis identifies which of the original environmental variables shows the highest correlation with shape variation in a given PLS covariance model. With the methodology described above, we compared four apomictic nothotaxa recently approved as allopolyploids by [51] and their progenitors, but also apomicts of clusters 1–3, sexuals and apomicts within the clusters 1–3, and eight apomictic taxa within cluster 2 (taxon-rich and with known progenitors).

To corroborate ecological (dis)similarity among and within clusters of sexual and apomictic populations, we performed a new non-linear, machine learning-based ordination technique, known as UMAP (uniform manifold approximation and projection for dimension reduction; [170]), using PAST. Ecological similarities were computed for 212 populations based on eight non-correlated variables (altitude, solar radiation, BIO1, BIO3, BIO4, BIO8, BIO9, and BIO18) and the Manhattan similarity index.

3. Results

3.1. Morphological Clustering with Genomic Background (RAD-Seq)

Comparing the clustering of the 220 populations (Figure 3A–F; Figure S7) according to the three morphological traits, BL (Figure 3A) and SL (Figure 3B) exhibited similar patterns with a well-separated cluster 1 and partly separated clusters 2 and 3. The cluster separations are significant for each trait and the concatenated dataset, respectively (Table S2). The RTs (Figure 3C) exhibited the lowest discriminant power to distinguish genomic clusters 1 and 2, and 1 and 3 (Table S2). The concatenated dataset (BL + SL + RT) consisting of sixty-six shape variables separated genomic clusters best (Figure 3D). Almost all sexual progenitor species (Figure 3E) clustered consistently throughout the analyses, except for *R. flabellifolius* which clustered differently in the BL and SL analyses. Considering the concatenated analysis, out of the sixty-six input (shape) variables, we observed the following most important vectors (Figure 3D,F): The first (BL_PC1) and fourth (BL_PC4) basal leaf principal components, describing shape variation between non-dissected BL and narrowly, three- to five-lobed BL (cluster 1—cluster 2 and 3), and between narrowly, three-lobed BL with roundish segments and broadly, three-lobed BL with acuminate segments (cluster 2—cluster 3), respectively; the first two SL principal components (SL_PC1, SL_PC2), describing shape variation between broadly lanceolate SL with teeth and linear segments SL (cluster 1 and 3—cluster 2), and between linear SL with sinuses and oval segments SL (cluster 2 and 3—cluster 1), respectively; and the first RT principal component (RT_PC1), describing shape variation between broad and long androclinium and oval and short gynoclinium RT on the one side and short and narrow androclinium and high gynoclinium RT on the other side (cluster 2—cluster 1 and 3).

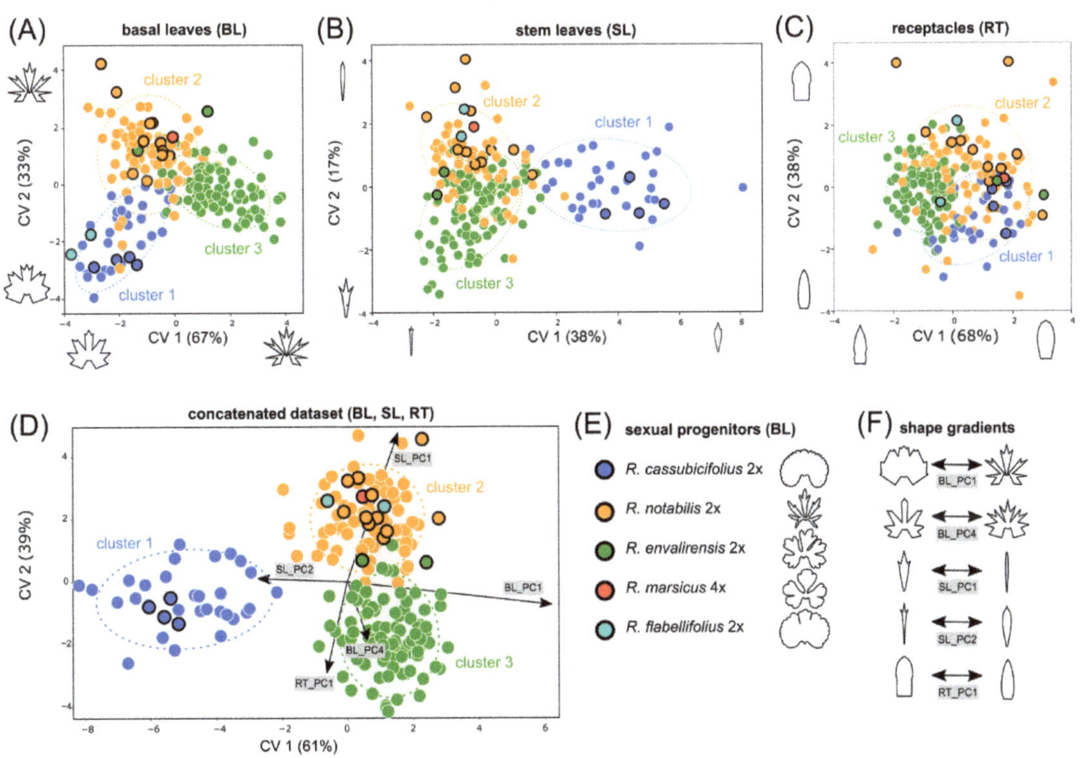

Figure 3. Morphological variation among sexual progenitors and polyploid apomictic derivative taxa with respect to genomic RAD-Seq (RADpainter) background. Canonical variate analyses (CVA) were applied for the clustering of 220 populations based on basal leaves (**A**), stem leaves (**B**), receptacles (**C**), and the concatenation of all traits (**D**). The concatenated analysis of all three traits (D) shows the five best separating morphometric trends illustrated in (**F**). Each dot in the CVA scatter plots (**A–D**) represents a single population, and the colors reflect assignments into the three RADpainter clusters

(i.e., genomic clusters 1–3). An overview of the five sexual species in (**E**) shows their characteristic basal leaf morphotype, the most important taxonomic trait. The five best-separating shape trends (shape changes along the relative warps) are visualized in (**F**). The coloring of sexual progenitors and clusters follows Figure 2 and [51]. BL = basal leaves, Cluster 1 = RAD-Seq (RADpainter) cluster 1 containing *R. cassubicifolius* and polyploid apomictic relatives, Cluster 2 = RAD-Seq (RADpainter) cluster 2 containing *R. flabellifolius*, *R. marsicus*, and *R. notabilis* and polyploid apomictic relatives, Cluster 3 = RAD-Seq (RADpainter) cluster 3 containing *R. envalirensis* and polyploid apomictic relatives, CV = canonical variate (explained percentages of shape variation), SL = stem leaves, RT = receptacles.

A detailed look at the morphospace trait occupation of the sexual progenitors and polyploid apomictic derivatives indicates the presence of transgressive apomictic phenotypes (compare Figure 3 and Figure S7). We detected a set of apomictic populations that grouped outside the range of sexual progenitors for all three traits and in the concatenated data analysis. Transgressive phenotypes were established along all three morphological traits and all three clusters but were most abundant in cluster 3.

3.2. Comparison of Morphological Clustering Concerning Different Genomic Backgrounds

We inferred significant morphological clustering according to genomic RAD-Seq (RADpainter, sNMF, NeighborNet clusters), nuclear gene TEG (Stacey clusters), and CP backgrounds (plastid clades) in CVA analyses (Table S2). Nevertheless, morphological clustering is best resolved by genomic RAD-Seq data showing the highest average differentiation value among clusters (F = 15.18 for RAD-Seq > F = 10.87 for CP > F = 8.29 for TEG; p values in Table S2; Figure 4A–E) inferred from the concatenated datasets. Though only representing a subset of sixty-six populations, the morphological groups correspond to those inferred from the analysis of 220 populations with RAD-Seq backgrounds (Figure 3D). The morphological clustering according to CP data was not able to distinguish *R. notabilis* (cluster 2) and *R. envalirensis* (cluster 4) and their respective polyploid apomicts from each other (Figure 4C). Concerning the TEG-guided clustering, morphological clusters 2–5 were highly overlapping, which was also indicated by the lowest average distance among the clusters (Table S2; Figure 4C). In general, *R. notabilis* and *R. cassubicifolius* clustered close to their polyploid apomictic relatives in RAD-Seq, TEG (Figure 4B), and CP (Figure 4C) analyses. In contrast, the morphological position of *R. envalirensis* and *R. flabellifolius* and their relationships to closely related polyploids were ambiguous throughout the analyses. The general morphological trends (Figure 4D), which best separated among the clusters, were identical to those inferred from the larger dataset of 220 populations (Figure 3, Table S3).

3.3. Covariation of Traits

The covariation analyses revealed different trait behaviors among the three genomic RAD-Seq (RADpainter) clusters. The strongest significant association between basal leaves and stem leaves was found in cluster 1 (Figure 5A), showing a covariance between plants characterized by non-dissected BL with narrow blade base and broad lanceolate, teethed SLs and plants characterized by nearly five-dissected BL and narrow SL with sinuses. Within cluster 1, the relationships between the BL and RT (Figure 5B), and between the SL and RT (Figure 5C) shapes were much weaker, though significant. Within cluster 2, again BL and SL exhibited the strongest covariation structure (Figure 5D) compared to BL and RT (Figure 5E) and SL and RT (Figure 5F). Shape changes from broadly three-lobed BL with a narrow blade base and broad lanceolate teethed SL to narrowly, up to five-dissected BL with a broad blade base and sinuses and lineal SL segments. The covariation of BL and SL in cluster 3 (Figure 5G) was statistically similar to that described for cluster 2 but showed shape changes from broadly three- to five-dissected BL with a narrow blade base and deep sinuses and SL with deep sinuses to narrowly three-dissected BL with broad blade base. However, other covariation structures (BL and RT, SL and RT) within this cluster were non-significant (Figure 5H,I).

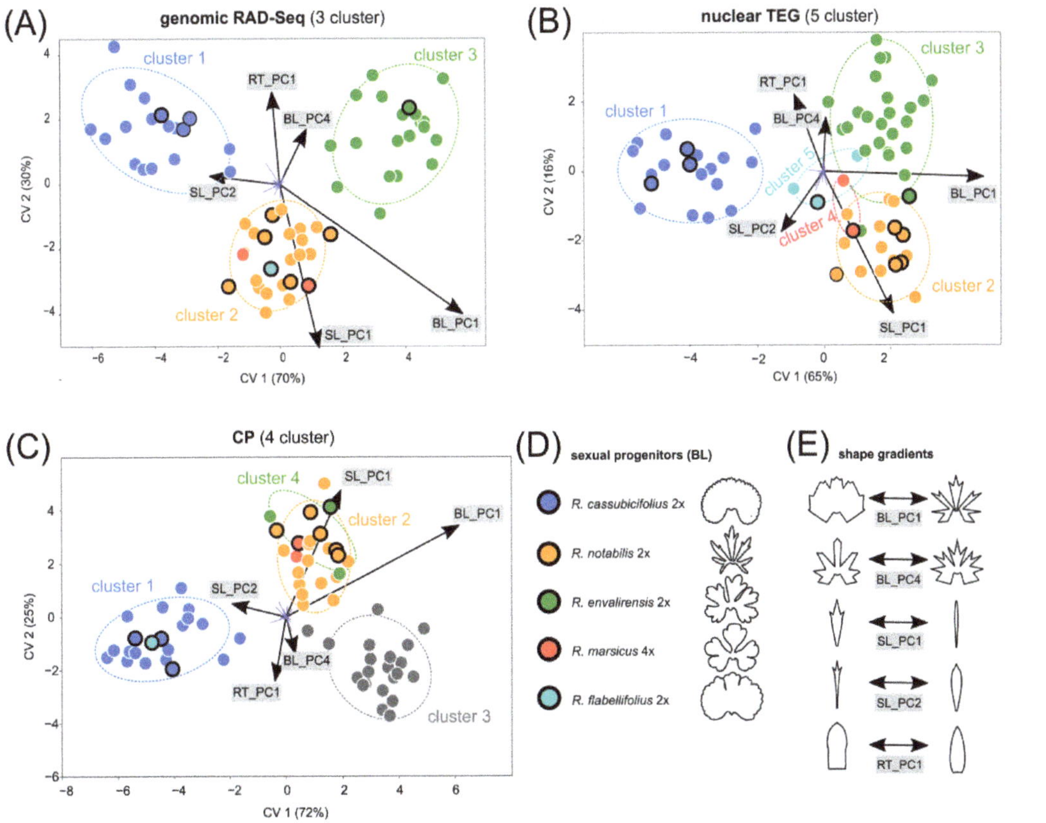

Figure 4. Morphological variation among sexual progenitors and polyploid apomictic derivative taxa with respect to different NGS backgrounds. Canonical variate analysis (CVA) was applied for the clustering of sixty-six populations based on the concatenated trait datasets (BL + SL + RT) and according to their assignment into clusters as inferred from genomic RAD-Seq (**A**), nuclear TEG (**B**), and CP (**C**) NGS-data (see also Materials and Methods for genomic cluster details). Each dot in the CVA scatter plots (**A**–**C**) represents a single population, and the colors reflect assignments into the RAD-Seq, TEG, and CP clusters (Table S1), respectively. Typical basal leaf morphotypes of the sexual species are shown in (**D**), and the best separating morphological trends are shown in (**E**). The coloring of sexual progenitors and clusters follows Figure 2 and [51]. BL = basal leaves; CP = plastid data; CV = canonical variate (explained percentages of shape variation); RAD-Seq = restriction-site associated DNA sequencing; RT = receptacles; SL = stem leaves; TEG = target enrichment nuclear genes.

3.4. Morphological Clustering of Polyploid Apomictic Nothotaxa

A detailed analysis of twenty-one polyploid apomictic nothotaxa and their respective sexual progenitor species pointed out similar patterns in all three genomic RAD-Seq (RADpainter) clusters. The sexual progenitors are clearly separated from the polyploid apomicts (nothotaxa) regarding all three morphological traits (BL, SL, RT; p values in Table S4). Within each RADpainter cluster, we observed that some polyploid nothotaxa were separated from each other across at least two different traits, but particularly in clusters 2 and 3, some polyploid nothotaxa strongly overlap in trait morphospace. We found a few examples of well-separated nothotaxa. For example, *R.* ×*platycolpoides* and *R.* ×*elatior* in cluster 1 (different in BL and SL, and partly in RT; Table S4), *R.* ×*fissifolius* and *R.* ×*obscurans* in cluster 2 (different in BL and SL, not in RT; Table S4), and *R.* ×*lucorum* and *R.* ×*reniger* in cluster 3 (different in BL and RT, not in SL; Table S4; Figure 6A–I). Nevertheless, the majority of polyploid apomictic nothotaxa overlap with another taxon across single or all traits. These are also often nothotaxa, which were hard to identify in the field, for example, *R.* ×*variabilis* and *R.* ×*phragmiteti* in cluster 2 (Figure 6D–F) or *R.* ×*alsaticus* and *R.* ×*vertumnalis* (Figure 6G–I) in Central Europe. In general, the weakest separation of nothotaxa was observed within cluster 3 (Figure 6G,H), showing highly overlapping trait variation.

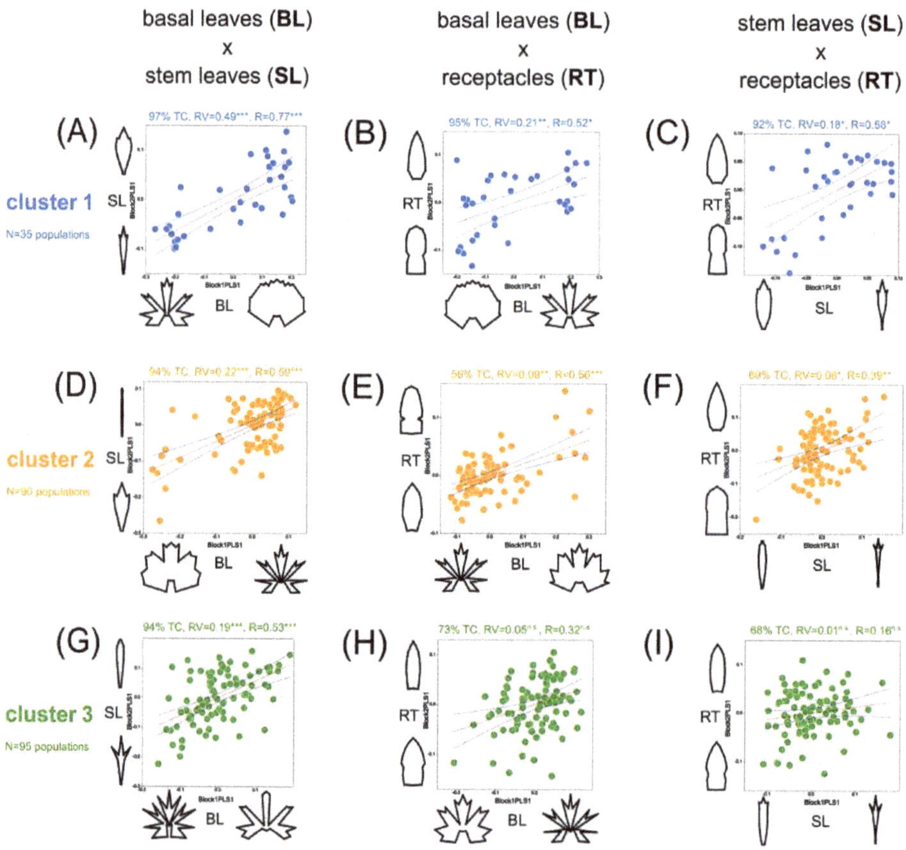

Figure 5. Covariation of the three taxonomically most informative traits is inferred for each RAD-Seq (RADpainter) cluster. The three morphological traits are plotted against each other in partial least-squares regression analyses of trait covariation. Within cluster 1 (inferred by RADpainter), the basal leaves are plotted against the stem leaves (**A**), against the receptacles (**B**), and the stem leaves against the receptacles (**C**). The same pairs of morphological traits were compared within clusters 2 (**D–F**) and 3 (**G–I**). Numbers above each plot give the amount of morphological covariation described by the first PLS axis (Block1PLS1 and Block2PLS1) as percentages of the total covariation (TC), a model fit statistic (RV) with its significance, and the correlation (R) of both PLS1 axes (each one representing one morphological trend). The coloring of sexual progenitors and clusters follows Figure 2 and [51], abbreviations as in Figures 3 and 4. R = correlation coefficient of PLS axes; RV = global correlation coefficient (multivariate analog of the squared correlation); * = $p < 0.05$; ** = $p < 0.01$; *** = $p < 0.001$; TC = total covariance.

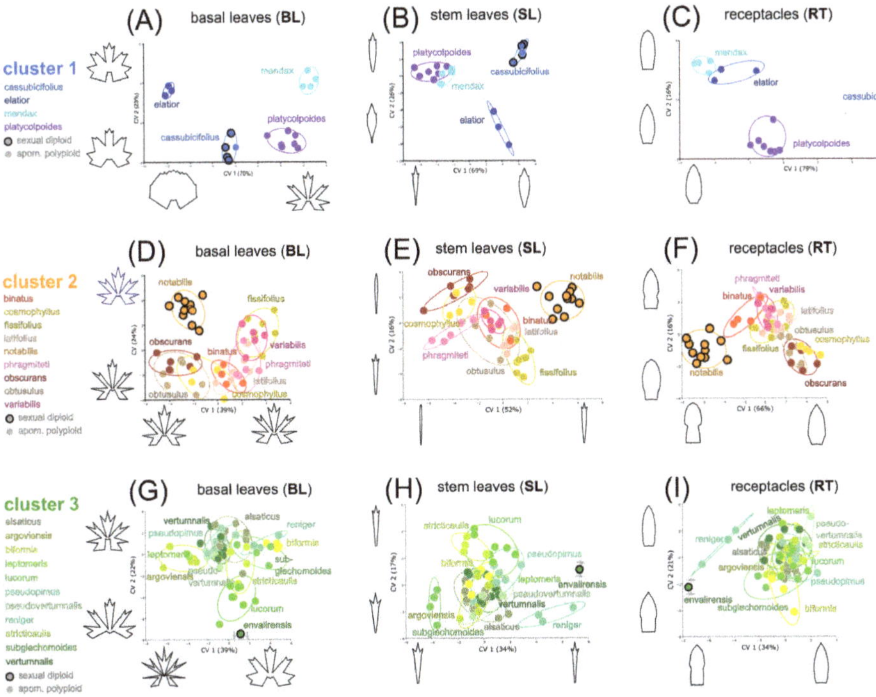

Figure 6. Morphological variation among sexual progenitors and polyploid apomictic derivative nothotaxa with respect to traits and genomic RAD-Seq (RADpainter) background. Canonical variate analyses (CVA) were applied for the clustering of twenty-four polyploid apomictic taxa based on the three morphological traits. Within cluster 1 (**A–C**), one sexual species and three apomictic polyploids were compared according to basal leaves (**A**), stem leaves (**B**), and receptacles (**C**). Within cluster 2 (**D–F**), one sexual species and eight apomictic polyploids were compared. In the case of cluster 3 (**G–I**), both sexual populations (*R. envalirensis*) were morphologically distant from all the polyploids, and their position within the plots was only graphically indicated by grey arrows. The coloring of sexual progenitors and clusters follows Figure 2 and [51]. BL = basal leaves, Cluster 1 = RAD-Seq (RADpainter) cluster 1 containing *R. cassubicifolius* and polyploid apomictic relatives, Cluster two = RAD-Seq (RADpainter) cluster 2 containing *R. flabellifolius*, *R. marsicus*, and *R. notabilis* and polyploid apomictic relatives, Cluster 3 = RAD-Seq (RADpainter) cluster 3 containing *R. envalirensis* and polyploid apomictic relatives; CV = canonical variate (explained percentages of shape variation); RT = receptacles; SL = stem leaves.

3.5. Subgenome Contributions from Sexual Progenitors with Associated Morphotypes of Polyploid Apomicts, and Ancestral Morphotype Reconstruction

The genomic contributions of the three analyzed sexual progenitor subgenomes (*R. cassubicifolius* 'C', *R. notabilis* 'N', *R. envalirensis* 'E') were significantly associated with BL, SL, and RT shape variation across 190 populations of polyploid apomictic taxa. The increasing contributions of subgenome 'C' were associated most strongly with BL and less strongly with SL shape changes ($p < 0.001$; Table S5; Figure 7A). With increasing subgenome C contribution, the associated morphological shape converts towards less dissected BL phenotypes and broad lanceolate SL phenotypes (Figure 7A). An association between subgenome C contribution and RT shape variation could not be determined (Table S5). Variable contributions of subgenome 'N' were associated with BL (strongest association), SL, and RT shape variation (Figure 7B). With increasing subgenome N contribution, the

BL shape of apomicts becomes more and more narrowly dissected, the SL shape becomes increasingly linear without sinuses, and the RT shape tends towards more roundish forms. Varying contribution of subgenome 'E' showed the strongest association with BL shape variation and a less pronounced association with SL and RT (Figure 7C; Table S5). With increasing subgenome E contribution, the BL shape of apomicts becomes more and more dissected with large sinuses, the SLs are narrower with sinuses, and the RT has a more roundish form with an elongated intervallum between the andro- and gynoclinium.

The reconstruction of the ancestral BL morphotype revealed an intermediary, three-lobed, and slightly dissected type at the root of the phylogenomic tree (Figure S8A), corroborated by the permutation test showing a significant phylogenetic signal in BL ($p < 0.001$). The evolutionary principal components analysis revealed that several shape changes (PC1-4) and not a single one (e.g., between non-dissected and dissected leaf types, as stressed out by several authors) played a role in the morphological evolution of the BL shape (Figure S8B). The two most important shape change components across the phylogenetic tree are concentrated at the incision and width of the middle segment and the blade base of a three-lobed to three-dissected BL (PC1/2; Figure S8B).

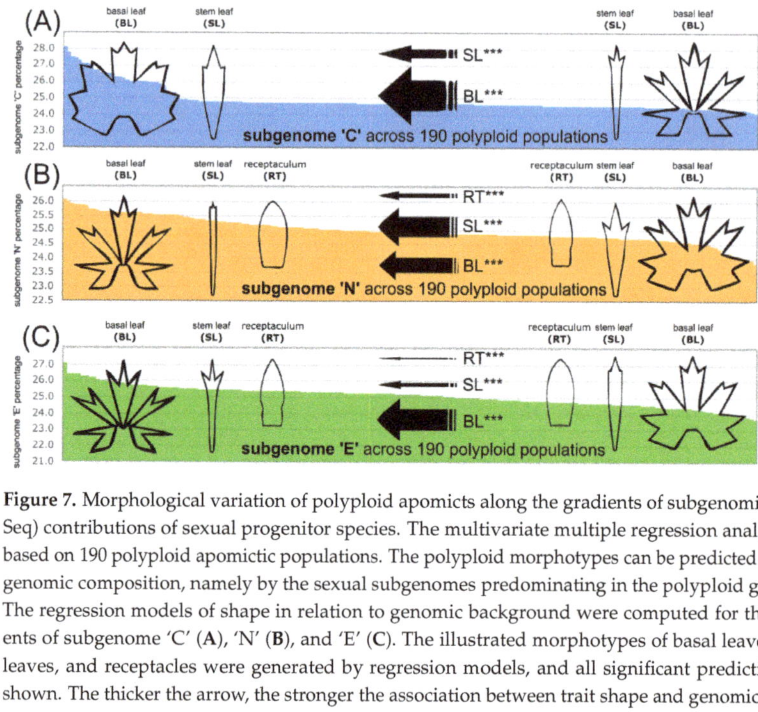

Figure 7. Morphological variation of polyploid apomicts along the gradients of subgenomic (RAD-Seq) contributions of sexual progenitor species. The multivariate multiple regression analyses are based on 190 polyploid apomictic populations. The polyploid morphotypes can be predicted by their genomic composition, namely by the sexual subgenomes predominating in the polyploid genomes. The regression models of shape in relation to genomic background were computed for the gradients of subgenome 'C' (**A**), 'N' (**B**), and 'E' (**C**). The illustrated morphotypes of basal leaves, stem leaves, and receptacles were generated by regression models, and all significant predictions are shown. The thicker the arrow, the stronger the association between trait shape and genomic composition. The coloring of sexual progenitors and clusters follows Figure 2 and [51]. BL = basal leaves; Cluster 1 = RAD-Seq (RADpainter) cluster 1 containing *R. cassubicifolius* and polyploid apomictic relatives; Cluster 2 = RAD-Seq (RADpainter) cluster 2 containing *R. flabellifolius*, *R. marsicus*, and *R. notabilis* and polyploid apomictic relatives; Cluster 3 = RAD-Seq (RADpainter) cluster 3 containing *R. envalirensis* and polyploid apomictic relatives; RT = receptacles; SL = stem leaves. *** = $p < 0.001$.

3.6. Environmental and Genomic Variables Associated with Phenotypic Variation

The calculated MRM revealed significant associations ($p < 0.001$, Table S6) between (i) overall shape distances (concatenated BL, SL, and RT datasets) among populations and their environmental distances (based on a set of eight environmental variables) and genomic distances (including their interaction), and (ii–iv) between morphological distances among populations (inferred from separate BL, SL, and RT datasets) and their environmental distances inferred from eight factors, and genomic distances. Overall, the shape distances among populations were strongly associated with their genomic distances (86%, $p = 0.001$), whereas environmental distance exhibited a relatively low explanatory power (14%, $p = 0.011$). When environmental and genomic distances were added to the same model, there was no change in the amount of explained variance. The BL shape distances among populations showed the strongest relationship with genomic distances (54%, $p = 0.001$) but also some associations with a couple of environmental factors (Figure 8A–C): precipitation of the warmest quarter (15%, $p = 0.001$; BIO18), isothermality (12%, $p = 0.001$; BIO3), mean temperature of the wettest quarter (11%, $p = 0.01$; BIO8), and altitude (7%, $p = 0.033$). In contrast to BL shape distances, which were associated mostly with genomic distances, SL and RT shape distances were more strongly associated with environmental factors than genomic distances (Figure 8A–C). SL shape distances exhibited associations with precipitation in the warmest quarter (32%, $p = 0.001$; BIO18), isothermality (31%, $p = 0.001$; BIO3), the mean temperature of the driest quarter (25%, $p = 0.001$; BIO9), and genomic distance (13%, $p = 0.001$). RT shape distances were associated with almost all environmental factors (precipitation in the warmest quarter > altitude > isothermality > mean temperature wettest quarter > mean temperature > temperature seasonality > solar radiation; in sum, 84%, $p < 0.05$) and genomic distances (16%, $p = 0.002$).

Regression models of shape variation (BL, SL, RT) in association with environmental variables (Table S7) showed a variety of morphological responses (Figure 8D–K). The precipitation of the warmest quarter (BIO18) was mostly associated with BL shape variation between finely dissected and more robustly dissected leaf phenotypes (Figure 8D). BIO18 has further associated with SL shape variation between finely dissected and more robustly dissected phenotypes (Figure 8D; Table S7) and RT shape variation between forms with a more pronounced intervallum and those without a distinct intervallum. The only other variable associated with the variation of all three traits (BL, SL, and RT) was isothermality (BIO3; BL > RT > SL; Figure 8E; Table S7). The BL shape variation correlated with BIO3 differs from that associated with BIO18, as the former varies not only with the fineness of the dissections but also with the blade base (Figure 8E). The SL shape variation correlated with BIO3 is opposite to that associated with BIO18, and RT varies between strongly pointed shapes and more rounded ones. The mean temperature of the wettest quarter (BIO8; Figure 8F) and altitude (Figure 8G) showed an association with shape variation in BL and RT, though the morphological trends differed between the two environmental predictors. The most pronounced morphological trend was the decreasing incision depth and blade base variation in BL associated with increasing altitude. The temperature seasonality (BIO4; Figure 8H), the mean annual air temperature (BIO1; Figure 8I), and solar radiation (Figure 8J) were correlated with the shape variation of RT (concerning mainly the appearance of the intervallum), with similar trends in BIO1 and solar radiation and an opposite trend in BIO4. The mean temperature of the driest quarter (BIO9) showed an association with SL variation between finer and more robustly dissected phenotypes (Figure 8K), similar to BIO3 (Figure 8E), and opposite to BIO18 (Figure 8D).

In a direct comparison of the apomicts from the tree clusters based on BL data, we observed morphological differentiation (Figure S9D–E). The jackknifed mean classification accuracy was 85% between clusters 1 and three and 79% for clusters 2 and 3. The morphological shift from more robust phenotypes of cluster 1 towards more gracile phenotypes of cluster 3 showed a significant covariation with four ecological variables, with the highest contribution being isothermality (BIO3; Figure S9F). Additionally, within the clusters, we observed differentiation between sexual and apomictic BL phenotypes (Figure S10). In cluster 1, the separation of sexuals and apomicts was the most pronounced (Figure S10A–C), while in cluster 2, it was the lowest among the three clusters (Figure S10D–F). In cluster 3, the differentiation among sexuals and apomicts was also significant (Figure S10G–I). The morphological differentiation exhibited covariances with ecological variables for cluster 1, we identified a shift towards drier climates by the apomicts (Figure S11A), while in clusters

2 and 3, most apomicts shifted towards more lowland climates than their sexual progenitors (Figure S11B,C). Among the apomictic taxa within the clusters, we could also identify morphological differentiation between more extreme BL morphotypes (e.g., *R.* ×*obscurans* and *R.* ×*binatus* in cluster 2; Figure S12A,B). The observed shift in BL phenotype showed a significant covariance with ecological variables (Figure S12C,D).

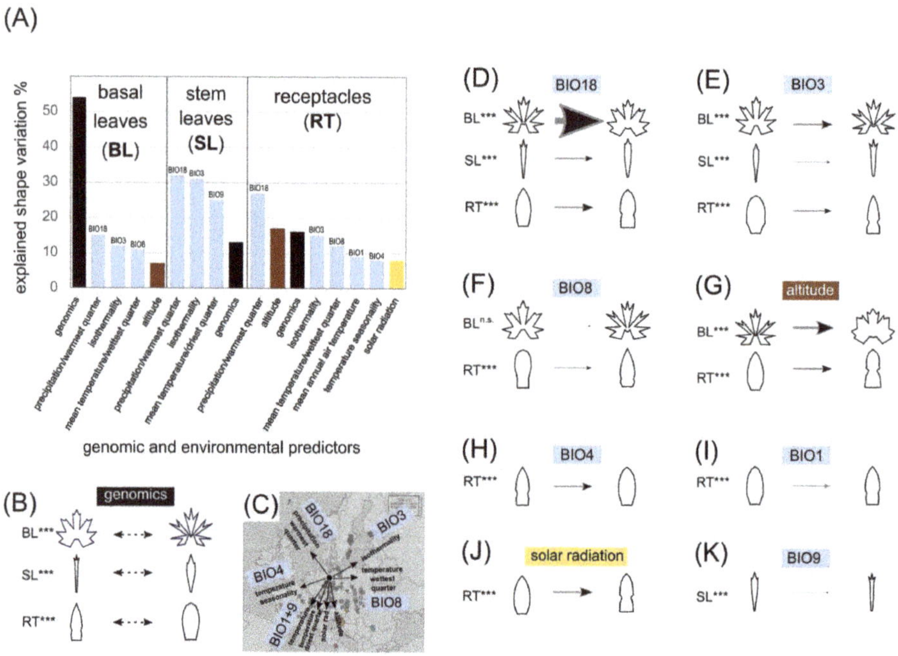

Figure 8. Shape variation of the three organs is associated with environmental variables and genomics. All variables with significant effects on shape variation are illustrated in (**A**), and the percentages give their relative importance for predicting the associated trait phenotypes. The genomic effects on shape (**B**) are visualized as shape variation at the first canonical variate, as shown in Figure 3. Environmental gradients affecting shape are plotted in the map of Europe (**C**) and the respective shape changes along the environmental variables were generated by regression models and visualized in (**D**–**K**). (**D**) The precipitation of the warmest quarter (BIO18) correlated to shape changes in BL, SL, and RT. (**E**) The isothermality (BIO3) is correlated to shape changes in B, SL, and RT. (**F**) The mean temperature of the wettest quarter (BIO8) correlated to shape changes in RT, (**G**) The altitude correlated to shape changes in basal leaves and receptacles. The RT shape changes were further correlated to the temperature seasonality (BIO4; **H**), the mean annual air temperature (BIO1; **I**), and solar radiation (**J**). The SL shape changes were also correlated to the mean temperature of the driest quarter (BIO9; **K**). The thicker the arrow, the stronger the association between trait shape variation and environmental predictors. *** = $p < 0.001$.

3.7. Morphologically and Ecologically Intermediate to Transgressive Polyploid Hybrids

R. ×*pseudocassubicus* from cluster 1 is an obligate apomictic polyploid hybrid of *R. cassubicifolius* (cluster 1) and *R. envalirensis* (cluster 3), and it exhibited an intermediate position between the parental species within the BL morphospace (Figure 9A,B). This taxon showed an ecological shift towards lowland climatic conditions, which are outside the parental range but more distant from *R. envalirensis* and closer to *R. cassubicifolius* (Figure 9C). *R.* ×*platycolpoides* from cluster 1 is an obligate apomictic polyploid hybrid of *R. cassubicifolius* (cluster 1) and *R. notabilis* (cluster 2), exhibiting an intermediate (but closer to *R. notabilis*) morphological position between its parental species (Figure 9D,E), and showing a pronounced ecological shift towards a drier climate outside the parental niche preferences (Figure 9F). *R. ×hungaricus* from cluster 1 is an obligate apomictic polyploid hybrid of *R. cassubicifolius* (cluster 1) and *R. flabellifolius* (cluster 2) with mainly transgressive BL phenotypes (Figure 9G,H) and mainly intermediate ecological preferences located inside the parental species niche space (Figure 9I,J). *R. ×leptomeris* from cluster three is an obligate apomictic polyploid hybrid of *R. envalirensis* (cluster 3) and *R. flabellifolius* (cluster 2), with pronounced transgressive BL phenotypes (Figure 9K,L). *R. ×leptomeris* exhibited a slight ecological shift outside the range of its parental species toward a more lowland climate with less extreme temperature fluctuations (Figure 9M,N). Moreover, the UMAP analysis of ecological similarity among sexuals and apomicts inferred substantial ecological shifts of the apomicts far outside of the range of the progenitor species (Figure S13).

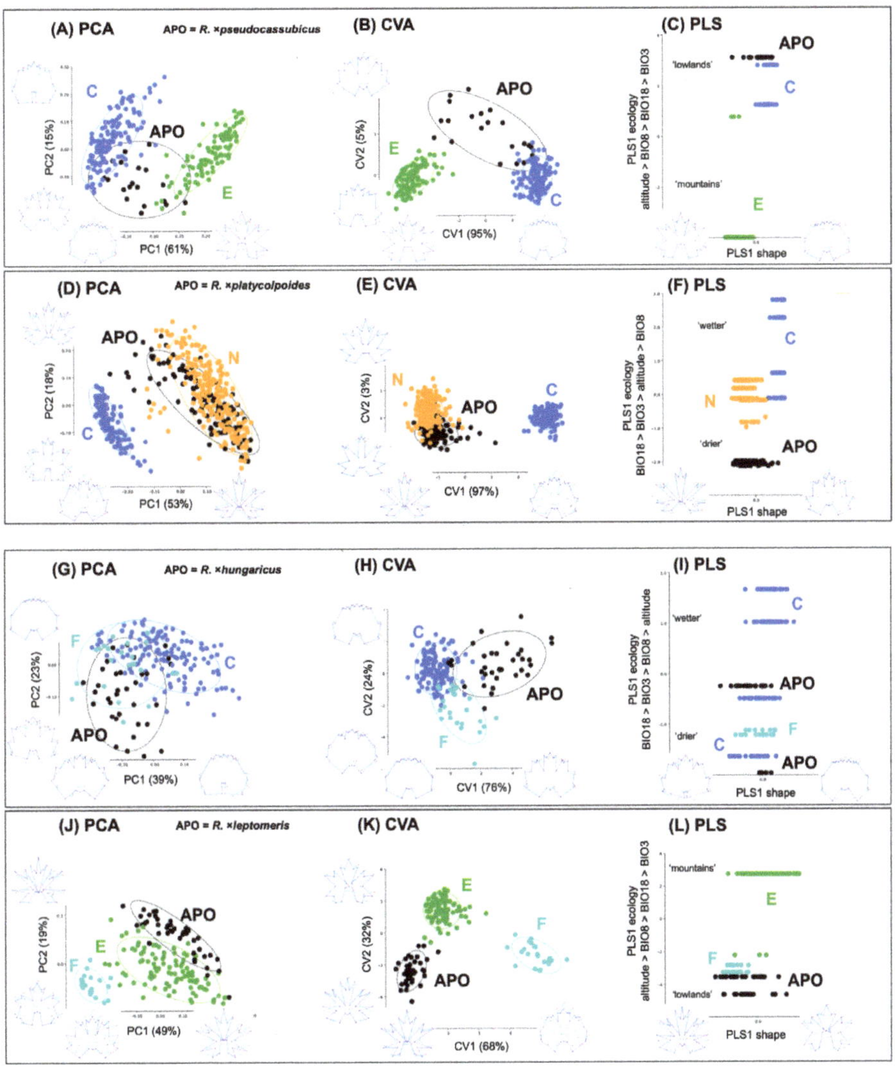

Figure 9. BL variation of sexual species and their selected allopolyploid apomictic (APO) derivatives. (**A–C**) *R.×pseudocassubicus*; (**D–F**) *R. ×platycolpoides*; (**G–J**) *R. ×hungaricus*; (**K,L**) *R. ×leptomeris*. PCA,

principal component analyses, CVA, canonical variate analyses; PLS, partial least square analyses, (**A**) PCA of *R.* ×*pseudocassubicus* (black dots) and two sexual progenitor species *R. cassubicifolius* (blue dots) and *R. envalirensis* (green dots). (**B**) CVA of the taxa in (A). (**C**) PLS of the taxa in (A). The first PLS axis described 99.6% of the covariance between BL shapes and the four ecological variables, and the covariance model had $p < 0.001$. The first ordination axis from the shape data (PLS1 shape) was significantly correlated (0.90; $p < 0.001$) with the first ordination axis from the ecological data (PLS1 ecology). The highest contribution to the ecological component was provided by altitude (−0.79). (**D**) PCA of *R.* ×*platycolpoides* (black dots) and two sexual species, *R. cassubicifolius* (blue dots) and *R. notabilis* (orange dots). (**E**) CVA of the taxa in (D); (**F**) PLS of the taxa in (D). The first PLS axis described 97% of the covariance between BL shapes and the four ecological variables, and the covariance model had $p < 0.001$. The first ordination axis from the shape data (PLS1 shape) was significantly correlated (0.63; $p < 0.001$) with the first ordination axis from the ecological data (PLS1 ecology). The highest contribution to the ecological component was provided by the precipitation of the warmest quarter (BIO18; 0.75). (**G**) PCA of *R.* ×*hungaricus* (black dots) and two sexual species, *R. cassubicifolius* (blue dots) and *R. flabellifolius* (turquoise dots). (**H**) CVA of the taxa in (G). (**I**) PLS of the taxa in (G). The first PLS axis described 55% of the covariance between BL shapes and the four ecological variables, and the covariance model had $p < 0.001$. The first ordination axis from the shape data (PLS1 shape) was significantly correlated (0.45; $p < 0.001$) with the first ordination axis from the ecological data (PLS1 ecology). The highest contribution to the ecological component was provided by the precipitation of the warmest quarter (BIO18; 0.93) (**J**) PCA of *R.* ×*leptomeris* (black dots) and two sexual species, *R. cassubicifolius* (blue dots) and *R. flabellifolius* (turquoise dots). (**K**) CVA of the taxa in (J). (**L**) PLS of the taxa in (J). The first PLS-axis described 97% of the covariance between BL shapes and the four ecological variables, and the covariance model had $p < 0.001$. The first ordination axis from the shape data (PLS1 shape) was significantly correlated (0.64; $p < 0.001$) with the first ordination axis from the ecological data (PLS1 ecology). The highest contribution to the ecological component was provided by the altitude (0.78).

4. Discussion

This study gathered and evaluated what is currently the largest landmark-based, multi-trait GM dataset available for an intricate polyploid plant species complex, comprising more than 11,000 trait measurements from 220 diploid sexual to polyploid apomictic populations from 80 *R. auricomus* taxa. The GM dataset was tested for congruence with groupings derived from genomic datasets (genomic RAD-Seq, nuclear TEG, plastid CP), morphological distinctiveness, and morphological and ecological novelty of polyploid apomicts, but also for the discriminating power of different traits and NGS datasets. Additional information on ploidy and reproduction modes, as well as ecological data, was included following an integrative taxonomic approach, as recommended by several authors [36,44,45,50,51]. Consequently, we were able to analyze for the first time, objectively and in detail, the phenotypic diversity of the polyploid apomictic *R. auricomus* complex under a comprehensive (phylo)genomic background [24,51]. We showed that (1) the three previously defined genomic clusters representing five sexual species and 75 apomictic *R. auricomus* taxa correspond to morphological groupings based on both basal leaves and all traits together, and genomic RAD-Seq, as opposed to TEG and CP datasets, best fits the morphological resolution; (2) the apomictic taxa usually overlap within the trait morphospace except for those taxa at the morphospace edges; (3) trait-based phenotypes are highly shaped by genomic composition and to a lesser extent by environmental factors; and (4) allopolyploid apomictic taxa, compared to sexual progenitors, resemble a mosaic of ecological and morphological intermediaries to novel (transgressive) biotypes.

4.1. GM Methodology

Both main directions of morphometric approaches, i.e., traditional morphometry [64,171] as well as geometric morphometrics [87,88], have already been applied in the *R. auricomus* complex using the taxonomically most informative leaf and fruit characters. The use of traditional morphometry has enabled, for example, the quantitative evaluation of a few closely related sexual and polyploid apomictic taxa [64,171]. Traditional morphometry has a methodological advantage in that it can measure things such as the length and width of a leaf blade, its ratio, stem height, or length of the carpellophore or fruits. What traditional morphometry cannot capture, however, is more complex, detailed shape information and changes within a single objective statistical analysis, which can be interpreted in an anatomical, ecological, and evolutionary context [95,97,98,172]. In the case of *R. auricomus*, particularly the basal and stem leaves appear to have seemingly infinite variation across taxa, which challenges quantitative taxonomic treatments. GM uses modern digitalization and mathematical approaches to turn landmarks or outlines into quantitative variables, which can subsequently be analyzed with up-to-date multivariate statistics. Currently, the most useful contribution of GM for species delimitation is to test morphological hypotheses within an evolutionary theory-based framework and to provide a metric that makes biological trait variation among taxa measurable and comparable.

The first GM application for *R. auricomus* by [87] quantified basal leaf variation in two sexual diploid progenitor species and one natural apomictic hybrid. The authors compared their variation to that of artificially produced hybrids generated from the same progenitor species. This study demonstrated that even a relatively small dataset and landmark scheme (14 landmarks; in the present study we analyzed twenty-six landmarks) can reveal fundamental phenotypic shape changes in basal leaves, particularly in the highly variable dissected forms (e.g., *R. notabilis* and *R. variabilis*). The follow-up study by [88] expanded the landmark scheme to morphometrically analyze not only the variable dissected leaf shapes but also the undissected types (e.g., *R. cassubicifolius*; Figures 1B and S1A). The new approach used 26 homologous landmarks to capture shape changes among the genetically and morphologically most distantly related sexual progenitor species and their crossings. Many artificially produced hybrids could be traced back to polyploid apomictic morphotypes found in nature, and thus GM further corroborated the idea about the hybrid origin of the *R. auricomus* complex. This study also showed that even an apparent continuum of forms can be decomposed into previously unknown (cryptic) morphological clusters [39], further extending the landmark approach to include two new traits, i.e., stem leaves (SL) and receptacles (RT). This phylogenomic-morphometric study on sexual progenitor species of the *R. auricomus* complex laid the foundation for the incorporation of GM into the taxonomy of this intricate plant group. For the first time, the significance of these three taxonomically most informative traits for separation among taxa could be quantified, which corroborated the final taxonomic treatment. The present study particularly makes progress in extending the multi-trait, landmark-based GM approach of [39] to polyploid apomicts and different genomic NGS datasets within an evolutionary framework, supported by the knowledge of previous studies (e.g., progenitor species circumscription, ploidy and reproduction modes, polyploid apomictic clusters, and genome evolution).

Our GM approaches are mainly aimed at unraveling complex-wide relationships in morphological trait variability. Therefore, we disregarded shape changes at the fine-grained level that could lead to more precise discrimination of apomictic taxa. Such additional traits concern the shape of early spring basal leaves, color of shoots, indumentum of basal leaves, shape of teeth at basal leaf blade margins, and indumentum of the receptacle. Although these fine-grained traits were often used by taxonomists to describe taxa (e.g., [121,128,142]), they are usually inconspicuous in the field, and some of these characters are available only at specific developmental stages (e.g., a reddish color on shoots appears in early spring during sprouting but disappears later) or are not stable in cultivation [161]. However, most fine-grained traits are related to phenotypic plasticity and do not discriminate genomic clusters or species (e.g., as shown for the indumentum of the receptacle in [39]).

4.2. Congruence of Genetic and Morphological Clustering, and Taxonomical Implications

The concatenated GM dataset revealed three significantly differentiated morphological groupings that are largely congruent with previously observed genomic clusters observed in [51] (Figure 3, Table S2). This finding is surprising because, in the light of field sampling and garden observations, no clear morphological groupings could be inferred due to seemingly endless phenotypic variation (Figure 1B–G). Each grouping comprises a single or few sexual progenitor species surrounded by polyploid apomictic derivatives. From the genomic perspective, this pattern is unique compared to other plant species complexes (e.g., [70,173]), where frequently several progenitors are found within clades or clusters or only polyploid descendants are observed. The origin of polyploid apomictic *Auricomi* is shaped by the hybridization of sexual progenitors, followed by early hybrid segregation, backcrossing to parents, polyploidization, and gene flow among apomicts due to facultative sexuality, leading to substantial genome evolution and consequently subgenome dominance [51,64,87,88]. The grouping of apomictic morphotypes around sexual progenitors probably represents the consequence of subgenome dominance and thus phenotypic trait expression rather similar to the dominant parent (Figure 7). Subgenome dominance is frequently observed in young allopolyploids [6,174]. However, effects on allopolyploid phenotypic trait expression based on exact subgenomic contributions have been so far less regarded in non-model plants, and hence, this study sheds new light on genomic-based changes in phenotypic and ecological features of naturally occurring allopolyploids (Figures 8, 9 and S13).

Although the current GM approach is labor- and cost-intensive (but see Section 4.4 on perspectives), it proved its value by unraveling and characterizing the morphological differentiation within a large part of the *R. auricomus* complex for the first time (Figure 3D). Cluster 1 (including progenitor *R. cassubicifolius*) is characterized by non-dissected to three-lobed BL with a narrow blade base, broadly lanceolate SL with teeth, and RT with short and narrow androclinium and high gynoclinium. Cluster 2 (including progenitors *R. flabellifolius* and *R. notabilis*) shows three-lobed to five-dissected BL with wide blade base and roundish leaf segments, SL with slightly to non-dissected, narrow to lineal segments, and RT with broad and long androclinium and oval and short gynoclinium. Finally, cluster 3 (including progenitor *R. envalirensis*) exhibits three-lobed to five-dissected BL with narrow blade base and strongly dissected leaf segments, SL with slightly to strongly dissected, narrow to broad segments, and RT with short and narrow androclinium and high gynoclinium.

In detail, the GM analysis of single and concatenated traits (BL, SL, RT) demonstrated that RAD-Seq genetic clusters show a quantifiable degree of morphological differentiation (Figures 3, 4 and S7; Table S2), and confirmed the usefulness of the GM approach in [39] also for a polyploid apomictic species complex. The RAD-Seq dataset is more effective in resolving morphological patterns because it provides magnitudes more information in coding and non-coding regions than the TEG single-copy genes applied here [51], and subgenome dominance seems to be more pronounced than maternal effects (CP data, Figure 4E). The last finding is supported by diploid crossing experiments, where F_2 hybrids showed equal ratios of maternal, intermediate, and paternal phenotypes [88]. Morphological differentiation is corroborated by an overall classification accuracy of 85% for assigning populations into the three clusters (concatenated trait dataset). The stem leaves mainly separated clusters 1 and 2 (91%), clusters one and three (94%), and the basal leaves mainly distinguished clusters 1 and 3 (91%). On average, the concatenated data separated all pairs of clusters slightly better than single best-separating traits (Table S2). Interestingly, the separation between clusters was only slightly affected by the inclusion or exclusion of sexual populations (Tables S2 and S3, Figure S9C–E), supporting the stability of inferred morphological groupings. The overall multi-trait morphological differentiation was highest between clusters 1 and 3, and the lowest between clusters 2 and 3 (Table S2), which is in congruence with genomic clusters inferred previously [51]. Cluster 1 is most distinct from all other clusters because of its unique non-dissected BL and broad lanceolate SL, which are largely congruent with previous classifications of "*R. cassubicus*" or "*R. cassubicus* group" of several authors [127,142,175]. In contrast to expectations, we can morphologically characterize the previously only genetically recognized clusters 2 and 3, providing a basis for an informal grouping concept for the species complex [120]. However, these clusters do not match previous taxonomic treatments of "*R. auricomus*" and an intermediate "*R. fallax*" group, but rather follow a geographical, and longitudinal differentiation of genetic clusters, as outlined in [24,51].

Basal leaves bear the most discriminative power, followed by significant contributions from stem leaves and receptacles. The importance of BL shape variation has been stressed by previous taxonomists but requires a careful comparison of leaves within their leaf cycles. The distinction of cluster 1 is alongside important differences in stem leaves also due to a basal leaf cycle without dissected leaves that appear in most individuals of clusters 2 and 3 on separate, adventitious shoots during anthesis [120,121,142,161]. Our approach of using the functionally equivalent final leaves of cluster 1 (fully developed during anthesis) for comparison between clusters follows taxonomic practice. However, except for the general dissection of BLs, other distinguishing characters are also present in the complex to-discriminate clusters. For example, BLs of clusters 2 and 3 differ mostly by the angle of the blade base and segment dissection of the BL and differences in the width and dissection of SL segments, as described above (Figures 3 and S14). The strong correlation between basal and stem leaf shape changes is expected from the shared developmental background of leaf organs. In Ranunculaceae, developmental studies [176] revealed that the ancestral leaf type has a trilobed to the ternate blade, from which either undivided leaves (by faster growth of blade than of the segmental meristems) or dissected leaves (by further secondary divisions of the segments) are independently derived. Our reconstruction of the ancestral BL shape of sexual species led to the same conclusion (Figure S8A), that the root of the Eurosiberian *Auricomi* possessed a three-lobed BL morphotype. The analysis also showed that the BL contained a phylogenetic signal. Interestingly, the shapeshift between non-dissected and dissected BL phenotypes (PC3) was not the transition dominating the morphological evolution of the species complex but rather two divergent traits from an ancestral intermediate (trilobed) leaf type (Figure S8B). Notably, the primary leaves of the BL cycle in clusters 2 and 3 are also often trilobed (e.g., Figure 1A, leaves 1–3) and might still reflect the ancestral shape, from which then the following leaves develop differentially. In general, the morphological clustering within the complex matches the background of a young, less than 1.0-Myr-old polyploid complex (sensu [177]) with a low degree of differentiation. The marked congruence of genetic and morphological patterns of allopolyploid clusters rather speaks for cluster criteria after [32] with regard to the entire *R. auricomus* complex, whereas only the sexual progenitors can be properly treated as species in the sense of evolutionary lineages and non-overlapping genetic/morphological clusters [39]. A modern ancestor-descendant lineage concept after [33] is hard to apply for the obligate to facultative apomictic allopolyploid nothotaxa due to multiple origins of the same morphotype (polyphyly, [51]), and even not at a cluster-wide scale, due to genetic and morphological instability of clusters in space and time caused by ongoing reticulate evolution.

4.3. Sexual Species and Apomictic Derivative Taxa in Relation to Morphospace and Ecology, and Taxonomic Implications

At a fine-grained morphological scale, and at first glance, clear morphological differentiation of polyploid apomictic taxa was not recognizable (Figure 6), especially when analyzing all available observations per trait (Figure S9B). Taxa of RAD-Seq clusters 1–3 usually overlap within single-trait morphospace (BL, SL, RT), except for sexual progenitor species and those polyploid apomictic taxa toward the morphospace edges (Figures 6 and S10). Closer inspection also revealed some polyploid apomictic taxa that consistently formed well-separated morphological groupings in all three traits (Figures 6 and S12). In cluster 1, e.g., *R.* ×*platycolpoides* and *R.* ×*elatior* are separated by all three traits. Nevertheless, one should keep in mind that several taxa from boreal Finland and Russia ([134], described under "*R. cassubicus*" and "*R. fallax*") were not sampled here and that the entire variability of cluster 1 is not yet fully documented. In clusters 2 and 3, our more comprehensive sampling for Central Europe shows that apomictic taxa appear largely intermingled, with a few exceptions exhibiting pairwise morphological differentiation. For example, the apomictic taxa *R.* ×*obscurans* and *R.* ×*binatus* (cluster 2) are distinguishable from each other in all three morphological traits (mostly by SL and RT) as well as in the concatenated trait dataset (Figure S12A–C). However, most polyploid apomicts seem morphologically intermingled due to the broad variability of characters and mosaic-like character combinations as typical for hybrids (e.g., [5,102]). The shape of the BL can even vary among shoots of the same clone [161,171], which could be explained by differential gene expression of subgenomes [174] in these allopolyploids. The allopolyploid origin of apomicts from at least four distinct sexual progenitors has resulted in hundreds of local and regional morphotypes [51]. These morphotypes lack the homogenizing effect of regular (obligate) sexuality and hence cannot form coherent lineages and frequently do not form phenotypic clusters that could be recognized as species (sensu [32]). Within genomic clusters, the sexuals separate from the apomicts due to reproductive barriers via different ploidy levels [178], which speaks against concepts that simply sink apomicts into sexual species (e.g., as for sexual autotetraploids in *R. cassubicifolius*, [179]). Moreover, agamospecies concepts are largely inapplicable because facultative sexuality is still present, especially in Central Europe [24]. A classification as nothotaxa [51], despite methodological issues (allopolyploids here are hybrids of hybrids), appears to be a pragmatic solution to link existing names to a morphotype and its type location and to separate apomicts formally from sexual species [120].

Morphological shape changes within the European *R. auricomus* complex are mainly associated with the subgenomic composition of the allopolyploids and overall genomic differentiation, and less so with abiotic environmental conditions (Figures 6–8). Results suggest a predominant heritable (epi)genetic control and a minor environmental regulation, particularly of BL features (e.g., [180–182]). BL leaf shape follows the pattern of an increased degree of basal leaf incisions under drier and hotter environments, but interestingly, under isothermal climatic conditions, the BL thus becomes more and more dissected towards temperature and precipitation stress conditions (same partly applies for SL, Figure 8A,D,E). These leaf traits probably reflect both climatic dependencies with a geographical west-east gradient (continentality) and altitude (Figure 8). Changes in BL leaf segments, margin incision, and the number of teeth usually influence leaf surface area, stomatal conductance, transpiration, and thus leaf energy balance and temperature, representing adaptation to water-limited and/or climatically variable environments [183,184]. Since shape differences are mainly attributed to genomic differences in the *R. auricomus* complex, these different leaf shapes probably represent selective advantages in their respective environments, for example, large non-dissected BL taxa along relatively water-rich but rather continental streamside habitats (e.g., *R. cassubicifolius*, *R.* ×*pseudocassubicus*), or strongly dissected BL taxa in less continental but rather dry anthropogenic meadows in Central-Eastern Europe (e.g., *R. notabilis*, *R.* ×*variabilis*). Interestingly, garden experiments with different levels of soil nutrients did not reveal changes in leaf shape, and different light treatments influenced only the size of plants and the number of leaves but not the shape of BL [161]. This indicates the predominant genomic fixation of BL features and thus supports the findings of this study. In addition, the shape of the receptacle has so far occasionally been utilized for descriptions of apomictic taxa (e.g., [121,137,142,185]), but not for the main groups. Contrary to expectation, results revealed that the RT shape separates clusters, specifically clusters 2 and 3 (Figures 3D and 7B,C), but not the taxa within these clusters (Figure 6F,I). The RT is probably shaped by a mix of genomic and climatic factors.

Allopolyploid apomictic taxa, compared to sexual progenitors, resemble a mosaic of ecological and morphological intermediate to novel (transgressive) biotypes (Figure 9). Intermediate biotypes in sympatry and the same ecological niche as their progenitors could create an unpleasant "smear" between distinct sexual species, making their circumscription in practice difficult (e.g., [64]). However, intermediate morphotypes in allopatry or in different ecological niches can be recognized separately from sexual species. For instance, *R.* ×*pseudocassubicus* showed an ecological shift towards lowland climatic conditions, which are more distant to its Pyrenean mountain progenitor *R. envalirensis* and closer to sympatric

R. cassubicifolius but outside of the parental range (Figure 9C). *R.* ×*platycolpoides* exhibited a pronounced ecological shift towards drier climates in southern Finland, which occurs not only allopatrically to the Central European parents, but also outside the parental ecological niche preferences (Figure 9F). Moreover, the UMAP analysis of ecological similarity among sexuals and apomicts inferred substantial ecological shifts of the apomicts far outside the range of the progenitor species (Figure S13). Ecological niche shifts can contribute significantly to the range expansions of allopolyploid apomicts compared to their progenitors ("geographical parthenogenesis") [9,24,186]. We also observed allopolyploids with intermediate ecology but transgressive morphology, i.e., with characters outside the parental's morphospace (e.g., *R.* ×*hungaricus* and *R.* ×*leptomeris*). In general, our results support a general hypothesis of genomic and phenotypic novelty in allopolyploids [3,7,55–57].

4.4. Perspectives of GM and Species Identification

Our results corroborated the most recent taxonomic treatments of the complex for Central Europe, to separate sexual progenitors as species and to treat the allopolyploid derivates as nothotaxa that can be grouped into three main informal clusters [39,51,120]. Identification of described or new taxa, however, is still a challenge because of the mosaic-like diversity of character combinations. Recent technological advances in automatically identifying plant species using machine learning (ML) can also be used in genetic modification (GM) and, by extension, in systematic biology. Similar to the landmark approach described in this study, automatic plant identification in the beginning also relied on manual feature extraction. From images of leaves or flowers, morphological features such as leaf shapes, leaf margins, leaf textures, flower shapes, or flower color were extracted [187]. The respective developed model refers to these features in the subsequent classification step. In recent years, so-called artificial neural networks (ANNs, a type of ML), and, more specifically, convolutional neural networks (CNNs), have made significant breakthroughs in automatic image classification [188]. They are already used in automatic plant identification [189,190] as well as in the extraction of plant features from herbaria [191,192]. The computer independently learns to recognize the structure of data, sometimes using up to millions of plant images.

Defining and setting landmarks as described in this study is labor-intensive, subjective, and a task for experts only. A species-specific machine learning approach to setting homologous landmarks automatically along the outline of a specific plant organ would be desirable. Attempts to combine GM and ML have been recently made in anthropology [193] and zoology [194–196]. However, similar to automatic plant identification, features extracted from ANNs [197] can be used for morphometric analysis of plant organs. It would be worth testing whether ANNs can similarly distinguish the different groups as unraveled in the GM approach applied herein. Structured image datasets as created in this study allow the visualization of self-learned features using, for example, Grad-CAM [198] to infer which leaf or receptacle region in the input image makes a large impact on species classification. Subsequently, it would be possible to investigate what the machine 'sees' across images and compare these results with the GM approach based on landmarks defined by experts. Such comparisons would make an important contribution to the explainability of features extracted by ML and provide important insights for species delimitation and final taxonomic decisions.

5. Conclusions

The polyploid apomictic *Ranunculus auricomus* complex exhibits enormous variability in morphological traits, which is often the case for predominantly hybridogenic species complexes and TCGs. After previous studies identified a structure of three genetic clusters within the *R. auricomus* complex, in the present study, we searched for morphological differentiation between clusters and apomictic nothotaxa. Morphological differentiation among the genetic clusters could indeed be detected by an extensive sampling of populations across Europe and using quantitative geometric morphometrics. The basal leaves as well as concatenated-traits data proved particularly useful for the morphological differentiation of the clusters. The hitherto confusing diversity of trait phenotypes and trait phenotype combinations thus received a basic structure on which future taxonomic treatments can build. However, most of the agamospecies described so far within these clusters could not be discriminated against. Moreover, it was also possible to figure out whether the genetic background alone or both the genetic background and the abiotic environment have a big effect on phenotypic diversity in the *R. auricomus* complex. We demonstrated that the hybridogenic phenotypic variability of polyploid apomicts is predominantly genetically determined, which means that the hybrid phenotypes are strongly shaped by parental subgenome contributions. Nevertheless, a couple of environmental parameters (e.g., temperature, precipitation, and temperature variability) could be identified, which influence phenotypic trait expression of leaves and receptacles. While most hybridogenic apomictic nothotaxa are morphologically within but ecologically outside the range of their progenitors, transgressive phenotypes have also evolved. ML techniques in combination with genomics and morphometrics promise new opportunities for future research on plant phenotypic variation and differentiation and integrative taxon-omics.

This study confirms a concept of classification proposed by [39], in which only sexual taxa represent well-defined species, while apomictic hybridogenic taxa are classified formally as nothotaxa [51,120]. The three big genetic and morphological clusters found here represent a geographical structure, but are not congruent with traditional taxonomic treatments of "main species" sensu [122,134]. These clusters will provide the foundation for a novel taxonomic treatment applying either a cluster criterion-based approach [38] or another infrageneric category, as soon as the whole complex has been analyzed. Our study highlights that detailed genomic and morphometric studies are needed to understand the evolution and structuring of agamic TCGs, which is required for a modern evolutionary classification. Traditional descriptive morphological treatments, however, failed to recognize taxa as natural entities on all levels of the hierarchy. Our study exemplifies a timely approach to the classification of TCGs.

Supplementary Materials: The following supporting information can be downloaded at: https://www.mdpi.com/article/10.3390/biology12030418/s1, Figure S1: Landmark digitization of the taxonomically most informative *Ranunculus auricomus* traits; Figure S2: Correlation plot concerning non-autocorrelated (r < 0.8), standardized (0 mean, unit variance) abiotic environmental factors; Figure S3: Correlation plot concerning standardized axis shape scores of all taxonomically informative traits; Figure S4: Correlation plot concerning standardized axis shape scores of basal leaves; Figure S5: Correlation plot concerning standardized axis shape scores of stem leaves; Figure S6: Correlation plot concerning standardized axis shape scores of receptacles; Figure S7: Morphological variation among sexual progenitors and polyploid apomictic derivative taxa with respect to genomic RAD-Seq (RADpainter) background; Figure S8: Ancestral basal leaf (BL) shape reconstruction using the sexual progenitors; Figure S9: Basal leaf (BL) differentiation among clusters: apomictic polyploids; Figure S10: Basal leaf (BL) differentiation within clusters: sexuals vs. apomicts; Figure S11: Basal leaf (BL) differentiation within clusters: ecological covariance; Figure S12: Morphological differentiation among apomictic polyploids; Figure S13: Ecological clustering of sexual and apomictic populations; Figure S14: Selected sexuals and apomicts showing consistent genetic+morphological clusterings; Table S1: Information on 220 populations with complete evidence of collection (GPS), morphometric (BL + SL + RT), and genomic/RAD-Seq clustering data; Table S2: Discriminant analysis of the differentiation among genomic (RAD-Seq), nuclear (TEG), and plastid (CP) clusters based on pairwise comparisons of single morphological traits and concatenated data; Table S3: Discriminant analysis of the differentiation among the three genomic (RAD-Seq) clusters based on pairwise comparisons of single morphological traits and concatenated data; Table S4: Discriminant analysis of the differentiation among selected nothotaxa within each of the three genomic (RAD-Seq) clusters based on pairwise comparisons of the three morphological traits; Table S5: Regression models of the association between genomic contributions of the sexual species and shape variation in apomictic polyploids (mostly allopolyploids); Table S6: Distance matrix-based multiple regression models (MRM) to explain the source of shape variation; Table S7: Regression models of the association between eight non-correlated ecological predictors and variation of all three morphological traits.

Author Contributions: Conceptualization, K.K. and E.H.; methodology, K.K., L.H., S.T. and J.W.; software, L.H., K.K., S.T. and J.W.; validation, K.K., L.H. and S.T.; formal analysis, L.H., K.K. and S.T.; investigation, L.H. and K.K.; resources, E.H. and J.W.; data curation, L.H and K.K.; writing—original draft preparation, K.K., L.H. and E.H.; writing—review and editing, E.H., J.W. and J.P.B.; visualization, L.H. and K.K.; supervision, E.H.; project administration, E.H.; funding acquisition, E.H. All authors have read and agreed to the published version of the manuscript.

Funding: This research was funded by the Deutsche Forschungsgemeinschaft (German Research Foundation) DFG, grant number Ho4395/10-1/2 to E.H. within the priority program "Taxon-Omics: New Approaches for Discovering and Naming Biodiversity" (SPP 1991). This study was further supported by the German Ministry of Education and Research (BMBF), grant number 01IS20062 to J.W. We acknowledge support from the Open Access Publication Funds of Göttingen University.

Institutional Review Board Statement: Not applicable.

Informed Consent Statement: Not applicable.

Data Availability Statement: The authors declare that basic data supporting the findings are available within the manuscript and Supporting Information. RAD-Seq and target enrichment reads are deposited on the National Center for Biotechnology Information Sequence Read Archive (SRA): BioProject ID PRJNA627796, https://www.ncbi.nlm.nih.gov/bioproject/627796 (accessed on 7 March 2023) and BioProject ID PRJNA628081, https://www.ncbi.nlm.nih.gov/bioproject/628081 (accessed on 7 March 2023), respectively. More detailed data, tables, and figures concerning (phylo)genomic analyses are deposited on FigShare (https://doi.org/10.6084/m9.figshare.14046305) (accessed on 7 March 2023). Flow cytometric (FC) and flow cytometric seed screening (FCSS) data (ploidy levels, reproduction modes) are also stored in Figshare (https://doi.org/10.6084/m9.figshare.13352429) (accessed on 7 March 2023). We deposited image data processed for geometric morphometric analyses on FigShare upon publication (https://doi.org/10.6084/m9.figshare.21393375) (accessed on 31 January 2023).

Acknowledgments: We acknowledge Franz G. Dunkel and Volker Melzheimer for inspiring discussions about the *R. auricomus* complex and Silvia Friedrichs and Gabriele Ließmann for garden work. We thank Esther Philine Zieschang, Anne-Sophie Burmeister, and Jennifer Krüger for processing leaf scans for geometric morphometric analyses, and Michael Kloster for providing scripts to automatically cut leaf scans into basal and stem leaf parts.

Conflicts of Interest: The authors declare no conflict of interest.

References

1. Hörandl, E. Novel Approaches for Species Concepts and Delimitation in Polyploids and Hybrids. *Plants* **2022**, *11*, 204. [CrossRef] [PubMed]
2. Soltis, P.S.; Soltis, D.E. The Role of Genetic and Genomic Attributes in the Success of Polyploids. *Proc. Natl. Acad. Sci. USA* **2000**, *97*, 7051–7057. [CrossRef] [PubMed]
3. Soltis, P.S.; Marchant, D.B.; Van de Peer, Y.; Soltis, D.E. Polyploidy and Genome Evolution in Plants. *Curr. Opin. Genet. Dev.* **2015**, *35*, 119–125. [CrossRef]
4. Leebens-Mack, J.H.; Barker, M.S.; Carpenter, E.J.; Deyholos, M.K.; Gitzendanner, M.A.; Graham, S.W.; Grosse, I.; Li, Z.; Melkonian, M.; Mirarab, S.; et al. One Thousand Plant Transcriptomes and the Phylogenomics of Green Plants. *Nature* **2019**, *574*, 679–685. [CrossRef]
5. Soltis, P.S.; Soltis, D.E. The Role of Hybridization in Plant Speciation. *Annu. Rev. Plant Biol.* **2009**, *60*, 561–588. [CrossRef]
6. Blischak, P.D.; Mabry, M.E.; Conant, G.C.; Pires, J.C. Integrating Networks, Phylogenomics, and Population Genomics for the Study of Polyploidy. *Annu. Rev. Ecol. Evol. Syst.* **2018**, *49*, 253–278. [CrossRef]
7. Comai, L. The Advantages and Disadvantages of Being Polyploid. *Nat. Rev. Genet.* **2005**, *6*, 836–846. [CrossRef]
8. Van de Peer, Y.; Ashman, T.-L.; Soltis, P.S.; Soltis, D.E. Polyploidy: An Evolutionary and Ecological Force in Stressful Times. *Plant Cell* **2021**, *33*, 11–26. [CrossRef]
9. Hörandl, E. The Complex Causality of Geographical Parthenogenesis. *New Phytol.* **2006**, *171*, 525–538. [CrossRef]
10. Rice, A.; Šmarda, P.; Novosolov, M.; Drori, M.; Glick, L.; Sabath, N.; Meiri, S.; Belmaker, J.; Mayrose, I. The Global Biogeography of Polyploid Plants. *Nat. Ecol. Evol.* **2019**, *3*, 265–273. [CrossRef]
11. Qiu, T.; Liu, Z.; Liu, B. The Effects of Hybridization and Genome Doubling in Plant Evolution via Allopolyploidy. *Mol. Biol. Rep.* **2020**, *47*, 5549–5558. [CrossRef]
12. Mallet, J. Hybridization as an Invasion of the Genome. *Trends Ecol. Evol.* **2005**, *20*, 229–237. [CrossRef] [PubMed]

13. Barke, B.H.; Karbstein, K.; Daubert, M.; Hörandl, E. The Relation of Meiotic Behaviour to Hybridity, Polyploidy and Apomixis in the *Ranunculus auricomus* Complex (Ranunculaceae). *BMC Plant Biol.* **2020**, *20*, 523. [CrossRef]
14. Izmailow, R. Observations in Embryo and Endosperm Development in Various Chromosomic Types of Apomictic Species *Ranunculus cassubicus* L. *Acta Biol. Cracoviensia Ser. Bot.* **1967**, *10*, 99–111.
15. Ainouche, M.L.; Baumel, A.; Salmon, A.; Yannic, G. Hybridization, Polyploidy and Speciation in Spartina (Poaceae). *New Phytol.* **2004**, *161*, 165–172. [CrossRef]
16. Barke, B.H.; Daubert, M.; Hörandl, E. Establishment of Apomixis in Diploid F_2 Hybrids and Inheritance of Apospory from F_1 to F_2 Hybrids of the *Ranunculus auricomus* Complex. *Front. Plant Sci.* **2018**, *9*, 1111. [CrossRef]
17. Hojsgaard, D.; Klatt, S.; Baier, R.; Carman, J.G.; Hörandl, E. Taxonomy and Biogeography of Apomixis in Angiosperms and Associated Biodiversity Characteristics. *CRC Crit. Rev. Plant Sci.* **2014**, *33*, 414–427. [CrossRef]
18. Zhang, Y.; Wu, H.; Hörandl, E.; de Oliveira Franca, R.; Wang, L.; Hao, J. Autonomous Apomixis in *Praxelis clematidea* (Asteraceae: Eupatorieae), an Invasive Alien Plant. *AoB Plants* **2021**, *13*, plab007. [CrossRef]
19. Pellino, M.; Hojsgaard, D.; Schmutzer, T.; Scholz, U.; Hörandl, E.; Vogel, H.; Sharbel, T.F. Asexual Genome Evolution in the Apomictic *Ranunculus auricomus* Complex: Examining the Effects of Hybridization and Mutation Accumulation. *Mol. Ecol.* **2013**, *22*, 5908–5921. [CrossRef]
20. Hojsgaard, D.; Hörandl, E. The Rise of Apomixis in Natural Plant Populations. *Front. Plant Sci.* **2019**, *10*, 358. [CrossRef]
21. Nogler, G.A. Genetics of Apospory in Apomictic *Ranunculus auricomus*. V. Conclusion. *Bot. Helv.* **1984**, *94*, 411–422.
22. Ozias-Akins, P.; van Dijk, P.J. Mendelian Genetics of Apomixis in Plants. *Annu. Rev. Genet.* **2007**, *41*, 509–537. [CrossRef]
23. Asker, S.; Jerling, L. *Apomixis in Plants*; CRC press: Boca Raton, FL, USA, 1992; ISBN 0849345456.
24. Karbstein, K.; Tomasello, S.; Hodač, L.; Lorberg, E.; Daubert, M.; Hörandl, E. Moving beyond Assumptions: Polyploidy and Environmental Effects Explain a Geographical Parthenogenesis Scenario in European Plants. *Mol. Ecol.* **2021**, *30*, 2659–2675. [CrossRef] [PubMed]
25. Stuessy, T.F.; Crawford, D.J.; Soltis, D.E.; Soltis, P.S. Plant Systematics: The Origin, Interpretation, and Ordering of Plant Biodiversity. In *Regnum Vegetabile*; Koeltz Scientific Books: Königsstein (Taunus), Germany, 2014; p. 425; ISBN 3874294528.
26. Miralles, A.; Bruy, T.; Wolcott, K.; Scherz, M.D.; Begerow, D.; Beszteri, B.; Bonkowski, M.; Felden, J.; Gemeinholzer, B.; Glaw, F.; et al. Repositories for Taxonomic Data: Where we are and what is Missing. *Syst. Biol.* **2020**, *69*, 1231–1253. [CrossRef] [PubMed]
27. Small, E. Systematics of Biological Systematics (Or, Taxonomy of Taxonomy). *Taxon* **1989**, *38*, 335–356. [CrossRef]
28. Bebber, D.P.; Carine, M.A.; Wood, J.R.I.; Wortley, A.H.; Harris, D.J.; Prance, G.T.; Davidse, G.; Paige, J.; Pennington, T.D.; Robson, N.K.B.; et al. Herbaria Are a Major Frontier for Species Discovery. *Proc. Natl. Acad. Sci. USA* **2010**, *107*, 22169–22171. [CrossRef] [PubMed]
29. IUCN. The IUCN Red List of Threatened Species. Version 2022-1. Available online: https://www.iucnredlist.org (accessed on 7 March 2023).
30. Larsen, B.B.; Miller, E.C.; Rhodes, M.K.; Wiens, J.J. Inordinate Fondness Multiplied and Redistributed: The Number of Species on Earth and the New Pie of Life. *Q. Rev. Biol.* **2017**, *92*, 229–265. [CrossRef]
31. Stuessy, T.F. *Plant Taxonomy: The Systematic Evaluation of Comparative Data*, 2nd ed.; Columbia University Press: New York, NY, USA, 2009; ISBN 0231518641.
32. Mallet, J. A Species Definition for the Modern Synthesis. *Trends Ecol. Evol.* **1995**, *10*, 294–299. [CrossRef]
33. De Queiroz, K. Species Concepts and Species Delimitation. *Syst. Biol.* **2007**, *56*, 879–886. [CrossRef]
34. Sukumaran, J.; Knowles, L.L. Multispecies Coalescent Delimits Structure, Not Species. *Proc. Natl. Acad. Sci. USA* **2017**, *114*, 1607–1612. [CrossRef]
35. Sukumaran, J.; Holder, M.T.; Knowles, L.L. Incorporating the Speciation Process into Species Delimitation. *PLOS Comput. Biol.* **2021**, *17*, e1008924. [CrossRef]
36. Ott, T.; Schall, M.; Vogt, R.; Oberprieler, C. The Warps and Wefts of a Polyploidy Complex: Integrative Species Delimitation of the Diploid *Leucanthemum* (Compositae, Anthemideae) Representatives. *Plants* **2022**, *11*, 1878. [CrossRef] [PubMed]
37. Freudenstein, J.V.; Broe, M.B.; Folk, R.A.; Sinn, B.T. Biodiversity and the Species Concept—Lineages Are Not Enough. *Syst. Biol.* **2017**, *66*, 644–656. [CrossRef] [PubMed]
38. Hörandl, E. The Classification of Asexual Organisms: Old Myths, New Facts, and a Novel Pluralistic Approach. *Taxon* **2018**, *67*, 1066–1081. [CrossRef]
39. Karbstein, K.; Tomasello, S.; Hodač, L.; Dunkel, F.G.; Daubert, M.; Hörandl, E. Phylogenomics Supported by Geometric Morphometrics Reveals Delimitation of Sexual Species within the Polyploid Apomictic *Ranunculus auricomus* Complex (Ranunculaceae). *Taxon* **2020**, *69*, 1191–1220. [CrossRef]
40. Vences, M.; Miralles, A.; Brouillet, S.; Ducasse, J.; Fedosov, A.; Kharchev, V.; Kostadinov, I.; Kumari, S.; Patmanidis, S.; Scherz, M.D.; et al. iTaxoTools 0.1: Kickstarting a Specimen-Based Software Toolkit for Taxonomists. *Megataxa* **2021**, *6*, 77–92. [CrossRef]
41. Irisarri, I.; Darienko, T.; Pröschold, T.; Fürst-Jansen, J.M.R.; Jamy, M.; De Vries, J. Unexpected Cryptic Species among Streptophyte Algae Most Distant to Land Plants. *Proc. R. Soc. B Biol. Sci.* **2021**, *288*, 20212168. [CrossRef]
42. Chambers, E.A.; Hillis, D.M. The Multispecies Coalescent Over-Splits Species in the Case of Geographically Widespread Taxa. *Syst. Biol.* **2020**, *69*, 184–193. [CrossRef]
43. Mallet, J. Hybridization, Ecological Races and the Nature of Species: Empirical Evidence for the Ease of Speciation. *Philos. Trans. R. Soc. B Biol. Sci.* **2008**, *363*, 2971–2986. [CrossRef]

44. Schlick-Steiner, B.C.; Steiner, F.M.; Seifert, B.; Stauffer, C.; Christian, E.; Crozier, R.H. Integrative Taxonomy: A Multisource Approach to Exploring Biodiversity. *Annu. Rev. Entomol.* **2010**, *55*, 421–438. [CrossRef]
45. Pinheiro, F.; Dantas-Queiroz, M.V.; Palma-Silva, C. Plant Species Complexes as Models to Understand Speciation and Evolution: A Review of South American Studies. *CRC Crit. Rev. Plant Sci.* **2018**, *37*, 54–80. [CrossRef]
46. Dayrat, B. Towards Integrative Taxonomy. *Biol. J. Linn. Soc.* **2005**, *85*, 407–415. [CrossRef]
47. Oberprieler, C. The Wettstein Tesseract: A Tool for Conceptualising Species-rank Decisions and Illustrating Speciation Trajectories. *Taxon* **2023**, *72*, 1–7. [CrossRef]
48. Ennos, R.; French, G.; Hollingsworth, P. Conserving Taxonomic Complexity. *Trends Ecol. Evol.* **2005**, *20*, 164–168. [CrossRef] [PubMed]
49. Federici, S.; Galimberti, A.; Bartolucci, F.; Bruni, I.; De mattia, F.; Cortis, P.; Labra, M. DNA Barcoding to Analyse Taxonomically Complex Groups in Plants: The Case of *Thymus* (Lamiaceae). *Bot. J. Linn. Soc.* **2013**, *171*, 687–699. [CrossRef]
50. Karbstein, K. Untying Gordian Knots: The Evolution and Biogeography of the Large European Apomictic Polyploid *Ranunculus auricomus* Plant Complex. Ph.D. Thesis, University of Göttingen, Göttingen, Germany, 2021.
51. Karbstein, K.; Tomasello, S.; Hodač, L.; Wagner, N.; Marinček, P.; Barke, B.H.; Paetzold, C.; Hörandl, E. Untying Gordian Knots: Unraveling Reticulate Polyploid Plant Evolution by Genomic Data Using the Large *Ranunculus auricomus* Species Complex. *New Phytol.* **2022**, *235*, 2081–2098. [CrossRef] [PubMed]
52. Sochor, M.; Vašut, R.J.; Sharbel, T.F.; Trávníček, B. How Just a Few Makes a Lot: Speciation via Reticulation and Apomixis on Example of European Brambles (*Rubus* Subgen. *Rubus*, Rosaceae). *Mol. Phylogenet. Evol.* **2015**, *89*, 13–27. [CrossRef]
53. Wagner, F.; Ott, T.; Zimmer, C.; Reichhart, V.; Vogt, R.; Oberprieler, C. 'At the Crossroads towards Polyploidy': Genomic Divergence and Extent of Homoploid Hybridization Are Drivers for the Formation of the Ox-Eye Daisy Polyploid Complex (*Leucanthemum*, Compositae-Anthemideae). *New Phytol.* **2019**, *223*, 2039–2053. [CrossRef]
54. Burgess, M.B.; Cushman, K.R.; Doucette, E.T.; Frye, C.T.; Campbell, C.S. Understanding Diploid Diversity: A First Step in Unraveling Polyploid, Apomictic Complexity in *Amelanchier*. *Am. J. Bot.* **2015**, *102*, 2041–2057. [CrossRef]
55. Chen, Z.J. Genetic and Epigenetic Mechanisms for Gene Expression and Phenotypic Variation in Plant Polyploids. *Annu. Rev. Plant Biol.* **2007**, *58*, 377–406. [CrossRef]
56. Wendel, J.F. The Wondrous Cycles of Polyploidy in Plants. *Am. J. Bot.* **2015**, *102*, 1753–1756. [CrossRef]
57. Spoelhof, J.P.; Soltis, P.S.; Soltis, D.E. Pure Polyploidy: Closing the Gaps in Autopolyploid Research. *J. Syst. Evol.* **2017**, *55*, 340–352. [CrossRef]
58. Liu, B.; Wendel, J.F. Epigenetic Phenomena and the Evolution of Plant Allopolyploids. *Mol. Phylogenet. Evol.* **2003**, *29*, 365–379. [CrossRef]
59. Abbott, R.; Albach, D.; Ansell, S.; Arntzen, J.W.; Baird, S.J.E.; Bierne, N.; Boughman, J.; Brelsford, A.; Buerkle, C.A.; Buggs, R.; et al. Hybridization and Speciation. *J. Evol. Biol.* **2013**, *26*, 229–246. [CrossRef] [PubMed]
60. Fox, D.T.; Soltis, D.E.; Soltis, P.S.; Ashman, T.L.; Van de Peer, Y. Polyploidy: A Biological Force from Cells to Ecosystems. *Trends Cell Biol.* **2020**, *30*, 688–694. [CrossRef] [PubMed]
61. Barker, M.S.; Arrigo, N.; Baniaga, A.E.; Li, Z.; Levin, D.A. On the Relative Abundance of Autopolyploids and Allopolyploids. *New Phytol.* **2016**, *210*, 391–398. [CrossRef] [PubMed]
62. Hörandl, E. Morphological Differentiation within the *Ranunculus cassubicus* Group Compared to Variation of Isozymes, Ploidy Levels, and Reproductive Systems: Implications for Taxonomy. *Plant Syst. Evol.* **2002**, *233*, 65–78. [CrossRef]
63. Turland, N.J.; Wiersema, J.H.; Barrie, F.R.; Greuter, W.; Hawksworth, D.L.; Herendeen, P.S.; Knapp, S.; Kusber, W.-H.; Li, D.-Z.; Marhold, K.; et al. (Eds.) *International Code of Nomenclature for Algae, Fungi, and Plants (Shenzhen Code) Adopted by the Nineteenth International Botanical Congress Shenzhen, China, July 2017*; Regnum Vegetabile 159; Koeltz Botanical Books: Glashütten, Germany, 2018; Volume 159, ISBN 9783946583165.
64. Hörandl, E.; Greilhuber, J.; Klímová, K.; Paun, O.; Temsch, E.; Emadzade, K.; Hodálová, I. Reticulate Evolution and Taxonomic Concepts in the *Ranunculus auricomus* Complex (Ranunculaceae): Insights from Analysis of Morphological, Karyological and Molecular Data. *Taxon* **2009**, *58*, 1194–1215. [CrossRef]
65. Beck, J.B.; Alexander, P.J.; Allphin, L.; Al-Shehbaz, I.A.; Rushworth, C.; Bailey, C.D.; Windham, M.D. Does Hybridization Drive the Transition to Asexuality in Diploid Boechera? *Evolution* **2012**, *66*, 985–995. [CrossRef]
66. Cushman, K.R.; Burgess, M.B.; Doucette, E.T.; Nelson, G.A.; Campbell, C.S. Species Delimitation in Tetraploid, Apomictic *Amelanchier* (Rosaceae). *Syst. Bot.* **2017**, *42*, 234–256. [CrossRef]
67. Rothfels, C.J. Polyploid Phylogenetics. *New Phytol.* **2021**, *230*, 66–72. [CrossRef]
68. Gehrke, B.; Bräuchler, C.; Romoleroux, K.; Lundberg, M.; Heubl, G.; Eriksson, T. Molecular Phylogenetics of *Alchemilla*, *Aphanes* and *Lachemilla* (Rosaceae) Inferred from Plastid and Nuclear Intron and Spacer DNA Sequences, with Comments on Generic Classification. *Mol. Phylogenet. Evol.* **2008**, *47*, 1030–1044. [CrossRef]
69. Wagner, N.D.; He, L.; Hörandl, E. Phylogenomic Relationships and Evolution of Polyploid Salix Species Revealed by RAD Sequencing Data. *Front. Plant Sci.* **2020**, *11*, 36–41. [CrossRef]
70. Carter, K.A.; Liston, A.; Bassil, N.V.; Alice, L.A.; Bushakra, J.M.; Sutherland, B.L.; Mockler, T.C.; Bryant, D.W.; Hummer, K.E. Target Capture Sequencing Unravels Rubus Evolution. *Front. Plant Sci.* **2019**, *10*, 1615. [CrossRef] [PubMed]

71. Wagner, F.; Ott, T.; Schall, M.; Lautenschlager, U.; Vogt, R.; Oberprieler, C. Taming the Red Bastards: Hybridisation and Species Delimitation in the *Rhodanthemum arundanum*-Group (Compositae, Anthemideae). *Mol. Phylogenet. Evol.* **2020**, *144*, 106702. [CrossRef]
72. Pätzold, C.; Wood, K.R.; Eaton, D.A.R.; Wagner, W.L.; Appelhans, M.S. Phylogeny of Hawaiian *Melicope* (Rutaceae): RAD-Seq Resolves Species Relationships and Reveals Ancient Introgression. *Front. Plant Sci.* **2019**, *10*, 1074. [CrossRef] [PubMed]
73. Šlenker, M.; Kantor, A.; Marhold, K.; Schmickl, R.; Mandáková, T.; Lysak, M.A.; Perný, M.; Caboňová, M.; Slovák, M.; Zozomová-Lihová, J. Allele Sorting as a Novel Approach to Resolving the Origin of Allotetraploids Using Hyb-Seq Data: A Case Study of the Balkan Mountain Endemic *Cardamine barbaraeoides*. *Front. Plant Sci.* **2021**, *12*, 659275. [CrossRef]
74. Gulyaev, S.; Cai, X.-J.; Guo, F.-Y.; Kikuchi, S.; Applequist, W.L.; Zhang, Z.-X.; Hörandl, E.; He, L. The Phylogeny of Salix Revealed by Whole Genome Re-Sequencing Suggests Different Sex-Determination Systems in Major Groups of the Genus. *Ann. Bot.* **2022**, *129*, 485–498. [CrossRef] [PubMed]
75. Stuessy, T.F.; Mayer, V.; Hörandl, E. (Eds.) *Deep Morphology: Toward a Renaissance of Morphology in Plant Systematics*; Koeltz Botanical Books; Gantner: Ruggel, Liechtenstein, 2003; ISBN 3-906166-07-4.
76. Jalas, J. Thymus L. In *Flora Europaea, 3*; Tutin, T.G., Heywood, V.H., Burges, N.A., Moore, D.M., Valentine, D.H., Eds.; Cambridge University Press: Cambridge, UK, 1972; pp. 172–182.
77. Morales, R. The History, Botany and Taxonomy of the Genus *Thymus*. In *Thyme—The genus Thymus*; Stahl-Biskup, E., Sáez, F., Eds.; Taylor & Francis: London, UK; New York, NY, USA, 2002; pp. 1–43. ISBN 978-0-203-29389-8.
78. Stace, C.A. Species Recognition in Agamosperms—The Need for a Pragmatic Approach. *Folia Geobot.* **1998**, *33*, 319–326. [CrossRef]
79. Haveman, R. Freakish Patterns—Species and Species Concepts in Apomicts. *Nord. J. Bot.* **2013**, *31*, 257–269. [CrossRef]
80. Scoppola, A.; Angeloni, D.; Franceschini, C. Classical Morphometrics in *V. arvensis* and *V. kitaibeliana* (Viola Sect. Melanium) Reveals Intraspecific Variation with Implications for Species Delimitation: Inferences from a Case Study in Central Italy. *Plants* **2022**, *11*, 379. [CrossRef] [PubMed]
81. Gratani, L. Plant Phenotypic Plasticity in Response to Environmental Factors. *Adv. Bot.* **2014**, *2014*, 208747. [CrossRef]
82. Karbstein, K.; Tomasello, S.; Prinz, K. Desert-like Badlands and Surrounding (Semi-)Dry Grasslands of Central Germany Promote Small-scale Phenotypic and Genetic Differentiation in *Thymus* praecox. *Ecol. Evol.* **2019**, *9*, 14066–14084. [CrossRef] [PubMed]
83. Klingenberg, C. Analyzing Fluctuating Asymmetry with Geometric Morphometrics: Concepts, Methods, and Applications. *Symmetry* **2015**, *7*, 843–934. [CrossRef]
84. Karbstein, K.; Gockel, S.; Frischbier, N.; Kahlert, K.; Konnert, M.; Profft, I. "High-Altitude Spruces" in Central Europe—A Summarizing Contribution to Phenotypic and (Epi)Genetic Differentiation within *Picea abies* (L.) H.KARST. *Allg. Forst- und Jagdzeitschrift* **2021**, *191*, 197–211. [CrossRef]
85. Wallace, M.; Bonhomme, V.; Russell, J.; Stillman, E.; George, T.S.; Ramsay, L.; Wishart, J.; Timpany, S.; Bull, H.; Booth, A.; et al. Searching for the Origins of Bere Barley: A Geometric Morphometric Approach to Cereal Landrace Recognition in Archaeology. *J. Archaeol. Method Theory* **2019**, *26*, 1125–1142. [CrossRef]
86. Cardini, A. Geometric Morphometrics. In *Encyclopedia of life support systems (EOLSS), Developed under the Auspices of the UNESCO*; Eolss Publishers: Paris, France, 2013; Available online: http://www.eolss.net (accessed on 31 January 2023).
87. Hodač, L.; Scheben, A.P.; Hojsgaard, D.; Paun, O.; Hörand, E. ITS Polymorphisms Shed Light on Hybrid Evolution in Apomictic Plants: A Case Study on the *Ranunculus auricomus* Complex. *PLoS ONE* **2014**, *9*, 28–30. [CrossRef]
88. Hodač, L.; Barke, B.H.; Hörandl, E. Mendelian Segregation of Leaf Phenotypes in Experimental F2 Hybrids Elucidates Origin of Morphological Diversity of the Apomictic *Ranunculus auricomus* Complex. *Taxon* **2018**, *67*, 1082–1092. [CrossRef]
89. Bookstein, F.L. *Morphometric Tools for Landmark Data: Geometry and Biology*; Cambridge University Press: Cambridge, UK, 1997; ISBN 0521585988.
90. Rohlf, F.J. Morphometric Spaces, Shape Components and the Effects of Linear Transformations. In *Advances in Morphometrics*; Springer: Boston, MA, USA, 1996; pp. 117–129.
91. Cooke, S.B.; Terhune, C.E. Form, Function, and Geometric Morphometrics. *Anat. Rec.* **2015**, *298*, 5–28. [CrossRef]
92. Jensen, R.J.; Ciofani, K.M.; Miramontes, L.C. Lines, Outlines, and Landmarks: Morphometric Analyses of Leaves of *Acer rubrum*, *Acer saccharinum* (Aceraceae) and Their Hybrid. *Taxon* **2002**, *51*, 475–492. [CrossRef]
93. van der Niet, T.; Zollikofer, C.P.E.; de León, M.S.P.; Johnson, S.D.; Linder, H.P. Three-Dimensional Geometric Morphometrics for Studying Floral Shape Variation. *Trends Plant Sci.* **2010**, *15*, 423–426. [CrossRef] [PubMed]
94. Leménager, M.; Burkiewicz, J.; Schoen, D.J.; Joly, S. Studying Flowers 3D Using Photogrammetry. *New Phytol.* **2022**, *237*, 1922–1933. [CrossRef]
95. Klingenberg, C.P. Evolution and Development of Shape: Integrating Quantitative Approaches. *Nat. Rev. Genet.* **2010**, *11*, 623–635. [CrossRef] [PubMed]
96. Viscosi, V. Geometric Morphometrics and Leaf Phenotypic Plasticity: Assessing Fluctuating Asymmetry and Allometry in European White Oaks (*Quercus*). *Bot. J. Linn. Soc.* **2015**, *179*, 335–348. [CrossRef]
97. Viscosi, V.; Fortini, P.; Slice, D.E.; Loy, A.; Blasi, C. Geometric Morphometric Analyses of Leaf Variation in Four Oak Species of the Subgenus *Quercus* (Fagaceae). *Plant Biosyst.* **2009**, *143*, 575–587. [CrossRef]
98. Stojnić, S.; Viscosi, V.; Marković, M.; Ivanković, M.; Orlović, S.; Tognetti, R.; Cocozza, C.; Vasić, V.; Loy, A. Spatial Patterns of Leaf Shape Variation in European Beech (*Fagus sylvatica* L.) Provenances. *Trees* **2022**, *36*, 497–511. [CrossRef]

99. Kilian, N.; Henning, T.; Plitzner, P.; Müller, A.; Güntsch, A.; Stöver, B.C.; Müller, K.F.; Berendsohn, W.G.; Borsch, T. Sample Data Processing in an Additive and Reproducible Taxonomic Workflow by Using Character Data Persistently Linked to Preserved Individual Specimens. *Database* **2015**, *2015*, bav094. [CrossRef]
100. Piedra-Malagón, E.M.; Albarrán-Lara, A.L.; Rull, J.; Piñero, D.; Sosa, V. Using Multiple Sources of Characters to Delimit Species in the Genus *Crataegus* (Rosaceae): The Case of the *Crataegus rosei* Complex. *Syst. Biodivers.* **2016**, *14*, 244–260. [CrossRef]
101. Lexer, C.; Joseph, J.; Van Loo, M.; Prenner, G.; Heinze, B.; Chase, M.W.; Kirkup, D. The Use of Digital Image-Based Morphometrics to Study the Phenotypic Mosaic in Taxa with Porous Genomes. *Taxon* **2009**, *58*, 349–364. [CrossRef]
102. Thompson, K.A.; Urquhart-Cronish, M.; Whitney, K.D.; Rieseberg, L.H.; Schluter, D. Patterns, Predictors, and Consequences of Dominance in Hybrids. *Am. Nat.* **2021**, *197*, E72–E88. [CrossRef]
103. Andrews, K.R.; Good, J.M.; Miller, M.R.; Luikart, G.; Hohenlohe, P.A. Harnessing the Power of RADseq for Ecological and Evolutionary Genomics. *Nat. Rev. Genet.* **2016**, *17*, 81–92. [CrossRef] [PubMed]
104. McKain, M.R.; Johnson, M.G.; Uribe-Convers, S.; Eaton, D.; Yang, Y. Practical Considerations for Plant Phylogenomics. *Appl. Plant Sci.* **2018**, *6*, e1038. [CrossRef]
105. Dauphin, B.; Grant, J.R.; Farrar, D.R.; Rothfels, C.J. Rapid Allopolyploid Radiation of Moonwort Ferns (Botrychium; Ophioglossaceae) Revealed by PacBio Sequencing of Homologous and Homeologous Nuclear Regions. *Mol. Phylogenet. Evol.* **2018**, *120*, 342–353. [CrossRef]
106. Eriksson, J.S.; de Sousa, F.; Bertrand, Y.J.K.; Antonelli, A.; Oxelman, B.; Pfeil, B.E. Allele Phasing Is Critical to Revealing a Shared Allopolyploid Origin of *Medicago arborea* and *M. strasseri* (Fabaceae). *BMC Evol. Biol.* **2018**, *18*, 9. [CrossRef]
107. Lautenschlager, U.; Wagner, F.; Oberprieler, C. AllCoPol: Inferring Allele Co-Ancestry in Polyploids. *BMC Bioinform.* **2020**, *21*, 441. [CrossRef] [PubMed]
108. Rannala, B.; Yang, Z. Bayes Estimation of Species Divergence Times and Ancestral Population Sizes Using DNA Sequences from Multiple Loci. *Genetics* **2003**, *164*, 1645–1656. [CrossRef]
109. Yang, Z.; Rannala, B. Unguided Species Delimitation Using DNA Sequence Data from Multiple Loci. *Mol. Biol. Evol.* **2014**, *31*, 3125–3135. [CrossRef]
110. Weitemier, K.; Straub, S.C.K.; Cronn, R.C.; Fishbein, M.; Schmickl, R.; McDonnell, A.; Liston, A. Hyb-Seq: Combining Target Enrichment and Genome Skimming for Plant Phylogenomics. *Appl. Plant Sci.* **2014**, *2*, 1400042. [CrossRef]
111. Folk, R.A.; Mandel, J.R.; Freudenstein, J.V. A Protocol for Targeted Enrichment of Intron-Containing Sequence Markers for Recent Radiations: A Phylogenomic Example from *Heuchera* (Saxifragaceae). *Appl. Plant Sci.* **2015**, *3*, 1500039. [CrossRef] [PubMed]
112. Stull, G.W.; Soltis, P.S.; Soltis, D.E.; Gitzendanner, M.A.; Smith, S.A. Nuclear Phylogenomic Analyses of Asterids Conflict with Plastome Trees and Support Novel Relationships among Major Lineages. *Am. J. Bot.* **2020**, *107*, 790–805. [CrossRef] [PubMed]
113. Huang, D.I.; Hefer, C.A.; Kolosova, N.; Douglas, C.J.; Cronk, Q.C.B. Whole Plastome Sequencing Reveals Deep Plastid Divergence and Cytonuclear Discordance between Closely Related Balsam Poplars, *Populus balsamifera* and *P. trichocarpa* (Salicaceae). *N. Phytol.* **2014**, *204*, 693–703. [CrossRef] [PubMed]
114. Paetzold, C.; Barke, B.H.; Hörandl, E. Evolution of Transcriptomes in Early-Generation Hybrids of the Apomictic *Ranunculus auricomus* Complex (Ranunculaceae). *Int. J. Mol. Sci.* **2022**, *23*, 13881. [CrossRef] [PubMed]
115. Paun, O.; Stuessy, T.F.; Hörandl, E. The Role of Hybridization, Polyploidization and Glaciation in the Origin and Evolution of the Apomictic *Ranunculus cassubicus* Complex. *New Phytol.* **2006**, *171*, 223–236. [CrossRef] [PubMed]
116. Emadzade, K.; Gehrke, B.; Peter Linder, H.; Hörandl, E. The Biogeographical History of the Cosmopolitan *Genus ranunculus* L. (Ranunculaceae) in the Temperate to Meridional Zones. *Mol. Phylogenet. Evol.* **2011**, *58*, 4–21. [CrossRef]
117. Emadzade, K.; Lebmann, M.J.; Hoffmann, M.H.; Tkach, N.; Lone, F.A.; Hörandl, E. Phylogenetic Relationships and Evolution of High Mountain Buttercups (*Ranunculus*) in North America and Central Asia. *Perspect. Plant Ecol. Evol. Syst.* **2015**, *17*, 131–141. [CrossRef]
118. Jalas, J.; Suominen, J. *Atlas Florae Europaeae, Vol. 8, Nymphaeaceae to Ranunculaceae*; The committee for mapping the flora of Europe and Societas Biologica Fennica Vanamo: Helsinki, Finland, 1989.
119. GBIF. Secretariat 2017. *Ranunculus auricomus* L; GBIF Backbone Taxonomy. Checklist Dataset. Available online: https://doi.org/10.15468/39omei (accessed on 21 November 2018).
120. Melzheimer, V.; Hörandl, E. *Die Ranunculaceae der Flora von Zentraleuropa: Ranunculus sect. Auricomus*; Universitätsbibliothek Johann Christian Senckenberg: Frankfurt, Germany, 2022.
121. Borchers-Kolb, E. *Ranunculus* sect. *Auricomus* in Bayern und den angrenzenden Gebieten II. Spezieller Teil. *Mitt. Bot. München* **1985**, *21*, 49–300.
122. Marklund, G. Der *Ranunculus auricomus*-Komplex in Finnland I. Diagnosen und Fundortslisten einiger Sippen des *R. auricomus* L. Coll. (*s. str.*). *Flora Fenn.* **1961**, *3*, 1–128.
123. Jalas, J.; Kvist, G. Atlas Florae Europaeae Notes. 7. Taxa of the *Ranunculus auricomus* Group Recognized as Species by Marklund in 1940. *Ann. Bot. Fenn.* **1987**, *12*, 221–223.
124. Hörandl, E.; Raab-Straube, E. Ranunculaceae. Available online: http://ww2.bgbm.org/EuroPlusMed (accessed on 7 March 2023).
125. Leslie, A.C. Ranunculaceae. Section 2. *Auricomus*. In *Flora of Great Britain and Ireland*; Sell, P., Murrell, G., Eds.; Cambridge University Press: Cambridge, UK, 2018; pp. 671–677.
126. Julin, E. *Ranunculus auricomus* L. in Södermanland, East-Central Sweden. *Opera Bot.* **1980**, *57*, 1–145.

127. Hörandl, E.; Gutermann, W. Der *Ranunculus auricomus*-Komplex in Österreich 2. Die *R. cassubicus-*, *R. monophyllus-* und *R. fallax*-Sammelgruppe. *Bot. Jahrb. Syst.* **1998**, *120*, 545–598.
128. Dunkel, F.G.; Gregor, T.; Paule, J. New Diploid Species in the *Ranunculus auricomus* Complex (Ranunculaceae) from W and SE Europe. *Willdenowia* **2018**, *48*, 227–257. [CrossRef]
129. Dunkel, F.G. Contribution to the Knowledge of the *Ranunculus auricomus* Complex (Ranunculaceae) in Spain. *Stapfia* **2021**, *112*, 5–59.
130. Linnaeus, C. *Species Plantarum*; Holmiae: Uppsala, Sweden, 1753.
131. Kvist, G. Identity of the Material of the *Ranunculus auricomus* Group in the Linnaean Herbarium. *Ann. Bot. Fenn.* **1987**, *24*, 73–77.
132. Sloboda, D. *Rostlinnictví: Čili návod k snadnému určení a pojmenování rostlin v Čechách, Moravě a jiných zemích Rakouského Mocnářství domácích*; V kommissí u Fr. Rivnáce: Prague, Bohemia (Austrian Empire), 1852.
133. Ovchinnikov, P. *Ranunculus monophyllus* Ovcz. *in Bot. Mater. Gerb. Glavn. Bot. Sada RSFSR.* **1922**, *3*, 54. Available online: https://www.europlusmed.org/cdm_dataportal/taxon/c2da9f70-7f7d-4320-95bd-425566191fa8 (accessed on 31 January 2023).
134. Marklund, G. Der *Ranunculus auricomus*-Komplex in Finnland II: Diagnosen und Fundortslisten einiger Sippen von *R. fallax* (W. et Gr.) Schur, *R. monophyllus* Ovcz. und *R. cassubicus* L. *Flora Fenn.* **1965**, *4*, 1–198.
135. Soó, R. Die *Ranunculus auricomus* L. Emend. Korsh. Artengruppe in der Flora Ungarns und der Karpaten I. *Acta Bot. Acad. Sci. Hung* **1964**, *10*, 221–237.
136. Schwarz, O. Beiträge zur Kenntnis kritischer Formenkreise im Gebiet der Flora von Thüringen. *Mitteilungen der Thüringischen Bot. Gesellschaft* **1949**, *1*, 120–143.
137. Hörandl, E.; Gutermann, W. Zur Kenntnis des *Ranunculus auricomus*-Komplexes in Österreich: Die Arten der *R. phragmiteti-* und *R. indecorus*-Gruppe. *Phyton* **1998**, *37*, 263–320.
138. Hörandl, E.; Gutermann, W. Der *Ranunculus auricomus*-Komplex in Österreich und benachbarten Gebieten 3. Die Arten der *R. latisectus-*, *R. puberulus-*, *R. stricticaulis*-und *R. argoviensis* (*R. auricomus*-Sammelgruppe). *Bot. Jahrb. Syst.* **1999**, *121*, 99–138.
139. Hörandl, E. Species Concepts in Agamic Complexes: Applications in the *Ranunculus auricomus* Complex and General Perspectives. *Folia Geobot.* **1998**, *33*, 335–348. [CrossRef]
140. Ericsson, S. The Microspecies of the *Ranunculus auricomus* Complex Treated at the Species Level. *Ann. Bot. Fenn.* **1992**, *29*, 123–158.
141. Jäger, E. Ranunculus L. In *Rothmaler Exkursionsflora von Deutschland. Gefäßpflanzen: Grundband 2*; Springer: Berlin/Heidelberg, Germany, 2017.
142. Borchers-Kolb, E. Ranunculus sect. Auricomus in Bayern und den angrenzenden Gebieten—I. Allgemeiner Teil. *Mitt. Bot. München* **1983**, *19*, 363–429.
143. Tomasello, S.; Karbstein, K.; Hodač, L.; Pätzold, C.; Hörandl, E. Phylogenomics Unravels Quaternary Vicariance and Allopatric Speciation Patterns in Temperate-montane Plant Species: A Case Study on the *Ranunculus auricomus* Species Complex. *Mol. Ecol.* **2020**, *29*, 2031–2049. [CrossRef]
144. Hörandl, E.; Gutermann, W. Die Bearbeitung der *Ranunculus auricomus*-Gruppe für die "Flora von Österreich"–Ein Werkstattbericht. *Fl. Austr. Novit* **1995**, *2*, 12–27. [CrossRef]
145. Paun, O.; Greilhuber, J.; Temsch, E.M.; Hörandl, E. Patterns, Sources and Ecological Implications of Clonal Diversity in Apomictic *Ranunculus carpaticola* (*Ranunculus auricomus* Complex, Ranunculaceae). *Mol. Ecol.* **2006**, *15*, 897–910. [CrossRef]
146. Reichenbach, L. *Flora Germanica excurso-ria [...]*; Cnobloch, C., Ed.; Carolum Cnobloch: Leipzig, Germany, 1832.
147. Grau, J. Vorläufige Übersicht der iberischen Vertreter von *Ranunculus* sect. *auricomus*. *Mitt. Bot. München* **1984**, *20*, 11–28.
148. Masci, S.; Miho, A.; Marchi, P. *Ranunculus auricomus* L. Aggr. (Ranunculaceae) in Italy. I. Sexual Tetraploids on the Apennines. *Caryologia* **1994**, *47*, 97–108. [CrossRef]
149. Rieseberg, L.H.; Willis, J.H. Plant Speciation. *Science* **2007**, *317*, 910–914. [CrossRef] [PubMed]
150. Dittrich-Reed, D.R.; Fitzpatrick, B.M. Transgressive Hybrids as Hopeful Monsters. *Evol. Biol.* **2013**, *40*, 310–315. [CrossRef]
151. Frichot, E.; Mathieu, F.; Trouillon, T.; Bouchard, G.; François, O. Fast and Efficient Estimation of Individual Ancestry Coefficients. *Genetics* **2014**, *196*, 973–983. [CrossRef]
152. Frichot, E.; François, O. LEA: An R Package for Landscape and Ecological Association Studies. *Methods Ecol. Evol.* **2015**, *6*, 925–929. [CrossRef]
153. Malinsky, M.; Trucchi, E.; Lawson, D.J.; Falush, D. RADpainter and FineRADstructure: Population Inference from RADseq Data. *Mol. Biol. Evol.* **2018**, *35*, 1284–1290. [CrossRef]
154. Jones, G. Algorithmic Improvements to Species Delimitation and Phylogeny Estimation under the Multispecies Coalescent. *J. Math. Biol.* **2017**, *74*, 447–467. [CrossRef] [PubMed]
155. Kozlov, A.M.; Darriba, D.; Flouri, T.; Morel, B.; Stamatakis, A. RAxML-NG: A Fast, Scalable and User-Friendly Tool for Maximum Likelihood Phylogenetic Inference. *Bioinformatics* **2019**, *35*, 4453–4455. [CrossRef] [PubMed]
156. Fick, S.E.; Hijmans, R.J. WorldClim 2: New 1-Km Spatial Resolution Climate Surfaces for Global Land Areas. *Int. J. Climatol.* **2017**, *37*, 4302–4315. [CrossRef]
157. Dormann, C.F.; Elith, J.; Bacher, S.; Buchmann, C.; Carl, G.; Carré, G.; Marquéz, J.R.G.; Gruber, B.; Lafourcade, B.; Leitão, P.J.; et al. Collinearity: A Review of Methods to Deal with It and a Simulation Study Evaluating Their Performance. *Ecography* **2013**, *36*, 27–46. [CrossRef]
158. Taiyun, W.; Simko, V. Corrplot—Visualization of a Correlation Matrix. 2022. Available online: https://cran.r-project.org/web/packages/corrplot/corrplot (accessed on 31 January 2023).

159. R Core Team. *R: A Language and Environment for Statistical Computing*; R Foundation for Statistical Computing: Vienna, Austria, 2022; Available online: http://www.r-project.org (accessed on 7 March 2023).
160. Volkova, P.A.; Kasatskaya, S.A.; Boiko, A.A.; Shipunov, A.B. Stability of Leaf Form and Size during Specimen Preparation of Herbarium Specimens. *Feddes Repert.* **2010**, *121*, 219–225. [CrossRef]
161. Lohwasser, U. Biosystematische Untersuchungen an *Ranunculus auricomus* L. (Ranunculaceae) in Deutschland. *Diss. Bot.* **2001**, *343*, 220.
162. Hörandl, E.; Gutermann, W. Der *Ranunculus auricomus*-Komplex in Österreich. 1. Methodik. Gruppierung der mitteleuropäischen Sippen. *Bot. Jahrb. Syst.* **1998**, *120*, 1–44.
163. Rohlf, F.J. The Tps Series of Software. *Hystrix* **2015**, *26*, 9–12. [CrossRef]
164. Klingenberg, C.P. MorphoJ: An Integrated Software Package for Geometric Morphometrics. *Mol. Ecol. Resour.* **2011**, *11*, 353–357. [CrossRef]
165. Hammer, Ø.; Harper, D.A.T.; Ryan, P.D. PAST: Paleontological Statistics Software Package for Education and Data Analysis. *Palaeontol. Electron.* **2001**, *4*, 9. [CrossRef]
166. Karbstein, K.; Prinz, K.; Hellwig, F.; Römermann, C. Plant Intraspecific Functional Trait Variation Is Related to Within-habitat Heterogeneity and Genetic Diversity in *Trifolium montanum* L. *Ecol. Evol.* **2020**, *10*, 5015–5033. [CrossRef] [PubMed]
167. Goslee, S.C.; Urban, D.L. The Ecodist Package for Dissimilarity-Based Analysis of Ecological Data. *J. Stat. Softw.* **2007**, *22*, 1–19. [CrossRef]
168. Maddison, W.P. Squared-Change Parsimony Reconstructions of Ancestral States for Continuous-Valued Characters on a Phylogenetic Tree. *Syst. Biol.* **1991**, *40*, 304–314. [CrossRef]
169. Klingenberg, C.P.; Gidaszewski, N.A. Testing and Quantifying Phylogenetic Signals and Homoplasy in Morphometric Data. *Syst. Biol.* **2010**, *59*, 245–261. [CrossRef] [PubMed]
170. McInnes, L.; Healy, J.; Melville, J. UMAP: Uniform Manifold Approximation and Projection for Dimension Reduction. *arXiv* **2018**, arXiv:1802.03426.
171. Hörandl, E.; Jakubowsky, G.; Dobeš, C. Isozyme and Morphological Diversity within Apomictic and Sexual Taxa of the *Ranunculus auricomus* Complex. *Plant Syst. Evol.* **2001**, *226*, 165–185. [CrossRef]
172. Jiang, X.; Hipp, A.L.; Deng, M.; Su, T.; Zhou, Z.; Yan, M. East Asian Origins of European Holly Oaks (*Quercus* Section Ilex Loudon) via the Tibet-Himalaya. *J. Biogeogr.* **2019**, *46*, 2203–2214. [CrossRef]
173. Kirschner, J.; Záveská Drábková, L.; Štěpánek, J.; Uhlemann, I. Towards a Better Understanding of the *Taraxacum* Evolution (Compositae–Cichorieae) on the Basis of nrDNA of Sexually Reproducing Species. *Plant Syst. Evol.* **2015**, *301*, 1135–1156. [CrossRef]
174. Alger, E.I.; Edger, P.P. One Subgenome to Rule Them All: Underlying Mechanisms of Subgenome Dominance. *Curr. Opin. Plant Biol.* **2020**, *54*, 108–113. [CrossRef]
175. Ericsson, S. Microspecies within the *Ranunculus auricomus* Complex. In *Flora Nordica 2*; Jonsell, B., Ed.; Bergius Foundation: Stockholm, Sweden, 2001; pp. 382–397.
176. Kürbs, S. Vergleichend-Entwicklungsgeschichtliche Studien an Ranunculaceaen-Fiederblättern. Teil I, II. *Bot. Jahrb. Syst.* **1973**, *93*, 130–167.
177. Grant, V. *Plant Speciation*; Columbia University Press: New York, NY, USA, 1981.
178. Hörandl, E. The Evolution of Self-Fertility in Apomictic Plants. *Sex. Plant Reprod.* **2010**, *23*, 73–86. [CrossRef] [PubMed]
179. Hörandl, E.; Greilhuber, J. Diploid and Autotetraploid Sexuals and Their Relationships to Apomicts in the *Ranunculus cassubicus* Group: Insights from DNA Content and Isozyme Variation. *Plant Syst. Evol.* **2002**, *234*, 85–100. [CrossRef]
180. Sailer, C.; Schmid, B.; Grossniklaus, U. Apomixis Allows the Transgenerational Fixation of Phenotypes in Hybrid Plants. *Curr. Biol.* **2016**, *26*, 331–337. [CrossRef]
181. Migicovsky, Z.; Li, M.; Chitwood, D.H.; Myles, S. Morphometrics Reveals Complex and Heritable Apple Leaf Shapes. *Front. Plant Sci.* **2018**, *8*, 2185. [CrossRef]
182. Tsukaya, H. Leaf Shape: Genetic Controls and Environmental Factors. *Int. J. Dev. Biol.* **2005**, *49*, 547–555. [CrossRef] [PubMed]
183. Peppe, D.J.; Royer, D.L.; Cariglino, B.; Oliver, S.Y.; Newman, S.; Leight, E.; Enikolopov, G.; Fernandez-Burgos, M.; Herrera, F.; Adams, J.M.; et al. Sensitivity of Leaf Size and Shape to Climate: Global Patterns and Paleoclimatic Applications. *New Phytol.* **2011**, *190*, 724–739. [CrossRef]
184. Wang, H.; Wang, R.; Harrison, S.P.; Prentice, I.C. Leaf Morphological Traits as Adaptations to Multiple Climate Gradients. *J. Ecol.* **2022**, *110*, 1344–1355. [CrossRef]
185. Halamski, A.T.; Dunkel, F.G.; Temsch, E.M. *Ranunculus kobendzae*, a New Species of the *R. auricomus* Complex (Ranunculaceae) from Poland. *BioOne* **2019**, *56*, 123–134. [CrossRef]
186. Hörandl, E. Geographical Parthenogenesis in Alpine and Arctic Plants. *Plants* **2023**, *12*, 844. [CrossRef]
187. Wäldchen, J.; Mäder, P. Plant Species Identification Using Computer Vision Techniques: A Systematic Literature Review. *Arch. Comput. Methods Eng.* **2018**, *25*, 507–543. [CrossRef]
188. LeCun, Y.; Bengio, Y.; Hinton, G. Deep Learning. *Nature* **2015**, *521*, 436–444. [CrossRef] [PubMed]
189. Rzanny, M.; Mäder, P.; Deggelmann, A.; Chen, M.; Wäldchen, J. Flowers, Leaves or Both? How to Obtain Suitable Images for Automated Plant Identification. *Plant Methods* **2019**, *15*, 77. [CrossRef] [PubMed]

190. Mäder, P.; Boho, D.; Rzanny, M.; Seeland, M.; Wittich, H.C.; Deggelmann, A.; Wäldchen, J. The Flora Incognita App—Interactive Plant Species Identification. *Methods Ecol. Evol.* **2021**, *12*, 1335–1342. [CrossRef]
191. Weaver, W.N.; Ng, J.; Laport, R.G. LeafMachine: Using Machine Learning to Automate Leaf Trait Extraction from Digitized Herbarium Specimens. *Appl. Plant Sci.* **2020**, *8*, e11367. [CrossRef] [PubMed]
192. Katal, N.; Rzanny, M.; Mäder, P.; Wäldchen, J. Deep Learning in Plant Phenological Research: A Systematic Literature Review. *Front. Plant Sci.* **2022**, *13*, 805738. [CrossRef]
193. Cintas, C.; Quinto-Sánchez, M.; Acuña, V.; Paschetta, C.; de Azevedo, S.; Cesar Silva de Cerqueira, C.; Ramallo, V.; Gallo, C.; Poletti, G.; Bortolini, M.C.; et al. Automatic Ear Detection and Feature Extraction Using Geometric Morphometrics and Convolutional Neural Networks. *IET Biom.* **2017**, *6*, 211–223. [CrossRef]
194. Bellin, N.; Calzolari, M.; Callegari, E.; Bonilauri, P.; Grisendi, A.; Dottori, M.; Rossi, V. Geometric Morphometrics and Machine Learning as Tools for the Identification of Sibling Mosquito Species of the Maculipennis complex (*Anopheles*). *Infect. Genet. Evol.* **2021**, *95*, 105034. [CrossRef]
195. Courtenay, L.A.; Yravedra, J.; Huguet, R.; Aramendi, J.; Maté-González, M.Á.; González-Aguilera, D.; Arriaza, M.C. Combining Machine Learning Algorithms and Geometric Morphometrics: A Study of Carnivore Tooth Marks. *Palaeogeogr. Palaeoclimatol. Palaeoecol.* **2019**, *522*, 28–39. [CrossRef]
196. Le, V.-L.; Beurton-Aimar, M.; Zemmari, A.; Marie, A.; Parisey, N. Automated Landmarking for Insects Morphometric Analysis Using Deep Neural Networks. *Ecol. Inform.* **2020**, *60*, 101175. [CrossRef]
197. Rzanny, M.; Wittich, H.C.; Mäder, P.; Deggelmann, A.; Boho, D.; Wäldchen, J. Image-Based Automated Recognition of 31 Poaceae Species: The Most Relevant Perspectives. *Front. Plant Sci.* **2022**, *12*, 3315. [CrossRef] [PubMed]
198. Selvaraju, R.R.; Cogswell, M.; Das, A.; Vedantam, R.; Parikh, D.; Batra, D. Grad-CAM: Visual Explanations from Deep Networks via Gradient-Based Localization. In Proceedings of the IEEE International Conference on Computer Vision, Venice, Italy, 22–29 October 2017; pp. 618–626. [CrossRef]

biology

MDPI

Article

Apomictic Mountain Whitebeam (*Sorbus austriaca*, Rosaceae) Comprises Several Genetically and Morphologically Divergent Lineages

Alma Hajrudinović-Bogunić [1], Božo Frajman [2,*], Peter Schönswetter [2], Sonja Siljak-Yakovlev [3] and Faruk Bogunić [1,*]

1 Faculty of Forestry, University of Sarajevo, Zagrebačka 20, 71000 Sarajevo, Bosnia and Herzegovina
2 Department of Botany, University of Innsbruck, Sternwartestrasse 15, 6020 Innsbruck, Austria
3 Ecologie Systématique Evolution, CNRS, Université Paris-Sud, AgroParisTech, Université Paris-Saclay, 91400 Orsay, France
* Correspondence: bozo.frajman@uibk.ac.at (B.F.); f.bogunic@sfsa.unsa.ba (F.B.)

Simple Summary: The genus *Sorbus* (whitebeams, rowans, and service trees) encompasses forest trees and shrubs characterised by exceptional diversity resulting from the interplay of polyploidisation, hybridization, and apomixis. The spatiotemporal processes driving *Sorbus* diversification remain poorly understood. This research aims to provide insights into the evolution and diversification patterns of mountain whitebeam (*S. austriaca*) covering most of its range in the mountains of Central and South-eastern Europe. Our molecular and morphometric data revealed pronounced cryptic diversity within the *S. austriaca* complex; it is composed of different lineages, that likely originated via multiple allopolyploidisations accompanied by apomixes, and these lineages exhibit different distribution patterns. Our results are particularly valuable from a biodiversity conservation perspective due to the continuing generation of novel diversity in sympatric populations of the parental taxa. Such derived diversity requires process-oriented conservation plans and measures.

Abstract: The interplay of polyploidisation, hybridization, and apomixis contributed to the exceptional diversity of *Sorbus* (Rosaceae), giving rise to a mosaic of genetic and morphological entities. The *Sorbus austriaca* species complex from the mountains of Central and South-eastern Europe represents an allopolyploid apomictic system of populations that originated following hybridisation between *S. aria* and *S. aucuparia*. However, the mode and frequency of such allopolyploidisations and the relationships among different, morphologically more or less similar populations that have often been described as different taxa remain largely unexplored. We used amplified fragment length polymorphism (AFLP) fingerprinting, plastid DNA sequencing, and analyses of nuclear microsatellites, along with multivariate morphometrics and ploidy data, to disentangle the relationships among populations within this intricate complex. Our results revealed a mosaic of genetic lineages—many of which have not been taxonomically recognised—that originated via multiple allopolyploidisations. The clonal structure within and among populations was then maintained via apomixis. Our results thus support previous findings that hybridisation, polyploidization, and apomixis are the main drivers of *Sorbus* diversification in Europe.

Keywords: apomixis; hybridisation; multiple origins; polyploidy; *Sorbus austriaca*

Citation: Hajrudinović-Bogunić, A.; Frajman, B.; Schönswetter, P.; Siljak-Yakovlev, S.; Bogunić, F. Apomictic Mountain Whitebeam (*Sorbus austriaca*, Rosaceae) Comprises Several Genetically and Morphologically Divergent Lineages. *Biology* **2023**, *12*, 380. https://doi.org/10.3390/biology12030380

Academic Editor: Valeria Terzi

Received: 30 January 2023
Revised: 22 February 2023
Accepted: 23 February 2023
Published: 27 February 2023

1. Introduction

The evolutionary importance of polyploidisation in flowering plants as a key mechanism shaping plant diversity is widely acknowledged [1–4]. Whole-genome multiplication via auto- or allopolyploidy can reshuffle genome structure, alter gene expression, induce phenotypic and physiological changes, and provide adaptive potential to polyploid plants [5,6]. Polyploidisation is sometimes connected to the breakdown of self-incompatibility systems and allows a transition from outcrossing to asexual (uniparental) reproduction [7]. Apomixis, which is asexual reproduction involving seed formation, has proven to be an effective strategy in the evolution of certain plant groups, as it preserves and maintains hybrid and heterozygous genetic lineages, as well as cytotypes with unbalanced chromosomes, enabling their long-term persistence and dispersal [8,9]. Apomictic polyploids produce clonal offspring genetically identical to the mother plant; most, however, retain residual sexuality (facultative apomixis [10]).

In most apomicts, the combination of factors such as clonality, residual sexuality, and multiple origins affects both geographic distribution and genotypic diversity within populations [11,12]. Apomicts show a better adaptive capacity and greater colonisation ability by often occupying extreme ecological niches or disturbed habitats. Furthermore, they may have larger distribution ranges than their sexual relatives, a phenomenon known as 'geographical parthenogenesis' [9,13–16]. While some apomictic species have broad ranges, there are numerous examples of apomict species with narrow ranges, such as *Rubus* L. [17], *Sorbus* L. [18,19], or *Taraxacum* F.H. Wigg [20]. The distribution ranges of apomictic species, however, largely depend on the applied species concept, either treating single or a few apomictic populations as distinct species [19,21–26] or lumping them into more broadly distributed morphospecies [27,28]. In addition, many hybridogenous apomictic complexes originated via polytopic allopolyploidisation events, e.g., in *Crataegus* L. [29], *Potentilla* L. [30], *Rubus* [31], or *Taraxacum* [12], but also in the intricate genus *Sorbus* [32,33].

Sorbus (whitebeams, rowans, and service trees, amongst others) encompasses trees and shrubs characterized by exceptional genetic and morphological diversity resulting from the interplay of polyploidisation, hybridisation, and apomixis [34–37]. It includes about 190 species inhabiting the Northern Hemisphere [19]. Diversification of European *Sorbus* has been primarily driven by hybridisation of four widely distributed diploids, namely *S. aria* (L.) Crantz, *S. aucuparia* L., *S. chamaemespilus* (L.) Crantz, and *S. torminalis* (L.) Crantz, and backcrossing of hybrids with their parental species, which led to the formation of many allopolyploid apomictic lineages [18,34,38]. Most hybrid derivatives are tri- and tetraploids that reproduce by apomixis and are restricted to geographical areas of varying sizes; many of them have been described as distinct species [18,26,39,40]. In addition, a small portion of the polyploids in *Sorbus* originated via autopolyploidisation [33,36].

Nevertheless, most of the *Sorbus* polyploids are facultative apomicts, and only a few are obligate apomicts [37,41]. The most common scenario for the formation of novel diversity involves mixed-ploidy communities consisting of species and cytotypes with different mating systems and their interaction through interploid gamete exchange, often resulting in polyploid offspring [41]. The recurrent interaction of hybridisation and polyploidisation generates hybrid swarms of intermediate morphology that are stabilised through apomixis and represent an inexhaustible source of *Sorbus* diversity. This, however, often results in complicated and unresolved taxonomies [42] vs. [43]. Treating morphologically recognizable and apomictically reproducing triploid and tetraploid entities of monophyletic origin as distinct species has been a species concept adopted by most recent authors [18]. These species often have restricted distributions, frequently confined to single localities in which they likely originated, thus representing single clones or related clone mates [42]. Such species concepts resulted in the description of numerous apomictic taxa in different parts of Europe [18,23,24,26,44,45].

Sorbus subgen. *Soraria* Májovský and Bernátova is one of the five hybridogenous subgenera, which include diploid and polyploid species that originated from hybridisation between *Sorbus* subgen. *Sorbus* and *Sorbus* subgen. *Aria* Pers. One of its members, *Sorbus austriaca* (Beck) Hedl. (Austrian whitebeam or mountain whitebeam [46]), is a tetraploid obligate pseudogamous apomict species [37,41]. It was described from the valley Rettenbachgraben, close to the village of Prein an der Rax (Niederösterreich, Austria [47]), but based on the morphological similarity of other populations considered to be distributed in the Eastern Alps, the Balkan Peninsula, and the Carpathians [19,48], it has a relatively large range compared with most other species of subgen. *Soraria*. However, the circumscription of this species and its differentiation from other species of *Sorbus* subgen. *Soraria* is unclear, and its precise distribution is thus unknown [19,49].

Initially, Austrian whitebeam was treated as a variety or subspecies of *S. mougeotii* Soyer-Willemeet and Godron [*S. mougeotii* var. *austriaca* (G. Beck) C.K. Schneider; *S. mougeotii* subsp. *austriaca* (G. Beck) Hedl.] or even as a subspecies of *S. aria* [*S. aria* subsp. *austriaca* (G. Beck) Bornm.]. After its recognition as a species, it was split into four subspecies: *S. austriaca* subsp. *austriaca*, *S. austriaca* subsp. *croatica* Kárpáti, *S. austriaca* subsp. *hazslinszkyana* (Soó) Kárpáti, and *S. austriaca* subsp. *serpentini* Kárpáti [46]. However, the most recent taxonomic treatments do not support the recognition of subspecific taxa [19,26,42]. Consequently, Kurto et al. [19] treated subsp. *hazslinszkyana*, distributed in Slovakia and northern Hungary, as a distinct species, *S. hazslinszkyana* (Soó) Boros. Furthermore, they included *S. austriaca* subsp. *serpentine*, distributed in Eastern Austria, in the *S. aria* species complex and excluded subsp. *croatica* distributed in Croatia, as an unresolved name. The *S. austriaca* species complex thus represents an allopolyploid apomictic system of populations that originated following hybridisation between *S. aria* and *S. aucuparia* in the mountains of Central and South-Eastern Europe, but with unclear relationships.

Besides *S. austriaca*, the only morphologically similar tetraploid species with a large distribution in Europe is *S. mougeotii*, ranging from Spain to western Austria, where it is adjacent to the western distribution margin of *S. austriaca* [19,46]. In addition, other morphologically similar tetraploid species with more narrow distributions have been described from Central Europe, e.g., *S. lippertiana* N. Mey. and Meierott, endemic to Germany and Austria, with main distribution in the north-eastern Limestone Alps, but ranging also to the south-eastern Limestone Alps [40], *S. pekarovae* Májovský and Bernátová, stenoendemic to the Žilina region in Slovakia [50], and *S. pauca* M. Lepší and P. Lepší, stenoendemic to the Doksy region in Czech Republic [25]. Furthermore, similar species described outside of Central Europe include *S. anglica* Hedl. (endemic to Great Britain and Ireland, mainly south-western England and Wales [18]), *S. cuneifolia* T.C.G. Rich (stenoendemic to Great Britain, distributed in Denbighshire county, Wales [22]), and *S. subsimilis* Hedl. (endemic to South Norway [51]). However, since many of the narrow endemic species were described solely on morphological and karyological grounds, nothing is known about their phylogenetic relationships, exact circumscription, and distribution patterns.

The aim of this study is to provide insights into relationships and diversification patterns within *S. austriaca* as currently circumscribed [19] as well as its relationships to other species belonging to the complex. We sampled 21 localities, covering most of the distribution of *S. austriaca* (*sensu* Kurtto et al. [19]), but also including four other described microspecies (one locality per microspecies) from Central and Western Europe, as well as three localities of *S. mougeotii* for comparison. More specifically, we aimed to (1) reconstruct the phylogenetic relationships among the sampled populations of *Sorbus* subgen and *Soraria* and in relation to their parents, *S. aria* and *S. aucuparia*, using amplified fragment length polymorphism (AFLP) fingerprinting, nuclear microsatellites, and plastid DNA sequencing. Furthermore, (2) we inferred the studied populations' ploidy and reproductive mode using flow cytometry; (3) we tested self-compatibility as a component of the reproductive system; and (4) we explored the geographic distribution and morphological variation of the apomictic lineages. We tested the hypothesis that different populations within the group represent independent evolutionary entities and thus represent hitherto cryptic diversity. We do not aim to provide a revised taxonomy of this group and name all evolutionary entities—which we consider premature—but rather discuss the taxonomic implications of our results.

2. Materials and Methods

2.1. Plant Material

Leaf material from 37 localities was collected and silica-dried for molecular analyses and herbarized for morphological analyses (Figure 1A, Table S1); localities 22, 27, 36, and 37 were sampled only for molecular analyses. The taxa were identified using *Flora*

Europaea [46], Euro + Med Plantbase [52], and national floras [18,53–56]. Voucher specimens are kept at the National Museum of Bosnia and Herzegovina (SARA).

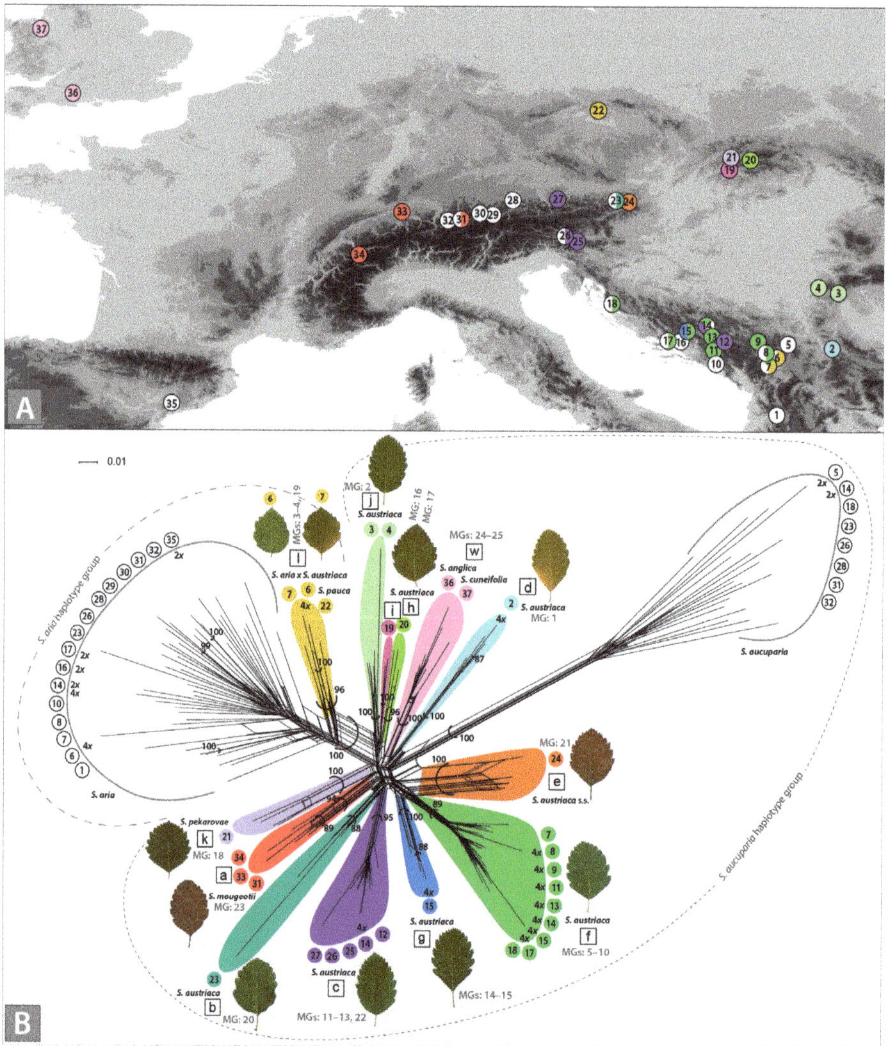

Figure 1. Geographic origin (**A**) and amplified fragment length polymorphism (AFLP) variation (**B**) of analysed *Sorbus* accessions. Sampled localities are numbered; details are in Table S1. AFLP clusters within *Sorbus* subgen. *Soraria* are colour-coded and labelled with the letters *a–l*; in (**A**) their presence in each sampled locality is indicated (the parental taxa *S. aria* and *S. aucuparia* are in white). The NeighborNet diagram is supplemented with bootstrap values > 85% derived from a Neighbour-joining analysis (Figure S1); multilocus genotypes (MG's) derived from nuclear microsatellites; and ploidy data obtained by flow cytometry; dashed lines denote the plastid haplotype affiliation according to the plastid *trnT–trnF* phylogenetic analysis. Note that only a subset of individuals was sequenced for plastid DNA variation.

2.2. *Amplified Fragment Length Polymorphism (AFLP)*

One to 13 individuals per locality, totaling 172 individuals from 37 sampled localities, were included in the AFLP analysis (Table S1). A modified CTAB-procedure [57] was used for extraction of total genomic DNA from c. 20 mg of silica-dried leaf material. The AFLP protocol was followed [58], with the modifications described in [39]. We used the following primer combinations for the selective PCR (fluorescent dye in brackets): *Eco*RI (6-FAM)-ACA/*Mse*I-CAC, *Eco*RI (VIC)-AAG/*Mse*I-CTG, and *Eco*RI (NED)-ACC/*Mse*I-CAG (*Mse*I- and *Eco*RI-primers: Sigma-Aldrich). Reproducibility was tested using fifteen replicated samples. Electropherograms were analysed with Peak Scanner 1.0 (Applied Biosystems, Waltham, MA, USA) with default peak detection parameters. The minimum fluorescent threshold was set at 50 relative fluorescence units (RFU). RawGeno 2.0 [59], a package for R [60], was used for automated data scoring with the following settings: 75–500 bp scoring range, 50 RFU minimum intensity, bin width 1.0–1.5. Fragments with reproducibility lower than 80% based on sample-replicate comparisons were excluded. A neighbour-joining analysis based on Nei-Li genetic distances [61] was conducted and bootstrapped (2000 pseudo-replicates) with TREECON 1.3 b [62]. A NeighborNet was produced from a matrix of uncorrected P distances using SplitsTree 4.12 [63]. A principal coordinate analysis (PCoA) based on Jaccard distances was conducted using PAST 2.15 [64].

2.3. *Analysis of Nuclear Microsatellites*

The amplification of six nuclear microsatellite-specific loci (CH01F02, MSS5, MSS13, MSS16, D11, and H10) was successfully performed for 171 individuals from 37 sampled localities (Table S1), following Robertson et al. [33,34]. An ABI PRISM 310 Genetic Analyzer (Applied Biosystems) was used for electrophoretic separation of the PCR products. Alleles were sized relative to the internal size standard TAMRA 500 (Applied Biosystems). Electropherograms were analysed using GeneMapper (Applied Biosystems). To study the genetic diversity, we determined the multilocus genotype (MG) for each individual on the basis of microsatellite alleles for each of the six loci using the software GenoType 1.2 [65]. Assignment of individuals to a particular clone was completed using the algorithm of Meirmans and Van Tienderen [65] according to the calculation of a genetic distance matrix and a threshold value (set to 2 after testing different thresholds as recommended) under the stepwise mutation model option. Clonal diversity was presented as the sum of the total number of multilocus genotypes (*Ng*), effective number of genotypes (*Eff*) and genotypic diversity (*Div*), calculated with GenoDive 1.2 [65]. Relationships among multilocus genotypes were visualised by principal coordinate analysis (PCoA) based on Jaccard distances using PAST 3.17 [64]. The maximum number of alleles per locus was used to infer the ploidy level.

2.4. *Plastid trnT–trnF Sequencing and Phylogenetic Analyses*

The plastid *trnT–trnF* region was sequenced for 39 individuals from 31 localities (GenBank number in Table S1), following the procedure described by Hajrudinović et al. [39] and using the primers TabA, TabC and TabF [66]; in addition, 15 sequences were included from Hajrudinović et al. [39]. Sequences were edited and aligned with Geneious Pro 5.5.9 [67]. We coded indels as binary characters using simple gap coding [68] with SeqState 1.25 [69].

Maximum parsimony (MP) and MP bootstrap (MPB) analyses of the concatenated plastid sequences (including gap codes) were performed using PAUP 4.0b10 [70]. The most parsimonious trees were searched for heuristically with 100 replicates of random sequence addition, TBR swapping, and MulTrees on. All characters were equally weighted and unsorted. The data set was bootstrapped using full heuristics with 1000 replicates, TBR branch swapping, MulTrees option off, and a random addition sequence with five replicates. Bayesian analyses of the same dataset were performed using MrBayes 3.2.1 [71], applying the HKY85 substitution model proposed by the Akaike information criterion implemented in MrAIC.pl 1.4 [72]. The alignment was partitioned into nucleotide and indel data sets, and the latter was treated as morphological data according to the model of Lewis [73]. Values for all parameters, such as the shape of the gamma distribution, were estimated during the analyses. The settings for the Metropolis-coupled Markov chain Monte Carlo process included four runs with four chains each (three heated ones using the default heating scheme), running simultaneously for 10,000,000 generations each, sampling trees every 1000th generation using default priors. The posterior probabilities (PP) of the phylogeny and its branches were determined from the combined set of trees, discarding the first 1001 trees of each run as burn-in.

2.5. Genome Size Estimation

Genome size estimation followed the protocol of Hajrudinović et al. [41], using flow cytometry. Briefly, fresh leaves of 45 individuals from 12 populations (Table S1) were co-chopped with a razor blade with fresh leaves of the internal standard *Medicago truncatula* Gaertn. cv. R108-1 (0.98 pg [74]) in 600 mL of cold Gif nuclear buffer. The suspension was filtered through a 50 µm nylon mesh (CellTrics, Partec), and RNAse (Roche) was added to 25 U mL^{-1}. The nuclei were stained with propidium iodide (Sigma-Aldrich) with a final concentration of 50 mg mL^{-1} and incubated on ice for ca. 15 min prior to analysis. The fluorescence of ~3000 nuclei was recorded for each sample using a Partec CyFlow SL3 (Partec, Münster, Germany) 532 nm laser cytometer or CyFlow Ploidy Analyser (Sysmex Europe SE) 532 nm laser. The 2C DNA values were obtained, and DNA ploidy levels [75] were inferred by comparison with the 2C DNA values of individuals of known chromosome counts, i.e., $2n = 2x = 34$ for diploids and $2n = 4x = 68$ for tetraploids [76].

For the purpose of reproduction mode identification, we conducted flow cytometric seed screening (FCSS) on 40 seeds from 10 locations (see Results), following Hajrudinović et al. [41]. Seeds were collected from previously cytotyped mother individuals (Table S1). Furthermore, to test for self-compatibility, the inflorescences of three *S. austriaca* trees from locality 14 were covered with pollination bags before anthesis. Seven seeds from pollination bags were collected and analysed. Only well-formed seeds, cleaned, shortly dried at room temperature, and kept in paper bags at 4 °C prior to analysis were used. Each seed was analysed separately. Endosperm ploidy was calculated using the inferred monoploid genome size of the embryo. Following Hajrudinović et al. [41], DNA ploidies of embryo and endosperm were compared with distinguish between sexual and apomictic origin of each seed.

2.6. Morphometric Analyses

Morphometric measurements were performed on 96 individuals from 24 localities (Table S1) of simple-leaved populations belonging to *S. aria* × *S. austriaca*, *S. austriaca*, *S. mougeotii* and *S. pekarovae* (all from the subgenus *Soraria*). Leaves of *S. anglica*, *S. cuneifolia*, and *S. pauca*, distributed outside the range of *S. austriaca*, were not available for measurements. The measurements included leaf characters that were previously shown to be informative [18,23,24,27,77]. The following 18 quantitative leaf characters were measured: lamina length (LLEAV), petiole length (LPET), length of the first, second, and third tooth (1SEINL, 2SEINL, 3SEINL), length of the first, second, and third nerve (1NERV, 2NERV, 3NERV), angle between the first, second, and third nerve compared with the primary nerve (1NANG, 2NANG, 3NANG), lamina width (WLEAV), distance of the leaf base to the line of maximal leaf width (MXWLEAV), leaf width 1 cm beneath the leaf apex (1ALEAV), leaf width 1 cm above the leaf base (1BLEAV), number of secondary nerve pairs (NNER), ratio of lamina length and lamina width (LLEAV/WLEAV) and ratio of lamina width and distance of the leaf base to the line of maximal leaf width (WLEAV/MXLEAV). The measurements were completed by hand, using millimeter paper and a digital calliper. The arithmetic means of three to five measurements per leaf character for each individual (from different mid-leaves of short sterile shots) were used for statistical analyses. Multivariate principal component analysis (PCA), canonical discriminant analysis (CDA), and classificatory discriminant analysis (DA) were performed for two data matrices. The first matrix encompassed data of all samples (96 individuals), while *S. pekarovae*, *S.*

mougeotii and three putative back-crossed individuals *S. aria* × *S. austriaca* (AFLP group *l* in Figure 1B) were excluded from the second matrix (84 individuals of the *S. austriaca* lineages, see Results). Two canonical discriminant analyses (CDA 1 and CDA 2) followed by two classificatory discriminant analyses (DA 1 and DA 2) were performed for the two data matrices [78]. Prior to the analyses, the data matrix was standardised due to the different measurement units used. PCA based on the correlation matrix characters of the first matrix was aimed at displaying a general pattern of variation and relationships among individuals/populations/taxa. CDA based on Mahalanobis' distances was used to analyse the morphology of *a priori* defined groups using the 12 AFLP groups with simple leaves marked as *a–l* in Figure 1B. The obtained results were validated using DA. The validation criterion in the identification of morphological groups was >70% of *a posteriori* correctly classified cases into the *a priori* defined groups in CDA [79]. Classificatory DA was performed using a leave-one-out cross-validation (jackknifing) procedure. Both analyses were performed using PAST 3.14 [64]. Furthermore, basic descriptive statistical parameters were calculated for the analysed taxa/genetic groups: the arithmetic mean (μ), standard deviation (SD), value range (Min–Max) and coefficient of variation (CV%).

3. Results

3.1. AFLP Fingerprinting

We obtained 435 high-quality and reproducible AFLP fragments from 172 individuals. The initial average error rate was 4.1%. The neighbour-joining analysis inferred 15 clusters with high bootstrap support (BS > 85%; Figure S1) that were also divergent in the Neighbor-Net (Figure 1B). The two most divergent clusters contained the parental species, *S. aucuparia* and *S. aria*. Whereas the former cluster was genetically more uniform, separated by a long split from all other samples, the latter cluster was more diverse. All other 13 clusters that were positioned intermediate between the parental taxa corresponded to *Sorbus* subgen. *Soraria*. Populations treated as *S. austriaca* were included in nine clusters, which included single populations scattered in the Northeastern Limestone Alps (Austria; clusters *b* and *e*, the latter including population 24 sampled ten kilometres away from the *locus classicus* of *S. austriaca*), the central Balkan Peninsula (Bosnia and Herzegovina, *g*; Serbia, *d*), and the Western Carpathians (Slovakia, *h*, *i*). Cluster *j* included two populations from the Southern Carpathians (Romania), cluster *c* included five populations from the Northern and Southern Limestone Alps (Austria, Slovenia) and the central Balkan Peninsula (Bosnia and Herzegovina), and cluster *f* included nine populations from the western Balkan Peninsula (Dinaric Mountains; Bosnia and Herzegovina, Croatia, Kosovo, and Serbia). Three populations of *S. mougeotii* from the Western Alps (Switzerland) and the Northern Limestone Alps (Austria) were included in cluster *a*. Likewise, the single population of *S. pekarovae* from the Western Carpathians (Slovakia) formed its own cluster *k*, whereas the single populations of *S. anglica* and *S. cuneifolia* from England and Wales, respectively, were included in cluster *w*. Finally, two populations from the central Balkan Peninsula (Kosovo and Serbia) and the single individual of *S. pauca* from the eastern Sudetes (Czech Republic) were genetically closest to *S. aria* and were included in cluster *l*; we assume these individuals are hybrids between *S. aria* and *S. austriaca*.

3.2. Nuclear Microsatellites

A total of 25 clonal multilocus genotypes (MGs) were found within 110 accessions of *Sorbus* subgen. *Soraria* (Figure 2, Table 1). Twenty-four out of 28 localities contained at least one MG shared by a different number of individuals within a locality, whereas localities 22 (Czech Republic), 33, and 34 (Switzerland) contained only one sampled individual, and locality 7 (Kosovo) contained three unique genotypes (Table S2). Nineteen populations with >1 sampled individuals were completely clonal (Ng = 1; Table S2). The effective number of genotypes (Eff) was higher than 1 in six populations due to the presence of unique genotypes conferring higher values of genotypic diversity (Div) in those populations. Nineteen MGs were limited to single populations, and MG 2 was present in geographically close populations (Table 1). However, several MGs occurred at multiple localities (MGs 2, 5, 12, 22, 23, and 24; Table 1). The most widespread clonal genotype (MG 5, Table 1) occurred at seven localities in the Dinaric Mountains (localities 7, 11, 13, 14, 15, 17, and 18; Table 1).

The relationships among MGs (Figure 2) were generally consistent with the AFLP data (Figure 1B). The two most divergent *Soraria* MG clusters along the first PCoA axis thus corresponded to AFLP clusters *c* and *f*, including populations from the Balkan Peninsula and the Eastern Alps. The populations of most other AFLP clusters were scattered between these two main MG clusters; clusters *a* and *w* were most divergent from cluster *c* along the second PCoA axis. The AFLP cluster *l* including putative hybrids *S. austriaca* × *S. aria* was most diverse. Individuals from locality 14 (Bosnia and Herzegovina) had the most heterogeneous genotypes. Individuals were included in the two most divergent clusters, in accordance with the AFLP results, where these individuals were included in clusters *c* and *f*. The populations from other areas were more uniform, however, without a clear geographic pattern. There were also several monoclonal populations, namely in locations 19, 20, and 21 from the Western Carpathians, 23 and 24 from the Eastern Alps, and 2 from the Central Balkan Peninsula. On the other hand, the populations from locations 3 and 4 from the Southern Carpathians and 31, 33, and 34 from the Alps all shared a single clone. In addition, the allele composition of the parental species, *S. aucuparia* and *S. aria*, is given in Table S3.

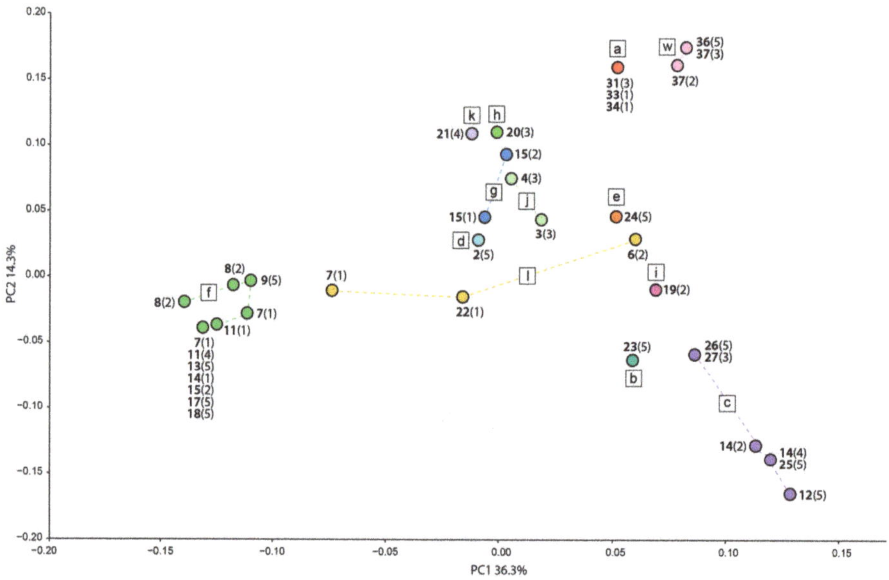

Figure 2. Principal coordinate analysis of Jaccard distances among the 25 multilocus genotypes found in 110 accessions of *Sorbus* subgen. *Soraria* accessions based on nuclear microsatellite data. Locality numbers correspond to Figure S1A and Table 1 and colours and letters to the AFLP clusters in Figure 1B. Multilocus genotypes belonging to the same AFLP cluster are connected with dashed lines. The numbers in parentheses denote the number of clones per locality.

Table 1. Multilocus genotypes (MGs) and allele composition of the studied 28 populations of *Sorbus* subgen *Soraria*.

MG	N	AFLP Cluster	N Per Locality	Loci						*Xo*	*Xe*
				MSS13	MSS5	CH01F02	D11	MSS16	H10		
1	5	d	**2**(5)	189, 193, 195	124, 126, 130	186, 200	143, 151	154, 158, 162	86	3	**4*x*** [a]
2	6	j	**3**(3), **4**(3)	193, 197	110, 124, 126	186, 190, 196, 200	137, 151	154, 158, 160, 162	92	**4**	-
3	2	l	**6**(2)	191, 193, 200	124, 134	194, 196	149, 151	154, 158, 160	82, 88, 92	3	**4*x*** [a]
4	1	l	**7**(1)	193, 195, 201	122, 124, 126, 130	190, 196, 198, 200	151, 157	156, 158	-	**4**	-
5	23	f	**7**(1), **11**(4), **13**(5), **14**(1), **15**(2), **17**(5), **18**(5)	191, 195, 201	124, 126, 150	186, 190, 196, 200	143, 151	156, 158	104, 106	**4**	**4*x*** [a]
6	1	f	**7**(1)	191, 195, 201	124, 126, 150	186, 190, 196, 200	143, 151	154, 158	104, 106	**4**	-
7	2	f	**8**(2)	193, 197, 201	124, 126	186, 196	143, 151	156, 158	104, 106	3	**4*x*** [a]
8	2	f	**8**(2)	191, 197, 201	124, 126	186, 190, 196	143, 151	156, 158	106	3	**4*x*** [a]
9	5	f	**9**(5)	191, 197, 201	124, 126	186, 190, 196	143, 151	154, 158, 166	104, 106	3	**4*x*** [a]
10	1	f	**11**(1)	191, 195, 201	124, 126, 150	186, 190, 196, 200	143, 151	156, 158	102, 104	**4**	**4*x*** [a]
11	5	c	**12**(5)	191, 193, 195	110, 112, 124	188, 196, 202	147, 153	154, 158, 162	92	3	-
12	9	c	**14**(4), **25**(5)	191, 193, 195	110, 112, 124	188, 196, 202	147, 151	154, 158, 162	92	3	**4*x*** [a]
13	2	c	**14**(2)	191, 193, 195	110, 112, 124	188, 196, 202	147, 151	154, 158, 162	80, 90	3	**4*x*** [a]
14	2	g	**15**(2)	191, 193	112, 120, 126	186, 190, 196, 200	151	154, 158, 164	98, 100	**4**	**4*x*** [a]
15	1	g	**15**(1)	191, 193	112, 120, 126	186, 190, 196, 200	151	154, 158, 164	102	**4**	**4*x*** [a]
16	2	i	**19**(2)	191, 193, 195	112, 120, 126	188, 190, 196, 200	147, 149, 151	154, 158, 160, 162	92	**4**	-
17	3	h	**20**(3)	-	110, 112, 126	186, 190, 196, 200	149, 151	154, 160	96	**4**	**4*x*** [d]
18	4	k	**21**(4)	193, 201	124, 126, 132	186, 190, 196	131, 151	154, 162	90	3	**4*x*** [d]
19	1	l	**22**(1)	191, 193, 201	112, 124, 126	190, 196, 200	149, 151	156, 158, 160, 162	92	**4**	**4*x*** [c]
20	5	b	**23**(5)	191, 193, 195	110, 112, 124	186, 196, 208	131, 151	154, 158, 162	102, 104	3	-
21	5	e	**24**(5)	191, 193, 195	124	190, 196	141, 147, 151	154, 158, 160	98, 100	3	-
22	8	c	**26**(5), **27**(3)	191, 193, 195	110, 112, 124	186, 196, 200	147, 151	154, 158, 162	92	3	-
23	5	a	**31**(3), **33**(1), **34**(1)	193, 197, 201	124, 132	186, 196	147, 151	154, 160, 162	100, 102	3	-
24	8	w	**36**(5), **37**(3)	191, 193, 195	118, 128, 134	186, 196, 200	147, 149, 151	154, 162	98, 100	3	**4*x*** [b]
25	2	w	**37**(2)	191, 193, 195	118, 128, 134	186, 196, 200	147, 149, 151	154, 162	96, 98	3	**4*x*** [b]
	110										

N—Number of individuals belonging to each multilocus genotype; N per locality—Number of individuals (in brackets) bearing a particular multilocus genotype per each locality (in bold), numbered as in Figure 1; *Xo*—Maximum number of alleles per locus; *Xe*—Expected ploidy level–obtained by flow cytometry for at least one individual per population from each MG group [a] or from published results for the same sampled locality [80] [b], [25] [c], [81] [d].

3.3. Plastid trnT–trnF Phylogenetic Relationships

The *trnT-trnF* alignment of the concatenated *trnT-trnL* intergenic spacer and the *trnL-trnF* partial sequence was 1943 bp long. The shortest sequences (1766 bp) were those of *S. anglica*, *S. aucuparia*, *S. cuneifolia*, *S. mougeotii*, *S. pekarovae*, and almost all of *S. austriaca*; they were all identical. Two individuals of *S. austriaca* from locality 23 were 8 bp longer. The sequences of *S. aria*, including also one of *S. pauca* and one of *S. aria* × *S. austriaca* varied much more in length, ranging from 1817 bp (*S. aria* from locality 41) over 1838 bp (*S. aria* 30, 32; *S. aria* × *S. austriaca* 6; *S. pauca* 22), 1840 bp (*S. aria* 14, 28, 39, 40, 42, 44, 46) to 1850 bp (*S. aria* 38), 1860 bp (*S. aria* 40) and 1864 bp (*S. aria* 14). Eight substitutions, of which two were outapomorphic, contributed to variability within the *S. aria* lineage. Twenty-three characters were parsimony-informative, and Bayesian and parsimony analyses of the *trnT*–

trnF sequences resulted in congruent phylogenies (Figure 3) that reflected the variation outlined above.

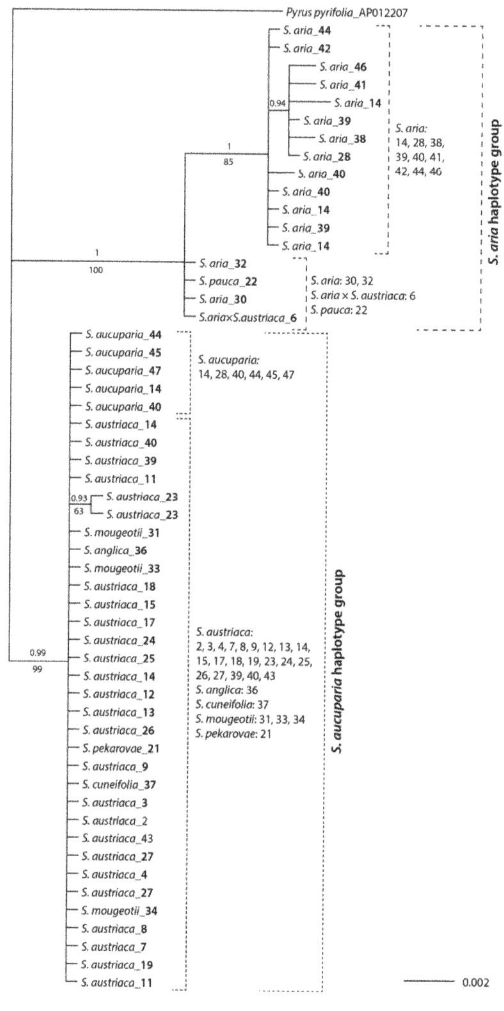

Figure 3. Bayesian consensus phylogram inferred from phylogenetic analyses of plastid *trnT–trnF* sequences. Numbers above branches are posterior probabilities > 0.90 and those below branches maximum parsimony bootstrap values > 50%. Locality numbers correspond to Table S1.

Two main clades were resolved: one (posterior probability, pp, 1; parsimony bootstrap, BS, 100%), named *S. aria* haplotype group, included two populations of *S. aria* (localities 30 and 32), one of *S. aria* × *S. austriaca* (locality 6), and one of *S. pauca* (locality 22) in a basal polytomy, and a clade (PP 1, BS 85%) including all other populations of *S. aria* (*S. aria* haplotype group; localities 14, 28, 38–42, 44, 46). The second main clade (PP 0.99, BS 99%), named *S. aucuparia* haplotype group, included all populations of *S. aucuparia* (localities 14, 44, 45, 47) and *S. austriaca* (localities 2–4, 7–9, 12–15, 17–19, 23–26), the single populations of *S. anglica* (locality 36), *S. cuneifolia* (locality 37), and *S. pekarovae* (locality 21), as well as the three populations of *S. mougeotii* (localities 31, 33, 34), with unresolved relationships.

3.4. Genome Size, Ploidy Level and Reproduction Mode

Flow cytometry of 45 individuals resulted in holoploid absolute genome sizes (GS, 2C value) corresponding to two ploidy levels, namely diploid and tetraploid (Figure 4).

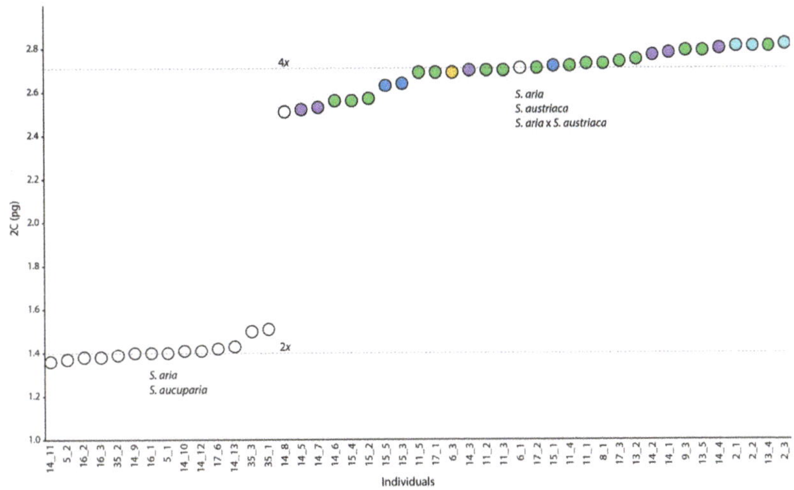

Figure 4. Scatterplot of absolute genome size values (2C pg) for analysed *Sorbus* accessions. Colours correspond to the AFLP clusters (Figure 1B) and locality numbers followed by individual numbers to Figure 1A and Table S1.

The GS was 1.36–1.43 pg in diploid *S. aucuparia*, 1.38–1.51 pg in diploid *S. aria*, 2.69 pg in tetraploid *S. aria* × *S. austriaca*, 2.52–2.82 pg in tetraploid *S. austriaca*, and 2.51–2.71 pg in tetraploid *S. aria* (Table 2).

Table 2. Flow cytometric results for nuclei from leaves and seeds of *Sorbus* taxa accompanied with the deduced origin of seeds/reproduction mode.

Locality No.	Affiliation in AFLP NNet	Taxon	*N*	Genome Size (2C pg)	Genome Size (1C*x* pg)	DNA Ploidy Level (2*n*)	*N* Seeds	Embryo (2C pg)	Endosperm (2C pg)	Embryo: Endosperm Ploidy	Seed Origin
2	*d*	*S. austriaca*	3	2.81–2.82	0.70–0.71	4*x*	-				
5	*S. aucuparia*	*S. aucuparia*	2	1.37–1.40	0.69–0.70	2*x*	2	1.36–1.41	2.07–2.14	2*x*:3*x*	Sexual
6	*l*	*S. aria* × *S. austriaca*	1	2.69	0.67	4*x*	1	2.67	8.13	4*x*:12*x*	Apomictic
	S. aria	*S. aria*	1	2.71	0.68	4*x*	1	2.77	8.15	4*x*:12*x*	Apomictic
8	*f*	*S. austriaca*	1	2.73	0.68	4*x*	1	2.73	8.08	4*x*:12*x*	Apomictic
9	*f*	*S. austriaca*	1	2.79	0.70	4*x*	1	2.73	8.10	4*x*:12*x*	Apomictic
11	*f*	*S. austriaca*	5	2.69–2.73	0.67–0.68	4*x*	1	2.72	8.32	4*x*:12*x*	Apomictic
13	*f*	*S. austriaca*	3	2.75–2.81	0.69–0.70	4*x*	2	2.73–2.74	8.09–8.23	4*x*:12*x*	Apomictic
14	*f*	*S. austriaca*	1	2.56	0.64	4*x*	1	2.81	8.46	4*x*:12*x*	Apomictic
	c	*S. austriaca*	6	2.52–2.80	0.63–0.70	4*x*	19 (7 *) 1	2.69–2.80 2.65	7.99–8.65 6.70	4*x*:12*x* 4*x*:10*x*	Apomictic Apomictic
	S. aria	*S. aria*	2	1.40–1.41	0.70–0.71	2*x*	2	1.41–1.43	2.12–2.15	2*x*:3*x*	Sexual
			1	2.51	0.63	4*x*	1	2.74	8.34	4*x*:12*x*	Apomictic
	S. aucuparia	*S. aucuparia*	3	1.36–1.43	0.68–0.72	2*x*	3	1.36–1.41	2.03–2.13	2*x*:3*x*	Sexual

Table 2. *Cont.*

Locality No.	Affiliation in AFLP NNet	Taxon	*N*	Genome Size (2C pg)	Genome Size (1Cx pg)	DNA Ploidy Level (2n)	*N* Seeds	Embryo (2C pg)	Endosperm (2C pg)	Embryo: Endosperm Ploidy	Seed Origin
15	*f, g*	*S. austriaca*	5	2.56–2.72	0.64–0.68	4*x*	1	2.67	8.12	4*x*:12*x*	Apomictic
16	*S. aria*	*S. aria*	3	1.38–1.40	0.69–0.70	2*x*	2	1.37–1.43	2.06–2.13	2*x*:3*x*	Sexual
17	*f*	*S. austriaca*	3	2.69–2.74	0.67–0.69	4*x*	1	2.66	8.11	4*x*:12*x*	Apomictic
	S. aria	*S. aria*	1	1.42	0.71	2*x*	-				
35	*S. aria*	*S. aria*	3	1.39–1.51	0.70–0.76	2*x*	-				
			45				40 (7 *)				

* Asterisks mark the number of seeds from inflorescences isolated in pollination bagss.

Flow cytometric seed screening of 40 seeds resulted in three different embryo:endosperm profiles, namely 2*x*:3*x*, 4*x*:10*x* and 4*x*:12*x* (Table 2). Diploid *S. aria* and *S. aucuparia* mother trees yielded seeds with 2*x* embryo and 3*x* endosperm, which represents a regular sexual profile. On the other hand, all seeds of analysed tetraploid *S. austriaca*, *S. aria* and *S. aria* × *S. austriaca* mother trees were of apomictic origin with 4*x* embryos and 12*x* endosperms, or, in one seed, 10*x* endosperm (Table 2). Moreover, several inflorescences of the three *S. austriaca* individuals from locality 14 that were covered with pollination bags to check for self-compatibility, yielded fruits. All seeds developed in pollination bags were of apomictic origin with 4*x* embryos and 12*x* endosperms (Table 2).

3.5. Morphology

Principal Component Analysis generated four significant principal components (two are displayed, Table S4). They accounted for 83.1% of the total variance (PC1 = 40.2%, PC2 = 19.6%, PC3 = 14.1%, and PC4 = 9.1%), with moderate correlation to the majority of corresponding morphological traits (Table S4). The PCA ordination diagram (Figure S2) showed a pattern in which most AFLP clusters overlapped among the four clusters (*c*, *d*, *f*, and *k*). The CDA 1 diagram of all samples (Figure S3A) showed that the overlapping AFLP clusters *c* (Dinaric Mountains, Southern Limestone Alps, Northern Limestone Alps), *h*, *i*, and *j* (all Carpathians) were morphologically divergent from the overlapping clusters *e* (Northern Limestone Alps), *f* and *g* (both Dinaric Mountains), and *l* (Dinaric Mountains, Bohemia), whereas the partly overlapping Alpine clusters *a* (Western Alps) and *b* (Northern Limestone Alps) were intermediate along DF1 (explaining 41.5% of variation). Along DF2 (explaining 20.6% of the variation), the populations from AFLP clusters *d* (Carpathians-Balkan Mountains) and *k* (Carpathians) were clearly divergent, whereas all other AFLP clusters were intermediate. Finally, along DF3 (11.3%; Figure S3B), some AFLP clusters that were overlapping along DF1 and DF2 were divergent, namely clusters *a*, *b*, *e*, and *j*. The characters that contributed mostly to the discrimination along DF1 were LLEAV/WLEAV, NNERV, 1NERV and 2NERV, those along DF2 were WLEAV, LLEAV/WLEAV, WLEAV, and those along DF3 1NANG, 2NANG. The classification matrix with the Jackknifed procedure for the CDA 1 dataset resulted in 82% of correctly classified individuals (Table S5).

The CDA 2 resulted in clearer morphological discrimination among the AFLP clusters of the *S. austriaca* lineages (Figure 5). Along with DF1 (45.2%) and DF2 (18.0%; Figure 5A), AFLP clusters *c*, *d*, and *j* were clearly divergent. On the other hand, *f*, *e*, and *g*, as well as *b*, *h*, and *i*, overlapped. Along DF3 (16.0%; Figure 5B) clusters *e*, *h*, *j* and *j* were divergent, whereas *c*, *d*, and *f*, as well as *f* and *g*, overlapped. The characters that contributed most to the discrimination along DF1 were LLEAV/WLEAV, 1NERV, NNER, and 2NERV; those along DF2 were WLEAV, LLEAV/WLEAV, NNERV, and 3NER; and those along DF3 were WLEAV/MXLEAV. The classification matrix with the Jackknifed procedure for the CDA 2 dataset resulted in 88% of correctly classified individuals (Table S6). Basic descriptive statistical parameters of measured leaf characters are given in Table S7.

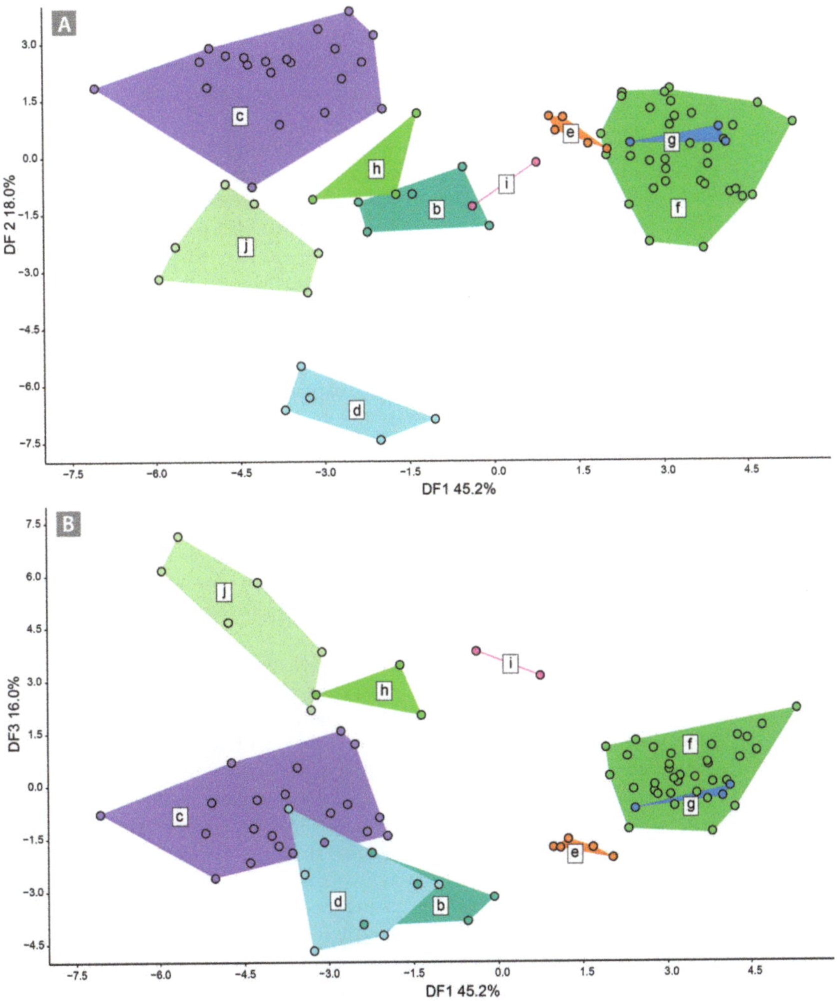

Figure 5. Canonical discriminant analysis ((**A**) DF1 vs. DF2; (**B**) DF1 vs. DF3) of nine predefined *Sorbus austriaca* groups (corresponding to the AFLP clusters shown in Figure 1) based on 18 morphological leaf characters.

4. Discussion

Our integrative approach combining AFLP fingerprinting, plastid DNA sequences, nuclear microsatellites, ploidy-level estimation, and morphometric analyses inferred intricate patterns of diversification within the *S. austriaca* complex. The genetic data revealed a clear divergence among (groups of) populations included in different allotetraploid apomictic lineages that originated in various parts of South-Eastern and Central Europe. Our results thus support previous findings that hybridisation, polyploidization, and apomixis are the main drivers of *Sorbus* diversification, at least in Europe [33,34,36,82].

4.1. Multiple Origins of *S. austriaca* Lineages

Our AFLP and nuclear microsatellite data suggest independent origins of the different *S. austriaca* lineages in different parts of the Alps, Dinaric Mountains, and the Carpathians, likely as a consequence of polytopic hybridisation between the parental diploid species *S.*

aria and *S. aucuparia* (Figures 1B and 2, Table 1). Hybridisation was followed by polyploidisation, as all individuals of *S. austriaca* for which we established the ploidy, were tetraploid (Figures 1B and 4, Tables 1 and 2). This pattern, along with the restriction of many lineages to single (lineages *b*, *d*, *e*, *g*, *h*, *i*, and *k*) or a few (lineages *a*, *j*, and *w*) localities, some of them in areas that were strongly glaciated during the Last Glacial Maximum (LGM) [83,84], may imply their recent origin. On the other hand, the geographically widespread lineages *c* and *f* included multiple populations and shared related multiclonal MGs differing in one or two alleles per locus (Figure 1A,B; Table 1). Cluster *f* includes populations solely from the Dinaric Mountains, which were much less affected by the Pleistocene glaciations than the Alps [85,86]. Cluster *c* also includes populations from this area and the southern margins of the Eastern Alps, which suggests that they might have originated earlier and thus had more time to disperse [87]. Their disjunct distributions could be a result of multiple dispersals and establishments in isolated localities, but also of a previously more continuous range disrupted by the climatic changes during the Pleistocene.

An important factor in the range expansion of apomicts is self-compatibility, a reproductive trait of the mating system of *S. austriaca* as determined in population 14 (Table 2). Self-compatibility most likely facilitated the range expansion of certain clonal genotypes, more specifically those from clusters *c* and *f* (Figure 1B). Due to self-compatibility, apomictic clones can establish populations via single individuals that use their own pollen for the required endosperm fertilisation to produce functional seeds [7]. In this way, apomictic genotypes promote range expansions to remote areas, where they can function as pioneer colonisers of new habitats [15]. The fleshy fruits of *Sorbus*, such as those of other *Malinae* (e.g., *Crataegus*), are adapted to dispersal by vertebrates, mainly birds, often over relatively long distances [16,88].

A certain level of genetic differentiation among the populations in the Dinaric cluster *f* (Figure 1) suggests their persistence in disjoint localities over a longer period rather than their recent dispersal. Such a differentiation is surprising, as it is not expected in an obligate apomictic such as *S. austriaca* [37,41]. Different mechanisms can facilitate genetic divergence even among obligate apomictic populations, namely residual sexuality [89], accumulation of mutations and chromosome rearrangements [12,90], recombination during restitutional meiosis [91], transposon activity [92], and heritable epigenetic variation [93]. On the other hand, in the Dinaric-Alpine cluster *c*, the level of diversification appears to be lower, suggesting more recent divergence of some populations. This is consistent with the extensive glaciation of the Alps during the LGM [94], rendering postglacial colonisation of the Alpine populations from the South likely. Nuclear microsatellite analysis confirmed that diploids from South-eastern Europe (Albania and Montenegro) and Central Europe (Slovenia) were involved in the formation of the Central European tetraploid populations of *S. aria* [87], which further supports the hypothesis of postglacial colonisation of the Alps from the South. In the same line, the Balkan Peninsula served as an important Pleistocene refugium and a source for post-glacial colonisation of Central Europe for other trees, e.g., alder buckthorn (*Frangula alnus* L. [95]), European beech (*Fagus sylvatica* L. [96]), and hornbeam (*Carpinus betulus* L. [97]).

The co-occurrence of the widespread genetic lineage *f* and the stenoendemic lineages *c* and *g* at the same localities in the central Dinaric Mountains (14 and 15, Figure 1) is an additional line of evidence for recurrent allopolyploidisations in sympatric populations of parental diploid sexuals. Weak reproductive barriers between *S. aucuparia* and *S. aria* facilitate gene flow and continuously generate novel hybrid derivatives, which are often polyploid [32]. Along the same line, the co-occurrence of AFLP group *f* in *S. austriaca* and *S. aria* at locality 7 in the south-eastern Dinaric Mountains could have led to hybridisation and, consequently, the origin of a novel genetic entity (i.e., cluster *l*).

4.2. Mostly *S. aucuparia*, but also *S. aria*, Served as Maternal Parents of Hybridogenous Lineages

As in most angiosperms [98], plastomes are maternally inherited in *Sorbus* [99]. Our plastid DNA phylogenies clearly show that hybridogenous populations of *S. anglica*,

S. austriaca, *S. cuneifolia*, *S. mougeotii*, and *S. pekarovae* all had *S. aucuparia* as their maternal parent (Figures 1A and 3), which is the most common pattern in the subgenus *Soraria* [32,34,39,81,100]. *Sorbus aria* acts as a pollen donor not only in hybrids with *S. aucuparia* but also with *S. torminalis* and *S. chamaemespilus* [32,34,39,100], albeit with some exceptions [88].

Also in our dataset, AFLP cluster *l* that was in the NeighborNet closest to *S. aria* (Figure 1B), shared its haplotype with *S. aria* (Figure 3). Due to their intermediate position between *S. aria* and the *S. austriaca* complex in the NeighborNet, we suggest that these populations represent backcrosses of the *S. austriaca* complex with *S. aria* as the maternal parent (*S. aria* × *S. austriaca*). Alternatively, direct hybridisation of *S. aria* with *S. aucuparia* as a pollen donor could be plausible, but we consider this less probable because such a scenario would result in an intermediate leaf morphology (semipinnate leaves [33]). To our best knowledge, *S. aucuparia* served as the maternal parent to all allopolyploid derivatives originating from hybridisation with *S. aucuparia*.

The individuals of cluster *l* in our study, including *S. pauca*, all shared a *S. aria* haplotype and were tetraploids with simple leaves (Figure 1B, Table 2). Whereas localities 6 and 7 are separated by only 27 km, and the clustering of their individuals is thus not surprising, the close relationship of the single analysed individual of *S. pauca* from a locality more than 970 km away, is unexpected. *Sorbus pauca* was recently described as a hybridogenous tetraploid apomictic endemic of the eastern Sudetes that most probably arose from two hybridisation events [25], but this hypothesis as well as the formal description of this new species were exclusively based on morphology and cytometry. It is likely that multiple recent hybridisations between genetically similar parents at different localities led to highly similar hybrid genotypes. Alternatively, cluster *l* has a single origin and is more common in intermediate areas, for instance the Carpathians.

The origin of *Sorbus* tetraploids may follow two different pathways. One involves an initial cross of a diploid sexual and a tetraploid apomict, leading to apomictic triploid offspring. Subsequently, unreduced triploid eggs are fertilised by reduced pollen of the parental diploid to produce new tetraploids [33,37,41]. Alternatively, crossing the reduced megagametophyte ($n = 2x$) of a sexual tetraploid with the reduced pollen ($n = 2x$) of an apomictic tetraploid can produce the same tetraploid offspring. In the case of *S. austriaca*, the first scenario would include fertilisation of a reduced diploid *S. aucuparia* megagametophyte ($n = x$) with reduced pollen of tetraploid *S. aria* ($n = 2x$) to produce triploid offspring; tetraploid *S. aria* is namely widespread in the Balkan Peninsula [39,41]. In the next step, the unreduced megagametophyte of such a triploid is fertilised by reduced pollen of *S. aucuparia* ($n = x$) to produce a tetraploid. Facultative sexuality of apomictic allopolyploids could allow backcrossing with parental species [101]. The second scenario would include fertilisation of a reduced allotetraploid megagametophyte (such as *S. bosniaca* with the same parental combination [39]) with reduced pollen of tetraploid *S. aria*. The problem with the latter scenario is that there is no evidence for sexuality in *S. bosniaca*. The third scenario presumes a direct cross via fertilisation of an unreduced *S. aucuparia* megagametophyte ($n = 2x$) with reduced pollen of tetraploid *S. aria* ($n = 2x$).

Therefore, different scenarios for the formation of *Sorbus* polyploids are plausible, at least in the Dinaric Mountains, where different cytotypes and taxa often coexist [39,41]. Despite recent methodological progress, reconstructing the origin of allopolyploids is still challenging due to their recurrent formation, recombination among homeologous chromosomes, different epigenetic expression, genome restructuring, or extinction of parental lineage [5,6,102,103]. It is particularly difficult to trace subgenomes of the *S. aria* group within allopolyploid complexes due to the enormous genetic and cytotypic variability of the members of this group [82].

4.3. Taxonomic Considerations

Taxonomic assessments are a serious challenge in apomictic groups, such as *Amelanchier* Medik. [104], *Antennaria* Gaertn. [105], *Crataegus* [106], *Taraxacum* [12], *Hieracium*

L. [107], *Ranunculus* L. [28], *Rubus* [17], and *Sorbus*. The taxonomic concept adopted by most European taxonomists in the past two decades in *Sorbus* has been the morphospecies concept, which implies a unique morphology coupled with distributional data [107]. Numerous apomictic species (microspecies) in Europe have been described based on unique morphology combined with cytometric or karyological and sometimes genotypic data [21,23,24,26,40,44,45,108,109].

In *Sorbus*, particularly ploidy data are considered important because in this genus, polyploidy confers apomictic reproduction, which is the key criterion in addition to morphology [18]. Apomictic reproduction *per se* does not necessarily imply uniform and distinctive morphology but may result in poorly differentiated individuals/populations arising from the same or related combinations of parental taxa/cytotypes, as shown for *S. aria* from Central Bavaria [82]. Our results show that genetically similar populations also tend to cluster together morphologically, regardless of their geographic origin (clusters *c* and *f*, Figures 1B, S2 and S3).

Our study reveals cryptic diversity within the *S. austriaca* complex. High congruence of molecular and morphological data demonstrates that the sampled populations of *S. austriaca* in fact form multiple evolutionary entities. Interestingly, both the level of genetic (Figure 1B) and morphological (Figures S2 and S3) divergence among described species of subgenus *Soraria*, namely *S. anglica* and *S. cuneifolia* (cluster *w*), *S. mougeotii* (*a*), and *S. pekarovae* (*k*), is similar to the divergence among different lineages of *S. austriaca* (clusters *b*–*g*). The *S. austriaca* clusters are differentiated from *S. pekarovae* and *S. mougeotii* based on leaf morphology (Figure S3), although some morphological overlap between *S. mougeotii* and *S. austriaca* (cluster *b*, population 23) is evident.

The strongest morphological differentiation is seen among *S. austriaca* lineages from the Balkan Peninsula (clusters *f* and *d*) and the Balkan-Alpine cluster *c* (Figure 5A). The Balkan populations have genetic affinities with both the Northern and Southern Limestone Alps (cluster *c*, Figures 1A and 5), which is also reflected in similar morphology. On the other hand, the pronounced genetic and morphological divergence of the Carpathian lineages of *S. austriaca* (the Southern Carpathian cluster *j* and the Western Carpathian clusters *i* and *h*) is in line with their geographical isolation relative to lineages from the Alps and the Dinaric Mountains. Their distinctiveness may be explained by a potentially different parental combination, i.e., a non-sampled paternal lineage of genetically variable *S. aria* [45,82].

Allopolyploidisation likely occurs polytopically, leading to the evolution of independent populations [81], and natural selection and chromosome rearrangements can then result in the formation of similar morphological forms [110]. On the other hand, epigenetic mechanisms coupled with polyploidy can produce different phenotypes regardless of the similarity of the genomic compositions of allopolyploids [111]. In our case, the morphological integrity of the *S. austriaca* lineages and their distinctiveness across the sampled area are likely maintained via apomixis and reproductive isolation. Even if some of the genetic clusters could be associated with existing species (e.g., cluster *c* could pertain to the recently described *S. lippertiana* [40]), the others likely represent undescribed cryptic microspecies. We here refrain from recognising the uncovered cryptic diversity as distinct taxonomic entities; additional studies, including a denser sampling and more detailed morphological analyses, are needed to taxonomically resolve this intricate complex.

5. Conclusions

Our molecular data revealed pronounced cryptic diversity within the *S. austriaca* complex; it is actually composed of different lineages, which likely originated at different time horizons and exhibit different distribution patterns. These data highlight the importance of genetic analyses on the one hand and including samples from a broader geographic area when taking taxonomic decisions and describing new species in *Sorbus* on the other hand. Apart from that, our results are particularly valuable from a biodiversity conservation perspective, because in the genus *Sorbus*, the interaction of hybridisation, polyploidy, and apomixis represents a powerful mechanism generating novel diversity in sympatric populations of the parental taxa. Traditional conservation efforts, based on clearly defined boundaries of taxonomic entities in order to specify appropriate action plans and measures, have limitations in preserving the diversity of taxonomic complex groups [112], such as the genus *Sorbus*, whose dynamic evolution requires a different approach. Therefore, a new conservation concept based on evolutionary processes was proposed (Process-Based Species Action Plan [113]). The goal of this concept is the conservation of processes that generate diversity, i.e., the preservation or increase of the number of individuals in a potential interaction. Studies such as the present one are timely, as they are setting the stage for such process-oriented biodiversity conservation.

Supplementary Materials: The following supporting information can be downloaded at: https://www.mdpi.com/article/10.3390/biology12030380/s1, Figure S1, Neighbor-Joining analysis of AFLP data for *Sorbus* subgen. *Soraria*, *S.* subgen. *Aria* and *S.* subgen. *Sorbus*. Localities' numbers and groups' colours follow the other Tables and Figures; Figure S2, PCA ordination of 18 morphological traits for all *Soraria* accessions. Colours and letters correspond to AFLP clusters in Figure 1; Figure S3, Canonical discriminant analysis (A: DF1 vs. DF2; B: DF1 vs. DF3) of 12 predefined groups (corresponding to AFLP clusters) based on individual plants and 18 morphological characters. Colours and letters correspond to AFLP clusters in Figure 1; Table S1, Geographic origin and number of individuals of *Sorbus* populations included in the analyses of AFLP, nuclear microsatellite, plastid DNA sequencing, flow cytometric and morphometric data, respectively; Table S2, Indices of population clonal diversity based on nuclear microsatellite data for the studied populations of *Sorbus* subgen. *Soraria*; Table S3, Alelle composition in six nuclear microsatellite loci for the presumed parental taxa of *Sorbus* subgen. *Soraria*; Table S4, Results of principal component analysis (PCA, see Figure S2) and canonical discriminant analysis (CDA, see Figure S3) of *Soraria* individuals based on the morphological characters of leaves; Table S5, Classification matrix with Jackknife procedure (82% of cases correctly classified) for the dataset of analysed *Sorbus* individuals based on morphometric leaf measurements (groups are defined according to AFLP clusters, Figure 1B); Table S6, Classification matrix with Jackknife procedure (88% of cases correctly classified) for the dataset of analysed *Sorbus austriaca* individuals based on morphometric leaf measurements (groups are defined according to AFLP clusters, Figure 1B); Table S7, Descriptive statistics of leaf morphometrics for *Sorbus* subgen. *Soraria* individuals presented for AFLP groups.

Author Contributions: Conceived and designed the study, B.F., P.S., A.H.-B. and F.B.; collected plant material, B.F., P.S., A.H.-B. and F.B.; performed molecular laboratory work, A.H.-B.; performed GS measurements, S.S.-Y. and F.B.; analysed the data A.H.-B., F.B., B.F. and P.S.; writing—original draft preparation A.H.-B., F.B. and B.F.; writing—review and editing, A.H.-B., B.F., P.S., S.S.-Y. and F.B.; visualisation, A.H.-B.; supervision, P.S.; project administration, F.B. and B.F.; funding acquisition, F.B., B.F. and P.S. All authors have read and agreed to the published version of the manuscript.

Funding: This research was funded by the Ministry of Science, Higher Education and Youth of Canton Sarajevo (project 27-02-11-41250-5/21 TreeCYTO) to F.B., and a Joint bilateral project between Austria and Bosnia and Herzegovina (Austrian Agency for International Cooperation ÖAD project BA 01/2023 to B.F; Ministry of Civil Affairs of Bosnia and Herzegovina project BA 10-33-11-7066/22 to F.B.). Additional support was provided by the Federal Fund for Environmental Protection (project 01-09-2-1476/19) to A.H.-B. and the International relations office of the University of Innsbruck to B.F. and F.B.

Institutional Review Board Statement: Not applicable.

Informed Consent Statement: Not applicable.

Data Availability Statement: The data presented in this study are available within article and Supplementary Materials.

Acknowledgments: We thank T. Rich, M. Lepší, Cs. Németh, A. Tribsch, R. Brus, M. Niketić, A. Medić, V. Kučerová, T. Knight, M. Falch, J. Theurillat, P. Pilsl, F. Gugerli, C. Pachschwöll, J. Vallès, W. Gutermann, D. Reich, R. Sander, M. Hofbauer, C. Gilli, B. Weis, M. Thalinger for their support with plant material. We thank M. Bourge for his assistance on the Imagerie-Gif Cytometry core facility of the Gif campus (https://www.i2bc.paris-saclay.fr/bioimaging/cytometry). We thank D. Pirkebner, M. Magauer and E. Silajdžić who supported the molecular genetic laboratory work.

Conflicts of Interest: The authors declare no conflict of interest.

References

1. Soltis, D.E.; Albert, V.A.; Leebens-Mack, J.; Bell, C.D.; Paterson, A.H. Polyploidy and angiosperm diversification. *Am. J. Bot.* **2009**, *96*, 336–348. [CrossRef]
2. Wood, T.E.; Takebayashi, N.; Barker, M.S.; Mayrose, I.; Greenspoon, P.B.; Rieseberg, L.H. The frequency of polyploid speciation in vascular plants. *Proc. Natl. Acad. Sci. USA* **2009**, *106*, 13875–13879. [CrossRef]
3. Madlung, A. Polyploidy and its effect on evolutionary success: Old questions revisited with new tools. *Heredity* **2013**, *110*, 99–104. [CrossRef] [PubMed]
4. Van de Peer, Y.; Ashman, T.-L.; Soltis, P.S.; Soltis, D.E. Polyploidy: An evolutionary and ecological force in stressful times. *Plant Cell* **2021**, *33*, 11–26. [CrossRef]
5. Hegarty, M.J.; Hiscock, S.J. Genomic clues to the evolutionary success of polyploid plants. *Curr. Biol.* **2008**, *18*, 435–444. [CrossRef]
6. Weiss-Schneeweiss, H.; Emadzade, K.; Jang, T.-S.; Schneeweiss, G.M. Evolutionary consequences, constraints and potential of polyploidy in plants. *Cytogenet. Genome Res.* **2013**, *140*, 137–150. [CrossRef] [PubMed]
7. Hörandl, E. The evolution of self-fertility in apomictic plants. *Sex. Plant Reprod.* **2010**, *23*, 73–86. [CrossRef]
8. Hojsgaard, D.; Klatt, S.; Baier, R.; Carman, J.G.; Hörandl, E. Taxonomy and biogeography of apomixis in angiosperms and associated biodiversity characteristics. *Crit. Rev. Plant Sci.* **2014**, *33*, 414–427. [CrossRef] [PubMed]
9. Coughlan, J.M.; Han, S.; Stefanović, S.; Dickinson, T.A. Widespread generalist clones are associated with range and niche expansion in allopolyploids of Pacific Northwest Hawthorns (*Crataegus* L.). *Mol. Ecol.* **2017**, *26*, 5484–5499. [CrossRef]
10. Ozias-Akins, P.; van Dijk, P.J. Mendelian genetics of apomixis in plants. *Ann. Rev. Genet.* **2007**, *41*, 509–537. [CrossRef]
11. Cosendai, A.C.; Wagner, J.; Ladinig, U.; Rosche, C.; Hörandl, E. Geographical parthenogenesis and population genetic structure in the alpine species *Ranunculus kuepferi* (Ranunculaceae). *Heredity* **2013**, *110*, 560–569. [CrossRef]
12. Majeský, L.; Vašut, R.J.; Kitner, M.; Trávníček, B. The pattern of genetic variability in apomictic clones of *Taraxacum officinale* indicates the alternation of asexual and sexual histories of apomicts. *PLoS ONE* **2013**, *7*, e41868. [CrossRef] [PubMed]
13. Paun, O.; Stuessy, T.F.; Hörandl, E. The role of hybridization, polyploidization and glaciation in the origin and evolution of the apomictic *Ranunculus cassubicus* complex. *New Phytol.* **2006**, *171*, 223–236. [CrossRef]
14. Hörandl, E. The complex causality of geographical parthenogenesis. *New Phytol.* **2006**, *171*, 525–538. [CrossRef] [PubMed]
15. Hojsgaard, D.; Hörandl, E. Apomixis as a facilitator of range expansion and diversification in plants. In *Evolutionary Biology: Biodiversification from Genotype to Phenotype*, 1st ed.; Pontarotti, P., Ed.; Springer: Cham, Switzerland, 2015; pp. 305–327. [CrossRef]
16. Dickinson, T.A.; Yan, B.X.; Han, S.; Zarrei, M. Niche shifts, hybridization, polyploidy and geographic parthenogenesis in western North American hawthorns (*Crataegus* subg. *Sanguineae*, Rosaceae). *Agronomy* **2021**, *11*, 2133. [CrossRef]
17. Kurtto, A.; Webber, H.E.; Lampinen, R.; Sennikov, A.N. (Eds.) *Atlas Florae Europaeae. Distribution of Vascular Plants in Europe. 15. Rosaceae (Rubus)*, 1st ed.; The Committee for Mapping the Flora of Europe & Societas Biologica Fennica Vanamo: Helsinki, Finland, 2010.
18. Rich, T.C.G.; Houston, L.; Robertson, A.; Proctor, M.C.F. *Whitebeams, Rowans and Service Trees of Britain and Ireland. A monograph of British and Irish Sorbus L.*, 1st ed.; B.S.B.I. Handbook No. 14; Botanical Society of the British Isles: London, UK, 2010.
19. Kurtto, A.; Sennikov, A.N.; Lampinen, R. (Eds.) *Atlas Florae Europaeae. Distribution of Vascular Plants in Europe. 17. Rosaceae (Sorbus s. lato)*, 1st ed.; The Committee for Mapping the Flora of Europe & Societas Biologica Fennica Vanamo: Helsinki, Finland, 2018.
20. Vašut, R.J. *Taraxacum* sect. *Erythrosperma in Moravia (Czech Republic): Taxonomic notes and the distribution of previously described species. Preslia* **2003**, *75*, 311–338.
21. Robertson, A.; Sydes, C. *Sorbus pseudomeinichii*, a new endemic *Sorbus* (Rosaceae) microspecies from Arran, Scotland. *Watsonia* **2006**, *26*, 85591923.
22. Rich, T.C.G.; Charles, C.; Houston, L.; Tillotson, A. The diversity of *Sorbus* L. (Rosaceae) in the Lower Wye Valley. *Watsonia* **2009**, *27*, 301–313.
23. Lepší, M.; Vit, P.; Lepší, P.; Boublik, K.; Suda, J. *Sorbus milensis*, a new hybridogenous species from northwestern Bohemia. *Preslia* **2008**, *80*, 229–244.
24. Lepší, M.; Vit, P.; Lepší, P.; Boublik, K.; Kolar, F. *Sorbus portae-bohemicae* and *Sorbus albensis*, two new apomictic endemic species recognized based on revision of *Sorbus bohemica*. *Preslia* **2009**, *81*, 63–89.
25. Lepší, M.; Lepší, P.; Sádlo, J.; Koutecký, P.; Vít, P.; Petřík, P. *Sorbus pauca* species nova, the first endemic species of the *Sorbus hybrida* group for the Czech Republic. *Preslia* **2013**, *85*, 63–80.
26. Lepší, M.; Lepší, P.; Koutecký, P.; Bílá, J.; Vít, P. Taxonomic revision of *Sorbus* subgenus *Aria* occurring in the Czech Republic. *Preslia* **2015**, *87*, 109–162.

27. Aldasoro, J.J.; Aedo, C.; Navarro, C.; Garmendia, F.M. The genus *Sorbus* (Maloideae, Rosaceae) in Europe and in North Africa: Morphological analysis and systematics. *Syst. Bot.* **1998**, *23*, 189–212. [CrossRef]
28. Karbstein, K.; Tomasello, S.; Hodač, L.; Dunkel, F.G.; Daubert, M.; Hörandl, E. Phylogenomics supported by geometric morphometrics reveals delimitation of sexual species within the polyploid apomictic *Ranunculus auricomus* complex (Ranunculaceae). *Taxon* **2020**, *69*, 1191–1220. [CrossRef]
29. Zarrei, M.; Stefanović, S.; Dickinson, T.A. Reticulate evolution in North American black-fruited hawthorns (*Crataegus* section *Douglasia*; Rosaceae): Evidence from nuclear ITS2 and plastid sequences. *Ann. Bot.* **2014**, *114*, 253–269. [CrossRef]
30. Paule, J.; Scherbantin, A.; Dobeš, C. Implications of hybridisation and cytotypic differentiation in speciation assessed by AFLP and plastid haplotypes—A case study of *Potentilla alpicola* La Soie. *BMC Evol. Biol.* **2012**, *12*, 132. [CrossRef]
31. Šarhanová, P.; Sharbel, T.F.; Sochor, M.; Vašut, R.; Dančák, M.; Trávníček, B. Hybridization drives evolution of apomicts in *Rubus* subgenus *Rubus*: Evidence from microsatellite markers. *Ann. Bot.* **2017**, *120*, 317–328. [CrossRef]
32. Nelson-Jones, E.B.; Briggs, D.; Smith, A.G. The origin of intermediate species of the genus *Sorbus*. *Theor. Appl. Genet.* **2002**, *105*, 953–963. [CrossRef]
33. Robertson, A.; Newton, A.C.; Ennos, R.A. Multiple hybrid origins, genetic diversity and population genetic structure of two endemic *Sorbus* taxa on the Isle of Arran, Scotland. *Mol. Ecol.* **2004**, *13*, 123–134. [CrossRef]
34. Robertson, A.; Rich, T.C.G.; Allen, A.M.; Houston, L.; Roberts, C.; Bridle, J.R.; Harris, S.A.; Hiscock, S.J. Hybridization and polyploidy as drivers of continuing evolution and speciation in *Sorbus*. *Mol. Ecol.* **2010**, *19*, 1675–1690. [CrossRef]
35. Ludwig, S.; Robertson, A.; Rich, T.C.G.; Djordjevic, M.; Cerovic, R.; Houston, L.; Harris, S.A.; Hiscock, S.J. Breeding systems, hybridisation and continuing evolution in Avon Gorge *Sorbus*. *Ann. Bot.* **2013**, *111*, 563–575. [CrossRef]
36. Hamston, T.J.; de Vere, N.; King, A.R.; Pellicer, J.; Fay, M.F.; Creswell, J.E.; Stevens, J.R. Apomixis and hybridization drives reticulate evolution and phyletic differentiation in *Sorbus* L.: Implications for conservation. *Front. Plant Sci.* **2018**, *9*, 1796. [CrossRef] [PubMed]
37. Lepší, M.; Koutecký, P.; Nosková, J.; Lepší, P.; Urfus, T.; Rich, T.C.G. Versatility of reproductive modes and ploidy level interactions in *Sorbus* s.l. (Malinae, Rosaceae). *Bot. J. Linn. Soc.* **2019**, *191*, 502–522. [CrossRef]
38. Jankun, A. Evolutive significance of apomixis in the genus *Sorbus* (Rosaceae). *Fragm. Florist. Geobot.* **1993**, *38*, 627–686.
39. Hajrudinović, A.; Frajman, B.; Schönswetter, P.; Silajdžić, E.; Siljak-Yakovlev, S.; Bogunić, F. Towards a better understanding of polyploid *Sorbus* (Rosaceae) from Bosnia and Herzegovina (Balkan Peninsula), including description of a novel, tetraploid apomictic species. *Bot. J. Linn. Soc.* **2015**, *178*, 670–685. [CrossRef]
40. Meyer, N.; Meierott, L. Ergänzende Beiträge zur *Sorbus*-Flora von Bayern. *Bayer. Bot. Ges.* **2021**, *91*, 21–48.
41. Hajrudinović, A.; Siljak-Yakovlev, S.; Brown, S.C.; Pustahija, F.; Bourge, M.; Ballian, D.; Bogunić, F. When sexual meets apomict: Genome size, ploidy level and reproductive mode variation of *Sorbus aria* s.l. and *S. austriaca* (Rosaceae) in Bosnia and Herzegovina. *Ann. Bot.* **2015**, *116*, 301–312. [CrossRef]
42. Sennikov, A.N.; Kurtto, A. A phylogenetic checklist of *Sorbus* s.l. (Rosaceae) in Europe. *Memo. Soc. Fauna. Flora. Fenn.* **2017**, *93*, 1–78.
43. Christenhusz, M.J.M.; Fay, M.F.; Byng, J.W. (Eds.) *The Global Flora. A Practical Flora to Vascular Plant Species of the World—Special Edition*, 1st ed.; GLOVAP Nomenclature Part 1; Plant Gateway Ltd.: Bradford, UK, 2018; Volume 4.
44. Meyer, N.; Meierott, L.; Schuwerk, H.; Angerer, O. *Beiträge zur Gattung Sorbus in Bayern*, 1st ed.; Berichte der Bayerischen Botanischen Gesellschaft zur Erforschung der heimischen Flora: Sonderband, Germany, 2005.
45. Velebil, J.; Lepší, M.; Nosková, J.; Lepší, P. Taxonomic assessment of *Sorbus* subgenus *Aria* in the Malé Karpaty Mountains. *Preslia* **2022**, *94*, 305–334. [CrossRef]
46. Warburg, E.F.; Kárpáti, Z.E. *Sorbus* L. In *Flora Europaea 2*; Tutin, T.G., Heywood, V.H., Burgess, N.A., Moore, D.M., Valentine, D.H., Walters, S.M., Webb, D.A., Eds.; Cambridge University Press: Cambridge, UK, 1968; pp. 67–71.
47. Beck, G.M. *Flora von Nieder-Österreich*; Druck und Verlag von Carl Gerold's Sohn: Wien, Austria, 1890; Volume 2, p. 714.
48. Kovanda, M. *Sorbus austriaca* (Beck) Prain in Italy. *Webbia* **1998**, *53*, 27–29. [CrossRef]
49. Rivers, M.C.; Beech, E. *Sorbus austriaca* (Beck) Hedl. *IUCN Red List. Threat. Species* **2017**. [CrossRef]
50. Májovský, J.; Bernátová, D. *Sorbus pekarovae* sp. nova: A new hybridogeneous species from the Veľká Fatra Mts. *Biologia* **1996**, *51*, 23–26.
51. Rich, T.G.C.; Hjertson, M.; Moberg, R. Typification of some T. Hedlund names in *Sorbus* L. (Rosaceae). *Taxon* **2006**, *55*, 203–206. [CrossRef]
52. Kurtto, A. Rosaceae (pro parte majore). In *Euro+Med Plantbase—The Information Resource for Euro-Mediterranean Plant Diversity, 2009+*; Available online: https://europlusmed.org/ (accessed on 9 November 2022).
53. Jovanović, B. Rod *Sorbus* L. In *Flora of SR Serbia 4*, 1st ed.; Josifović, M., Ed.; Srpska akademija Nauka i Umetnosti, Odeljenje Prirodno-Matematičkih Nauka: Beograd, Srbija, 1972; pp. 150–160.
54. Fischer, M.A.; Oswald, K.; Adler, W. *Exkursionsflora für Österreich, Liechtenstein und Südtirol*, 1st ed.; Biologiezentrum der Oberösterreichischen Landesmuseen: Linz, Austria, 2008.
55. Lauber, K.; Wagner, G.; Gygax, A. *Flora Helvetica–Illustrierte Flora der Schweiz*, 2nd ed.; Haupt Verlag: Bern, Switzerland, 2018.
56. Nikolić, T. *Flora Croatica 3—Vaskularna Flora Republike Hrvatske*, 1st ed.; ALFA d.d.: Zagreb, Croatia, 2020.
57. Tel-zur, N.; Abbo, S.; Myslabodski, D.; Mizrahi, Y. Modified CTAB procedure for DNA isolation from epiphytic cacti of genera *Hylocereus* and *Selenicereus* (Cactaceae). *Plant Mol. Biol. Rep.* **1999**, *17*, 249–254. [CrossRef]

58. Vos, P.; Hogers, R.; Bleeker, M.; Reijans, M.; van de Lee, T.; Hornes, M.; Frijters, A.; Pot, J.; Peleman, J.; Kuiper, M.; et al. AFLP: A new technique for DNA fingerprinting. *Nucleic Acids Res.* **1995**, *23*, 4407–4414. [CrossRef]
59. Arrigo, N.; Tuszynski, J.W.; Ehrich, D.; Gerdes, T.; Alvarez, N. Evaluating the impact of scoring parameters on the structure of intra-specific genetic variation using RawGeno, an R package for automating AFLP scoring. *BMC Bioinform.* **2009**, *10*, 33. [CrossRef]
60. R Core Team. *R: A Language and Environment for Statistical Computing*; R Foundation for Statistical Computing: Vienna, Austria, 2022; Available online: http://www.r-project.org/ (accessed on 9 November 2022).
61. Nei, M.; Li, W.H. Mathematical model for studying genetic variation in terms of restriction endonucleases. *Proc. Natl. Acad. Sci. USA* **1979**, *76*, 5269–5273. [CrossRef]
62. Van de Peer, Y.; De Wachter, R. Construction of evolutionary distance trees with TREECON for Windows: Accounting for variation in nucleotide substitution rate among sites. *Comput. Appl. Biosci.* **1997**, *13*, 227–230. [CrossRef]
63. Huson, D.H. SplitsTree: A program for analyzing and visualizing evolutionary data. *Bioinformatics* **1998**, *14*, 68–73. [CrossRef]
64. Hammer, Ø.; Harper, D.A.T.; Ryan, P.D. PAST: Paleontological statistics software package for education and data analysis. *Palaeontol. Electron.* **2001**, *4*, 1–9.
65. Meirmans, P.G.; Van Tienderen, P.H. GENOTYPE and GENODIVE: Two programs for the analysis of genetic diversity of asexual organisms. *Mol. Ecol. Res.* **2004**, *4*, 792–794. [CrossRef]
66. Taberlet, P.; Gielly, L.; Pautou, G.; Bouvet, J. Universal primers for amplification of three non-coding regions of chloroplast DNA. *Plant Mol. Biol.* **1991**, *17*, 1105–1109. [CrossRef]
67. Kearse, M.; Moir, R.; Wilson, A.; Stones-Havas, S.; Cheung, M.; Sturrock, S.; Buxton, S.; Cooper, A.; Markowitz, S.; Duran, C.; et al. Geneious Basic: An integrated and extendable desktop software platform for the organization and analysis of sequence data. *Bioinformatics* **2012**, *28*, 1647–1649. [CrossRef]
68. Simmons, M.P.; Ochoterena, H. Gaps as characters in sequence-based phylogenetic analyses. *Syst. Biol.* **2000**, *49*, 369–381. [CrossRef]
69. Müller, K. SeqState: Primer design and sequence statistics for phylogenetic DNA datasets. *Appl. Bioinform.* **2005**, *4*, 65–69. [CrossRef]
70. Swofford, D. *PAUP*. Phylogenetic Analysis Using Parsimony (*and Other Methods), Version 4.0b10*; Sinauer Associates: Sunderland, MA, USA, 2002. [CrossRef]
71. Ronquist, F.; Teslenko, M.; van der Mark, P.; Ayres, D.L.; Darling, A.; Höhna, S.; Larget, B.; Liu, L.; Suchard, M.A.; Huelsenbeck, J.P. MrBayes 3.2: Efficient Bayesian phylogenetic inference and model choice across a large model space. *Syst. Biol.* **2012**, *61*, 539–542. [CrossRef] [PubMed]
72. Nylander, J.A.A. *MrAIC.pl. Program Distributed by the Author*; Evolutionary Biology Centre, Uppsala University: Uppsala, Sweden, 2004.
73. Lewis, P.O. A likelihood approach to estimating phylogeny from discrete morphological character data. *Syst. Biol.* **2001**, *50*, 913–925. [CrossRef]
74. Marie, D.; Brown, S.C. A cytometric exercise in plant DNA histograms, with 2C values for 70 species. *Biol. Cell* **1993**, *78*, 41–51. [CrossRef] [PubMed]
75. Suda, J.; Krahulcová, A.; Trávníček, P.; Krahulec, F. Ploidy level versus DNA ploidy level: An appeal for consistent terminology. *Taxon* **2006**, *55*, 447–450. [CrossRef]
76. Siljak-Yakovlev, S.; Pustahija, F.; Šolić, M.E.; Bogunić, F.; Muratović, E.; Bašić, N.; Catrice, O.; Brown, S.C. Towards a genome size and chromosome number database of Balkan flora: C-values in 343 taxa with novel values for 242. *Adv. Sci. Lett.* **2010**, *3*, 190–213. [CrossRef]
77. Aldasoro, J.J.; Aedo, C.; Garmendia, F.M.; de la Hoz, F.P.; Navarro, C. Revision of Sorbus Subgenera Aria and Torminaria (Rosaceae-Maloideae). *Syst. Bot. Monogr.* **2004**, *69*, 1–148. [CrossRef]
78. McGarigal, K.; Cushman, S.; Stafford, S. *Multivariate Statistics for Wildlife and Ecology Research*, 1st ed.; Springer Science+Business Media Inc.: New York, NY, USA, 2000. [CrossRef]
79. Valcárcel, V.; Vargas, P. Quantitative morphology and species delimitation under the general lineage concept: Optimization for *Hedera* (Araliaceae). *Am. J. Bot.* **2010**, *97*, 1555–1573. [CrossRef]
80. Pellicer, J.; Clermont, S.; Houston, L.; Rich, T.C.G.; Fay, M.F. Cytotype diversity in the *Sorbus* complex (Rosaceae) in Britain: Sorting out the puzzle. *Ann. Bot.* **2012**, *110*, 1185–1193. [CrossRef]
81. Uhrinová, V.; Zozomova-Lihova, J.; Bernatova, D.; Paule, J.; Paule, L.; Gömöry, D. Origin and genetic differentiation of pink-flowered *Sorbus* hybrids in the Western Carpathians. *Ann. Bot.* **2017**, *120*, 271–284. [CrossRef]
82. Feulner, M.; Weig, A.; Paule, J.; Gregor, T.; Schott, L.F.; Aas, G. Genetic variability and morphology of tri- and tetraploid members of the *Sorbus aria* complex in northern Bavaria. *Preslia* **2017**, *89*, 275–290. [CrossRef]
83. van Husen, D. *Die Ostalpen in den Eiszeiten*, 1st ed.; Geologische Bundesanstalt: Wien, Austria, 1987.
84. Stepišnik, U.; Grlj, A.; Radoš, D.; Žebre, M. Geomorphology of Blidinje, Dinaric Alps (Bosnia and Herzegovina). *J. Maps* **2016**, *12*, 163–171. [CrossRef]
85. Turril, W.B. *The Plant Life of the Balkan Peninsula. Phytogeographical Study*, 1st ed.; Oxford: Clarendon, UK, 1929.

86. Horvat, I.; Glavač, V.; Ellenberg, H. *Vegetation Südosteuropas*, 1st ed.; Gustav Fischer Verlag: Stuttgart, Germany, 1974.
87. Feulner, M.; Weig, A.; Voss, T.; Schott, L.F.; Aas, G. Central European polyploids of *Sorbus* subgenus *Aria* (Rosaceae) recurrently evolved from diploids of central and south-eastern Europe: Evidence from microsatellite data. *Bot. J. Linn. Soc.* **2019**, *191*, 315–324. [CrossRef]
88. Oddou-Muratorio, S.; Guesnet, D.; Ozdemir, E.; Petit, R.J.; Demesure, B. Patterns of seed dispersal in a scattered forest tree species (*Sorbus torminalis*) based on multi-scale investigations of populations genetic structure for chloroplast DNA. In *Genetic Response of Forest Systems to Changing Environmental Conditions*, 1st ed.; Muller-Starck, G., Schubert, R., Eds.; Kluwer Academic Publisher: Dodrecht, The Netherlands, 2001; pp. 271–280.
89. Šarhanová, P.; Vašut, R.J.; Dančák, M.; Bureš, P.; Trávniček, B. New insight into the variability of reproduction modes in European populations of *Rubus* subg. *Rubus*: How sexual are polyploid brambles? *Sex. Plant Reprod.* **2012**, *25*, 319–335. [CrossRef] [PubMed]
90. Richards, A.J. Apomixis in flowering plants: An overview. *Phil. Trans. R. Soc. Lond. B* **2003**, *358*, 1085–1093. [CrossRef] [PubMed]
91. Spillane, C.; Steimer, A.; Grossniklauss, U. Apomixis in agriculture: The quest for clonal seeds. *Sex. Plant Reprod.* **2001**, *14*, 179–187. [CrossRef] [PubMed]
92. Ferreira de Carvalho, J.; Oplaat, C.; Pappas, N.; Derks, M.; Ridder, D.; de Verhoeven, J.F. Heritable gene expression differences between apomictic clone members in *Taraxacum officinale*: Insight into early stages of evolutionary divergence in asexual plants. *BMC Genom.* **2016**, *17*, 203. [CrossRef]
93. Wilschut, R.A.; Oplaat, C.; Snoek, L.B.; Kirschner, J.; Verhoeven, K.J.F. Natural epigenetic variation contributes to heritable flowering divergence in a widespread asexual dandelion lineage. *Mol. Ecol.* **2016**, *25*, 1759–1768. [CrossRef]
94. Ehlers, J.; Gibbard, P.L.; Hughes, P.D. *Quaternary Glaciations–Extent and Chronology: A Closer Look*, 1st ed.; Elsevier: Amsterdam, The Netherlands, 2011.
95. Hampe, A.; Arroyo, J.; Jordano, P.; Petit, R.J. Rangewide phylogeography of a bird-dispersed Eurasian shrub: Contrasting Mediterranean and temperate glacial refugia. *Mol. Ecol.* **2003**, *12*, 3415–3426. [CrossRef] [PubMed]
96. Magri, D.; Vendramin, G.G.; Comps, B.; Dupanloup, I.; Geburek, T.; Gömöry, D.; Latałowa, M.; Litt, T.; Paule, L.; Roure, J.M.; et al. A new scenario for the Quaternary history of European beech populations: Palaeobotanical evidence and genetic consequences. *New Phytol.* **2006**, *171*, 199–221. [CrossRef]
97. Grivet, D.; Petit, R.J. Chloroplast DNA phylogeography of the hornbeam in Europe: Evidence for a bottleneck at the outset of postglacial colonization. *Conserv. Genet.* **2003**, *4*, 47–56. [CrossRef]
98. Mogensen, H.L. Invited special paper: The hows and whys of cytoplasmic inheritance in seed plants. *Am. J. Bot.* **1996**, *83*, 383–404. [CrossRef]
99. Raspé, O. Inheritance of the chloroplast genome in *Sorbus aucuparia* L. (Rosaceae). *J. Hered.* **2001**, *92*, 507–509. [CrossRef] [PubMed]
100. Chester, M.; Cowan, R.S.; Fay, M.F.; Rich, T.C.G. Parentage of endemic *Sorbus* L. (Rosaceae) species in the British Isles: Evidence from plastid DNA. *Bot. J. Linn. Soc.* **2007**, *154*, 291–304. [CrossRef]
101. Karbstein, K.; Tomasello, S.; Hodač, L.; Wagner, N.; Marinček, P.; Barke, B.H.; Paetzold, C.; Hörandl, E. Untying Gordian knots: Unraveling reticulate polyploid plant evolution by genomic data using the large *Ranunculus auricomus* species complex. *New Phytol.* **2022**, *235*, 2081–2098. [CrossRef] [PubMed]
102. Soltis, P.S.; Soltis, D.E. Ancient WGD events as drivers of key innovations in angiosperms. *Curr. Opin. Plant Biol.* **2016**, *30*, 159–165. [CrossRef]
103. Melichárková, A.; Španiel, S.; Marhold, K.; Hurdu, B.-I.; Drescher, A.; Zozomová-Lihová, J. Diversification and independent polyploid origins in the disjunct species *Alyssum repens* from the Southeastern Alps and the Carpathians. *Am. J. Bot.* **2019**, *106*, 1499–1518. [CrossRef]
104. Cushman, K.R.; Burgess, M.B.; Doucette, E.T.; Nelson, G.A.; Campbell, C.S. Species delimitation in tetraploid, apomictic *Amelanchier* (Rosaceae). *Syst. Bot.* **2017**, *42*, 234–256. [CrossRef]
105. Thapa, R.; Bayer, R.J.; Mandel, J.R. Genetic diversity, population structure, and ancestry estimation in the *Antennaria rosea* (Asteraceae: Gnaphalieae) polyploid agamic complex. *Taxon* **2021**, *70*, 139–152. [CrossRef]
106. Liston, A.; Weitmier, K.A.; Letelier, L.; Podani, J.; Zong, Y.; Liu, L.; Dickinson, T.A. Phylogeny of *Crataegus* (Rosaceae) based on 257 nuclear loci and chloroplast genomes: Evaluating the impact of hybridization. *PeerJ* **2021**, *9*, e12418. [CrossRef]
107. Majeský, L.; Krahulec, F.; Vašut, R.J. How apomictic taxa are treated in current taxonomy: A review. *Taxon* **2017**, *66*, 117–1140. [CrossRef]
108. Levin, J.; Fay, M.F.; Pellicer, J.; Hedrén, M. Multiple independent origins of intermediate species between *Sorbus aucuparia* and *S. hybrida* (Rosaceae) in the Baltic region. *Nord. J. Bot.* **2018**, *35*. [CrossRef]
109. Rich, T.C.G.; Gree, D.; Houston, L.; Lepší, M.; Ludwig, S.; Pellicer, J. British *Sorbus* (Rosaceae): Six new species, two hybrids and a new subgenus. *New J. Bot.* **2014**, *4*, 2–12. [CrossRef]
110. Rieseberg, L.H. Hybrid origins of plant species. *Ann. Rev. Ecol. Syst.* **1997**, *28*, 359–389. [CrossRef]
111. Hegarty, M.J.; Batstone, T.O.M.; Barker, G.L.; Edwards, K.J.; Abbott, R.J.; Hiscock, S.J. Nonadditive changes to cytosine methylation as a consequence of hybridization and genome duplication in *Senecio* (Asteraceae). *Mol. Ecol.* **2011**, *20*, 105–113. [CrossRef]

112. Ennos, R.A.; French, G.C.; Hollingsworth, P.M. Conserving taxonomic complexity. *Trends Ecol. Evol.* **2005**, *20*, 164–168. [CrossRef]
113. Ennos, R.A.; Whitlock, R.; Fay, M.F.; Jones, B.; Neaves, L.E.; Payne, R.; Taylor, I.; De Vere, N.; Hollingsworth, P.M. Process-based species action plans: An approach to conserve contemporary evolutionary processes that sustain diversity in taxonomically complex groups. *Bot. J. Linn. Soc.* **2012**, *168*, 194–203. [CrossRef]

biology

MDPI

Article

Disentangling *Crocus* Series *Verni* and Its Polyploids

Irena Raca [1], Frank R. Blattner [2,3], Nomar Espinosa Waminal [2], Helmut Kerndorff [4], Vladimir Ranđelović [1] and Dörte Harpke [2,*]

1 Department of Biology and Ecology, University of Niš, 18000 Niš, Serbia
2 Leibniz Institute of Plant Genetics and Crop Plant Research (IPK), D-06466 Seeland-Gatersleben, Germany
3 German Centre of Integrative Biodiversity Research (iDiv) Halle-Jena-Leipzig, D-04103 Leipzig, Germany
4 São Romão, Cx 548 A, 8150-058 São Brás de Alportel, Portugal
* Correspondence: harpke@ipk-gatersleben.de

Simple Summary: In plants, the occurrence of polyploid lineages, which are plants with multiple instead of two sets of chromosomes, is quite common. Polyploids can originate as autopolyploids within a species or by combining the genomes of different species resulting in allopolyploids. Within the group of spring crocuses, a polyploid complex exists where it is unclear how it evolved and which species eventually contributed to polyploid formation. Among *Crocus* species, evolutionary analyses are further complicated by widely varying chromosome numbers that do not clearly correlate with di- or polyploidy. To reconstruct the evolution of these polyploids, we combine chromosome counts, genome size estimations, phylogenetic analyses based on maternally and bi-parentally inherited genomes, co-ancestry analysis, and morphometric data for all species potentially involved in polyploid formation. Through this approach, we show that polyploids in the *Crocus heuffelianus* group are allopolyploids that originated multiple times involving different parental genotypes and reciprocal crosses. Chromosome numbers partly changed after polyploidization. Numbers found in polyploids are therefore no longer in all cases additive values of their parents' chromosomes. We conclude that in crocuses, only an approach combining evidence from different analysis methods can uncover the evolutionary history of species if polyploidization is involved.

Abstract: Spring crocuses, the eleven species within *Crocus* series *Verni* (Iridaceae), consist of di- and tetraploid cytotypes. Among them is a group of polyploids from southeastern Europe with yet-unclear taxonomic affiliation. Crocuses are generally characterized by complex dysploid chromosome number changes, preventing a clear correlation between these numbers and ploidy levels. To reconstruct the evolutionary history of series *Verni* and particularly its polyploid lineages associated with *C. heuffelianus*, we used an approach combining phylogenetic analyses of two chloroplast regions, 14 nuclear single-copy genes plus rDNA spacers, genome-wide genotyping-by-sequencing (GBS) data, and morphometry with ploidy estimations through genome size measurements, analysis of genomic heterozygosity frequencies and co-ancestry, and chromosome number counts. Chromosome numbers varied widely in diploids with $2n$ = 8, 10, 12, 14, 16, and 28 and tetraploid species or cytotypes with $2n$ = 16, 18, 20, and 22 chromosomes. *Crocus longiflorus*, the diploid with the highest chromosome number, possesses the smallest genome (2C = 3.21 pg), while the largest diploid genomes are in a range of 2C = 7–8 pg. Tetraploid genomes have 2C values between 10.88 pg and 12.84 pg. Heterozygosity distribution correlates strongly with genome size classes and allows discernment of di- and tetraploid cytotypes. Our phylogenetic analyses showed that polyploids in the *C. heuffelianus* group are allotetraploids derived from multiple and partly reciprocal crosses involving different genotypes of diploid *C. heuffelianus* ($2n$ = 10) and *C. vernus* ($2n$ = 8). Dysploid karyotype changes after polyploidization resulted in the tetraploid cytotypes with 20 and 22 chromosomes. The multi-data approach we used here for series *Verni*, combining evidence from nuclear and chloroplast phylogenies, genome sizes, chromosome numbers, and genomic heterozygosity for ploidy estimations, provides a way to disentangle the evolution of plant taxa with complex karyotype changes that can be used for the analysis of other groups within *Crocus* and beyond. Comparing these results with morphometric analysis results in characters that can discern the different taxa currently subsumed under *C. heuffelianus*.

Citation: Raca, I.; Blattner, F.R.; Waminal, N.E.; Kerndorff, H.; Ranđelović, V.; Harpke, D. Disentangling *Crocus* Series *Verni* and Its Polyploids. *Biology* **2023**, *12*, 303. https://doi.org/10.3390/biology12020303

Academic Editor: Lorenzo Peruzzi

Received: 30 December 2022
Revised: 6 February 2023
Accepted: 7 February 2023
Published: 14 February 2023

Keywords: chromosome numbers; *Crocus heuffelianus* group; *Crocus* series *Verni*; dysploidy; genome size; genotyping-by-sequencing (GBS); morphometry; phylogeny; polyploidy

1. Introduction

Polyploidization or whole-genome duplication (WGD) is a common process in plants, resulting in individuals with different ploidy levels. Two major mechanisms are discerned: if WGD happens within a species, the resulting polyploid is termed autopolyploid, whereas hybridization between two different species followed by WGD is termed allopolyploidy [1]. Autopolyploids might suffer at least initially from reduced fertility due to distorted chromosome distribution during meiosis, while allopolyploids usually undergo normal meiosis. However, one has to understand auto- and allopolyploids as the endpoints of a continuum. Chromosome pairing and distribution among daughter cells depends on the overall similarity of the chromosomes. Thus, even within a species, chromosomes can be relatively different or rather similar in an allopolyploid if closely related species are involved. Polyploidization is often a driver of evolution, as doubling of genes releases one of the homeologous copies from purifying selection, allowing it to obtain new functions [2]. This event should be advantageous, particularly in stressful habitats or during changing environmental conditions [3]. Through time, polyploid genomes will accumulate differences not only at the level of allelic differences but, through karyotype changes, also regarding the overall structure of the genomes [1]. This results in diploidization, i.e., former homeologous chromosomes are no longer recognizable as such. Ancient polyploidization events are therefore not easy to detect and most often need in-depth genome analysis to be revealed [2,4]. For more recent WGD events, chromosome numbers and genome sizes can be indicative. While there are taxa where chromosome numbers correlate clearly with ploidy levels and genome size [5,6], dysploid chromosome number changes can blur such correlations through breaking and/or fusion of chromosomes [7]. In addition, downsizing or enlarging of genomes [4,8–10] through the loss of DNA or activation of transposable elements can hinder ploidy level recognition. However, these latter processes are normally acting at a slower pace than changes in chromosome numbers [11,12].

Crocus series *Verni* B.Mathew is a group of mostly spring-flowering crocuses from Central and South Europe, some of them being important ornamentals. The series consists of eleven species with unclear phylogenetic relationships, as earlier approaches with molecular markers arrived only at badly resolved species groups and species identification was also partly uncertain [13–15] or sampling incomplete [16]. Chromosome numbers range from $2n = 8$ to $2n = 28$ [16–23], often with uncertain ploidy levels, as high chromosome numbers do not necessarily correlate with larger genome sizes and, therefore, higher ploidy level [16]. Particular populations from the Balkan Peninsula, thought to taxonomically belong to *C. heuffelianus* Herb., exhibit highly diverse chromosome numbers with $2n = 10$, 18, 19, 20, 22, and 23 chromosomes [16,18]. Harpke et al. [15], in their account of series *Verni*, concluded from karyotype analysis that certain populations of *C. heuffelianus* (as well as *C. neglectus* Peruzzi and Carta) might have resulted from polyploidization but were not able to confirm this further. We refer to these potential polyploids throughout this article as "*C.* cf. *heuffelianus*", "*C.* cf. *tommasinianus*", and "*C.* cf. *vernus*", as their taxonomic status is unclear and name changes seem still premature to us. *Crocus neglectus*, in contrast, was already recognized as a separate species [15].

Here we intend to understand the evolution of *Crocus* series *Verni* with particular reference to the origin of the polyploid species and cytotypes. To arrive at this goal, we use molecular phylogenetic approaches based on (a) genotyping-by-sequencing (GBS [24]) to obtain highly informative genome-wide single-nucleotide polymorphisms (SNPs) for a robust phylogeny, (b) nuclear rDNA internal transcribed spacers (ITS) plus 14 single-copy gene sequences for tracing bi-parentally inherited genome parts and chloroplast DNA sequences for inferring maternal lineages within the study group, and (c) morpho-anatomical analyses to find traits that can discern the taxa, and combine these data with (d) chromosome counts and (e) genome size estimations of diverse populations in order to infer di- and polyploid taxa and cytotypes and reveal their parental contributors and geographic distribution.

2. Materials and Methods

2.1. Plant Materials

Our study includes plants from 63 populations: 24 *C. heuffelianus*/*C.* cf. *heuffelianus*, 13 *C. vernus* (L.) Hill/*C.* cf. *vernus*, nine *C. tommasinianus* Herb., three *C. bertiscensis* Raca, Harpke, Shuka, and V.Randjel., three *C. neapolitanus* (Ker Gawl.) Loisel., two *C. neglectus*, two *C. kosaninii* Pulević, one of each of the other series *Verni* species, two populations of outgroup *C. malyi* Vis., and one ornamental cultivar (Table 1, Supplementary Table S1). To differentiate the $2n = 18$ karyotypes of *C.* cf. *heuffelianus*, we labeled them Western Carpathian clade (WCC), Pannonian-Illyric clade (PIC), and Southern Carpathian clade (SCC). The sampling covers the whole distribution area of the series except for the westernmost *C. vernus* populations from the Pyrenees, which could only be included in the chloroplast and nuclear single-copy marker dataset (Table 1). To encumber poaching on the wild populations we here provide only rather general locations for the studied materials instead of populations' GPS coordinates. Chloroplast and nuclear single-copy markers were investigated for a smaller number of representatives of series *Verni* while the GBS analyses were based on the most exhaustive number of individuals (Table 1).

Table 1. Materials from *Crocus* ser. *Verni* used in this study (for an extended list see Supplementary Table S1).

Species	Individuals in Chloroplast/ GBS/NSCG [1]/Morpho-Anatomical Analyses	Origin	*x*/Chromosome Number/Average 2C Genome Size
Crocus bertiscensis Raca, Harpke, Shuka, and V.Randjel.	4/6/1/0	Albania (ALB), Montenegro (MNE)	2*x*/12/6.66 pg
Crocus etruscus Parl.	2/3/0/0	Italy (ITA)	2*x*/8/7.58 pg
Crocus ilvensis Peruzzi and Carta	4/5/1/0	Italy (ITA)	2*x*/8/7.88 pg
Crocus kosaninii Pulević	2/3/2/0	Kosovo (XKX), Serbia (SRB)	2*x*/14/7.95 pg
Crocus longiflorus Raf.	2/1/2/0	Italy (ITA)	2*x*/28/3.21 pg
Crocus neglectus Peruzzi and Carta	1/1/1/0	Germany (GER)	4*x*/16/12.24 pg
Crocus neapolitanus (Ker Gawl.) Loisel.	3/2/1/0	Italy (ITA)	2*x*/8/–
Crocus siculus Tineo	0/1/0/0	Italy (ITA)	2*x*/8/–
Crocus tommasinianus Herb.	5/25/2/0	Bosnia and Herzegovina (BIH), Italy (ITA), Montenegro (MNE), Serbia (SRB)	2*x*/16/5.53 pg
Crocus vernus (L.) Hill	14/27/4/100	Albania (ALB), Bosnia and Herzegovina (BIH), France (FRA), Montenegro (MNE), Slovenia (SLO), Switzerland (CHE)	2*x*/8/5.78 pg
Crocus heuffelianus Herb.	8/22/4/70	Romania (ROU), Slovakia (SVK), Ukraine (UKR)	2*x*/10/7.73 pg
Crocus cf. *heuffelianus* (SCC)	7/19/1/70	Romania (ROU)	4*x*/18/12.84 pg
Crocus cf. *heuffelianus* (PIC)	5/26/1/40	Bosnia and Herzegovina (BIH), Slovenia (SLO)	4*x*/18/10.88 pg
Crocus cf. *heuffelianus* (WCC)	3/9/2/35	Slovakia (SVK)	4*x*/18/12.75 pg
Crocus cf. *heuffelianus*	12/23/0/80	Montenegro (MNE), Serbia (SRB)	4*x*/20/11.82 pg
Crocus cf. *heuffelianus*	3/14/1/40	Albania (ALB), Kosovo (XKX)	4*x*/22/11.95 pg
Crocus cf. *vernus*	5/9/1/0	Albania (ALB)	4*x*/16/12.38 pg
Crocus cf. *tommasinianus*	0/1/0/0	Montenegro (MNE)	4*x*/-/-
Crocus malyi Vis. (outgroup)	2/2/2/0	Croatia (HRV)	2*x*/30/–

[1] NSCG = nuclear single-copy genes plus rDNA ITS.

2.2. DNA Extraction and Sanger Sequencing

Total genomic DNA was extracted from silica-gel-dried leaf tissue with the DNeasy Plant Mini Kit (Qiagen) according to the instructions of the manufacturer. After DNA extraction, we checked DNA quality and concentration on 1% agarose gels. For the amplification of the two chloroplast regions, we used the primers matKf, rpS16in1_r, rpS16in1_f, trnQr, ycf1bF, and ycf1bR [25]. PCR amplification protocols for all markers followed Harpke and Kerndorff [25]. Forward and reverse strands of both regions were directly Sanger sequenced on an ABI 3730 XL using the amplification primers, edited where necessary, and assembled into single sequences in GENEIOUS Prime 2022.1.1 [26]. Afterward, the sequences were aligned using MAFFT version 1.5.0 [27] within GENEIOUS and manually corrected.

2.3. Genotyping-by-Sequencing

To obtain genome-wide SNPs, GBS analyses [24] were conducted for 91 di- and 102 tetraploid individuals, with one of the latter included twice as a replicate. For the library preparation, 200 ng of genomic DNA was used and cut with the two restriction enzymes *Pst*I-HF (NEB) and *Msp*I (NEB). Library preparation, individual barcoding, and single-end sequencing on the Illumina NovaSeq were performed following Wendler et al. [28].

Barcoded reads from the 194 samples were de-multiplexed using the CASAVA pipeline 1.8 (Illumina). Adapter trimming of GBS sequence reads was performed with CUTADAPT [29] within IPYRAD v.0.9.58 [30] and reads shorter than 60 bp after adapter removal were discarded. GBS reads were clustered using the IPYRAD 0.7.5 [30] pipeline with a clustering threshold of 0.85. We tested diverse IPYRAD settings but at the end the default settings of parameter files generated with IPYRAD were optimal for the other parameters. We generated one output that included the outgroup *C. malyi*, which was used for phylogenetic analyses, and a second output without *C. malyi*, which was used for principal component analysis (PCA) and population assignment analyses.

2.4. Analyses of Population Structure

Principal component analysis (PCA) was conducted in IPYRAD. For model-based Bayesian population assignment analysis we used the R package LEA [31]. Population assignment was performed for K = 1–15 with 20 repetitions each and ploidy set to four. Additionally, FASTSTRUCTURE [32] was used with "simple" as prior for K = 1–15 with 20 repetitions each for cross-validation. The optimal K was then determined with the function "chooseK" in FASTSTRUCTURE. The Q-matrices obtained with LEA (for K = 5, K = 8) and FASTSTRUCTURE (for K = 4), which include the ancestral assignment frequencies, were sorted using the R package tidyverse [33] and plotted with ggplot2 [34], discerning different ancestral clusters with color-coding. The ggplot2 package was also used for plotting the PCA results.

2.5. Heterozygosity and Fst Determination

The allelic constitution of SNP positions was checked in the vcf file obtained with IPYRAD to infer if and to what extent at these positions more than two alleles for a heterozygous SNP were present in the diploid individuals. As overall only less than 3% of SNPs were not biallelic, DNASP v. 6 [35] was used to infer the heterozygosity as well as the fixation index (F*st*) of the data based on the vcf files generated with IPYRAD for di- and polyploid individuals. The DNASP output was used to calculate the ratio of heterozygous sites to the total number of sites of the samples.

2.6. Next-Generation Sequencing of Nuclear Markers

Fourteen nuclear single-copy markers were amplified (see Supplementary Table S2) using the Phusion High Fidelity DNA polymerase (ThermoFisher Scientific). To obtain sequences of the nuclear rDNA internal transcribed spacer region (ITS), we used the primers ITS-A and ITS-B [36] following the protocol of Blattner [37]. PCR products of each amplicon of one sample were pooled together regardless of their concentration, purified using a NucleoFast plate (Marcherey-Nagel), and finally diluted in 34 µL triple-distilled water. Sixteen microliters of the purified amplicon pool were digested using the NEBNext™ dsDNA Fragmentase kit (NEB) for an incubation time of 1.5 min and another 16 µL of the amplicon pool was digested for 4.5 min following the manufacturer's instructions (NEB). Both were pooled, 100 µL triple distilled water was added, and a NucleoFast plate (Marcherey-Nagel) was used to remove too-small fragments and contaminants. Fifty microliters were subjected to size-selection targeting fragment sizes of 400–600 bp using BluePippin (Sage Science), blunt-end repaired, and used for sequencing library preparation according to the Illumina TruSeq DNA library protocol. Adaptors and barcodes were ligated to the samples. The libraries were size-selected with BluePippin. Fragment size distribution and DNA concentration were evaluated on an Agilent BioAnalyzer High Sensitivity DNA Chip and using the Qubit DNA Assay Kit in a Qubit 2.0 Fluorometer (Life Technologies). Finally, the DNA concentration of the libraries was checked with a quantitative PCR run. Cluster generation on Illumina cBot and paired-end sequencing 2 × 250 bp on the Illumina HiSeq 2000/2500 and NovaSeq6000 platform, respectively, followed Illumina's recommendation and included 1% Illumina PhiX library as internal control. The targeted output per sample was 300,000 reads. Reads were initially iteratively mapped against the forward primer as reference or already existing sequences of the marker in GENEIOUS Prime 2022.1.1 using the GENEIOUS mapping tool. In a second step, the pre-defined reads were assembled into haplotypes, i.e., representing the different alleles present in the sequenced marker region. Finally, three out of 14 nuclear single-copy markers (*orcf, rcf2, topo6*) were informative and were used for the investigation of haplotype differences of di- and polyploids to determine the parental species involved in allotetraploid formation.

2.7. Chromosome Counts

Chromosome counts were either obtained from the literature, obtained from direct observations in this study, or extrapolated for a few individuals based on the genome size data together with published chromosome counts of nearby populations. For direct observations, roots were cut about 2 cm from the tips, pretreated with 2 mM 8-hydroxyquinoline for 5 h at room temperature, and then kept in cold water overnight in a refrigerator. The roots were fixed in Carnoy's solution (3:1 ethanol:acetic acid) for 24 h and stored in 70% ethanol until use. Slide preparation was carried out according to Waminal et al. [38] and Rodríguez-Domínguez et al. [39]. Slides were fixed in 2% formaldehyde solution (47608, Sigma-Aldrich) for 3 min and dehydrated in an ethanol series (70%, 90%, 99%). Chromosomes were stained with 1 µg/µL DAPI in 2× SSC. Images were captured using a 100× objective of an Olympus BX61 fluorescence microscope (Olympus).

2.8. Genome Size Measurements

Genome sizes were measured for 134 individuals. Due to a lack of material, we could not measure the genome sizes of *C. neapolitanus* and *C. siculus*. Genome sizes for *C. bertiscensis* were partially taken from Raca et al. [16].

Genome size was determined using propidium iodide (PI) as a stain in flow cytometry with a Cyflow Space (Sysmex Partec) flow cytometer, following essentially the procedure described in Jakob et al. [6]. We mainly used rye (*Secale cereale*; 16.01 pg/2C) or pea (*Pisum sativum*; 9.09 pg/2C) as internal size standards and the buffer CyStain PI Absolute P (Sysmex Partec). Genome size measurements aimed at identifying diploids and polyploids. To link the genome sizes with the molecular data, we used silica-gel-dried leaves from the same individual used for DNA extraction whenever possible. Initial tests showed that fresh and silica-gel-dried materials arrived at the same genome size estimations in *Crocus*. However, the quality of data obtained is slightly lower for dried leaves. A detailed overview of the material measured and standards used is given in Supplementary Table S3.

2.9. Phylogenetic Inferences and Origin of Allopolyploids

Maximum parsimony (MP) analyses were conducted in PAUP* 4.0a169 [40] using a two-step heuristic search as described in Blattner [37] with 1000 initial random addition sequences (RAS) restricting the search to 25 trees per replicate. The resulting trees were afterwards used as starting trees in a search with maxtree set to 10,000. To test clade support, bootstrap analyses were run on all datasets with re-sampling 1000 times with the same settings as before, except that we did not use the initial RAS step. PAUP* was also used to infer the best-fitting model of sequence evolution for the sequence datasets using the Bayesian information criterion (Table 2). Analyses were run for (a) a dataset consisting of the combined sequences of the two chloroplast marker regions *mat*K-*trn*Q and *ycf*1; for the GBS-derived sequences including (b) the diploid species and (c) the diploid plus all tetraploid individuals of the *C. heuffelianus*/*C. vernus* complex plus *C. neglectus*, the only tetraploid species formally recognized to date; (d) for three nuclear single-copy genes; and (e) for the rDNA ITS region. The nuclear single-copy and ITS datasets were used to identify homeotic alleles in tetraploids and their diploid parents instead of deriving detailed phylogenetic relationships. The MP analysis of GBS data derived from di- and tetraploid individuals was used to infer the closest diploid parent of allopolyploids and to see if allopolyploids are monophyletic or originated multiple times.

Table 2. Characteristics of the analyzed molecular datasets for *Crocus* ser. *Verni*.

	Chloroplast (*mat*K–*trn*Q + *ycf*1)	**GBS Data 2*x*/2*x* + 4*x***	**Nuclear Gene Regions *orcp*/*rcf*2/*topo*6/ITS**
Number of sequences	81	93/194	37/42/39/22
Alignment lengths	4123	187,846	1160/568/819/646
Constant characters	3989	178,748/176,958	982/490/737/599
Variable characters	134	9098/10,888	178/78/82/47
Parsimony-informative characters	114	5899/6954	125/47/38/33
Number of MP trees	4	45/5400	142/10k/10k/10k [1]
MP tree length	156	16,298/25,300	333/99/89/56
Consistency index (CI)	0.87	0.58/0.45	0.57/0.83/0.96/0.88
Retention index (RI)	0.98	0.83/0.77	0.66/0.88/0.97/0.93
Model of sequence evolution (BIC)	GTR + I	GTR + I + Γ	HKY + I + Γ/HKY + I + Γ/TrN + I/HKY + I

[1] 10,000 trees were set as maxtree in MP analyses.

For the GBS-derived data, we also calculated SVDquartets in PAUP*, evaluating all quartets for the dataset consisting of 92 diploid individuals running 500 bootstrap re-samples. Individuals were partitioned according to their species affiliation, and trees were selected using QFM quartet assembly and the multispecies coalescent (MSC) as tree model.

Bayesian phylogenetic inference (BI) was conducted in MRBAYES 3.2.7 [41] for the chloroplast, the nuclear single-copy genes, and the ITS datasets. In BI, two times four chains were run for 5 million generations specifying the respective model of sequence evolution. A tree was sampled every 500 generations. Converging log-likelihoods, potential scale reduction factors for each parameter, and inspection of tabulated model parameters in MRBAYES suggested that stationary had been reached in all cases. The first 25% of trees of each run were discarded as burn-in.

In addition to the phylogeny-based inference of parental species of allotetraploids we also used STACKS v2.55 [42] to generate an input file for RADPAINTER [43]. A locus needed to be present in 80% of the individuals of a population and in 50% of all individuals to be processed. Population structure was inferred using 10,000 burn-in steps in the Monte Carlo Markov Chain (MCMC) analysis with 100,000 further iterations, keeping every 1000th sample. This run was continued, adding an additional 100,000 steps, treating the original run as burn-in. To obtain the best posterior state for the tree, 1000 attempts were used. Results were visualized in R with the functions provided with RADPAINTER.

The GBS-related Supplementary Datasets S1–S4 are available online through the e!DAL PGP data repository (https://doi.org/10.5447/ipk/2023/5, accessed on 10 February 2022).

2.10. Morpho-Anatomical Analyses

The morphological analysis was performed on fresh material, including 435 individuals in total (*C. vernus*: five populations; *C. heuffelianus*: two populations; mixed *C. heuffelianus*/*C.* cf. *heuffelianus* $2n = 18$ SCC: two populations; *C.* cf. *heuffelianus* $2n = 18$ SCC: three populations; *C.* cf. *heuffelianus* $2n = 18$ WCC: two populations; *C.* cf. *heuffelianus* $2n = 18$ PIC: two populations; *C.* cf. *heuffelianus* $2n = 20$: four populations; *C.* cf. *heuffelianus* $2n = 22$: two populations). Leaf cross-sections were made using a manual microtome [44]. The leaf sections of all 435 individuals were double-stained with safranin (1 g of safranin dissolved in 100 mL of 50% ethanol) and alcian blue (1 g of dye dissolved in 100 mL of distilled water, with a couple of crystals of phenol and three drops of glacial acetic acid). Stained sections were then dehydrated through an alcohol series (50%, 70%, 96%, 100%), examined, and photographed with a Leica DM 1000 microscope (Leica Microsystems) [16,45,46]. Anatomical features were measured in IMAGEJ [47]. A list of 42 characters from the literature relevant for this group [46,48,49] related to morphology and leaf anatomy was taken into consideration (see Supplementary Table S4). The qualitative characters were standardized as states represented by numbers (see Supplementary Table S4).

Principal component (PCA) and discriminant (CDA) analyses for morpho-anatomical characters were computed using STATISTICA 7.0 [50]. Due to unbalanced sample numbers (two populations of diploid vs. 13 populations of polyploid *C.* cf. *heuffelianus*), the set of differential characters was defined based on representative populations for each taxon/cytotype derived from type localities (140 individuals; Supplementary Table S4), computing a PCA. PCA served as a tool to point out the significant traits. The characters highlighted as important with the PCA were furthermore used for CDA. Moreover, eight a priori-defined groups of parental species and polyploids were included in CDA: *C. vernus* (100 individuals); *C. heuffelianus* (40); mixed *C. heuffelianus*/*C.* cf. *heuffelianus* $2n = 18$ SCC (40); *C.* cf. *heuffelianus* $2n = 18$ SCC (60); *C.* cf. *heuffelianus* $2n = 18$ WCC (35); *C.* cf. *heuffelianus* $2n = 18$ PIC (40); *C.* cf. *heuffelianus* $2n = 20$ (80); *C.* cf. *heuffelianus* $2n = 22$ (40) (435 individuals; Supplementary Table S4). The CDA results were plotted with ggplot2 [34].

3. Results

3.1. Determination of Ploidy Levels

Chromosome counts—Most chromosome counts obtained in this study coincided with previous reports (Supplementary Table S3). The lowest chromosome count was in *C. etruscus*, *C. ilvensis*, and *C. vernus* ($2n = 8$) followed by *C. heuffelianus* ($2n = 10$); *C. bertiscensis* ($2n = 12$); *C. kosaninii* ($2n = 14$); *C. neglectus*, *C. tommasinianus*, and *C.* cf. *vernus* ($2n = 16$); PIC, SCC, and WCC populations of *C.* cf. *heuffelianus* ($2n = 18$); *C.* cf. *heuffelianus* from Montenegro and Serbia ($2n = 20$); *C.* cf. *heuffelianus* from Albania and Kosovo ($2n = 22$); and *C. longiflorus* ($2n = 28$). The chromosome count of a *C. vernus*-derived polyploid population from Central Albania, herein referred to as *C.* cf. *vernus* ($2n = 16$), is reported here for the first time (Supplementary Table S3, Supplementary Figure S1).

Genome sizes—The smallest genome size was observed in *C. longiflorus* (2C = 3.21 pg), followed by *C. tommasinianus* (2C = 5.53 pg) and *C. vernus* (2C = 5.78 pg). *Crocus bertiscensis* was observed with an average genome size of 2C = 6.66 pg. *Crocus etruscus* (2C = 7.58 pg), *C. ilvensis* (2C = 7.88 pg), *C. heuffelianus* (2C = 7.73 pg), and *C. kosaninii* (2C = 7.95 pg) had similar genome sizes (Supplementary Table S3). Populations of *C. heuffelianus* with higher chromosome counts ($2n = 18, 20, 22$) were measured with average genome sizes per cytotype ranging between 2C = 10.88 and 2C = 12.84 pg (Table 1). Similar genome sizes were measured for *C. neglectus* (2C = 12.24 pg) and *C.* cf. *vernus* (2C = 12.23 pg).

Plotting of genome sizes against the chromosome counts showed a general negative relationship between genome size and chromosome number in both diploid and tetraploid taxa. This relationship is weaker when *C. longiflorus* is excluded (Supplementary Figure S1).

Crocus longiflorus is sister to the core series *Verni* taxa and has a seemingly polyploid chromosome count ($2n = 28$), although genome size indicates a diploid genome (Supplementary Table S3).

GBS-derived heterozygosity to ploidy—Most SNPs in the datasets were bi-allelic even in polyploids. The heterozygosity H_0 ranged between 0.0115 and 0.0628 for the GBS data set used (Supplementary Table S5). Individuals can be divided into two groups: samples with a H_0 of 0.0115 to 0.0278 and samples with a H_0 of 0.0359 to 0.0628 (Supplementary Table S5). If heterozygosity is considered in the context of known genome sizes, the group with the higher H_0 possesses the larger genome sizes (2C = >10 pg). The group with lower H_0 also has smaller genome sizes (2C = <8 pg) and generally lower chromosome numbers (except for *C. longiflorus*) and comprises *C. vernus* (2C = 5.78 pg, $2n = 8$), *C. etruscus* (2C = 7.58 pg, $2n = 8$), *C. ilvensis* (2C = 7.88 pg, $2n = 8$), *C. heuffelianus* (2C = 7.73 pg, $2n = 10$), *C. bertiscensis* (2C = 6.66 pg, $2n = 12$), *C. kosaninii* (2C = 7.95 pg, $2n = 14$), *C. tommasinianus* (2C = 5.53 pg, $2n = 16$), and *C. longiflorus* (2C = 3.21 pg, $2n = 28$). As a consequence, all samples with a H_0 below 0.03 are considered to represent diploids, while all samples with a H_0 above 0.035 are considered polyploids. The latter group includes *C.* cf. *heuffelianus* with higher chromosome counts and genome sizes (2C = 10.88 to 12.84 pg, $2n = 18, 20, 22$), *C. neglectus* (2C = 12.24 pg, $2n = 16$), and *C.* cf. *vernus* (2C = 12.23 pg, $2n = 16$).

3.2. Phylogenetic Inference and Origin of Allopolyploids

GBS-derived data—Initially, we created a single dataset for all GBS-derived analyses in IPYRAD, i.e., including all 194 sequences in one alignment (2009 loci, 187,846 bp alignment length, 20.88% missing sites). From this, we derived the dataset that includes only the diploid accessions by excluding all polyploid individuals from the data matrix (Table 2). The SVDquartets analysis of the diploid species (Figure 1) using the multispecies coalescent was used to infer the phylogenetic relationships among the diploid species, which are the basic taxonomic units in this group. Here *C. longiflorus* ($2n = 2x = 28$) from Sicily is the sister species to all other taxa within series *Verni*. The next two consecutively branching clades consist of the species with $2n = 2x = 8$ with *C. ilvensis* grouping together with *C. etruscus* followed by the clade of *C. neapolitanus*, *C. siculus*, and *C. vernus*. These species all occur in Italy or the Alps. The latter group is sister to a clade that harbors the eastern species *C. heuffelianus*, *C. bertiscensis*, *C. kosaninii*, and *C. tommasinianus* with the higher chromosome numbers of $2n = 2x = 10$, 12, 14, and 16, respectively.

The MP analysis of diploid taxa (Figure 2, Supplementary Figure S2) differs strongly from the diploid's SVDquartets tree topology. *Crocus longiflorus* is in both analyses sister to all other series *Verni* species. In the next clade the positions of the $2n = 8$ and $2n \geq 10$ taxa are, however, reversed. Although the species in the tree received very high bootstrap support, for the clades along the backbone of the tree, support values are low (Supplementary Figure S2), so that the topology of this tree has no strong support. We used MP mainly to infer the topology of the polyploids in relation to their diploid progenitors. In MP allopolyploids mostly group within or as sister to the parental species where they share higher genetic similarity and, as MP is sensitive to reticulate data structure, indicates if a polyploid is monophyletic or might consist of different subgroups. In the analysis of the combined di- and tetraploid GBS data (Figure 3, Supplementary Figure S3), *C. longiflorus*, *C. kosaninii*, and *C. bertiscensis* are the first species branching off, with the latter being sister to a large clade consisting of three subclades: (a) *C. etruscus*, *C. ilvensis*, and *C. neglectus* being sister to *C. tommasinianus*, (b) *C. neapolitanus*, *C. siculus*, and *C. vernus* as sister group of (c) *C. heuffelianus*. The polyploids were grouped in between the diploid taxa (Figure 3, Supplementary Figure S3).

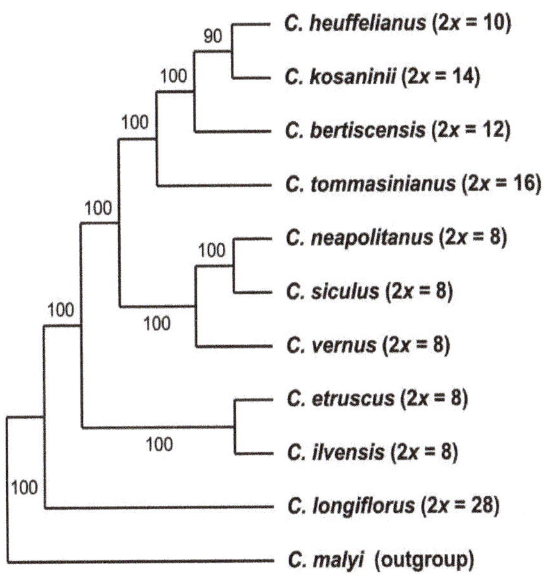

Figure 1. Phylogenetic tree of diploid species and accessions of *Crocus* ser. *Verni* based on the GBS dataset analyzed with SVDquartets analysis using the multispecies coalescent as tree model. Numbers along branches indicate bootstrap support values; chromosome numbers are provided in brackets.

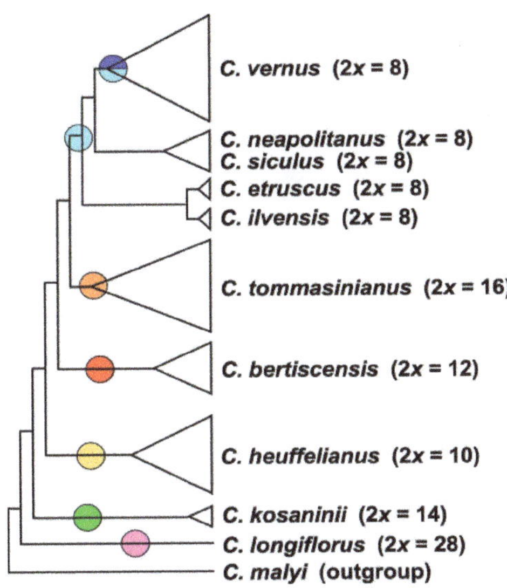

Figure 2. Schematic representation of the strict consensus tree topology of 45 most parsimonious trees derived from an MP analysis of the GBS dataset including only diploid accessions of *Crocus* ser. *Verni* (for original data see Supplementary Figure S2). In brackets, the ploidy levels and chromosome numbers are given. Colored circles refer to the chloroplast types present in the respective clades (below).

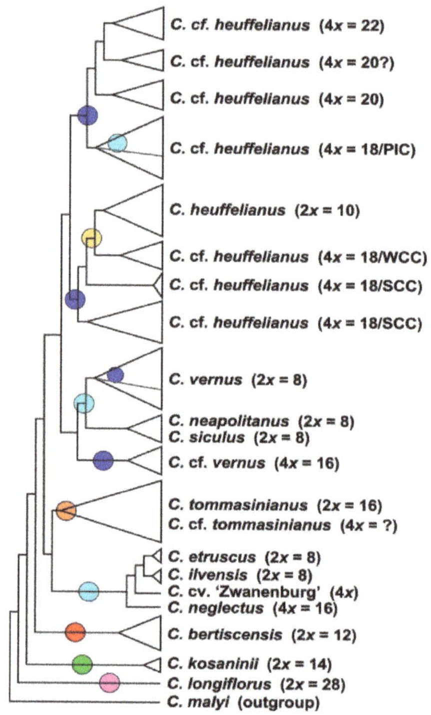

Figure 3. Schematic representation of the strict consensus tree topology of 5400 most parsimonious trees derived from an MP analysis of the GBS dataset including di- and tetraploid accessions of *Crocus* ser. *Verni* (for original data see Supplementary Figure S3). In brackets the ploidy levels and chromosome numbers are given. For tetraploids with $4x = 18$ the geographic affiliation is also provided, as Pannonian-Illyric Clade (PIC), Southern Carpathian Clade (SCC), and Western Carpathian Clade (WCC). Colored circles refer to the chloroplast types present in the respective clades (below).

Chloroplast-derived data—Series *Verni* diploids are split into two major groups in the phylogenetic tree of combined chloroplast data. The first group comprises *C. heuffelianus* ($2n = 2x = 10$), *C. bertiscensis* ($2n = 2x = 12$), *C. kosaninii* ($2n = 2x = 14$), and *C. tommasinianus* ($2n = 2x = 16$) (Figure 4, Supplementary Figure S4). The second group consists of two subgroups: (a) a group comprising the $2n = 2x = 8$ species *C. etruscus*, *C. ilvensis*, *C. neapolitanus*, *C. siculus*, and *C. vernus* from its northern and western distribution range (Alps to Pyrenees, NW type), and (b) *C. vernus* from the southeastern distribution range (Dinaric Alps; SE type), which has *C. longiflorus* ($2n = 2x = 28$) as its sister group.

The allotetraploids were grouped with their maternal parents. *Crocus neglectus* ($2n = 4x = 16$) possesses a chloroplast similar to *C. ilvensis*. *Crocus* cf. *heuffelianus* (WCC; $2n = 4x = 18$) from Slovakia groups with *C. heuffelianus*. Most of the *C.* cf. *heuffelianus* (PIC; $2n = 4x = 18$) individuals from Bosnia and Herzegovina as well as Slovenia grouped in a clade with the western $2n = 2x = 8$ species and are partly identical with the NW chloroplast type of *C. vernus* (e.g., *C. vernus* from Slovenian Alps is identical with *C.* cf. *heuffelianus* from Bosnia and Herzegovina). However, we also found *C.* cf. *heuffelianus* PIC grouping with the SE chloroplast type of *C. vernus*, indicating an independent origin. *Crocus* cf. *heuffelianus* (SCC; $2n = 4x = 18$) from Romania, $2n = 4x = 20$ from Montenegro and Serbia, $2n = 4x = 22$ from Northern Albania and Kosovo, as well as *C.* cf. *vernus* ($2n = 4x = 16$) from Central Albania were found in the clade with the southeastern *C. vernus* populations (SE type).

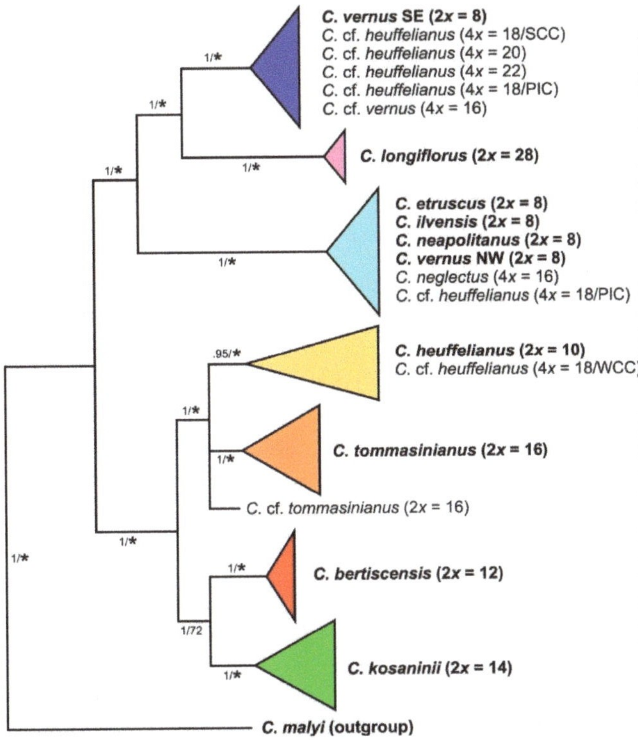

Figure 4. Schematic representation of relationships of species and cytotypes of *Crocus* ser. *Verni* obtained through Bayesian phylogenetic inference of sequences from two chloroplast regions (for original data see Supplementary Figure S4). Diploid accessions, as basic units in the series, are given in boldface. Numbers along branches indicate Bayesian posterior probabilities/maximum parsimony bootstrap values (≥50%), with asterisks for bootstrap support values >80%.

Nuclear single-copy markers—Three variable nuclear single-copy regions (*orcp*, *rcf*2, *topo*6; Supplementary Table S2) were chosen to identify the position of alleles of the allo-ploids by the criterion that the marker regions showed differences between the potential diploid parental species. Up to four alleles could be found within the allopolyploids (Figure 5, Supplementary Figures S5–S7). *Crocus* cf. *heuffelianus* $2n = 4x = 18$ (WCC) and *C.* cf. *heuffelianus* $2n = 4x = 18$ (PIC) were found to be grouping with *C. vernus* from its western distribution range and/or with *C. heuffelianus*. *Crocus* cf. *heuffelianus* $2n = 4x = 18$ (SCC) shared similar alleles with *C. vernus* from its eastern distribution range and/or *C. heuffelianus*. The same applies for *C.* cf. *heuffelianus* $2n = 4x = 20$ and $2n = 4x = 22$. *Crocus neglectus* ($2n = 4x = 16$) grouped with *C. ilvensis* or within a clade comprising *C. neapolitanus* as well as other allotetraploids in two of the selected nuclear single-copy genes. Generally, the gene-tree topologies were different from the topologies in chloroplast or GBS-derived datasets and differences also occurred among the single-copy genes. For example, in *topo*6, *C. longiflorus* occupies a similar position as in the chloroplast marker tree where it groups in one clade with other Italian taxa such as *C. neapolitanus*, *C. ilvensis*, and *C. vernus* (Figure 5). In the *rcf*2-derived tree its position is similar to the GBS results, where it groups as sister to the core series *Verni* taxa. *Crocus tommasinianus*, or one of its alleles, was found in one clade with *C. vernus* (NW) in *topo*6 and *rcf*2 (Figure 5), while it groups with *C. bertiscensis*, *C. kosaninii*, and *C. heuffelianus* in the SVDquartets of GBS and the chloroplast trees. The *orcp* data set had the highest number of parsimony-informative sites of the three closely examined nuclear single-copy genes, but also the highest homoplasy and was mostly characterized by polytomies (Table 2, Supplementary Figure S7).

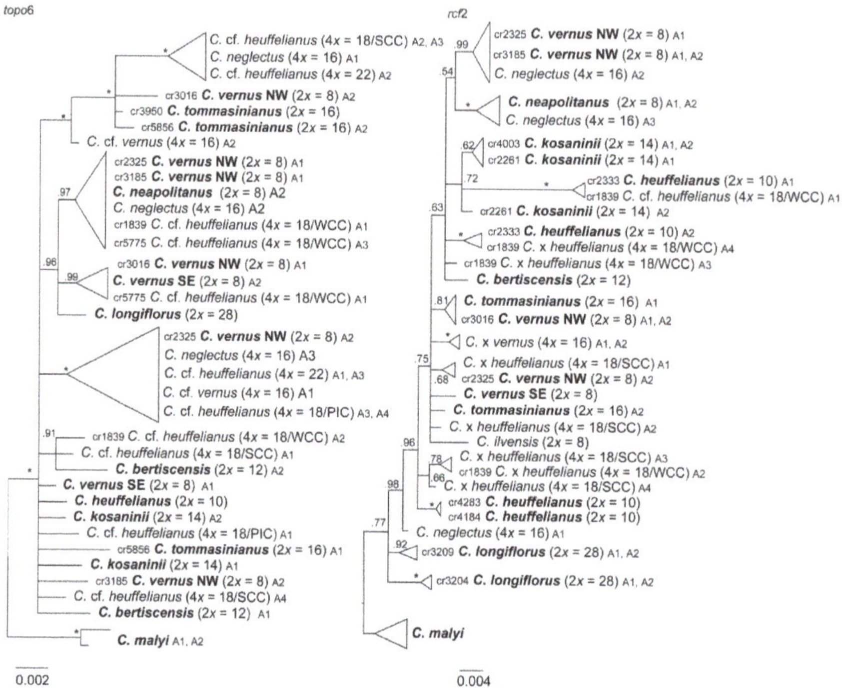

Figure 5. Phylogenetic trees obtained through Bayesian phylogenetic inference based on haplotype sequences of nuclear single-copy genes in *Crocus* ser. *Verni*. Numbers along branches indicate BI posterior probabilities (pp). Support values of 1.0 are indicated by asterisks. Allelic differences (A1–A4) in these markers were used to track the bi-parental contributions of diploids to allotetraploids. If more than one individual per species or cytotype was included, their DNA identifier (cr number) is provided. For detailed information, see Supplementary Figures S5 and S6.

ITS—In the ITS tree (Supplementary Figure S8) *C. longiflorus* is sister to the other series *Verni* taxa (pp 1.0). In the sister clade of *C. longiflorus*, *C. kosaninii* is separated from the remaining species of the series but with very low support (pp 0.61). The majority of series *Verni* species is found in a large polytomy with only a few subclades. In one of the subclades, *C. ilvensis* groups with *C. etruscus*, while most of the other subclades are formed by samples of the same species, with the exception of the subclade comprising *C. vernus*. Here, *C. neapolitanus*, *C. siculus*, *C. vernus*, and all allotetraploids included in the dataset can be found (pp 1.0; Supplementary Figure S8), albeit their relationships remained unresolved.

3.3. Phylogenomic Analysis

The 192 GBS sequences of series *Verni* (excluding the outgroup *C. malyi*) with a threshold number of 120 samples sharing a locus resulted in a dataset comprising 2207 loci with 18.38% missing sites. The data matrix of unlinked SNPs included 2172 SNPs. The sample with the lowest number of reads (559,922) and loci (866) was *C. siculus*.

The GBS-based PCA (Figure 6) placed *C. tommasinianus* clearly separate from all other samples in the negative part of the PC1 axis. In between *C. tommasinianus* and the majority of all other taxa, in the positive part of the PC1, *C. heuffelianus* and *C. vernus* can be found. However, *C. heuffelianus* in the positive part of the PC2 axis (5 or higher) and *C. vernus* in the negative part of the PC2 axis (−4 or lower) were distinct from each other. The two individuals of *C. neapolitanus* were close to *C. vernus*. *Crocus etruscus* and *C. ilvensis* were placed together in close proximity to *C. bertiscensis*, *C. kosaninii*, and *C. longiflorus* representatives, all between −1 and 1 on the PC1 axis and 0 to 4 on the PC2 axis. *Crocus siculus* was found in the lower negative part of the PC2 axis (ca. −2), partly overlapping with the polyploids *C.* cf. *heuffelianus*, *C. neglectus*, and *C.* cf. *vernus*. They were placed in between *C. heuffelianus* and *C. vernus* but always in the positive part of the PC1 axis.

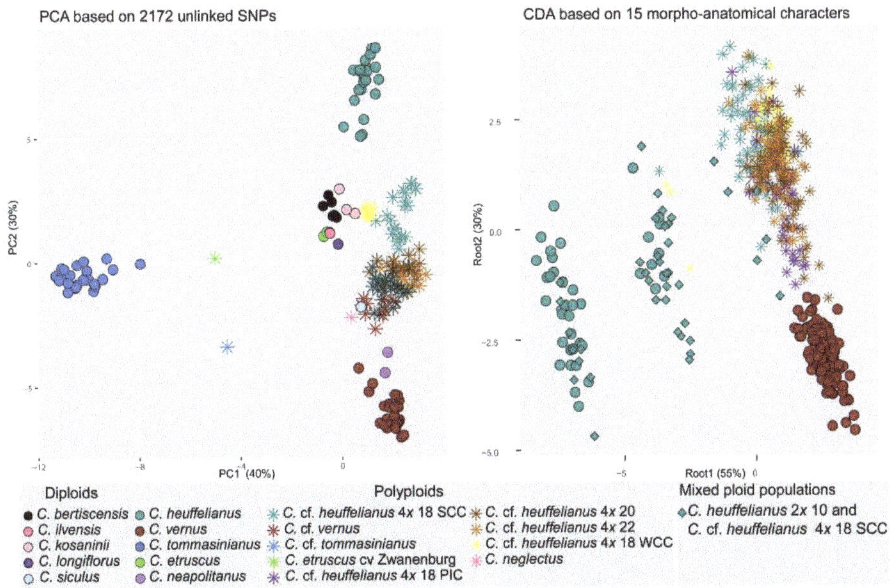

Figure 6. PCA based on 2127 unlinked GBS-derived SNPs (**left panel**) comprising all species of *Crocus* ser. *Verni*, and CDA based on 15 morpho-anatomical characters (**right panel**) of *C. vernus*, *C. heuffelianus*, and *C.* cf. *heuffelianus*.

Co-ancestry analysis with RADPAINTER showed admixture for *C. neglectus* (Supplementary Figure S9) with both *C. etruscus* (co-ancestry 234.2, 241.8) and *C. ilvensis* (co-ancestry 217.9, 221.8) as well as with *C. neapolitanus* (co-ancestry 99.2, 103.7). *Crocus etruscus* cv. 'Zwanenburg' shares a relatively high co-ancestry with *C. etruscus* (380.6, 386.3) followed by *C. ilvensis* (301.3, 301.1) and *C. tommasinianus* from Italy (110.5). *Crocus* cf. *tommasinianus* was found to be admixed with *C. vernus* (181.7) and *C. tommasinianus* (135.1) growing at the same location (Montenegro, Mt. Lovcen). The highest co-ancestry of *C.* cf. *heuffelianus* $2n = 4x = 18$ (PIC) and diploid series *Verni* species was found with *C. vernus* (52.8–86.9), while it was lower with *C. heuffelianus* (44.0–51.6). The level of co-ancestry for *C.* cf. *heuffelianus* $2n = 4x = 18$ (SCC) was high in the mixed-ploidy populations with *C. heuffelianus* (127.5–128.8), while it was between 50 and 101.5 with other *C. heuffelianus* populations and between 49.2 and 59.4 with *C. vernus*. *Crocus* cf. *heuffelianus* $2n = 4x = 18$ from the Western Carpathians (WCC) had its highest co-ancestry level with *C. heuffelianus* (49.4–93.4). Its co-ancestry shared with *C. vernus* ranged between 51.6 and 62.0. The co-ancestry levels of *C.* cf. *heuffelianus* $2n = 4x = 20$ and $2n = 4x = 22$ shared with *C. vernus* were lower than in the other allotetraploid *C.* cf. *heuffelianus* (48.0–58.2 and 47.8–58.5). The same applies to the shared co-ancestry with *C. heuffelianus* (46.1–54.0 and 45.9–52.2). In addition, they are even lower than the co-ancestry shared between diploid species such as *C. bertiscenesis* and *C. kosaninii* (57.1–62.9).

In the population assignment analysis (K with the lowest entropy was K = 8), most of the alleles present in *C. tommasinianus* samples were assigned to one ancestral population (Supplementary Figure S10). Moreover, *C. vernus* alleles were mostly assigned to one ancestral population, and representatives of *C. heuffelianus* were mostly assigned to their own ancestral population as well. However, *C. bertiscensis*, *C. etruscus*, *C. ilvensis*, *C. kosaninii*, and *C. longiflorus* appeared admixed, partly sharing *C. heuffelianus* and *C. vernus* (*C. etruscus*, *C. ilvensis*, and *C. longiflorus*) patterns or additionally having alleles assigned to *C. tommasinianus* and/or to *C.* cf. *heuffelianus*. The different *C.* cf. *heuffelianus* groups were partly assigned to their own ancestral population showing no admixture with LEA (K = 5, K = 8) or to *C. vernus* (K = 4) with FASTSTRUCTURE. *Crocus* cf. *vernus* and *C. neglectus* were partly assigned to the *C. vernus* ancestral population but also to different ancestral populations of *C.* cf. *heuffelianus*. *Crocus etruscus* cv. 'Zwanenburg' and *C.* cf. *tommasinianus* showed admixture, with parts of their alleles derived from *C. tommasinianus*. In the case of *C.* cf. *tommasinianus*, *C. vernus* contributed genomic materials, and *C. etruscus* cv. 'Zwanenburg' was complemented by *C. etruscus*.

Assigning the several polyploid groups mostly to one ancestral population while showing diploids as admixed was also observed for lower K or other SNP subsampling and/or other ancestral assignment methods (Supplementary Figures S9 and S10) regardless of the analysis program used.

Crocus cf. *heuffelianus* had the lowest F*st* with *C. heuffelianus* and *C. vernus* ($2n = 4x = 18$ WCC: F*st* = 0.24 and F*st* = 0.22; $2n = 4x = 18$ SCC: F*st* = 0.14 and F*st* = 0.17; $2n = 4x = 18$ PIC: F*st* = 0.17 and F*st* = 0.12; $2n = 4x = 20$: F*st* = 0.18 and F*st* = 0.15; $2n = 4x = 22$: F*st* = 0.22 and F*st* = 0.19) or, in cases of *C.* cf. *heuffelianus* (2*n* = 4x = 18 PIC), with *C. neapolitanus* (F*st* = 0.15; Supplementary Table S6). The lowest F*st* for *C.* cf. *vernus* was observed towards *C. vernus* (F*st* = 0.15), followed by *C. neapolitanus* (F*st* = 0.20) and *C. heuffelianus* (F*st* = 0.23). *Crocus neglectus* had the lowest F*st* towards *C. etruscus* (F*st* = 0.07), *C. vernus* (F*st* = 0.08), *C. neapolitanus* (F*st* = 0.10), and *C. ilvensis* (F*st* = 0.10). The cultivar 'Zwanenburg' had the lowest F*st* (0.00) with *C. etruscus*, followed by *C. ilvensis* (F*st* = 0.09) and *C. tommasinianus* (F*st* = 0.14). The lowest F*st* for *C.* cf. *tommasinianus* was found with *C. tommasinianus* (F*st* = 0.14) and *C. vernus* (F*st* = 0.13).

3.4. Morpho-Anatomical Analyses

The PCA of the morpho-anatomical dataset highlighted 14 characters with PC scores >0.70: outer and inner perigone segment length and width (Outer_ps_l, Outer_ps_w, Inner_ps_l, Inner_ps_w), anther length (Anther_l), throat hair (Th_hair), leaf section height and width (Sectioh_h, Section_w), arm length (Arm_l), central parenchyma area (Parenchyma_a), palisade cell and tissue height (Pal_cell_h, Pal_tissue_h), spongy tissue height (Sp_tissue_h), and xylem area (Xy_a) (Supplementary Table S7). This set was extended with stigma/anther ratio (S/a_r) (PC1 = 0.68) as the most important discriminative feature for *C. vernus* confirmed by previous research [15]. Finally, the CDA of the complete dataset (435 individuals of *C. vernus*, *C. heuffelianus*, mixed *C. heuffelianus*/*C.* cf. *heuffelianus* $2n = 18$ SCC, *C.* cf. *heuffelianus* $2n = 18$ SCC, *C.* cf. *heuffelianus* $2n = 18$ WCC, *C.* cf. *heuffelianus* $2n = 18$ PIC, *C.* cf. *heuffelianus* $2n = 20$, and *C.* cf. *heuffelianus* $2n = 22$) was computed based on 15 previously mentioned characters. The clear separation of *C. heuffelianus* (Figure 6) in the negative part of the CDA of both axes was caused by the absence of throat hair (Figure 7, Supplementary Table S8). The mixed populations of the diploid *C. heuffelianus* and its polyploid $2n = 18$ SCC cytotype overlapped with these two taxa (Figure 6), while all other polyploid populations were grouped in the positive part of both axes (Figure 6). The characters responsible for differentiation along the second axis were leaf cross-section width and arm length (Supplementary Table S8).

Figure 7. Flower, throat, and cross-section of leaf (from top to bottom) of parental species and a *C.* cf. *heuffelianus* polyploid: *C. vernus* (**A**), *C. heuffelianus* (**B**), and polyploid *C.* cf. *heuffelianus* representatives (**C**).

4. Discussion

4.1. Recognition of Recent Polyploids and Their Parents

Crocus heuffelianus represents one of the biggest taxonomical challenges within *Crocus* due to its high morphological variability. This morphological diversity seems mainly caused by the allotetraploid origin of the karyotypes of taxa possessing higher chromosome counts ($2n = 18, 20, 22$), which became evident in a former molecular study [15]. Through incongruences between the GBS tree (Figures 1–3) and the chloroplast tree (Figure 4) as well as to the ITS tree (Supplementary Figure S8) and/or the affiliation of different alleles of variable nuclear single-copy markers (Figure 5, Supplementary Figures S5–S7) we identified at least seven independent hybridization events involving *C. heuffelianus* ($2n = 10$) and *C. vernus* ($2n = 8$), mostly with *C. vernus* as maternal parent. We summarize these reticulate relationships in Figure 8, a tree based on the diploid's topology derived from the SVDquartets analysis. *Crocus vernus* was found to possess two different chloroplast types depending on the geographical distribution (Eastern Alps to Pyrenees: NW type; Dinaric Alps: SE type). *Crocus vernus* from the Alps hybridized *C. heuffelianus* resulting in allotetraploid Western Carpathian populations (WCC; $2n = 18$) with *C. heuffelianus* as the maternal parent. In the Pannonian-Illyric group of *C.* cf. *heuffelianus* (PIC; $2n = 18$), we found two different chloroplast haplotypes stemming from the SE and NW *C. vernus* types indicating two crosses involving *C. vernus* as maternal parents (Figure 8). Furthermore, we also observed differences among the chloroplast haplotypes stemming from the NW type of *C. vernus* (Figure 4). One differed by at least two substitutions and two indels from any other NW type (see also branch lengths in Supplementary Figure S4). This indicates multiple hybridization events between *C. vernus* as maternal and *C. heuffelianus* as paternal parent creating the Pannonian-Illyric *C.* cf. *heuffelianus*. The SE type-carrying *C. vernus* was identified as the maternal species for Southern Carpathian populations of *C.* cf. *heuffelianus* (SCC; $2n = 18$), as well as for the cytotypes $2n = 4x = 20$ and $2n = 4x = 22$ (Figure 8).

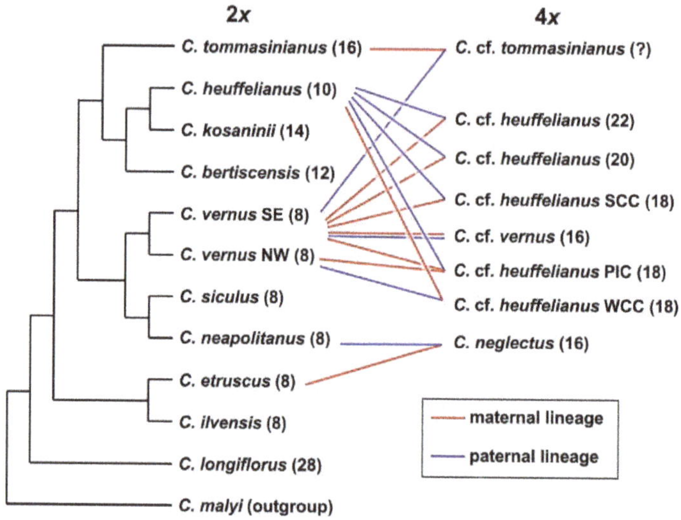

Figure 8. Schematic representation of species relationships in *Crocus* ser. *Verni* based on GBS, chloroplast, and nuclear single-copy gene data. For the tetraploids the maternal and paternal parents are indicated by lines connecting them to the respective diploid taxa. In brackets, chromosome numbers are provided.

Considering that the Southern Carpathian diploid cytotype of *C. heuffelianus* ($2n = 10$) is ancestral to all polyploid forms and that according to morphological and chorological characteristics it corresponds to the original description, it represents *C. heuffelianus* s. str. [16,49]. It comprises populations with darker perigones and predominantly glabrous throats or hardly visible sparse and short hairs, which makes it distinct from all other series *Verni* species. Several authors reported different distribution ranges for *C. heuffelianus* [51–53]. This confusion is likely caused by confusing *C. heuffelianus* s. str. with its morphologically similar allopolyploids. For instance, some authors reported *C. heuffelianus* s. str. to occur in regions that are, according to our investigations, only inhabited by allopolyploid *C.* cf. *heuffelianus*. (e.g., throughout Bosnia and Herzegovina [54] or northeastern Italy [55]).

As a result of our study, we confirm a Carpathian distribution of *C. heuffelianus* s. str. in Slovakia, Romania, and Ukraine. *Crocus* cf. *heuffelianus* allopolyploid cytotypes have partly sympatric distributions with one of their parents or are growing in between the parental distribution areas (Supplementary Figure S11).

A new polyploid, *C.* cf. *vernus* ($2n = 4x = 16$), was found in Central Albania, having *C. vernus* as the maternal parent. Its hybrid origin is indicated by its sister position in the GBS phylogeny, where it did not group within *C. vernus* as it would if it were an autopolyploid [5]. A (segmental) allotetraploid origin is also evident by its position in the SNP-based PCA (Figure 6). In the three closely examined variable nuclear single-copy genes (Figure 5, Supplementary Figures S5–S7), some of *C.* cf. *vernus*'s alleles were unique or usually grouped close to those of other taxa with $2n = 8$ chromosomes. Therefore, the paternal parent could either be an extinct *C. vernus*-like genotype, which probably had eight chromosomes, or stems from a *C. vernus* population that we have not yet collected.

Allopolyploids in *C.* cf. *heuffelianus* and *C.* cf. *vernus* are also genetically differentiated, as indicated by the Fst values, which were usually higher than 0.15. This genetic differentiation might explain the difficulty in assigning ancestral populations (Supplementary Figure S10) or in the inference of co-ancestry (Supplementary Figure S9), where some allotetraploids were not shown as admixed.

Crocus neglectus ($2n = 4x = 16$) could be confirmed here as an allotetraploid with *C. ilvensis* or *C. etruscus* as the maternal parent [15]. The fixation index (*C. etruscus* $Fst = 0.07$; *C.*

ilvensis Fst = 0.10) points to a more likely contribution of *C. etruscus*. However, *C. neglectus* shares its chloroplast haplotype with *C. ilvensis*, which was not found in *C. etruscus*. This discrepancy may be explained by the possibility that genetic drift eliminated this type of chloroplast in *C. etruscus* while it persisted in the geographically isolated *C. ilvensis* or that it was just not discovered in the individuals analyzed up to now. Seed size and germination are also more similar to *C. etruscus* [56], while flower bouquet is more similar to *C. ilvensis* [57]. *Crocus neapolitanus* is likely the paternal parent of *C. neglectus*. *Crocus neapolitanus* is genetically very similar to *C. vernus* but has a species-specific allele in one of the nuclear single-copy markers (*rcf*2; Figure 5) found in *C. neglectus*. The relatively high co-ancestry shared between *C. neglectus* and *C. neapolitanus* further supports *C. neapolitanus* as paternal parent.

4.2. General Results Regarding Phylogeny and Systematics

Until the advent of high-throughput sequencing technologies, phylogenetic studies were often restricted to a limited set of markers. Consequently, relationships could not be resolved, as in the case of series *Verni* [15]. In Raca et al. [16], we already successfully increased the resolution by using genome-wide SNP data obtained from GBS. However, since this study aimed to show the phylogenetic affiliation of *C. bertiscensis*, we included neither all the species in *Crocus* ser. *Verni* (*C. siculus* was lacking) nor a higher number of individuals per species and did not analyze species relationships in detail. Here we added additional samples, a chloroplast marker dataset, as well as a dataset comprising nuclear single-copy genes. The latter mainly served to identify the parental origin of the recent allotetraploids. Excluding the allotetraploids in the analysis increased the support values of the tree backbone of the GBS-based MP tree, similar to the BI-based analysis in Raca et al. [16]. However, our SVDquartets species tree (Figure 1) showed a different topology. The relatively high degree of homoplasy in the GBS dataset (Table 2) indicates incomplete lineage sorting (ILS) and/or hybridization that could result in genomic introgression. Incongruences of single-gene trees, as well as the chloroplast tree, support this hypothesis. For example, *C. vernus* today has two different chloroplast haplotypes (SE type and NW type). The NW type is shared with other closely related species such as *C. etruscus*, *C. ilvensis*, *C. neapolitanus*, and *C. siculus*, which all occur in Italy (Supplementary Figure S11). The SE type is found only in the southeastern range of *C. vernus* and groups with *C. longiflorus* as sister to the NW type clade. A possible interpretation could be that these two chloroplast types were once both present in Italy, where introgression of *C. vernus* or its ancestor with *C. longiflorus* occurred. Subsequently, the SE type was sorted out and persisted only in the eastern distribution of *C. vernus*. The position of *C. longiflorus* as sister to *C. vernus* was also observed in one of the nuclear single-copy markers (*topo*6), which supports the hypothesis of an ancient introgression event between *C. vernus* or its ancestor and *C. longiflorus*. There are also other examples where species (or one of their alleles) group with species to which they are only distantly related according to the GBS trees (SVDquartets as well as MP), such as *C. tommasinianus* grouping together with *C. vernus* in *topo*6 and *rcf*2 (Figure 5).

Single gene trees generally showed a poor resolution, which was the main reason we show only *rcf*2 and *topo*6, and even in data sets with a relatively high number of parsimony-informative sites, phylogenetic relationships remained largely unresolved (Supplementary Figure S7) due to the young age of the group and homoplasy in the data. Since multispecies coalescent methods consider ILS and introgression, the SVDquartets tree should more closely reflect the true species relationships although its topology differs from trees derived with concatenation approaches [58].

4.3. Chromosome and Genome Size Evolution

All angiosperms essentially have undergone polyploidization [59]. Dysploidy, chromatin elimination or expansion, nested polyploidizations, introgression, and hybridization can confound the evolutionary dynamics between chromosome number and genome size over time [60–62]. Consequently, these two genomic parameters show independent evolution and no clear correlation, especially in older polyploids, such as meso- and paleopolyploids [61,63,64]. Nevertheless, some correlation can still be observed in neopolyploids [6,61].

Neopolyploids often have roughly additive genome sizes and chromosome numbers of the progenitors, as in C. cf. *heuffelianus* ($2n = 18$, 2C = 10.88–12.84 pg), *C. neglectus* ($2n = 16$, 2C = 12.24 pg), and C. cf. *vernus* ($2n = 16$, 2C = 12.38 pg). However, although genome sizes could remain roughly additive in the absence of considerable chromatin loss [65], the additive pattern in chromosome number is often blurred when chromosome fusion (descending dysploidy) or fission (ascending dysploidy) occurs [61,66,67]. Considering the chromosome numbers and genome sizes of the parental genomes of C. cf. *heuffelianus*, *C. heuffelianus* ($2n = 10$), and *C. vernus* ($2n = 8$), it is highly likely that the C. cf. *heuffelianus* cytotypes from Albania, Kosovo, Montenegro, and Serbia ($2n = 20$ and $2n = 22$, 2C = 11.95 pg) have undergone ascending dysploidy. Indeed, a few shorter chromosomes have been observed in these cytotypes, as ascending dysploidy entails chromosome breakage resulting in an increase in chromosome number in general and shorter chromosomes in particular (Supplementary Figure S1).

The cryptic relationship between chromosome number and genome size increases over time in polyploids. This relationship can be observed between *C. longiflorus* and the core series *Verni* species. In general, series *Verni* species with more chromosomes have smaller genomes, thus showing a negative relationship between genome size and chromosome number, with or without *C. longiflorus* (Supplementary Figure S1), which is sister to all other taxa of series *Verni* and was recently moved to this series from the now-defunct series *Longiflori* B.Mathew [15]. *Crocus longiflorus* has the highest chromosome number but the lowest genome size ($2n = 28$, 2C = 3.21 pg) in the series. Likewise, *C. tommasinianus* ($2n = 16$), which has twice as many chromosomes as *C. vernus* ($2n = 8$), has a roughly similar genome size to *C. vernus*. Taking into account only chromosome counts, *C. longiflorus* and *C. tommasinianus* would likely be deemed polyploids. However, they show very low heterozygosity scores (Supplementary Table S5), indicating extensive chromatin elimination and essentially diploidized genomes. This chromosome number–genome size relationship between *C. longiflorus* and the core species of series *Verni* can indicate an ancient whole-genome duplication event prior to the divergence of series *Verni*, likely even before the diversification of *Crocus* [68].

A burst of certain repetitive DNA element families may also have promoted genome size expansion in the core series *Verni*. Thus, the genome sizes in the core taxa of series *Verni* have increased despite chromosome number reduction relative to *C. longiflorus*. Nevertheless, this hypothesis can only be supported when fused chromosome blocks [59,69] and expansion of lineage-specific repeat families [70,71] are observed within the core of series *Verni*. This can become possible by comparative genome and repeat analyses between *C. longiflorus* and the other species from series *Verni*. Combined with cytogenetic analyses involving all major groups within the genus, this should allow an understanding of karyotype evolution within the genus.

5. Conclusions

Our study was designed to disentangle the *C. heuffelianus* complex using GBS data and chloroplast markers and combine them with morphology, genome sizes, and chromosome counts. This strategy generally proved to be successful when (a) a broad sampling of allotetraploids and potential parental species are included and (b) the allotetraploids group as sister to their paternal parent in the phylogenetic GBS tree. In cases where our sampling was restricted to just a few samples, such as for some of the Italian species and allotetraploid *C. neglectus*, conclusions remain a bit uncertain. The combination of chloroplast data with only GBS failed to reveal the paternal contributor of C. cf. *heuffelianus* (SCC; $2n = 18$). Here, additional nuclear markers (Figure 5, Supplementary Figure S7) were necessary to identify the NW type of *C. vernus* as the paternal parent. Algorithms specifically designed to detect hybridization signals in GBS data were only partly able to recover the allopolyploids within series *Verni*, while in several cases co-ancestry values for their parents were lower than even between independent diploids. Thus, such an approach alone is not sufficient to infer polyploidization or indicate the diploid progenitors of polyploids in crocuses.

By linking molecular results and genome sizes with morphology, a clear differentiation of allopolyploids and parental species was possible. In the taxonomically confusing *C. heuffelianus* complex, a circumscription of the diploid taxon and its distinction from the allotetraploids is now possible. *Crocus heuffelianus* s. str. is the diploid cytotype ($2n = 10$) with mostly glabrous throats and darker perigone segments. Together with *C. vernus*, it represents the parental species for all the *C.* cf. *heuffelianus* allotetraploids. The cytotype $2n = 18$ of *C.* cf. *heuffelianus* is split into three groups: Western Carpathian (WCC), Pannonian-Illyric (PIC), and Southern Carpathian (SCC). *Crocus heuffelianus* s. str. is the mother of WCC only, while the NW and SE types of *C. vernus* are maternal lineages of PIC. The SE type of *C. vernus* is the only maternal parent of SCC, as well as for the cytotypes with $2n = 20$ and $2n = 22$ chromosomes. All analyzed *C.* cf. *heuffelianus* polyploids represent morphologically intermediate forms between their parental species, but currently, they cannot be distinguished based on the investigated morphological characters. Instead, chromosome counts are necessary.

While it is possible to unravel more recent polyploidization events, the detection of paleo-polyploidization remains difficult. Our incongruent gene trees indicate past hybridization events, which might have triggered genome size and chromosome number changes. However, while the methods applied here work well in this very young taxon group, they are not satisfactory in uncovering ancient and complex evolutionary histories, particularly those involving highly dynamic genome size and chromosome number changes. In series *Verni*, and in *Crocus* in general, the future availability of genome assemblies will enable comparative cytogenomic analyses to detect potential ancient polyploidization and to trace chromosomal rearrangements resulting in changing karyotypes.

Supplementary Materials: The following supporting information can be downloaded at: https://www.mdpi.com/article/10.3390/biology12020303/s1, Figure S1: Chromosome counts for diploid *C. vernus* and *C. heuffelianus* (A and B), tetraploid *C.* cf. *vernus* (C), and tetraploid *C.* cf. *heuffelianus* (D–F). Asterisks show relatively shorter chromosomes. (G) The chromosome number and genome size in diploid and tetraploid *C.* ser. *Verni* taxa show a negative relationship, which was only significant when *C. longiflorus* was included; Figure S2: Strict consensus MP tree based on 2009 GBS loci including only diploid accessions of *Crocus* ser. *Verni*; Figure S3: Strict consensus MP tree based on 2009 GBS loci including both diploid and tetraploid accessions and cytotypes of *Crocus* ser. *Verni*; Figure S4: BI phylogenetic tree of *Crocus* ser. *Verni* based on chloroplast markers. Numbers at branches provide BI posterior probabilities/bootstrap values from MP analysis. Asterisks indicate bootstrap support values $\geq$80%; Figure S5: Strict consensus MP tree of *topo6* in *Crocus* ser. *Verni*. Allelic differences (A1–A4) in these markers were used to track the bi-parental contributions of diploids to allotetraploids; Figure S6: Strict consensus MP tree of *rcf2* in *Crocus* ser. *Verni*. Allelic differences (A1–A4) in these markers were used to track the bi-parental contributions of diploids to allotetraploids; Figure S7: Strict consensus MP tree of *orcp* in *Crocus* ser. *Verni*. Allelic differences (A1–A4) in these markers were used to track the bi-parental contributions of diploids to allotetraploids; Figure S8: Phylogenetic tree of *Crocus* ser. *Verni* obtained through Bayesian phylogenetic inference based on rDNA ITS sequences. Numbers along branches indicate BI posterior probabilities (pp), pp supports of 1.0 are indicated by asterisks; Figure S9: FINERADSTRUCTURE co-ancestry matrix of the study species based on 10,591 loci (without outgroup species *C. malyi*; *C. siculus* was excluded due to low coverage). Black indicates maximum levels of co-ancestry between two individuals, white the minimum (scale on the right). Numbers below the plots indicate the sample ID; Figure S10: Population structure analysis in *Crocus* ser. *Verni* based on 2207 GBS loci using FASTSTRUCTURE at K = 4 (30,661 SNPs) and LEA K = 5 and 8 (2172 unlinked SNPs). Each vertical line represents one individual, while each color shows the genetic composition that is assigned into a distinct genetic cluster; Figure S11: Map with the approximate distributions of species in *Crocus* ser. *Verni*. Distribution areas of different species are indicated by different colors or patterns; Table S1: Detailed information about the individuals included in the chloroplast (CP), genotyping-by-sequencing (GBS), nuclear single-copy genes (NSCG), and morpho-anatomical analyses (MA) of *Crocus* ser. *Verni* separately (No.) and in total (No. total), accompanied with voucher numbers; Table S2: Nuclear single-copy marker PCR information in *Crocus* ser. *Verni;* Table S3: Genome size measurements and chromosome counts in *Crocus* ser. *Verni;* Table S4: The dataset for morpho-anatomical analyses (consisting of the parental diploid species *C. vernus* and *C. heuffelianus* and all polyploid cytotypes of *C.* cf. *heuffelianus*); Table S5: Information on GBS data, loci, and heterozygosity in *Crocus* ser. *Verni.;* Table S6: Fixation index (F*st*) of allotetraploids versus diploids in *Crocus* ser. *Verni;* Table S7: The differential characters based on representative populations dataset of *Crocus* ser. *Verni* highlighted by the principal component analysis (PCA); Table S8: Discriminant analysis (CDA) conducted on the dataset including all populations of diploid accessions of *C. vernus* and *C. heuffelianus* and all polyploid accessions of *C.* cf. *heuffelianus;* Dataset S1: Alignment of concatenated GBS loci used for phylogenetic analysis; Dataset S2: Variant calling format (vcf) file of the series *Verni* dataset used for FASTSTRUCTURE and LEA; Dataset S3: Unlinked SNP matrix in .geno format used for PCA. Dataset S4: RADPAINTER input file. References [72,73] are cited in Supplementary Materials.

Author Contributions: Conceptualization, D.H., V.R. and I.R.; methodology, data analysis, I.R., D.H., F.R.B. and N.E.W.; resources, I.R., V.R., D.H., and H.K.; data curation, D.H.; writing—original draft preparation, F.R.B., I.R. and D.H.; writing—review and editing, all authors; visualization, I.R., F.R.B. and D.H.; funding acquisition, D.H. and I.R. All authors have read and agreed to the published version of the manuscript.

Funding: This research was funded by grants of the Ministry of Education, Science and Technological Development of the Republic of Serbia (project No. 173030, contract No. 451-03-68/2020-14/200124) to I.R., Deutscher Akademischer Austauschdienst (DAAD) to I.R., International Association for Plant Taxonomy (IAPT) to I.R., and Deutsche Forschungsgemeinschaft (DFG grant HA 7550/2 and HA 7550/4) to D.H. Costs for open access publishing were partially funded by the Deutsche Forschungsgemeinschaft (DFG grant 491250510) and the Leibniz Open Access Fund.

Institutional Review Board Statement: Not applicable.

Informed Consent Statement: Not applicable.

Data Availability Statement: The analyzed DNA sequences and raw reads are available through ENA study accession number PRJEB57934, Datasets S1–S4 can be downloaded from e!DAL through https://doi.org/10.5447/ipk/2023/5, accessed on 10 February 2022.

Acknowledgments: We would like to thank I. Faustmann, C. Koch, B. Kraenzlin, and P. Oswald for help with plant cultivation and lab work; A. Himmelbach and S. König for performing Illumina sequencing; and L. Peruzzi, K. Hegedüšová, P. Vantara, T. Huber, S. Milanović, M. Slovák, J. Rukšāns, N. Sarajlić, B. Štulović, M. Mihajlović, D. Jakšić, and L. Shuka for providing *Crocus* materials for molecular and morpho-anatomical analysis and/or assistance during field work. We thank the authorities of the respective countries for supporting our work with collection permits.

Conflicts of Interest: The authors declare no conflict of interest.

References

1. Soltis, D.E.; Visger, C.J.; Marchant, D.B.; Soltis, P.S. Polyploidy: Pitfalls and paths to a paradigm. *Am. J. Bot.* **2016**, *103*, 1146–1166. [CrossRef] [PubMed]
2. Van de Peer, Y.; Fawcett, J.A.; Proost, S.; Sterck, L.; Vandepoele, K. The flowering world: A tale of duplications. *Trends Plant Sci.* **2009**, *14*, 680–688. [CrossRef]
3. Van de Peer, Y.; Ashman, T.L.; Soltis, P.S.; Soltis, D.E. Polyploidy: An evolutionary and ecological force in stressful times. *Plant Cell* **2021**, *33*, 11–26. [CrossRef] [PubMed]
4. Vu, G.T.H.; Schmutzer, T.; Bull, F.; Cao, H.X.; Fuchs, J.; Tran, T.D.; Jovtchev, G.; Pistrick, K.; Stein, N.; Pecinka, A.; et al. Comparative genome analysis reveals divergent genome size evolution in a carnivorous plant genus. *Plant Genome* **2015**, *8*, plantgenome2015-04. [CrossRef]
5. Brassac, J.; Blattner, F.R. Species-level phylogeny and polyploid relationships in *Hordeum* (Poaceae) inferred by next-generation sequencing and in silico cloning of multiple nuclear loci. *Syst. Biol.* **2015**, *64*, 792–808. [CrossRef] [PubMed]
6. Jakob, S.S.; Meister, A.; Blattner, F.R. The considerable genome size variation of *Hordeum* species (Poaceae) is linked to phylogeny, life form, ecology, and speciation rates. *Mol. Biol. Evol.* **2004**, *21*, 860–869. [CrossRef] [PubMed]
7. Albach, D.C.; Greilhuber, J. Genome size variation and evolution in *Veronica*. *Ann. Bot.* **2004**, *94*, 897–911. [CrossRef] [PubMed]
8. Leitch, I.J.; Bennett, M.D. Genome downsizing in polyploid plants. *Biol. J. Linn. Soc.* **2004**, *82*, 651–663. [CrossRef]

9. Leitch, I.J.; Fay, M. Plant genome horizons: Michael Bennett's contribution to genome research. *Ann. Bot.* **2008**, *101*, 737–746. [CrossRef]
10. Meudt, H.M.; Rojas-Andrés, B.M.; Prebble, J.M.; Low, E.; Garnock-Jones, P.J.; Albach, D.C. Is genome downsizing associated with diversification in polyploid lineages of *Veronica*? *Bot. J. Linn. Soc.* **2015**, *178*, 243–266. [CrossRef]
11. Petrov, D.A. Evolution of genome size: New approaches to an old problem. *Trends Genet.* **2001**, *17*, 23–28. [CrossRef] [PubMed]
12. Vu, G.T.H.; Cao, H.X.; Reiss, B.; Schubert, I. Deletion-bias in DNA double-strand break repair differentially contributes to plant genome shrinkage. *New Phytol.* **2017**, *214*, 1712–1721. [CrossRef] [PubMed]
13. Petersen, G.; Seberg, O.; Thorsøe, S.; Jørgensen, T.; Mathew, B. A phylogeny of the genus *Crocus* (Iridaceae) based on sequence data from five plastid regions. *Taxon* **2008**, *57*, 487–499. [CrossRef]
14. Harpke, D.; Meng, S.; Rutten, T.; Kerndorff, H.; Blattner, F.R. Phylogeny of *Crocus* (Iridaceae) based on one chloroplast and two nuclear loci: Ancient hybridization and chromosome number evolution. *Mol. Phylogenet. Evol.* **2013**, *66*, 617–627. [CrossRef]
15. Harpke, D.; Carta, A.; Tomović, G.; Ranđelović, V.; Ranđelović, N.; Blattner, F.R.; Peruzzi, L. Phylogeny, karyotype evolution and taxonomy of *Crocus* series *Verni* (Iridaceae). *Plant Syst. Evol.* **2015**, *301*, 309–325. [CrossRef]
16. Raca, I.; Harpke, D.; Shuka, L.; Ranđelović, V. A new species of *Crocus* ser. *Verni* (Iridaceae) with 2*n* = 12 chromosomes from the Balkans. *Plant Biosyst.* **2022**, *156*, 36–42. [CrossRef]
17. Karasawa, K. Karyological studies in *Crocus*. III. *Jap. J. Bot.* **1943**, *12*, 475–503.
18. Brighton, C.A. Cytological problems in the genus *Crocus* (Iridaceae): I. *Crocus vernus* aggregate. *Kew Bull.* **1976**, *31*, 33–46. [CrossRef]
19. Baldini, R.M. Karyological observations on two *Crocus* species (Iridaceae) from Tuscany (Italy). *Caryologia* **1990**, *43*, 341–345. [CrossRef]
20. Micevska, D. Caryological Analysis of the Species of the Genus *Crocus*. Master's Thesis, University of Skopje, Skopje, North Macedonia, 2000.
21. Mitic, D. Caryological Analysis of Some Species of the Genus *Crocus* L. Master's Thesis, University of Skopje, Skopje, North Macedonia, 2001.
22. Peruzzi, L.; Carta, A. *Crocus ilvensis* sp. nov. (sect. *Crocus*, Iridaceae), endemic to Elba Island (Tuscan Archipelago, Italy). *Nord. J. Bot.* **2011**, *29*, 6–13. [CrossRef]
23. Peruzzi, L.; Carta, A.; Garbari, F. Lectotypification of the name *Crocus sativus* var. *vernus* L. (Iridaceae) and its consequences within *Crocus* ser. *Verni*. *Taxon* **2013**, *62*, 1037–1040. [CrossRef]
24. Elshire, R.J.; Glaubitz, J.C.; Sun, Q.; Poland, J.A.; Kawamoto, K.; Buckler, E.S.; Mitchell, S.E. A robust, simple genotyping-by-sequencing (GBS) approach for high diversity species. *PLoS ONE* **2011**, *6*, e19379. [CrossRef]
25. Kerndorff, H.; Harpke, D. *Crocus cobbii* Kerndorff, Pasche & Harpke species nova (Liliiflorae, Iridaceae) and its relatives. *Stapfia* **2022**, *113*, 5–32. [CrossRef]
26. Kearse, M.; Moir, R.; Wilson, A.; Stones-Havas, S.; Cheung, M.; Sturrock, S.; Buxton, S.; Cooper, A.; Markowitz, S.; Duran, C.; et al. Geneious Basic: An integrated and extendable desktop software platform for the organization and analysis of sequence data. *Bioinformatics* **2012**, *28*, 1647–1649. [CrossRef]
27. Katoh, K.; Standley, D.M. MAFFT multiple sequence alignment software version 7: Improvements in performance and usability. *Mol. Biol. Evol.* **2013**, *30*, 772–780. [CrossRef] [PubMed]
28. Wendler, N.; Mascher, M.; Nöh, C.; Himmelbach, A.; Scholz, U.; Ruge-Wehling, B.; Stein, N. Unlocking the secondary gene-pool of barley with next-generation sequencing. *Plant Biotechnol. J.* **2014**, *12*, 1122–1131. [CrossRef] [PubMed]
29. Martin, M. Cutadapt removes adapter sequences from high-throughput sequencing reads. *EMBnet. J.* **2011**, *17*, 10. [CrossRef]
30. Eaton, D.A.R.; Overcast, I. ipyrad: Interactive assembly and analysis of RADseq datasets. *Bioinformatics* **2020**, *36*, 2592–2594. [CrossRef] [PubMed]
31. Frichot, E.; François, O. LEA: An R package for landscape and ecological association studies. *Methods Ecol. Evol.* **2015**, *6*, 925–929. [CrossRef]
32. Raj, A.; Stephens, M.; Pritchard, J.K. fastSTRUCTURE: Variational inference of population structure in large SNP data sets. *Genetics* **2014**, *197*, 573–589. [CrossRef]
33. Wickham, H.; Wickham, M.H. *Package 'Tidyverse'*, 2019; pp. 1–5. Available online: http://tidyverse.tidyverse.org(accessed on 20 July 2022).
34. Wickham, H. *ggplot2: Elegant Graphics for Data Analysis*; Springer: New York, NY, USA, 2016; pp. 1–189.
35. Rozas, J.; Ferrer-Mata, A.; Sánchez-DelBarrio, J.C.; Guirao-Rico, S.; Librado, P.; Ramos-Onsins, S.E.; Sánchez-Gracia, A. DnaSP 6: DNA sequence polymorphism analysis of large data sets. *Mol. Biol. Evol.* **2017**, *34*, 3299–3302. [CrossRef] [PubMed]
36. Blattner, F.R. Direct amplification of the entire ITS region from poorly preserved plant material using recombinant PCR. *Biotechniques* **1999**, *27*, 1180–1186. [CrossRef] [PubMed]
37. Blattner, F.R. Phylogenetic analysis of *Hordeum* (Poaceae) as inferred by nuclear rDNA ITS sequences. *Mol. Phylogenet. Evol.* **2004**, *33*, 289–299. [CrossRef]
38. Waminal, N.E.; Park, H.M.; Ryu, K.B.; Kim, J.H.; Yang, T.J.; Kim, H.H. Karyotype analysis of *Panax ginseng* C.A.Meyer, 1843 (Araliaceae) based on rDNA loci and DAPI band distribution. *Comp. Cytogenet.* **2012**, *6*, 425–441. [CrossRef]
39. Rodríguez-Domínguez, J.M.; Ríos-Lara, L.L.; Tapia-Campos, E.; Barba-Gonzalez, R. An improved technique for obtaining well-spread metaphases from plants with numerous large chromosomes. *Biotech. Histochem.* **2017**, *92*, 159–166. [CrossRef]

40. Swofford, D.L. *PAUP*: Phylogenetic Analysis Using Parsimony (* and Other Methods)*; Sinauer Assoc.: Sunderland, MA, USA, 2002.
41. Ronquist, F.; Teslenko, M.; van der Mark, P.; Ayres, D.L.; Darling, A.; Höhna, S.; Larget, B.; Liu, L.; Suchard, M.A.; Huelsenbeck, J.P. MrBayes 3.2: Efficient Bayesian phylogenetic inference and model choice across a large model space. *Syst. Biol.* **2012**, *61*, 539–542. [CrossRef]
42. Rochette, N.C.; Rivera-Colón, A.G.; Catchen, J.M. Stacks 2: Analytical methods for paired-end sequencing improve RADseq-based population genomics. *Mol. Ecol.* **2019**, *28*, 4737–4754. [CrossRef]
43. Malinsky, M.; Trucchi, E.; Lawson, D.J.; Falush, D. RADpainter and fineRADstructure: Population inference from RADseq data. *Mol. Biol. Evol.* **2018**, *35*, 1284–1290. [CrossRef] [PubMed]
44. Gligorijević, S.; Pejčinović, D. Contribution to the methodology of anatomical sections preparation. *Acta Biol. Med. Exp.* **1983**, *8*, 43–45.
45. Raca, I.; Ljubisavljević, I.; Jušković, M.; Ranđelović, N.; Ranđelović, V. Comparative anatomical study of the taxa from series *Verni* Mathew (*Crocus* L.) in Serbia. *Biol. Nyssana* **2017**, *8*, 15–22. [CrossRef]
46. Raca, I.; Jovanovic, M.; Ljubisavljevic, I.; Juskovic, M.; Randelovic, V. Morphological and leaf anatomical variability of *Crocus* cf. *heuffelianus* Herb. (Iridaceae) populations from the different habitats of the Balkan Peninsula. *Turk. J. Bot.* **2019**, *43*, 645–658. [CrossRef]
47. Schneider, C.A.; Rasband, W.S.; Eliceiri, K.W. NIH Image to ImageJ: 25 years of image analysis. *Nat. Methods* **2012**, *9*, 671–675. [CrossRef] [PubMed]
48. Mosolygo-L, A.; Sramko, G.; Barabas, S.; Czegledi, L.; Javor, A.; Molnar, V.A.; Suranyi, G. Molecular genetic evidence for allotetraploid hybrid speciation in the genus *Crocus* L. (Iridaceae). *Phytotaxa* **2016**, *258*, 121–136. [CrossRef]
49. Raca, I. Taxonomy and Phylogeny of Series Verni Mathew (*Crocus* L.) in Southeastern Europe—Morpho-Anatomical, Cytological and Molecular Approach. Ph.D. Thesis, University of Niš, Niš, Serbia, 2021.
50. Statsoft Inc. *STATISTICA (Data Analysis Software System)*, Version 7; Statsoft Inc.: Tulsa, OK, USA, 2004.
51. Herbert, W.; Lindley, J. A history of the species of *Crocus*. *J. R. Hortic. Soc.* **1847**, *2*, 249–293.
52. Meusel, H.; Jäger, E.J.; Weinert, E. *Vergleichende Chorologie der zentraleuropäischen Flora*; VEB Fischer: Jena, Germany, 1965.
53. Malynovsky, K.A. *Roslynnisty vysokogirja Ukrainskykh Karpat*; Naukova dumka: Kyiv, Ukraine, 1980.
54. Beck-Mannagetta, G. *Flora von Bosnien, der Herzegowina und des Sandzaks Novipazar*; C. Gerold: Vienna, Austria, 1904.
55. Peruzzi, L. *Crocus heuffelianus* (Iridaceae), a new record for the Italian flora. *Phytotaxa* **2016**, *261*, 291–294. [CrossRef]
56. Carta, A.; Probert, R.; Moretti, M.; Peruzzi, L.; Bedini, G. Seed dormancy and germination in three *Crocus* ser. *Verni* species (Iridaceae): Implications for evolution of dormancy within the genus. *Plant Biol.* **2014**, *16*, 1065–1074. [CrossRef]
57. Carta, A.; Flamini, G.; Cioni, P.L.; Pistelli, L.; Peruzzi, L. Flower bouquet variation in four species of *Crocus* ser. *Verni*. *J. Chem. Ecol.* **2015**, *41*, 105–110. [CrossRef]
58. Molloy, E.K.; Warnow, T. To include or not to include: The impact of gene filtering on species tree estimation methods. *Syst. Biol.* **2017**, *67*, 285–303. [CrossRef]
59. Soltis, P.S.; Soltis, D.E. Ancient WGD events as drivers of key innovations in angiosperms. *Curr. Opin. Plant Biol.* **2016**, *30*, 159–165. [CrossRef]
60. Murat, F.; Armero, A.; Pont, C.; Klopp, C.; Salse, J. Reconstructing the genome of the most recent common ancestor of flowering plants. *Nat. Genet.* **2017**, *49*, 490–496. [CrossRef]
61. Mandakova, T.; Lysak, M.A. Post-polyploid diploidization and diversification through dysploid changes. *Curr. Opin. Plant Biol.* **2018**, *42*, 55–65. [CrossRef] [PubMed]
62. Glombik, M.; Bačovský, V.; Hobza, R.; Kopecký, D. Competition of parental genomes in plant hybrids. *Front. Plant Sci.* **2020**, *11*, 200. [CrossRef]
63. Lysak, M.A.; Cheung, K.; Kitschke, M.; Bures, P. Ancestral chromosomal blocks are triplicated in *Brassiceae* species with varying chromosome number and genome size. *Plant Physiol.* **2007**, *145*, 402–410. [CrossRef] [PubMed]
64. El-Shehawi, A.M.; Elseehy, M.M. Genome size and chromosome number relationship contradicts the principle of Darwinian evolution from common ancestor. *J. Phylogenet. Evol. Biol.* **2017**, *5*, 179. [CrossRef]
65. Fedoroff, N.V. Transposable elements, epigenetics, and genome evolution. *Science* **2012**, *338*, 758–767. [CrossRef] [PubMed]
66. Guerra, M. Agmatoploidy and symploidy: A critical review. *Genet. Mol. Biol.* **2016**, *39*, 492–496. [CrossRef]
67. Schubert, I.; Lysak, M.A. Interpretation of karyotype evolution should consider chromosome structural constraints. *Trends Genet.* **2011**, *27*, 207–216. [CrossRef] [PubMed]
68. Tank, D.C.; Eastman, J.M.; Pennell, M.W.; Soltis, P.S.; Soltis, D.E.; Hinchliff, C.E.; Brown, J.W.; Sessa, E.B.; Harmon, L.J. Nested radiations and the pulse of angiosperm diversification: Increased diversification rates often follow whole genome duplications. *New Phytol.* **2015**, *207*, 454–467. [CrossRef]
69. Murat, F.; Xu, J.-H.; Tannier, E.; Abrouk, M.; Guilhot, N.; Pont, C.; Messing, J.; Salse, J. Ancestral grass karyotype reconstruction unravels new mechanisms of genome shuffling as a source of plant evolution. *Genome Res.* **2010**, *20*, 1545–1557. [CrossRef]
70. Kelly, L.J.; Renny-Byfield, S.; Pellicer, J.; Macas, J.; Novák, P.; Neumann, P.; Lysak, M.A.; Day, P.D.; Berger, M.; Fay, M.F.; et al. Analysis of the giant genomes of *Fritillaria* (Liliaceae) indicates that a lack of DNA removal characterizes extreme expansions in genome size. *New Phytol.* **2015**, *208*, 596–607. [CrossRef]

71. Macas, J.; Novak, P.; Pellicer, J.; Cizkova, J.; Koblizkova, A.; Neumann, P.; Fukova, I.; Doležel, J.; Kelly, L.J.; Leitch, I.J. In depth characterization of repetitive DNA in 23 plant genomes reveals sources of genome size variation in the legume tribe Fabeae. *PLoS ONE* **2015**, *10*, e0143424. [CrossRef] [PubMed]
72. Brighton, C.A.; Mathew, B.; Marchant, C.J. Chromosome counts in the genus Crocus (Iridaceae). *Kew Bull.* **1973**, *28*, 451–464. [CrossRef]
73. Thiv, M.; Wörz, A. Die neue Identität des Zavelsteiner Krokus als Crocus neglectus nach DNA-Untersuchungen. *Jahresh. Ges. Nat. Württemberg* **2015**, *171*, 163–172.

biology MDPI

Article

Picks in the Fabric of a Polyploidy Complex: Integrative Species Delimitation in the Tetraploid *Leucanthemum* Mill. (Compositae, Anthemideae) Representatives

Christoph Oberprieler [1,*,†], Tankred Ott [1,†] and Robert Vogt [2]

1 Evolutionary and Systematic Botany Group, Institute of Plant Biology, University of Regensburg, Universitätsstr. 31, D-93053 Regensburg, Germany
2 Botanic Garden & Botanical Museum Berlin, Freie Universität Berlin, Königin-Luise-Str. 6-8, D-14191 Berlin, Germany
* Correspondence: christoph.oberprieler@ur.de
† Christoph Oberprieler and Tankred Ott have contributed equally to this work.

Simple Summary: The delimitation of species as the most important rank in biological classification is an essential contribution of taxonomy to biodiversity research, with all of its evolutionary, ecological, political, and legislative ramifications. Species delimitation is extremely tricky in plant groups evolving by polyploidisation (multiplication of chromosome sets) because the rapid formation of new, reproductively isolated lineages (species) is often not paralleled by conspicuous genetic, morphological, physiological, and/or ecological differentiation. Having clarified the taxonomy of diploid (2*x*) representatives of the genus *Leucanthemum* (marguerites, ox-eye daisies) in a previous contribution, the present study aims at an objective and reproducible delimitation of evolutionarily significant units (species) at the tetraploid (4*x*) level. We used DNA-based fingerprinting and statistical analyses of leaf shapes, ecological niches, and distribution ranges for eight predefined morphotaxa to judge their ranks as species or subspecies and propose a taxonomical treatment for the surveyed group with six species (two of them with two subspecies). Having clarified the taxonomic structure of the ancestral diploid (the 'warps and wefts') and the subsequent tetraploid layer (the 'picks of the fabric'), we will be able to provide a taxonomy for the remainder of this well-known plant group and study its reticulate evolutionary history.

Abstract: Based on the results of a preceding species-delimitation analysis for the diploid representatives of the genus *Leucanthemum* (Compositae, Anthemideae), the present study aims at the elaboration of a specific and subspecific taxonomic treatment of the tetraploid members of the genus. Following an integrative taxonomic approach, species-level decisions on eight predefined morphotaxon hypotheses were based on genetic/genealogical, morphological, ecological, and geographical differentiation patterns. ddRADseq fingerprinting and SNP-based clustering revealed genetic integrity for six of the eight morphotaxa, with no clear differentiation patterns observed between the widespread *L. ircutianum* subsp. *ircutianum* and the N Spanish (Cordillera Cantábrica) *L. cantabricum* and the S French *L. delarbrei* subsp. *delabrei* (northern Massif Central) and *L. meridionale* (western Massif Central). The inclusion of differentiation patterns in morphological (leaf dissection and shape), ecological (climatological and edaphic niches), and geographical respects (pair-wise tests of sympatry vs. allopatry) together with the application of a procedural protocol for species-rank decisions (the 'Wettstein tesseract') led to the proposal of an acknowledgement of the eight predefined morphotaxon hypotheses as six species (two of them with two subspecies). Nomenclatural consequences following from these results are drawn and lead to the following new combinations: *Leucanthemum delarbrei* subsp. *meridionale* (Legrand) Oberpr., T.Ott & Vogt, comb. nov. and *Leucanthemum ruscinonense* (Jeanb. & Timb.-Lagr.) Oberpr., T.Ott & Vogt, comb. et stat. nov.

Citation: Oberprieler, C.; Ott, T.; Vogt, R. Picks in the Fabric of a Polyploidy Complex: Integrative Species Delimitation in the Tetraploid *Leucanthemum* Mill. (Compositae, Anthemideae) Representatives. *Biology* **2023**, *12*, 288. https://doi.org/10.3390/biology12020288

Academic Editor: Lorenzo Peruzzi

Received: 20 January 2023
Revised: 7 February 2023
Accepted: 9 February 2023
Published: 10 February 2023

Keywords: Asteraceae; ecology; ddRADseq; geography; leaf morphology; nomenclature; polyploidy; taxonomy

1. Introduction

The delimitation of species as the most paramount rank in biological classification is an essential contribution of taxonomy to biodiversity research, with all of its evolutionary, ecological, political, and legislative ramifications and corollaries. Following Stuessy [1] and Zachos et al. [2], this alpha-taxonomy procedure has a twofold nature: while in the first step (the 'grouping' step), 'species discovery' and 'species validation' methods are used to infer and subsequently test species-group hypotheses (the 'species taxa' of Zachos et al. [2]) by detecting genealogical, morphological, or ecological discontinuities [3,4], the second step (the 'ranking' step) constitutes 'an executive decision that the species taxon warrants recognition at the species level' [2]. However, the latter—the decision whether 'species taxa' should be ranked as species under the Linnaean classification system—is clearly subjective, due to its dependence on the acceptance of a species concept. Of these, a broad array exists, though without any cognisable chance for the unrestricted applicability of a single one throughout the realm of organismic diversity.

The 'unified species concept' proposed by De Queiroz [5] was a game-changer in the futile search for a generally applicable species concept because it altered the perspective that the properties entertained by the plethora of concepts (e.g., reproductive isolation, genealogy, morphology, ecology, geography, etc.) are helpful in species conceptualisation. Instead, the 'unified species concept' defined species as hypothetical independently evolving metapopulation lineages, for which the above-mentioned properties could be made to subserve as indicators or proxies. This conceptual shift paved the way for the renaissance of 'biosystematics' or 'experimental taxonomy' approaches to species delimitation used in the second half of the 20th century as 'integrative taxonomy' [6,7], which makes use of all available sources of empirical evidence for the conceptualisation of species rank. This is nowadays carried out either by using computational tools, e.g., GENELAND [8] for the joint analysis of morphology, genetics, and geography; "multivariate normal mixtures and tolerance regions" analysis [9,10] for morphology and geography; iBPP [11] for genealogy and morphology; regression analysis [12] for genetics and geography; or by entertaining procedural protocols, e.g., [13–16]. Most recent approaches also try to incorporate the speciation process itself into species-delimitation software programs (DELINEATE [17]).

Species formation by polyploidy—a speciation mode realised in significant numbers of pteridophyte and angiosperm groups—poses considerable problems for species delimitation [18]: while polyploidisation will instantly lead to postzygotic reproductive isolation between parental taxa and their polyploid derivatives (biological species concept), detectable trait differences entertained by a morphological or a physiological/ecological species concept may only be realised in an allopolyploid speciation scenario and not (or at least not immediately or obviously) in an autopolyploid one. Additionally, a strict phylogenetic species concept (with species defined as monophyletic evolutionary units) is violated both in the auto- and allopolyploid speciation mode due to (a) the parental species becoming paraphyletic relative to the newly formed polyploid species and (b) the potentially polyphyletic nature of a polyploid species caused by its multiple (and sometimes reciprocally parented) origin or (c) gene flow between independently formed allo- and/or autopolyploid populations/lineages. Finally, at least in some plant groups, the switch of polyploid lineages toward asexual reproduction (agamospermy) makes the application of a biological species concept senseless. As a consequence, these peculiarities of polyploid species formation make the application of a 'unified species concept' sensu De Queiroz [5] (species as metapopulation lineages) and the integrative approach to species delimitation indispensable for plant groups diversifying through this speciation mode.

The genus *Leucanthemum* Mill. (Compositae, Anthemideae; marguerites or ox-eye daisies) comprises 39 [19] to 42 species [20], with ploidy levels ranging from diploid (2*x*) to dodecaploid (12*x*), and one species [*L. lacustre* (Brot.) Samp.) from Portugal even showing a chromosome number of $2n = 22x = 198$ (docosaploid level). *Leucanthemum* is distributed over the whole European continent and extends into Northern Asia (the tetraploid *L. ircutianum* DC. is found in Siberia), while some species were also introduced into temperate regions of the Northern and Southern Hemispheres [21]. The current species delimitation of the genus is mostly based on differences in morphology, especially in general leaf shape and leaf dissection, as the flower characters are relatively invariant, and ploidy level, as well as geographical distribution [22,23] and, more recently, genetic differentiation [19].

Previous studies addressed the species delimitation and phylogeny of the diploid *Leucanthemum* representatives based on multicopy nuclear markers (nrDNA ETSs), AFLP fingerprinting, and single-copy nuclear markers [24–27]. Cloning of nrDNA ETS amplicons revealed that some diploid taxa exclusively possess a plesiomorphic ETS ribotype cluster closely related to ETS ribotypes of the outgroup and that others are characterised by the exclusive possession of an apomorphic ETS ribotype cluster, while a third group of taxa exhibit an additive pattern of the two types [24]. This finding was supported by Konowalik et al. [25] using AFLP fingerprinting and multilocus species-tree reconstructions based on low-copy markers of the nuclear genome, and it was demonstrated that the species with the plesiomorphic ETS ribotypes form an early-diverging paraphyletic grade, in which the monophyletic group of taxa with the apomorphic ETS ribotypes are nested. Furthermore, the authors applied coalescent-based simulations to distinguish between hybridisation and incomplete lineage sorting (ILS), revealing that for most of the diploid taxa involved (and especially for those in the second group) incongruence among gene trees could not be explained by ILS alone and that recent hybridisation or even homoploid hybrid speciation events must be assumed. Consequently, some infraspecific taxa were raised to species level to account for their assumed independent formation through homoploid hybrid speciation (i.e., *L. cacuminis*, *L. eliasii*, and *L. pyrenaicum*).

Species delimitation in a morphologically close-knit group of taxa in the clade with the apomorphic nrDNA ETS ribotypes was subsequently carried out by Wagner et al. [26], who used AFLP fingerprinting, sequence information from plastid and nuclear low-copy markers, and coalescent-based Bayesian delimitation methods to infer species boundaries in the *L. ageratifolium* group, despite the frequent presence of hybrid individuals in this group. Finally, Wagner et al. [27] presented a multilocus phylogenetic reconstruction of the subtribe Leucantheminae, in which the diversification among diploid *Leucanthemum* species was dated to the last 1.9 (1.1–2.9) Ma, arguing for the strong influence of Pleistocene oscillations on species formation in this genus. Most recently, Ott et al. [19] combined restriction-site-associated DNA sequencing (RADseq), ecological niche modelling (ENM), geographical patterns, and geometric morphometrics for integrative species delimitation among the diploid *Leucanthemum* representatives.

In contrast to the extensive investigation of species delimitation and the phylogenetic relationships of diploids, the evolutionary histories of only a few tetraploid *Leucanthemum* species have been the subject of previous studies. Oberprieler et al. [28] found indications of an allopolyploid origin of the tetraploid *L. ircutianum* subsp. *ircutianum* based on AFLP markers. One year later, Greiner et al. [29] showed that the tetraploid *L. pseudosylvaticum* is able to form fertile offspring when crossed with its diploid relative *L. pluriflorum*, suggesting a relatively recent diversification. Another study of the *L. pluriflorum* group by Greiner et al. [30] produced evidence for the allopolyploid and autopolyploid origins of *L. pseudosylvaticum* and *L. corunnense*, respectively, and raised *L. pseudosylvaticum*, which was formerly a subspecies of *L. ircutianum*, to species rank. Finally, the most recent contribution to the taxonomy of the tetraploid *Leucanthemum* representatives by Oberprieler et al. [31] suggested infraspecific ranks for the widely distributed *L. ircutianum* subsp.

ircutianum and its amphi-Adriatic counterpart *L. ircutianum* subsp. *leucolepis* based on AFLP fingerprinting and sequence variations in nuclear and plastid DNA.

The present contribution aims at a comprehensive and integrative assessment of all tetraploid *Leucanthemum* taxa. Applying the conceptual framework for species-rank decisions proposed by Oberprieler [16], morphological, ecological, geographical, and genealogical/genetic evidence is used to devise a taxonomic treatment for eight tetraploid taxa that have been hitherto accepted as occupying specific or infraspecific ranks in Central and Southern Europe, the only exception being the NW Spanish *L. corunnenese* Lago, for which an autotetraploid origin based on the diploid *L. pluriflorum* has been shown in a previous study [30].

2. Material and Methods

2.1. Taxon Selection

Besides the widespread *L. ircutianum* subsp. *ircutianum*, the following seven geographically more restricted tetraploid taxa were considered as entities, for which an integrative taxonomic treatment was envisioned in the present contribution: these comprise the three species endemic to the Iberian Peninsula, *L. cantabricum* Sennen, *L. crassifolium* (Lange) Lange, and *L. pseudosylvaticum* (Vogt) Vogt & Oberpr., along with the NE Spanish and SW French *L. delarbrei* Timb.-Lagr. subp. *delarbrei*, subsp. *ruscinonense* (Jeanb. & Timb.-Lagr.) Vogt et al. and *L. meridionale* Legrand and the amphi-Adriatic *L. ircutianum* subsp. *leucolepis* (Briq. & Cavill.) Vogt & Greuter (see Figure 1 for a distribution map).

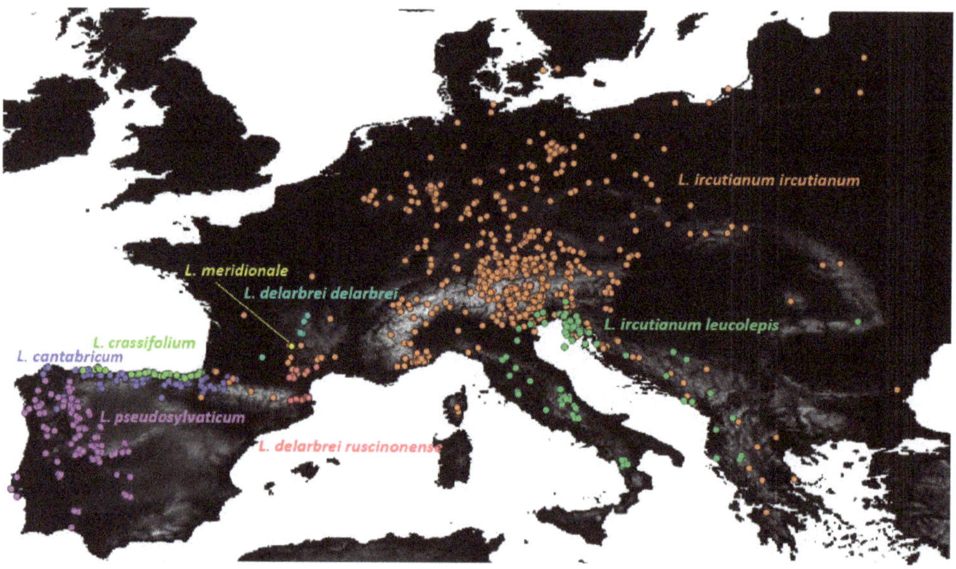

Figure 1. Distribution of the eight tetraploid *Leucanthemum* morphotaxa in Southern and Central Europe based on locality information from revised herbarium specimens which was used in the ecological niche modelling and geographical overlap analyses of the present study (see Supplementary Material ES02 for the list of georeferenced accessions).

2.2. Morphological Analyses

Differences in leaf shape, an important delimitation criterion for *Leucanthemum* taxa, have been demonstrated in the diploids of the genus [19] and were assessed here by measuring the general leaf shape and the degree of dissection of the leaves by applying elliptic Fourier analysis (EFA; [32]) and calculating leaf dissection indices (LDIs), respectively. For this purpose, 285 images of digitised herbarium specimens were provided by the herbarium of the Berlin Botanical Museum (B; see Supplementary Material ES01). Using these images, we manually annotated 1,338 intact leaves with polygons around the leaves' outlines and polylines along the leaves' main veins using the Computer Vision Annotation Tool (CVAT; https://github.com/openvinotoolkit/cvat (accessed on 31 January 2023)) according to Ott et al. [19]. Subsequently, we straightened the leaves to reduce the influence of deformations either caused by developmental irregularities or introduced by the drying process and extracted the leaf contours as binary masks.

Using the binary masks, we conducted EFA and calculated LDIs using the Python packages scikit-image [33], numpy [34], scikit-learn [35], and PyEFD (https://github.com/hbldh/pyefd (accessed on 31 January 2023)). For the EFA, we used 20 harmonics, normalised the descriptors, and applied principal component analysis (PCA) for decorrelation and feature extraction to the non-constant descriptors (normalisation causes the three descriptors A1, B1, and C1 to be constant).

To find differences in general leaf shape, we used the first 15 principal components (PCs), explaining 50% of the total variance with two different testing strategies: (1) the permutation-based test for differences in Euclidean distance in PC space as applied by [19]; and (2) non-parametric multivariate analysis of variance (NPMANOVA; [36]) implemented in the R package 'vegan' v2.6 (function 'adonis'; [37]). The number of permutations was set to 5000 and 100 for the former and the latter, respectively. To assess differences in leaf dissection, we subjected the LDI values to Welch's tests. All tests were corrected using Bonferroni´s method.

2.3. Ecological Niche Modelling

Ecoclimatological and edaphic niches of the tetraploid *Leucanthemum* taxa were reconstructed using ecological niche modelling (ENM) and compared using permutation-based statistical tests. For this purpose, we retrieved the collection locations of 1470 individuals from several herbaria (Supplementary Material ES02) and obtained rasters of 19 bioclimatic and 10 edaphic variables at depth levels of 0–5 cm, 5–15 cm, and 15–30 cm (Supplementary Material ES03) from Worldclim Bioclim [38] and SoilGrids [39], respectively. The rasters were cropped to encompass Central Europe (longitude -10.0°–26.0°; latitude 35.9°–52°) and scaled to a resolution of 2.5 arc minutes using the R package 'raster' v3.5.15 [40] (https://github.com/rspatial/raster (accessed on 31 January 2023)). In addition to the climate rasters for presence, we retrieved paleoclimate rasters for the last glacial maximum (LGM; CCSM4) and the last interglacial (LIG; lig_30s) from Worldclim Bioclim and subjected them to the same preprocessing, but with additional recoding of temperature, since LGM and LIG datasets used Bioclim 1 temperature encoding. For the edaphic variables, we averaged the raster values of the three mentioned depth levels to obtain a single raster for each soil variable. Since it was computationally not tractable to work with all 29 rasters, we applied principal component analysis (PCA) with standardisation implemented in the R package 'ENMtools' v1.0.6 [41] for feature extraction to the recent Bioclim and SoilGrids rasters, separately. The first three principal-component (PC) rasters for each of the datasets were selected for the ENM.

To compare the ecological niches of the different taxa, we reconstructed potential distribution ranges for all tetraploid taxa except *L. meridionale* using MaxEnt v3.4.4 [42], with 5000 iterations and 6-fold cross-validation, and applied niche-equivalency tests implemented in 'ENMTools' for all combinations of tetraploid taxa with MaxEnt as species distribution, 200 replicates, a species range of 50 km, and 1000 background points with a range of 20 km. *Leucanthemum meridionale* was excluded from these comparisons because this species is endemic to a very limited area, causing all collection points to fall into the same raster cell and thus rendering ENM impossible. Potential niches at LGM and LIG were also reconstructed using MaxEnt with the same parameters as for the PC rasters, but this time with LGM and LIG as projection rasters and including collection data for the diploid species provided by Ott et al. [19].

2.4. Geographical Distribution

The geographic co-distribution of the tetraploid taxa, excluding *L. meridionale* again due to its restricted sampling area as a point endemic, was assessed by evaluating the overlap of spatial distributions using the same data as for the ENM analyses. Unfortunately, there are no comprehensive distribution rasters available for the tetraploid *Leucanthemum* representatives, which is why we had to resort to approximating the geographic distribution from the available sampling points. We applied the method proposed by Ott et al. [19], which approximates the true geographical distribution by reconstructing the potential area using an ENM (in this case, MAXENT v3.4.4; [42]) and subsequently removes unconnected and unsampled regions (i.e., regions that are separated from collection points by areas of low probability). To test for sympatry, we applied the permutation approach of Ott et al. [19] with 400 simulated datasets and a threshold of 0.25; the resulting *p*-values were corrected for multiple testing using Bonferroni´s method.

2.5. RADseq Assembly

Double-digest RADseq (ddRADseq; [43]) was conducted based on an accession set comprising 51 individuals from all presently accepted 17 diploid *Leucanthemum* species (20 taxa; [19]) and 35 individuals from 8 tetraploid *Leucanthemum* taxa (see the table in Supplementary Material ES04), the latter being the focal group of the present study. ddRADseq reads of the diploid samples were taken from Ott et al. [19], while for the tetraploid individuals, genomic DNA was extracted from silica-dried specimens according to the CTAB DNA extraction protocol of Doyle and Dickson [44]. For assembly-quality assessment, replicates were generated by repeating the extraction for two samples (*L. ircutianum* DC. subsp. *ircutianum*: accession L055-03/-031; *L. ircutianum* subsp. *leucolepis*: accession 170-02/-021). ddRAD Illumina sequencing (2 x 150 bp; NextSeq 500, Illumina Inc., San Diego, CA, USA) using the restriction enzymes *Pst1* and *ApeK1*, including demultiplexing and adapter clipping, was conducted by LGC Genomics (Berlin, Germany).

Demultiplexed and adapter clipped reads from the diploid and tetraploid samples were assembled using IPYRAD v0.9.81 [45] against the reference provided by Ott et al. [19]; the minimum number of samples per locus (min_samples_locus) was set to 20, while the remaining parameters were kept at default values. For detection of assembly errors, locus, allele, and SNP error rates were calculated (see [19,46]).

2.6. RADseq Network Analysis

Neighbor-nets based on SNP-based Nei distances ([47]; https://github.com/simjoly/pofad (accessed on 31 January 2023)) were calculated using SPLITSTREE4 v4.15.1 [48] for (a) the complete dataset and (b) for the tetraploid accessions exclusively. Distances were calculated based on the variant output (VCF file) of IPYRAD using a custom Python and C tool (https://github.com/TankredO/nei_vcf (accessed on 31 January 2023)). With SNP-based Nei distances, differences in variant (SNP) frequencies are directly included in distance calculation, rendering this kind of distance metric particularly useful for comparing diploid and polyploid samples.

2.7. RADseq Consensus Clustering

Weighted ensemble of random *(k)k*-means (WKM) clustering [49] was applied to detect clusters of genetically similar individuals within the group of tetraploid *Leucanthemum* representatives. WKM clustering is a semi-supervised consensus (ensemble) clustering technique, allowing for a priori-defined fuzzy pairwise must-link and must-not-link constraints. Roughly, WKM clustering fits a number of *k*-means clusterings (e.g., 1000), each for a random subset of features ('variables') and data points ('samples'), and with a random number of clusters *(k)*. For each *k*-means run, clustering-level and cluster-level consistencies are calculated, which measure deviations from the fuzzy constraints concerning the whole clustering and single clusters, respectively. In addition, the mean silhouette index is calculated as an internal clustering-quality metric. The three values are non-linearly combined to obtain a clustering weight (i.e., a value determining the clustering quality). The clustering co-association matrix, i.e., the matrix determining which samples were clustered together (this can be thought of as a similarity matrix with only 1 and 0 entries; 1 if the samples were placed in the same cluster and 0 otherwise), is multiplied by the clustering weight. All weighted co-association matrices are summed (and optionally scaled) to obtain a consensus (co-association) matrix. Finally, a consensus clustering is calculated using this consensus matrix via hierarchical or spectral clustering.

For clustering of the tetraploid *Leucanthemum* representatives, the previously calculated SNP-based Nei distances (see above, under *Network analysis*) were subjected to a principal coordinate analysis (PCoA) with Lingoes negative eigenvalue correction, and the first principal coordinates (PCos) explaining at least 50% of the variance were selected. Based on the PCo scores, WKM clustering was applied with must-link constraints for the individuals of each of the taxa *L. cantabricum*, *L. crassifolium*, *L. delarbrei* subsp. *delarbrei*, *L. delarbrei* subsp. *ruscinonense*, *L. ircutianum* subsp. *leucolepis*, and L. *meridionale* and must-not-link constraints among individuals from *L. delarbrei* subsp. *delarbrei* and *L. meridionale*. The fraction of random features and samples was set to 0.60 and 0.80, respectively. The number of *k*-means runs was set to 5000, and *k* was allowed to take values from 2 to 14, including the lower and upper bounds. Finally, the consensus matrix was subjected to an average-linkage hierarchical clustering for each of the *k*-values, and the clustering scoring that was best according to the Bayesian information criterion (BIC) and the Calinski–Harabasz criterion (CH) was selected. All analyses were performed using pyckmeans v0.9.4 (https://github.com/TankredO/pyckmeans (accessed on 31 January 2023)).

2.8. Genealogical Species Delimitation

To find the potential diploid parental species of the tetraploid taxa under study, we applied a custom version of SNIPLOID [50], as proposed by Wagner et al. [51]. Very similar to the original SNIPLOID algorithm, our script compares two parent individuals (i.e., diploids) with one child individual (i.e., a tetraploid) on a per-SNP basis, distinguishing and counting the frequencies of five different SNP categories: categories 1 and 2 count so-called inter-specific SNPs, where exactly one parental SNP is identical to the child SNP and the other not. Categories 3 and 4 are so-called derived SNPs, meaning that an SNP is unique to the child individual, while the parents are homozygous for the same variant; patterns 3 and 4 cannot be distinguished based on unphased SNP data. Category 5 SNPs are homeo-SNPs, where the child is heterozygous (polymorphic), combining both parents' homologous and monomorphic alleles. For all categories, only SNPs where both parents were homozygous (monomorphic) were considered.

We applied the SNIPLOID approach to look for signs of allopolyploid speciation and expected recently formed allotetraploids to express a high number of category 5 SNPs, while longer established allotetraploid species should express a higher frequency of derived SNPs (i.e., category 3 and 4 SNPs). Finally, category 1 and 2 SNPs may indicate gene flow to the putative parents or alternatively be an additional signature of recent polyploidisation. For each tetraploid taxon and each possible pair of diploid parent species, therefore, we applied our custom Python implementation of the SNIPLOID algorithm to the concatenated SNP output returned by IPYRAD (.snps ouput). For each triplet, we tested all individuals and calculated the arithmetic means of SNP patterns 1 through 5.

3. Results

3.1. Morphology

The degree of leaf dissection varied significantly at an alpha level of p-value < 0.01 for all but three taxon pairs (Table 1, upper triangle). Differences in general leaf shape, measured as Euclidean distances among principal-component (PC)-transformed Fourier descriptors, were significant for 14 of the 28 taxon pairs (Table 1, lower triangle, left number). When applying NPMANOVA to the PC scores, 16 of the 28 taxon pairs showed significant differences in leaf-outline shape (Table 1, lower tringle, right number).

Table 1. Bonferroni-corrected *p*-values for pairwise tests of morphological similarity. The upper triangle represents *p*-values of the LDI-based Welch's tests, determining divergence in the dissection of the leaves. The lower triangle comprises *p*-values of the permutation tests of Euclidean distances in PC space (left) and NPMANOVA results (right), quantifying differences in general leaf shape. Corrected *p*-values are truncated to 1.0.

	cantabricum	*crassifolium*	*delarbrei*	*ruscinonense*	*ircutianum*	*leucolepis*	*meridionale*	*pseudosylvaticum*
cantabricum		<0.01	0.21	<0.01	<0.01	<0.01	<0.01	<0.01
crassifolium	1.0/1.0		<0.01	<0.01	<0.01	<0.01	<0.01	<0.01
delarbrei	1.0/1.0	0.44/1.0		<0.01	<0.01	<0.01	1.0	<0.01
ruscinonense	0.04/<0.01	0.02/<0.01	1.0/1.0		<0.01	<0.01	<0.01	<0.01
ircutianum	<0.01/<0.01	0.17/<0.01	0.02/0.83	<0.01/<0.01		1.0	<0.01	<0.01
leucolepis	<0.01/<0.01	<0.01/<0.01	<0.01/<0.01	<0.01/<0.01	1.0/<0.01		<0.01	<0.01
meridionale	1.0/1.0	1.0/1.0	1.0/0.83	1.0/1.0	1.0/<0.01	0.38/<0.01		<0.01
pseudosylvaticum	<0.01/<0.01	<0.01/<0.01	0.06/<0.01	0.01/<0.01	<0.01/<0.01	<0.01/<0.01	1.0/1.0	

3.2. Ecological Niche Modelling and Geographical Range Overlap

All pairwise niche-equivalency tests except those for the two taxon pairs *L. cantabricum*–*L. crassifolium* (both N Spain) and *L. cantabricum* (N Spain)–*L. delarbrei* subsp. *ruscinonense* (NE Spain, SW France) were significant at an alpha level of 0.01 (see Table 2, where *p*-values of 0 indicate no niche overlap at all). Maps depicting predicted potential distribution ranges of the diploids and tetraploids based on recent, LGM, and LIG bioclimatic variables are provided in the supporting information (Supplementary Material ES05, ES06).

Table 2. Bonferroni-corrected *p*-values for pairwise niche equivalency tests based on the two test statistics, D (upper triangle) and I (lower triangle).

	cantabricum	*crassifolium*	*delarbrei*	*ruscinonense*	*ircutianum*	*leucolepis*	*meridionale*	*pseudosylvaticum*
cantabricum		0.52	0	0.32	0	0	n/a	0
crassifolium	0.31		0	0	0	0	n/a	0
delarbrei	0	0		0	0	0	n/a	0
ruscinonense	0.52	0	0		0	0	n/a	0
ircutianum	0	0	0	0		0	n/a	0
leucolepis	0	0	0	0	0		n/a	0
meridionale	n/a	n/a	n/a	n/a	n/a	n/a		n/a
pseudosylvaticum	0	0	0	0	0	0	n/a	

In the tests on geographical range overlap (Table 3), all taxon pairs except for *L. cantabricum*–*L. crassifolium* (both N Spain) and *L. delarbrei* subsp. *ruscinonense* (NE Spain, SW France)–*L. ircutianum* subsp. *ircutianum* (large parts of Europe) were significant, suggesting a deviation from sympatric distributions for most of the tetraploid taxa.

Table 3. Bonferroni-corrected *p*-values for pairwise tests of geographical overlap, i.e., sympatry. Corrected *p*-values are truncated to 1.0.

	cantabricum	*crassifolium*	*delarbrei*	*ruscinonense*	*ircutianum*	*leucolepis*	*meridionale*	*pseudosylvaticum*
cantabricum		1.0	0	0	0	0	n/a	0
crassifolium			0	0	0	0	n/a	0
delarbrei				0	0	0	n/a	0
ruscinonense					0.05	0	n/a	0
ircutianum						0	n/a	0
leucolepis							n/a	0
meridionale								n/a

3.3. RADseq Assembly and Analysis

Of 216,551,342 raw reads, 81,517,011 demultiplexed, adapter-clipped, and restriction-enzyme-filtered reads were mapped against the reference of Ott et al. [19]. The resulting assembly comprised 7,342 loci with 156,057 SNPs, of which 83,375 were parsimony-informative. The percentages of missing data for sequence and SNP matrices were 36.9% and 34.5%, respectively. The trustworthiness of the RADseq fingerprinting procedure was confirmed by the high degree of similarity between re-extracted and reanalysed accessions (L055-031, 170-021) and their counterparts (L055-03, 170-02) in the subsequent analyses.

The network reconstruction based on SNP-based Nei distances for the complete dataset (i.e., all diploid and tetraploid individuals) placed all tetraploid taxa into the so-called *L. vulgare* group (Figure 2A). While all diploid samples were found to be clustered according to species membership, the majority of the tetraploid accessions followed this pattern except representatives of *L. ircutianum* subsp. *ircutianum* and *L. delarbrei* subsp. *ruscinonense*. Regarding the former, the accessions were observed to form two independent clusters comprising two and five accessions (the latter including L055-03 and its replicate L055-031), and a single individual (437-01) was ungrouped with any cluster. For *L. delarbrei* subsp. *ruscinonense*, two clusters with two (accessions 139-01 and 355-03) and three (accessions 101-01, 110-01, and 349-01) individuals were observed in quite distant positions in the network. When subjecting the tetraploid accessions alone to Neighbor-net clustering (Figure 2B), sample 437-01 was found to be grouped with the larger of the two *L. ircutianum* subsp. *ircutianum* clusters. As a consequence, this cluster comprises accessions of the taxon from Austria (L062-04), Corsica (437-01), Germany (L052-02, L055-03), Italy (87-01), and Montenegro (177-01), while the second one, closer to *L. cantabricum*, contains the two accessions from SW France (106-01 and 343-01). Additionally, all accessions of *L. delarbrei* subsp. *ruscinonense* now cluster together, and the network reconstruction indicates that there are closer genetic similarities between (a) *L. delarbrei* subsp. *delarbrei* and *L. meridionale* (both S France) and (b) *L. crassifolium* (N Spain), *L. pseudosylvaticum* (NW Spain), and *L. delarbrei* subsp. *ruscinonense* (SW France, NE Spain), while accessions of *L. ircutianum* subsp. *leucolepis* (Italy, Balkan Peninsula) form a cluster relatively isolated from these.

The optimal number of clusters according to the Bayesian information criterion (BIC; 'lower is better') was found to be six (Figure 3B), while the Calinksi–Harabasz (CH; 'higher is better') score was optimal for $k = 2$, but also had a local optimum at six clusters. The consensus clustering with the co-association matrix at $k = 6$ (Figure 3A) merged *L. meridionale* and *L. delarbrei* subsp. *delarbrei* (pink cluster) and *L. ircutianum* subsp. *ircutianum* and *L. cantabricum* (yellow cluster), while the remaining assignments of accessions to clusters followed their taxonomic classifications.

3.4. Genealogical Species Delimitation

Plots of the category proportions for all taxon triplets surveyed in the SNiPloid approach to infer parental diploids are provided in the supporting information (Supplementary Material ES07). We found that SNP categories 1 and 2 increased with the genetic distances of the corresponding parental species from the child taxa. The sum of categories 3 and 4 generally decreased with genetic distance. For the analysed taxa, we could not find any irregularity in this pattern. The frequency of category 5 was relatively constant, but also slightly decreased with the genetic distances of the parental species. There seemed to be slight shifts in category frequencies, with at least one parent outside of the *L. vulgare* group (i.e., *L. vulgare*, *L. gaudinii*, *L. pluriflorum*, *L. ageratifolium*, *L. monspeliense*, *L. legraenum*, and *L. ligusticum*) included as a parental species.

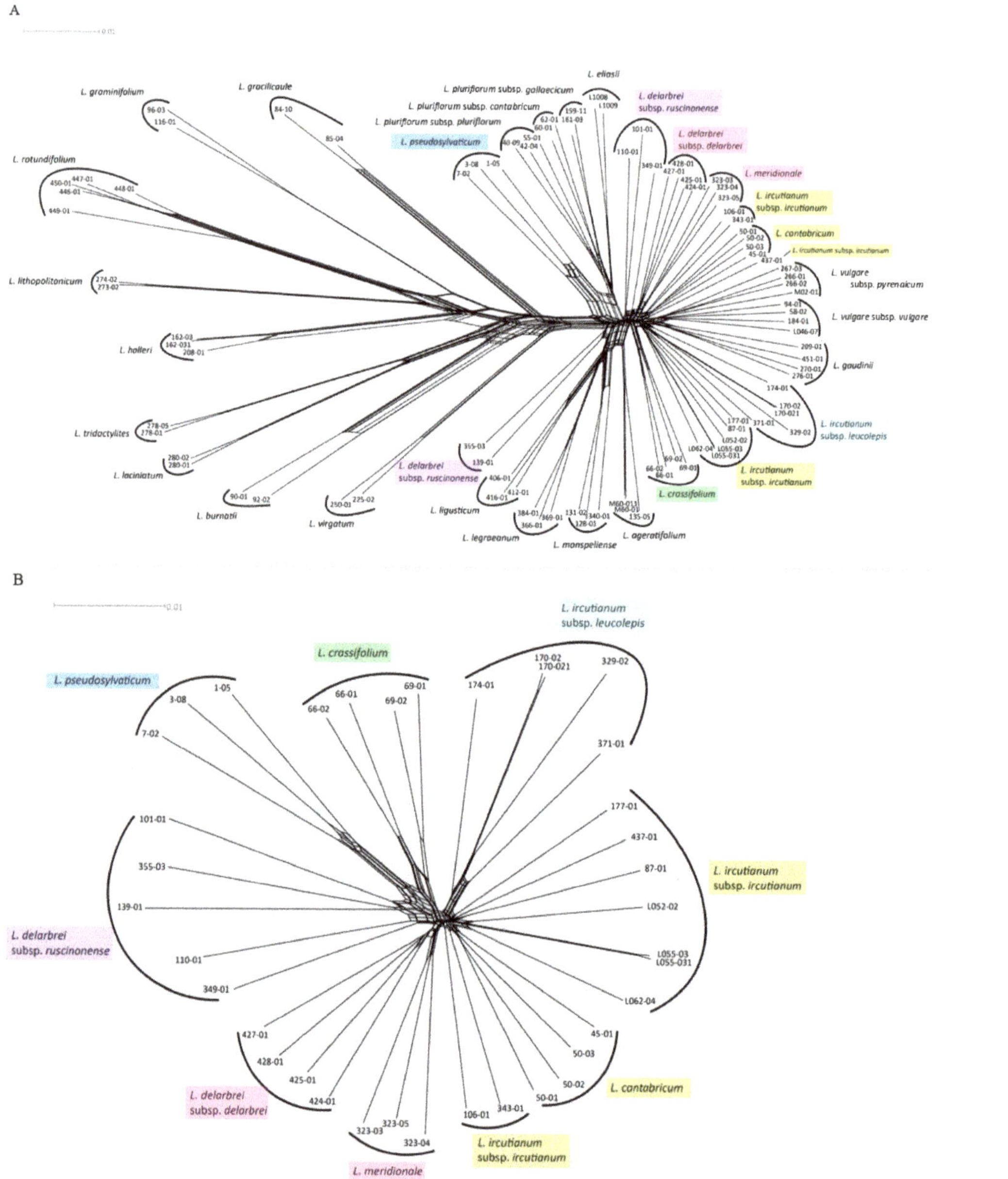

Figure 2. NeighborNet network reconstructions based on SNP-level Nei distances of the full dataset (**A**), comprising individuals from all 17 diploid species and eight tetraploid taxa (in coloured boxes), and of the tetraploids only (**B**). For the full dataset, all tetraploid taxa are found within the so-called *L. vulgare*-group (right). The *L. vulgare*-group is separated from the remaining diploid species by relatively long branches (center). All diploid samples are clustered according to their species membership (**A**). The same is the case for the tetraploid taxa, except for *L. ircutianum* subsp. *ircutianum*, which is separated into two groups ((**B**), lower right).

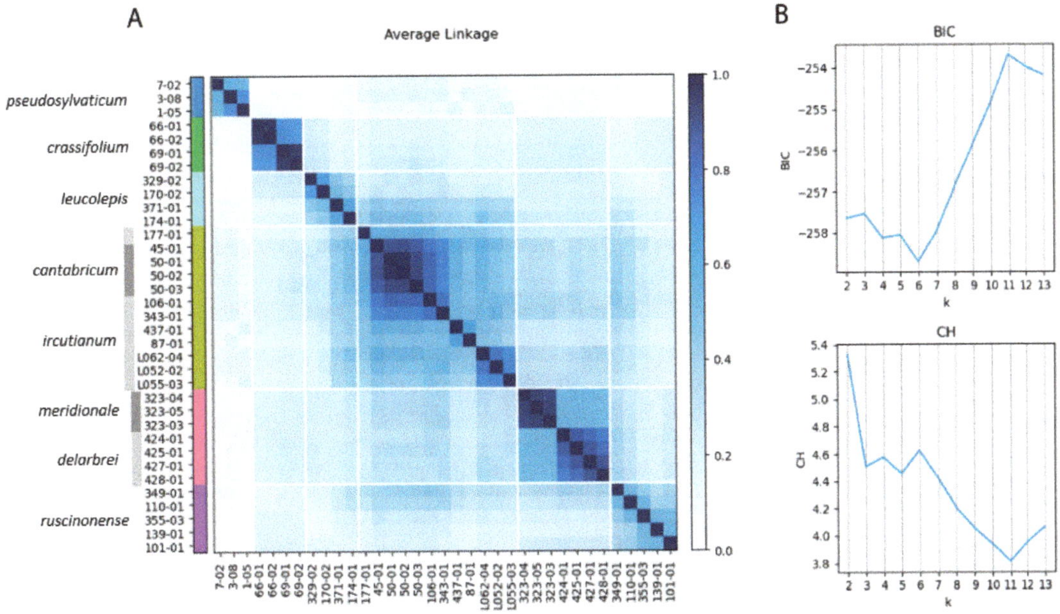

Figure 3. (**A**) Co-association matrix for the tetraploid *Leucanthemum* taxa based on weighted ensemble of random *(k)k*-Means (WKM) clustering and SNP-based Nei distances. Consensus clusters for $k = 6$ are shown, with colour bars depicting consensus cluster membership. (**B**) BIC and CH metrics for number of clusters ranging from $k = 2$ to $k = 13$.

4. Discussion

Having provided an integrative classification scheme for the diploid representatives of the genus [19], the present contribution aims at a comparable taxonomic treatment of tetraploid *Leucanthemum* taxa based on the same sources of evidence for species delimitation: genetic diversification, morphological discontinuities, ecological differentiation, and information on the geographical distribution of taxa. In addition to the methodological approach taken in the diploid case, we additionally tried to incorporate genealogical aspects into this integrative taxonomy of *Leucanthemum* tetraploids: by trying to infer the parentage of tetraploid taxa from RADseq-based single-nucleotide polymorphism (SNP) data, we hoped for additional arguments for a classification scheme that incorporates the evolutionary history of the study group. This follows the rationale that, by elucidating the combination of diploid genomes in the polyploids, the disentangling of auto- and alloploid formations of these taxa and the knowledge of their independent vs. non-independent evolutionary trajectories could be used in taxonomic decisions.

4.1. Genealogical and Genetic Patterns

Unfortunately, the present analyses provide us only with a quite limited notion about the evolutionary origins of the tetraploid *Leucanthemum* taxa under study. This disappointing result is somewhat unexpected in consideration of the huge amount of genomic information received from the GBS fingerprinting, with over 150,000 SNPs from over 7000 loci. However, our present analyses simultaneously point towards the probable explanation for the unresolved evolutionary patterns in this plant group: all eight surveyed tetraploid taxa appear to be closely related to the close-knit species group of diploids around *L. vulgare* (Figure 2A), for which already at the diploid level species delimitation and reconstruction of phylogenetic relationships were found to be problematic [19]. Owing to the fact that the radiation of the whole genus *Leucanthemum* was dated to the last 1.9

(1.1–2.9) Ma [27] and the diversification of the *L. vulgare* group is presumably not older than 400 ka, polyploid species formation from these diploid ancestors has to be assumed to have occurred not earlier than the Middle (Chibanian, 0.77–0.13 Ma) or Late Pleistocene (0.13–0.01 Ma) with its glaciation cycles of the Mindel, Elster, Riss, and Würm eras.

It is comprehensible, therefore, that our present efforts to trace the parenthood of diploid species for the formation of tetraploid taxa by application of the SNIPLOID-based strategy proposed by Wagner et al. [51] for *Salix* polyploids revealed no patterns in *Leucanthemum*. We think that the observed failure of the mentioned strategy, in contrast to its successful application in *Salix* tetraploids, is not due to a lack of sufficient loci and sampled SNPs (23,393 loci and 320,010 SNPs in *Salix* vs. 7342 loci and 156,057 SNPs in *Leucanthemum*) but due to the fact that the putative parental diploid species in *Salix* are much older than in *Leucanthemum*. He et al. [52] provide a dated phylogeny for the *Salix* subg. *Chamaetia/Vetrix* clade, in which the two diploids *S. purpurea* L. and *S. repens* L.—the two putative parents of the tetraploid *S. caesia* Vill. analysed by [51] using SNIPLOID—are dated as having diverged from each other in the Early Miocene, around 20 Ma ago. It appears obvious that in a constellation like this, with the long evolutionary independence of diploid precursor lineages and a relatively recent formation (Pleistocene) of a tetraploid taxon, the SNIPLOID-based strategy proposed by Wagner et al. [51] will reveal a clear signal. However, this cannot be expected in the *Leucanthemum* case, with its comparatively young diploid lineages.

There may be methodological concerns that the SNIPLOID approach is strongly influenced by the filtering of RADseq reads gained from polyploids. This is due to the fact that, when parameters (cluster thresholds) are optimised in IPYRAD to avoid under- and oversplitting of loci and to minimise paralogy, paralogy caused by whole-genome duplication (i.e., homoeology) will be also reduced and this could diminish potential additive allelic signals of allopolyploidy. However, we circumvented this problem by applying a reference-based assembly of reads from tetraploids (mapped against the diploid reference from Ott et al. [19]) instead of performing a de novo assembly of reads from both diploid and tetraploid accessions. As a consequence, we are confident that our negative SNIPLOID results are mainly due to the young age of polyploidisation events in *Leucanthemum*.

Nevertheless, despite the failure to pinpoint potential diploid parental taxa of the tetraploids using the SNIPLOID-based strategy, some relationships among taxa at the two ploidy levels were revealed by the SNP-based network reconstruction (Figure 2). The most obvious connection of a tetraploid taxon with diploid species is the case of the NW Iberian *L. pseudosylvaticum*, which was found to cluster with the group comprising the N Spanish *L. eliasii* and the three subspecies of the NW Spanish *L. pluriflorum*. This position supports former reconstructions based on AFLP fingerprinting and sequence variation in cpDNA intergenic spacers and nrDNA ETSs [24,30] that all pointed towards the contribution of *L. pluriflorum* subsp. *pluriflorum* to an alleged allopolyploid origin of *L. pseudosylvaticum*, while they remained equivocal with respect to the donor of the other diploid genome to the latter. The allopolyploid nature of *L. pseudosylvaticum* receives considerable support from our present SNP-based analysis due to the obvious position of the three accessions of the species (1-05, 3-08, and 7-02) at the vertex of a parallelogram at their base in the NeighborNet network of Figure 2A. Following an interpretation of this parallelogram as an indication for *L. pseudosylvaticum* sharing one edge (genome) with *L. pluriflorum*/*L. eliasii*, one may feel justified in hypothesising the sharing of the other edge (genome) with one of the diploid species making up the left side of the network. Owing to the fact that all other members of this subgroup are presently not found in the Iberian Peninsula, *L. gracilicaule*—a diploid endemic to the region around Valencia in SE Spain—could then be the most probable candidate for the other parental taxon of *L. pseudosylvaticum*, if an extinct diploid may not have acted as such. The latter scenario receives plausibility from studies on other polyploid complexes that have encountered so-called 'ghost (sub)genomes' of extinct diploids in polyploid taxa (e.g., in *Viola* [53], in *Fragaria* [54], and in *Brachypodium* [55]).

In contrast to the evolutionary history of *L. pseudosylvaticum*, hypotheses about the formation of the other seven tetraploid taxa included in the present study remain even more obscure due to the nearly star-like structure in the right part of the NeighborNet network of Figure 2A. The position of the closely related diploid species *L. gaudinii* and *L. vulgare* (with its two subspecies *L. vulgare* subsp. *vulgare* and subsp. *pyrenaicum*) amongst accessions of the tetraploid taxa *L. ircutianum* (with its two subspecies *L. ircutianum* subsp. *ircutianum* and subsp. *leucolepis*) and *L. cantabricum* may point towards the participation of either of these two diploids (or a joint ancestor of both) in the formation of the latter. In the case of *L. ircutianum* subsp. *ircutianum*, it has been shown by Oberprieler et al. [28] that *L. vulgare* subsp. *vulgare* may have acted as the paternal partner in this allopolyploidisation and *L. virgatum* as the maternal partner [56]. At least the latter parentage receives little support from our present SNP-based reconstructions that locate *L. virgatum* quite distantly from all members of the *L. vulgare* cluster. Strangely enough, however, the cpDNA-based analysis of Greiner et al. [56] demonstrated that chloroplast haplotypes of *L. virgatum* (or haplotypes closely related to these) could not only be found in accessions of *L. ircutianum* subsp. *ircutianum* but also in some or even all representatives of *L. ircutianum* subsp. *leucolepis*, *L. cantabricum*, and *L. crassifolium* included in the mentioned study. Finally, chloroplast haplotypes characterised for *L. delarbrei* subsp. *delarbrei* and subsp. *ruscinonense* (sub *L. monspeliense*) were found to be closely related to *L. halleri* and *L. vulgare*, respectively [56]. In summary, these findings may either indicate real maternal parentages in the allopolyploidisation events (with subsequent assimilation of the nuclear genome towards the paternal parent) or events of chloroplast capture caused by hybridisation at the tetraploid level or between diploids and polyploids (tetraploids or taxa with even higher ploidy levels), indeed questioning the possibility of a comprehensive reconstruction of phylogenetic relationships among all taxa of the ploidy complex of *Leucanthemum*.

In genetic terms, SNP-based species-delimitation analyses carried out by weighted ensemble of random *(k)k*-means (WKM) clustering [49] and determination of the optimal number of clusters according to the Bayesian information criterion (BIC) and Calinski–Harbaz (CH) scores revealed significant discontinuities among six of the eight morphotaxon hypotheses (Figure 3). A lack of sufficient genetic differentiation was inferred for *L. cantabricum* and *L. ircutianum* subsp. *ircutianum* and for *L. meridionale* and *L. delarbrei* subsp. *delarbrei*, respectively. In both cases, this was found to correspond to the close positions of accessions from these taxon pairs in the NeighborNet network for the tetraploids (Figure 2B). In methodological respects, WKM clustering is a pattern-based, phenetic species-delimitation procedure not equivalent to process-based species-delimitation methods resting on the multispecies coalescent (MSC), which are used in many present-day delimitation studies based on molecular data in diploid taxon groups (e.g., [19,26,27] in *Leucanthemum* and [57] in *Rhodanthemum*). The lack of an MSC-based model adapted for polyploids hampers coalescent-based species delimitation here. However, since we have demonstrated in *Leucanthemum* diploids that pattern-based methods, such as consensus *k*-means (CKM; [58]) clustering, revealed genetic differentiation patterns equivalent to those derived according to a coalescent-based species-delimitation approach [19], we are confident that the differentiation patterns among tetraploids inferred with WKM clustering are trustworthy approximations to their genealogical structures.

It is obvious from both the NeighborNet network (Figure 2) and the WKM clustering (Figure 3) that *L. delarbrei* subsp. *delarbrei* and *L. meridionale* are closely related to each other. Despite the lack of significant genetic differentiation between them, as indicated by the BIC and CH scores, the two taxa appear to be genetically homogenous. This is remarkable due to the dispersed origins of accessions of the former taxon from four different locations on two volcanic mountain massifs in the southern Massif Central, France (Puy de Sancy, Monts du Cantal), while the latter is endemic to Puy de Wolf, a serpentine mountain c. 65 km SW of the Monts du Cantal. The situation is different for the other taxon pair, for which genetic differentiation was found to be not significant according to BIC and CH scores. In this case, *L. cantabricum*, with four accessions from two populations in NW Spain

(Galicia), is also distinct but nested in a highly diverse cluster representing *L. ircutianum* subsp. *ircutianum*, with eight accessions from eight populations sampled throughout the European range of the taxon.

Two circumstances, however, are considered noteworthy here in evaluating the genetic/genealogical relationship of the two taxa: (a) The two populations sampled for *L. cantabricum* in the present analysis are the westernmost ones in the distributional range of the species that is mainly distributed in the W Pyrenean mountains and the Cordillera Cantábrica in N Spain [23], the *locus classicus* of the taxon being around 250 km to the east of the sampled populations. (b) In his revision of the genus for the Iberian Peninsula, Vogt [22] considered the stronger dissected leaves of *L. cantabricum* (sub *L. ircutianum* subsp. *cantabricum*) as being the main difference between this taxon and *L. ircutianum* subsp. *ircutianum*. The author, however, reports also that the morphological variation observed in the former taxon is considerable and that morphologically intermediate individuals with respect to the latter are observed frequently. Sampling of accessions from the centre of the distributional range of *L. cantabricum* (especially from the *locus classicus*) may reveal more pronounced genetic differences to *L. ircutianum* subsp. *ircutianum*, and the two populations sampled for the present study may turn out to be intermediate forms rather than pure representatives of *L. cantabricum*. Nevertheless, the undeniable closeness of the two taxa and their connectedness via intermediates will remain an unequivocal genealogical pattern.

4.2. Morphological Patterns

Leaf morphology—especially leaf outline and leaf dissection—is of paramount importance for taxon delimitation and taxon determination in *Leucanthemum*. As demonstrated by many taxonomic treatments of the genus in floras of European countries (e.g., *Flora Iberica* [23], *Flora Gallica* [59], and *Flora d 'Italia* [60]), other morphological features, such as the colour of margins of involucral bracts, plant size, number of capitula per stalk, and achene size, come only second to these characters. Therefore, we felt justified in utilising leaf morphology as the sole proxy for morphological variation in the study group, especially because leaf-morphological features could be easily and objectively inferred using the applied machine-learning techniques and images of digitised herbarium specimens. These techniques allowed us to analyse 285 representative specimens and more than 1300 individual leaves across the study group in terms of leaf dissection and leaf shape in a time-effective and reproducible manner and to test for significant differences between pairs of the eight morphotaxa. We are fully aware of the fact that significant differences in these tests do not indicate the suitability of these characteristics as sole taxon-diagnostic features, as all taxa show considerable variation and overlap broadly in leaf characteristics. However, the main motivation for the inclusion of morphological characteristics in species-delimitation analyses is not for practical reasons of applicability (determinability), but—as in the case of genetic/genealogical features—for the assessment of morphological discontinuities as potential proxies for the evolutionary independence of lineages and their reciprocal reproductive isolation.

When compared across all possible pair-wise test combinations, leaf dissection (as measured by LDI) discriminates among the eight morphotaxa more profoundly than leaf outline (measured in elliptic Fourier analyses and tested for significance either with a permutation or an NPMANOVA approach): while only three (11%) of the 28 possible pair-wise comparisons of LDIs revealed no significant differences between taxa, 14 (50%) and 12 (43%) of the tests addressing leaf-shape differences showed non-significant results. We think that these contrasting results do not only have a biological reason, but also a methodological one: in its present and here-applied formulation (comparison of the length of leaf outline with leaf area), LDI does not only measure the intensity of leaf incision, but also the deviation of the leaf shape from a perfect circle. That means that it also captures leaf shape. In contrast, elliptic Fourier analysis (EFA) not only exclusively captures the outline of leaves, but also (with higher harmonics) the subdivision of the leaf lamina. However, since parameters gained with the latter harmonics of the EFA are down-weighted against parameters from earlier ones in a principal component analysis (PCA), it may be justified to consider both measures applied here (i.e., LDI and EFA) as overlapping with respect to what they capture in terms of leaf morphology, but with a tendency towards assessment of leaf dissection with the former and leaf outline with the latter.

Discrepancies between LDIs and EFA descriptors of leaf morphology were found to be especially pronounced when *L. meridionale* was included in the pair-wise comparisons: while six of the seven comparisons concerning LDIs revealed significance, the majority of EFA-based comparisons resulted in non-significant results. The former is quite reasonable when the habit of the species is studied both in its natural habitat and in a herbarium: compared with all other tetraploid taxa of *Leucanthemum*, *L. meridionale*, with its reduced scape size and its filigree leaves, resembles more the diploid representatives of the genus than the tetraploid ones. This led Tison and de Foucault [59] among others, who did not know about the tetraploid nature of this taxon, to speculate on the close (and allegedly) hybridogenic relationship of the species to *L. vulgare* and *L. graminifolium*. The somewhat reduced habit of *L. meridionale* may be a consequence of the exceptional habitat preference of the species for serpentine soils; the lack of a significant difference in LDIs with respect to *L. delarbrei* subsp. *delarbrei*, which is found on volcanic soils further north and is genetically closely related with it, however, may also point towards a close phylogenetic relationship between these two tetraploid *Leucanthemum* taxa.

A further noteworthy similarity in terms of leaf-morphological descriptors was observed between *L. ircutianum* subsp. *ircutianum* and subsp. *leucolepis*, for which all three tests remained non-significant (Table 1). As described by Oberprieler et al. [31], the former wide-spread taxon and the latter amphi-Adriatic one (including the S Italian *L. ircutianum* subsp. *asperulum* as a synonym) are allopatrically distributed in the Apennine Peninsula but sympatrically distributed in the Balkan Peninsula and form morphologically intermediate forms through hybridisation in overlapping regions of their distribution ranges. The main difference between the two taxa is in the colouration of the margins of involucral bracts, which are black to dark-brown in subsp. *ircutianum* and hyaline in subsp. *leucolepis*. In genetic respects, the two taxa represent two significantly distinct (Figure 3), albeit closely related (Figure 2), clusters.

4.3. Ecological and Geographical Patterns

As in the case of pair-wise statistical testing of leaf-morphological aspects, testing for significantly non-overlapping distribution ranges (Table 3) and ecological niches (Table 2) among the tetraploid *Leucanthemum* morphotaxa did not reveal complete and therefore diagnostic separation in these respects (strict allopatry or non-overlapping ecological niches); it rather led to the detection of patterns of bimodality in the spatial and ecoclimatological distributions of the species pairs compared. Therefore, the overwhelming number of significant test results (even after Bonferroni correction for multiple testing) for both geographical (2 of 21 tests) or ecological (also 2 of 21 tests) differences between species pairs may just represent tendencies in spatial or ecoclimatological differentiation and not complete discontinuities. Nevertheless, when looking at the distribution ranges of the eight tetraploid morphotaxa under study (Figure 1) and adding information from taxon descriptions concerning their ecological behaviour (e.g., [22] for the taxa of the Iberian Peninsula), it becomes obvious that the mentioned test results understate rather than hyperbolise differentiation patterns.

In geographical respects, the distribution ranges of the eight morphotaxa clearly follow an allopatric pattern, with exceptions being the observed non-significant differentiation between *L. ircutianum* subsp. *ircutianum* and *L. delarbrei* subsp. *ruscinonense* on the one hand and *L. cantabricum* and *L. crassifolium* on the other. As in the case of the point-endemic *L. meridionale*, for which statistical testing for geographical and ecological differences was not possible due to the restricted distribution range of the species, we think, however, that these test results are strongly scale-dependent and more on the conservative than on the oversensitive side. In the case of the taxon pair of *L. cantabricum* and *L. crassifolium*, the two species show a clear allopatric distribution on a small scale, with the former being restricted to the mountains of the Cordillera Cantábrica and adjacent regions [22,23] and the latter to littoral habitats of the N Spanish coast. In the other cases, the wide-spread (and constantly further spreading) *L. ircutianum* subsp. *ircutianum* encloses the geographically restricted *L. delarbrei* subsp. *ruscinonense* and *L. meridionale* in a more parapatric than allopatric pattern, which may be responsible for the non-significant test result. For all three taxon pairs, however, ecological differences were either revealed by our tests on niche overlap (*L. ircutianum* subsp. *ircutianum* vs. *L. delarbrei* subsp. *ruscinonense*) or by small-scale habitat differences—especially in edaphic factors—that have not been captured by the geography-based ecoclimatological niche modelling underlying the test-setup, *L. meridionale* being adapted to serpentine soils and *L. crassifolium* to saliferous coastal habitats.

4.4. Integration of Sources of Evidence

Species delimitation in plants—especially in hybridising plant groups (syngameons) and polyploid complexes—is a problematic matter [18]. The applicability of a strict biological species concept (BSC) has been denied by the majority of botanists owing to the sheer frequency of hybridisation in the plant kingdom and the numerous groups with agamospermic (apomictic) reproduction. Additionally, the usage of actual (hybridisation) or potential interbreeding (crossability) as a proxy for the evolutionary independence of lineages is deceptive. The lack of correlation between hybrid-formation capability and phylogenetic proximity in plants is exemplified by many examples of old and well-characterised lineages (species, sometimes even from different genera) that easily hybridise when brought into contact on the one hand and extremely young lineages (like autopolyploids) that are reproductively isolated instantly on the other, demonstrating that interfertility is a plesiomorphic character state [61]. Other species concepts, as enumerated by Zachos [62], share the problem that the criteria entertained in the different concepts differ in temporal sequence and relative importance across the tree of life due to the fact that speciation is a continuous process "over a timeframe that is too long to study from start to finish" (the 'speciation continuum'; [63]). With these unsatisfactory consequences for biological classification, taxonomic decisions following a 'unified species concept' sensu De Queiroz [5] have considerable attraction because this concept shifts the focus away from properties being used for defining species towards their usage as indicators for independently evolving metapopulation lineages (species). Additionally, it allows for and demands the integration of multiple sources of evidence for taxonomic ranking.

Here, we follow a procedural protocol for species-rank decisions designed by Oberprieler [16] that is based on the integration of morphology, ecology, and geography, as proposed by von Wettstein [64], and expanded for the inclusion of an additional genetic/genealogical axis (the 'Wettstein tesseract'). Conceptually, it follows the evolutionary species concept (EvoSC) of Wiley [65], who defined species as "a single lineage of ancestral-descendent populations of organisms which maintains its identity from other such lineages and which has its own evolutionary tendencies and historical fate", but addresses the major drawback of that definition, namely, that genealogy-based, multispecies-coalescent species-delimitation methods tend to mistake population structures for species boundaries [66]. By adding geographical, ecological, and morphological sources of evidence, the 'Wettstein tesseract' provides a tool for conceptualising decisions on species or subspecific ranks and has been successfully applied to the diploids of *Leucanthemum*—the "warps and wefts" of this polyploid complex [19].

Despite the somewhat pessimistic outlook on the phylogenetics of the *Leucanthemum* polyploidy complex, the present study is extremely helpful in terms of providing a pattern-based species delimitation at the tetraploid level and its integrative taxonomical conceptualisation. Aiming at a reproducible, objective classification of eight morphotaxon hypotheses in terms of delimitation and ranking, we have analysed genetic, ecological, and morphological variation together with information from taxon distributions, summarised in Figure 4. In the diagram, statistically significant differences found between pairs of taxon hypotheses are depicted in red, while the lack of significant discontinuities is shown in green. Owing to the restricted distribution range of *L. meridionale*, which is a point endemic of a serpentine mountain in S France (Aveyron, Rodez, Decazeville, Firmi, and Puy de Wolf), this taxon has been omitted from pair-wise comparisons that are dependent on modelling procedures based on spatial distribution data comprising more than a single raster cell (scaled to a resolution of 2.5 arc minutes). Due to its described endemicity, *L. meridionale* (accessions 323-03, -04, and -05) is allopatrically distributed with all other tetraploid *Leucanthemum* taxa except *L. ircutianum* subsp. *ircutianum*, which is found in parapatry on non-serpentine soils surrounding Puy de Wolf (and sampled with accession 343-01). Its ecological niche is uniquely determined by edaphic factors connected to serpentine soils (only paralleled by the diploid *L. pluriflourum* subsp. *gallaecicum* in NW Spain and the decaploid *L. pachyphyllum* in N Italy) rather than climatic ones.

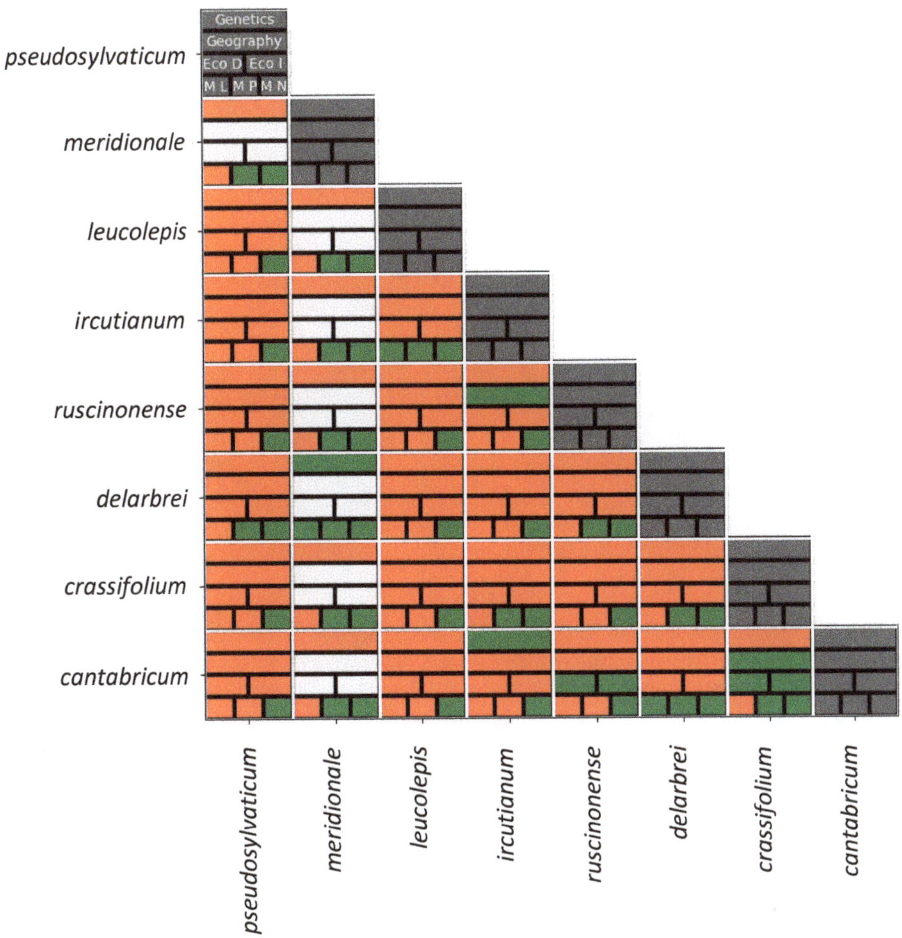

Figure 4. Taxon pair-wise differences in genealogy, geography, ecology (EcoD: Schoener´s D; EcoI: Warren´s I), and morphology [ML: Welch's test on leaf-dissection indices (LDI); MP: permutation test on leaf shape (Elliptic Fourier Analysis, EFA), MN: NPMANOVA on leaf shape (EFA)]. Significant differences are marked in red, non-significant ones in green; white cells indicate lack of testability due to the restricted geographical range of *L. meridionale*.

Application of the 'Wettstein tesseract' tool to the present study group of *Leucanthemum* tetraploids in seeking the most reasonable ranking for closely related species hypotheses (morphotaxa) leads to the following reasoning:

(a) *Lecanthemum crassifolium* (N Spain), *L. delarbrei* subsp. *ruscinonense* (E Pyrenees and northeastern foothills), and *L. pseudosylvaticum* (W Iberian Peninsula) are best considered as being genetically closely related but independent lineages (Figures 2 and 3) that merit species ranking due to their significant ecological (Table 2) and leaf-morphological differences (Table 1) that have evolved in strict allopatry. While for *L. pseudosylvaticum* at least one parental diploid species is known (*L. pluriflorum* subsp. *pluriflorum*; [24,30]), the matter is unsettled as yet for the other two tetraploid lineages. Following haplotype-network reconstructions for the whole genus [56], *L. pluriflorum* subsp. *pluriflorum*, as the maternal diploid ancestor of these, could be excluded definitively, and *L. virgatum* appears to be the most probable candidate for this role in *L. crassifolium*, while *L. delarbrei* subsp. *ruscinonense* (sub *L. monspeliense* in [56]) shares its chloroplast haplotype with a large number of diploid taxa: *L. ageratifolium* (NE Spain, SW France; sub *L. vulgare* subsp. *pujiulae* in [56]), *L. burnatii* (S France), *L. eliasii* (N Spain; sub *L. vulgare* subsp. *eliasii* in [56]), *L. gaudinii* (Alps, Carpathian Mountains), *L. graminifolium* (S France), *L. pluriflorum* subsp. *cantabricum* (N Spain; sub *L. gaudinii* subsp. *cantabricum* in [56]), and both subspecies of *L. vulgare* (subsp. *vulgare*, widespread; subsp. *barrelieri*, Pyrenees).

(b) *Leucanthemum delarbrei* subsp. *delarbrei* (northern Massif Central, France) and *L. meridionale* (western Massif Central, France) are allopatrically distributed sister groups (Figure 2) that lack sufficient genetic differentiation to merit species ranking (Figure 3) and show non-significantly different leaf morphology (Table 1) but exhibit edaphic differences (the former growing on siliceous rocks of old volcano cones, the latter on serpentine soils). Therefore, subspecific ranking appears appropriate for these two taxa. The chloroplast haplotype found in one of two accessions of *L. delarbrei* subsp. *delarbrei* surveyed by Greiner et al. [56] was found to be closely related to the diploid *L. halleri* and may point towards a phylogenetic relationship with this Alpine species, while information on a chloroplast haplotype in *L. meridionale* is still lacking.

(c) Finally, *L. cantabricum* and the two subpecies of *L. ircutianum* also show close genetic relationships in the NeighborNet network of Figure 2. Here, however, genetic differentiation patterns (Figure 3) do not correspond to the hitherto proposed classification because *L. ircutianum* subsp. *ircutianum* shows a closer genetic similarity with *L. cantabricum* than with *L. ircutianum* subsp. *leucolepis*. When following the conceptual framework of species-rank decision making with the 'Wettstein tesseract', the significant differentiations between *L. ircutianum* subsp. *ircutianum* and subsp. *leucolepis* in genetic, geographical, and ecoclimatological aspects would argue for an acknowledgement of the two taxa at species level (the lack of leaf-morphological differences is counterbalanced by differences in the colours of the margins of the involucral bracts). The main argument for treating the two entities as independent species, following the logic of von Wettstein [64], is that the ecoclimatiological differences between the two would allow them to keep their lineage identity when allopatry would change into sympatry in the future. Oberprieler et al. [31] have shown, however, that there are mixed stands of the two taxa in Central Italy, where hybridisation and backcrossing leads to introgressive hybrid swarms, with a complete blurring of taxon limits. Since the situation seems comparable in the Western Balkan Peninsula, where the two taxa grow sympatrically and intermediate forms are found [31], we propose to keep the two entities at subspecific rank for conservative reasons aiming at minimizing taxonomic and nomenclatural disruptions if not unequivocally demanded by the underlying data and analyses.

Subspecific ranking under *L. ircutianum*, on the other hand, is less equivocal for *L. cantabricum* (Cordillera Cantábrica in N Spain) following the here-presented results: the two taxa are allopatrically distributed and show ecoclimatological and morphological differences (more strongly dissected leaves in *L. cantabrica*) but lack genetic differentiation

(Figure 4). The observation of morphologically intermediates by Vogt [22] further argues for hybridisation between the two taxa and the appropriateness of subspecific ranking.

5. Taxonomic Treatment

With the described conceptual framework at hand, species delimitation in the group of tetraploid *Leucanthemum* taxa under study could be put into effect as follows:

(1) ***Leucanthemum crassifolium*** (Lange) Lange in Willk. & Lange, Prodr. Fl. Hispan. 2: 96. 1865 ≡*Leucanthemum pallens* var. *crassifolium* Lange in Vidensk. Meddel. Dansk Naturhist. Foren. Kjøbenhavn, ser. 2, 3: 77. 1861 ≡*Leucanthemum ircutianum* subsp. *crassifolium* (Lange) Vogt in Ruizia 10: 127. 1991—Lectotype (Vogt, Ruizia 10: 128. 1991): In rupibus maritimis ad Portugalete, Cantabria, Oct. 1851, Herb. Joh. Lange (C! (C10007128)).

Notes.—Endemic to the Cantabrian coast between Asturias and the Basque region (NE Spain and SW France). Habitats are coastal rocks and salt-influenced coastal slopes from sea level to 20 m. *Leucanthemum crassifolium* is characterised by succulent leaves (Figure 5) and involucral bracts with dark-brown hyaline margins of involucral bracts. It was first described as a variety of *L. pallens* by Lange [67] and subsequently raised to subspecific [22] and specific rank [23].

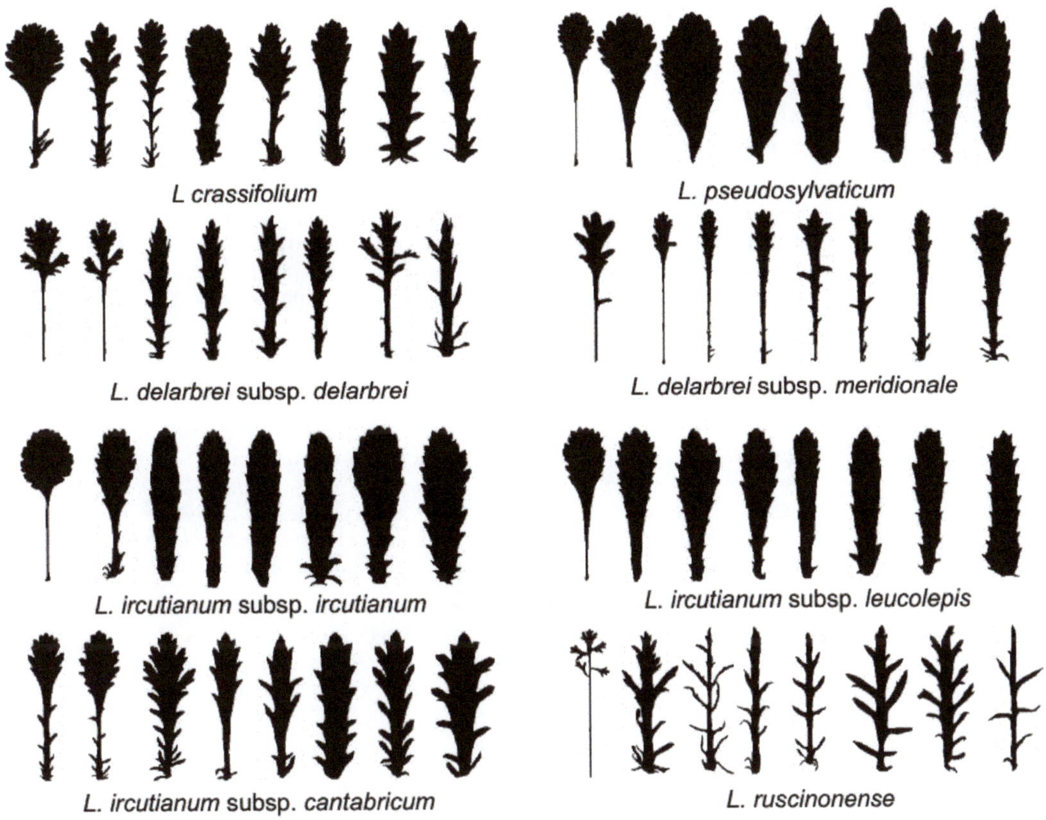

Figure 5. Silhouettes of basal and cauline leaves from different accessions of the eight tetraploid *Leucanthemum* taxa under study.

(2a) ***Leucanthemum delarbrei*** Timb.-Lagr. (**subsp. *delarbrei***) in Mém. Acad. Sci. Clermont-Ferrand 20: 508. 1878.—Lectotype: Not yet designated. =*Chrysanthemum leucanthemum* var. *pinnatifidum* Lecoq. & Lamotte, Cat. Pl. Plateau Central: 227. 1847 ≡*Leucanthemum vulgare* var. *pinnatifidum* (Lecoq. & Lamotte) Briq. & Cavill. in Burnat, Fl. Alpes Marit. 6: 91. 1916 ≡*Leucanthemum ircutianum* var. *pinnatifidum* (Lecoq. & Lamotte) D. Löve & J.-P. Bernard in Svensk. Bot. Tidskr. 53: 444. 1959.—Ind. loc.: "AR.- Mont Dore; pâturages et pentes herbeuses de Chaudefour, bords du chemin de Sancy à Vassivière. Bozat! AR."—**Lectotype (designated here by Vogt & Oberprieler):** Mont Dore, Sancy, 15 (..) 1844, leg. *H. Lecoq & M. Lamotte* (P! (P00729975)).

Notes.—Endemic to the siliceous rocks of the old volcano cones of the central Massif Central in France (Monts Dore, Monts du Cantal). Its habitats are siliceous rocks and meadows between 1550 m and 1750 m. *Leucanthemum delarbrei* subsp. *delarbrei* is characterised by strongly dissected leaves, from pinnatifid to bipinnatisect (Figure 5), and dark- to light-brown hyaline margins of involucral bracts.

(2b) ***Leucanthemum delarbrei* subsp. *meridionale*** (Legrand) Oberpr., T. Ott & Vogt, **comb. nov.** ≡*Leucanthemum meridionale* Legrand in Bull. Soc. Bot. France 28: 56. 1881 (basionym) ≡*Leucanthemum vulgare* var. ('χ') *meridionale* (Legrand) Rouy, Fl. France 8: 274. 1903 ≡*Chrysanthemum leucanthemum* f. ('g') *meridionale* (Legrand) Fiori in Fiori & Béguinot, Fl. Italia 3: 239. 1903 ≡*Leucanthemum vulgare* subsp. *meridionale* (Legrand) Nyman, Consp. Fl. Eur., Suppl. 2: 169. 1889.—Ind loc.: "Habite dans les interstices des rochers serpentineuses du puy de Wolf, près de Firmy (Aveyron); fleurit de fin mai à juillet. Je reçus cette plante en 1879 ... leg. *F(rère) Saltel* (Baenitz, Herbarium eur. n° 4184)."—**Lectotype (designated here by Vogt & Oberprieler):** Puy de Wolf, pr. Firmy; mai, juin 1879 (Aveyron).—Gallia merid., Saltel.—Comm. Le Grand (P! (P00729957)).

Notes.—Endemic to the Puy de Wolf (France, Aveyron, Rodez, Decazeville, Firmi), a serpentine mountain in the western Massif Central. It is found in dry and open, south-facing grassland on serpentine soil between 400 m and 600 m. *Leucanthemum meridionale* is characterised by its lanky habitus resembling the diploid *L. vulgare* or a member of the genus *Leucanthemopsis* with quite narrow leaves and cuneate leaf bases (Figure 5).

(3a) ***Leucanthemum ircutianum*** DC. (**subsp. *ircutianum***), Prodr. 6: 47. 1838.—Lectotype (Vogt, Ruizia 10: 119. 1991): In pratis, 1828, Turczaninoff a Irkoutsk, Turcz.: 1830 (G-DC! (G00451151)).

Notes.—Besides the diploid *L. vulgare* Lam., this taxon is the most widely distributed species of the genus [22]. It is present in nearly all countries of the Euro + Med region [20] but has also been introduced into all continents except Antarctica. Habitats are anthropogenetically influenced and include meadows and roadsides. *Leucanthemum ircutianum* subsp. *ircutianum* is morphologically very similar to *L. ircutianum* subsp. *leucolepis* (see Figure 5) but can be differentiated by the dark-brown hyaline margins of involucral bracts.

(3b) ***Leucanthemum ircutianum* subsp. *cantabricum*** (Sennen) Vogt in Ruizia 10: 121. 1991 ≡*Leucanthemum cantabricum* Sennen, Diagn. Nouv.: 50. 1936 ≡*Leucanthemum vulgare* subsp. *cantabricum* Sennen, Diagn. Nouv. 50. 1936 (nom. altern.).—Lectotype (Vogt, Ruizia 10: 121. 1991): Santander: La Calda de Besaya, rochers silliceux humides, 5.6.1927, *E. Leroy* (BC-Sennen!).

Notes.—Distributed between the western foothills of the Pyrenees (Spain and France), along the northern slopes of the Cantabrian Mountains to Asturias and Galicia in the west. Habitats include road and meadow margins, meadows, and pastures; the elevational distribution ranges from sea level to 800 m. *Leucanthemum cantabricum* is characterised by dissected (pinnatisect to pinnatipartite), non-succulent leaves (Figure 5) and dark-brown hyaline margins of involucral bracts. It was described as a species by Sennen [68] but reduced to subspecific rank under *L. ircutianum* by Vogt [22] before being reacknowledged at species rank by Vogt [23].

(3c) ***Leucanthemum ircutianum* subsp. *leucolepis*** (Briq. & Cavill.) Vogt & Greuter in Willdenowia 33: 41. 2003 ≡*Leucanthemum leucolepis* (Briq. & Cavill.) Gajić in Josifović, Fl. SR Srbije 9: 185. 1977 ≡*Leucanthemum vulgare* subsp. *leucolepis* Briq. & Cavill. in Burnat, Fl. Alpes Marit. 6: 93. 1916 ≡*Chrysanthemum leucanthemum* subsp. *leucolepis* (Briq. & Cavill.) Schinz & Thell., Fl. Schweiz, ed. 4, 1: 685. 1923 ≡*Leucanthemum leucolepis* (Briq. & Cavill.) Horvatić in Acta Bot. Croat. 22: 214. 1963, nom. inval. (Turland et al. 2018: Art. 41.5) ≡*Leucanthemum pallens* subsp. *leucolepis* (Briq. & Cavill.) Faverger in Anales Inst. Bot. Cavanilles 32: 1236. 1975.—Lectotype (Oberprieler et al., Bot. J. Linn. Soc. 199: 844. 2022): Flora Italica Exsiccata, curantibus Adr. Fiori, A. Béguinot, R. Pampanini, number 175—Etruria, Prov. di Firenze, Vallombrosa, in pratis, alt. 900–1000 m., solo pingui, siliceo, 19.6.1904, *A. Fiori* (G-BU! (G00848032)).

Notes.—A taxon with an amphi-Adriatic distribution, with populations throughout Italy and along the Adriatic coast of the Balkan Peninsula. The taxon was described as a subspecies of the diploid *L. vulgare* but was subsequently considered a subspecies of the hexaploid *L. pallens* due to its white hyaline margins of involucral bracts or seen as an independent species. The observation of an allopatric distribution with *L. ircutianum* subsp. *ircutianum* as more Mediterranean facies of the species, together with the observation of hybrid individuals and hybrid swarms in areas of joint occurrence of the two tetraploid taxa in the Apennine and Balkan Peninsulas by Oberprieler et al. [31] argued for treatment as a subspecies of *L. ircutianum*. In contrast to *L. ircutianum* subsp. *ircutianum*, it has paler hyaline margins of the involucral bracts.

(4) ***Leucanthemum pseudosylvaticum*** (Vogt) Vogt & Oberpr. in Ann. Bot. (Oxford), n.s., 111: 1121. 2013 ≡*Leucanthemum ircutianum* subsp. *pseudosylvaticum* Vogt in Ruizia 10: 134. 1991.—Holotype: Portugal, Distrito Porto, Serra do Marão, Mesão Frio-Amarante, feuchter Hang südlich der Paßhöhe, ca. 800 m, 20.07.1986, *R. Vogt* 4711 & *E. Bayón* (M! (M-0030173)).

Notes.—Distributed throughout the western part of the Iberian Peninsula (Portugal and Spain). Its habitats comprise road margins and slopes, margins of creeks, and ditches, at an elevational range from 100 m to 1600 m. *Leucanthemum pseudosylvaticum* is characterised by leaves with proximally (sub)entire margins (Figure 5) and involucral bracts with pallid to light-brown hyaline margins. It was described by Vogt [22] as a subspecies of *L. ircutianum* but shown to merit species rank due to its evolutionary independence from the latter species by Oberprieler et al. [24] and Greiner et al. [29,30].

(5) ***Leucanthemum ruscinonense*** (Jeanb. & Timb.-Lagr.) Oberpr., T.Ott & Vogt, **comb. et stat. nov.** ≡*Leucanthemum palmatum* var. *ruscinonense* Jeanb. & Timb.-Lagr. in Mém. Acad. Sci. Toulouse, ser 8, 1(2): 192. 1879 (basionym) ≡*Leucanthemum monspeliense* var. *ruscinonense* (Jeanb. & Timb.-Lagr.) O. Bolòs & Vigo in Collect. Bot. (Barcelona) 17: 91. 1988 ≡*Leucanthemum cebennense* var. *ruscinonense* (Jeanb. & Timb.-Lagr.) Gaut., Catal. Rais. Fl. Pyr. Orient.: 232. 1898 ≡*Leucanthemum delarbrei* subsp. *ruscinonense* (Jeanb. & Timb.-Lagr.) Vogt, Florian Wagner & Oberpr., Fl. Iber. 16(3): 1866. 2019—Ind loc.: "... les Alberes ...".—Holotype: ... la tour de la Massane dans la altura, Pyr. Orient., 21.5.1877, *E. Timbal-Lagrave* (TL-Timbal-Lagrave!).

Notes.—Endemic to the eastern Pyrenees (Spain and France) and the SW parts of the Massif Central (Haut Languedoc, Montagne Noire). Habitats comprise road margins, creeks, and stony slopes, between 250 m and 1900 m. It is characterised by strongly dissected leaves, pinnatifid to bipinnatisect (Figure 5), and dark- to light-brown hyaline margins. High variability in terms of leaf dissection is speculated to be the result of potential hybridisation. The taxon was, for a long time, considered part of *L. monspeliense* L. (e.g., [22]), which is also distributed in the Massif Central (Cevennes), but has a diploid chromosome number.

Supplementary Materials: The following supporting information can be downloaded at: https://www.mdpi.com/article/10.3390/biology12020288/s1. File ES01: List of herbarium specimens of *Leucanthemum* tetraploids housed at the Berlin Botanical Museum (B) and used for the morphological analyses of the present study. File ES02: List of collection localities of *Leucanthemum* tetraploids used in the ecological niche modelling and geographical overlap analyses of the present study. File ES03: Descriptions of the bioclimatological and edaphic variables used for ecological niche modelling. File ES04: List of samples used for the ddRAD analysis with information on voucher specimens in the Botanical Museum Berlin (B) and ploidy, collection localities, coordinates, and collectors. File ES05: Maps depicting predicted potential distribution ranges of the *Leucanthemum* diploids based on recent, last glacial maximum (LGM), and last interglacial (LIG) bioclimatic variables. File ES06: Maps depicting predicted potential distribution ranges of the *Leucanthemum* tetraploids based on recent, last glacial maximum (LGM), and last interglacial (LIG) bioclimatic variables. File ES07: Plots of the SNIPLOID category proportions for all triplets (two diploid potential parent species and one tetraploid daughter species).

Author Contributions: Conceptualisation, C.O., T.O. and R.V.; methodology, T.O.; software, T.O.; validation, T.O. and C.O.; formal analysis, T.O.; investigation, T.O. and C.O.; resources, C.O. and R.V.; data curation, C.O. and R.V.; writing—original draft preparation, C.O., T.O. and R.V.; writing—review and editing, C.O. and R.V.; visualisation, C.O. and T.O.; supervision, C.O.; project administration, C.O.; funding acquisition, C.O. and R.V. All authors have read and agreed to the published version of the manuscript.

Funding: This work was funded by grants from the German Research Foundation (DFG) in the frame of the SPP 1991 "Taxon-omics—New Approaches for Discovering and Naming Biodiversity" to C.O. (OB 155/13-1) and R.V. (VO 1595/3-1).

Institutional Review Board Statement: Not applicable.

Informed Consent Statement: Not applicable.

Data Availability Statement: The raw ddRAD reads were deposited at NCBI (Bioproject PRJNA884628).

Acknowledgments: We thank Anja Heuschneider for preparing the samples for ddRADseq sequencing. The suggestions from Agnes Scheunert and Ulrich Lautenschlager, considering the analyses and the manuscript, were, as always, welcome and are very much appreciated. Points raised by the three reviewers of the present contribution are thankfully acknowledged and improved the manuscript considerably.

Conflicts of Interest: The authors declare no conflict of interest.

References

1. Stuessy, T.F. Ultrastructural data for the practicing plant systematist. *Am. Zool.* **1979**, *19*, 621. [CrossRef]
2. Zachos, F.E.; Christidis, L.; Garnett, S.T. Mammalian species and the twofold nature of taxonomy: A comment on Taylor et al. *Mammalia* **2020**, *84*, 1. [CrossRef]
3. Ence, D.D.; Carstens, B.C. SpedeSTEM: A rapid and accurate method for species delimitation. *Mol. Ecol. Resour.* **2011**, *11*, 473. [CrossRef] [PubMed]
4. Smith, M.L.; Carstens, B.C. Species delimitation using molecular data. In *Species Problems and Beyond–Contemporate Issues in Philosophy and Practice*; Wilkins, J.S., Zachos, F.E., Pavlinov, I.Y., Eds.; CRC Press: Boca Raton, FL, USA, 2022; pp. 145–159.
5. De Queiroz, K. Species concepts and species delimitation. *Syst. Biol.* **2007**, *56*, 879. [CrossRef] [PubMed]
6. Dayrat, B. Towards integrative taxonomy. *Biol. J. Linn. Soc.* **2005**, *85*, 407. [CrossRef]
7. Will, K.W.; Mishler, B.D.; Wheeler, Q.D. The perils of DNA barcoding and the need for integrative taxonomy. *Syst. Biol.* **2005**, *54*, 844. [CrossRef]
8. Guillot, G.; Renaud, S.; Ledevin, R.; Michaux, J.; Claude, J. A unifying model for the analysis of phenotypic, genetic, and geographical data. *Syst. Biol.* **2012**, *61*, 897. [CrossRef] [PubMed]
9. Zapata, F.; Jiménez, I. Species delimitation: Inferring gaps in morphology across geography. *Syst. Biol.* **2012**, *61*, 179. [CrossRef]
10. Vásquez-Cruz, M.; Vovides, A.P.; Sosa, V. Disentangling species limits in the *Vauquelinia corymbosa* complex (Pyreae, Rosaceae). *Syst. Bot.* **2017**, *42*, 1. [CrossRef]
11. Solís-Lemus, C.; Knowles, L.L.; Ané, C. Bayesian species delimitation combining multiple genes and traits in a unified framework. *Evolution* **2015**, *69*, 492. [CrossRef]
12. Hausdorf, B.; Hennig, C. Species delimitation and geography. *Mol. Ecol. Resour.* **2020**, *20*, 950. [CrossRef] [PubMed]
13. Doyen, J.T.; Slobodchikoff, C.N. An operational approach to species classification. *Syst. Zool.* **1974**, *23*, 239. [CrossRef]

14. Schlick-Steiner, B.C.; Steiner, F.M.; Seifert, B.; Stauffer, C.; Christian, E.; Crozier, R.H. Integrative taxonomy: A multisource approach to exploring biodiversity. *Annu. Rev. Entomol.* **2010**, *55*, 421. [CrossRef] [PubMed]
15. Padial, J.M.; Miralles, A.; De la Riva, I.; Vences, M. The integrative future of taxonomy. *Front. Zool.* **2010**, *7*, 16. [CrossRef]
16. Oberprieler, C. The Wettstein tesseract: A tool for conceptualising species-rank decisions and illustrating speciation trajectories. *Taxon*. [CrossRef]
17. Sukumaran, J.; Holder, M.T.; Knowles, L.L. Incorporating the speciation process into species delimitation. *PLoS Comput. Biol.* **2021**, *17*, e1008924. [CrossRef] [PubMed]
18. Hörandl, E. Novel approaches for species concepts and delimitation in polyploids and hybrids. *Plants* **2022**, *11*, 204. [CrossRef]
19. Ott, T.; Schall, M.; Vogt, R.; Oberprieler, C. The warps and wefts of a polyploid complex: Integrative species delimitation of the diploid *Leucanthemum* (Compositae, Anthemideae) representatives. *Plants* **2022**, *11*, 1878. [CrossRef]
20. Euro+Med (2006): Euro+Med PlantBase-the information resource for Euro-Mediterranean plant diversity. Available online: http://ww2.bgbm.org/EuroPlusMed/ (accessed on 24 March 2022).
21. Meusel, H.; Jäger, E.J. *Vergleichende Chorologie der Zentraleuropäischen Flora*; Gustav Fischer Verlag: Jena, Germany; Stuttgart, Germany; New York, NY, USA, 1992; Volume 3.
22. Vogt, R. Die Gattung *Leucanthemum* Mill. (Compositae—Anthemideae) auf der Iberischen Halbinsel. *Ruizia* **1991**, *10*, 1–261.
23. Vogt, R. *Leucanthemum* Mill. In *Flora Iberica*; Benedí, C., Buira, A., Rico, E., Crespo, M.B., Quintanar, A., Aedo, C., Eds.; Real Jardín Botánico, CSIC: Madrid, Spain, 2019; Volume 16, pp. 1848–1880.
24. Oberprieler, C.; Greiner, R.; Konowalik, K.; Vogt, R. The reticulate evolutionary history of the polyploid NW Iberian *Leucanthemum pluriflorum* clan (Compositae, Anthemideae) as inferred from nrDNA ETS sequence diversity and eco-climatological niche-modelling. *Mol. Phylogenet. Evol.* **2014**, *70*, 478. [CrossRef]
25. Konowalik, K.; Wagner, F.; Tomasello, S.; Vogt, R.; Oberprieler, C. Detecting reticulate relationships among diploid *Leucanthemum* Mill. (Compositae, Anthemideae) taxa using multilocus species tree reconstruction methods and AFLP fingerprinting. *Mol. Phylogenet. Evol.* **2015**, *92*, 308. [CrossRef] [PubMed]
26. Wagner, F.; Härtl, S.; Vogt, R.; Oberprieler, C. 'Fix Me Another Marguerite!': Species delimitation in a group of intensively hybridizing lineages of ox-eye daisies (*Leucanthemum* Mill., Compositae-Anthemideae). *Mol. Ecol.* **2017**, *26*, 4260. [CrossRef] [PubMed]
27. Wagner, F.; Ott, T.; Zimmer, C.; Reichhart, V.; Vogt, R.; Oberprieler, C. 'At the crossroads towards polyploidy': Genomic divergence and extent of homoploid hybridization are drivers for the formation of the ox-eye daisy polyploid complex (*Leucanthemum*, Compositae-Anthemideae). *New Phytol.* **2019**, *223*, 2039. [CrossRef]
28. Oberprieler, C.; Eder, C.; Meister, J.; Vogt, R. AFLP fingerprinting suggests an allopolyploid origin of two members of the *Leucanthemum vulgare* aggregate (Compositae, Anthemideae) in central Europe. *Nord. J. Bot.* **2011**, *29*, 370. [CrossRef]
29. Greiner, R.; Oberprieler, C. The role of inter-ploidy block for reproductive isolation of the diploid *Leucanthemum pluriflorum* Pau (Compositae, Anthemideae) and its tetra- and hexaploid relatives. *Flora* **2012**, *207*, 629. [CrossRef]
30. Greiner, R.; Vogt, R.; Oberprieler, C. Evolution of the polyploid north-west Iberian *Leucanthemum pluriflorum* clan (Compositae, Anthemideae) based on plastid DNA sequence variation and AFLP fingerprinting. *Ann. Bot.* **2013**, *111*, 1109. [CrossRef]
31. Oberprieler, C.; Conti, F.; Dorfner, M.; Eder, S.M.; Heuschneider, A.; Ott, T.; Scheunert, A.; Vogt, R. The taxonomy of *Leucanthemum ircutianum* (Asteraceae, Anthemideae) in the Apennine Peninsula based on AFLP fingerprinting, plastid DNA sequence variation and eco-climatological niche reconstruction. *Bot. J. Linn. Soc.* **2022**, *194*, 830. [CrossRef]
32. Kuhl, F.P.; Giardina, C.R. Elliptic Fourier features of a closed contour. *Comput. Graph. Image Process.* **1982**, *18*, 236. [CrossRef]
33. Van der Walt, S.; Schönberger, J.L.; Nunez-Iglesias, J.; Boulogne, F.; Warner, J.D.; Yager, N.; Gouillart, E.; Yu, T. scikit-image: Image processing in Python. *PeerJ* **2014**, *2*, e453. [CrossRef] [PubMed]
34. Harris, C.R.; Millman, K.J.; van der Walt, S.J.; Gommers, R.; Virtanen, P.; Cournapeau, D.; Wieser, E.; Taylor, J.; Berg, S.; Smith, N.J.; et al. Array programming with NumPy. *Nature* **2020**, *585*, 357. [CrossRef]
35. Pedregosa, F.; Varoquaux, G.; Gramfort, A.; Michel, V.; Thirion, B.; Grisel, O.; Blondel, M.; Prettenhofer, P.; Weiss, R.; Dubourg, V.; et al. Scikit-learn: Machine learning in Python. *J. Mach. Learn. Res.* **2011**, *12*, 2825–2830.
36. Anderson, M.J. A new method for non-parametric multivariate analysis of variance. *Austral Ecol.* **2001**, *26*, 32. [CrossRef]
37. Oksanen, J.; Simpson, G.; Blanchet, F.; Kindt, R.; Legendre, P.; Minchin, P.; O'Hara, R.; Solymos, P.; Stevens, M.; Szoecs, E.; et al. Vegan: Community ecology package–R package version 2.6-4, 2022. Available online: https://CRAN.R-project.org/package=vegan (accessed on 31 January 2023).
38. Fick, S.E.; Hijmans, R.J. WorldClim 2: New 1-km spatial resolution climate surfaces for global land areas. *Int. J. Climatol.* **2017**, *37*, 4302. [CrossRef]
39. Hengl, T.; Jesus, J.M.; de Heuvelink, G.B.M.; Gonzalez, M.R.; Kilibarda, M.; Blagotić, A.; Shangguan, W.; Wright, M.N.; Geng, X.; Bauer-Marschallinger, B.; et al. SoilGrids250m: Global gridded soil information based on machine learning. *PLoS ONE* **2017**, *12*, e0169748. [CrossRef]
40. Hijmans, R. Raster: Geographic Data Analysis and Modeling–R package Version 3.6-11. 2022. Available online: https://CRAN.R-project.org/package=raster (accessed on 31 January 2023).
41. Warren, D.L.; Matzke, N.J.; Cardillo, M.; Baumgartner, J.B.; Beaumont, L.J.; Turelli, M.; Glor, R.E.; Huron, N.A.; Simões, M.; Iglesias, T.L.; et al. ENMTools 1.0: An R package for comparative ecological biogeography. *Ecography* **2021**, *44*, 504. [CrossRef]

42. Phillips, S.; Dudík, M.; Schapire, R. Maxent Software for Modeling Species Niches and Distributions, Version 3.4.4. 2021. Available online: http://biodiversityinformatics.amnh.org/open_source/maxent/ (accessed on 31 January 2023).
43. Poland, J.A.; Brown, P.J.; Sorrells, M.E.; Jannink, J.L. Development of high-density genetic maps for barley and wheat using a novel two-enzyme genotyping-by-sequencing approach. *PLoS ONE* **2012**, *7*, e32253. [CrossRef] [PubMed]
44. Doyle, J.J.; Dickson, E.E. Preservation of plant samples for DNA restriction endonuclease analysis. *Taxon* **1987**, *36*, 715. [CrossRef]
45. Eaton, D.A.R.; Overcast, I. ipyrad: Interactive assembly and analysis of RADseq datasets. *Bioinformatics* **2020**, *36*, 2592. [CrossRef]
46. Mastretta-Yanes, A.; Arrigo, N.; Alvarez, N.; Jorgensen, T.H.; Piñero, D.; Emerson, B.C. Restriction site-associated DNA sequencing, genotyping error estimation and de novo assembly optimization for population genetic inference. *Mol. Ecol. Resour.* **2015**, *15*, 28. [CrossRef]
47. Joly, S.; Bruneau, A. Incorporating allelic variation for reconstructing the evolutionary history of organisms from multiple genes: An example from *Rosa* in North America. *Syst. Biol.* **2006**, *55*, 623. [CrossRef]
48. Huson, D.H.; Bryant, D. Application of phylogenetic networks in evolutionary studies. *Mol. Biol. Evol.* **2006**, *23*, 254. [CrossRef] [PubMed]
49. Lai, Y.; He, S.; Lin, Z.; Yang, F.; Zhou, Q.; Zhou, X. An adaptive robust semi-supervised clustering framework using weighted consensus of random *k*-means ensemble. *IEEE Trans. Knowl. Data Eng.* **2019**, *33*, 1877–1890. [CrossRef]
50. Peralta, M.; Combes, M.C.; Cenci, A.; Lashermes, P.; Dereeper, A. SNiPloid: A utility to exploit high-throughput SNP data derived from RNA-Seq in allopolyploid species. *Int. J. Plant Genom.* **2013**, *2013*, 890123. [CrossRef] [PubMed]
51. Wagner, N.D.; He, L.; Hörandl, E. Phylogenomic relationships and evolution of polyploid *Salix* species revealed by RAD sequencing data. *Front. Plant Sci.* **2020**, *11*, 1077. [CrossRef]
52. He, L.; Wagner, N.D.; Hörandl, E. Restriction-site associated DNA sequencing data reveal a radiation of willow species (*Salix* L., Salicaceae) in the Hengduan Mountains and adjacent areas. *J. Syst. Evol.* **2021**, *59*, 44. [CrossRef]
53. Marcussen, T.; Heier, L.; Brysting, A.K.; Oxelman, B.; Jakobsen, K.S. From gene trees to a dated allopolyploid network: Insights from the angiosperm genus *Viola* (Violaceae). *Syst. Biol.* **2015**, *64*, 84. [CrossRef]
54. Kamneva, O.K.; Syring, J.; Liston, A.; Rosenberg, N.A. Evaluating allopolyploid origins in strawberries (*Fragaria*) using haplotypes generated from target capture sequencing. *BMC Evol. Biol.* **2017**, *17*, 1. [CrossRef]
55. Sancho, R.; Inda, L.A.; Dáz-Pérez, A.; Des Marais, D.L.; Gordon, S.; Vogel, J.P.; Lusinska, J.; Hasterok, R.; Contreras-Moreira, B.; Catalán, P. Tracking the ancestry of known and 'ghost' homeologous subgenomes in model grass *Brachypodium* polyploids. *Plant J.* **2022**, *109*, 1535. [CrossRef]
56. Greiner, R.; Vogt, R.; Oberprieler, C. Phylogenetic studies in the polyploid complex of the genus *Leucanthemum* (Compositae, Anthemideae) based on cpDNA sequence variation. *Plant Syst. Evol.* **2012**, *298*, 1407. [CrossRef]
57. Wagner, F.; Ott, T.; Schall, M.; Lautenschlager, U.; Vogt, R.; Oberprieler, C. Taming the Red Bastards: Hybridisation and species delimitation in the *Rhodanthemum arundanum*-group (Compositae, Anthemideae). *Mol. Phylogenet. Evol.* **2020**, *144*, 106702. [CrossRef]
58. Monti, S.; Tamayo, P.; Mesirov, J.; Golub, T. Consensus clustering: A resampling-based method for class discovery and visualization of gene expression microarray data. *Mach. Learn.* **2003**, *52*, 91. [CrossRef]
59. Tison, J.M.; de Foucault, B. *Flora Gallica: Flore de France*; Biotope: Mèze, France, 2014.
60. Lo Presti, R.M.; Oberprieler, C.; Vogt, R. Tribus VI–*Anthemideae* (gen. 56, 58-73, 75-79). In *Flora d 'Italia*, 2nd, 3rd ed.; Pignatti, S., Ed.; Edagricole: Milano, Italy, 2018; pp. 815–881.
61. Seberg, O. Genome analysis, phylogeny, and classification. *Plant Syst. Evol.* **1989**, *166*, 159. [CrossRef]
62. Zachos, F.E. *Species Concepts in Biology*; Springer International Publishing: Cham, Switzerland, 2016.
63. Stankowski, S.; Ravinet, M. Defining the speciation continuum. *Evolution* **2021**, *75*, 1256. [CrossRef] [PubMed]
64. Von Wettstein, R. *Grundzüge der Geographisch-Morphologischen Methode der Pflanzensystematik*; Gustav Fischer Verlag: Jena, Germany, 1898.
65. Wiley, E.O. The evolutionary species concept reconsidered. *Syst. Biol.* **1978**, *27*, 17. [CrossRef]
66. Sukumaran, J.; Knowles, L.L. Multispecies coalsecent delimits structure, not species. *Proc. Natl. Acad. Sci. USA* **2017**, *114*, 1607. [CrossRef] [PubMed]
67. Lange, J.M. Pugillus plantarum imrimis hispanicarum quas in itinere 1851-52 legit Joh. *Lange, Vidensk. Meddel. Dan. Naturhist. Kjöbenhaven* **1861**, *3*, 33–116.
68. Sennen, F. *Diagnoses des Nouveautés Parues dans les Exsiccata Plantes d 'Espagne et du Maroc de 1928 à Imp*; Anglada: Vic, Spain, 1936.

biology

MDPI

Article

Cytogenetics, Typification, Molecular Phylogeny and Biogeography of *Bentinckia* (Arecoideae, Arecaceae), an Unplaced Indian Endemic Palm from Areceae

Suhas K. Kadam [1,†], Rohit N. Mane [2,3,†], Asif S. Tamboli [1], Sandip K. Gavade [4], Pradip V. Deshmukh [2], Manoj M. Lekhak [2], Yeon-Sik Choo [1] and Jae Hong Pak [1,*]

1 Research Institute for Dok-do and Ulleung-do Island, Department of Biology, School of Life Sciences, Kyungpook National University, Daegu 41566, Republic of Korea
2 Angiosperm Taxonomy Laboratory, Department of Botany, Shivaji University, Kolhapur 416004, India
3 Department of Botany, Rayat Shikshan Sanstha's, Balwant College, Vita 415311, Sangli, India
4 Department of Botany, Shri Swami Vivekanand Shikshan Santha's, Dattajirao Kadam Arts, Science and Commerce College, Ichalkaranji 416115, Maharashtra, India
* Correspondence: jhpak@knu.ac.kr
† These authors contributed equally to this work.

Simple Summary: *Bentinckia* is an Indian endemic genus belonging to the tribe Areceae (Arecaceae). This genus contains two species, *B. condapanna* and *B. nicobarica*, and both need to be conserved as they come under the threatened category. *Bentinckia*, along with nine genera, remains unplaced in Areceae. The members of the unplaced Areceae show characteristics corresponding to all the subtribes. Therefore, morphologically it is difficult to assign any subtribes to these genera. Many molecular phylogenetic analyses have reported the relationships within Areceae. However, all of these are unable to show confident position and support for the species. In the present article, we constructed the molecular phylogeny of Areceae based on an appropriate combination of chloroplast and nuclear loci that satisfactorily depicts the phylogenetic positions of all species from Areceae. Phylogeny and evolutionary history disclose that *Bentinckia* together with unplaced *Clinostigma* and *Cyrostachys* show a close relationship with the subtribe Arecinae and might have originated in Eurasia and India. In addition, this study reports a taxonomic revision of *Bentinckia*. In addition, it provides a new chromosome number (cytotype), i.e. $2n = 30$ for *B. condapanna*. This study will form the very basis for assessing and refining the systematic position of all the species from the tribe Areceae.

Abstract: *Bentinckia* is a genus of flowering plants which is an unplaced member of the tribe Areceae (Arecaceae). Two species are recognized in the genus, viz. *B. condapanna* Berry ex Roxb. from the Western Ghats, India, and *B. nicobarica* (Kurz) Becc. from the Nicobar Islands. This work constitutes taxonomic revision, cytogenetics, molecular phylogeny, and biogeography of the Indian endemic palm genus *Bentinckia*. The present study discusses the ecology, morphology, taxonomic history, distribution, conservation status, and uses of *Bentinckia*. A neotype was designated for the name *B. condapanna*. Cytogenetical studies revealed a new cytotype of *B. condapanna* representing $2n = 30$ chromosomes. Although many phylogenetic reports of the tribe Areceae are available, the relationship within the tribe is still ambiguous. To resolve this, we carried out Bayesian Inference (BI) and Maximum Likelihood (ML) analysis using an appropriate combination of chloroplast and nuclear DNA regions. The same phylogeny was used to study the evolutionary history of Areceae. Phylogenetic analysis revealed that *Bentinckia* forms a clade with other unplaced members, *Clinostigma* and *Cyrostachys*, and together they show a sister relationship with the subtribe Arecinae. Biogeographic analysis shows *Bentinckia* might have originated in Eurasia and India.

Keywords: Arecaceae; *Bentinckia*; biogeographic analysis; karyomorphology; molecular phylogeny; typification

Citation: Kadam, S.K.; Mane, R.N.; Tamboli, A.S.; Gavade, S.K.; Deshmukh, P.V.; Lekhak, M.M.; Choo, Y.-S.; Pak, J.H. Cytogenetics, Typification, Molecular Phylogeny and Biogeography of *Bentinckia* (Arecoideae, Arecaceae), an Unplaced Indian Endemic Palm from Areceae. *Biology* **2023**, *12*, 233. https://doi.org/10.3390/biology12020233

Academic Editor: Lorenzo Peruzzi

Received: 18 December 2022
Revised: 28 January 2023
Accepted: 29 January 2023
Published: 1 February 2023

1. Introduction

Arecaceae (Palmae), a monocotyledonous family, is classified into five subfamilies, namely Arecoideae Burnett, Coryphoideae Griff., Ceroxyloideae Drude, Calamoideae Griff., and Nypoideae Griff. Arecoideae is the largest and the most diverse subfamily [1]. Approximately 60% of the palm genera (107 out of 188) and more than 50% of species (ca. 1300 out of 2585) belong to this group. The members of Arecoideae are distributed throughout the tropics and subtropics, occurring mainly in rainforests and to a lesser extent in some seasonally dry habitats. Some species belonging to Arecoideae have economic importance, for example, oil palm, coconut, betel nut, and peach palm; many species are cultivated as ornamental.

The Areceae is the largest tribe among the palms considering the genera and species number [1]. Areceae represents the following characters: diminutive to robust, acaulescent, erect or rarely climbing, unarmed or armed; pinnate leaves or entire bifid, the leaflet tips entire; sheaths forming a crownshaft or crownshaft absent; infrafoliar or interfoliar inflorescences, spicate to highly branched; inflorescence bracts usually but not always comprising a prophyll and a single peduncular bract; flowers borne in triads at the base, paired or solitary staminate flowers or pits in sunk; pistillate flowers with petals distinct or connate basally, valvate distally; staminodes distinct, or very rarely connate in a conspicuous ring; pseudomonomerous gynoecium; fruit remains with basal or apical stigmatic; epicarp smooth [1]. Areceae consists of 11 subtribes, 61 genera and ca. 660 species. Among these 61 genera, 10 genera with ca. 150 species have not been placed in any subtribes. These species show representative characters of Areceae and have little anatomical diversity, and their general level of variations corresponds to place them in all the subtribes [2]. These unplaced Areceae genera include monotypic *Dransfieldia*, *Dictyosperma*, *Loxococcus* and polytypic *Rhopaloblaste*, *Iguanura*, *Hydriastele*, *Cyrtostachys*, *Bentinckia*, *Clinostigma*, and *Heterospathe*. Although consisting of a considerable number of species, morphological diversity within Areceae is limited, and considerable uncertainty of the phylogenetic relationships among the tribe exists. Few Areceae members can be described as massive trees, whereas many species of the genera *Iguanura* and *Pinanga*, etc., are qualified as 'palmlets' of the forest understorey [1].

Bentinckia Berry ex Roxb. is a small genus characterised by its distinct crownshafts and much-branched inflorescence with rachillae bearing flowers in laterally compressed pits [1]. It is endemic to India and comprises only two species, viz. *B. condapanna* (Figure 1) and *B. nicobarica* (Figure 2) [3] and shows disjunct distribution [4]. *Bentinckia condapanna* is only found on mountain cliffs of the southernmost Western Ghats at 1200–1900 m in the forests of Kerala and Tamil Nadu. It is a strong light demander, fog resilient, fire and drought loving, and a good colonizer. It is a very sensitive species in its regeneration process and prefers to grow in open places. The species occur well in shallow, porous soil with good drainage in first or second order streams where a continuous soil moisture regime is ensured frequent precipitation [5]. This species is medicinally important and used in the Siddha system of medicine [6]. Its terminal buds and young leaves are edible. It is cultivated as an ornamental palm in botanical gardens and parks due to its slender stem and feather-like leaves [5]. *Bentinckia nicobarica* grows at low altitudes in the humid forests of the Kamorta, Katchal, Great Nicobar, Nancowry, and Trinket Islands [3,7]. It has a small number of populations in the Nicobar group of islands and is at risk of extinction [8].

Figure 1. *Bentinckia condapanna* (**a**) habitat, (**b**) habit, (**c**) inflorescence, (**d**) infructescence, and (**e**) fruits.

Figure 2. *Bentinckia nicobarica* (**a**) habit, (**b**) spathe, (**c**) inflorescence, and (**d**) infructescence.

The generic name *Bentinckia* was validly published in Roxburghs' Flora Indica [9] with one species, *B. condapanna*. The name was given after Lord William Henry Cavendish Bentick (1774–1839), who was the Governor General of India between 1828 and 1835. The second species of the genus was first described as *Orania nicobarica* Kurz in the Journal of Botany, British and Foreign [10]. Beccari [11] transferred *O. nicobarica* to the genus *Bentinckia*, i.e. *B. nicobarica* (Kurz) Becc.

Bentinckia condapanna has brightly coloured fruits and a floral axis. Although its new population grows drastically by seeds, these species are rarely reported from accessible forests as these are cut down for terminal shoots that native people and wild elephants eat. Deforestation for tea plantations is a reason for localised habitat on cliffs of mountains. Habitat loss and overuse are significant threats to the survival of *B. condapanna*. Its Populations in the Western Ghats are decreasing and are much restricted in distribution [12]. *B. condapanna* is particular in its habitat (on cliffs of hills) and thereby restricted in distribution. The distribution is restricted to Kerala and Tamil Nadu, including Agastyamala, Pachakkanaum, Kulathupuzha, Uppupara, Peerumedu, Peppara, Moozhiar, South Travancore, and Tirunelveli [12,13]. Basu et al. [5] reported the distribution in Tirunelveli and Travancore Hills. They examined the habitations of the species in Kalakad Mundanthurai Tiger Reserve (KMTR) (Tamil Nadu) using GPS, Gr5IS, and stratified random sampling techniques. Further, the authors studied the growth habit, silvicultural characters, ethnobotany, places of endemism, the phytogeography parameters of field, and its phytosociological layout. Based on these outcomes, the authors concluded that *B. condapanna* is an endangered species that needs to be conserved immediately.

Bentinckia nicobarica grows with other palms such as *Areca triandra*, *A. catechu*, *Rhopaloblaste augusta*, and *Pinanga manii* at low altitudes in moist forests of Katchal Island. *B. nicobarica* was declared as a threatened species in its natural habitat [14]. The leading causes are habitat alteration, human intervention, expansion of agriculture, annual burning, cutting, and the depletion of natural resources. Due to restricted distribution and probable habitat loss, this palm is categorized as an endangered species in the wild population. There is an urgent need to develop some means of protecting it.

Cytogenetical data have considerable significance in plant taxonomy, and many researchers have studied the cytogenetics of palms. Although recent information on the comparative cytogenetics of the Arecaceae remains limited, no other large family of tropical woody angiosperms has been studied in better detail. To date, chromosome numbers have been published for approximately 330 species in 126 genera [1]. Chromosome morphology and genome size have been studied in only a few of these species. All reliable chromosome counts clearly show that diploid chromosome numbers in palms range from $2n = 26$–606 (*Chamaedorea pumila* $2n = 26$ and *Voanioala gerardii* $2n = 606$) [1]. Supernumerary chromosomes have also been recorded in *Chelyocarpus*, *Chamaerops*, *Trachycarpus*, *Pritchardia*, and *Desmoncus* [15–17]. Chromosome numbers are typically uniform within genera and sometimes within larger groups. However, there are well-documented cases of chromosome number variations within *Phoenix*, *Chamaedorea*, *Ravenea*, and *Dypsis*. In the subfamily Arecoideae, chromosome counts span the entire range of palm diploid chromosome numbers ($2n = 26$–606); however, $2n = 32$ appears to be the most common number. Although the subfamily Arecoideae is large and morphologically diverse, many studies typically reported $2n = 32$ chromosome numbers. Arecoideae chromosomes were dominantly metacentric to submetacentric. Sharma [18] and Read [19] reported $2n = 32$ chromosomes in *Bentinckia condapanna* and *B. nicobarica*, respectively.

In addition, even after being the oldest monocots to appear in fossil records around 95 million years ago, the relationship within the tribe Areceae is ambiguous [1]. Numerous molecular studies on Areceae have been reported; however, all these reports have been unable to obtain well-resolved molecular phylogeny [1,2]. Moreover, the main focus of these studies has been classification at the family and subfamily level and not at the tribe level of Areceae. Hence, the relationships within the subtribes are poorly established and the monophyly of some of the subtribes remains uncertain. In addition, the position of unplaced members of Areceae has also remained obscure. The basic morphological variation and anatomical diversity among unplaced Areceae correspond to all the subtribes. Therefore, incorporating these species into any particular subtribe is a challenging task. Strong morphological and molecular phylogenetic support is necessary to assign the subtribes to unplaced Areceae. Considering all these aspects, it is clear that there is a need to construct a well-resolved molecular phylogeny of Areceae to resolve taxonomic issues.

The current paper provides taxonomic revision, cytogenetics, and molecular phylogeny of *Bentinckia*. Molecular phylogeny of Areceae is reported based on PRK, RPB2, *acc*D, *rpo*C1, *rbc*L, *rps*16, *trn*L-F, *mat*K, and *ndh*F regions. The phylogenetic position and biogeography of unplaced members of Areceae are also discussed.

2. Materials and Methods

2.1. Taxon Sampling

Specimens for both species of *Bentinckia*, viz. *B. condapanna* and *B. nicobarica*, were collected from the Kerala and Kolkata Botanical Gardens, India, respectively. Data for species such as *Arenga wightii* Griff., *A. pinnata* (Wurmb) Merr., *Trachycarpus takil* Becc., and *Hyphaene dichotoma* (D.White bis ex Nimmo) Furtado. were adapted from our previous study [20]. The collection locality of *Bentinckia* is shown in Figure 3. The voucher specimens were collected and submitted to SUK (Department of Botany, Shivaji University, Kolhapur, Maharashtra, India). Herbarium preparation of all specimens followed the protocol reported by Jain and Rao [21]. Plant identification was based on consultation of the relevant literature [3] and the description provided in protologue and type specimens. We constructed a dataset of 67 taxa representing 63 species of the tribe Areceae and 4 of Coryphoideae (outgroup) (Supplementary File S1). This dataset was created by combining sequences produced in the present study and extracted sequences from the NCBI database. The created dataset covers the sampling of all the subtribes and unplaced members of the tribe Areceae.

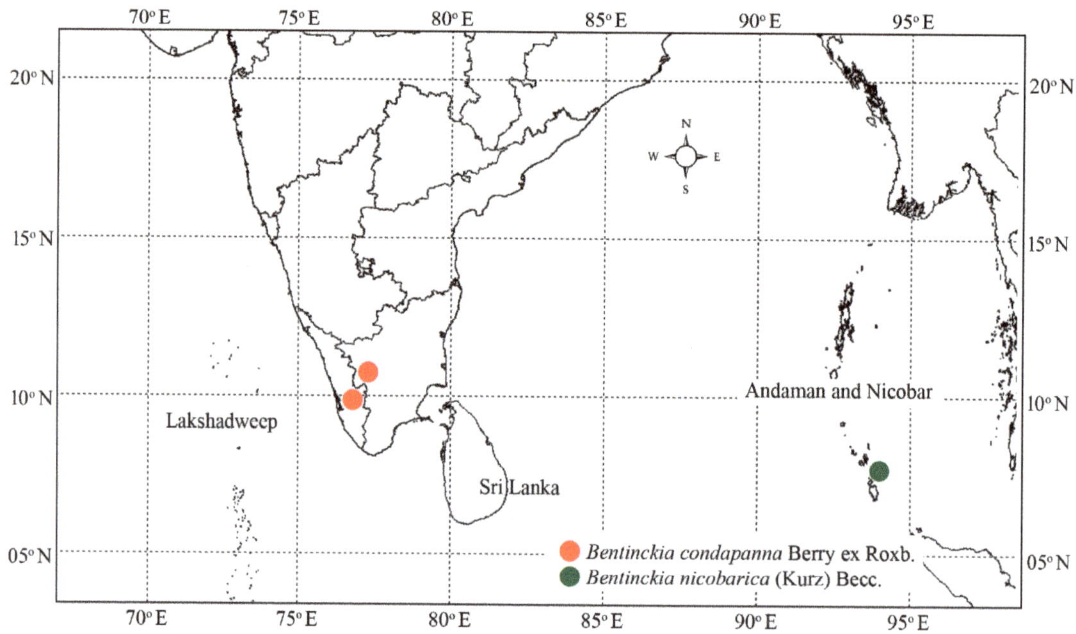

Figure 3. Distribution map of *Bentinckia condapanna* and *B. nicobarica*.

2.2. Cytogenetics

A wild population of sampled *Bentinckia* (fruits and seeds) was used for cytological study. The chromosome preparation was carried out using methods described by Mane and Yadav [22]. Well-spread chromosomes were photographed using a Leica DM 2000. Ten plates of well-segregated metaphase chromosomes were assigned for karyotype analysis as described by Levan et al. [23]. Chromosome morphology was determined by the centromeric index as: short arm $\times$ 100/total length of the chromosome. Homologous chromosomes were paired by centromeric index and length. The length of a chromosome was estimated from the mean of the total length of the chromosome. Mean chromosome length (MCL) and the sum of lengths of all chromosomes of the complement (THL) were also calculated. Comparative karyograms were prepared for both species. The degree of karyotype asymmetry was determined following the categories of Stebbins [24] and parameters proposed by Peruzzi & Eroglu [25], viz. the Coefficient of Variation of Chromosome Length (CV_{CL}), Coefficient of Variation of Centromeric Index (CV_{CI}) and Mean Centromeric Asymmetry (M_{CA}).

2.3. DNA Extraction, PCR, and Sequencing

DNA from the fresh green leaves of *Bentinckia condapanna* and *B. nicobarica* was extracted using a modified CTAB method reported by Paterson et al. [26]. The PCR amplification of *mat*K, *ndh*F, *rbc*L, RPB2, *rps*16, and PRK genes was carried out as mentioned in a previous study [20]. The sequencing of amplified loci were carried out at Macrogen, Inc. (Republic of Korea). The obtained accession numbers of *acc*D, *mat*K, PRK, *ndh*F, *rbc*L, RPB2, *rpo*C1, *rps*16, and *trn*L-F gene sequences used in the construction of Areceae phylogeny are mentioned in Table S1.

2.4. Phylogenetic Analysis

The sequence analysis was done using Sequencher v. 5.1. Multiple sequence alignment of individual genes was carried out in MEGA 11 [27] using the MUSCLE [28] program. All the alignments were refined using a Gblocks server [29].

We used ML and BI methods to construct phylogenetic trees of combined (nrDNA + cpDNA) datasets. The jModelTest 2 program [30] was used to select best-fit nucleotide substitution models under AIC. The suggested best-fit model (TVM + I+G) was not present in the MrBayes. Hence, we chose the second model GTR + I+G for the construction of phylogenies. BI phylogeny was constructed in MrBayes v.3.2.7 [31] with similar parameters described in our previous study [20]. ML analysis was also carried out based on the same best-fit model using the IQ tree via IQ tree web server [32]. The number of bootstrap replications was kept at 1000 replicates to assess the robustness of the nodes.

2.5. Biogeographic Analysis

Biogeographic areas were defined considering the earlier biogeographic studies and distribution of all Areceae species [33,34]. The distribution range of Areceae was coded as follows: (A) Eurasia up to Wallace's Line and the Andaman and Nicobar Islands, (B) India and Sri Lanka, (C) Indian Ocean Islands and Madagascar, and (D) the Pacific (areas east of Wallace's Line and Australia). The S-DIVA analysis was performed on an all-compatible Bayesian tree in RASP v 4.2 [35]. To obtain trustworthy results of biogeographic analysis, 1332 binary trees were used to run S-DIVA.

3. Results

3.1. Cytogenetics

3.1.1. Bentinckia Condapanna Berry Ex Roxb

Our study showed that *B. condapanna* collected from Chemunjii, Thiruvananthapuram, Kerala had $2n = 30$ chromosomes (Figure 4a) (Table 1). This report of *B. condapanna* with $2n = 30$ chromosomes forms a new cytotype for the species. The CV_{CL} and CV_{CI} were observed at 18.18 and 8.05, respectively. The length of the shortest chromosome was 1.48 µm and the longest chromosome was 3.00 μm. Haploid chromosome length was 34.04 μm. The karyotype formula of this species consisted of 15 median pairs. The karyotype of this species was classified as Stebbins 4B asymmetry class. M_{CA} was 14.05. The karyogram is depicted in Figure 4c.

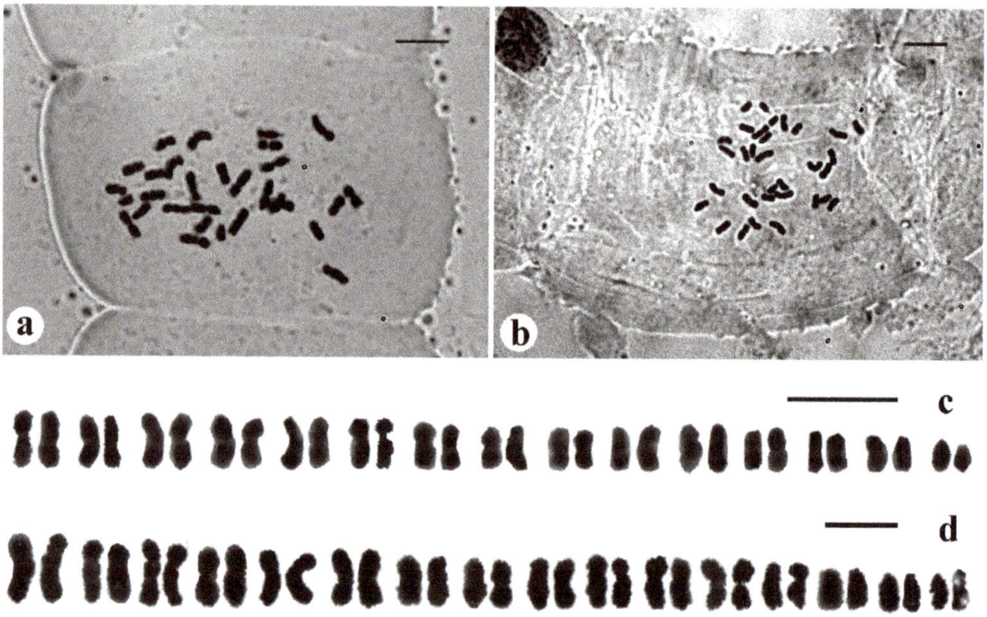

Figure 4. Mitotic metaphase chromosomes and karyograms. (**a**) and (**c**) show *Bentinckia condapanna* ($2n$ = 30); (**b**) and (**d**) show *B. nicobarica* ($2n$ = 32). Scale bars = 5 μm.

Table 1. Comparative karyotypes of *Bentinckia condapanna* and *B. nicobarica.*

	B. condapanna	*B. nicobarica*
$2n$	30	32
Total haploid chromosome length (THCL ± SE (μm))	34.04 ± 0.59	25.83 ± 0.40
Haploid karyotype formula	15m	9m + 7sm
Coefficient of Variation of Chromosome Length (CV_{CL})	18.18	27.17
Coefficient of Variation of Centromeric Index (CV_{CI})	8.05	11.17
Mean Centromeric Asymmetry (M_{CA})	14.05	22.02
Shortest chromosome (S ± SE (μm))	1.48 ± 0.34	0.91 ± 0.25
Longest chromosome (L ± SE (μm))	3.00 ± 0.25	2.30 ± 0.16
Mean chromosome length (MCL ± SE (μm))	1.62 ± 0.59	1.61 ± 0.11
Longest to shortest chromosome ratio (R)	2.03	2.52
Stebbins category (St)	4B	3B

3.1.2. *Bentinckia Nicobarica* (Kurz) Becc

Our study showed that *B. nicobarica* had a diploid chromosome number of $2n$ = 32 (Figure 4b) (Table 1). The CV_{CL} and CV_{CI} were observed at 27.17 and 11.17, respectively. The shortest chromosome length was 0.91 μm, and the longest chromosome was 2.30 μm in length. Haploid chromosome length was 25.83 μm. The karyotype formula of this species consisted of nine median pairs and seven submedian pairs. The karyotype of this species was classified as Stebbins 3B symmetry class. M_{CA} was 22.02. The karyogram is depicted in Figure 4d.

3.2. Molecular Phylogeny of Areceae

The combined matrix of plastid and nuclear loci was used to construct the Maximum Likelihood and Bayesian Inference molecular phylogeny of the tribe Areceae. The aligned sequence dataset of combined nuclear + chloroplast includes 67 genera covered of 7026 characters (Supplementary Material S1). The constructed phylogeny based on the combined dataset resolves nine subtribes with strongly supported clades, including Archontophoenicinae (PP = 1 and BS = 92), Ptychospermatinae (PP = 1 and BS = 100), Laccospadicinae (PP = 1 and BS = 100), Clinospermatinae (PP = 1 and BS = 100), Carpoxylinae (PP = 1 and BS = 100), Verschaffeltiinae (PP = 1 and BS = 96), Dypsidinae (PP = 1 and BS = 100), Arecinae (PP = 1 and BS = 100), and Oncospermatinae (PP = 1 and BS = 100) (Figure 5). However, Basseliniinae and Rhopalostylidinae were clustered in a single group. In addition, both sampled *Bentinckia* species grouped together and formed a clade with other unplaced Areceae.

3.3. Ancestral Area Reconstruction

Ancestral area reconstruction of the tribe Areceae was performed using S-DIVA method. The reconstruction based on the combined datasets (Nuclear + Chloroplast) showed a Maximal S-DIVA value of 5099.00 (Figure 6). Node 129 signifies an equal probability of Eurasia (A), India and Sri Lanka (B), Indian Ocean Islands and Madagascar (C), and the Pacific (D) to be an origin of the tribe Areceae. Biogeography analysis showed that Eurasia might be a place of origin of the sampled genus *Bentinckia*. The group of unplaced Areceae containing *Bentinckia*, *Clinostigma*, and *Cyrtostachys* also originated in Eurasia. The other genus, *Iguanura*, originated in Eurasia and the Indian Ocean with 100% probability. The origin of the genus, *Hydriastele*, was found to be in the Indian Ocean and the Pacific. The group of unplaced Areceae IV containing *Dictyosperma* and *Rhopaloblaste* diverged from the rest of the Indian Ocean clade at the Indian Ocean, Eurasia, and the Pacific and distributed into the Indian Ocean. The origin of *Dransfieldia* and *Heterospathe* might be the Pacific and the Indian Ocean and the Pacific, respectively. The genus *Loxococcus* might have originated from India and the Pacific (Figure 6).

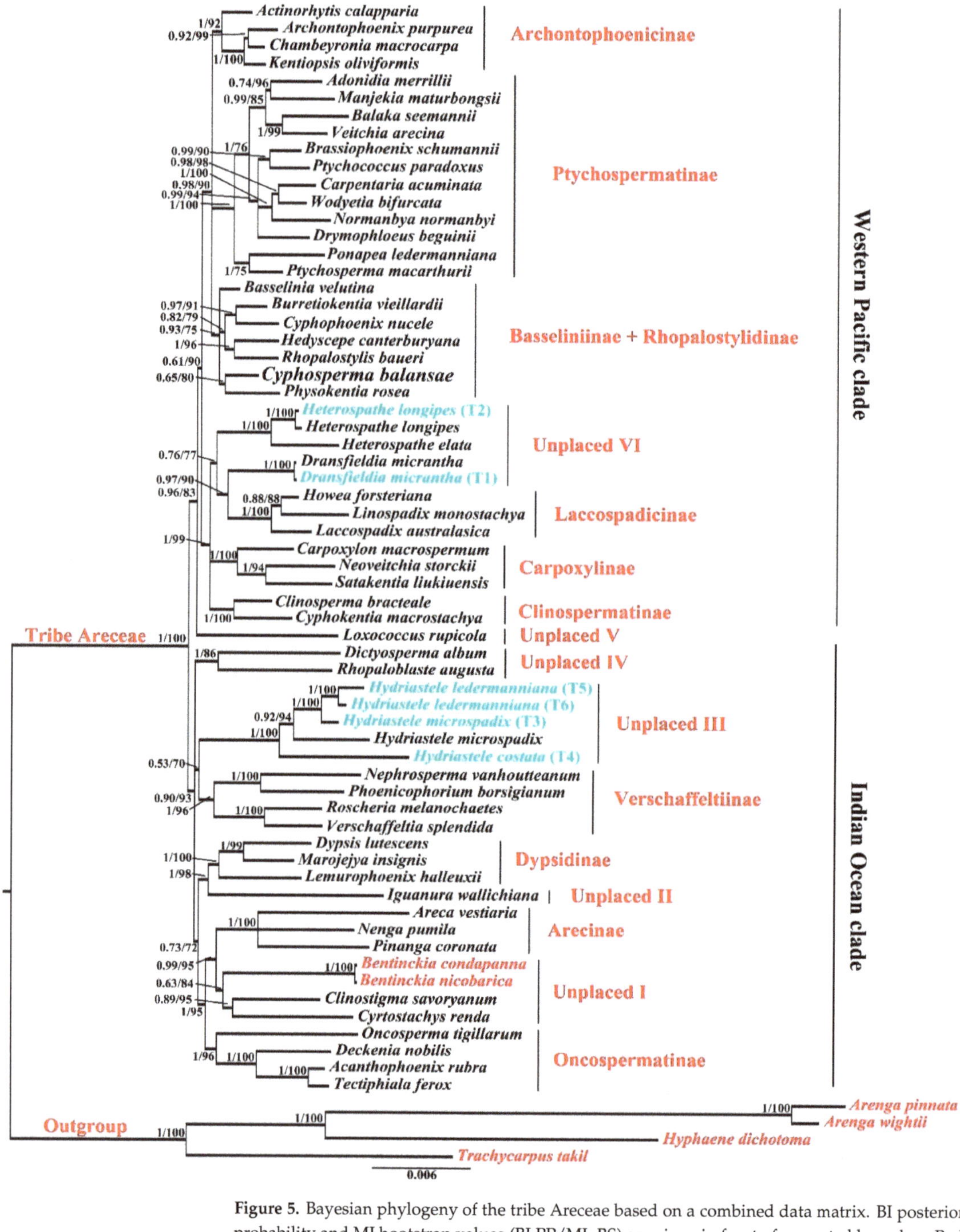

Figure 5. Bayesian phylogeny of the tribe Areceae based on a combined data matrix. BI posterior probability and MI bootstrap values (BI PP/ML BS) are given in front of respected branches. Red colour represents sampled species whereas blue colour represents revised names of species.

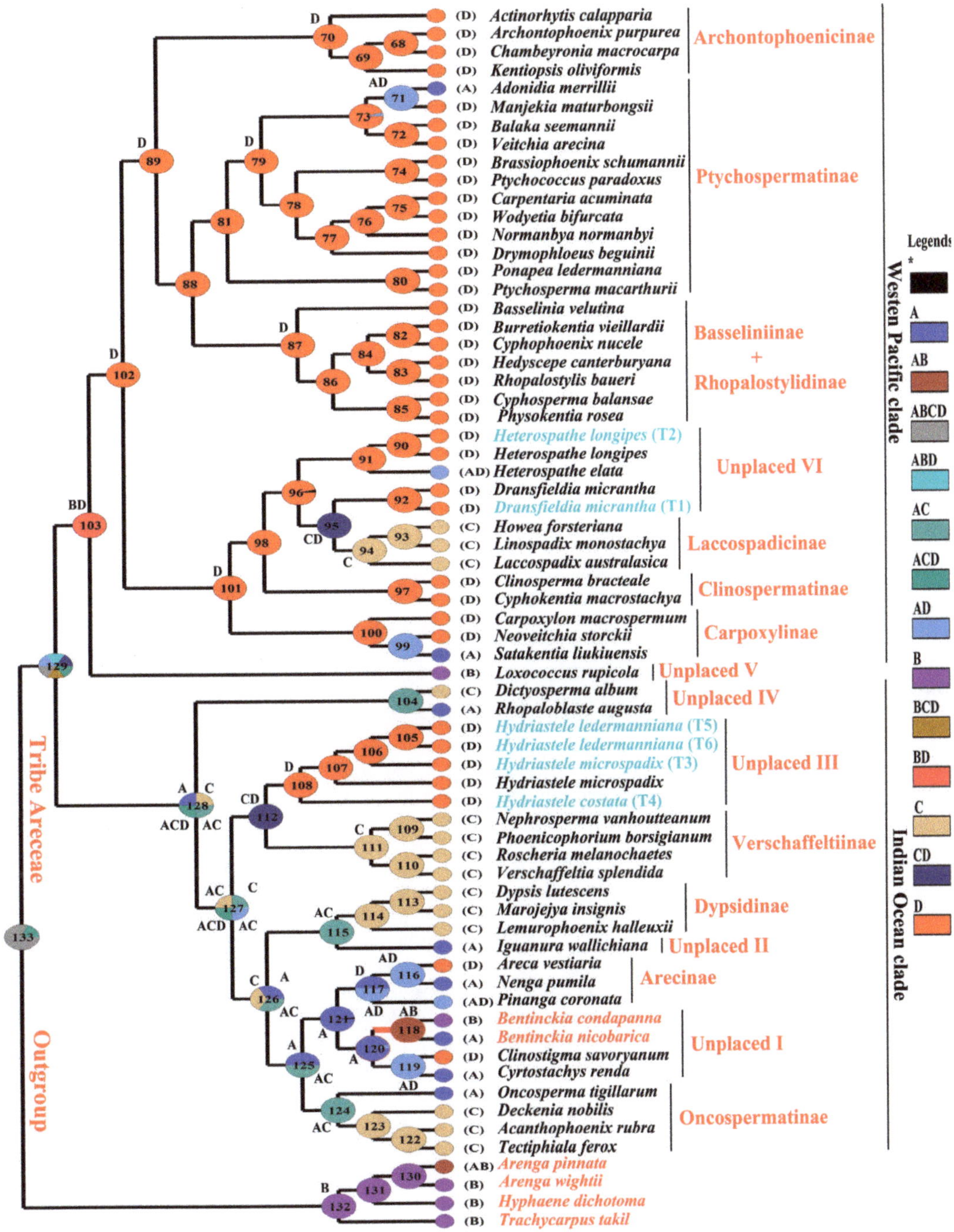

Figure 6. Biogeographic analysis of Areceae based on the Bayesian all compatible groups tree. * (Black colour), ranges with probabilities < 5% are hidden and lumped together and reported as *.

4. Discussion

The Phylogenetic analysis of Areceae was carried out on nuclear and chloroplast regions. The tribe Areceae is one of the largest and most important tribes from Arecaceae. Several attempts were carried out to find the relationships between the tribe, but it remains poorly understood. In the current study, we have discovered the correct combination of the molecular marker to achieve a better phylogenetic resolution in the tribe Areceae. By using the same dataset, we studied on the phylogenetic placement and evolutionary history of unplaced Areceae members. This study also gives phylogenetic support to the previous systematic revisions carried out in unplaced Areceae. In addition, it will help to resolve the question of unplaced Areceae and future doubts in the tribe Areceae. In addition, this study reports a taxonomic revision and a new cytotype with $2n = 30$ chromosomes for *Bentinckia*. This study is the first report of $2n = 30$ chromosomes from the genus *Bentinckia* and the tribe Areceae.

4.1. *Taxonomic Treatment of Bentinckia*

Bentinckia Berry ex Roxb., Fl. Ind. 3: 621 (1832).

Type: *Bentinckia condapanna* Berry ex Roxb.

Keppleria Mart. ex Endl., Gen. pl. 251 (1837).

Type: *Keppleria tigillaria* (Jack) Meisn.

Unarmed palms. Leaves terminal, equally pinnate, spathes numerous, two lower short incompletes, upper 2-fid. Spadix interfoliar, branched; flowers minute, monoecious or polygamous, solitary or 3-nate with the intermediate female clustered in spirally arranged form. Pits on the branches, bracts forming a 2-lipped mouth to each pit; bracteoles 2. Male flower sub symmetric, glumaceous, often reduced to ciliate scales; sepals oblong, obtuse, connate below, imbricate; petals longer, connate bellow into a stipes, valvate; stamens 6, anthers versatile; pistillode conical. Female flower ovoid; sepals broad, obtuse, imbricate; petals longer, convolute; staminodes 6. Ovary 3-celled, 1-ovuled; stigmas 3, recurved. Fruit 1.3–1.5 cm in diameter, subspherical. Seeds pendulous from the top of the cavity, sinuately grooved or ridged; albumen equable.

Key to the species of *Bentinckia*

1. Stem slender, up to 10 m tall, flowering branches light pink around pits, ripened fruits deep scarlet; Southern India. *B. condapanna*.

2. Stem robust, up to 20 m tall, flowering branches yellowish white, ripened fruits deep brown; Nicobar Islands. *B. nicobarica*.

Bentinckia condapanna Berry in Roxb., Fl. Ind. 3: 621. 1832; Griff., Calcutta J. Nat. Hist. 5 467. 1845; Griffith, Palms Brit. Ind. 160. 1850; Mart., Hist. Nat. Palm. 3: 165, 228. t. 1–39. 1823–1853; Becc. & Hook. f. in Hook. f., Fl. Brit. India 6: 418. 1892; Hook., Fl. Brit. India 6: 418. 1894; Fischer in Gamble. F1. Madras. 1555–1556. 1931; Basu & Chakraverty, Man. Cult. Palms India, 128. 1994; David, Palms Throughout the World, 142. 1995; Renuka & Sreekumar, A field guide to the palms of India, 32–33. 2012.

Neotype (Designated here): India, Peninsular India or Travancore, s.d., N. Wallich s.n. (M0208636) (Figure 7).

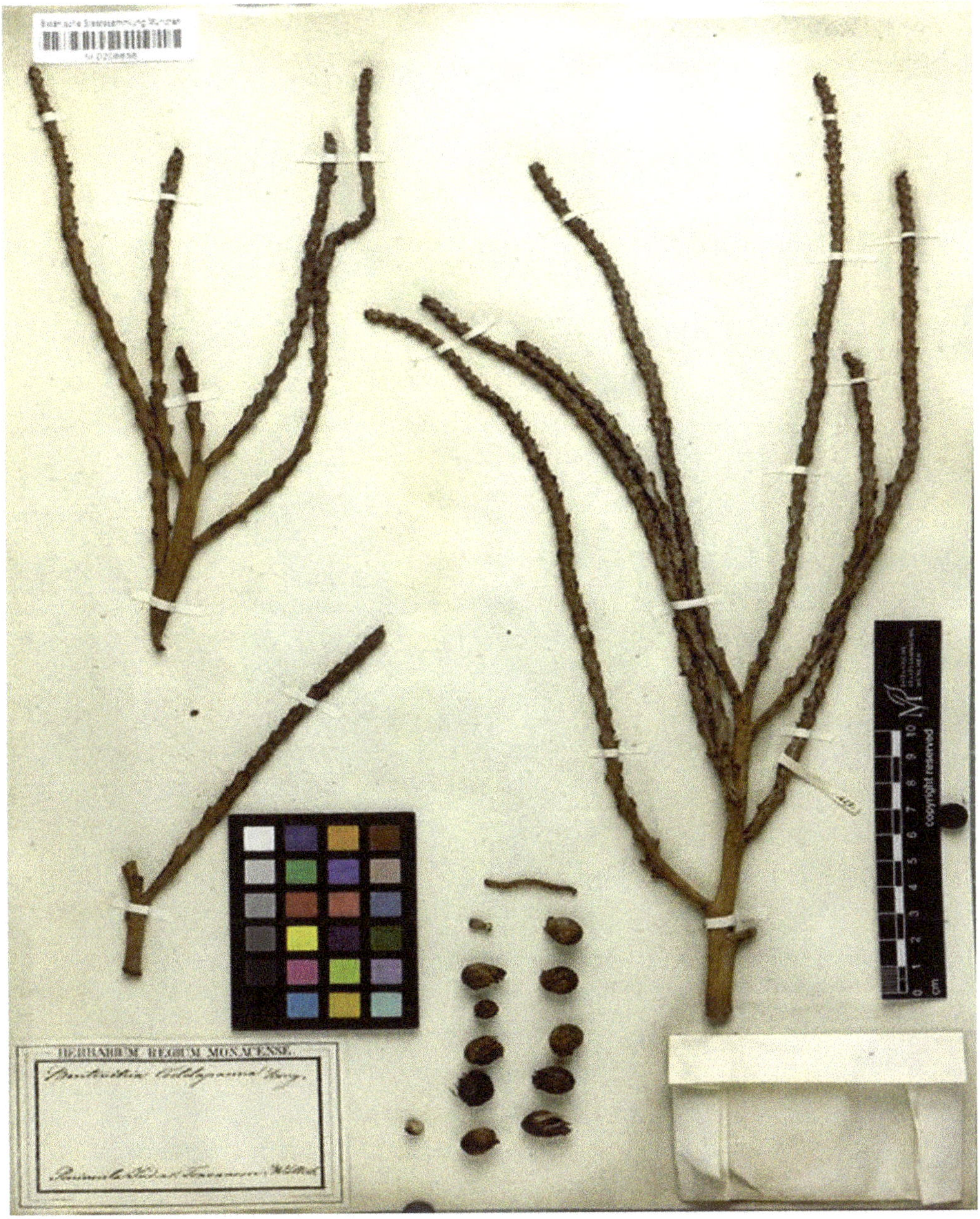

Figure 7. Neotype of *Bentinckia condapanna* Berry ex Roxb. (M0208636). © Botanische Staatssammlung München (M).

Solitary slender stemmed monoecious palm, stem erect, up to 8 m long, ca. 20 cm diameter near base; crownshaft cylindrical ca. 1 m long. Leaves pinnate, 1–1.5 m long, ascending to spreading in all directions; leaflets linear, acuminate, deep green in colour, up to 80 cm long, to 4 cm broad at middle, duplicately folded near the point of attachment; midnerve conspicuous on upper side bifurcating into long narrow lobes. Inflorescence infrafoliar, decompound; prophyll and peduncular bract large bicarinate, 25–30 cm long, fall off after emergence of flower branches; peduncle flattened, deep green in colour, approximately 4.5 cm long; basal flower branches bracteate, divided into fourth order. Fruit globose to ovoid, bright chocolate coloured when ripe, 1.3–1.5 cm in diameter, seed shining brown, conspicuously grooved adaxially and laterally; endosperm homogenous. Seeds fleshy pendulous in fruit cavity, suspended from the top, grooved and ringed; albumen is horny. Seeds ovate to oblong, conspicuously grooved, convex on one side, ribbed. Embryo is closer to apex, slightly lateral (Figure 8a–c).

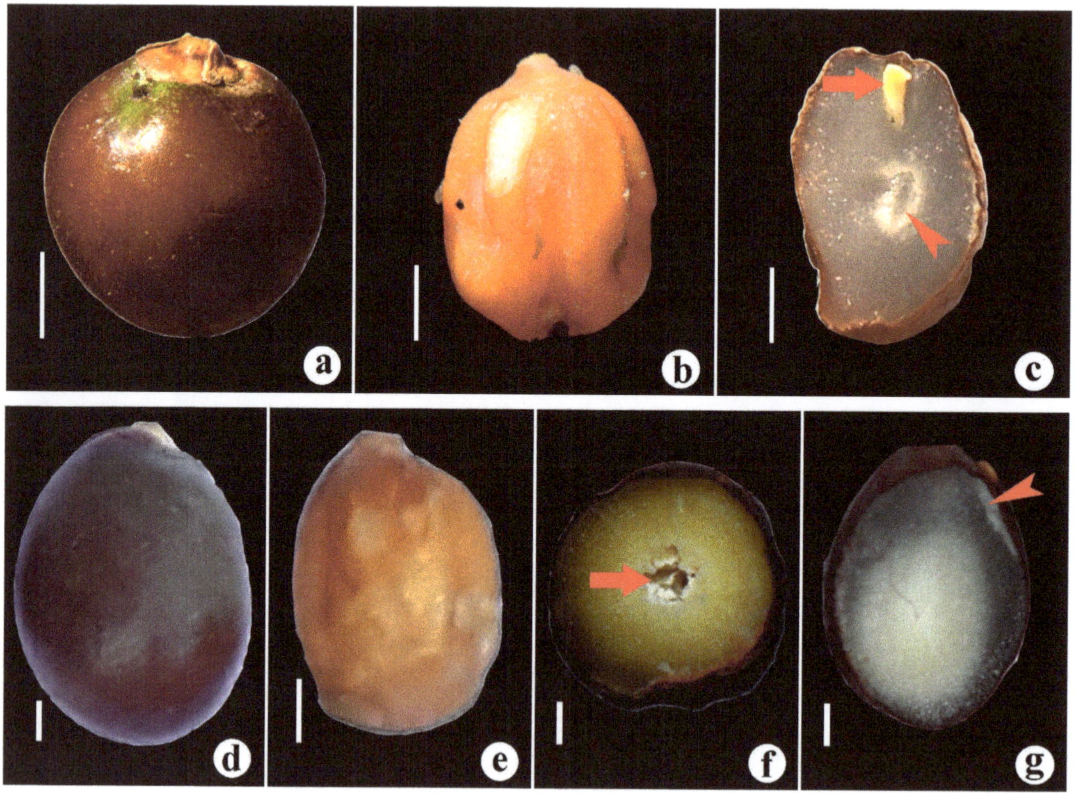

Figure 8. Fruits and seeds in *Bentinckia* species. *B. condapanna* (**a**) fruit, (**b**) seed, and (**c**) cross section of seed. Arrow shows apical embryo and arrowhead horny albumen. Scale bars = 0.2 cm. *B. nicobarica* (**d**) fruit (**e**) seeds, and (**f**,**g**) cross section of seed. Arrow shows ribbed albumen and arrowhead apical embryo. Scale bars = 0.2 cm.

Nomenclatural Note: The name *Bentinckia condapanna* Berry ex Roxb. was initially proposed by Berry [9] and was validated by Roxburgh [9] in his *Flora Indica*. In the protologue, Roxburgh [9] stated that Dr. Berry found this plant species in the mountains of Travancore. Berry was a surgeon in the East India Company in Madras and was superintendent of the Company's Cactus Garden at Marmalong in 1790. It is unclear whether Berry sent the plant to Roxburgh at Calcutta Botanical Garden, and whether it grew well there and Wallich made the specimens or not. We could not locate Berry's specimens anywhere. However, while searching, we could locate six specimens of *B. condapanna* in M (M0208631, M0208632, M0208633, M0208636, M0208635, M0208634). All these specimens bear a label 'Peninsula Ind. or. Travancore Wallich' but are not an original material for the name *Bentinckia condapanna*. Since no original material appears to be extant, the Wallich specimen (M0208636) in M is chosen here as the neotype.

Etymology: The specific epithet 'condapanna' was derived from the local terms 'conda', used to describe the characteristic casual hairstyle in the local language that resembles opened inflorescence of the palm, and 'pana', which designates a colloquial term for palm.

Distribution: India (Kerala and Tamil Nadu).

Habitat: Grows only in the steep slopes of evergreen forests. Locally common on rocky cliffs between 1000 and 1900 m above sea level.

Local Names: Malayalam (Kantal, Kanthakamugu, Kantha-kamugu, Kanthal, Parapakku, Vareikamuku); Tamil (Kantha Panai, Varei Kamugu, Varukamuvu).

Common Name: Hill Areca nut.

Uses: The palm heart is edible; inflorescences are used in religious ceremonies by the tribal people; the trunk is used for construction purposes; and planted as an ornamental species.

Conservation Status: Vulnerable [3,36].

Note: This species is reported to be rare; distribution in the Western Ghats is restricted to the south Palakkad gap; mainly in the mountains of Agasthymala, Peerumedu, and the Palani Hills [3]. Basu [14] has enlisted this plant under the rare category in his Red Data Book on Indian plants. The World Conservation Monitoring Centre (1996) also considered the species in the rare category. However, a recent field survey assessed the species as Endangered (EN)-A1c, B2a, bi, ii, iii, CI, E. [5].

Specimens examined: INDIA. Kerala: Quilon district, Ponnambala medu, 15 December 1981, C. N. Mohanan 72826 (CAL); Vallakkadavu, Peerumedu, Idukki, 19 October 1996, P. V. Anto 7325 (KFRI); Pullupara, Vallakadav, 20 November 2001, V. B. Sreekumar & V. V. Rangan 7606 (KFRI); Tamil Nadu: Kanyakumari district, Upper kodayar, 7 August 1977, A. N. Henry 49651 (CAL); Salem district, botanical garden, Botanical Survey of India, Yercaud, 10 August 2019, R. N. Mane 144 (SUK).

Bentinckia nicobarica (Kurz) Becc. Annales du Jardin Botanique de Buitenzorg 2: 165. 1885; Becc. & Hook. f. in Hook. f. Fl. Brit. India 6:418. 1892; Blatt. Palms Brit. Ind. & Ceyl. 376, t. 67. 1978 (Repr. ed.); Hook., Fl. Brit. India 6: 418. 1894; Basu & Chakraverty, Man. Cult. Palms India, 129. 1994; David, Palms Throughout the World, 142. 1995; Renuka & Sreekumar, A field guide to the palms of India, 34–35. 2012.

Orania nicobarica Kurz, J. Bot. 13: 331 (1875).

Lectotype (designated by Mane and Lekhak, [37]): INDIA, Nicobar Islands: Kamorta, February 1875, W. S. Kurz, s.n. (K000736204) (Figure 9).

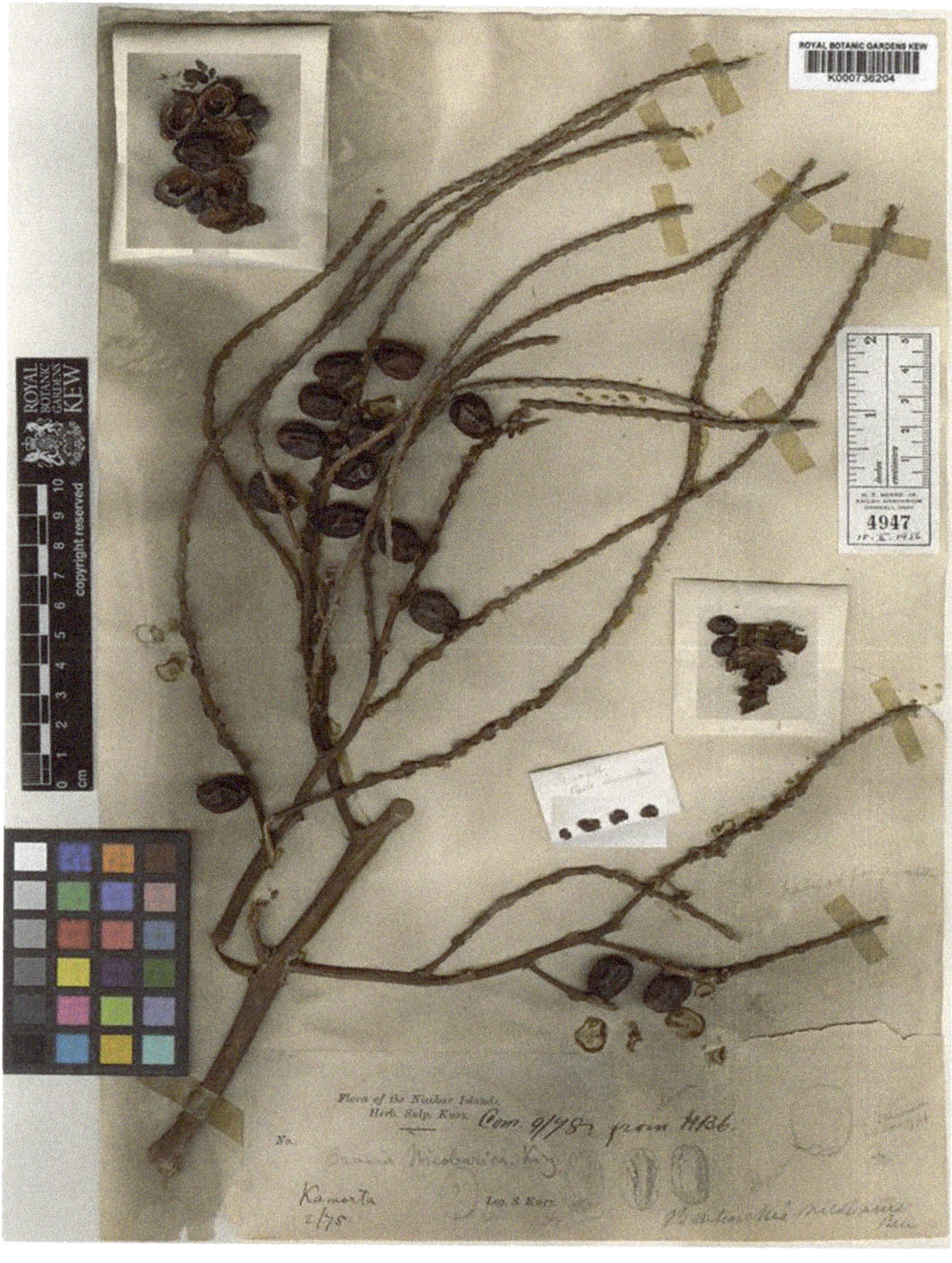

Figure 9. Lectotype of *Orania nicobarica* Kurz (≡*Bentinckia nicobarica* (Kurz) Becc.) (K000736204). © Royal Botanic Gardens, Kew.

Solitary tall palm, stem columnar, distinctly annulate, up to 20 m long, up to 40 cm in diameter near base; crownshaft cylindrical, green, ca. 1 m long. Leaves ascending to arching, approximately 2.5 m long; leaflets closely packed, linear lanceolate, acuminate, alternate to subopposite in adult trees; laterally jointed in younger plants, 50–60 cm long with conspicuous midnerve on upper side; terminal leaflets jointed. Inflorescence infrafoliar, decompound; prophyll and peduncular bracts large, green bicarinate, spatuliform; flower branches greenish yellow; ultimate flower branches slightly inserted at the point of attachment; flowers bracteolate. Fruit subglobose to ellipsoid, deep brown in colour; middle portion fibrous; inner portion brittle; seed ovoid 0.8–0.9 cm long, endosperm white, homogenous. Showing ovoid-oblong seed, ventrally flat, dorsally convex and rugosely ribbed, seed showing apical embryo, cross section of seed showing ribbed albumen (Figure 8d–g).

Nomenclatural Note: The binomial *Bentinckia nicobarica* (Kurz) Becc. was based on basionym *Orania nicobarica* Kurz. In search of type, Mane and Lekhak [37] could locate two specimens of *O. nicobarica* at K (K000736204 and K000736203) and four specimens at CAL. (CAL0000001211, CAL0000001212, CAL0000001213 and CAL0000001214). All specimens serve as syntypes, and Mane and Lekhak [37] designated a specimen from K (K000736204) collected by Kurz from the Nicobar Islands as a lectotype.

Etymology: The specific epithet 'nicobarica' was given after the type locality, i.e. the Nicobar Islands.

Distribution: INDIA (Nicobar Islands). Known only from the Nicobar Islands, being more common in the northern islands according to Kurz [38]. Widely cultivated throughout South East Asia.

Habitat: Lowland evergreen forests at 100–150 m elevation.

Common Names: Bentinck palm, Nicobar palm.

Uses: The tree trunk of this species is used for construction. This species is grown in gardens as an ornamental palm.

Conservation Status: Endangered [39], Critically Endangered [3].

Specimens examined: INDIA, Nicobar Islands: Arong, Car Nicobar, 3 October 2008, E. L. Linto 10724 (KFRI); West Bengal: Acharya Jagadish Chandra Bose Indian Botanic Garden, Kolkata, 5 February 2018, R. N. Mane 110 (SUK).

4.2. Cytogenetics

A new cytotype with $2n = 30$ was reported for *B. condapanna*. In contrast, Sharma [18] reported $2n = 32$ from the same species. The diploid chromosome ranges from $2n = 30$ in *B. condapanna* to $2n = 32$ in *B. nicobarica*. Read [19] reported $2n = 32$ in the species, and it has been confirmed in the present study; karyotype analysis is also provided. The earlier authors studied chromosomes, and the most common somatic chromosome number was $2n = 32$ which suggests that the base number (x) for the genus is 16. The intrachromosomal index (M_{CA}) is due to the centromeric position while the interchromosomal index (CV_{CL}) depicts heterogeneity among chromosome sizes in a complement. Higher values of M_{CA} and CV_{CL} for *B. nicobarica* indicate more asymmetry in its karyotype. The *B. nicobarica* karyotype shows two types of chromosomes, m and sm, whereas in *B. condapanna*, only m type chromosomes are present. In addition, the high R ratio in *B. nicobarica* reflects more heterogeneity in its chromosome complement.

The sequence of basic chromosomes in palms ($n = 13$–18) is known as the dysploid series [1]. The chromosomes in Arecaceae show a great variability in length and chromosome number (*Voanioala gerardii* $2n = 550, 596, 606$) [1]. The family Arecaceae is large and morphologically diverse. Metacentric to submetacentric (m to sm) chromosomes are dominant in Areceae [1]. Our study also showed that metacentric and submetacentric chromosomes are present in *Bentinckia*.

4.3. Molecular Phylogeny of Areceae

The palm family is considered among the oldest and most well-studied families. Arecoideae and Areceae are the largest subfamily and tribe from Arecaceae, respectively.

Many studies have reported phylogenetic analysis of Areceae; however, the relationship within the tribe is still ambiguous. Several reports strongly support the monophyly of Areceae [40–44] while others recover it with less support [45–48]. In addition, a few other studies also reported the phylogenetic relationship of Areceae [49–52], but all of them were unable to report a well-resolved molecular phylogeny of Areceae. Many authors reported phylogeny based on single plastid or nuclear DNA (PRK and RPB2) regions. However, the role of low-copy nuclear and chloroplast DNA becomes very important while studying the phylogeny of palm. Therefore, the right combination of molecular markers is inevitable for a well-resolved phylogeny [20]. Areceae contains 11 subtribes, 61 genera (10 unplaced), and 660 species, even though the phylogenetic positions of several genera are not well understood. Here, we are reporting a well-resolved phylogeny of Areceae built using the appropriate combination of chloroplast and nuclear regions. Based on this phylogeny, it is clear that Areceae is divided into two main groups: the Western Pacific clade and the Indian Ocean clade. The Indian Ocean clade consists of four subtribes: Oncospermatinae, Arecinae, Dypsidinae, and Verschaffeltiinae while the Western Pacific clade consists of seven subtribes: Archontophoenicinae, Ptychospermatinae, Basseliniinae, Rhopalostylidinae, Laccospadicinae, Clinospermatinae, and Carpoxylinae. Based on PRK and RPB2, Norup et al. [43] found that the Western Pacific and Indian Ocean clades are monophyletic. Nevertheless, the interrelationship within the members of the subtribes remained unresolved on account of polytomy.

Unplaced Areceae

The ten unplaced members of Areceae are divided into six groups and they are distributed among the Western Pacific and the Indian Ocean clade. The sampled Indian endemic *Bentinckia* groups together and forms a clade with other unplaced *Clinostigma* and *Cyrostachys*. This clade of unplaced Areceae shows a sister relationship with the subtribe Arecinae. This relationship was recovered in both the supertree and the supermatrix analysis conducted by Baker et al. [49]. Supertree and supermatrix analyses were carried out using the same 16 partitions; however, the phylogenetic relationships within Areceae are in contrast with each other. *Iguanura wallichiana* shows a sister relationship with subtribe Dypsidinae with strong support. The clade of unplaced *Hydriastele* stands confidently showing a sister relation with the subtribe Verschaffeltiinae. Additionally, Petoe et al. [53] recognized both the former species *Gronophyllum chaunostachys* (Burret) H.E.Moore and *Hydriastele chaunostachys* (Burret) W.J.Baker & Loo as *Hydriastele ledermanniana* (Becc.) W.J.Baker & Loo [54], and these two former species group together and show close relationships with other *Hydriastele* species in our phylogeny. Baker and Loo [55] synonymized *Gulubia macrospadix* (Burret) H.E.Moore as *Hydriastele microspadix* (Burret) W.J.Baker & Loo [54], which is also shown by our findings, as *G. macrospadix* nests within other accessions of *Hydriastele microspadix*. Similarly, *Gulubia costata* (Becc.) Becc., a synonym of *Hydriastele costata* F.M.Bailey [55], shows a close relationship with other *Hydriastele* species. Therefore, our phylogeny of Areceae supports all the revised circumscriptions of *Gronophyllum*, *Gulubia*, and *Hydriastele*. The paraphyletic clade of unplaced *Dictyosperma* and *Rhopaloblaste* shows a close relationship with the remaining species of the Indian Ocean clade. Baker et al. [44] also recovered this clade representing *Dictyosperma* and *Rhopaloblaste*. In addition, this study resolved the phylogenetic position of a few of the subtribes but with very low support. Most of them resolved as a polytomy.

In the Western Pacific clade, the unplaced member of Areceae *Loxococcus* stands confidently as a sister to the western Pacific clade. Other unplaced members, *Dransfieldia* and *Heterospathe*, show a close relation with the subtribe Laccospadicinae. *Dransfieldia* shows a sister relationship with the subtribe Laccospadicinae and is placed in the same clade while *Heterospathe* stands as a sister to that clade. In addition, Baker et al. [56] revised the former species *Ptychosperma micranthum* Becc. as *Dransfieldia micrantha* (Becc.) W.J.Baker & Zona [54] and both species show a close relationship in our proposed phylogeny. Further, Norup [57] revised former species *Alsmithia longipes* H.E.Moore as *Heterospathe longipes* (H.E.Moore) Norup [54], and our phylogeny supports this. Loo et al. [42] also showed molecular support for the inclusion of some species of *Gronophyllum* and *Gulubia* in *Hydriastele*. Our phylogeny supports this inclusion. In addition, it also provides molecular support to the revisions carried out in *Ptychosperma*, *Alsmithia*, and *Hydriastele* species concerned with unplaced Areceae.

4.4. Ancestral Area Reconstruction of Areceae

Studying evolutionary history based on molecular phylogeny is obligatory to understand precise biogeographical evolution [58]. To date, no robust biogeographical work has been carried out on the tribe Areceae. Previous reports [33,34] have studied the biogeography of the palm; however, the relationship within Areceae, as well as Arecaceae, was not resolved [20]. The main focus of those studies was the family or subfamily-level relationship. Our S-DIVA analysis suggests Eurasia (A), India and Sri Lanka (B), Indian Ocean Islands and Madagascar (C), and the Pacific (D) may have been the centre of origin of the tribe Areceae. The unplaced members of Areceae, including *Bentinckia*, *Clinostigma*, and *Cyrtostachys*, show Eurasia might be the centre of origin of this group. *Bentinckia* diverged from *Clinostigma* and *Cyrtostachys* in Eurasia and was distributed in Indian territory. *Bentinckia condapanna* is spread in Kerala and Tamil Nadu while *Bentinckia nicobarica* remains endemic to the Nicobar Islands. Baker and Couvreur [33] reported a contrasting result to our study. These authors showed *Cyrtostachys* diverged from *Bentinckia* and *Clinostigma*. *Iguanura* diverged from the subtribe Dypsidine in Eurasia and the Indian Ocean while Baker and Couvreur [33] reported the diversion of *Iguanura* from the rest of Areceae. Our biogeography analysis shows that *Hydriastele* originated in the Indian Ocean and the Pacific and diverged from the subtribe Verschaffeltiinae. However, the previous study reported that *Hydriastele* diverged from the subtribe Oncospermatinae [33,34]. *Dictyosperma* and *Rhopaloblaste* might have originated in Eurasia, the Indian Ocean, and the Pacific and then diverged from the rest of the Indian Ocean clade. The genus *Loxococcus* diverged from the Western Pacific clade in India and the Pacific. The origin of *Dransfieldia* and *Heterospathe* might be the Pacific and the Indian Ocean and the Pacific, respectively. *Heterospathe*, *Dransfieldia*, and the subtribe Laccospadicinae diverged like *Heterospathe*—(*Dransfieldia*—Laccospadicinae). All the results of the biogeography reported by Baker and Couvreur [33] were based on a supertree of the palm of Baker et al. [49]. This study could not resolve the relationship within the tribe Areceae as the relationship within the tribe was not well understood at that time.

5. Conclusions

The present study reports a revision of *Bentinckia* and assigns a neotype for *B. condapanna*. In addition, it documents a new cytotype for *B. condapanna* with $2n = 30$ chromosomes, which is the first different chromosome number reported from the genus *Bentinckia* and the tribe Areceae. The genus *Bentinckia* needs to be conserved because both species fall in a threatened category. As per the IUCN, *B. condapanna* is considered in the vulnerable (A1c, B2a ver 3.1) [5] category, and *B. nicobarica* is listed in the endangered (C2a ver 2.3) [39] category. This study provides the first well-resolved phylogeny of Areceae built using the appropriate combination of chloroplast and nuclear regions that supports all previous systematic revisions from unplaced Areceae. Moreover, molecular phylogeny and biogeographic analysis give the phylogenetic position and evolutionary history of all unplaced Areceae concerning closely related subtribes. However, to assign respective subtribes to unplaced genera, a strong morphological support is necessary. In addition, for precise evolutionary history, more sampling from Areceae is necessary. Nevertheless, this study confirms the phylogenetic placement and evolutionary history of *B. condapanna* and *B. nicobarica*. Phylogenetic analysis revealed that both species of *Bentinckia* form a clade with the other unplaced members, *Clinostigma* and *Cyrostachys*, and together they show a sister relationship with the subtribe Arecinae. Biogeography analysis shows *Bentinckia* might have originated in Eurasia and India.

Supplementary Materials: The following supporting information can be downloaded at: https://www.mdpi.com/article/10.3390/biology12020233/s1, Supplementary Material S1: Aligned combined sequence data (PRK + RPB2 + *acc*D + *rpo*C1 + *rbc*L + *rps*16 + *trn*L-F + *mat*K + *ndh*F) matrix. Supplementary Table S1: Voucher information and GenBank numbers (PRK, RPB2, *rbc*L, *rps*16, *mat*K, *ndh*F, *rpo*C1 *acc*D, and *trn*L-F) for all accessions used in this study. The sequences generated in this study are marked with * and — represent missing sequences.

Author Contributions: Conceptualization and methodology, S.K.K., R.N.M. and J.H.P.; fieldwork, collection of specimens, cytogenetical, and morphological analyses, R.N.M., S.K.G. and P.V.D.; molecular work and data analyses, S.K.K. and A.S.T.; writing—original draft preparation, S.K.K., R.N.M. and S.K.G.; writing—review and editing, Y.-S.C., M.M.L. and J.H.P.; supervision, M.M.L. and J.H.P.; funding acquisition, J.H.P. All authors have read and agreed to the published version of the manuscript.

Funding: This research was supported by the Basic Science Research Program through the National Research Foundation of Korea (NRF) funded by the Ministry of Education (NRF-2016R1A6A1A05011910).

Institutional Review Board Statement: Not applicable.

Data Availability Statement: Data supporting the findings of this study are available within the article and its Supplementary Materials.

Acknowledgments: R.N.M., P.V.D. and M.M.L. thank the Head, Department of Botany, Shivaji University, Kolhapur for providing necessary facilities.

Conflicts of Interest: The authors declare no conflict of interest.

References

1. Dransfield, J.; Uhl, N.W.; Asmussen, C.B.; Baker, W.J.; Harley, M.M.; Lewis, C.E. *Genera palmarum: The Evolution and Classification of Palms*; Royal Botanic Gardens, Kew: London, UK, 2008.
2. Tomlinson, P.B.; Horn, J.W.; Fisher, J.B. Subfamily Arecoideae. In *The Anatomy of Palms: Arecaceae–Palmae*; Tregear, J., Ed.; Oxford Academic: London, UK, 2012; pp. 185–240.
3. Renuka, C.; Sreekumar, V.A. *Field Guide to the Palms of India*; Kerala Forest Research Institute: Peechi, India, 2012.
4. Irwin, S.J.; Narasimhan, D. Endemic Genera of Angiosperms in India: A review. *Rheedea* **2011**, *21*, 87–105.
5. Basu, A.; Sarkar, M.K.; Muthukrishnan, A.; Arockiasamy, D.I. Recovery Plan for *Bentinckia Condapanna* Berry—An Endemic and Threatened Tree Species (palm) of Tropical Semi-Evergreen forests in Southern Western Ghats, India. *Indian For.* **2014**, *140*, 439–447.
6. Ved, D.K. *List of Medicinal Plants in Tamil Nadu*; Foundation for Revitalisation of Local Health Traditions (FRLHT): Bangalore, India, 2000.
7. Kulkarni, A.R.; Mulani, R.M. Indigenous palms of India. *Curr. Sci.* **2004**, *86*, 1598–1603.
8. Mathew, S.P.; Abraham, S. The vanishing palms of the Andaman and Nicobar Islands, India. *Principes* **1994**, *38*, 100–104.
9. Roxburgh, W. *Flora Indica*; Carey, W., Ed.; Cambridge University Press: Calcutta, India, 1832; Volumes 1–3.
10. Trimen, H.; Baker, J.G. Description of new plants from Nicobar Islands. *J. Bot. Brit. Foreign* **1875**, *13*, 331–332.
11. Beccari, O. Reliquiae Schefferianae illustrazione di alcune palme viventi nel Giardino Botanico di Buitenzorg. *Ann. Jard. Bot. Buitenzorg* **1885**, *2*, 77–171.
12. Renuka, C. *Palms of Kerala*; Kerala Forest Research Institute: Thrissur, India, 1999; p. 72.
13. Varghese, A.O.; Menon, A.R. Ecological niches and amplitudes of rare, threatened and endemic trees of Peppara Wildlife Sanctuary. *Curr. Sci.* **1997**, *76*, 1204–1208.
14. Basu, S.K. *Red Data Book of Indian Plants*; Nayar, M.P., Sastry, A.R.K., Eds.; B.S.I.: Calcutta, India, 1988; Volume 32, p. 268.
15. Read, R.W. More chromosomes counts by mail. *Principes* **1967**, *11*, 77.
16. Roser, M. Variation and evolution of karyotype characters in palm subfamily Coryphoideae s.l. *Bot. Acta* **1993**, *106*, 170–182. [CrossRef]
17. Roser, M. Pathways of karyological differentiation in Palms (*Arecaceae*). *Plant Syst. Evol.* **1994**, *189*, 83–122. [CrossRef]
18. Sharma, A.K. Annual report, 1967–1968. *Res. Bull. Univ. Calcutta. Cytogenet. Lab.* **1970**, *2*, 1–50.
19. Read, R.W. New chromosome counts in the Palmae. *Principes* **1966**, *10*, 55–61.
20. Kadam, S.K.; Tamboli, A.S.; Mane, R.N.; Yadav, S.R.; Choo, Y.-S.; Burgos-Hernández, M.; Pak, J.H. Revised molecular phylogeny, global biogeography, and diversification of palms subfamily Coryphoideae (*Arecaceae*) based on low copy nuclear and plastid regions. *J. Plant Res.* **2022**. [CrossRef] [PubMed]
21. Jain, S.K.; Rao, R.R. *Handbook of Field and Herbarium Methods*; Today & Tommorrow Printers and Publishers: New Delhi, India, 1977; pp. 50–80.
22. Mane, R.N.; Yadav, S.R. Karyomorphological Analysis of *Syzygium laetum* (Myrtaceae) from India. *Cytologia* **2022**, *87*, 227–230. [CrossRef]

23. Levan, A.; Fredge, K.; Sandberg, A. Nomenclature for centromeric positions on chromosomes. *Hereditas* **1964**, *52*, 201. [CrossRef]
24. Stebbin, G.L. *Chromosomal Evolution in Higher Plants*; Edward Arnold: London, UK, 1971.
25. Peruzzi, L.; Eroğlu, H.E. Karyotype asymmetry: Again, how to measure and what to measure? *Comp. Cytogenet.* **2013**, *7*, 1–9. [CrossRef]
26. Paterson, A.; Brubaker, C.; Wendel, J. A rapid method for extraction of cotton (*Gossypium* spp.) genomic DNA suitable for RFLP or PCR analysis. *Plant Mol. Biol. Rep.* **1993**, *11*, 122–127. [CrossRef]
27. Tamura, K.; Stecher, G.; Kumar, S. MEGA11: Molecular Evolutionary Genetics Analysis version 11. *Mol. Biol. Evol.* **2021**, *38*, 3022–3027. [CrossRef] [PubMed]
28. Edgar, R.C. MUSCLE: A multiple sequence alignment method with reduced time and space complexity. *BMC Bioinform.* **2004**, *5*, 113. [CrossRef] [PubMed]
29. Talavera, G.; Castresana, J. Improvement of phylogenies after removing divergent and ambiguously aligned blocks from protein sequence alignments. *Syst. Biol.* **2007**, *56*, 564–577. [CrossRef]
30. Darriba, D.; Taboada, G.L.; Doallo, R.; Posada, D.J. Model Test 2: More models, new heuristics and parallel computing. *Nat. Methods* **2012**, *9*, 772–775. [CrossRef]
31. Ronquist, F.; Huelsenbeck, J.P. MRBAYES 3: Bayesian phylogenetic inference under mixed 451 models. *Bioinformatics* **2012**, *19*, 1572–1574. [CrossRef] [PubMed]
32. Trifinopoulos, J.; Nguyen, L.T.; von Haeseler, A.; Minh, B.Q. W-IQ-TREE: A fast online phylogenetic tool for maximum likelihood analysis. *Nucl. Acids Res.* **2016**, *44*, W232–W235. [CrossRef] [PubMed]
33. Baker, W.J.; Couvreur, T.L.P. Global biogeography and diversification of palms sheds light on the evolution of tropical lineages. I. historical biogeography. *J. Biogeogr.* **2013**, *40*, 274–285. [CrossRef]
34. Baker, W.J.; Couvreur, T.L.P. Global biogeography and diversification of palms sheds light on the evolution of tropical lineages. II. Historical biogeography. *J. Biogeogr.* **2013**, *40*, 286–298. [CrossRef]
35. Yu, Y.; Harris, A.J.; Blair, C.; He, X. RASP (Reconstruct Ancestral State in Phylogenies): A tool for historical biogeography. *Mol. Phylogenet. Evol.* **2015**, *87*, 46–49. [CrossRef] [PubMed]
36. Johnson, D. *Bentinckia condapanna. IUCN Red List Threat. Species* **1998**, e.T38449A10119906. [CrossRef]
37. Mane, R.N.; Lekhak, M.M. Lectotypification of *Corypha taliera* Roxburgh and *Orania nicobarica* Kurz (*Arecaceae*). *Phytotaxa* **2021**, *489*, 236–239. [CrossRef]
38. Kurz, W.S. Descriptions of new plants from the Nicobar Islands (including a few from the Andaman Islands). *J. Bot.* **1875**, *13*, 321–326.
39. Johnson, D. *Bentinckia nicobarica. IUCN Red List Threat. Species* **1998**, e.T38450A10120169. [CrossRef]
40. Hahn, W.J. A phylogenetic analysis of the Arecoid Line of palms based on plastid DNA sequence data. *Mol. Phylogenet. Evol.* **2002**, *23*, 189–204. [CrossRef]
41. Lewis, C.E.; Doyle, J.J. A phylogenetic analysis of tribe Areceae (Arecaceae) using two low-copy nuclear genes. *Plant Syst. Evol.* **2002**, *236*, 1–17. [CrossRef]
42. Loo, A.H.B.; Dransfield, J.; Chase, M.W.; Baker, W.J. Low copy nuclear DNA, phylogeny and the evolution of dichogamy in the betel nut palms and their relatives (Arecinae, Arecaceae). *Mol. Phylogenet. Evol.* **2006**, *39*, 598–618. [CrossRef] [PubMed]
43. Norup, M.V.; Dransfield, J.; Chase, M.W.; Barfod, A.S.; Fernando, E.S.; Baker, W.J. Homoplasious character combinations and generic delimitations: A case study from the Indo-Pacific arecoid palms (Arecaceae: Areceae). *Am. J. Bot.* **2006**, *93*, 1065–1080. [CrossRef] [PubMed]
44. Baker, W.J.; Norup, M.V.; Clarkson, J.J.; Couvreur, T.L.P.; Dowe, J.L.; Lewis, C.E.; Pintaud, J.C.; Savolainen, V.; Wilmot, T.; Chase, M.W. Phylogenetic relationships among arecoid palms (Arecaceae: Arecoideae). *Ann. Bot.* **2011**, *108*, 1417–1432. [CrossRef] [PubMed]
45. Uhl, N.W.; Dransfield, J.; Davis, J.I.; Luckow, M.A.; Hansen, K.S.; Doyle, J.J. Phylogenetic relationships among palms: Cladistic analyses of morphological and chloroplast DNA restriction site variation. In *Monocotyledons: Systematics and Evolution*; Rudall, P.J., Cribb, P.J., Cutler, D.F., Humphries, C.J., Eds.; Royal Botanic Gardens, Kew: London, UK, 1995; pp. 623–661.
46. Asmussen, C.B.; Chase, M.W. Coding and noncoding plastid DNA in palm systematics. *Am. J. Bot.* **2001**, *88*, 1103–1117. [CrossRef]
47. Lewis, C.E.; Doyle, J.J. Phylogenetic utility of the nuclear gene malate synthase in the palm family (Arecaceeae). *Mol. Phylogenet. Evol.* **2001**, *19*, 409–420. [CrossRef] [PubMed]
48. Asmussen, C.B.; Dransfield, J.; Deickmann, V.; Barfod, A.S.; Pintaud, J.C.; Baker, W.J. A new subfamily classification of the palm family (Arecaceae): Evidence from plastid DNA phylogeny. *Bot. J. Linn. Soc.* **2006**, *151*, 15–38. [CrossRef]
49. Baker, W.J.; Savolainen, V.; Asmussen-Lange, C.B.; Chase, M.W.; Dransfield, J.; Forest, F.; Harley, M.M.; Uhl, N.W.; Wilkinson, M. Complete generic-level phylogenetic analyses of palms (Arecaceae) with comparisons of supertree and supermatrix approaches. *Syst. Biol.* **2009**, *58*, 240–256. [CrossRef]
50. Comer, J.R.; Zomlefer, W.B.; Barrett, C.F.; Davis, J.I.; Stevenson, D.W.; Heyduk, K.; Leebens-Mack, J.H. Resolving relationships within the palm subfamily Arecoideae (Arecaceae) using plastid sequences derived from next-generation sequencing. *Am. J. Bot.* **2015**, *102*, 888–899. [CrossRef]
51. Comer, J.R.; Zomlefer, W.B.; Barrett, C.F.; Stevenson, D.W.; Heyduk, K.; Leebens-Mack, J.H. Nuclear phylogenomics of the palm subfamily Arecoideae (Arecaceae). *Mol. Phylogenet. Evol.* **2016**, *97*, 32–42. [CrossRef] [PubMed]

52. Faurby, S.; Eiserhardt, W.L.; Baker, W.J.; Svenning, J. An all-evidence species-level supertree for the palms (Arecaceae). *Mol. Phylogenet. Evol.* **2016**, *100*, 57–69. [CrossRef] [PubMed]
53. Petoe, P.; Heatubun, C.D.; Baker, W.J. A monograph of *Hydriastele* (Areceae, Arecaceae) in New Guinea and Australia. *Phytotaxa* **2018**, *370*, 1–92. [CrossRef]
54. Plants of the World. Kew Science. 2022. Available online: www.plantssoftheworldonline.org (accessed on 12 December 2022).
55. Baker, W.J.; Loo, A.H.B. A synopsis of genus *Hydriastele*. *Kew Bull.* **2004**, *59*, 61–68. [CrossRef]
56. Baker, W.J.; Zona, S.; Heatubun, C.D.; Lewis, C.E.; Maturbongs, R.A.; Norup, M.V. *Dransfieldia* (Arecaceae)—A New Palm Genus from Western New Guinea. *Syst. Bot.* **2006**, *31*, 61–69. [CrossRef]
57. Norup, M.V. *Alsmithia Subsumed in Heterospathe (Arecaceae, Arecoideae)*; Missouri Botanical Garden Press: St. Louis, MI, USA, 2005; Volume 15, pp. 455–457.
58. Ali, S.S.; Yu, Y.; Pfosser, M.; Wetschnig, W. Inferences of biogeographical histories within subfamily Hyacinthoideae using S-DIVAand Bayesian binary MCMC analysis implemented in RASP (Reconstruct Ancestral State in Phylogenies). *Ann. Bot.* **2012**, *109*, 95–107. [CrossRef] [PubMed]

biology

MDPI

Article

Integrative Taxonomy of *Armeria arenaria* (Plumbaginaceae), with a Special Focus on the Putative Subspecies Endemic to the Apennines

Manuel Tiburtini [1,*], Giovanni Astuti [2], Fabrizio Bartolucci [3], Gabriele Casazza [4], Lucia Varaldo [4], Daniele De Luca [5], Maria Vittoria Bottigliero [5], Gianluigi Bacchetta [6], Marco Porceddu [6], Giannantonio Domina [7], Simone Orsenigo [8] and Lorenzo Peruzzi [1]

1 Department of Biology, University of Pisa, 56126 Pisa, Italy; lorenzo.peruzzi@unipi.it
2 Botanic Garden and Museum, University of Pisa, 56126 Pisa, Italy; giovanni.astuti@unipi.it
3 Floristic Research Center of the Apennine, University of Camerino—Gran Sasso Laga National Park, San Colombo, 67021 Barisciano, Italy; fabrizio.bartolucci@unicam.it
4 Department for the Earth, Environment and Life Sciences (DISTAV), University of Genoa, 16132 Genoa, Italy; gabriele.casazza@unige.it (G.C.); lucia.varaldo@edu.unige.it (L.V.)
5 Department of Biology, University of Naples Federico II, 80100 Naples, Italy; daniele.deluca@unina.it (D.D.L.); ma.bottigliero@studenti.unina.it (M.V.B.)
6 Centre for Conservation of Biodiversity (CCB), Department of Life and Environmental Sciences, University of Cagliari, 09123 Cagliari, Italy; bacchet@unica.it (G.B.); porceddu.marco@unica.it (M.P.)
7 Department of Agricultural, Food and Forest Sciences, University of Palermo, 90128 Palermo, Italy; giannantonio.domina@unipa.it
8 Department of Earth and Environmental Science, University of Pavia, Via S. Epifanio 14, 27100 Pavia, Italy; simone.orsenigo@unipv.it
* Correspondence: manuel.tiburtini@phd.unipi.it

Citation: Tiburtini, M.; Astuti, G.; Bartolucci, F.; Casazza, G.; Varaldo, L.; De Luca, D.; Bottigliero, M.V.; Bacchetta, G.; Porceddu, M.; Domina, G.; et al. Integrative Taxonomy of *Armeria arenaria* (Plumbaginaceae), with a Special Focus on the Putative Subspecies Endemic to the Apennines. *Biology* **2022**, *11*, 1060. https://doi.org/10.3390/biology11071060

Academic Editor: Frank H. Hellwig

Received: 3 June 2022
Accepted: 11 July 2022
Published: 14 July 2022

Publisher's Note: MDPI stays neutral with regard to jurisdictional claims in published maps and institutional affiliations.

Simple Summary: *Armeria arenaria* is a highly variable Western European species, for which three subspecies are recorded in Italy. *Armeria arenaria* subsp. *arenaria* has been reported from Northern Italy, while *A. arenaria* subsp. *marginata* and *A. arenaria* subsp. *apennina* are considered endemic to the Apennines. The taxonomic value of these two latter taxa is unclear and the actual occurrence of *A. arenaria* s.str. in Italy has never been addressed. Following an integrated taxonomic approach, in this study we show that all the Italian records of *A. arenaria* s.str. should be actually referred to *A. arenaria* subsp. *praecox* and that only one Northern Apennine endemic taxon can be recognized, namely, *A. arenaria* subsp. *marginata*.

Abstract: Three subspecies of *Armeria arenaria* are reported from Italy, two of which are considered endemic to the Apennines. The taxonomic value of these two taxa (*A. arenaria* subsp. *marginata* and *A. arenaria* subsp. *apennina*) is unclear. Moreover, the relationships between *A. arenaria* subsp. *praecox* and Northern Italian populations—currently ascribed to *A. arenaria* subsp. *arenaria*—have never been addressed. Accordingly, we used an integrated taxonomic approach, including morphometry, seed morpho–colorimetry, karyology, molecular systematics (*psbA–trnH*, *trnQ–rps16*, *trnF–trnL*, *trnL–rpl32*, and *ITS* region), and comparative niche analysis. According to our results, French–Northern Italian populations are clearly distinct from Apennine populations. In the first group, there is evidence which allows the recognition of *A. arenaria* s.str. (not occurring in Italy) and *A. arenaria* subsp. *praecox*. In the second group, the two putative taxa endemic to the Northern Apennines cannot be separated, so a single subspecies is here recognized: *A. arenaria* subsp. *marginata*.

Keywords: endemism; morphometrics; image analysis; molecular analysis; niche similarity; nomenclature

1. Introduction

Most of our biological knowledge of plant diversity comes from the foundations laid by alpha taxonomy, which played a crucial role in discovering and documenting plant diversity around the world. Nevertheless, although science is progressing, taxonomists seem to struggle to keep pace with novel methods and approaches. Indeed, hundreds of putative new species are described annually [1], but most of them are still described on qualitative grounds. In such approaches, the information that a taxonomist collects to shape his/her idea about the "species" in question is often obscure [2], so that biases [3] in taxonomist decisions [4] can dramatically affect taxonomic treatment. For instance, the number of species in the genus *Armeria* varies dramatically under different taxonomic circumscriptions elaborated by different taxonomists [5–7]. The subjectiveness of these processes may have contributed to what has been called taxonomic anarchy [8]. Integrated taxonomic approaches aim to address this problem with the consilience principle [2], according to which multiple and complementary approaches (morphology, phylogenetics, cytology, etc.) [9] are used to try falsifying taxonomic hypotheses in a Popperian sense [10]. This represents a step towards an omega taxonomy [11,12] that needs the integration of different skills [13,14].

The genus *Armeria* Willd. (Plumbaginaceae, Limonioideae) includes up to 95 accepted, mostly Holarctic, perennial species [15]. In Italy, the current knowledge on the taxonomy and systematics of this genus is largely derived from traditional alpha–taxonomic revisions [16,17], which indicate the existence of 23 taxa in 18 species [18]. However, the taxonomic value of some of these taxa is still debated [18], and the picture is further complicated by the fact that species boundaries within *Armeria* are difficult to establish [6,19] and weak [20–22]. In this scenario, it has been demonstrated that homoploid hybrid speciation [21] can play a crucial role in the emergence of new species [23,24], given that all species tested so far are diploid, with $2n = 2x = 18$ chromosomes [16,25,26]. The use of nrDNA and (maternally inherited) cpDNA markers helps to elucidate the phylogenetic relationships even under hybridization scenarios [27,28]. *Armeria arenaria* (Pers.) F. Dietr. complex currently includes 13 subspecies in its whole range [29] and, according to Arrigoni [17], three subspecies occur in Italy: *A. arenaria* subsp. *arenaria*, distributed across the Central–Western Alps [18]; *A. arenaria* subsp. *apennina* Arrigoni, endemic to the Tuscan–Aemilian Apennines; and *A. arenaria* subsp. *marginata* (Levier) Arrigoni, also endemic to the Northern and up to the Central Apennines. *Armeria arenaria* subsp. *praecox* (Jord.) Kerguélen ex Greuter, Burdet & G. Long, described from south–eastern France, is reported as doubtfully occurring in Italy. Arrigoni [17] considers *A. arenaria* subsp. *apennina* as intermediate between *A. arenaria* s.str. and *A. arenaria* subsp. *marginata*. The same author [17] also claims that there is a series of unclear intermediate forms distinguished by the transition of some putatively diagnostic character states. However, the circumscription of these subspecies is based only on a qualitative morphological approach. All these factors led to the consideration of *A. arenaria* subsp. *apennina* and *A. arenaria* subsp. *marginata* as two subspecies of uncertain taxonomic value [18].

For these reasons, there is need to use an integrated approach to address the taxonomy [9] of these putative subspecies. To achieve a sound taxonomic circumscription, we performed morphometric analyses, including living populations from type localities, complemented by seed morpho–colorimetry, karyotype asymmetry estimation, molecular systematics, and comparative niche analysis (for similar integrative approaches, see [27,28]). In this study we aim: (1) to test the current taxonomic circumscription; (2) to verify the occurrence in Italy of *A. arenaria* subsp. *praecox*; and (3) to clarify the nomenclature of the group.

2. Materials and Methods

2.1. Sampling

In total, we selected 12 populations (Table 1) across the Northern Apennines and Western Alps up to Central France. The populations studied were selected based on three criteria: (1) to include all the type localities of the four taxa putatively occurring in Italy (FB, LA, LL, and MB—acronyms as in Table 1); (2) to include other populations explicitly cited in [19]: AA, BO, BR, MC, MP, and TV; and (3) to also include a lowland (GA) and the easternmost (PS) populations in Italy.

Table 1. Taxa and populations of *Armeria arenaria* sampled in this study, according to the current taxonomic hypothesis [17]. "Code" corresponds to population acronyms used elsewhere in the manuscript. * = type locality. "Voucher" refers to the specimens stored at Herbarium Horti Botanici Pisani (PI) and freely available for consultation at http://erbario.unipi.it/, accessed on 10 July 2022. See also Figure 6 for the geographical localisation of the sampled populations.

Code	Current Taxonomic Hypothesis	Population	Voucher	*n*
MB *	*A. arenaria* subsp. *apennina*	Italy, Emilia–Romagna, between M. Marmagna & M. Braiola (WGS84: 44.401250 N, 9.994250 E)	*M. Tiburtini* et *S. Quitarrá*; 27 June 2020; PI53990–PI54009	20
MC	*A. arenaria* subsp. *apennina*	Italy, Emilia–Romagna, Monte Cusna (WGS84: 44.288194 N, 0.390055 E)	*M. Tiburtini* et *S. Quitarrá*; 26 June 2020; PI54653–PI54672	20
MP	*A. arenaria* subsp. *apennina*	Italy, Emilia–Romagna, Monte Prinzera (WGS84: 44.640833 N, 10.083772 E)	*M. Tiburtini* et *G. Astuti*; 12 June 2020; PI53010–PI54029	20
BO	*A. arenaria* subsp. *arenaria*	Italy, Emilia–Romagna, Monte Tre Abati, loc. Bobbio (WGS84: 44.752425 N, 9.436694 E)	*S. Orsenigo*; 3 June 2020; PI54050–PI54063	14
BR	*A. arenaria* subsp. *arenaria*	Italy, Val d'Aosta, Brusson (WGS84: 45.738166 N, 7.754111 E)	*S. Orsenigo*; 16 June 2020; PI53970–PI53989	20
FB *	*A. arenaria* subsp. *arenaria*	France, Île-de-France, Fontainebleau (WGS84: 48.437194 N, 2.690166 E)	*M. Tiburtini* et *F. Losacco*; 18 June 2020; PI56593–PI56615	23
GA	*A. arenaria* subsp. *arenaria*	Italy, Lombardia, Gambolò (WGS84: 45.268611 N, 8.961027 E)	*S. Orsenigo*; 26 May 2020; PI54032–PI54049	18
PS	*A. arenaria* subsp. *arenaria*	Italy, Lombardia, Piana di Salmezza (WGS84: 45.782583 N, 9.732277 E)	*S. Orsenigo*; 12 June 2020; PI53950–PI53969	20
TV	*A. arenaria* subsp. *arenaria*	Italy, Piemonte, Terme di Valdieri (WGS84: 44.204083 N, 7.265000 E)	*S. Orsenigo*; 26 June 2020; PI55569–PI56573	20
AA	*A. arenaria* subsp. *marginata*	Italy, Toscana, Alpi Apuane (WGS84: 44.124166 N, 10.212000 E)	*M. Tiburtini* et *S. Quitarrá*; 28 June 2020; PI54673–PI54686	14
LA *	*A. arenaria* subsp. *marginata*	Italy, Emilia–Romagna, Libro Aperto (WGS84: 44.157402 N, 10.712021 E)	*M. Tiburtini* et *S. Quitarrá*; 25 June 2020; PI56573–PI56592	20
LL *	*A. arenaria* subsp. *praecox*	France, Hautes Alpes, Le Lauzet (WGS84: 44.980166 N, 6.499611 E)	*M. Tiburtini* et *F. Losacco*; 16 June 2020; PI55549–PI55568	20

For each population, about 20 flowering individuals were sampled. The number of flowering scapes was counted in the field, whereas pictures were taken to assess the colour of the flowers and involucres of each plant. In total, 229 specimens were collected, and herbarium vouchers were prepared. All vouchers are stored at Herbarium Horti Botanici Pisani (PI), and high-resolution images are freely available for consultation at http://erbario.unipi.it/, accessed on 10 July 2022 (codes in Table 1). Concerning molecular systematics, dried leaves were picked from a subset of three individuals for each population and put in a paper bag with silica gel. Ripe fruits were also collected from the same populations. Seeds were dried at room temperature for two months and cleaned in the Germplasm Bank, Department of Biology of the University of Pisa, using sieves and Agriculex CB-1 Column Seed Cleaner complemented by manual cleaning.

2.2. Morphometric Analysis

In total, 49 qualitative and quantitative morphological characters (Table 2, see also Supplementary Materials for details concerning the calyx) were studied, with a resulting dataset of 223 individuals × 49 variables. Macroscopic measures were taken with a digital calliper (error ± 0.1 mm), whilst microscopic and calyx measurements [30] (Table 2 and Figure S1) were taken through bar-scaled pictures with a Fiji 2.1.0 [31]. To provide a more objective means of counting the number of leaf veins, free-hand transversal sections of leaves were prepared. We considered as a "vein" each fascicule composed of the xylem and phloem surrounded by sclerenchyma. The anatomy of summer leaves was surveyed under a Leitz Diaplan light microscope at 40× (Figure S2). We considered as "involucral bracts" those from the capitulum involucre, as "spikelet bracts" those subtending each spikelet, and as "bracteoles" those under each flower. To take into account the internal variability of the capitulum, we measured a spikelet collected from the middle of the capitulum ("inner spikelet") and a spikelet in contact with the inner involucral bract ("outer spikelet").

Table 2. Morphological characters studied in *Armeria arenaria*. QC = quantitative continuous, QD = quantitative discrete, CN = nominal, BI = binary, CO = ordered factor.

Character Name	Description of the Character	Type	Tool
SCAB	Calyx awns scabrous (yes/no)	BI	Stereo
AWN	Awn presence on the calyx's limb (yes/no)	BI	Stereo
CALYX_HAIRINESS	Calyx hairiness (holotrichous/pleurotrichous)	BI	Stereo
CALYX_VEINS	Number of calyx veins with hairs (10/5)	BI	Stereo
CAP_SHAPE	Shape of capitulum (hemispherical/subspherical)	BI	Stereo
DIMORP	Leaf dimorphism (yes/no)	BI	Stereo
INNER_SPI_BRACT_HAIR	Presence of hairs on inner spikelet bract (yes/no)	BI	Stereo
MAR_SUM_LEAF	Margin of the summer leaf (hyaline/dentate)	BI	Stereo
MAR_WIN_LEAF	Margin of the winter leaf (hyaline/dentate)	BI	Stereo
OUTER_SPI_BRACT_HAIR	Presence of hairs on the outer spikelet bract (yes/no)	BI	Stereo
INNER_SPI_BRACT_APEX	Shape of the inner spikelet bract apex (crenulate/undulate)	BI	Stereo
OUT_SPI_BRACT_APEX	Shape of the outer spikelet bract apex (crenulate/undulate)	BI	Stereo
SUM_LEAF_APEX	Shape of the summer leaf apex (acute/cucullate)	BI	Stereo
VEINS_HAIRS	Presence of hairs along the leaf veins (yes/no)	BI	Stereo
WIN_LEAF_APEX	Shape of the winter leaf apex (acute/cucullate)	BI	Stereo
COL_INV	Involucre colour (green/variegated/reddish/dry)	CN	Stereo
COL_PET	Petal colour (white/pink/fuchsia)	CN	Stereo
GEN_SHAPE_OUT_INV_BRACT	Shape of the outer involucral bract (deltate < triangular < strictly triangular)	CO	Stereo
SHAPE_APEX_INN_INV_BRACT	Shape of the inner involucral bract apex (round < acute < mucronate < apiculate)	CO	Stereo
SHAPE_APEX_INT_INV_BRACT	Shape of the intermediate involucral bract apex (round < acute < mucronate < apiculate)	CO	Stereo
SHAPE_APEX_OUT_INV_BRACT	Shape of the outer involucral bract apex (acute < mucronate < apiculate < acuminated < subulate < long subulate)	CO	Stereo
ANG_SUM_TIP	Summer leaf tip angle (°)	QC	Fiji
ANG_WIN_TIP	Winter leaf tip angle (°)	QC	Fiji
AWN_LENG	Awn length (mm)	QC	Fiji
DIAM_CAP	Capitulum diameter (mm)	QC	Calliper
HEIGTH	Plant height (mm)	QC	Ruler
LENG_CAL _PED	Calyx pedicel length (mm)	QC	Fiji
LENG_CAL_TUBE	Calyx tube length (mm)	QC	Fiji
LENG_INNER_INV_BRACT	Inner involucral bract length (mm)	QC	Calliper
LENG_INNER_SPI_BRACLE	Inner spikelet bracteole length (mm)	QC	Calliper
LENG_INNER_SPI_BRACT	Inner spikelet bract length (mm)	QC	Calliper
LENG_INTER_INV_BRACT	Intermediate involucral bract length (mm)	QC	Calliper
LENG_OUT_INV_BRACT	Outer involucral bract length (mm)	QC	Calliper
LENG_OUTER_SPI_BRACLE	Outer spikelet bracteole length (mm)	QC	Calliper
LENG_OUTER_SPI_BRACT	Outer spikelet bract length (mm)	QC	Calliper
LENG_SUM_LEAF	Summer leaf length (mm)	QC	Ruler
LENG_WIN_LEAF	Winter leaf length (mm)	QC	Ruler
LIMB_LENG	Limb length (mm)	QC	Fiji
SCA_DIAM	Scape diameter at 1 cm from the base (mm)	QC	Calliper
SCA_LENG	Scape length (mm)	QC	Ruler
SHEATH_LENG	Sheath length (mm)	QC	Calliper
WIDTH_CAL_TUBE	Calyx tube width below the limb (mm)	QC	Fiji
WIDTH_IAL_SUM	Width of the hyaline margin in summer leaf (mm)	QC	Fiji
WIDTH_IAL_WIN	Width of the hyaline margin in winter leaf (mm)	QC	Fiji

Table 2. *Cont.*

Character Name	Description of the Character	Type	Tool
WIDTH_INNER_INV_BRACT	Inner involucral bract width at the middle (mm)	QC	Calliper
WIDTH_INNER_SPI_ BRACT	Inner spikelet bract width at the middle (mm)	QC	Calliper
WIDTH_INNER_SPI_BRACLE	Inner spikelet bracteole width at the middle (mm)	QC	Calliper
WIDTH_INTER_INV_BRACT	Intermediate involucral bract width at the middle (mm)	QC	Calliper
WIDTH_OUT_INV_BRACT	Outer involucral bract width at the base (mm)	QC	Calliper
WIDTH_OUTER_SPI_ BRACT	Outer spikelet bract length at the middle (mm)	QC	Calliper
WIDTH_OUTER_SPI_BRATLE	Outer spikelet bracteole width at the middle (mm)	QC	Calliper
WIDTH_SUM_LEAF	Width of the summer leaf at the middle (mm)	QC	Calliper
WIDTH_WIN_LEAF	Width of the winter leaf at the middle (mm)	QC	Calliper
N_ WIN_VEINS	Number of veins (with sclerenchyma) of winter leaf	QD	Microscope
N_INV_BRACT	Number of involucral bracts	QD	—
N_SUM_VEINS	Number of veins (with sclerenchyma) of summer leaf	QD	Microscope
SCAP_NUM	Number of scapes	QD	—

All statistical analyses were conducted in R Studio (version 3.6.2) [32]. To test the suitability of the data for factor analysis, the Kaiser–Meyer–Olkin test (MSA = 0.86, *psych* package [33] and Bartlett sphericity test ($p < 0.001$, *REdaS* package [34]) were performed successfully on the correlation matrix. Since there were mixed variables, Gower distance in the *FD* package [35] with Podani correction [36] was used, whilst Cailliez correction [37] was applied due to the violation of the triangle inequality (i.e., the matrix was not Euclidean). On such a dissimilarity matrix, Principal Coordinate Analysis (PCoA) in the *ape* package [38] was used to explore the dataset. Graphs were plotted with the *ggplot2* package [39]. One–way ANOSIM in PAST (version 4.09) [40] was used to test the null hypothesis of no difference between groups in the Gower dissimilarity matrix. To test the current taxonomic hypothesis and other alternative groupings based on our results, we applied jackknifed Linear Discriminant Analysis (LDA) in the *MASS* (for plotting) and *Predpsych* (to obtain the confusion matrix [41]) packages. Qualitative variables were converted into numbers with integer encoding. Using the *PredPsych* package, Cohen's Kappa coefficient was estimated for each grouping hypothesis. K coefficient is a measure of how the classification results compare with the values assigned and is generally thought to be a more robust measure than simple percentage agreement calculation, since it considers the possibility of agreement occurring by chance [42]. It ranges from 0 to 1 and K values greater than 0.75 may be taken to represent excellent agreement beyond chance [43].

Each character was statistically tested. For all the quantitative characters, normality was tested with the Shapiro–Wilk test. Normal and non–normal data were checked for homoscedasticity with the Bartlett test and Levene–Brown–Forsythe test, respectively. After checking statistical assumptions, normal and log–normal quantitative characters were compared among groups with one–way ANOVA and the post hoc Tukey–Kramer test (homoscedastic data) or Welch's ANOVA and the post hoc Games–Howell test (heteroscedastic data). On the contrary, non–normal characters were tested with the Kruskal–Wallis and multiple comparisons test of Wilcoxon–Mann–Whitney (homoscedastic data) or a permutation test implemented using the pairwisePermutationTest function of the *rcompanion* package [44] (heteroscedastic data). To control the family–wise error rate of multiple comparisons, Holm's correction was applied to all the tests. Qualitative nominal and ordinal characters were tested with the R function pairwiseNominalIndependence and pairwiseOrdinalIndependence based on Fisher's exact test and implemented in the *rcompanion* package [44]. All the tests were considered significant with $\alpha < 0.01$. The number of statistically significant differences for a variable among population pairs was counted and a pairwise triangular matrix was built. Descriptive statistics for each group were calculated using the describeBy function in the *psych* package [33].

2.3. Seed Morpho–Colorimetric Analysis

For a sample of 100 seeds per accession (cleaned from the fruiting calyx and the membranous pericarp), digital images were acquired using a flatbed scanner (Epson Perfection V550) with a digital resolution of 1200 dpi. When an accession had fewer than 100 seeds, the analysis was carried out on the whole batch available. The system worked with 2D images: seeds were randomly disposed on the scanner tray, so that they did not touch one another, and covered using a box with white paper followed by a box with black paper to avoid interference from environmental light. The images were processed using the software package ImageJ (version 1.52b) (Available online: http://rsb.info.nih.gov/ij. (accessed on 11 March 2022), and the descriptors of seed–size, shape, and colour features were measured and analysed. A plugin, Particles8 [45] (Available online: https://blog.bham.ac.uk/intellimic/g-landini-software/. (accessed on 11 March 2022), was used to measure 20 colorimetric and 26 morphometric features. This plugin was further enhanced by adding algorithms that can compute the Elliptic Fourier Descriptors (EFDs) for each analysed seed, thus increasing the number of independent variables. Following Terral et al. [46] and Sarigu et al. [47], to minimize measurement errors and optimize the efficiency of shape reconstruction, 20 harmonics were used to define the seed boundaries, obtaining 78 additional variables that were useful to discriminate between the studied seeds. In total, 124 morphometric and colorimetric characters were measured for each seed [48]. Statistical analyses were performed with the software SPSS release 16 (SPSS 16.0 for Windows; SPSS Inc., Chicago, IL, USA) by applying stepwise Linear Discriminant Analysis (LDA).

2.4. Karyological Analysis

Seeds were germinated in Petri dishes with 1% agar at 25 °C in an alternating 12/12 h dark/light photoperiod. After about 4 days, radicles emerged, and seedlings were removed from the seed incubator and kept at 4 °C for 24 h in a fridge, then we followed the Feulgen staining protocol. Root tips were pre–treated with 0.4% colchicine for 3 h and then fixed in Carnoy fixative solution for 1 h. After hydrolysis in HCl 1 N at 60 °C for 8 min, the root tips were stained in leuco–basic fuchsine for 2 h; root tips were squashed in a solution of aceto–orcein on a microscope slide.

Chromosomes were observed with a Leitz Diaplan microscope at 100× and pictures were taken with a Leica MC–170HD camera using Leica LAS–EZ 3.0 imaging software. At least four good metaphase plates were measured for each population. Lastly, chromosome numbers and karyological variables, such as THL (Total Haploid Length), M_{CA} (Mean Centromeric Asymmetry), CV_{CL} (Coefficient of Variation of Chromosome Length), and CV_{CI} (Coefficient of Variation of Centromeric Index) were obtained from each plate with MATO 1.1 (version 20210101) [49]. Since all the karyological variables were normal and homoscedastic, they were statistically tested with One–way ANOVA and the post hoc Tukey–Kramer test for more than 3 groups or with two sample *t*–tests when comparing 2 groups.

2.5. DNA Extraction and Molecular Systematics

Total DNA was extracted using the GeneAll® Exgene™ Plant SV mini kit (GeneAll Biotechnology, Seoul, Korea), following the manufacturer's protocol for dried material. About 25 mg of leaf tissue was ground to powder using the Mixer Mill 300 (Retsch®, Verder Scientific, Haan, Germany). The quality and quantity of extracted DNA was evaluated by 0.8% gel electrophoresis using the high–molecular weight marker HyperLadder™ 1 Kb (Bioline, Meridian Bioscience, Cincinnati, OH, USA). The internal transcribed spacers *ITS1* and *ITS2(+5.8S)* and four chloroplast intergenic spacers (*trnF–trnL, trnH–psbA, trnL–rpl32, trnQ–rps16*) were amplified in a final volume of 25 μL containing: 10 ng DNA, 2X Kodaq PCR MasterMix (ABM®, Richmond, BC, Canada), 400 nM forward and reverse primers, and water to volume. The list of primers [50–52] and PCR conditions is reported in Tables S7 and S8. Amplification products were visualized by 1.5% gel electrophoresis and purified using 15–20% polyethylene glycol (PEG), according to the size of the fragment. The purified amplicons were sequenced at one (chloroplast markers) or both ends (ITS region) using the BrightDye® Terminator Cycle Sequencing Kit (MCLAB, Harbor Way, San Francisco, CA, USA). Capillary electrophoresis was carried out using the Applied Biosystems® 3130 Genetic Analyzer (Applied Biosystems, Thermo Fisher Scientific, Foster City, CA, USA). *ITS* sequences were submitted to GenBank (accession numbers: ON512680–ON512715), while the chloroplast intergenic spacers were submitted to DDBJ (*trnF–trnL*: LC710463–LC710498; *psbA–trnH*: LC710671–LC710706; *trnL–rpl32*: LC710707–LC710742; *trnQ–rps16*: LC710743–LC710778).

Sequences were visually inspected and aligned using the ClustalW algorithm [53] implemented in BioEdit (version 7.2.5) [54] with the default values. An incongruence length difference (ILD) test was carried out in Nona (version 2.0) [55], as a daughter process of Winclada (version 1.00.08) [56], to test the putative incongruence of nuclear and chloroplast partitions prior to combination; default values were used for the analysis. A nucleotide evolution model was calculated for each of the five sequenced regions using jModelTest (version 2.1.10) [57], and the best fitting model was chosen over the others using the Bayesian Information Criterion (BIC) [58]. A Bayesian phylogenetic tree was inferred in MrBayes (version 3.2.6) [59] in two simultaneous, independent runs with the following settings: 2,000,000 generations of MCMC sampling every 2000 generations, and four runs (three cold and one hot). Convergence and mixing were evaluated in Tracer (version 1.7.2) [60]. The consensus Bayesian tree was visualized in FigTree (version 1.4.2) [61]. The best evolution models were K80 [62] for *ITS* and F81 [63] for chloroplast markers.

2.6. Comparative Niche Analysis

Occurrence data for the studied taxa were retrieved directly in the field, from SILENE (French National Mediterranean Botanical Conservatory of Porquerolles) (Available online: http://flore.silene.eu/index.php?cont=accueil. accessed on 17 May 2022), GBIF (Available online: https://www.gbif.org. accessed on 17 May 2022), and Wikiplantbase #Italia [64], with a total of 496 points. To test for differentiation in environmental space, we represented and quantified niche overlap using the PCA-based method developed by Broennimann et al. [65]. The Schoener's D index, which ranges from 0 (no overlap) to 1 (full overlap), was used to measure niche overlap [66]. We used niche similarity tests [67] to assess whether the ecological niches of the taxa were more similar than expected at random from their geographical ranges. Niche similarity tests compare the environmental conditions occupied by taxa, taking into account the environmental conditions that are available in the geographic area occupied by each taxon. Briefly, the observed climatic niche overlap between two taxa was compared, with the overlap measured between the niche of one taxon and the randomized niche of the other taxon. This randomized niche was obtained by randomly sampling occurrence points in buffer areas of 10 km around occurrences (the 'background area').

2.7. Nomenclature and Distribution

Currently accepted names, basionyms, and homotypic synonyms within *Armeria arenaria* and its subspecies studied here were taken from the Med-Checklist [68]. Information about the herbaria in which the original material could be stored was derived from [69]. Accordingly, we digitally examined the following herbaria: B, FI, L, LY, M, MPU, P, and SLA (herbarium acronyms follow Thiers [70]). Once we had elaborated the new taxonomic scheme, we used our identification key to assess the geographical distribution of the recognised taxa by checking the herbarium materials stored at ANC, APP, B, CAME, FI, HLUC, MJG, MW, P, and RO. This material was then georeferenced and used to build the map in Figure 6.

3. Results

3.1. Morphometry

The first two axes of the PCoA explain 49% of the total variance. Along the first axis, there is a clear separation of four Apennine populations (AA, LA, MB, and MC, on the right side of Figure 1). Hereafter, we will refer to this group of four populations as "marginatoid". Another group uniting populations from northern Italy (BO, BR, GA, MP, PS, and TV) and France (FB and LL) emerged. We will refer to this group hereafter as "arenarioid" (on the left side of Figure 1). The MP population, initially attributed to *A. arenaria* subsp. *apennina*, clearly falls among arenarioid plants.

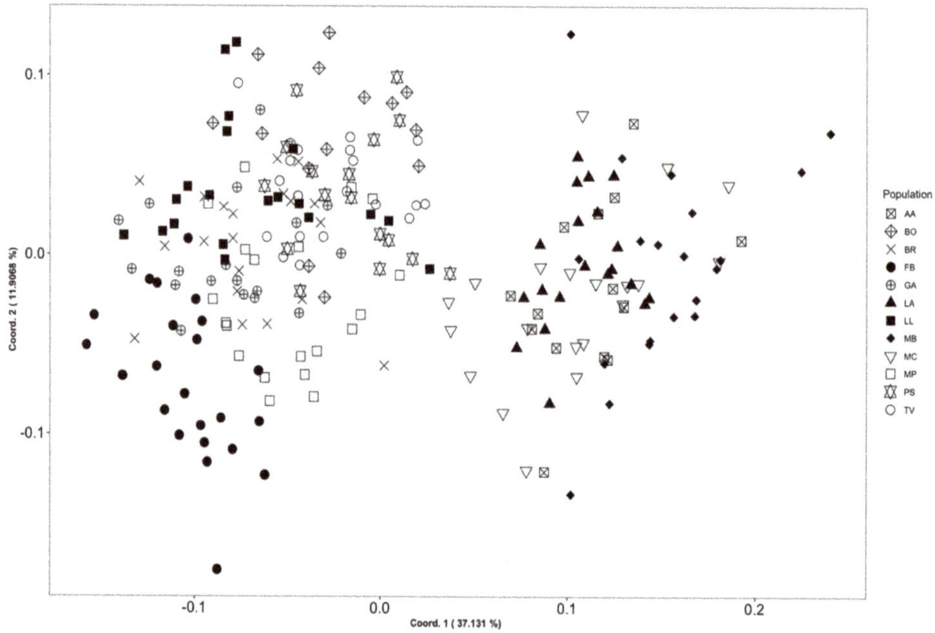

Figure 1. PCoA based on the 49 morphological characters measured in *Armeria arenaria* populations. Solid symbols represent individuals from type localities of the four taxa studied. AA = Apuan Alps, N Apennines; BO = Bobbio, N Apennines; BR = Brusson, Pennine Alps; FB = Fontainebleau, Île–de–France; GA = Gambolò, West Po Valley; LA = Libro Aperto; N Apennines; LL = Le Lauzet, Dauphiné Alps; MB = Marmagna–Braiola, N Apennines; MC = Monte Cusna, N Apennines; MP = Monte Prinzera, N Apennines; PS = Piana di Salmezza, Lombard Prealps; TV = Terme di Valdieri, Maritime Alps. Further population details are provided in Table 1 and Figure 6.

Along the second axis, the topotypical population of *A. arenaria* subsp. *arenaria* (FB) shows a slight separation from the other arenarioid populations (Figure 2). One–way ANOSIM showed that there was, indeed, a significant difference between FB and the rest of the arenarioid populations (BO, BR, GA, MP, PS, TV, and LL) ($R = 0.6573$, $p = 0.001$), confirming the separation shown along the second axis of PCoA. LDA performed on the current taxonomic hypothesis (Table 3) obtained an 87% correct classification and $K = 0.8$. The lowest value of sensibility was scored by *A. arenaria* subsp. *marginata* (77.7%), followed by *A. arenaria* subsp. *apennina* (85.4%). The percentage of correct classifications and K increased to 99% and 0.9696, respectively, when comparing arenarioid with marginatoid plants.

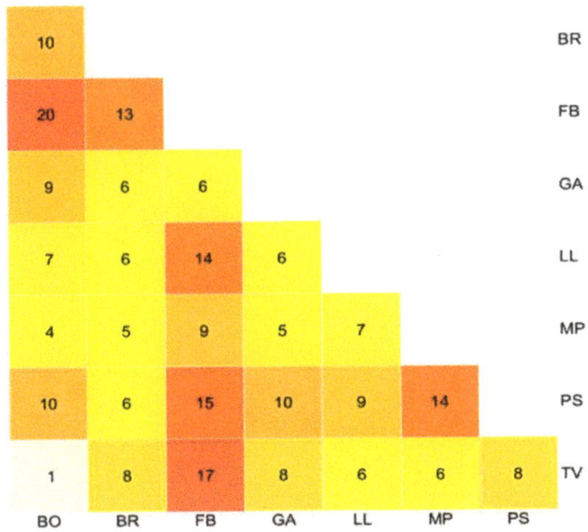

Figure 2. Heatmap of the pairwise comparisons of the 49 morphological characters for which we found statistically significant differences between population pairs in *Armeria arenaria*. Numbers inside the cells indicate the sum of statistically different characters. Colours are a function of the number of characters showing significant differences: whitish–yellow colours indicate that the pair is almost identical, whereas orange–red colours indicate that the pair shows several differences. AA = Apuan Alps, N Apennines; BO = Bobbio, N Apennines; BR = Brusson, Pennine Alps; FB = Fontainebleau, Île-de-France; GA = Gambolò, West Po Valley; LA = Libro Aperto; N Apennines; LL = Le Lauzet, Dauphiné Alps; MB = Marmagna–Braiola, N Apennines; MC = Monte Cusna, N Apennines; MP = Monte Prinzera, N Apennines; PS = Piana di Salmezza, Lombard Prealps; TV = Terme di Valdieri, Maritime Alps. Further population details are provided in Table 1 and Figure 6.

Table 3. Confusion matrix of the LDA based on the 49 morphometric characters, assuming the current taxonomic hypothesis of *Armeria arenaria* subspecies as a priori groups, as proposed by Arrigoni [17]. Rows show the membership of each a priori established group, whereas columns show the membership predicted by the classification model.

A. arenaria subsp.	*apennina*	*arenaria*	*marginata*	*praecox*	Total
apennina	46	5	8	0	59
arenaria	2	105	1	2	110
marginata	6	0	28	0	34
praecox	0	5	0	15	20
Total	54	115	37	17	223

To further investigate the morphological variation within arenarioid plants, we carried out a pairwise comparison using univariate statistical analyses on single characters. Figure 2 shows that the highest number of pairwise differences (94) was found between FB and all the other arenarioid populations. The population that shows the second most number of differences is PS (72).

Accordingly, we set up two new alternative grouping hypotheses in both of which marginatoid plants (AA, LA, MB, and MC) were combined in a single group. In the first grouping hypothesis (I), we tested FB, together with all the Northern Italian populations, as belonging to the same taxon (as in the current taxonomic hypothesis) against the single population LL, which is the topotypical population of *A. arenaria* subsp. *praecox*. In the second grouping hypothesis (II), we tested LL as belonging to the same taxon as all the other Northern Italian arenarioid populations against the single population of FB (which corresponds to *A. arenaria* s.str.). The performance of LDA was 96% (K = 0.925) under grouping hypothesis I and 98% (K = 0.968) under grouping hypothesis II. The two most important qualitative characters are provided in Tables S1–S3, whereas mean (± standard deviation) values of the quantitative morphological characters for each population are provided in Tables S4 and S5.

3.2. Seed Morpho–Colorimetry

The LDA performed on the current taxonomic hypothesis of a priori groups gave an overall cross–validated classification performance of 51.7% (Table 4). *Armeria arenaria* subsp. *marginata* showed the highest percentage of discrimination, with values of 71.5%, while the lowest (36.3 %) was detected in *A. arenaria* subsp. *apennina* (Table 4).

Table 4. Confusion matrix of the LDA based on the seed morpho–colorimetric dataset (percentages of correct classification), assuming the current taxonomic hypothesis of *Armeria arenaria* subspecies as a priori groups, as proposed by Arrigoni [17]. Rows show the membership of each a priori established group, whereas columns show the membership predicted by the classification model.

A. arenaria subsp.	*apennina*	*marginata*	*arenaria*	*praecox*	Total
apennina	36.3	30	15	18.7	100
marginata	18.5	71.5	7.5	2.5	100
arenaria	15.5	8.2	48.7	27.7	100
praecox	20	—	15	65	100

The second LDA, contrasting arenarioid and marginatoid plants, provides an overall percentage of 86% correct classification, with high discrimination performance for the two groups (Table 5).

Table 5. Confusion matrix of the LDA based on the seed morpho–colorimetric dataset (percentages of correct classification), assuming "arenarioid" and "marginatoid" groups for *Armeria arenaria* populations. Rows show the membership of each a priori established group, whereas columns show the membership predicted by the classification model.

Group	Arenarioid	Marginatoid	Total
Arenarioid	87.4	12.6	100
Marginatoid	16.3	83.3	100

According to two alternative grouping hypotheses derived from the morphometric analysis, FB was tested, together with all Northern Italian populations, as belonging to the same group, against the single population LL (hypothesis I), and LL was tested as belonging to the same group as all other Northern Italian arenarioid populations against the single population FB (hypothesis II). The discriminant analysis provided an overall percentage of classification of 77.3% and 84.4% for hypotheses I and II, respectively (Table 6, Figure 3). In hypothesis I, high discrimination performance was obtained for LL (78.8%) and marginatoid plants (81.3%). Concerning hypothesis II, higher performances, ranging from 91.0% (in FB) to 81.5% (in marginatoid plants), were detected.

Table 6. Confusion matrix of the LDA based on the seed morpho–colorimetric dataset (percentages of correct classification), according to the alternative grouping hypotheses I and II for *Armeria arenaria* populations. Rows show the membership of each a priori established group, whereas columns show the membership predicted by the classification model.

Groupings		Marginatoid	Arenarioid (LL Excluded)	LL	Total
Grouping hypothesis I	Marginatoid	81.3	12.3	6.5	100
	Arenarioid (LL excluded)	13.1	52.3	34.6	100
	LL	1	21	78.0	100

Table 6. *Cont.*

		Marginatoid	Arenarioid (FB Excluded)	FB	Total
Grouping hypothesis II	Marginatoid	81.5	18.5	—	100
	Arenarioid (FB excluded)	13.3	84.1	2.6	100
	FB	—	9	91.0	100

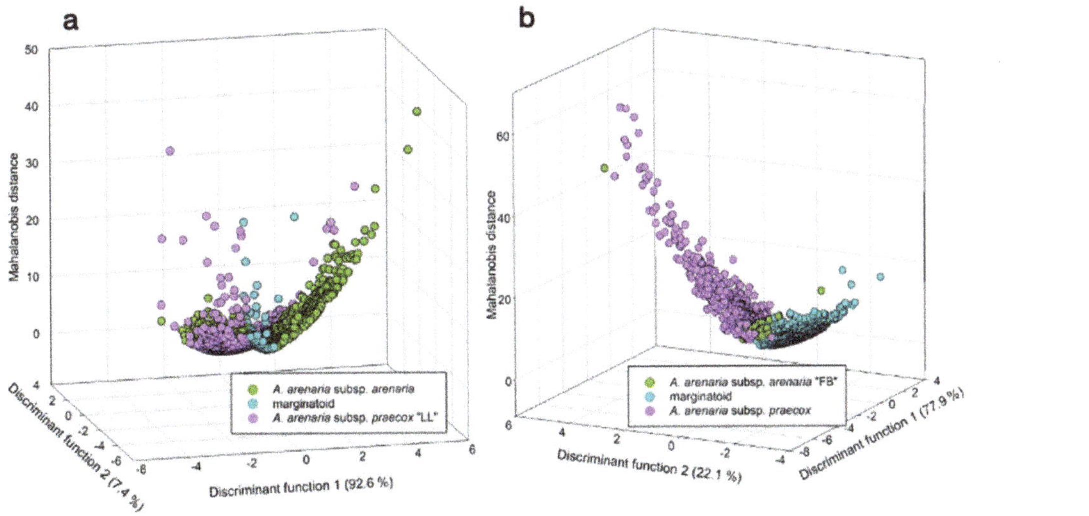

Figure 3. Graphical representation of the linear discriminant analysis (LDA) for the alternative grouping hypothesis for *Armeria arenaria* populations. (**a**) Grouping hypothesis I; (**b**) Grouping hypothesis II. FB = Fontainebleau, Île–de–France; LL = Le Lauzet, Dauphiné Alps.

3.3. *Karyotype Structure and Asymmetry*

All the studied populations were diploid, with $2n = 2x = 18$ chromosomes. They showed medium–sized (4.68 ± 0.64 μm), mostly metacentric (48.6%) or submetacentric (50.7%), chromosomes (see also Figures S3 and S4). One–way ANOVA revealed that all four karyological indices showed no significant differences among the four subspecies as circumscribed according to the current taxonomic hypothesis. However, the arenarioid plants ($n = 45$) showed significantly lower M_{CA} ($t = -4.52$, df = 59, $p < 0.001$) and THL ($t = -4$, df = 59, $p < 0.001$) values when compared to marginatoid plants ($n = 16$). Lower, but not significantly different, values were also observed in CV_{CL} and CV_{CI}. Mean (± standard deviation) values for the karyological indices at population level are provided in Table S6.

Under grouping hypothesis I, keeping all the Italian arenarioid populations as *A. arenaria* subsp. *arenaria* and contrasting them with LL and with marginatoid plants, one–way ANOVA revealed that THL ($F = 8.056$; $p < 0.001$) and M_{CA} ($F = 10.52$, $p < 0.001$) were significantly different, but not CV_{CL} and CV_{CI}. A post hoc Tukey–Kramer test showed that M_{CA} and THL values differed significantly ($p < 0.001$) between marginatoid and Italian arenarioid plants (+FB). In contrast, M_{CA} and THL were not significantly different between LL and the marginatoid group, or between the two arenarioid groups.

Under grouping hypothesis II, grouping all the Italian arenarioid populations with LL, contrasting them with FB and with marginatoid plants, One–way ANOVA revealed that THL ($F = 8.158$; $p < 0.001$), M_{CA} ($F = 11.03$; $p < 0.001$), and CV_{CI} ($F = 5.221$; $p < 0.01$) were significantly different among the three groups, while no difference was found in CV_{CL} values.

A post hoc Tukey test showed that M_{CA} and CV_{CI} values differed significantly between marginatoid plants and FB at $p < 0.01$, whereas THL differs significantly at $p < 0.001$ between marginatoid plants and all Italian arenarioid plants (+ LL) (Figure 4). In contrast, there was no significant difference between Italian arenarioid plants (+ LL) and FB in any of the studied karyological indices.

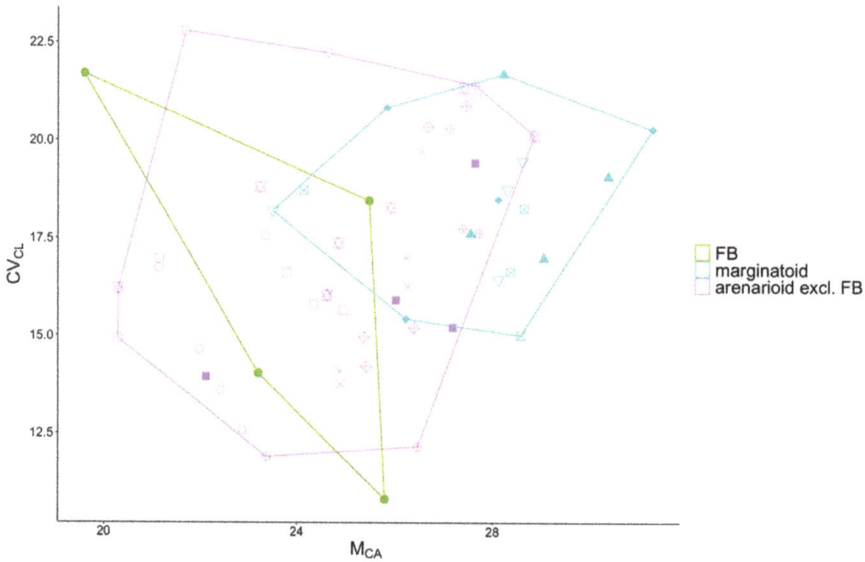

Figure 4. Scatterplot of the two karyotype asymmetry indices M_{CA} vs. CV_{CL} in *Armeria arenaria*. Accessions are enclosed by convex hulls according to grouping hypothesis II derived from the morphometric and seed morpho-colorimetric analysis, which sees LL as belonging to the same group as all the other Northern Italian arenarioid populations against the single population FB (which corresponds to *A. arenaria* s.str.). Symbols of populations as in Figure 1. FB = Fontainebleau, Île-de-France.

3.4. Molecular Systematics

The number of phylogenetically informative characters obtained from the amplification of the five markers was 36, corresponding to approximately 1.5% of the entire alignment. The markers that showed the highest number of informative characters were the intergenic spacers *trnL–rpl32* and *ITS*, with 13 and 11 phylogenetically informative characters (Table S9). The results of the ILD test showed that all plastid markers were congruent ($p > 0.05$, Table S10). On the contrary, increasing the number of replicates (up to 100), all the pairwise combinations were congruent except for *ITS* and *trn*L–*rpl*32, which turned incongruent at $p = 0.01$ (Table S10). Indeed, removing the *trn*L–*rpl*32 marker, the *ITS* and the resulting concatenated plastid matrix become congruent ($p = 0.09$). However, since the topology of the concatenated trees with and without *trn*L–*rpl*32 were not in conflict, we decided to retain the full matrix, which was 2337 bp long.

The Bayesian concatenated consensus unrooted tree is shown in Figure 5. Arenarioid populations are split into two main clades but are collectively well distinct from marginatoid (+ arenarioid PS) populations. The former main clade is more variable and encompasses accessions from the French populations FB and LL (forming a clade), accessions from MP, a clade with the accessions from GA, TV, BO (the latter in a separate clade), as well as those from BR, which do not form a monophyletic group. The second main clade contains two clades and the accessions from PS with an unresolved position. Separate ITS and plastid phylogenies are provided in Figures S5 and S6.

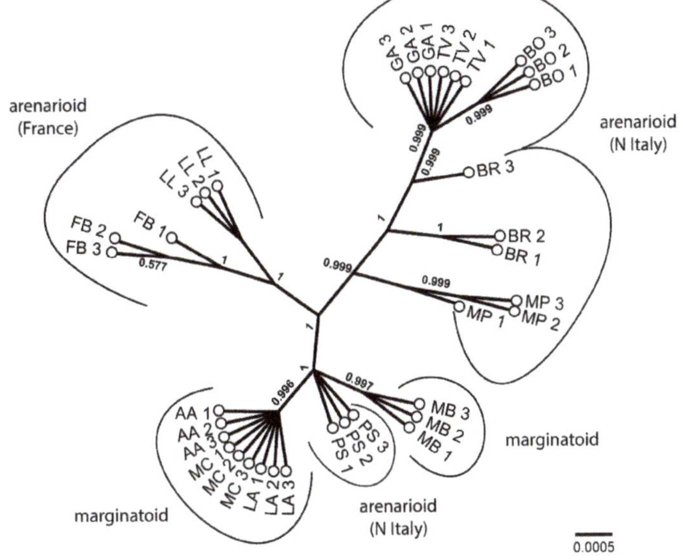

Figure 5. Bayesian unrooted consensus phylogenetic tree (concatenated dataset) of *Armeria arenaria* populations. AA = Apuan Alps, N Apennines; BO = Bobbio, N Apennines; BR = Brusson, Pennine Alps; FB = Fontainebleau, Île-de-France; GA = Gambolò, West Po Valley; LA = Libro Aperto; N Apennines; LL = Le Lauzet, Dauphiné Alps; MB = Marmagna–Braiola, N Apennines; MC = Monte Cusna, N Apennines; MP = Monte Prinzera, N Apennines; PS = Piana di Salmezza, Lombard Prealps; TV = Terme di Valdieri, Maritime Alps. Further population details are provided in Table 1 and Figure 6.

3.5. Comparative Niche Analysis

Schoener's D values were generally low, ranging from 0 to 0.208. In particular, *A. arenaria* subsp. *praecox* was the subspecies showing a niche that overlapped less with those of the other subspecies, according to the current taxonomic hypothesis (Table 7). The lack of significance in the similarity test indicated that the low niche overlap values were due to habitat availability in the background areas rather than an effect of habitat selection. Taken together, these results suggest differences in optimal niche positions without niche shift.

Table 7. Results of niche similarity tests in environmental spaces among the different taxa and circumscription hypotheses of *Armeria arenaria*. Backgrounds were defined by applying 10 km buffer zones around the occurrence points. Current taxonomic hypothesis as stated by Arrigoni [17], (I) = first alternative grouping hypothesis, (II) = second alternative grouping hypothesis. ns = not significant.

Armeria arenaria	subsp.	*arenaria*	*praecox*	*marginata*
Current taxonomic hypothesis	*praecox*	0.036 ns/ns		
	marginata	0.041 ns/ns	0.044 ns/ns	
	apennina	0.006 ns/ns	0.000 ns/ns	0.208 ns/ns
Grouping hypothesis I	*praecox*	0.036 ns/ns		
	marginata	0.033 ns/ns	0.012 ns/ns	
Grouping hypothesis II	*praecox*	0.005 ns/ns		
	marginata	0.000 ns/ns	0.107 ns/ns	

4. Discussion

All our results concur in highlighting that the current taxonomic hypothesis available for *Armeria arenaria* is no longer supported. Starting from the marginatoid plants, there is no morphometric support at all for distinguishing the two taxa as proposed by Arrigoni [17]. Moreover, the idea that *A. arenaria* subsp. *apennina* represents a taxon somehow intermediate between *A. arenaria* subsp. *marginata* and *A. arenaria* subsp. *arenaria* [17] is only supported by their climatic requirements. Nevertheless, it also should be noticed that the two putative marginatoid taxa show the highest values of niche overlap detected. There is no karyological difference between the two marginatoid taxa, but together they show higher M_{CA} and THL values with respect to the arenarioid plants. Phylogenetically, all marginatoid plants form a highly supported clade, in which the accessions of the two putative subspecies are intermingled. A single alpine arenarioid population (PS) is placed phylogenetically close to the marginatoid plants, suggesting that the genetic differentiation between arenarioid and marginatoid plants occurred only recently and may be derived from incomplete lineage sorting or gene flow. Despite this, morphological evidence fully places PS among arenarioid plants. Accordingly, we deem that the maintenance of the subspecific rank for marginatoid with respect to arenarioid plants is appropriate. Concerning arenarioid plants, they share a set of morphological and karyological features. Altogether, our data also support the maintenance of the subspecific rank for *Armeria arenaria* subsp. *praecox* with respect to *A. arenaria* subsp. *arenaria*, albeit with different circumscriptions, since all the Italian arenarioid populations agree much better with *A. arenaria* subsp. *praecox* than with *A. arenaria* subsp. *arenaria*. Indeed, from a morphometric point of view, the FB population (*A. arenaria* s.str.) shows the highest number of pairwise differences among all the other arenarioid populations, and it also shows the smallest seeds.

As a consequence, we exclude *Armeria arenaria* subsp. *arenaria* from the Italian flora, in favor of *A. arenaria* subsp. *praecox*, so that the range of the former subspecies is now reduced to Portugal, Spain, and France [68]. We cannot rule out that the range of that subspecies could be further narrowed in the future, given that this taxon is *"conceived as a mixed bag that includes the variability of the rest of the populations"* [25,71]. *Armeria arenaria* subsp. *praecox* has been only doubtfully recorded for Italy so far [17]. However, we clearly show that Italian arenarioid plants have a morphology highly overlapping that of the typical *A. arenaria* subsp. *praecox* (Figure 1). Indeed, the highest values of correct classification and K obtained by the discriminant analyses conducted for morphology and seed morpho–colorimetry were found when all the Italian arenarioid populations were grouped with LL and not with FB (which corresponds to the typical *A. arenaria* s.str.). Italian arenarioid populations are also phylogenetically more closely related to LL than to FB in the plastid tree (Figure S6). The possible occurrence in Italy of other subspecies occurring in Southern France can be excluded based on the comparison of our data with those published by Baumel et al. [72], Tison et al. [73], and Tison et De Foucault [74] (data not shown).

Geographically and climatically, the marginatoid plants from the Northern Apennines are replaced in the central–western alpine and perialpine areas by *A. arenaria* subsp. *praecox*, which is in turn replaced by *A. arenaria* subsp *arenaria* in Central–Northern France (Figure 6).

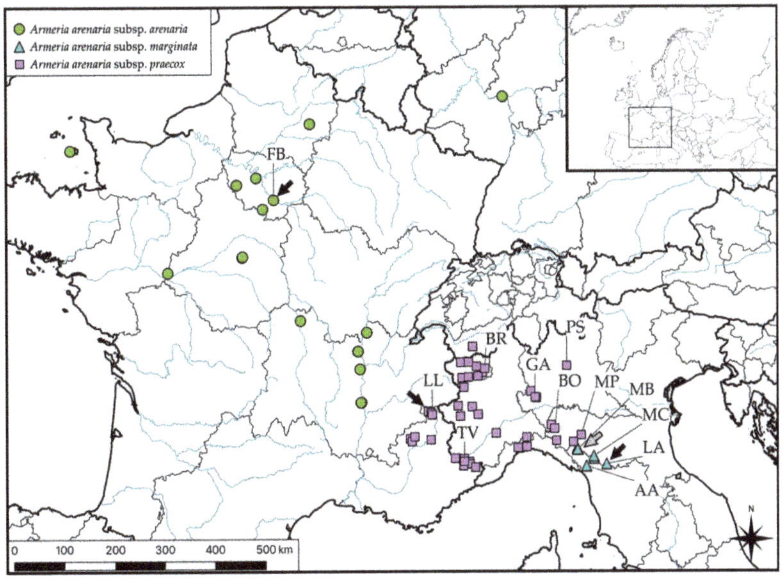

Figure 6. Distribution based on 81 herbarium specimens, including the localities sampled in this study, of *Armeria arenaria* subsp. *arenaria*, *A. arenaria* subsp. *marginata*, and *A. arenaria* subsp. *praecox*, as newly circumscribed. Solid arrows indicate type localities of the three taxa listed in the legend at the top–left corner on the map, whereas the crosshatched arrow indicates the type locality of *A. arenaria* subsp. *apennina*. Population codes as in Table 1.

5. Taxonomic and Nomenclatural Scheme

5.1. Identification Key

1. Leaves with flat apex, middle vein translucid or nearly so, veins (section!) usually $\leq$5, bracts of the inner spikelet 8.78 $\pm$ 0.95 mm long, capitula 18.3 $\pm$ 1.5 mm ... ... **Armeria arenaria subsp. marginata**
1. Leaves with cucullate apex, middle vein not translucid, veins (section!) usuall >5, bracts of the inner spikelet 6.67 $\pm$ 0.91 mm long, capitula 16.6 $\pm$ 2 mm wide **2**
2. Capitulum intermediate scales 3.87 $\pm$ 0.81 mm long; capitula 13.91 $\pm$ 1.5 mm wide; calyx 0.7 $\pm$ 0.15 mm wide, seeds 2.38 $\pm$ 0.07 mm long ... ... **Armeria arenaria subsp. arenaria**
2. Capitulum intermediate scales 5.98 $\pm$ 0.89 mm long; capitula 16.57 $\pm$ 2 mm wide; calyx 1 $\pm$ 0.16 mm wide, seeds 2.89 $\pm$ 0.25 mm long ... ... **Armeria arenaria subsp. praecox**

5.2. Nomenclature and Distribution

Armeria arenaria (Pers.) F.Dietr., Nachtr. Vollst. Lex. Gärtn. 1: 313 (1815) subsp. ***arenaria*** ≡ *Statice arenaria* Pers., Syn. Pl. 1: 332 (1805) ≡ *Armeria arenaria* (Pers.) Schult. in Roem. & Schult., Syst. Veg., ed. nov. (15), 6: 771 (1820) isonym ≡ *Armeria arenaria* (Pers.) Ebel, Armeriae Gen. Diss. 35 (1840) isonym.

Type: (neotype, here designated):—FRANCE. *Statice arenaria*, freq. rura Parisiis et alibi, s.d., *Persoon s.n.* (L2648462!).

In the protologue, a short diagnosis ("*caul. scapo longo, bract 2–3 capitulo longiorib., fol. linearib. rigidulis glabris*"), the habitat ("*in arenosis*"), and the provenance ("*Copiose prope Fontainebleau*") are provided. No original material occurs at L (M. Scherrenberg, pers. comm.), where the Persoon's herbarium and types are deposited [75].

Specimens seen. GERMANY. Gonsenheimer Wald bei Mainz, 8 July 1876, *A. Vigener* (FI!). BAILIWICK OF JERSEY. The Quennevais Jersey, 1860, *H.L.* (FI!). FRANCE. Sables aux Aulnois–sous–Laon (Aisne), 10 July 1873, *Favre* (FI!); Forêt de Rambouillet (Seine et Oise), sables au champ de manoeuvres le long de la route de Saint–Léger, 13 August 1932, *B. de Retz* (P05086601!); Prairies sèches, bois secs et clairs sur l'alluvion, près du Plessis–Piquet, aux environs de Paris, July 1841, *Kralik* (P05386780!); Nei dintorni de l'Hippodrome de la Solle, Fontainebleu (Seine–et–Marne) nelle schiarite sabbiose del bosco (WGS84: 48°26′13.9″ N 2°41′24.588″ E), 17 June 2020, *F. Losacco et M. Tiburtini* (PI56593–PI56615!); Buthiers (Seine–et–Marne) friches sablonneuses sur le coteau dominant la route de Malasherbes, 27 June 1943, *B. de Retz* (P05086604!); Gallia—La Sologne. In arenosis, 22 July 1924, *G. Lacaita* (FI!); La Sologne, entre Gien et Orléans: in arenosis, 22 July 1924, *G. Lacaita* (FI!); Saint-Nicolas, grande route de Thours (Tours), July 1845, *L. d'Espianay* (P00707605!); Moulis, sables de l'Allier, Jul 1890, *H. Bourdot* (FI!); Feillens près Mâcon, 7 July 1872, *H. Lareniq* (FI!); Arnas, par Villefranche (Rhone), France, in arenosis, 7 September 1876, *M. Gandoger* (FI!); Gall. Lyon, August 1900, *M. Gandoger* (FI!); Serras (Ardèche): prairies sablonneuses, 6 May 1878, *E. Chabert* (FI!).

Armeria arenaria subsp. ***marginata*** (Levier) Arrigoni, Fl. Medit. 25 (Special Issue): 15 (2015) ≡ *Armeria majellensis* var. *marginata* Levier, Atti Soc. Tosc. Sci. Nat. Pisa Processi Verbali 6 (11 novembre 1888): 157 (1888) ≡ *Armeria marginata* (Levier) Bianchini, Giorn. Bot. Ital. n.s., 111: 49 (1977)

Type (lectotype, designated by [17]: 15):—ITALY. In monti Libro Aperto, Apennini Pistoriensis supra Boscolungo, 1700 m, July 1881, *Levier s.n.* (FI barcode FI002438!) = *Armeria arenaria* subsp. *apennina* Arrigoni, Fl. Medit. 25 (Special Issue): 13, Figure 1 (2015). Type (holotype): ITALY. Emilia–Romagna, Corniglio (Parma). Vaccinieti e rupi della cresta rocciosa tra il M.te Marmagna e M.te Braiola, m 1600–1800, substr. Arenaria, 21 July 1986, *Arrigoni, Foggi et Ricceri s.n.* (FI barcode FI007466!)

The two names were published simultaneously at subspecies level by Arrigoni [17] on November 20, 2015. We opt here for *Armeria arenaria* subsp. *marginata* as having priority over the competing *A. arenaria* subsp. *apennina* (Art. 11.5 of the ICN [76]).

Specimens seen. ITALY. Emilia–Romagna: Lago santo—Sotto il M. Orsaio versante parmigiano, 28 June 1902, *S. Sommier* (FI!); Vetta del Monte Cusna (Rescadore—Reggio Emilia) a 2100 m s.l.m. (WGS84: 44°17′17.538″ N 10°23′24.192″ E), 26 June 2020, *M. Tiburtini et S. Quitarrá* (PI54653–PI54672!); Monte Cusna—prati rocciosi della vetta, 10 August 1988, *B. Foggi et C. Ricceri* (FI!); Ligonchio M.te Prado. Cresta rocciosa tra lo sperone di Prado e la vetta Prati e Vaccinieti, Alt. 1955–2054, substrato: arenaria, 28 July 1987, *B. Foggi et C. Ricceri* (FI!); Vetta del Libro Aperto (Fiumalbo—Modena) a 1930 m s.l.m. (WGS84: 44°9′25,64″ N 10°42′45,078″ E), 25 June 2020, *M. Tiburtini et S. Quitarrá* (PI56573–PI56592!); Toscana: Salendo da Pracchiola al M. Orsaio Pascoli verso 1900 m, 28 June 1903, *S. Sommier* (FI!); *ibidem*, 1700 m, 28 June 1903, *S. Sommier* (FI!); *ibidem*, 1200 m, 28 June 1903, *S. Sommier* (FI!); Pascoli alpini al M. Orsaio presso la Foce di Catelea e la cima, 21 July 1838, *F. Parlatore* (FI!); Lungo il sentiero sulla Sella del Braiola (Lagdei, Parma) a circa 1600 m s.l.m. (WGS84: 44°24′4.512″ N 9°59′39.318″ E), 27 June 2020, *M. Tiburtini et S. Quitarrá* (PI53990–PI54009!); Appennino lucchese reggiano M. Prado erbosi su macigno esposti a sud vicino alla vetta alt. 2000 m, 20 August 1992, *E. Ferrarini* (FI!); Sommità di Monte Prado nelle Alpi di Mommio, July 1851, *F. Calderini* (FI!); Libro Aperto Appennino Pistoiese, 7 August 1898, *S. Sommier* (FI!); Alpi Apuane, Serenaia Minucciano sotto l'Orto di Donna, 28 May 1960, *B. Lanza* (FI!); *ibidem*, 1000 m, 28 May 1960, *B. Lanza* (FI!); Ambienti prativi rocciosi lungo il sentiero nei pressi del Masso del Gigante (Località Altare) sotto Foce di Cardeto, Alpi Apuane (Minucciano—Lucca) (WGS84: 44°7′26.98″ N 10°12′43.188″ E), 26 June 2020, *M. Tiburtini et S. Quitarrá* (PI54673–PI54686!).

Armeria arenaria subsp. ***praecox*** (Jord.) Kerguélen ex Greuter, Burdet & G.Long, Med-Checkl. 4: 309 (1989) ≡ *Armeria praecox* Jord. In Boreau, Fl. Centre France, ed. 3, 2: 537 (1857) ≡ *Armeria arenaria* subsp. *praecox* (Jord.) Kerguélen, Lejeunia nov. ser., 120: 49 (1987), comb. inval. (Art. 41.5 of the ICN [72]).

Type (lectotype, here designated):— FRANCE. Hautes–Alpes, Monêtier–les–Bains, 1839, *Jordan s.n.* (LY barcode LY0421392!).

In the protologue, the name *Armeria praecox* is published in a note, reporting a short diagnosis and the provenance ("dans les Alpes"). Both the name ("*A. praecox* Jord.!") and description are clearly attributed to Claude Thomas Alexis Jordan (Boreau, Fl. Centre France, ed. 3, 1: 12. 1857: "*Quant aux espèces que M. Jordan m'a communiquées avant de les avoir publiées, je me suis efforcé d'en saisir les caractères, et s'ils ne sont pas convenablement mis en lumière, c'est mon insuffisance seule qui devra être mise en cause*"). We traced one specimen at LY (barcode LY0421392), where Jordan's herbarium and original material are preserved [75,77,78]. This specimen bears five parts of the same plant and the label "*Armeria praecox* Jord. | Hautes Alpes Monetier 1839 | (Herbier Jordan)". LY0421392 is part of the original material, is congruent with the Boreau's diagnosis, and is here designated as the lectotype for this name.

Specimens seen. FRANCE. Prairies de Serras—Ardèche, 8 May 1869, *Chabert* (FI!); Bords du le Mont du Lautaret, 23 July 1899, *P. Favre* (FI!); Le Lauzet (Hautes–Alpes): Lieux secs le long des chemins, 8 August 1875, *P. Favre* (FI!); Tra Le Lauzet e Le Monêtier–les–Bains (Hautes–Alpes) lungo i bordi della strada carrabile (WGS84: 44°58′48.588″ N 6°29′58.65″ E), 16 June 2020, *F. Losacco et M. Tiburtini* (PI55549–PI55568!); Col Bayard près Gaps: H.tes Alpes 1300 m, 23 June 1900, *L. Girod* (FI!); Le Roche–des–Arnauds près Gap, Jul 1886, *Serres* (FI!); La Freyssinouse (H.tes Alpes) 1000 m alt., 13 June 1898, *L. Girod* (FI!); Lieux incultes au Lauzet (H.tes Alpes), France, 2 August 1889, *E. Neyraut* (FI!); Terrains incultes au Lauzet—Hautes Alpes, 5 May 1883, *E. Neyraut* (FI!); Saint–Martin Vésubie, vallon de Salèses, 28 July 1910, *R. Pampanini* (FI!); Alpes de Tende, July 1843, *G.F. Reuter* (FI!). SWITZERLAND. Helvetia: Vallesia centralis in pratis saxosis aridis Vallis Hérens "Intericos la Sage et al. Forclas", in consortio *Dianthus carthusianorum, Aster alpini, Galium borealis* etc. solo silic., 1700 m, 12 August 1926, *A. Remaud* (FI!). ITALY. Valle d'Aosta: Près de Saint Rémy en Aoste (Italy) pelouses sèches, bords des champs 1630 m, 2 July 1875, *F. Tripet* (FI!); Tra Ollomont e Valpelline 1000–1400 m, 25 June 1902, *L. Vaccari* (FI!); Saint Marcel valle inf. fino a Prabornaz, 7 August 1902, *L. Vaccari* (FI!); Cogne salita al M. Herban, 1500–2000 m, s.d., *L. Vaccari* (FI!); Valpelline et Oyace, 17 July 1914, *P. Bolzon* (FI!); Aosta tra Arpuille e Plau de Dian—1300–1500 m, 24 July 1899, *L. Vaccari* (FI!); Valle di Champorcher a 700 m, 1 June 1899, *L. Vaccari* (FI!); Balze su substrato ofiolitico e prati sfalciati circostanti, Castello di Graines (Brusson, Aosta) (WGS84: 45°44′17.4″ N 07°45′14.8″ E), 16 June 2020, *S. Orsenigo* (PI53970–PI53989!); Bassa Val d'Ayas, tra Barme e Carogne, sopra il castello di Verrés, in prato secco, 810 m, 23 May 2006, *M. Bovio* (FI!); Degioz—Valsavarenche Valle d'Aosta, 26 July 1935, *U. Losacco* (FI!). Piemonte: Ceresole Reale nei prati in fondo alla valle sotto la chiesa 1400 m, 25 July 1910, *L. Vaccari et Wileyk* (FI!); Regione Valensana, 20 May 1914, *P. Bolzon* (FI!); Mt. Musiné Piedmont, 21 May 1870, *A. Chamber* (FI!); Stupinigi, dans le bois Piedmont, April 1854, *A. Chevalier* (FI!); Perosa, erbosi sopra Pomaretto, 4 July 1937, *G. Negri* (FI!); Env. d'Alba et de Turin, 1868, *Borguais* (FI!); Laghi della Lavagnina (Ovada), s.d., *I. Vagge* (ANC6837); Vallone di San Bernoni–Bernolfo (Alpi Marittime), 23 July 1889, *A. Ferrari* (FI!); Vallone della Meris: erbosi aridi a 1500 m, Val Gesso (A.M.), 1 August 1961, *P.G. Bono* (FI!); Luoghi silvestri alla confluenza del torrente Vallasco nel Gesso, alle Terme di Valdieri m. 1370 frequentissima nei luoghi aridi attorno alle terme, 12 July 1897, *O. Boggiani* (FI!); Prati e affioramenti rocciosi lungo la mulattiera che porta al rifugio Valasco (Terme di Valdieri, Cuneo) (WGS84: 44°12′14.7″ N 07°15′54.0″ E), 29 June 2020, *S. Orsenigo* (PI55569–PI56573!); Alpi Marittime, Valle del Gesso Tra Entracque e San Giacomo, 11 August 1887, *T. Caruel* (FI!); Erbosi aridi in Vallone di M.te Colombo presso Prà del Bosur (1350–1400 m), Val Gesso (A.M.), 17 July 1961, *P.G. Bono* (FI!); Lombardia: Prati sfalciati (Salmezza, Bergamo) (WGS84: 45°46′57.3″ N 09°43′56.2″ E), 12 June 2020, *S. Orsenigo* (PI53950–PI53969!); Abbiategrasso Valle del Ticino = Siti sabbiosi secchi, 4 June 1895, *C. Camperio* (FI!); Radure aride, su substrato sabbioso/ghiaioso, Parco del Ticino (Gambolò, Pavia) (WGS84: 45°16′07″ N 8°57′39.7″ E), 26 May 2020, *S. Orsenigo* (PI54032–PI54049!); Zelata, Pavia, 1844, *L. Rota* (FI!); Prov. di Pavia Sassi Neri (Penice) serpentini 600–700 m,

1 August 1916, *Mafra* (FI!); Liguria: Monte Maggiorasca, s.d., *s.coll.* (FI!); Monte Beigua (Lig. occ.) alt. 1200 m, 26 July 1885, *N. Mezzana* (FI!); Appennino a Voltri, 2 July 1871, *Baglietto* (FI!); Arenzano M.te Tardia, 2 July 1871, *Grey* (FI!); Liguria—Varazze M. Sciguello 1160 m, 23 May 1928, *S. Fresino* (FI!); Emilia–Romagna: Su detrito e roccia, affioramento ofiolitico, Monte Tre Abati (Bobbio, Piacenza), 3 June 2020, *S. Orsenigo* (PI54050–PI54063!); Monte Prinzera (Fornovo di Taro, Parma) su serpentini (WGS84: 44°38′27.048″ N 10°5′1.58″ E), 12 June 2020, *G. Astuti et M. Tiburtini* (PI53010–PI54029!); Aemilia—Prov. di Parma, abunde in rupium fissuris montis Prinzera, solo siliceo, 20 May 1905, *P. Bolzon* (MW0786588!); Emilia—Appennino Parmense Val Taro Roccamurata in loc. Groppo di Gorro substr. Serpentinoso, 1 June 1980, *F. Roffi* (FI!).

6. Conclusions

In this work, we conducted an integrative taxonomic study of *Armeria arenaria* in Italy. On the basis of nomenclatural, morphometric, seed morpho–colorimetric, karyological, molecular, and comparative niche evidence we were able to demonstrate that the current taxonomic setting for this species is no longer supported. Specifically, we proved that *A. arenaria* subsp. *apennina* is a heterotypic synonym of *A. arenaria* subsp. *marginata* and that all the previous records of *A. arenaria* subsp. *arenaria* for Italy should be attributed to *A. arenaria* subsp. *praecox*. Finally, we also provided an identification key for dried herbarium specimens to facilitate the identification of these taxa. The same key was used to reconstruct the distribution of the three subspecies based on 81 herbarium specimens.

Supplementary Materials: The following supporting information can be downloaded at: https://www.mdpi.com/article/10.3390/biology11071060/s1, Table S1, Shape of the leaf apex, according to the second grouping hypothesis; Table S2, Presence of teeth on summer leaf margins, according to the second grouping hypothesis; Table S3, Number of veins on summer leaves based on a cross section, according to the second grouping hypothesis; Table S4, Mean and standard deviation values of quantitative characters for the twelve populations studied; Table S5, Median and inter–quartile range of quantitative discrete characters for the twelve populations studied; Table S6, Means and standard deviations for the karyological indices in *Armeria arenaria* from the twelve studied populations; Table S7, List of molecular markers and their primers; Table S8, PCR settings used; Table S9, Number of phylogenetically informative characters per marker and for the concatenated matrix; Table S10, IDL test results; Figure S1, Schematic drawing of a calyx and related morphometric characters; Figure S2, Schematic drawings of selected summer leaf cross sections from the type localities of the four putative *Armeria arenaria* subspecies; Figure S3, Haploid idiograms ($x = 9$) of the twelve populations studied; Figure S4, Selected metaphasic plates of the twelve populations studied; Figure S5. Bayesian unrooted consensus phylogenetic tree of ITS matrix; Figure S6. Bayesian unrooted consensus phylogenetic tree of plastid matrix.

Author Contributions: Conceptualization, L.P. and G.A.; methodology, L.P., G.A. and M.T.; sampling and data collection, M.T., S.O., F.B., G.A. and L.P.; morphometry and karyology, M.T.; seed morpho–colorimetry, M.P. and G.B.; molecular systematics, D.D.L. and M.V.B.; niche analysis, L.V. and G.C.; nomenclature, G.D., F.B. and L.P.; data curation, M.T., D.D.L. and G.C.; writing—original draft preparation, M.T. and L.P.; writing—review and editing, M.T. and L.P.; supervision, L.P.; project administration, L.P.; funding acquisition, L.P. All authors have read and agreed to the published version of the manuscript.

Funding: This research was funded by Progetto di Ricerca di Rilevante Interesse Nazionale" (PRIN) "PLAN.T.S 2.0—towards a renaissance of PLANt Taxonomy and Systematics" led by the University of Pisa, grant number 2017JW4HZK (Principal Investigator: Lorenzo Peruzzi).

Institutional Review Board Statement: Not applicable.

Informed Consent Statement: Not applicable.

Data Availability Statement: The data presented in the current study are available within the article and supplementary material.

Acknowledgments: We thank François Thevenon and Jacopo Franzoni for collecting/sending seeds from Fontainebleau (France) and Libro Aperto (Italy), respectively. We also thank Federica Losacco and Sergio Quitarrá for helping during some field activities. We thank, too, Alexander Baumel for providing data about his studies. We also thank Marco Sarigu and Ludovica Dessì for helping during seed image acquisition and morpho–colorimetric analysis. We thank Riccardo Pennesi, Roberta Vallariello, Robert Philipp Wagensommer, Simonetta Fascetti, and Agnese Tilia for floristic information provided and Gabriele Galasso for nomenclatural advices. We are grateful to Chiara Nepi and Simona Casavecchia for the authorization provided to directly study the plant material housed in FI and ANC, respectively. We thank Mélanie Thiébaut and Martijn Scherrenberg for their help in searching Jordan's and Persoon's materials stored in LY and L, respectively.

Conflicts of Interest: The authors declare no conflict of interest.

References

1. Christenhusz, M.; Byng, J. The number of known plant species in the world and its annual increase. *Phytotaxa* **2016**, *261*, 201–217. [CrossRef]
2. Mayo, S.J. Plant taxonomic species and their role in the workflow of integrative species delimitation. *Kew Bull.* **2022**, *77*, 1–26. [CrossRef]
3. Klayman, J. Varieties of confirmation bias. In *Psychology of Learning and Motivation*; Elsevier: Amsterdam, The Netherlands, 1995; Volume 32, pp. 385–418, ISBN 978-0-12-543332-7.
4. Ariely, D.; Carmon, Z. Gestalt characteristics of experiences: The defining features of summarized events. *J. Behav. Decis. Mak.* **2000**, *13*, 191–201. [CrossRef]
5. Bernis, F. El genero *Armeria* Willd. en Portugal. *Bol. Soc. Brot.* **1949**, *23*, 225–263.
6. Bernis, F. Revisión del género *Armeria* Willd. con especial referencia a los grupos ibéricos. *Anales Inst. Bot. Cavanilles* **1954**, 5–287.
7. Bernis, F. Revisión del género *Armeria* Willd. con especial referencia a los grupos ibéricos. Parte Secunda (Conclusión). *Anales Inst. Bot. Cavanilles* **1957**, *14*, 259–432.
8. Garnett, S.T.; Christidis, L. Taxonomy anarchy hampers conservation. *Nature* **2017**, *546*, 25–27. [CrossRef]
9. Dayrat, B. Towards integrative taxonomy. *Biol. J. Linn. Soc.* **2005**, *85*, 407–415. [CrossRef]
10. Popper, K. *The Logic of Scientific Discovery*; Routledge: Abingdon-on-Thames, UK, 2005; ISBN 0-203-99462-0.
11. Turrill, W.B. The investigation of plant species. *Proc. Linn. Soc. London* **1935**, *147*, 1–144. [CrossRef]
12. Turrill, W.B. The expansion of taxonomy with special reference to spermatophyta. *Biol. Rev. Biol. Proc. Camb. Phil. Soc.* **1938**, *13*, 342–373. [CrossRef]
13. Knapp, S. Taxonomy as a team sport. In *The New Taxonomy*; Systematics Association Special Volumes; CRC Press: Broken Sound Park, NY, USA, 2008; pp. 33–53, ISBN 978-1-4200-0856-2.
14. Pupulin, F. Why we have no serious alternatives but cooperative taxonomy. *Lankesteriana* **2016**, *16*, 279–291. [CrossRef]
15. WFO, World Flora Online. Available online: http://www.worldfloraonline.org (accessed on 4 December 2021).
16. Arrigoni, P.V. Contributo alla conoscenza delle Armerie Sardo-Corse. *Webbia* **1970**, *25*, 137–182. [CrossRef]
17. Arrigoni, P.V. Contribution to the study of the genus *Armeria* (Plumbaginaceae) in the Italian peninsula. *Fl. Medit.* **2015**, *25*, 7–32. [CrossRef]
18. Bartolucci, F.; Peruzzi, L.; Galasso, G.; Albano, A.; Alessandrini, A.; Ardenghi, N.; Astuti, G.; Bacchetta, G.; Ballelli, S.; Banfi, E.; et al. An updated checklist of the vascular flora native to Italy. *Plant Biosyst.* **2018**, *152*, 179–303. [CrossRef]
19. Nieto Feliner, G.; Aguilar, F.A.; Rosselló, J.A. A new species of *Armeria* (Plumbaginaceae) from southern spain with molecular and morphometric evidence on its origin. *Bot. J. Linn. Soc.* **2001**, *135*, 71–84. [CrossRef]
20. Nieto Feliner, G. Natural and experimental hybridization in *Armeria* (Plumbaginaceae): *Armeria salmantica*. *Int. J. Plant Sci.* **1997**, *158*, 585–592. [CrossRef]
21. Nieto Feliner, G.; Álvarez, I.; Fuertes-Aguilar, J.; Heuertz, M.; Marques, I.; Moharrek, F.; Piñeiro, R.; Riina, R.; Rosselló, J.A.; Soltis, P.S.; et al. Is homoploid hybrid speciation that rare? An empiricist's view. *Heredity* **2017**, *118*, 513–516. [CrossRef]
22. Nieto Feliner, G.; Rosato, M.; Alegre, G.; San Segundo, P.; Rosselló, J.A.; Garnatje, T.; Garcia, S. Dissimilar molecular and morphological patterns in an introgressed peripheral population of a sand dune species (*Armeria pungens*, Plumbaginaceae). *Plant Biol.* **2019**, *21*, 1072–1082. [CrossRef]
23. Tauleigne-Gomes, C.; Lefèbvre, C. Natural hybridisation between two coastal endemic species of *Armeria* (Plumbaginaceae) from Portugal. 1. Populational in situ investigations. *Plant Syst. Evol.* **2005**, *250*, 215–230. [CrossRef]
24. Tauleigne-Gomes, C.; Lefèbvre, C. Natural hybridisation between two coastal endemic species of *Armeria* (Plumbaginaceae) from Portugal. 2. Ecological investigations on a hybrid zone. *Plant Syst. Evol.* **2008**, *273*, 225–236. [CrossRef]
25. Nieto Feliner, G. *Armeria* Willd. In *Flora Ibérica: Plantas Vasculares de la Península Ibérica e Islas Baleares*; Castroviejo, S., Laínz, M., López, G., Montserrat, P., Muñoz Garmendia, F., Paiva, J., Villar, L., Eds.; Consejo Superior de Investigaciones Científicas: Madrid, Spain, 1990; Volume 2, pp. 642–721.
26. Donadille, P. Contribution à l'étude du genre *Armeria* Willd. (Plumbaginaceae). III. Clé générale des taxons français. *Bull. Soc. Bot. France* **1969**, *116*, 511–521. [CrossRef]

27. De Giorgi, P.; Giacò, A.; Astuti, G.; Minuto, L.; Varaldo, L.; De Luca, D.; De Rosa, A.; Bacchetta, G.; Sarigu, M.; Peruzzi, L. An integrated taxonomic approach points towards a single-species hypothesis for *Santolina* (Asteraceae) in Corsica and Sardinia. *Biology* **2022**, *11*, 356. [CrossRef]
28. Liu, L.; Astuti, G.; Coppi, A.; Peruzzi, L. Different chromosome numbers but slight morphological differentiation and genetic admixture among populations of the *Pulmonaria hirta* complex (Boraginaceae). *Taxon* **2022**, *in press*. [CrossRef]
29. POWO—Plants of the World Online, Facilitated by the Royal Botanic Gardens, Kew. Available online: http://www.plantsoftheworldonline.org/ (accessed on 15 November 2021).
30. Lawrence, G.H.M. Armerias, native and cultivated. *Gentes Herbarum* **1940**, *11*, 391–418.
31. Schindelin, J.; Arganda-Carreras, I.; Frise, E.; Kaynig, V.; Longair, M.; Pietzsch, T.; Preibisch, S.; Rueden, C.; Saalfeld, S.; Schmid, B.; et al. Fiji: An open-source platform for biological-image analysis. *Nat. Methods* **2012**, *9*, 676–682. [CrossRef]
32. RStudio Team. *RStudio: Integrated Development for R.*; RStudio, Inc.: Boston, MA, USA, 2020.
33. Revelle, W. *Psych: Procedures for Psychological, Psychometric, and Personality Research*; CRAN, 2020.
34. Johannes Maier, M. *R: Einführung Durch Angewandte Statistik*, Wiley series Scientific Tools; 2nd ed.; Pearson Studium: Munchen, Germany, 2015.
35. Laliberté, E.; Legendre, P. A distance-based framework for measuring functional diversity from multiple traits. *Ecology* **2010**, *91*, 299–305. [CrossRef]
36. Podani, J. Multivariate Exploratory Analysis of Ordinal Data in Ecology: Pitfalls, Problems and Solutions. *Journal of Vegetation Science* **2005**, *16*, 497–510. [CrossRef]
37. Cailliez, F. The analytical solution of the additive constant problem. *Psychometrika* **1983**, *48*, 305–308. [CrossRef]
38. Paradis, E.; Schliep, K. ape 5.0: An Environment for modern phylogenetics and evolutionary analyses in R. *Bioinformatics* **2019**, *35*, 526–528. [CrossRef] [PubMed]
39. Wickham, H. *ggplot2: Elegant Graphics for Data Analysis*; Springer: New York, NY, USA, 2016; ISBN 978-3-319-24277-4.
40. Hammer, O.; Harper, D.; Ryan, P. PAST: Paleontological statistics software package for education and data analysis. *Palaeontol. Electron.* **2001**, *4*, 1–9.
41. Koul, A. *PredPsych: Predictive Approaches in Psychology*; CRAN, 2019. Available online: https://cran.r-project.org/web/packages/PredPsych/PredPsych.pdf (accessed on 10 July 2022).
42. Cohen, J. A coefficient of agreement for nominal scales. *Educ. Psychol. Meas.* **1960**, *20*, 37–46. [CrossRef]
43. Fleiss, J.L.; Levin, B.; Paik, M.C. The measurement of interrater agreement. In *Statistical Methods for Rates and Proportions*; Wiley Series in Probability and Statistics; Wiley: Hoboken, NJ, USA, 2003; ISBN 978-0-471-45861-6.
44. Mangiafico, S. *Rcompanion: Functions to Support Extension Education Program Evaluation*; CRAN, 2020. Available online: https://cran.r-project.org/web/packages/rcompanion/rcompanion.pdf (accessed on 10 July 2022).
45. Landini, G. Quantitative analysis of the epithelial lining architecture in radicular cysts and odontogenic keratocysts. *Head Face Med.* **2006**, *2*, 4. [CrossRef]
46. Terral, J.-F.; Tabard, E.; Bouby, L.; Ivorra, S.; Pastor, T.; Figueiral, I.; Picq, S.; Chevance, J.-B.; Jung, C.; Fabre, L.; et al. Evolution and history of grapevine (*Vitis vinifera*) under domestication: New morphometric perspectives to understand seed domestication syndrome and reveal origins of ancient european cultivars. *Ann. Bot.* **2010**, *105*, 443–455. [CrossRef]
47. Sarigu, M.; Grillo, O.; Bianco, M.L.; Ucchesu, M.; d'Hallewin, G.; Loi, M.C.; Venora, G.; Bacchetta, G. Phenotypic identification of plum varieties (*Prunus domestica* L.) by endocarps morpho-colorimetric and textural descriptors. *Comput. Electron. Agric.* **2017**, *136*, 25–30. [CrossRef]
48. Sarigu, M.; Porceddu, M.; Schmitt, E.; Camarda, I.; Bacchetta, G. Taxonomic discrimination of the *Paeonia mascula* group in the tyrrhenian islands by seed image analysis. *Syst. Biodiv.* **2019**, *17*, 801–810. [CrossRef]
49. Altınordu, F.; Peruzzi, L.; Yu, Y.; He, X. A Tool for the analysis of chromosomes: KaryoType. *Taxon* **2016**, *65*, 586–592. [CrossRef]
50. Aceto, S.; Caputo, P.; Cozzolino, S.; Gaudio, L.; Moretti, A. Phylogeny and evolution of *Orchis* and allied genera based on ITS DNA variation: Morphological gaps and molecular continuity. *Mol. Phylog. Evol.* **1999**, *13*, 67–76. [CrossRef]
51. Shaw, J.; Lickey, E.B.; Beck, J.T.; Farmer, S.B.; Liu, W.; Miller, J.; Siripun, K.C.; Winder, C.T.; Schilling, E.E.; Small, R.L. The tortoise and the hare II: Relative utility of 21 noncoding chloroplast DNA sequences for phylogenetic analysis. *Amer. J. Bot.* **2005**, *92*, 142–166. [CrossRef]
52. Shaw, J.; Lickey, E.B.; Schilling, E.E.; Small, R.L. Comparison of whole Chloroplast genome sequences to choose noncoding regions for phylogenetic studies in angiosperms: The tortoise and the hare III. *Amer. J. Bot.* **2007**, *94*, 275–288. [CrossRef]
53. Thompson, J.D.; Higgins, D.G.; Gibson, T.J. CLUSTAL W: Improving the sensitivity of progressive multiple sequence alignment through sequence weighting, position-specific gap penalties and weight matrix choice. *Nucleic Acids Res. Spec. Publ.* **1994**, *22*, 4673–4680. [CrossRef]
54. Hall, T. BioEdit: A user-friendly biological sequence alignment editor and analysis program for windows 95/98/NT. *Nucleic Acids Sympos. Ser.* **1999**, *41*, 95–98.
55. Goloboff, P. NONA ver. 2. Published by the author: Tucumán, Argentina, 1999.
56. Nixon, K. WinClada Ver. 1.0000. Published by the author: Ithaca, Greece, 2002.
57. Darriba, D.; Taboada, G.L.; Doallo, R.; Posada, D. JModelTest 2: More models, new heuristics and parallel computing. *Nat. Methods* **2012**, *9*, 772. [CrossRef] [PubMed]
58. Schwarz, G. Estimating the dimension of a model. *Ann. Stat.* **1978**, *6*, 461–464. [CrossRef]

59. Huelsenbeck, J.P.; Ronquist, F. MRBAYES: Bayesian inference of phylogenetic trees. *Bioinformatics* **2001**, *17*, 754–755. [CrossRef] [PubMed]
60. Rambaut, A.; Drummond, A.J.; Xie, D.; Baele, G.; Suchard, M.A. Posterior summarization in bayesian phylogenetics using tracer 1.7. *Syst. Biol.* **2018**, *67*, 901–904. [CrossRef]
61. Rambaut, A. *FigTree 1.4. 2 Software*; Institute of Evolutionary Biology, University Edinburgh: Edinburgh, UK, 2014.
62. Kimura, M. A simple method for estimating evolutionary rates of base substitutions through comparative studies of nucleotide sequences. *J. Mol. Evol.* **1980**, *16*, 111–120. [CrossRef]
63. Felsenstein, J. Evolutionary trees from DNA sequences: A maximum likelihood approach. *J. Mol. Evol.* **1981**, *17*, 368–376. [CrossRef]
64. Peruzzi, L.; Roma-Marzio, F.; Pinzani, L.; Bedini, G. Wikiplantbase #Italia v1.0. Available online: http://bot.biologia.unipi.it/wpb/italia/index.html (accessed on 4 December 2021).
65. Broennimann, O.; Fitzpatrick, M.C.; Pearman, P.B.; Petitpierre, B.; Pellissier, L.; Yoccoz, N.G.; Thuiller, W.; Fortin, M.; Randin, C.; Zimmermann, N.E. Measuring ecological niche overlap from occurrence and spatial environmental data. *Glob. Ecol. Biogeogr.* **2012**, *21*, 481–497. [CrossRef]
66. Schoener, T.W. Nonsynchronous spatial overlap of lizards in patchy habitats. *Ecology* **1970**, *51*, 408–418. [CrossRef]
67. Warren, D.L.; Glor, R.E.; Turelli, M. ENMTools: A toolbox for comparative studies of environmental niche models. *Ecography* **2010**, *33*, 607–611. [CrossRef]
68. Euro+Med PlantBase. Available online: https://www.emplantbase.org/home.html (accessed on 20 May 2021).
69. Stafleu, F.A.; Cowan, R.S. *Taxonomic Literature: A Selective Guide to Botanical Publications and Collections with Dates, Commentaries and Types*, 2nd ed.; Bohn, Scheltema & Holkema: Utrecht, The Netherlands, 1976.
70. Thiers, B. Index Herbariorum: A Global Directory of Public Herbaria and Associated Staff. Available online: http://sweetgum.nybg.org/science/ih/ (accessed on 4 December 2021).
71. Nieto Feliner, G. El género *Armeria* Willd. (Plumbaginaceae) en la Península Ibérica: Aclaraciones y novedades para una síntesis. *Anal. Jard. Bot. Madr.* **1987**, *44*, 319–348.
72. Baumel, A.; Auda, P.; Torre, F.; Médail, F. Morphological polymorphism and RDNA internal transcribed spacer (ITS) sequence variation in *Armeria*(Plumbaginaceae) from South-Eastern France. *Bot. J. Linn. Soc.* **2009**, *159*, 255–267. [CrossRef]
73. Tison, J.-M.; Jauzein, P.; Michaud, H. *Flore de la France Méditerranéenne Continentale*; Naturalia Publications: Turriers, France, 2014; ISBN 978-2-909717-90-6.
74. Tison, J.-M.; de Foucault, B. *Flora Gallica: Flore de France*; Biotope Editions: Mèze, France, 2014; Volume 1, ISBN 2-36662-012-8.
75. HUH Index of Botanists. Available online: https://kiki.huh.harvard.edu/databases/botanist_search.php?mode=details&id=47839 (accessed on 4 April 2022).
76. Turland, N.J.; Wiersema, J.H.; Barrie, F.R.; Greuter, W.; Hawksworth, D.L.; Herendeen, P.S.; Knapp, S.; Kusber, W.-H.; Li, D.-Z.; Marhold, K.; et al. *International Code of Nomenclature for Algae, Fungi, and Plants (Shenzhen Code) Adopted by the Nineteenth International Botanical Congress Shenzhen, China, July 2017. Regnum Vegetabile*; Koeltz Botanical Books: Glashütten, Germany, 2019; p. 159. [CrossRef]
77. Giacò, A.; Astuti, G.; Peruzzi, L. Typification and nomenclature of the names in the *Santolina chamaecyparissus* species complex (Asteraceae). *Taxon* **2021**, *70*, 189–201. [CrossRef]
78. Domina, G.; Astuti, G.; Bacchetta, G.; Barone, G.; Rešetnik, I.; Terlević, A.; Thiébaut, M.; Peruzzi, L. Typification of 14 names in the *Dianthus virgineus* group (Caryophyllaceae). *Phytokeys* **2021**, *187*, 1–14. [CrossRef]

biology

MDPI

Article

Adonis fucensis (*A.* sect. *Adonanthe*, Ranunculaceae), a New Species from the Central Apennines (Italy)

Fabio Conti [1], Christoph Oberprieler [2], Marco Dorfner [2], Erik Schabel [2], Roxana Nicoară [3] and Fabrizio Bartolucci [1,*]

[1] Floristic Research Center of the Apennine, University of Camerino—Gran Sasso Laga National Park, San Colombo, 67021 Barisciano, Italy
[2] Evolutionary and Systematic Botany Group, Institute of Plant Biology, University of Regensburg, Universitätsstr. 31, D-93053 Regensburg, Germany
[3] Department of Taxonomy, Ecology and Nature Conservation, Institute of Biology Bucharest—Romanian Academy, Splaiul Independenței 296, District 6, 060031 Bucharest, Romania
* Correspondence: fabrizio.bartolucci@unicam.it

Simple Summary: *Adonis* sect. *Adonanthe* is characterized by species with strongly gibbous abaxial side of achenes, reticulate-venation on its surface, with short and recurved style and includes four series: ser. *Amurenses*, ser. *Coeruleae*, ser. *Apenninae*, ser. *Vernales*. In the Euro-Mediterranean area three species belonging to *A.* sect. *Adonanthe* are currently recognized: *A. apennina* (ser. *Apenninae*), *A. volgensis* (incl. *A. transsilvanica*; ser. *Vernales*), *A. vernalis* (ser. *Vernales*). In 2021 was discovered in the Central Apennines (Italy) a yellow-flowered *Adonis* population belonging to sect. *Adonanthe* similar to *A. volgensis*. Following an integrated taxonomic approach, we have shown that the newly discovered population should be regarded as a new species, named *A. fucensis*, endemic to Abruzzo (Central Apennines, Italy).

Abstract: *Adonis fucensis* is herein described as a new species based on morphological and molecular analyses. It is endemic to one locality of the Central Apennines between Amplero and Fucino plains within the NATURA 2000 network in the SAC IT7110205 (Central Italy). The only discovered population is composed of 65 individuals and is at risk of extinction. The conservation status assessment according to IUCN categories and criteria is proposed and discussed. The new species belongs to *A.* sect. *Adonanthe* and is morphologically similar to *A. volgensis* (incl. *A. transsilvanica*), a species distributed in Hungary, Romania, Bulgaria, and Turkey as well as eastward to SW Siberia and Central Asia. *Adonis fucensis* can be distinguished from *A. volgensis* by larger cauline leaves, pentagonal with lobes lanceolate, larger stipules with more lobes and teeth, and larger flowers. Finally, an analytical key to *Adonis* species belonging to sect. *Adonanthe* distributed in Europe is presented.

Keywords: *Adonanthe*; endemism; molecular phylogeny; nomenclature; taxonomy; steppic plant

Citation: Conti, F.; Oberprieler, C.; Dorfner, M.; Schabel, E.; Nicoară, R.; Bartolucci, F. *Adonis fucensis* (*A.* sect. *Adonanthe*, Ranunculaceae), a New Species from the Central Apennines (Italy). *Biology* **2023**, *12*, 118. https://doi.org/10.3390/biology12010118

Academic Editor: Eduardo Blumwald

Received: 15 December 2022
Revised: 9 January 2023
Accepted: 9 January 2023
Published: 11 January 2023

1. Introduction

The genus *Adonis* L. (Ranunculaceae) comprises 38 accepted, annual and perennial, species and subspecies, distributed in the northern hemisphere and native to Asia, Europe, northern Africa, and Mediterranean region [1]. According to Wang [2,3], based on a morphological study, the genus *Adonis* should be divided into two subgenera, six sections, and six series: subg. *Adonis* (divided into three sections and two series) and subg. *Adonanthe* (Spach) W.T.Wang (divided into three sections and four series). Recent molecular studies [4,5] do not fully support the taxonomic treatment based on morphological features proposed by Wang [2,3], whereas a phylogenetic classification has not yet been established. In Italy, the genus *Adonis* is represented by 10 taxa (species and subspecies): the annual and red-flowered *A. annua* L., *A. flammea* Jacq. (with two subspecies), *A. aestivalis* L. (with two subspecies), *A. microcarpa* DC., and the perennial and yellow-flowered *A. distorta* Ten.

and *A. vernalis* L. *Adonis distorta* is the only species of the genus endemic to Italy [6,7], growing in the alpine belt on limestone screes and less frequently on more stabilized rocky slopes between 1845 and 2675 m a.s.l. of the Central Apennines [8]. The Abruzzo administrative region, in Central Italy, hosts seven *Adonis* taxa, the highest number among the Italian administrative regions, including the rare endemic *A. distorta*, and the only Italian populations of the steppe species *A. vernalis* [9]. In recent years extensive field surveys have been carried out for floristic and vegetation research in the National Park of Abruzzo, Lazio, and Molise [10–14], thanks to which some plants typical of the Alpine continental valleys or even of the E-European steppes have been discovered (i.e., [15,16]), confirming that the inner basins of the Central Apennine mountains have a pronounced steppic character. In March 2021 a group of hikers, discovered within the buffer area of the National Park, a yellow-flowered *Adonis* population. After analyzing the photographic material sent to us by Marina Buschi, who discovered the plant, we immediately realized that we were dealing with a very interesting population of *Adonis* morphologically very different from the two yellow-flowered species that already occurred in Italy, which are *A. vernalis* and *A. distorta*. From March to June 2021, we performed field surveys in the discovery locality, close to Amplero in Collelongo municipality (L'Aquila, Abruzzo, Central Italy), to evaluate the numerical consistency of the population and look for new stations. According to the classification proposed by Wang [2,3], the new discovered *Adonis* population belongs to *A.* sect. *Adonanthe* W.T.Wang ser. *Vernales* Bobr. ex Poschk. The section is characterized by species with strongly gibbous abaxial side of achenes, reticulate-venation on its surface, with short and recurved style and includes four series [2,3]: ser. *Amurenses* Poschk. with petiolate lower cauline leaves, ovoid, triangular or elliptic, and yellow or white petals; ser. *Coeruleae* Poschk. with petiolate lower cauline leaves, oblong or ovoid-oblong, 3–4 pinnatisect, and white or purple petals; ser. *Apenninae* Bobr. ex Poschk. with sessile or sub-sessile cauline leaves, pinnately compound, segments 2–3 pinnatisect, and yellow petals; ser. *Vernales* with sessile cauline leaves, palmately compound, segments 3 pinnatisect, and yellow petals. In the Euro-Mediterranean area three species belonging to *A.* sect. *Adonanthe* are currently recognized [1,17]: *A. apennina* L. (ser. *Apenninae*), *A. volgensis* DC. (incl. *A. transsilvanica* Simonov.; ser. *Vernales*), and *A. vernalis* L. (ser. *Vernales*). All other currently accepted species [1] within *A.* sect. *Adonanthe* are distributed exclusively in Asia: *A. amurensis* Regel & Radde (ser. *Amurenses*), *A. davidi* Franch. (ser. *Amurenses*), *A. multiflora* Nishikawa & Koji Ito (ser. *Amurenses*), *A. pseudoamurensis* W.T.Wang (ser. *Amurenses*), *A. ramosa* Franch. (ser. *Amurenses*), *A. shikokuensis* Nishikawa & Koji Ito (ser. *Amurenses*), *A. sutchuenensis* Franch. (ser. *Amurenses*), *A. coerulea* Maxim. (ser. *Coeruleae*), *A. bobroviana* Simonov. (ser. *Apenninae*), *A. mongolica* Simonov. (ser. *Apenninae*), *A. tianschanica* (Adolf) Lipsch. (ser. *Apenninae*), *A. turkestanica* (Korsh.) Adolf (ser. *Apenninae*), and *A. villosa* Ledeb. (ser. *Apenninae*).

The closest species within *A.* sect. *Adonanthe* ser. *Vernales* based on morphology is *A. volgensis*, a typical plant of the E-European and Asiatic steppes, distributed in Hungary, Romania, Bulgaria, and Turkey, as well as eastward to SW Siberia and Central Asia [1,17,18]. The populations from Romania and Hungary were regarded by Wang [3] as a different species with the name *A. transsilvanica* Simonov. Instead, others authors have considered *A. transsilvanica* as synonym of *A. volgensis* (i.e., [1]), as a name ambiguous [19], or have not listed it at all [17,18].

An extensive morphological and molecular investigation has been carried out providing evidence about the differentiation between *A. volgensis* and the new discovered Apennines' population. Our results, and the disjunct and isolated geographical distribution of the population occurring in the Central Apennines, allowed us to describe it as a species new to science, named *A. fucensis*.

2. Materials and Methods

2.1. Plant Material

This study is based on field surveys, an extensive analysis of the relevant literature, and examination of herbarium specimens (including the original material of *A. volgensis* and *A. transsilvanica*; Supplementary File S1) kept in APP, B, BP, BRNU, CL, G, K, LD, LE, MW, S, UPS, and US (acronyms follow [20]).

Morphological characters, recognized as taxonomically discriminant in *Adonis* [2,3,18,21] were scored in the herbarium specimens kept in APP, BP, CL, K, MW, UPS, and US. All morphological characters were observed and measured under a Leica MZ16 stereoscopic microscope, using a digital caliper with 0.1 mm precision. Digital images of herbarium specimens from online databases were measured with IC Measure version 2.0.0.245.

Regarding the new species, having found only one small population, we have not collected whole individuals, but only some parts, such as petals, sepals, leaves, etc., and then dried them. Only two individuals, without rhizomes, probably displaced by wild fauna, were collected. For the description of the new species some characters, i.e., length of the sepals and petals, were also scored in the field on fresh material.

2.2. Morphometric Analyses

A total of 18 morphological characters were selected and scored in 87 dried individuals belonging to *A. volgensis* (63) from Romania, Moldavia, Russia and to the new population from the Central Apennines (Italy), named *A. fucensis* (24). Two characters, i.e., height (H) and number of petals (NP), were scored for *A. fucensis* on the field. Among the morphological characters studied, 14 are quantitative, 1 is calculated ratio, and 3 are qualitative (Table 1). Samples with missing data were not included in the multivariate analysis (resulting dataset of 50 individuals × 18 variables). For each quantitative character an independent sample *t*-test was carried out with SPSS v25 software (IBM Corp., Armonk, NY, USA) [22]. A non-metric multidimensional scaling (NMDS) and Cluster Analysis (CA) using the average linkage method (UPGMA), were performed with PAST package v4.11 software (Natural History Museum, Oslo, Norway) [23]. The similarity matrix was calculated using the Gower coefficient, suitable for mixed data [24]. Furthermore, the variability of the analyzed morphological characters was described by standard statistical parameters (mean, standard deviation, minimum, maximum, and 25th and 75th percentiles). Boxplots were built through SPSS v25.

Table 1. Morphological characters studied.

Abbreviation	Description of the Character	Type
H	Height (cm)	quantitative continuous
MLL	Middle leaf length (mm)	quantitative continuous
MLW	Middle leaf width (mm)	quantitative continuous
SML	Shape of middle leaf (0: pentagonal; 1: triangular-ovate)	qualitative
NLML	Number of lobes and teeth of the middle leaf	quantitative discrete
STL	Shape of terminal lobe (0: lanceolate; 1: narrowly lanceolate; 2: linear)	qualitative
ATL	Angle of terminal apex lobe (°)	quantitative continuous
LMW	Subterminal lobe max width (mm)	quantitative continuous
LWB	Subterminal lobe width at base (mm)	quantitative continuous
LMW/LWB	Ratio lobe max width/lobe width at base	calculated ratio
SL	Stipule length (mm)	quantitative continuous
SW	Stipule width (mm)	quantitative continuous
NSL	Number stipule lobes and teeth	quantitative discrete
SLD	Sepal length (mm)	quantitative continuous
SWD	Sepal width (mm)	quantitative continuous
PLD	Petal length (mm)	quantitative continuous
PWD	Petal width (mm)	quantitative continuous
NP	Number of petals	quantitative discrete

2.3. DNA Sequencing and AFLPseq Fingerprinting

Genomic DNA for both sequencing and genetic fingerprinting was extracted according to the CTAB DNA extraction protocol of Doyle and Dickson [25] and Doyle and Doyle [26]. Amplification of the two internal transcribed spacer regions (ITS1, ITS2) of the nuclear ribosomal repeat (nrDNA) was carried out with primers ITS-18SF [27] and ITS2 [28] for ITS1 and ITS-D [29] and ITS-SR [30] for ITS2, respectively. After purification of PCR amplicons with AmpliClean (Nimagen, Nijmegen, The Netherlands) magnetic beads, Sanger sequencing was carried out by a contract sequencing company (Macrogen Europe, Amsterdam, The Netherlands). Electropherograms were manually edited with CHROMAS v2.6.6 [31]; polymorphisms observed in accession A1251 were resolved manually and the two resulting sequences were independently included in the alignment together with sequences of other species of *Adonis* sect. *Adonanthe* and an outgroup sequence (from *Trollius ranunculoides* Hemsl.). We used PAUP* v4.0a169 [32] to calculate distances among the aligned sequences based on the Kimura-2-Parameter model and constructed a Neighbor-joining tree. A bootstrap analysis was performed with 1000 replicates.

The AFLPseq fingerprinting method has been proposed by [33] and combines the genome-complexity reducing AFLP approach [34] with the next-generation sequencing (NGS) of resulting AFLP bands using the Nanopore sequencer MinION from Oxford Nanopore Technologies (Oxford, UK). It provides sequence and single-nucleotide polymorphism (SNP) information for hundreds of anonymous loci from across the whole genome and could be used for both population genetic, phylogenetic, and species delimitation studies. It is suited for both well-preserved DNA from silica-gel dried leaf material and degraded DNA from herbarium specimens.

The present AFLPseq study comprised 12 *Adonis* accessions (Supplementary Table S1), either recently collected, silica-gel dried material (five accessions from Italy and Romania), or well-preserved herbarium material housed in the herbaria B and BRNO (seven accessions from Romania, the Russian Federation, and Kazakhstan). The accessions were selected (a) to cover large parts of the distribution range of *A. volgensis* and (b) to include only plant material in the fingerprinting procedure, for which extracts of unfragmented genomic DNA was expectable. The AFLPseq procedure followed the protocol given in [33] with the following modifications: in the restriction-ligation step, we used a double-digestion procedure with restriction enzymes MseI and EcoRI. After ligation of MseI and EcoRI adapters (MseI adapter: 5′-GACGATGAGTCCTGAG-3′ + 5′-TACTCAGGACTCAT-3′; EcoRI adapter: 5′-CTCGTAGACTGCGTACC-3′ + 5′-AATTGGTACGCAGTCTAC-3′), we continued with the AFLP genome-reduction protocol using primers with 1bp-overhangs (MseI-C: 5′-GATGAGTCCTGAGTAAC-3′; EcoRI-A: 5′-GACTGCGTACCAATTCA) in the pre-selective amplification step and in the selective amplification step with additional 1bp- (EcoRI side) or 2bp-overhangs (MseI side), respectively. The two primers used in the latter amplification step, however, were additionally tailored to include Nanopore barcode adapter sequences at the 5′ end of the primers (Mse_CTG_Nanopore_fw: 5′-TTTCTGTTGGTGCTGATATTGCGATGAGTCCTGAGTAACTG-3′; Eco_AA_Nanopore_rv: 5′-ACTTGCCTGTCGCTCTATCTTCGACTCCGTACCAATTCAA-3′), as suggested in the 'Ligation sequencing amplicons—PCR barcoding (SQK-LSK109 with EXP-PBC001)' protocol by Oxford Nanopore Technologies, substituting a subsequent ligation of the Nanopore barcode adapter with an additional barcoding PCR. To ensure specific binding with long and tailed primers, a two-step variation of the selective PCR was conducted (94 °C for 2 min; followed by 30 cycles of 94 °C for 20 s and 72 °C for 2 min; and a final step at 72 °C for 2 min). To every 2 μL of 1:10 diluted preselective PCR product, 5 μL Taq DNA Polymerase Master Mix RED, 0.25 μL of each 10 μM tailed selective primer, and 2.5 μL H2O were added. After the selective PCR, the length of the fragments ranged from 200–500 bp. All subsequent steps (Nanopore barcode PCR, sample multiplexing, size selection, preparation of Nanopore sequencing library) followed [33]. The resulting library was sequenced with the MinION using a Flongle flow cell. Read data processing, de novo locus assembly, identification of orthologous loci, and reference-based SNP calling with the SLANG pipeline, and the final calculation of frequency-sensitive SNP-based Nei distances followed the protocol described by [33]. Based on these pairwise distances both a phylogenetic network reconstruction using the Neighbor-joining method in SPLITSTREE v4.16.1 [35] and a principal co-ordinate analysis (PCoA) with a custom R v4.0.5 script using the 'phangorn' library to read the distance matrices and the 'ape' package to calculate and plot the PCoA was carried out.

3. Results

3.1. Morphometric Analyses

The NMDS, performed with three dimensions, yielded an ordination with a stress value of 0.09224. The scatterplot shows on the first two axes a clear distinction between *A. volgensis* and *A. fucensis*, and no overlapping areas among individuals were found (Figure 1). The UPGMA dendrogram (Figure 2) yielded two well-defined clusters, one including all individuals of *A. volgensis* and the other all individuals of *A. fucensis*.

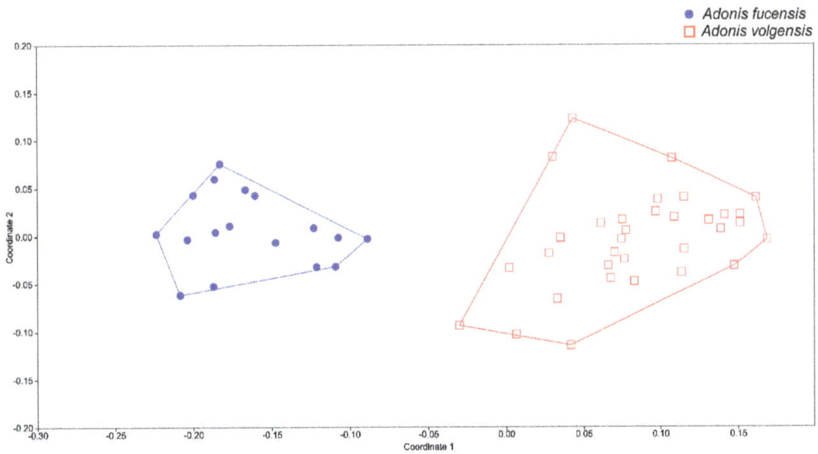

Figure 1. Non-metric multidimensional scaling scatterplot showing the first two dimensions of the analysis.

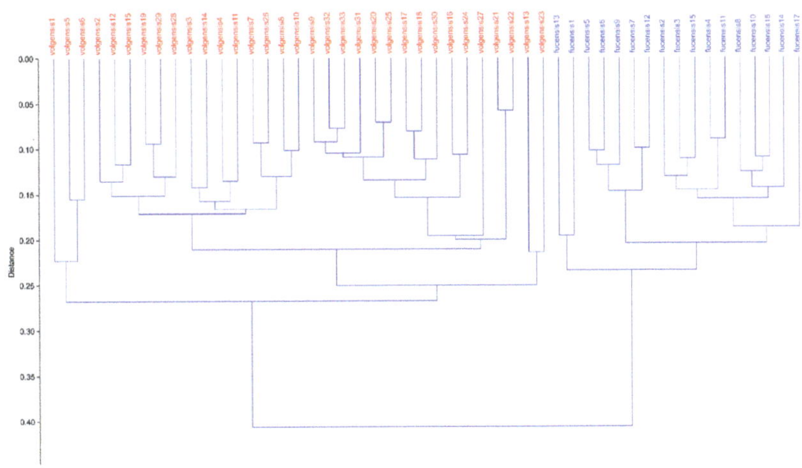

Figure 2. Hierarchical clustering of individuals of *A. volgensis* and *A. fucensis* using paired group algorithm (UPGMA) and Gower Similarity Index. Cophenetic correlation coefficient is 0.8566.

Comparisons of morphological characters between *A. volgensis* and *A. fucensis* (Figure 3) are summarized in Table 2. The states of 13 characters (H, MLL, MLW, NMLN, ATL, LMW, LWB, LMW/ LWB, SL, SW, NSL, CD, SLD, SWD, PLD, and PWD) show significant differences between the two species ($p < 0.01$). Boxplots of relevant characters are showed in Figure 4.

Figure 3. Comparison of *A. fucensis* and *A. volgensis*: (**A**) *A. fucensis* from Mt. Annamunna locality (Italy, Abruzzo, photo by F. Bartolucci); (**B**) *A. volgensis* from Murfatlar locality (Romania, Constanța County, photo by R. Nicoară); (**C**) dry cauline leaf of *A. fucensis*, pentagonal leaf blade; (**D**) dry cauline leaf of *A. volgensis*, triangular-ovate leaf blade. Scale bar 1 cm.

Table 2. Morphological comparison of *A. volgensis* and *A. fucensis*. Quantitative continuous characters are expressed in mm and are reported as mean ± standard deviation and 25–75 percentiles (extreme values in brackets). For quantitative discrete cardinal characters, 25–75 percentiles are given (extreme values in brackets). Significantly different character states are shown in bold ($p < 0.01$).

	Adonis volgensis		*Adonis fucensis*	
Height (cm)	**22.80 ± 7.21**	**(10.49)17.42–27.39(39.61)**	**12.06 ± 2.46**	**(8)10.25–13.75(16)**
Middle leaf length (mm)	**50.36 ± 16.86**	**(24.12)38.70–58.35(109.66)**	**74.28 ± 12.39**	**(56.70)65.75–86.22(96.06)**
Middle leaf width (mm)	**35.01 ± 13.75**	**(10.00)27.00–42.60(81.30)**	**64.31 ± 10.83**	**(45.16)57.01–75.06(79.30)**
Shape of middle leaf		triangular-ovate, rarely pentagonal		pentagonal
Number of lobes and teeth of the middle leaf		**(12)32–85(176)**		**(84)123–187(280)**
Shape of terminal lobe		linear to narrowly lanceolate		lanceolate
Angle of terminal apex lobe (°)	**31.95 ± 11.21**	**(15.04)23.73–37.55(64.96)**	**50.45 ± 11.76**	**(31.73)43.36–54.32(74.4)**
Subterminal lobe max width (mm)	**1.13 ± 0.39**	**(0.53)0.84–1.34(2.39)**	**1.76 ± 0.48**	**(1.14)1.41–1.91(2.83)**
Subterminal lobe width at base (mm)	**0.94 ± 0.28**	**(0.51)0.70–1.10(1.75)**	**1.36 ± 0.38**	**(0.90)1.08–1.49(2.29)**
Ratio lobe max width/lobe width at base	1.21 ± 0.27	(0.91)1.02–1.32(2.52)	1.29 ± 0.11	(1.13)1.21–1.37(1.57)
Stipule length (mm)	**12.63 ± 4.36**	**(4.64)9.31–15.88(22.00)**	**22.41 ± 4.53**	**(14.13)19.83–25.23(31.00)**
Stipule width (mm)	**8.45 ± 4.82**	**(1.69)4.85–10.60(24.63)**	**22.06 ± 5.56**	**(9.59)16.40–23.24(32.87)**
Number of stipule lobes and teeth		**(2–)3–6(–12)**		**(7)10–21(–28)**
Sepal length (mm)	**12.23 ± 3.67**	**(6.40)9.26–14.51(21.07)**	**18.81 ± 5.80**	**(11.00)12.25–23.63(27.00)**
Sepal width (mm)	**4.98 ± 1.71**	**(2.50)3.57–6.05(9.46)**	**10.61 ± 1.82**	**(7.50)9.63–12.00(14.00)**
Petal length (mm)	**17.2 ± 6.04**	**(7.61)12.12–20.69(29.28)**	**23.48 ± 5.54**	**(11.00)21.75–27.00(30.00)**
Petal width (mm)	**5.59 ± 2.15**	**(2.85)3.95–6.83(11.68)**	**9.37 ± 1.96**	**(6.00)8.00–10.50(13.00)**
Number of petals		(10)12–13(16)		(9)12–14(18)

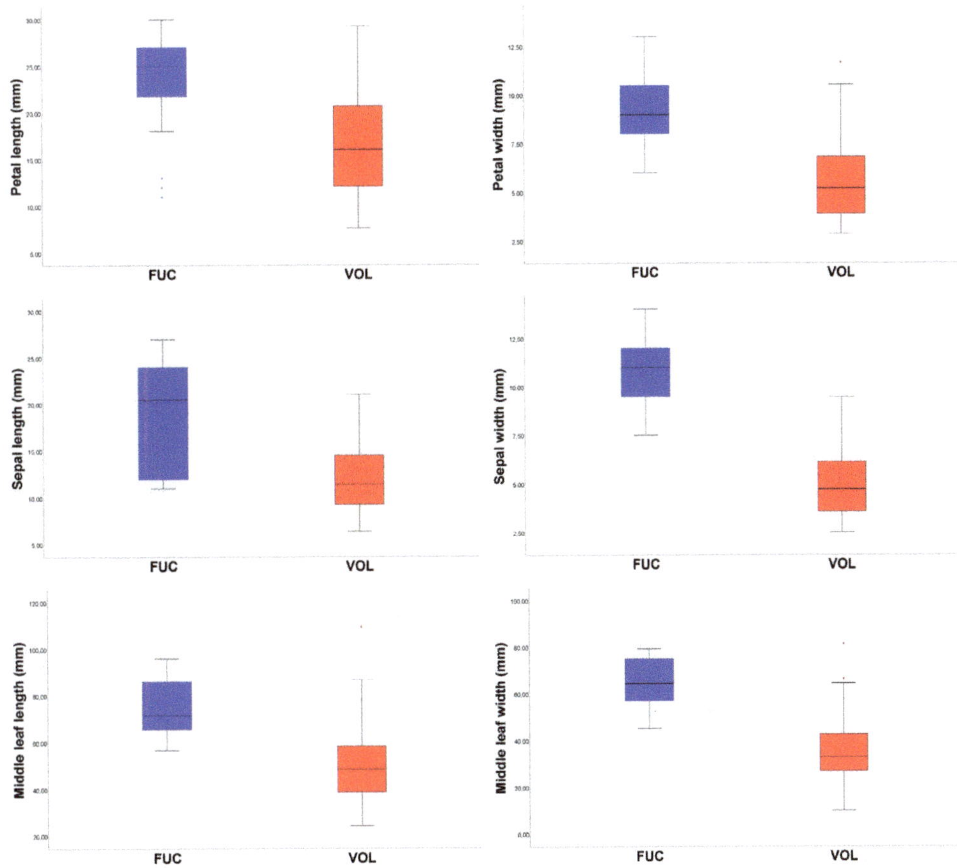

Figure 4. *Cont.*

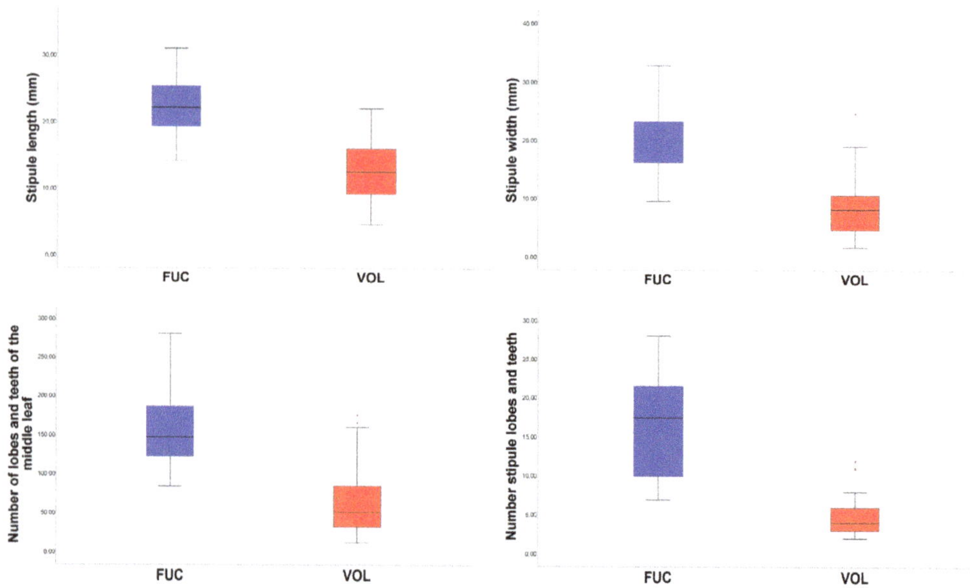

Figure 4. Boxplots expressing morphological variation between *A. fucensis* (FUC) and *A. volgensis* (VOL). Outlined central box depicts middle 50% of data, extending from 25th and 75th percentiles, and horizontal bar is the median. Ends of vertical lines (or "whiskers") indicate minimum and maximum data values, unless outliers are present, in which case whiskers extend to a maximum of 1.5 times inter-quartile range. Circles indicate outliers.

3.2. *nrDNA Sequence Variation*

The Neighbor-joining tree based on Kimura-2-parameter distances among nrDNA ITS sequences of 15 *Adonis* accessions is shown in Figure 5. The central Italian *Adonis fucensis* accession (A1252) is found being closely related with *A. volgensis* and *A. vernalis* in the monophyletic group of *A.* sect. *Adonanthe* ser. *Vernales.* As also found by a more comprehensive phylogenetic analysis of section *Adonanthe* performed by [4], series *Amurenses* did not form a monophyletic group.

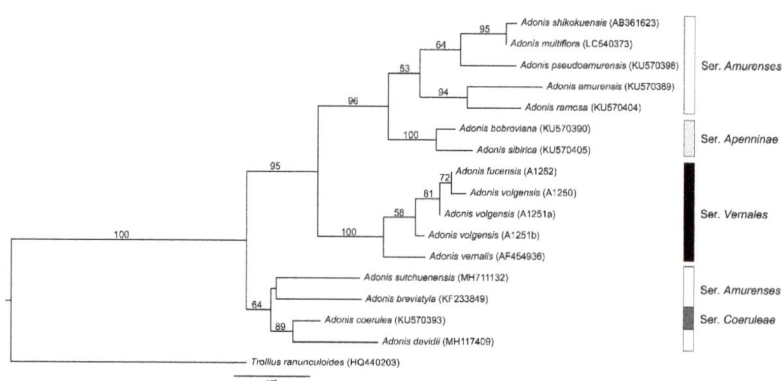

Figure 5. Neighbor-joining tree of 15 accessions of *Adonis* sect. *Adonanthe* based on nrDNA ITS sequence variation and Kimura-2-parameter distances. GenBank accession numbers and probe numbers of the present study (*A. fucensis, A. volgensis*) are given in brackets. Numbers above branches are bootstrap values based on 1000 replicates.

3.3. AFLPseq Fingerprinting

In total, 731,698 reads and 243.72 Mbp were sequenced for the 12 *Adonis* accessions. After read preprocessing, 592,432 reads with lengths between 10 bp and 614 bp passed the Q5 quality filter. With the SLANG pipeline (cluster thresholds optimized to values of 0.85 and 0.95 for the first and second cluster step, respectively), 486 orthologous loci were inferred, containing 2944 SNPs. After calculation of pairwise Nei distances, the resulting Neighbor-joining tree (Supplementary File S2) and the PCoA plot were received (Figure 6). While in the first, the *Adonis* accession from the Central Apennines (A1252) is connected with the remaining *A. volgensis* representatives without any exceptionally longer branch than the other accessions, the PCoA plot demonstrates the clear separation between the two taxa; with accessions of the latter on the left and accession A1252 on the right side of principal co-ordinate PCo axis 1, which account for 20.8% of the total variation in the data set. Additionally, PCo axis 2 (accounting for additional 15.0% of the total variation) shows a clear geographical separation within *A. volgensis*, with accessions of this species from Romania (sometimes considered as being an independent species, *A. transsilvanica*) on the positive and accessions from Russia and Kazakhstan on the negative side of the axis.

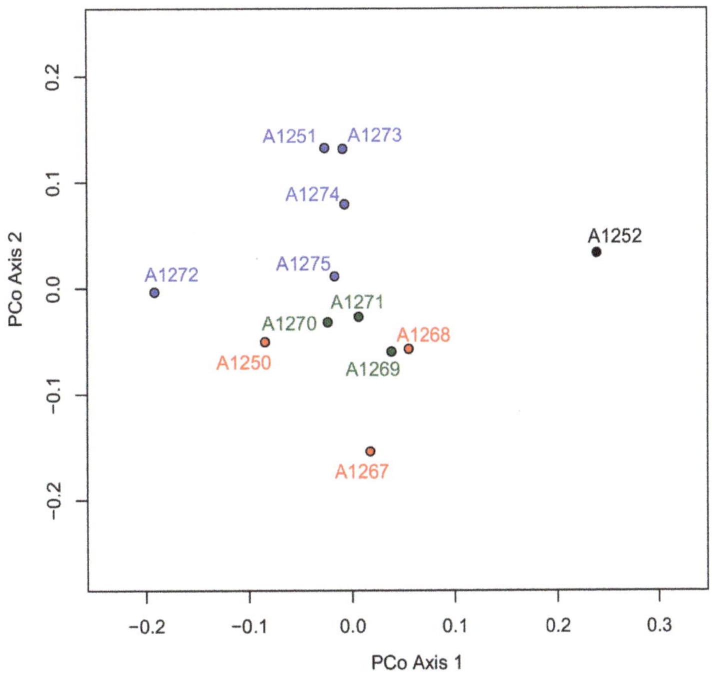

Figure 6. Ordination of accessions of *A. fucensis* (black) and *A. volgensis* (blue: Romania; red: Russia; green: Kazakhstan) based on pairwise Nei distances from 2944 single-nucleotide polymorphisms (SNPs) from 486 AFLPseq loci in a Principal Co-ordinate Analysis (PCoA), with axis 1 explaining 20.8% and axis 2 explaining 15.0% of the total variance, respectively.

An additional result of the analysis is worth mentioning in methodological respects: accessions A1251, A1273, and A1274 are very similar to each other in spite of the fact that the three probes come from the same locality (Romania, Constanta, Cotu Văii), but were recently collected as silica-gel dried leaf material (the latter two) or as an herbarium specimen (A1251) twenty years ago. This observation adds to the trustworthiness of the AFLPseq protocol and the comparability of differently preserved DNAs in terms of sequence information retrieved through this process.

4. Discussion

Morphological and molecular analyses provide evidence that *A. fucensis* should be regarded as a new species, endemic to Abruzzo (Central Apennines, Italy). It is similar to *A. volgensis*, a typical plant of the E-European and Asiatic steppes, distributed in Hungary, Romania, Bulgaria, and Turkey, as well as eastward to SW Siberia and Central Asia, but it can be distinguished by several quantitative and qualitative morphological characters, as shown in Table 2. The new species lives in shrub-steppe habitat in contact environments between bushes dominated by *Prunus spinosa* L. subsp. *spinosa* and steppe grasslands with the presence of *Festuca valesiaca* Schleich. ex Gaudin subsp. *valesiaca*. Abruzzo is the Italian administrative region with the highest number of taxa belonging to the genus *Adonis*, and also hosts the only Italian populations of the extremely rare steppe species *A. vernalis*.

The dry sub-continental climate of internal basins of the Central Italy, together with wild herbivore disturbance and prehistoric anthropogenic fires [36], may have reduced the post-glacial reforestation. Subsequently sheep grazing and the practice of transhumance, dating back to the 6th century BC or earlier in Abruzzo and widely practiced until the 1950s [37,38], has probably favored the spread of grasslands [39]. Around Amplero, close to the locality of *A. fucensis*, lies an archaeological site inhabited since the VI century B.C. The area hosted, from the Bronze age until Medieval times and beyond, important shepherd settlements and was located on transhumance routes [40]. These causes explain the persistence of steppe species in the internal areas of the Central Apennines.

In the internal basins of Abruzzo such as the Fucino and the L'Aquila plains, there is a consistent number of grassland taxa featuring a disjunction with E-European steppes, giving these areas a pronounced steppic character: *Alyssum desertorum* Stapf., *Androsace maxima* L., *Astragalus danicus* Retz., *A. exscapus* L. subsp. *exscapus*, *A. onobrychis* L., *A. vesicarius* L. subsp. *vesicarius*, *Ceratocephala falcata* (L.) Cramer, *Crocus variegatus* Hoppe & Hornsch, *Falcaria vulgaris* Bernh., *Festuca valesiaca* Schleich. ex Gaudin subsp. *valesiaca*, *Pulsatilla montana* (Hoppe) Rchb. subsp. *montana*, *Poa perconcinna* J.R.Edm., *Salvia aethiopis* L., and *Stipa capillata* L. [12,15,16]. In addition, some Italian endemics living in the same area should be considered of steppic origin such as *Goniolimon tataricum* (L.) Boiss. subsp. *italicum* (Tammaro, Pignatti & Frizzi) Buzurović [41,42], and *Astragalus aquilanus* Anzal. [43]. The presence of these plants in the central Apennines is due to different migrations from east to west. Plants with similar morphological features, with respect to the Northern and Eastern populations, e.g., *A. exscapus* subsp. *exscapus*, or *F. valesiaca* subsp. *valesiaca*, probably arrived in the Central Apennines during late-Pleistocene [15]. An initial wave of steppic plants probably occurred during the Messinian Salinity Crisis, such as the spread of an ancient *G. tataricum* lineage throughout south-eastern Europe [37]. This could be a hypothesis on the origin of the presence of *A. fucensis* in the Central Apennines.

Alternatively, a recent study on the phylogeography of the closely related *A. vernalis* [5] revealed that this plant species expanded its range from SE Europe into the Euro-Siberian steppe, with a Spanish population of the species being the earliest-diverging lineage. Whether members of our present study group parallel this migration pattern and *A. fucensis* constitutes the earliest-diverging remnant of an eastwards expanding *A. volgensis* could be hypothesized here but must await a much denser sampling of the latter species. Due to the restricted number of accessions analyzed in the present contribution, the biogeographical history of *A. fucensis* and *A. volgensis* remain unresolved.

Adonis fucensis is a very rare species, consisting of a very small population of 65 individuals, assessed here as critically endangered. In the two years (2021–2022) in which we were able to study the population we observed that although plants have a large number of flowers, they produce few fruits (we have observed many abortive achenes), and its survival is probably related to vegetative reproduction with consequent loss of genetic diversity. It will be absolutely necessary to undertake a dialogue with the National Park of Abruzzo, Lazio, and Molise to plan correct in situ and ex situ conservation strategies, to try to save this new species from extinction.

5. Taxonomic Treatment

Adonis subg. ***Adonanthe*** (Spach) W.T.Wang, Bull. Bot. Res., Harbin 14(1): 22 (1994) ≡ *Adonanthe* Spach, Hist. Nat. Vég. 7: 227 (1838).

Type: *A. vernalis* L. [Lectotype, Herb. Linn. No. 714.4, LINN [digital photo!]); image of the lectotype is available at https://linnean-online.org (accessed on 28 November 2022)].

Adonis sect. ***Adonanthe*** W.T.Wang, Bull. Bot. Res., Harbin 14(1): 26 (1994).

Type: A. vernalis L.

Adonis ser. ***Vernales*** Bobr. ex Poschk., Novosti Sist. Vyssh. Rast. 14: 83 (1977).

Type: A. vernalis L.

Adonis volgensis DC., Syst. Nat. [Candolle] 1: 545 (1817) ≡ *Adonanthe volgensis* (DC.) Chrtek & Slavíková, Preslia 50(1): 24 (1978).

Holotype: *Ad Volgam*, 1817, *Steven s.n.* [G barcode G00144834 [digital photo!]; image of the holotype is available at http://www.ville-ge.ch/musinfo/bd/cjb/chg (accessed on 28 November 2022)].

= *Adonis transsilvanica* Simonovich, Dokl. Akad. Nauk Belorusski. S.S.R., IX(6): 396 (1965).

Holotype: Hungaria. Transsilvania. *In collibus herbidis ad "Szénafü" prope "Kolozsvár"*, May and June 1910, *A. Richter 5201* [LE barcode LE00012366 [digital photo!], isotypes LD No. 2196730 [digital photo!], S No. 7488 [digital photo!]; image of the holotype is available at https://en.herbariumle.ru (accessed on 28 November 2022)].

Adonis fucensis F.Conti & Bartolucci, sp. *nov.* (Figure 7).

Figure 7. *Adonis fucensis* F.Conti & Bartolucci [Italy, Abruzzo, Mt. Annamunna, photo by F. Conti (**A–E**,**G**) and (**F**). Bartolucci (F)]. (**A**) Habitat and flowering plants of *A. fucensis*; (**B**) whole plants; (**C**) flowering plants; (**D**) cauline leaf; (**E**) flower, dorsal view; (**F**) flower, front view; (**G**) not mature aggregate fruit.

Type: ITALY. Abruzzo, Valle Lupara alla base del Monte Annamunna (Collelongo, L'Aquila; WGS84 33T 41°55′21″ N, 13°38′8″ E), radura, margini e cespuglieti a *Prunus spinosa* L. subsp. *spinosa*, 1038 m, 9 April 2021, *F. Conti s.n.* (holotype APP No. 66211).

Diagnosis: it is similar to *A. volgensis* but can be distinguished by larger cauline leaves, leaf blade pentagonal vs. triangular-ovate, lobes lanceolate vs. linear to narrowly lanceolate, larger stipules more divided, and larger sepals and petals.

Description: Herbs perennial. Rhizomatous. Roots numerous, fibrous. Stems erect, branched, pubescent, (8)10.25–13.75(16) cm tall. Scales membranous, alternate at lower parts of stem. Leaves stipulate, alternate, palmately compound with segments 3-pinnatifid, sessile, pentagonal, (56.7)65.75–86.22(96.06) mm long, (45.16)57.01–75.06(79.30) mm wide, green, hairy, n. of lobes and teeth (84)123–187(280), terminal lobes lanceolate, dentate, mucronate to acute at apex with angle (31.73)43.36–54.32(74.4)°, subterminal lobes (3.3)3.5–5.7(6.7) mm long, (1.14)1.41–1.91(2.83) mm max wide, (0.90)1.08–1.49(2.29) mm wide at base; stipules pinnatifid, (14.13)19.83–25.23(31.00) mm long, (9.59)16.40–23.24(32.87) mm wide, n. of stipules lobes and teeth (7)10–21(–28). Flower solitary, (40)44.75–60(66) mm in diameter in vivo, yellow. Sepals 5–8, ovate, obovate to elliptic, rarely truncate or dentate, (17)19–27.8(33) mm long (fresh), (11)12.25–23.63(27) mm long (dry), (9)10.1–14.9(18) mm wide (fresh), (7.5)9.63–12(14) mm wide (dry), olive-green, yellowish, brownish, purplish abaxially, hairy. Petals (9)12–14(18), yellow, obovate to narrowly obovate, (19)21–32(33) mm long (fresh), (11)21.75–27(30) mm long (dry), (7–)8–12.9(–14) mm wide (fresh), (6)8–10.50(13) mm wide (dry), obtuse, entire or rarely with rounded notches at apex, glabrous. Stamens numerous, basifixed; anther 2-loculed, oblong, 2–2.2 mm long, 0.5–0.7 mm wide; filaments filiform ca. 6 mm. Pistil numerous, ovary ovate, puberulous; styles 0.9–1.5 mm long, recurved. Aggregated fruit subglobular to ellipsoidal, 9–11 mm long, 8–10 mm wide. Receptacle elliptical pubescent. Achenes numerous, obovoid to ellipsoidal, 3.5–4.3 mm long, 2.5–3.2 mm wide, pubescent; style 0.9–1.5 mm long, recurved at base, ± appressed.

Etymology: *Adonis fucensis* is named after Fucino Plain located nearby to the north and affected by the presence of the third largest Italian lake drained in 1878.

Habitat: The species grows in the contact zone between bushes dominated by *Prunus spinosa* subsp. *spinosa* and steppic grasslands with *Festuca valesiaca* subsp. *valesiaca*, *Achillea setacea* Waldst. & Kit., *Koeleria splendens* C.Presl, *Centaurea jacea* L. subsp. *gaudinii* (Boiss. & Reut.) Gremli, and *Galium verum* L. subsp. *verum*.

Phenology: Flowering from March to April; fruiting from May to June.

Distribution: Endemic to one locality of Abruzzo (Central Italy) within the SAC IT7110205 "Parco Nazionale d'Abruzzo". The species grows in a small flat clearing on the slopes of Mt. Annamunna, between Amplero and Fucino plains (Supplementary File S1).

Conservation status: *Adonis fucensis* is known only by one location (locus classicus) where, during 2021, we counted only 65 individuals (genets). It is located within the NATURA 2000 network in the SAC IT7110205 "Parco Nazionale d'Abruzzo". The area of occupancy (AOO) is 4 km^2 (cell grid 2 × 2 km), calculated with GeoCAT (Geospatial Conservation Assessment Tool) software (http://geocat.kew.org/about (accessed on 10 October 2022)) [44]. We observed pressure due to the grazing of wild animals (especially wild boars that dig up single plants). Observing the aerial photos of the 1980s it is evident how in the *A. fucensis* habitat the shrub and tree vegetation increased by reducing the surface of the pastures probably due to a decrease in grazing by livestock. The natural succession of vegetation is a pressure and a threat for the population of *A. fucensis*. It is not possible to be certain of the decline of the species even if it is reasonable to assume that it was more common in the past. According to IUCN criterion B2ab(iii) [45], the species is assessed as Critically Endangered (CR).

Key to Adonis Species Belonging to sect. Adonanthe Distributed in Europe

1. Leaves pinnately compound .... ***A. apennina***
1. Leaves palmately compound 2
2. Leaves glabrous, with narrowly linear, entire lobes, rarely few-dentate .... ***A. vernalis***

2. Leaves pubescent, rarely glabrescent, with linear-lanceolate to lanceolate, dentate lobes 3

3. Middle cauline leaves triangular-ovate, rarely pentagonal, (24.12)38.70–58.35(109.66) mm long × (10.00)27.00–42.60(81.30) mm wide, with (12)32–85(176) lobes and teeth linear to narrowly lanceolate; stipule (4.64)9.31–15.88(22.00) mm long, with (2–)3–6(–12) lobes and teeth; sepal (6.40)9.26–14.51(21.07) mm long × (2.50)3.57–6.05(9.46) mm wide, petal (7.61)12.12–20.69(29.28) mm long × (2.85)3.95–6.83(11.68) mm wide .... ***A. volgensis***

3. Middle cauline leaves pentagonal, (56.70)65.75–86.22(96.06) mm long × (45.16)57.01–75.06(79.30) mm wide, with (84)123–187(280) lobes and teeth lanceolate; stipule (14.13)19.83–25.23(31.00) with 10–22.4(–26) lobes and teeth; sepal (11.00)12.25–23.63(27.00) mm long × (7.50)9.63–12.00(14.00) mm wide, petal (11.00)21.75–27.00(30.00) mm long × (6.00)8.00–10.50(13.00) mm wide .... ***A. fucensis***

Supplementary Materials: The following supporting information can be downloaded at: https://www.mdpi.com/article/10.3390/biology12010118/s1, File S1: List of the herbarium specimens examined; distribution map of *Adonis fucensis*. File S2: Neighbor-joining tree of accessions of *Adonis fucensis* and *A. volgensis*. Table S1: *Adonis* populations sampled for the present study with information on localities, voucher specimens, and GenBank accession numbers for nrDNA ITS.

Author Contributions: Conceptualization and methodology, F.C., F.B. and C.O.; field investigations, F.C., F.B. and R.N.; morphological measurements, F.C.; morphological analyses, F.B.; molecular analyses, C.O., M.D. and E.S.; writing—original draft preparation, F.C., F.B. and C.O.; writing—review and editing, F.C., F.B., C.O. and R.N.; supervision, F.C. and F.B.; funding acquisition, F.C. and F.B. All authors have read and agreed to the published version of the manuscript.

Funding: This research was funded by Abruzzo, Lazio, and Molise National Park (grant number BVI400031).

Institutional Review Board Statement: Not applicable.

Informed Consent Statement: Not applicable.

Data Availability Statement: The data presented in the current study are available within the article and Supplementary Materials.

Acknowledgments: The authors wish to thank Directors and Curators of B, BP, CL, K, MW, UPS, and US for providing us digital images herbarium specimens. We sincerely thank our friend Marina Buschi, who was the first to discover this new species during a CAI (Club Alpino Italiano) excursion, as well as Valeria Giacanelli and Biagio Sucapane for accompanying us during field investigations. Maximilian Schall is thankfully acknowledged for his technical help in the molecular systematic laboratory of C.O. (Regensburg University). Finally, we would like to thank the President, the Director and the Staff at Abruzzo, Lazio and Molise National Park Agency for encouraging and supporting this research. This work was supported by the "Progetto di Ricerca di Rilevante Interesse Nazionale" (PRIN) "PLAN.T.S. 2.0—towards a Renaissance of PLANt Taxonomy and Systematics" led by the University of Pisa, under grant number 2017JW4HZK.

Conflicts of Interest: The authors declare no conflict of interest.

References

1. POWO. Plants of the World Online, Facilitated by the Royal Botanic Gardens, Kew. Available online: http://www.plantsoftheworldonline.org/ (accessed on 10 October 2022).
2. Wang, W.T. Revision of *Adonis* (Ranunculaceae) I. *Bull. Bot. Res.* **1994**, *14*, 1–31.
3. Wang, W.T. Revision of *Adonis* (Ranunculaceae) II. *Bull. Bot. Res.* **1994**, *14*, 105–138.
4. Son, D.C.; Park, B.K.; Ko, S.C. Phylogenetic study of the section Adonanthe of genus *Adonis* L. (Ranunculaceae) based on ITS sequences. *Korean J. Plant Taxon.* **2016**, *46*, 1–12. [CrossRef]
5. Seidl, A.; Tremetsberger, K.; Pfanzelt, S.; Lindhuber, L.; Kropf, M.; Neuffer, B.; Blattner, F.R.; Király, G.; Smirnov, S.V.; Friesen, N.; et al. Genotyping-by-sequencing reveals range expansion of *Adonis vernalis* (Ranunculaceae) from Southeastern Europe into the zonal Euro-Siberian steppe. *Sci. Rep.* **2022**, *12*, 19074. [CrossRef] [PubMed]
6. Peruzzi, L.; Conti, F.; Bartolucci, F. An inventory of vascular plants endemic to Italy. *Phytotaxa* **2014**, *168*, 1–75. [CrossRef]

7. Bartolucci, F.; Peruzzi, L.; Galasso, G.; Albano, A.; Alessandrini, A.; Ardenghi, N.M.G.; Astuti, G.; Bacchetta, G.; Ballelli, S.; Banfi, E.; et al. An updated checklist of the vascular flora native to Italy. *Plant Biosyst.* **2018**, *152*, 179–303. [CrossRef]
8. Bartolucci, F.; Stinca, A.; Conti, F. Typification of the name *Adonis distorta* (Ranunculaceae). *Phytotaxa* **2021**, *523*, 264–268. [CrossRef]
9. Conti, F.; Bartolucci, F. The vascular flora of Gran Sasso and Monti della Laga National Park (Central Italy). *Phytotaxa* **2016**, *256*, 1–119. [CrossRef]
10. Conti, F.; Bartolucci, F. *The Vascular Flora of the National Park of Abruzzo, Lazio and Molise (Central Italy): An Annotated Checklist*; Geobotany Studies; Springer: Cham, Switzerland, 2015; 254p.
11. Conti, F.; Bartolucci, F. *La Flora Vascolare del Parco Nazionale d'Abruzzo, Lazio e Molise*; FastEdit: Acquaviva Picena, Italy, 2022.
12. Filibeck, G.; Cancellieri, L.; Sperandii, M.G.; Belonovskaya, E.A.; Sobolev, N.; Tsarevskaya, N.; Becker, T.; Berastegi, A.; Bückle, C.; Che, R.; et al. Biodiversity patterns of dry grasslands in the Central Apennines (Italy) along a precipitation gradient: Experiences from the 10th EDGG Field Workshop. *Bull. Eurasian Dry Grassl. Group* **2018**, *36*, 25–41. [CrossRef]
13. Filibeck, G.; Cancellieri, L.; Rosati, L. *Carta dei Tipi di Habitat di Interesse Comunitario (Direttiva 92/43/CEE) del Parco Nazionale d'Abruzzo, Lazio e Molise e dei siti Natura 2000: ZSC IT7110205, ZPS IT7120132, ZSC IT6050018, ZSC IT6050020, ZSC IT7212121—Note Illustrative*; Università degli Studi della Tuscia: Viterbo, Italy, 2022; 114p.
14. Conti, F.; Cancellieri, L.; Cangelmi, G.; Filibeck, G.; Rosati, L.; Bartolucci, F. New records of native and alien vascular plants from Abruzzo, Lazio and Molise National Park (Italy)—And additions to the flora of Abruzzo and Molise administrative regions. *Ann. Bot.* **2022**, *12*, 23–34. [CrossRef]
15. Cancellieri, L.; Sperandii, M.G.; Filibeck, G. First record of the steppic relict *Astragalus exscapus* L. subsp. *exscapus* in the Apennines (Abruzzo, Italy), and biogeographic implications. *Plant Biosyst.* **2017**, *151*, 944–948. [CrossRef]
16. Filibeck, G.; Cancellieri, L.; Bartolucci, F.; Becker, U.; Conti, F.; Maestri, S.; Mürz, M.; Schommer, E.; Sperandii, M.G.; Becker, T. *Festuca valesiaca* Schleich. ex Gaudin newly discovered in the Central Apennines (Italy): A further example of steppe relict in the Abruzzo "dry valleys". *Plant Biosyst.* **2020**, *154*, 593–600. [CrossRef]
17. Raab-Straube, E.V.; Hand, R.; Hörandl, E.; Nardi, E. Ranunculaceae. Available online: https://europlusmed.org/ (accessed on 27 November 2022).
18. Akeroyd, J.R. *Adonis* L. In *Flora Europaea*; Tutin, T.G., Heywood, V.H., Burges, N.A., Valentine, D.H., Walters, S.M., Webb, D.A., Eds.; Cambridge University Press: Cambridge, UK, 1993; Volume 1, pp. 267–269.
19. WFO. *Adonis transsilvanica* Simonovich. Available online: http://www.worldfloraonline.org/taxon/wfo-0000521284 (accessed on 15 December 2022).
20. Thiers, B. Index Herbariorum: A Global Directory of Public Herbaria and Associated Staff. New York Botanical Garden's Virtual Herbarium. Available online: http://sweetgum.nybg.org/science/ih/ (accessed on 10 October 2022).
21. Son, D.C.; Park, B.K.; Chang, K.S.; Choi, K.; Shin, C.H. Cladistic analysis of the section *Adonanthe* under genus *Adonis* L. (Ranunculaceae) from East Asia. *J. Asia-Pac. Biodivers.* **2017**, *10*, 232–236. [CrossRef]
22. IBM. *IBM SPSS Statistics for Windows, Version 25.0*; IBM Corp.: Armonk, NY, USA, 2017.
23. Hammer, Ø.; Harper, D.A.; Ryan, P.D. PAST: Paleontological statistics software package for education and data analysis. *Palaeontol. Electron.* **2001**, *4*, 1–9.
24. Gower, J.C. A general coefficient of similarity and some of its properties. *Biometrics* **1971**, *27*, 857–874. [CrossRef]
25. Doyle, J.; Dickson, E. Preservation of plant samples for DNA restriction endonuclease analysis. *Taxon* **1987**, *36*, 715–722. [CrossRef]
26. Doyle, J.; Doyle, J. A rapid DNA isolation procedure for small quantities of fresh leaf tissue. *Phytochem. Bull.* **1987**, *19*, 11–15.
27. Rydin, C.; Pedersen, K.R.; Friis, E.M. On the evolutionary history of *Ephedra*: Cretaceous fossils and extant molecules. *Proc. Natl. Acad. Sci. USA* **2004**, *101*, 16571–16576. [CrossRef]
28. White, T.J.; Bruns, T.; Lee, S.; Taylor, J. Amplification and direct sequencing of fungal ribosomal RNA genes for phylogenetics. In *PCR Protocols: A Guide to Methods and Applications*; Gelfand, D.H., Sninsky, J.J., White, T.J., Innis, M.A., Eds.; Academic Press: San Diego, CA, USA, 1990; pp. 315–322. [CrossRef]
29. Blattner, F.R. Direct amplification of entire ITS region from poorly preserved plant material using recombinant PCR. *BioTechniques* **1999**, *27*, 1180–1186. [CrossRef]
30. Blattner, F.R.; Weising, K.; Bänfer, G.; Maschwitz, U.; Fiala, B. Molecular Analysis of Phylogenetic Relationships among Myrmecophytic *Macaranga* Species (Euphorbiaceae). *Mol. Phylogenet. Evol.* **2001**, *19*, 331–344. [CrossRef]
31. Technelysium Pty Ltd. CHROMAS (Version 2.6.6). Available online: http://technelysium.com.au/wp/chromas (accessed on 15 October 2022).
32. Swofford, D.L. PAUP* (Version PAUP* v.4.0a169). Phylogenetic Analysis Using Parsimony (*and Other Methods). Available online: http://phylosolutions.com/paup-test (accessed on 15 October 2022).
33. Dorfner, M.; Ott, T.; Ott, P.; Oberprieler, C. Long-read genotyping with SLANG (Simple Long-read loci Assembly of Nanopore data for Genotyping). *Appl. Plant Sci.* **2022**, *10*, e11484. [CrossRef] [PubMed]
34. Vos, P.; Hogers, R.; Bleeker, M.; Reijans, M.; van de Lee, T.; Hornes, M.; Friters, A.; Pot, J.; Paleman, J.; Kuiper, M.; et al. AFLP: A new technique for DNA fingerprinting. *Nucleic Acids Res.* **1995**, *23*, 4407–4414. [CrossRef] [PubMed]
35. Huson, D.H.; Bryant, D. Application of phylogenetic networks in evolutionary studies. *Mol. Biol. Evol.* **2006**, *23*, 254–267. [CrossRef] [PubMed]
36. Barker, G. *Landscape and Society: Prehistoric Central Italy (Studies in Archaeology)*; Academic Press: London, UK, 1981.

37. Manzi, A. *Storia Dell'ambiente nell'Appennino Centrale: La Trasformazione della Natura in Abruzzo Dall'ultima Glaciazione ai Giorni nostri*; Meta: Treglio, Italy, 2012; 316p.
38. Primi, R.; Filibeck, G.; Amici, A.; Bückle, C.; Cancellieri, L.; Di Filippo, A.; Gentile, C.; Guglielmino, A.; Latini, R.; Mancini, L.D.; et al. From Landsat to leafhoppers: A multidisciplinary approach for sustainable stocking assessment and ecological monitoring in mountain grasslands. *Agric. Ecosyst. Environ.* **2016**, *234*, 118–133. [CrossRef]
39. Hejcman, M.; Hejcmanová, P.; Pavlů, V.; Beneš, J. Origin and history of grasslands in Central Europe—A review. *Grass Forage Sci.* **2013**, *68*, 345–363. [CrossRef]
40. Grossi, G. Il territorio di Casali d'Aschi dall'antichita al medioevo. In *Casali d'Aschi Ieri e Oggi*; Grossi, G., Palozzi, L., Amendola, V., Melchiorre, A., Eds.; Edizioni dell'Urbe: Rome, Italy, 1983; pp. 11–57.
41. Morretti, F.; Puppi, G.; Giuliani, C.; Conti, F. Heterostyly in *Goniolimon italicum* (Plumbaginaceae), endemic to Abruzzo (central Apennines, Italy). *An. Jard. Bot. Madr.* **2015**, *72*, e014. [CrossRef]
42. Buzurović, U.; Tomović, G.; Niketić, M.; Bogdanović, S.; Aleksić, J.M. Phylogeographic and taxonomic considerations on *Goniolimon tataricum* (Plumbaginaceae) and its relatives from south-eastern Europe and the Apennine Peninsula. *Plant Syst. Evol.* **2020**, *306*, 29. [CrossRef]
43. Anzalone, B. Su un nuovo *Astragalus* scoperto in Abruzzo. *Webbia* **1970**, *24*, 723–734. [CrossRef]
44. Bachman, S.; Moat, J.; Hill, A.W.; de la Torre, J.; Scott, B. Supporting Red List threat assessments with GeoCAT: Geospatial conservation assessment tool. *ZooKeys* **2011**, *150*, 117–126. [CrossRef]
45. IUCN. IUCN Guidelines for Using the IUCN Red List Categories and Criteria. Version 15.1. Available online: https://www.iucnredlist.org/resources/redlistguidelines (accessed on 23 September 2022).

biology

MDPI

Article

Morphological Continua Make Poor Species: Genus-Wide Morphometric Survey of the European Bee Orchids (*Ophrys* L.)

Richard M. Bateman * and Paula J. Rudall

Jodrell Laboratory, Royal Botanic Gardens Kew, Richmond, Surrey TW9 3DS, UK
* Correspondence: r.bateman@kew.org

Simple Summary: Our frequently deployed approach to optimally circumscribing species requires large-scale field sampling within and between populations for large numbers of morphometric characters, followed by multivariate ordinations to objectively seek discontinuities (or, failing that, zones of limited overlap) among sets of populations considered to represent species. Corresponding boundaries are sought in DNA-based outputs, either phylogenies or preferably ordinations based on population genetic data. Herein, we analyse within a molecular phylogenetic framework detailed morphometric data for the charismatic bee orchids (*Ophrys*), seeking a 'mesospecies' species concept that might provide a compromise between the nine 'macrospecies' recognised primarily through DNA barcoding and the several hundred 'microspecies' recognised primarily through perceived pollinator specificity. Our analyses failed to find robust groupings that could be regarded as credible mesospecies, instead implying that each macrospecies constitutes a morphological continuum. This problematic result encouraged us to reappraise both our morphometric approach and the relative merits of the contrasting macrospecies and microspecies concepts, and to reiterate the key role played by genetics in species circumscription.

Abstract: Despite (or perhaps because of) intensive multidisciplinary research, opinions on the optimal number of species recognised within the Eurasian orchid genus *Ophrys* range from nine to at least 400. The lower figure of nine macrospecies is based primarily on seeking small but reliable discontinuities in DNA 'barcode' regions, an approach subsequently reinforced and finessed via high-throughput sequencing studies. The upper figure of ca. 400 microspecies reflects the morphological authoritarianism of traditional taxonomy combined with belief in extreme pollinator specificity caused by reliance on pollination through pseudo-copulation, enacted by bees and wasps. Groupings of microspecies that are less inclusive than macrospecies are termed mesospecies. Herein, we present multivariate morphometric analyses based on 51 characters scored for 457 individual plants that together span the full morphological and molecular diversity within the genus *Ophrys*, encompassing 113 named microspecies that collectively represent all 29 mesospecies and all nine macrospecies. We critique our preferred morphometric approach of accumulating heterogeneous data and analysing them primarily using principal coordinates, noting that our conclusions would have been strengthened by even greater sampling and the inclusion of data describing pseudo-pheromone cocktails. Morphological variation within *Ophrys* proved to be exceptionally multidimensional, lacking strong directional trends. Multivariate clustering of plants according to prior taxonomy was typically weak, irrespective of whether it was assessed at the level of macrospecies, mesospecies or microspecies; considerable morphological overlap was evident even between subsets of the molecularly differentiable macrospecies. Characters supporting genuine taxonomic distinctions were often sufficiently subtle that they were masked by greater and more positively correlated variation that reflected strong contrasts in flower size, tepal colour or, less often, plant size. Individual macrospecies appear to represent morphological continua, within which taxonomic divisions are likely to prove arbitrary if based exclusively on morphological criteria and adequately sampled across their geographic range. It remains unclear how much of the mosaic of subtle character variation among the microspecies reflects genetic versus epigenetic or non-genetic influences and what proportion of any contrasts observed in gene frequencies can be attributed to the adaptive microevolution that is widely considered to dictate speciation in the genus. Moreover, supplementing

Citation: Bateman, R.M.; Rudall, P.J. Morphological Continua Make Poor Species: Genus-Wide Morphometric Survey of the European Bee Orchids (*Ophrys* L.). *Biology* **2023**, *12*, 136. https://doi.org/10.3390/biology12010136

Academic Editor: Lorenzo Peruzzi

Received: 29 November 2022
Revised: 3 January 2023
Accepted: 9 January 2023
Published: 16 January 2023

weak morphological criteria with extrinsic criteria, typically by imposing constraints on geographic location and/or supposed pollinator preference, assumes rather than demonstrates the presence of even the weakest of species boundaries. Overall, it is clear that entities in *Ophrys* below the level of macrospecies have insufficiently structured variation, either phenotypic or genotypic, to be resolved into discrete, self-circumscribing ("natural") entities that can legitimately be equated with species as delimited within other less specialised plant genera. Our search for a non-arbitrary (meso)species concept competent to circumscribe an intermediate number of species has so far proven unsuccessful.

Keywords: demographic systematics; ethology; evolution; morphometrics; natural selection; next-generation sequencing; ordination; phylogeny; reproductive isolation; sexual deceit; speciation; species circumscription

1. Introduction

1.1. Bee Orchids as an Evolutionary Case-Study

Few plant genera have gained as much academic attention as *Ophrys* L. (bee orchids). When attracting researchers, orchids per se have the advantage of being renowned for their complexity and diversity, as well as being readily recognisable (they are almost unique in developing their 'male' and 'female' reproductive organs as a single unified structure, the gynostemium [1–3]). But even among orchid genera, bee orchids are exceptionally charismatic, largely as a result of their near-ubiquitous reliance on pollination through sexual deception—specifically, through attempts made by (typically) bees and wasps to mate with their intricate flowers. Passing insects are attracted by first biochemical, then visual and finally tactile cues. Their combined effect must be sufficiently convincing to trick a male insect into twice believing that he is interacting with a conspecific female—collecting at least one of the two pollinaria from one flower and soon depositing it on the stigma of another flower. Pollination via pseudo-copulation is a high-risk strategy that achieves considerably less success than deceitfully promising and far less success than actually providing, a nectar reward [4–6]. Although pseudo-copulatory pollination is gradually being detected in other groups of plants, both within and outwith the orchid family, *Ophrys* remains the most intensively researched system of sexual deception and therefore acts as the archetypal textbook case-study [7,8].

Initial studies of evolution within the genus focused on behavioural ecology and functional morphology, therefore concentrating on visual and tactile cues [9–11]. Even among orchids, the labellum (the median petal, evolutionarily modified to act as a combined landing stage and sex-doll) is exceptionally complex in its overall three-dimensional topography, outline, markings and in the distribution of trichomes, papillae and glandular cells across its surface [12–14]. Technological advances enabled later studies to reveal biochemical aspects of the flower's phenotype to be equally extraordinary [15–18]. Complex cocktails of volatile exudates were shown to function as reputedly species-specific pseudo-pheromones, constituting the earliest-acting and arguably the most important of the three categories of pollinator cues. Early genetic work focused on determining relationships among taxa within the genus [19–21], but more recently, far larger datasets have begun to allow the identification of genes responsible for some of these intriguing phenotypic features [18,22,23].

Unsurprisingly, *Ophrys* has become a textbook case of what are widely regarded as numerous accumulated adaptations aiding pseudo-copulation, assumed by most authors to be largely the product of speciation through natural selection. In the eyes of many observers, *Ophrys* represents a classic example of recent, rapid and ongoing adaptive radiation, a position that, by definition, requires an exceptionally high speciation rate [24–26]. However, this undeniably attractive interpretation is valid only if the many supposed species that are the end-product of the radiation are genuine, rather than the consequence of a seriously flawed species concept that gives greater emphasis to evolutionary process than evolutionary product [27–29].

1.2. The Explicit Taxonomic Controversy: Microspecies, Mesospecies and Macrospecies

As well as becoming a classic case-study of pseudo-copulatory pollination and adaptive radiation, the genus *Ophrys* has also become a classic case-study of taxonomic controversy—a controversy of sufficient gravitas that it has now extended beyond the realm of systematic biology [27,30], even attracting the attention of philosophers of science [31]. In contrast with today, for much of the 20th century, the taxonomy of *Ophrys* was remarkably stable. Pursued through species concepts that were rooted in perceived (rather than quantified) degrees of morphological difference, several authoritative (but also inevitably authoritarian) classifications were published between 1960 and 1990 that recognised between 19 and 21 species, most of them congruent among these studies, together with a larger number of subspecies [32–34]. That traditional view of relatively few species plus many subspecies extended into some 21st century treatments; for example, the *Ophrys* monograph by Pedersen and Faurholdt [35] listed just 19 species and gave rise a decade later to a European orchid flora that listed only 22 species [36].

However, the early 1990s witnessed the first of two radical revolutions that impacted heavily on the taxonomy of *Ophrys*. Two 1994 publications elevated to species level many taxa that had been regarded by previous authors as subspecies or varieties: the technical monograph of Devillers and Devillers-Terschuren [37] listed 150 species, and the first edition of Delforge's European orchid flora recognised 142 species, a figure that presaged linear increases through three subsequent editions to 215, 252 and 353 species, respectively [38,39]. Many of these putative species were credited with only very restricted geographic distributions. We contend that this order-of-magnitude increase in recognised species was driven largely by an increasingly widely held belief that the unusual pseudo-copulatory pollination mechanism of *Ophrys* typically yielded a single preferred (or even sole) insect pollinator for each orchid species. This assumption led logically to the dubious conclusion that whatever phenotypic differences were apparent among taxa should be considered sufficient to justify their status as different species [40–42]. This concept yielded an increasing number of supposed local endemics, a mind-set that in turn encouraged recognition as different species of morphologically similar populations inhabiting contrasting geographic locations (it is no accident that *Ophrys* biodiversity is judged to peak among the numerous islands of the Aegean [25]). The species recognised in these high-diversity classifications were later termed '**microspecies**' by Bateman [28,29,43].

The second taxonomic revolution began in 2008, when Devey et al. [21] (see also [28,44]) applied DNA barcoding techniques (nuclear ribosomal ITS plus two plastid regions) to a wide range of named *Ophrys* taxa but were able to recognise with reasonable confidence only ten groupings (labelled A–J). Moreover, many of the individual plants analysed yielded multiple ITS ribotypes, implying recent and most probably ongoing gene-flow among microspecies. The subsequent application of next-generation sequencing techniques to a more restricted range of microspecies confirmed earlier DNA 'barcode' results but further reduced the number of groupings that could reliably be recognised to just nine (Figure 1). These nine essentially self-circumscribing ('natural') taxonomic units were termed '**macrospecies**' by Bateman [28,29,43,45].

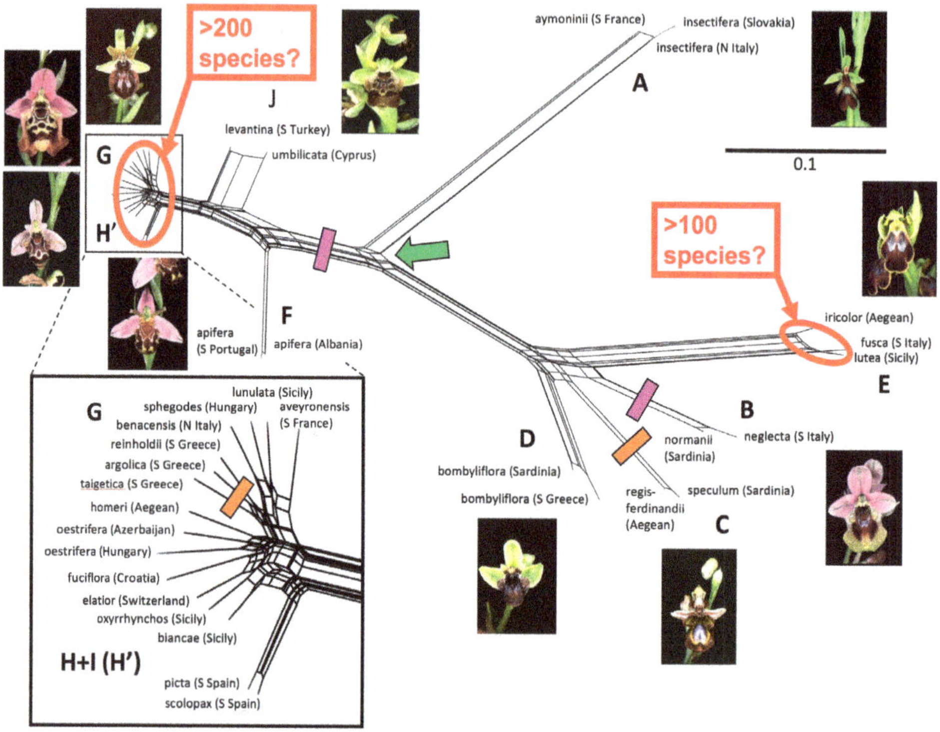

Figure 1. Unrooted *SplitsTree* network based on 4059 RAD-seq-derived single nucleotide polymorphisms (SNPs) for 32 plants that together represent the nine *Ophrys* macrospecies (**A–H',J**) circumscribed by Bateman et al. [43]. Inset: Magnified view of that portion of the topology that represents groups G and H'. The root of the tree remains equivocal but certainly lies close to the green-arrowed node. Red ovals emphasise the poverty of genetic variation present in the microspecies-rich groups E, G and H'. Pink bars indicate two predicted origins of the ability to infuse sepals and petals with pink–purple anthocyanin pigments; brown bars indicate two predicted origins of the ability to generate longitudinal stripes of red–brown pigment on the lateral sepals.

Many observers (including, we admit, ourselves) felt intuitively that the genus *Ophrys* should contain more than nine species but considerably fewer than 353 species. Such intermediate classifications have persisted into the 21st century. At the time of writing (November 2022), the authoritarian (but by no means authoritative) *World Checklist of Selected Plant Families* (WCSP) lists an impressive 2356 formal epithets (including those of hybrid 'nothospecies') for the genus *Ophrys* but nonetheless conservatively recognises only 29 embarrassingly heterogeneous species. Both Kreutz [46] and Baumann et al. [47] listed 65 species plus roughly three times that number of subspecies, thereby encompassing approximately as many formal epithets as did the exclusively microspecies-based classification of Delforge [39]. However, Delforge [39] did usefully organise his 353 microspecies into 29 intermediate groups. Herein, those supraspecific groups of Delforge, together with the species listed in any other classification of *Ophrys* that recognises between 20 and 65 species, are termed '**mesospecies**', again following Bateman [28,29,43,45]. The mesospecies is the most heterogeneous of the three categories of putative species; indeed, it is unclear whether any of the available classifications that includes mesospecies [37–39] was constructed on an explicit underlying species concept. Thus, a crucial outstanding question is whether a credible example of such a concept can be developed.

1.3. The General Taxonomic Controversy: Natural Versus Artificial Species

The extensive literature on *Ophrys* systematics is matched by that discussing the theory and philosophy of species concepts [26–29,31,48–57]. Most philosophically informed taxonomic debates are initially couched in terms of "artificial" versus "natural" species. Members of the various "natural" schools believe that species are fundamental biological entities made cohesive by shared biological processes. Most subscribe to "*the* biological species concept" (harkening back to Ernst Mayr's [48,49] oft-repeated belief that "species are groups of actually or potentially interbreeding populations which are reproductively isolated from other such groups"). This regrettable but persistent use ignores at least a dozen competing biological species concepts (reviewed by [50,51]). Members of the competing "artificial" schools fall into two broad classes—idealistic and pragmatic. Idealistic members have seriously considered and then rejected the idea that species have independent evolutionary fates, whereas pragmatic members argue that simply creating convenient pigeon-holes for organisms is the most rapid and effective approach to the urgent practical task of categorising the Earth's biodiversity.

Unfortunately, such elevated philosophical debates tend to ignore the reality of current taxonomic practice, at least as applied to higher plants. Most formal species-level epithets are rooted in the floras, monographs and taxonomic notes generated through 270 years of traditional herbarium taxonomy. Even today, such outputs rarely make any mention of which species concept is being applied, while also routinely side-stepping the fact that science, by definition, requires the formulation of explicit questions that can subsequently be addressed through the gathering and quantitative analysis of appropriate data. In the consistent absence of such cycles of analysis, the species thus circumscribed are, by definition, artificial. Those among the traditional practitioners who take seriously the idea that species should have biological reality hope that their output of morphologically circumscribed alpha-taxonomic species will in the future be proven to be what is euphemistically termed "natural" (a term now so often abused in a taxonomic framework that it arguably no longer has meaning). The most obvious way to test their species hypotheses is through the acquisition of additional categories of data, permitting reciprocal illumination through potential congruence among data-sets.

Returning to the three categories of *Ophrys* species established in Section 1.3—microspecies, mesospecies and macrospecies—are they best viewed as natural or artificial? The strength of the nine macrospecies is that they are unequivocally natural, irrespective of which of the many definitions of "natural" is applied. Their genetic profiles demonstrate clearly that they are bona fide clades that have enjoyed independent evolutionary fates for considerable periods of time, and they are also widely regarded as being readily distinguished through their morphologies. Our use of the term "self-circumscribing" was criticised by reviewers of this paper, but we stand by the concept; members of the same plant species "recognise" each other (albeit unconsciously) irrespective of whether the boundaries that they perceive match those ascribed to them by taxonomists.

In contrast, two radically different taxonomic world-views have combined to generate the current plethora of *Ophrys* microspecies. Although classifications containing hundreds of *Ophrys* species are a recent phenomenon, many of the formal epithets were coined in earlier eras that preceded not only molecular systematics but also Mayr's biological species concept and the recognition of universal pseudo-copulatory pollination among bee orchids. When first established, the epithets were at that stage undeniably artificial, but it could be argued that their artificiality was recognised by deploying most of them at infraspecific levels. It is only recently that belief in extreme pollinator specificity within *Ophrys* appears to have biologically legitimised the subsequent elevation of these epithets to species level, even though such taxonomic changes are still often performed in the absence of any explicit data analysis. In other cases, it is explicit studies of pollination that prompt circumscription of new *Ophrys* microspecies bearing novel epithets. Thus, *Ophrys* microspecies are hybrid progeny, generated when traditional artificial taxa were reappraised via taxonomic concepts that are rooted explicitly in reproductive biology and could therefore be regarded as strongly natural.

Similar ambiguity inevitably pervades the search for a mesospecies concept that would yield a number of species intermediate to those of the current macrospecies and microspecies concepts. Herein, we have used as our initial framework the 29 mesospecies of Delforge [39], which we suspect primarily reflect pragmatic motives. Although unsupported by a conceptual definition, they offer the great advantage of having an explicit relationship with the 363 microspecies recognised by an experienced European orchidologist. Our present search is focused on finding mesospecies that are both "natural" and can be accommodated within a single conceptual definition of species.

1.4. Aims of the Present Study

When preparing to write this paper, we first surveyed 55 papers that (a) compared *Ophrys* taxa, (b) were published during the last quarter-century and (c) presented at least some quantitative data. Among these papers, only 20% included analyses of morphological data, compared for example with 35% that quantified the composition of pseudo-pheromone cocktails. The majority of these data-sets were subjected to various forms of multivariate ordination, but in most cases, the data gathered were confined to sampling between two and five microspecies, typically putatively closely related within a single macrospecies; also, in most cases, only a few morphological characters considered a priori to be of particular significance (termed 'traits') were scored [22,58–65]. Consequently, these highly focused studies could only explore the credibility of supposed boundaries separating microspecies; higher-level mesospecies and macrospecies were not tested. The exception to this rule was our own recent morphometric ordination of 124 plants collectively representing 33 microspecies within macrospecies Sphegodes, which primarily sought mesospecies boundaries, employing modern approaches to analysing genotype as well as phenotype [45].

The taxonomic scope of the present paper is far broader. We examined morphometrically a much larger number of plants (457) that collectively span 113 of the 353 microspecies and all 29 of the mesospecies recognised by Delforge [39]; these plants adequately represent the variation found within all nine of the molecularly circumscribed macrospecies recognised by Bateman et al. [43]. Our sampling strategy was designed to explore every level in the demographic hierarchy, downwards from macrospecies > mesospecies > microspecies > population > individual plant within population, in order to test Bateman's [28,29,45] continuum hypothesis of variation within macrospecies. Given that species are the most fundamental units in our attempts to understand evolution and ecology, to conserve nature and to predict the consequences of future environmental changes, it is our contention that species should be self-circumscribing "natural" entities rather than mere conveniences of classification.

Taking as read the nine essentially self-circumscribing molecular macrospecies of *Ophrys* (Figure 1), we sought evidence that might suggest the existence of self-circumscribing entities within those macrospecies—in other words, taxa that could allow a more finely resolved natural classification based on some kind of mesospecies concept. We also sought repeated patterns of correlation among the 51 characters scored—patterns that could credibly be interpreted as adaptive trends resulting from directional or disruptive selection, critically reappraising evidence that *Ophrys* is presently in the midst of an evolutionary radiation. We used this case-study to review more broadly the strengths and weaknesses of our preferred morphometric approach, which we have applied to seven Eurasian orchid genera since it was devised by Bateman and Denholm [56] more than 40 years ago, before returning to a popular recent debate—the most appropriate role to award evolutionary mechanisms when attempting to optimise taxonomy.

Throughout the text, epithets representing the macrospecies of Bateman et al. [43] are presented in roman script with a capitalised first letter. In contrast, the mesospecies and microspecies of Delforge [39] are italicised; mesospecies epithets have a capitalised first letter, whereas microspecies epithets are presented entirely in lower case (they are usually also preceded by '*O.*').

2. Materials and Methods

2.1. Plant Materials

Fieldwork for our broader research programme targeted specifically at the genus *Ophrys* began in 2004 in the Peloponnese and ended in 2019 on Rhodes; populations were sampled across most of the geographic range of the genus, excepting only the extreme East in the Levant and the Caucasus. Initially, morphometric analyses were confined to macrospecies Fuciflora and Sphegodes, but we later expanded our taxonomic coverage; the last macrospecies to be included in our study was Fusca, sampled from 2008 onward.

Our standard procedure has for long been to randomly select study plants within field populations and to measure their vegetative characters in situ, also obtaining 1:1-scale perpendicular and lateral images of a representative flower taken from each plant as a record of its morphology and potential source of coordinates for use in geometric morphometrics [66–68]. Sampling is thus confined to excising one flower for mounting and morphometric study later in the same day while placing a second flower in a sachet of fine-ground silica gel to permit subsequent DNA analyses. However, our aims in most previous morphometric studies of European orchids have been to typically sample ten plants per population and ten or more populations per putative species [66,69]. In this study, several factors, most notably the vast number of microspecies described within the genus *Ophrys*, necessitated cruder, more pragmatic sampling that tolerated not only much smaller sample sizes per putative microspecies but also greater heterogeneity of sampling among microspecies.

2.2. Morphometric Data Collection

Morphometric characters employed in the present study are listed in Appendix A (for terminology see also Figures 2 and 3). Our initial list of 53 characters included two microscopic characters describing marginal bract cells, but these characters quickly proved to be insufficiently informative relative to the considerable amount of time consumed in recording them; hence, they were soon discarded. The remaining 51 characters contributing to the statistical analyses describe the stem and inflorescence (5), leaves and bracts (7), gynostemium and ovary (3), labellum (20) and lateral petals and sepals (16). They can alternatively be categorised as metric (33), meristic (3), multistate-scalar (13) and bistate (2). Flower colour was recorded by matching the colour of the lower half of the labellum (excluding the speculum), the sepals and the lateral petals to the closest colour block(s) of the Royal Horticultural Society Colour Chart for subsequent conversion into three quantified variables long recognised by the Commission Internationale de l'Eclairage.

Data for individual plants were summarised in an Excel v15.4 spreadsheet. Two rounds of multivariate data analysis were performed. The first, stand-alone analysis involved the complete matrix of 457 individuals (summarised in Table 1), together encompassing all 29 of the mesospecies and 113 (32%) of the 353 microspecies listed by Delforge [39].

The complete morphometric matrix of 457 plants × 51 characters (total 23,307 cells) contained 4.4% (1026) missing values—a figure that fell to just 0.7% (120) if only the 39 floral characters were considered. The most frequent cause of missing values among vegetative characters was the premature desiccation of the leaves, a phenomenon that affects plants growing in an unusually arid environment or sampled during an unusually arid spring (most commonly affecting macrospecies Sphegodes and Apifera).

Table 1. Summary of plants subjected to morphometric analysis in the present study, organised according to nine molecularly circumscribed macrospecies.

RAD-Seq (and ITS) Groups	Number of Sampled:				
	Plants	**Microspecies**	**Mesospecies**	**Localities**	**Countries**
Insectifera (A)	14	1	1	6	3
Speculum (C)	13	2	1	12	3
Bombyliflora (D)	14	1	1	14	5
Tenthredinifera (B)	31	11	1	27	5
Fusca (E)	51	33	8	46	7
Apifera (F)	15	1	1	15	6
Umbilicata (J)	52	8 *	1 *	44	3
Fuciflora (H + I)	143	23	6	68	8
Sphegodes (G)	124	33	9	104	7
TOTAL	*457*	*113*	*29*		

* Two of the analysed microspecies from Cyprus placed in mesospecies *Bornmuelleri* by Delforge [39] were molecularly assigned to macrospecies *Umbilicata* by Devey et al. [21] and Bateman et al. [43].

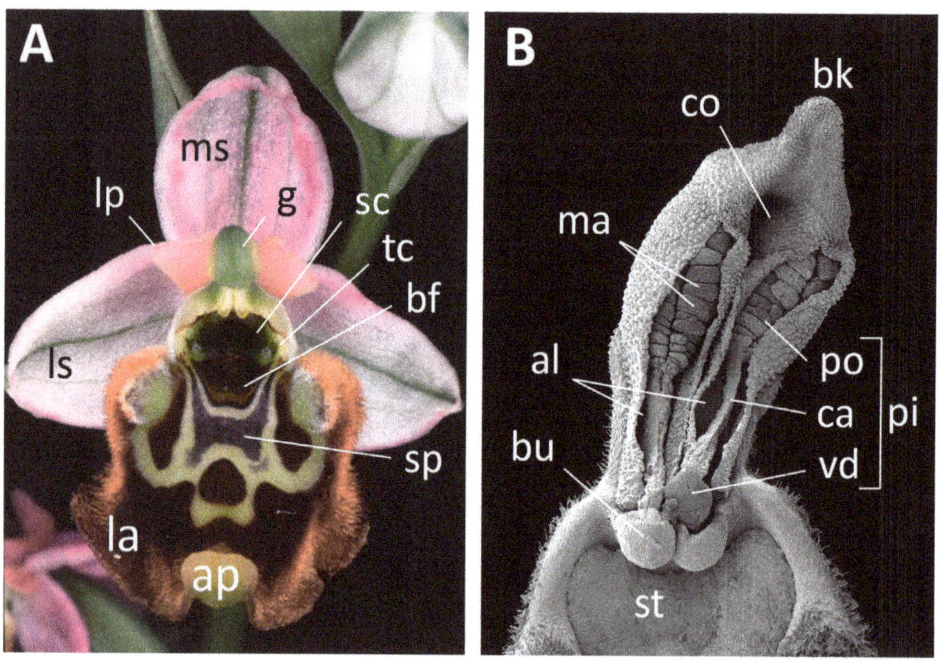

Figure 2. Terminology of a bee orchid flower. (**A**) Floral features, exemplified by microspecies *O. episcopalis* (mesospecies *Fuciflora*, macrospecies Fuciflora). (**B**) Gynostemium features, exemplified by *O. cretensis* (mesospecies *Mammosa*, macrospecies Sphegodes). Labels on (**A**): la, labellum (lip); lp, lateral petal; ms, median sepal; ls, lateral sepal; g, gynostemium (column); sc, stigmatic cavity; tc, temporal callosity ("pseudoeye"); bf, basal field; sp, speculum; ap, appendix. Labels on (**B**): al, anther locules; bu, bursicle (enclosing viscidial disc); bk, beak; co, connective; st, stigmatic surface; po, pollinium; ma, massulae; ca, caudicle; v, viscidial disc (these features are collectively termed the pollinarium–pi). Images: A = Richard Bateman, B = Paula Rudall.

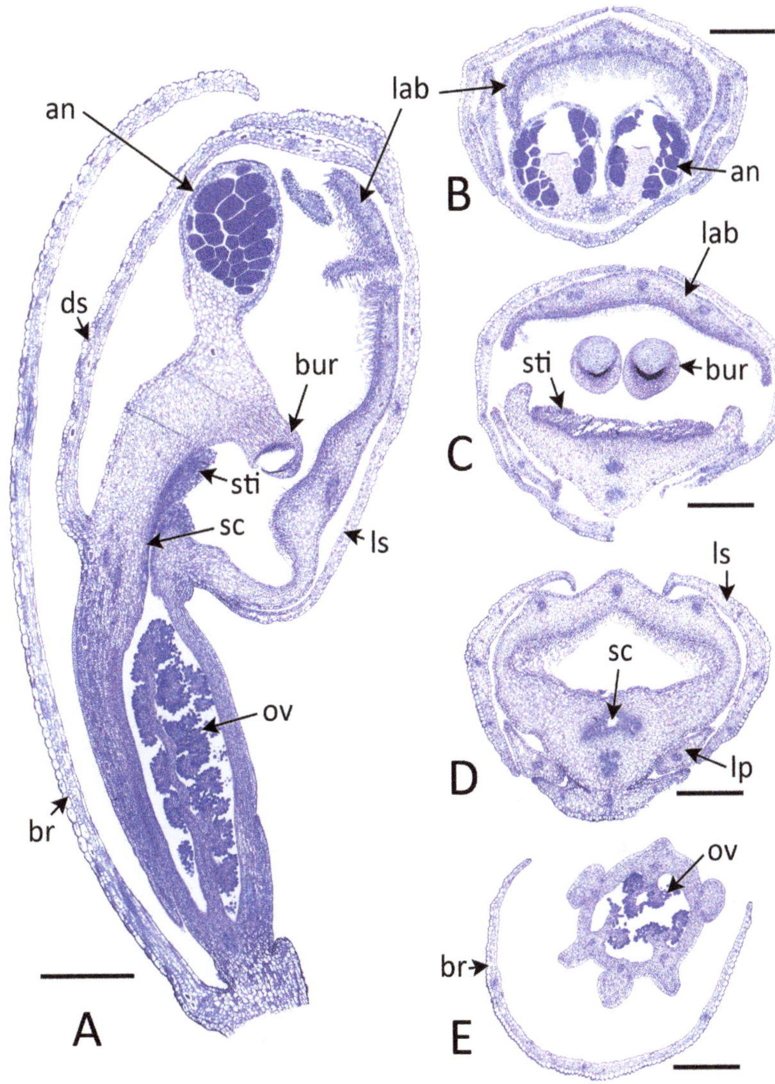

Figure 3. Anatomical sections of pre-anthetic flower bud of *O. pallida* (mesospecies *Obaesa*, macrospecies Fusca). (**A**) Median longitudinal section, showing entire bud subtended by bract. The anther, stigma and prominent bursicles are enclosed on one side by the labellum (which become strongly recurved once the flower has opened); it bears localised projecting trichomes. Note the exceptionally narrow stylar canal below the stigmatic papillae—a constriction encouraging pollen-tube competition. The elongate ovary contains numerous tiny ovules borne on branched placentae. Nectaries are absent. (**B–E**) Transverse sections. (**B**) Distal region, showing the anther, which contains dark-staining massulae, and the labellum, which bears outward- and backward-projecting trichomes. (**C**) Central region, showing the stigma and the lower parts of the prominent bursicles. (**D**) Proximal region, showing the narrow stylar canal. (**E**) Unilocular ovary containing many tiny ovules borne on branched parietal placentae. Labels: an, anther; bur, bursicle; br, bract; lab, labellum; lp: lateral petal; ls, lateral sepal; ov, ovules; sc, stylar canal; sti, stigma. All scales = 1 mm. All images: Paula Rudall.

2.3. Morphometric Data Analysis

For all matrices and submatrices, the assembled data were analysed via multivariate methods using Genstat v14 [70]. They were employed to compute a symmetrical matrix that quantified the similarities of pairs of datasets (i.e., plants) using the Gower Similarity Coefficient [71] on unweighted datasets scaled to unit variance. The matrix was in turn used to construct a dendrogram and a minimum spanning tree [72] and subsequently to calculate principal coordinates [73,74]—compound vectors that incorporate positively or negatively correlated characters that are most variable and therefore potentially diagnostic of putative taxa. As discussed in greater detail in Section 4.1, principal coordinates are especially effective for simultaneously analysing heterogeneous suites of morphological characters and can comfortably accommodate missing values. They have proven invaluable for assessing relationships among orchid species and populations throughout the last three decades (reviewed by Bateman [29,67,68]) and are the crux of the morphometric element of the present study.

For each multivariate analysis, the first four principal coordinates (PC1–4) were plotted together in pairwise combinations to assess the degree of the morphological separation of individuals (and thereby of populations and taxa) in these dimensions, and pseudo-F statistics were obtained to indicate the relative contributions to each coordinate of each of the original variables. The resulting ordinations were presented using Deltagraph v7.1 (SPSS/Red Rock software). In our many previous morphometric studies of orchids, we have also presented data based on overall similarity values (i.e., dendrograms and/or minimum spanning trees), but herein, we have focused exclusively on ordinations (a) because they have reliably proved to be more taxonomically discriminatory, (b) because we wished to compare ordinations resulting from alternative combinations of characters and ratios, and (c) because we wished to label some of the ordinations according to three contrasting categories: geographic origin, mesospecies assignment and microspecies identity.

Each of the 447 plants measured was attributed to one of the nine molecularly circumscribed macrospecies. Wherever possible, individuals were assigned to macrospecies on the basis of DNA data presented in previous studies [19–21,24,43,44,75], but where this was not possible, they were assigned on the basis of the closest morphological match that had benefited from molecular analysis. We recognise that this system is not wholly infallible, as was demonstrated by the supposed Greek local endemic microspecies *O. delphinensis*; when analysed using RAD-seq, this microspecies proved to be of recent hybrid origin, having a phenotype closer to that of macrospecies Sphegodes but a maternally inherited plastid genotype typical of macrospecies Fuciflora [45].

Each of the nine macrospecies was analysed separately. For the four macrospecies that have not (yet) been split taxonomically into numerous microspecies (Insectifera, Speculum, Bombyliflora, Apifera) and hence are represented herein by only small numbers of individuals, only two analyses were conducted: the full suite of 51 characters and a more restricted character-set consisting only of the 39 floral characters. The motivations for experimenting with removing the 12 vegetative characters (i.e., characters C40–51) were the facts that (a) these characters incurred higher frequencies of missing values, and (b) in terrestrial orchids, vegetative organs consistently show considerably greater epigenetic and ecophenotypic variation [76]. For each of the five macrospecies that have been split into numerous microspecies—and in most cases into several mesospecies (i.e., macrospecies Tenthredinifera, Fusca, Umbilicata, Fuciflora, Sphegodes)—and hence are represented herein by between 31 and 137 individuals (Table 1), further analyses were conducted wherein all metric floral measurements had been converted to ratios, a transformation enacted in order to minimise the impact of overall size differences on the resulting ordinations. All ordinations were labelled according to both microspecies identity and geographic origin, but only those groups rich in microspecies required additional categorisation according to mesospecies assignment.

Finally, similar analytical approaches were applied to a master matrix combining data for all nine macrospecies (also a submatrix consisting of the three most closely related macrospecies), with the aim of identifying the strongest patterns of morphological variation evident across the entire genus *Ophrys*. This ordination was labelled according to macrospecies, mesospecies and geographic origin. In addition, a second genus-wide ordination was conducted after reducing all metric measurements in the matrix to ratios, aiming to maximise contrasts in the shape of various organs while minimising the influence of plant size and vigour.

2.4. Scanning Electron Microscopy and Anatomy

Flowers of 12 selected microspecies collectively representing seven of the nine macrospecies were either collected by us in the field (most in Crete in 2017) or, in five cases, selected from among samples previously collected by knowledgeable individuals and deposited in the Spirit Collection of the Royal Botanic Gardens Kew.

Preparation for scanning electron microscopy involved selecting flowers from each inflorescence for dehydration through an alcohol series to 100% ethanol. They were then stabilised using an Autosamdri 815B critical-point drier, mounted onto stubs using double-sided adhesive tape, coated with platinum using an Emtech K550X sputter-coater, and examined under a Hitachi cold-field emission SEM S-4700-II at 2 kV. The resulting images were recorded digitally for subsequent aggregation in Adobe Photoshop.

In addition, flowers of a representative member of macrospecies Fusca from Sicily were prepared for anatomical light microscopy. Alcohol-fixed flowers were transferred through an ethanol series to 100% ethanol, followed by an ethanol–LR-White resin series, then embedded in LR-White resin using a vacuum oven set at 60 °C. Semi-thin sections were cut using a Reichert-Jung Ultracut ultramicrotome and a glass knife before mounting on glass slides. Sections were stained with Alcian Blue and imaged using a Leica DM6000B light microscope.

3. Results

3.1. Scanning Electron Microscopy and Anatomy

Two morphological extremes within the genus *Ophrys* are illustrated in Figures 2A and 3, together showing all of the relevant features of bee orchids. Both flowers show the classic orchid architecture of two closely spaced, alternating whorls of sepals and petals, the median petal forming a larger and more elaborate labellum that, along with the fused gynostemium into which it grades, is responsible for much of the functional morphology of the flower. The labellum is spur-less and dark brown; it varies in outline and three-dimensional topography, in the size and distribution of trichomes, and in the size and complexity of the paler trichome-free region termed a speculum. A distal appendix and lateral 'horns' may be present (Figure 2A) or absent (Figure 3). The lateral petals are considerably shorter and narrower than the sepals; all five tepals are typically coloured green or, less often, pink–purple.

Figures 3 and 4 show that, in most of its characteristics, the gynostemium of *Ophrys* is typical of other closely related genera within the subtribe Orchidinae. The paired, elongate anther-sacs are near-parallel and closely juxtaposed, their length typically exceeding considerably the diameter of the underlying stigmatic surface. Each loculus contains a single tripartite, club-shaped pollinarium. The distal pollinium typically bears 50–100 massulae, each consisting of many tightly packed pollen tetrads; it is linked to the proximal adhesive viscidial disc by a caudicle of similar length (Figure 4B). The viscidium is enclosed in a protective, desiccation-resistant bursicle. More distinctive of the genus *Ophrys* is the fact that, although closely juxtaposed and rather scrotal in appearance, the bursicles are actually separate, allowing (and perhaps even encouraging) a visiting insect to remove only one of the two pollinaria. The stigmatic surface is more concave, equidimensional and simpler in outline than those of many other Orchidinae. The narrow stylar canal links the stigmatic surface to the unilocular ovary (Figure 3A,D), guiding innumerable pollen tubes towards similarly large numbers of minute ovules. *Ophrys* massulae are unusually cohesive, as demonstrated by the remains of a disaggregated pollinium that are attached to the stigmatic surface of the *O. leochroma* flower shown in Figure 4K.

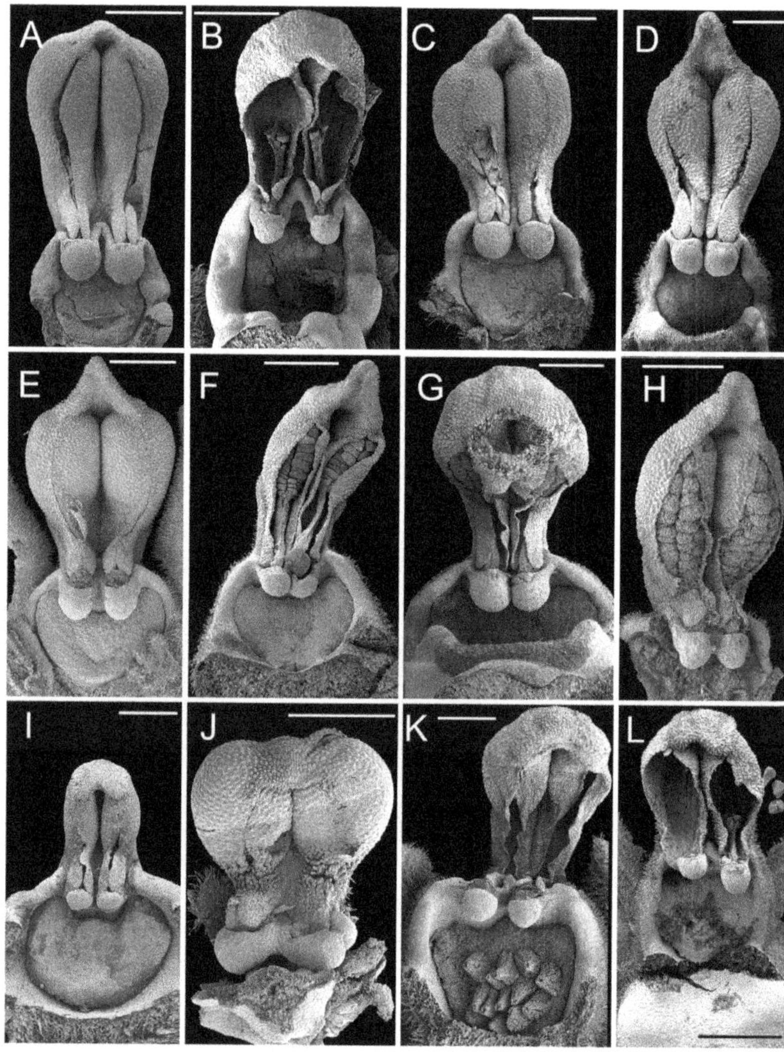

Figure 4. Scanning electron micrographs of gynostemia of 12 *Ophrys* microspecies, collectively representing seven of the nine macrospecies. (**A**) *O. vernixia*: mesospecies *Speculum*, macrospecies Speculum. (**B**) *O. regis-ferdinandii*: mesospecies *Speculum*, macrospecies Speculum. (**C**) *O. ferrum-equinum*: mesospecies *Mammosa*, macrospecies Sphegodes. (**D**) *O. heldreichii*: mesospecies *Heldreichii*, macrospecies Fuciflora. (**E**) *O. mammosa*: mesospecies *Mammosa*, macrospecies Sphegodes. (**F**) *O. cretensis*: mesospecies *Mammosa*, macrospecies Sphegodes. (**G**) *O. lacaena*: mesospecies *'Bornmuelleri'*, macrospecies Fuciflora. (**H**) *O. sphegodes*: mesospecies *Sphegodes*, macrospecies Sphegodes. (**I**) *O. phryganae*: mesospecies *Lutea*, macrospecies Fusca. (**J**) *O. bombyliflora*: mesospecies *Bombyliflora*, macrospecies Bombyliflora. (**K**) *O.* cf. *leochroma*: mesospecies *Tenthredinifera*, macrospecies Tenthredinifera. (**L**) *O. insectifera*: mesospecies *Insectifera*, Macrospecies Insectifera. Scale in all images =1 mm. Images: Paula Rudall.

However, there are relatively few differences among the viscidia of contrasting macrospecies within the genus *Ophrys* (Figure 4). It has long been known [43] that a clade consisting of five macrospecies (groups F–J in Figure 1) is characterised by the possession of a triangular beak extending upwards from the connective (Figure 4C–H, though almost severed in G). This extension reaches its maximum expression in the "flying duck" gynostemium of macrospecies Apifera (not shown). However, Figure 4 usefully increases the number of autapomorphic states recognised in macrospecies Insectifera, specifically by revealing the presence of a thickened, strongly papillose connective that resembles a toupé (Figure 4L). The anthers appear comparatively short in the macrospecies Insectifera (Figure 4L), Fusca (Figure 4I) and especially Bombyliflora (Figure 4J) but more elongated in the macrospecies Speculum (Figure 4A,B). The stigmatic surface is unusual in being slightly higher than wide in two macrospecies, Insectifera (Figure 4L) and Speculum (Figure 4A,B), and is especially deeply recessed—and hence particularly well-defined—in the Fusca group (Figure 4I).

3.2. *Morphometrics: Analyses Involving Multiple Macrospecies*

3.2.1. All Nine Macrospecies

Seeking a preliminary overview, our initial analysis used all 457 individuals and all 51 characters (Figure 5). The first two principal coordinates encompassed 34% of the total variation, the first coordinate being considerably stronger than the second. As expected, given the extreme multi-directionality of character variation within the genus, the analysis failed to yield the complete separation of any of the nine macrospecies from the remainder, though it did helpfully suggest polarisation between two morphological extremes. The first coordinate gave almost complete separation of a group of four comparatively cohesive macrospecies (Insectifera, Fusca, Speculum, Bombyliflora) that reliably possess green (rather than pink) lateral petals and sepals and a labellum that shows little or no expression of an appendix, bears a relatively simple speculum and (with the exception of Bombyliflora) tends to be comparatively two-dimensional. Fuciflora and Apifera constitute the opposing pole of PCo1, leaving in intermediate positions Tenthredinifera, Umbilicata and Sphegodes; the latter macrospecies appears especially incohesive.

The second coordinate yielded the almost complete separation of Tenthredinifera from the remainder, based mainly on its trichome-rich, apically notched labellum, featuring a yellow marginal zone and a small and simple but prominent speculum. These characters also allow some differentiation among macrospecies Insectifera, Bombyliflora, Speculum and Fusca, though plants of the latter are more widely spread due to their considerable variation in flower size and trichome development.

Surprisingly, the third coordinate offered no taxonomic separation—unfortunately, it was dominated by a character (C49) that was conceived with the intention of describing leaf shape but proved to be overly crude. The fourth coordinate (not shown) used characters such as petal curvature, speculum length and the presence versus absence of brown stripes on the lateral sepals to establish polarity between Speculum at one extreme and Insectifera at the other, but otherwise this axis was not taxonomically discriminatory.

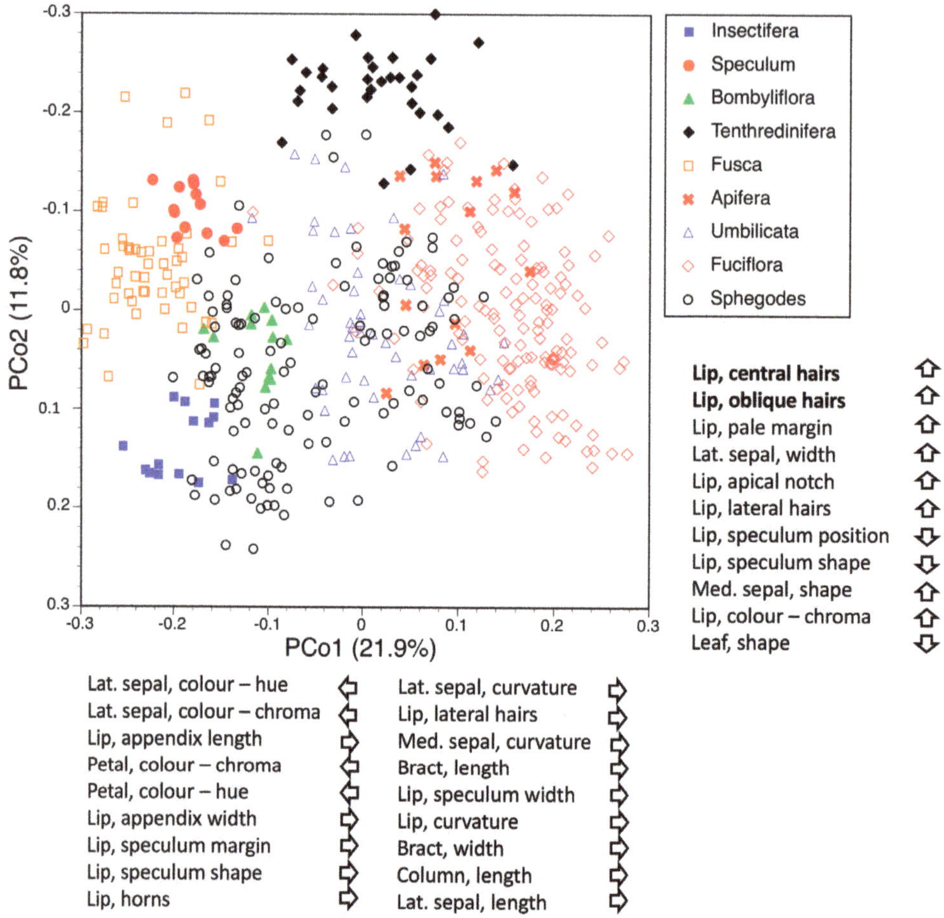

Figure 5. Plot of principal coordinates 1 and 2 for the genus *Ophrys*, labelled according to nine macrospecies groups: 447 individuals from ca. 300 localities, 51 variable characters. Characters contributing to each coordinate are listed in descending order, with arrows indicating the direction of increase in the value of each character (boldface characters dominate that coordinate).

3.2.2. The Umbilicata-Fuciflora-Sphegodes Clade

In the hope of reducing dimensionality in the data, we therefore attempted an analysis of three of the four macrospecies most intensively sampled here (Figure 6), which (as determined in our RAD-seq tree; Figure 1) together form a discrete, relatively tight-knit clade: Umbilicata, Fuciflora and Sphegodes. Based on analysis of 319 plants, the first coordinate largely separated most Fuciflora from the majority of Sphegodes plants, with Umbilicata intermediate. However, Sphegodes plants were split into two groups of approximately equal numbers of plants, based largely on whether their petals and sepals were green or pink. Similarly, green-flowered Fuciflora plants (a minority within this macrospecies) were consequently placed closer to Sphegodes.

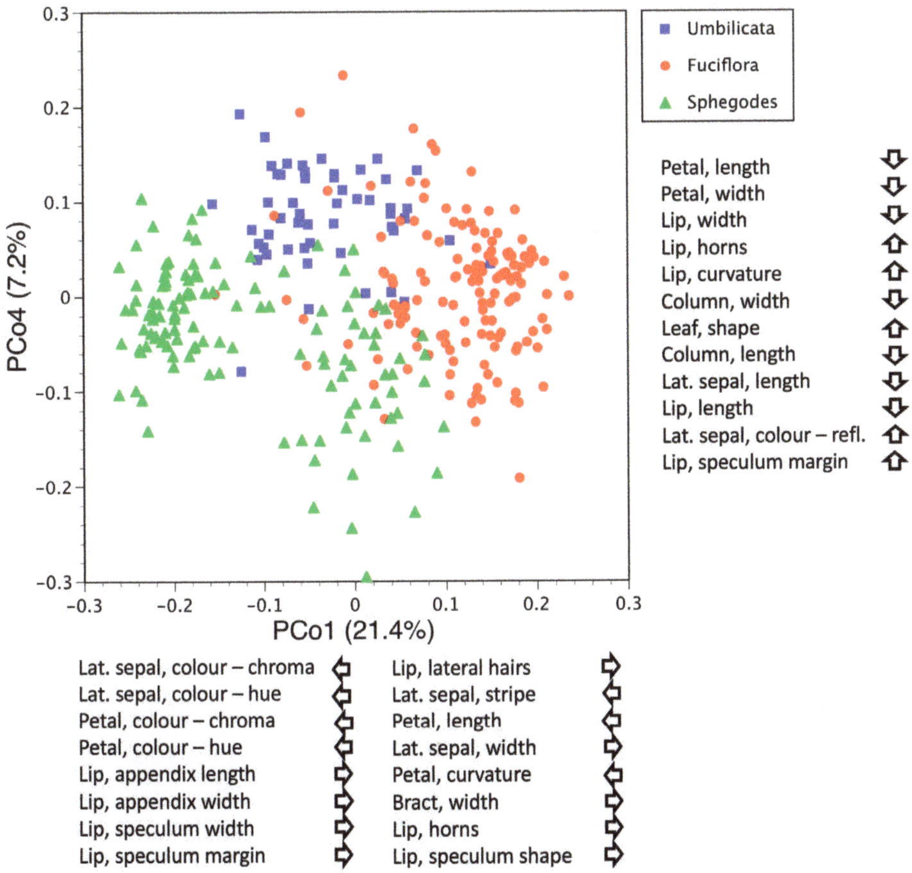

Figure 6. Plot of principal coordinates 1 and 4 for the macrospecies Umbilicata plus Fuciflora plus Sphegodes, labelled according to three macrospecies groups: 319 individuals from ca. 200 localities and 51 variable characters. Characters contributing to each coordinate are listed in descending order, with arrows indicating the direction of increase in the value of each character.

We anticipated that re-analysing this matrix after removing the six characters that represent petal and sepal colour would unify the two groups of Sphegodes, which it did, and would thereby reduce the overall morphological overlap of Sphegodes with Fuciflora, which it did not (results not shown). In fact, the amount of overlap increased. Evidently, flower colour is in fact important for distinguishing the macrospecies, but it largely operates at a level that is more subtle than simply contrasting green with pink. Clearly, the presence of colour dimorphism within each of these macrospecies is problematic with regard to taxon circumscription at finer scales. Nonetheless, additional characters contributing to the first coordinate, such as the presence and size of a labellar appendix, nature of the speculum, petal length and sepal width, also contribute considerably to distinguishing the three macrospecies in this analysis.

The second and third coordinates were much weaker than the first and provided little taxonomic separation. Only when reaching down to the fourth coordinate (Figure 6) was Umbilicata partially separated from Fuciflora and Sphegodes, on the basis of its smaller petals and columns, and (for many Umbilicata microspecies) also the narrower, more three-dimensional labella bearing prominent horns.

3.3. *Morphometrics: Analyses Involving Single Macrospecies but Multiple Mesospecies*

3.3.1. Macrospecies Sphegodes

The first two principal coordinates for the morphometric matrix of 124 plants of macrospecies Sphegodes are plotted together in Figure 7, individual plants being labelled according to mesospecies assignment sensu Delforge [39]. Although totalling only 32% of the total variance, the first two coordinates are considerably stronger than the remainder. The first coordinate is dominated by the contrast between plants with green (left) versus pink (right) lateral petals, which again divides the plants into two ill-defined clusters. Subordinate contributory characters show that plants located toward the right of the plot typically possess labellar appendixes and have comparatively long columns, large sepals and labella that are comparatively hairy but tend to lack forward-pointing horns. The more egalitarian second coordinate represents several characters that contribute approximately equally, together reflecting comparatively large plant size and, to a lesser degree, large flower size—effectively constituting a vigour coordinate. The third coordinate combines other vegetative dimensions and leaf number with a mixed bag of labellum characters. All coordinates of fourth order and below reflect very few characters and offer little if any taxonomic discrimination.

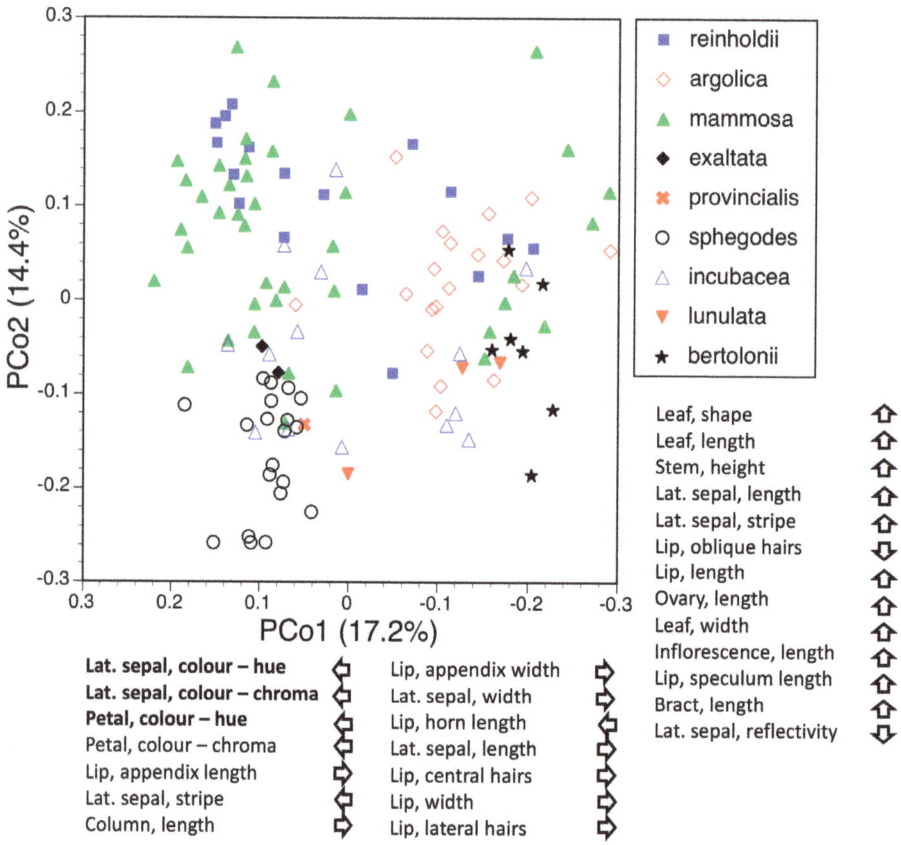

Figure 7. Plot of principal coordinates 1 and 2 for the macrospecies Sphegodes, labelled according to nine mesospecies groups: 124 individuals from 104 localities and 51 variable characters. Characters contributing to each coordinate are listed in descending order, with arrows indicating the direction of increase in the value of each character (boldface characters dominate that coordinate).

Viewed at the level of mesospecies, the impact of the first coordinate (Figure 7) is overly dependent on whether the mesospecies in question encompasses a mixture of green- and pink-petaled plants; consequently, those mesospecies considered capable of exhibiting both colours (*Mammosa, Incubacea, Reinholdii*) are spread more widely along the first coordinate than are those that are either reliably green (all lack anthocyanins, e.g., *Sphegodes*) or reliably pink (all possess anthocyanins, e.g., *Bertolonii*). The second coordinate gives the almost complete separation of the large-bodied, large-flowered mesospecies *Mammosa* and *Reinholdii* from the more modestly sized mesospecies *Sphegodes* and *Provincialis*. The third coordinate serves only to partially separate mesospecies *Incubacea* from the remainder (not shown).

In an additional experiment, morphological variation within populations of a single microspecies was estimated through analysis of 15 pairs of con-microspecific plants that together encompassed the full morphological range exhibited by macrospecies Sphegodes. Distances separating these paired individuals on Figure 7 varied greatly from less than 0.01 to 0.21, averaging 0.050 ± 0.046 for PCo1 and 0.079 ± 0.061 for PCo2. The comparatively large mean value for this comparison on the second coordinate relative to the first coordinate is readily explained by contrasts in plant size encapsulated by PCo2 that are likely to reflect differences in the development (ontogeny) and environment of growth (ecophenotypy) at least as much as any direct genetic influence (see also [45]).

3.3.2. Macrospecies Fuciflora

The first two coordinates for the 143 plants of Fuciflora represent just 27% of the total variance (Figure 8), and neither is effective at distinguishing among the six sampled mesospecies. As with Sphegodes, the first coordinate is dominated by sepal and petal colour, supported by various flower size characters plus leaf shape. Most of the characters contributing to the second coordinate are also positively correlated dimensions of various floral parts, suggesting that much of the morphological variation within macrospecies Fuciflora reflects differences in flower size.

The third and fourth coordinates proved to be more discriminatory at the mesospecies level. The third coordinate almost completely separated *Fuciflora* and *Tetraloniae* from most of the remaining mesospecies on the basis of their poorly developed labellar horns and shoulder hair, larger petals and modest vegetative size. The fourth coordinate partially separated mesospecies *Heldreichii* and *Scolopax* from the remainder on the basis of their laterally reflexed labella with prominent lateral lobes and their comparatively low vegetative vigour.

A finer-scale analysis of this relatively well-sampled macrospecies is presented in Section 3.5.

3.3.3. Macrospecies Umbilicata

The eastern Mediterranean macrospecies Umbilicata was analysed for 52 plants representing two mesospecies and eight microspecies. This plot differed from almost all of the other analyses performed in that the plants were resolved by the first coordinate into two discrete groups separated by a genuine discontinuity (Figure 9). Specifically, the Cypriot microspecies *levantina* and *aphrodite* were distinguished from the remainder primarily as a result of their simpler, less striking specula, supported by characters demonstrating that their labella were also more hirsute and less three-dimensional, with little if any development of lateral lobes, and were more likely to develop apical notches; they were also more likely to be presented parallel to the stem.

The second coordinate organised the remaining six microspecies into a near-linear arrangement of three crude clusters: *umbilicata* plus *lapethica*, *attica* plus *flavomarginata* plus *rhodia*, and *kotschyi*. This cline represents two correlated trends: from strong pink through pink-washed green to strong green sepals and petals, and from smaller to larger flowers, especially with regard to labellum length.

The third axis almost completely separated microspecies *aphrodite*, *attica* and *rhodia* from the remainder on the basis of their wider sepals and often a lack of a brownish–pink wash on the petals, whereas in contrast the fourth axis reflected vegetative vigour, thereby separating *aphrodite* from the more modestly proportioned *levantina* (not shown).

3.3.4. Macrospecies Tenthredinifera

Macrospecies Tenthredinifera is represented by 31 plants encompassing 11 (mostly recently conceived) microspecies. The first coordinate largely separated *normanii*, *grandiflora* and *ficalhoana* from the remaining microspecies (Figure 10). It reflected their combination of vegetative vigour, notably larger bracts and leaves, and floral characters such as lateral sepal size, labellum width and the presentation of the flowers parallel to the stem. The considerably weaker second coordinate provided little discrimination, other than partially separating the Sardinio–Corsican microspecies *neglecta* according to its simple, basally concentrated, brightly margined speculum, short petals and comparatively narrow column.

The third coordinate, mainly reflecting the reflectivity and hue of the lateral sepals, offered no taxonomic discrimination, whereas the fourth coordinate largely separated the Sardinian microspecies *grandiflora* and *normanii* from the remaining nine microspecies analysed, primarily because they possessed ovate rather than obovate median sepals.

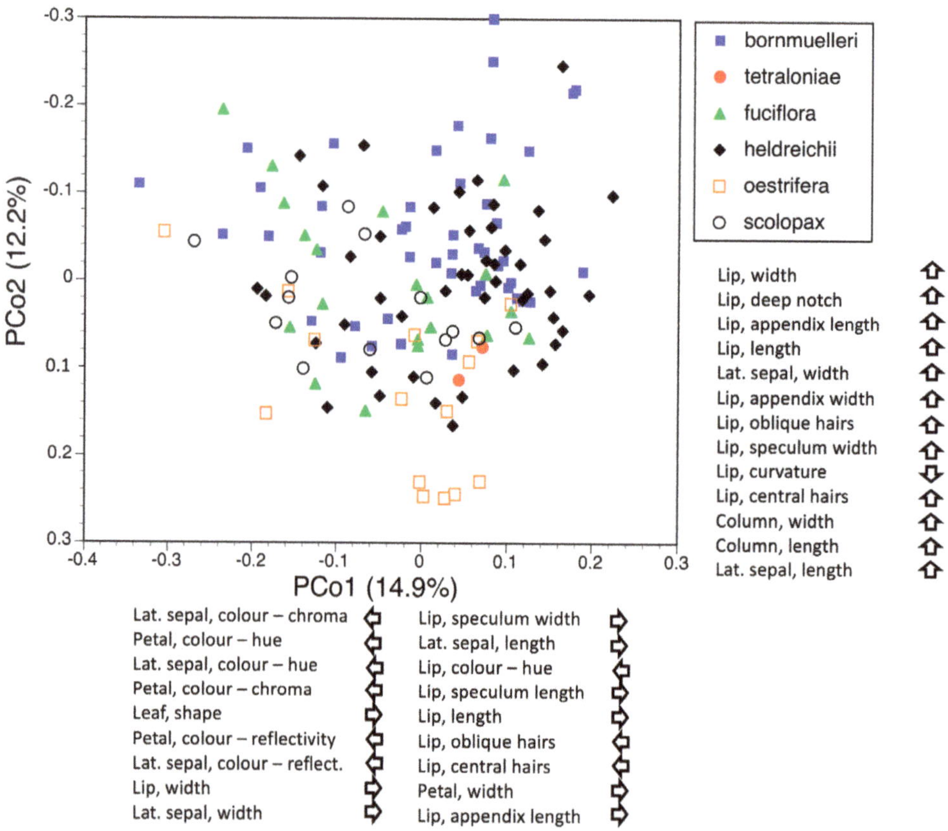

Figure 8. Plot of principal coordinates 1 and 2 for the macrospecies Fuciflora, labelled according to six mesospecies groups: 143 individuals from 68 localities and 51 variable characters. Characters contributing to each coordinate are listed in descending order, with arrows indicating the direction of increase in the value of each character (see also Figure 16).

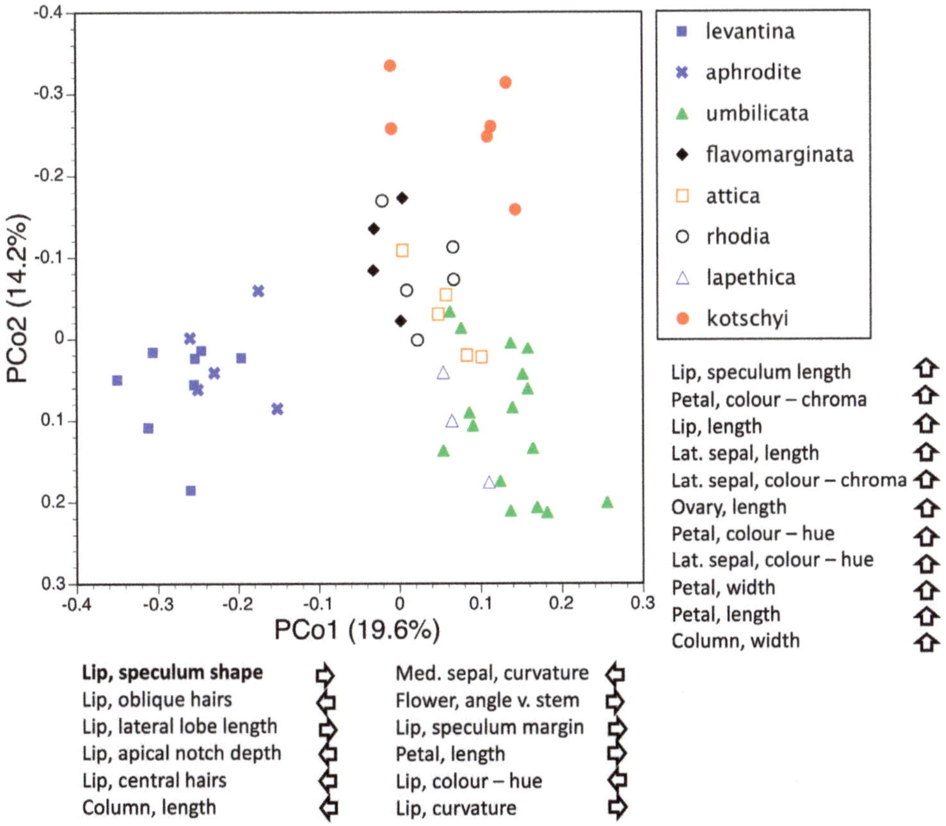

Figure 9. Plot of principal coordinates 1 and 4 for the macrospecies Umbilicata, labelled according to eight microspecies groups: 52 individuals from 44 localities and 51 variable characters. Characters contributing to each coordinate are listed in descending order, with arrows indicating the direction of increase in the value of each character (boldface characters dominate that coordinate).

3.3.5. Macrospecies Fusca

The 51 plants (and 33 microspecies) of macrospecies Fusca yielded a first coordinate that represented correlated size contrasts in all floral organs (Figure 11), mesospecies *Lutea*, *Attaviria* and *Obaesa* having the smaller flowers. The second coordinate represented a negative correlation between increases in several plant size characters versus increased hairiness of labellum and possession of an obovate median sepal; short, few-flowered plants with relatively trichome-rich labella characterise mesospecies *Omegaifera* and *Obaesa*. Neither the third coordinate, dominated by leaf shape, nor the fourth coordinate, dominated by the width, colour (chroma) and pale margin of the labellum, provided meaningful distinction among mesospecies.

Because the first two coordinates relied so heavily on labellar dimensions, we re-analysed the matrix after replacing 18 metric measurements with 10 ratios derived from those metric measurements (not shown), aiming to emphasise the shapes of structures rather than their relative sizes. This seemingly radical modification imposed on the underlying data had surprisingly little impact on the revised distribution of plants across the first two coordinates; moreover, it neither strengthened nor weakened the clustering of individuals according to mesospecies assignment.

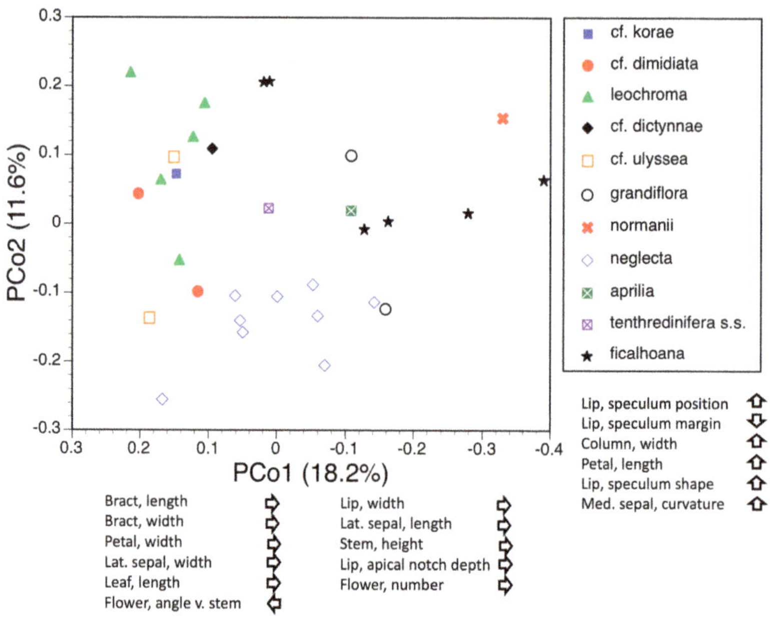

Figure 10. Plot of principal coordinates 1 and 2 for the macrospecies Tenthredinifera, labelled according to nine mesospecies groups: 31 individuals from 27 localities and 47 variable characters. Characters contributing to each coordinate are listed in descending order, with arrows indicating the direction of increase in the value of each character.

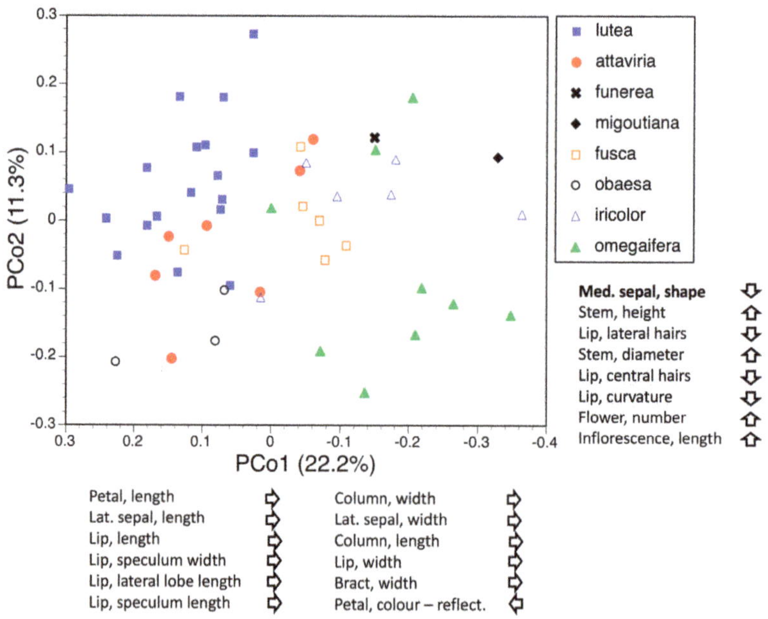

Figure 11. Plot of principal coordinates 1 and 2 for the macrospecies Fusca, labelled according to eight mesospecies groups: 51 individuals from 46 localities and 46 variable characters. Characters contributing to each coordinate are listed in descending order, with arrows indicating the direction of increase in the value of each character (boldface characters dominate that coordinate).

3.4. Morphometrics: Analyses Involving Single Macrospecies and Single Mesospecies

3.4.1. Macrospecies Speculum

Although dominated by *O. speculum* s.s., this small-scale analysis of 13 plants from 12 localities spanning the Mediterranean also included two plants of the Eastern microspecies *O. regis-ferdinandii* (Figure 12). The two microspecies were readily separated on the first coordinate, *regis-ferdinandii* tending to produce a larger number of flowers that possessed a lip with strongly recurved lateral margins and lateral sepals that curved more strongly forwards and bore broader brown longitudinal stripes. All lower-order coordinates failed to distinguish among plants of *O. speculum* s.s. from contrasting regions of the Mediterranean Basin, the two plants sampled in Andalusian Spain being especially widely separated from each other on the second coordinate. This result discourages the division of *O. speculum* into further microspecies.

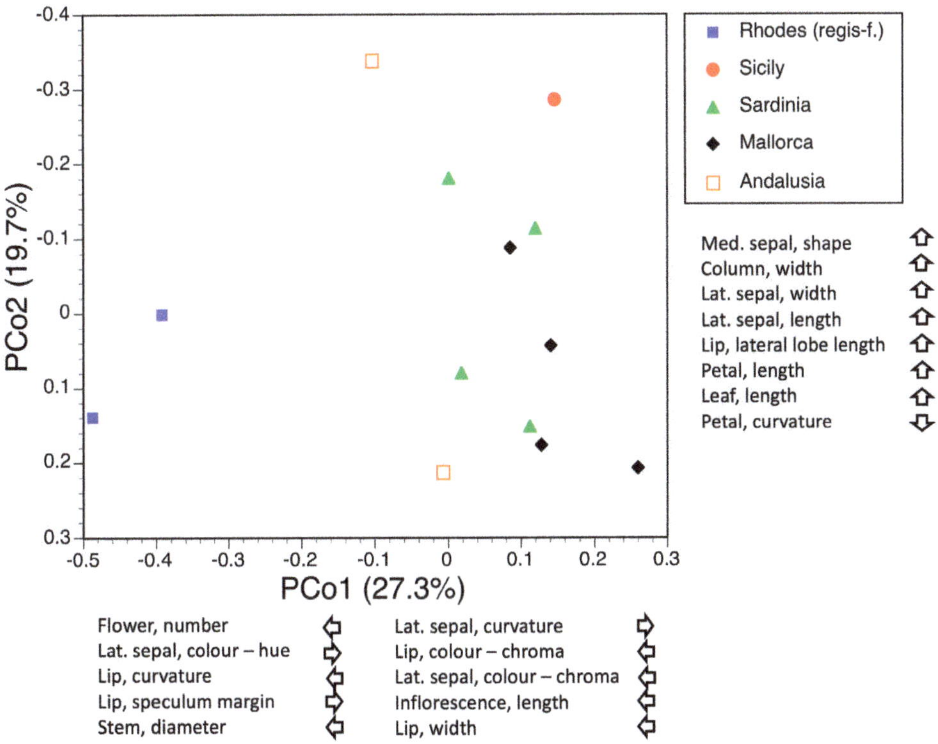

Figure 12. Plot of principal coordinates 1 and 2 for the macrospecies Speculum, labelled according to five regions of origin: 13 individuals from 12 localities and 42 variable characters. Characters contributing to each coordinate are listed in descending order, with arrows indicating the direction of increase in the value of each character.

3.4.2. Macrospecies Bombyliflora

Thus far, macrospecies Bombyliflora has escaped the taxonomic fragmentation into microspecies that has afflicted most other macrospecies of *Ophrys*. Nonetheless, we were interested to see whether any correlated trends might emerge from our small but geographically extensive set of plants: 14 plants from 14 localities, sampled on seven islands that together span the entire Mediterranean. The four islands that yielded multiple datasets suggest that, as expected, morphological variation on particular islands is less than that encompassed by the macrospecies as a whole. Nonetheless, no geographical clines are evident, epitomised by the fact that plants from the easternmost and westernmost of the islands sampled are placed together in the top-left region of the ordination (Figure 13). Again, division into microspecies appears unjustified.

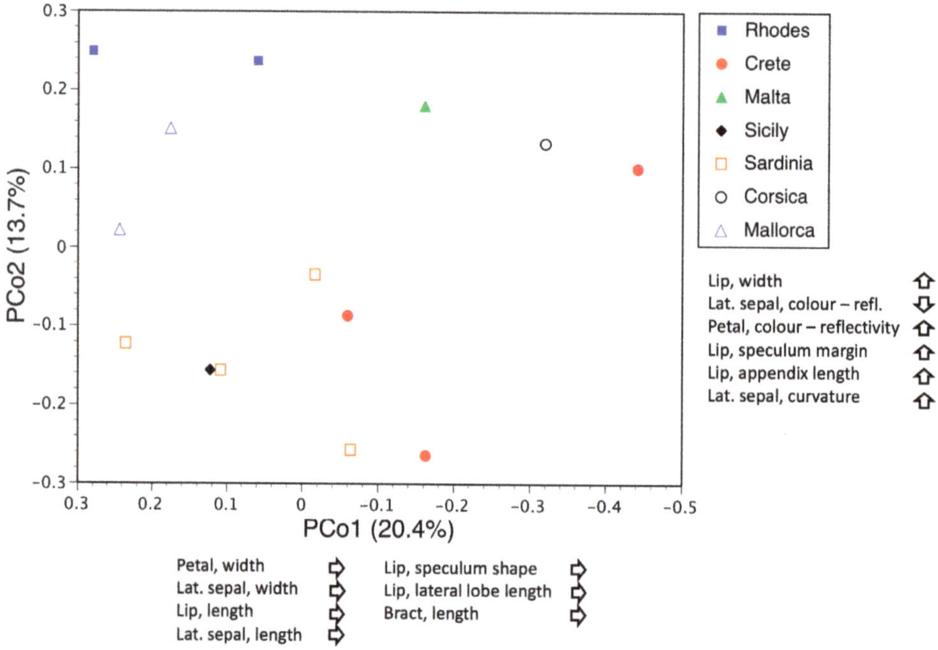

Figure 13. Plot of principal coordinates 1 and 2 for the macrospecies Bombyliflora, labelled according to four regions of origin: 14 individuals from 14 localities and 47 variable characters. Characters contributing to each coordinate are listed in descending order, with arrows indicating the direction of increase in the value of each character.

The first coordinate largely reflects flower size. The second coordinate is less thematic, but examination of the underlying data showed that the heterogeneous suite of floral characters vary independently rather than in concert. The third and fourth coordinates are uninformative, being dictated by single characters: leaf shape and flower number, respectively.

3.4.3. Macrospecies Apifera

Flowers of macrospecies Apifera are dominantly self-pollinating, thus encouraging the maintenance of a wide spectrum of named phenotypic variants; happily, those variants are most commonly described as infraspecific taxa rather than as different microspecies (presumably because they share the same primary pollinators, uniquely within *Ophrys*—wind and rain). Our analysis is based on 15 plants from 15 localities spanning eight geographic regions. Unusually, the first and third coordinates (Figure 14) provided a more informative plot than the first and second coordinates. Multiple samples from the same geographical region are reliably placed fairly close together.

Characters contributing to the first coordinate show that the British and Tuscan plants had sepals that were significantly smaller than those of plants from the remaining regions, but with lateral petals that were slightly longer and labella that were less hirsute. The second coordinate was dominated by subtle differences in the degree of recurvation of the petals, together with the more extensive speculum shown by the single plant sampled in northern Greece. The third coordinate was polarised between single plants from Tuscany and Epirus toward the negative end of the coordinate and from Cyprus toward the positive end, the former having toothed petals and minimal appendices, and the latter featuring lateral sepals that were oriented slightly forwards rather than being swept back in the posture more typical of macrospecies Apifera. The fourth coordinate, dominated by flower number, offered no obvious discrimination.

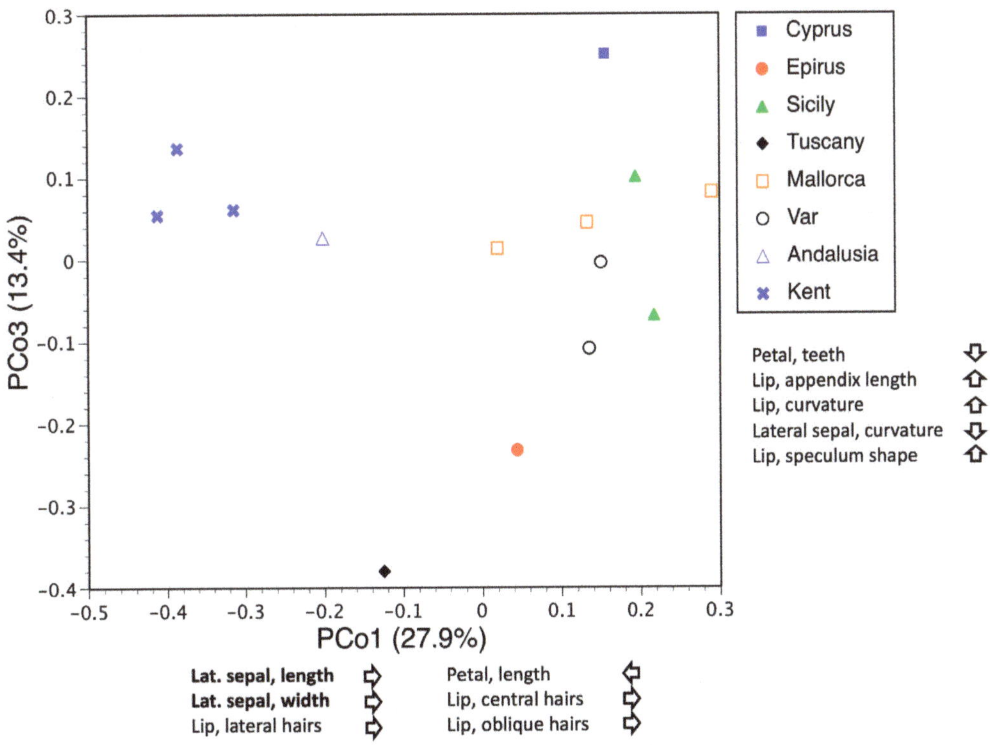

Figure 14. Plot of principal coordinates 1 and 3 for the macrospecies Apifera, labelled according to eight regions of origin: 15 individuals from 15 localities and 49 variable characters. Characters contributing to each coordinate are listed in descending order, with arrows indicating the direction of increase in the value of each character (boldface characters dominate that coordinate).

3.4.4. Macrospecies Insectifera

Sampling of macrospecies Insectifera was especially poor; single plants from Hungary and the Swedish island of Gotland were compared with trios of plants sampled from each of four localities that together formed a west–east transect across southern England. Multiple plants from single localities group together on the plot of the first two coordinates (Figure 15). Plants scoring negative values on the first coordinate (including Buckinghamshire and Gotland) possess discernible trichomes; their labella are moderately laterally recurved, and their median sepals are oriented vertically rather than forwards; the Gotland plant also had unusually small lateral labellar lobes. The second coordinate separated populations primarily according to lateral sepal curvature and colour, plants from Buckinghamshire and Hungary having somewhat paler green sepals that curve slightly forwards rather than being held vertically.

The third coordinate showed the non-British (i.e., Hungarian and Swedish) plants to have slightly paler petals and a comparatively shallow apical notch in the labellum. The fourth coordinate identified the Hungarian plant as having somewhat paler sepals and a narrower column; it also possessed the smallest labellum.

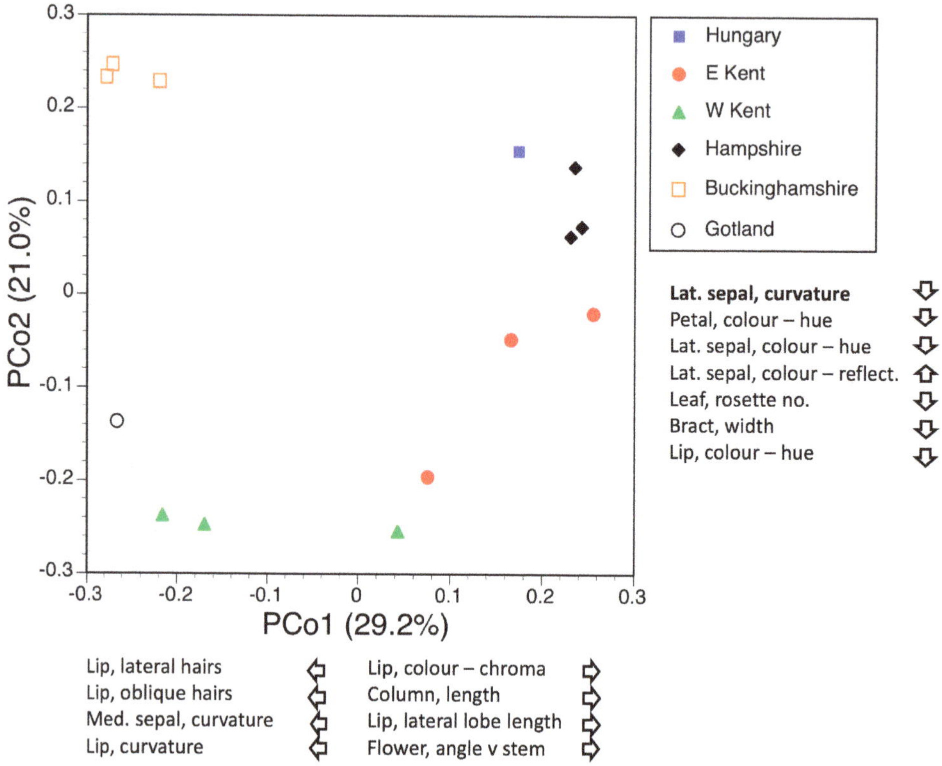

Figure 15. Plot of principal coordinates 1 and 2 for the macrospecies Insectifera, labelled according to six regions of origin: 14 individuals from six localities and 40 variable characters. Characters contributing to each coordinate are listed in descending order, with arrows indicating the direction of increase in the value of each character (boldface characters dominate that coordinate).

3.5. Deeper Analysis of Macrospecies Fuciflora

The above descriptions of the principal coordinates plots resulting from analyses of single macrospecies are qualitative and so vulnerable to accusations of subjectivity. In an attempt to explore a selected macrospecies more deeply and also to consider the microspecies level as seriously as the mesospecies level, we quantified the morphospace of both several mesospecies and several microspecies within the macrospecies Fuciflora—a taxon chosen because herein it is the most intensively sampled macrospecies in terms of plants analysed, is rich in microspecies and proved to be the only macrospecies wherein both the third and fourth principal coordinates appeared to offer as much taxonomic discrimination as the first two.

The approach taken was to quantify the morphospace occupied by each mesospecies and microspecies on four planes that each represented different pairwise combinations of the first four coordinates. For each such plane, the areal extent (i.e., morphospace) of the single macrospecies, its component mesospecies and their component microspecies were simply calculated as a convex hull (given a set of points distributed across a single plane, the convex hull of the set is the smallest convex polygon that contains all of its component points). The morphospace of each mesospecies and microspecies was then expressed as a percentage of that occupied by the entire macrospecies. Comparison was limited to those nine of the 23 sampled microspecies that were represented by five or more individuals and to those five of the six sampled mesospecies that were represented by more than 15 individuals. An exemplar plot, featuring the second and third coordinates and showing plants resolved at the microspecies level, is given as Figure 16, and the vital statistics of the four planes under comparison are given in Table 2.

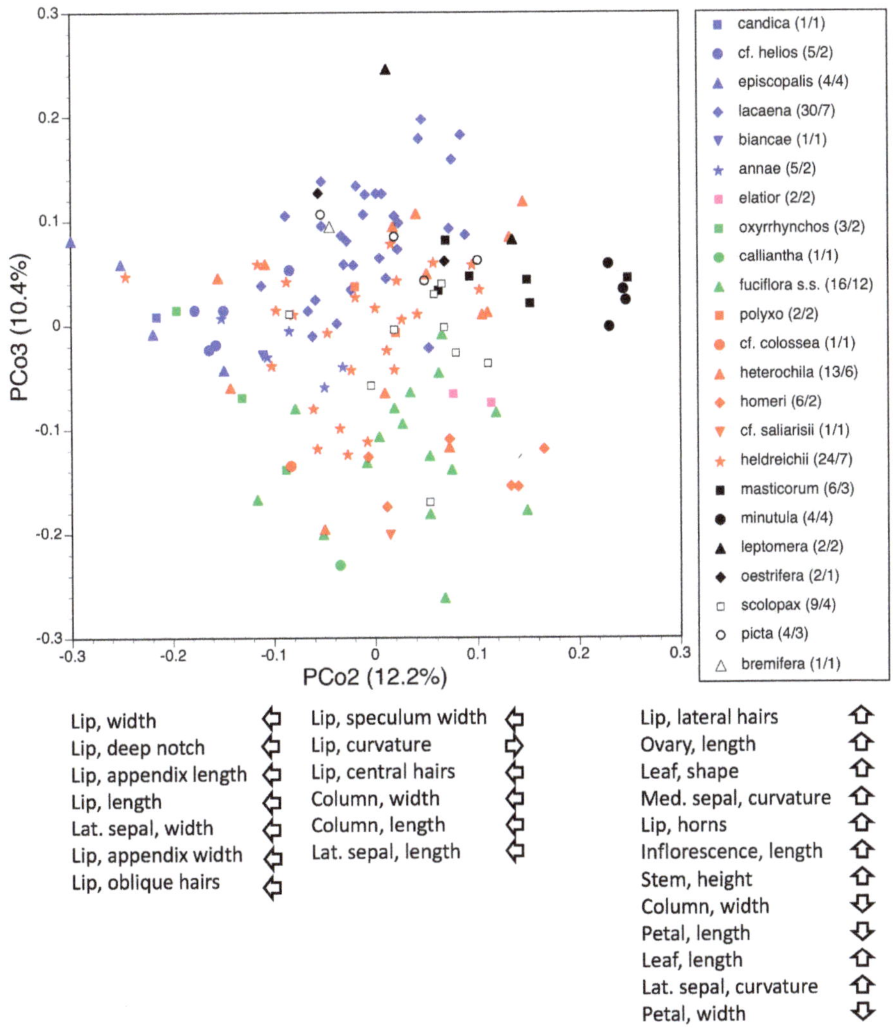

Figure 16. Plot of principal coordinates 2 and 3 for the macrospecies Fuciflora, labelled according to 23 microspecies (and six mesospecies, represented by identically coloured symbols): 143 individuals from 68 localities and 51 variable characters. Parenthetic numbers in the taxonomic key indicate the number of plants and localities respectively included in the analysis. Characters contributing to each coordinate are listed in descending order, with arrows indicating the direction of increase in the value of each character.

Table 2. Comparison of relative degrees, as measured via convex hulls, of congruence with (a) taxonomic circumscriptions of nine microspecies and five mesospecies and (b) sampling effort for microspecies at both the levels of individual plants and populations, in the three highest planes of variation evident in the principal coordinates resulting from the morphometric analysis of macrospecies *Fuciflora* (see also Figure 16). r^2 was assessed against negative exponential curves for nine microspecies, each represented by between five and 30 individuals.

Coordinates Plotted (and % of Total Variance Represented by Plot)	Mean ± SD (and Range) of Proportion of Group-Wide Variation (PGWV: %)		PGWV vs. Sampling Effort, Microspecies (–ve expl. r^2)	
	Microspecies	Mesospecies	Individuals	Populations
PCo1 vs. 2 (27.1)	12.8 ± 14.1 (2.6–25.3)	40.7 ± 21.4 (20.9–66.6)	0.82	0.92
PCo1 vs. 3 (25.3)	13.6 ± 10.8 (2.1–29.1)	38.5 ± 14.3 (24.3–54.2)	0.73	0.76
PCo2 vs. 3 (22.6)	13.4 ± 11.8 (1.2–35.8)	32.7 ± 13.6 (18.6–52.3)	0.53	0.67
PCo3 vs. 4 (18.5)	12.9 ± 11.2 (1.0–27.5)	33.4 ± 12.8 (19.4–44.9)	0.72	0.83

Focusing initially on the plane constructed of the first and second coordinates, the average microspecies occupies 13%, and the average mesospecies 41%, of the morphospace occupied by the entire macrospecies (Table 2). However, these average figures disguise the facts that (a) there is a ten-fold difference between the most compact and most diffuse microspecies, and (b) there is a three-fold difference between the most compact and most diffuse mesospecies. Morphometric variation was greatest in mesospecies *Fuciflora*, at 67% of the total observed for macrospecies Fuciflora. We anticipated from first principles that we would find a strong positive correlation between morphospace occupied and sampling intensity, irrespective of whether this was assessed as the number of individuals or number of populations sampled per taxon. Surprisingly, for a plane constructed using the first two coordinates, this expectation was fulfilled for microspecies but not for mesospecies.

Comparison of the four planes analysed revealed both surprising trends and a surprising lack of trends, depending on the precise question being asked. The average morphospace occupied by a particular microspecies varied remarkably little between plots. Perhaps because it was already poor when viewed at the microspecies level and abysmal when viewed at the mesospecies level, the cohesion of taxonomic groups did not decrease in plots employing lower-level coordinates. Indeed, there was a modest decrease in the average morphospace occupied by mesospecies in those plots that lacked the first coordinate (Table 2)—in other words, a slight increase in their perceived cohesion.

The most obvious explanation for the radical differences in the proportion of total morphospace occupied by individual microspecies and mesospecies is the equally radical contrasts in sample size. First principles correctly predicted that curve-fitting morphospace against sample size would approximate a negative exponential curve (Figure 17). This assumption proved to be correct, though regressing morphospace against sample size for microspecies yielded moderately contrasting r^2 values among plots—these values ranged from 0.82 for individuals and 0.92 for populations on the plot of the first and second coordinates, through to 0.53 for individuals and 0.67 for populations on the corresponding plot derived from the second and third coordinates (Table 2, Figure 17). The mesospecies deviating most from the negative exponential curves was *Heldreichii*, which offered unusually tight cohesion relative to sample size. The availability for comparison of only five mesospecies meant that r^2 values were statistically unreliable at this higher taxonomic level but nonetheless indicated that the relationship between sample size and morphological variation of mesospecies was particularly poor for the plot of the first two coordinates, irrespective of whether sample size was represented by the number of plants or number of populations. Thus, the combination of the first two coordinates yielded the strongest positive correlations between morphological variation and sample size at the level of microspecies but apparently the weakest at the level of mesospecies, hinting that different suites of characters may experience maximum variation at contrasting demographic levels.

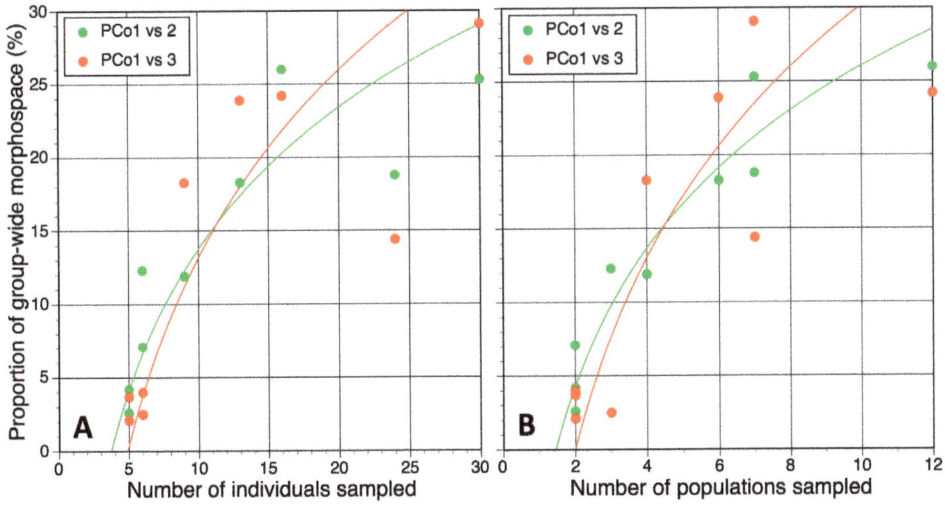

Figure 17. Relationship between the morphospace occupied by the nine best-sampled microspecies of macrospecies Fuciflora versus sampling effort, as measured in numbers of individual plants (**A**) and source populations (**B**) and compared for two morphometric planes (PCo1 vs. 2 and 1 vs. 3: see also Table 2).

4. Discussion

4.1. Strengths and Weaknesses of Our Chosen Morphometric Approach: General Principles

Before drawing specific conclusions regarding phenotypic variation across the genus *Ophrys*, we will first critically appraise the morphometric approach that we have employed here, in order to establish limits to the strength of the conclusions that can reasonably be drawn from this study. We should begin by noting that this is the 22nd such empirical study of European native orchids that one of us (R.B.) has published using this morphometric approach, together spanning seven genera: *Dactylorhiza* (6 papers [66,77]), *Gymnadenia* (2 papers [69,78]), *Platanthera* (4 papers [79,80]), *Pseudorchis* (1 paper), *Orchis* (4 papers [76,81]), *Himantoglossum* (3 papers [82]) and *Ophrys* (1 paper [45]). The strengths and weaknesses of our approach have therefore been amply demonstrated empirically.

Our objective is to characterise rapidly but accurately the macromorphology and 'mesomorphology' of the above-ground organs of each plant studied through minimal intervention by leaving that plant in situ and removing only one or two flowers for later morphological and possibly DNA-based examination. This approach is intentionally low-tech, requiring only a $\times 8/\times 10$ loupe, an RHS colour chart and a well-designed proforma data-sheet. Once placed under such close scrutiny, a terrestrial orchid genus can typically be resolved into between 40 and 55 credible numerical characters, the total depending on its relative morphological complexity. The initial aim is to characterise all above-ground organs with similar levels of resolution, though this principle does not yield similar numbers of characters per organ; the greater complexity of the flowers in general and labellum in particular inevitably means that they are divisible into larger numbers of valid characters than are the vegetative organs. In order to employ the widest feasible range of characters representing, at contrasting scales, size, shape, texture and colour, it is necessary to establish a mixture of characters that are genuinely quantitative (i.e., metric measurements and meristic counts) alongside characters that are semi-quantitative (i.e., multistate and bistate), in some cases requiring the imposition of potentially arbitrary state boundaries on variables that are arguably better described as qualitative.

Dividing organs into characters (and, in the case of semi-quantitative characters, dividing each such character into a linear sequence of alternative character states) requires much decision making that can only realistically be conducted under self-imposed prior constraints. These are, of necessity, guidelines, as attempts to develop fixed unbreakable rules generally prove counter-productive. Many of the most difficult decisions faced during this process concern maintaining a balance between the desire to maximise overall character number versus the desire to avoid duplications of similar characters that will likely cause spurious positive correlations—correlations capable of dominating the ensuing algorithmic analyses (see examples in *Ophrys* discussed below in Section 4.2). Also challenging is choosing the optimal number of character states to represent semi-quantitative characters such as the shapes of particular organs or features. Increasing the number of states per character tends to decrease the influence of the relevant feature on the subsequent analysis, but conversely, recognising too few states risks over-weighting a boundary between states that proves to be arbitrary.

One issue that requires post-hoc resolution is that, given the many life-challenges experienced by a typical plant, the completed morphological matrix for a large number of study plants will contain at least a small proportion of missing values, representing characters that could not be scored for a minority of individuals. Possible causes of missing values include damage to that feature of the plant (e.g., through herbivory or even carelessness on the part of the analyst) or programmed/environmentally induced decay prior to sampling (e.g., death through desiccation of the basal rosette of leaves prior to anthesis). Although such undesirably empty matrix cells can be filled through the application of arbitrary rules (for example, simply by inserting the group mean value), obviously it is better if the chosen analytical algorithm can instead readily accommodate missing values.

Popular multivariate algorithms such as principal components analysis (PCA) are designed to analyse matrices that are both complete and that consist of homogenous, fully quantitative characters. In contrast, we exclusively employ principal coordinates analysis (PCoA) based on the Gower similarity coefficient [73,74] because this approach (a) successfully accommodates heterogeneous suites of both fully quantitative and semi-quantitative characters, and (b) yields results that are unaffected by missing values. The price paid for these clear advantages is that it is more difficult to ascertain the relative contributions of the original characters to the resulting trees or ordinations; fortunately, this goal can be achieved in Genstat [70] via pseudo-F statistics derived through the 'RELATE' command. The most obvious alternative to PCA and PCoA is canonical variates analysis (CVA), but in our opinion, this approach is unnecessarily subjective, at least when attempting circumscription—it requires that the algorithm should be informed a priori of the assignment of individuals to groups and then maximises perceived distances between groups at the expense of distances within them.

Once the similarities among the study plants have been quantified, the choice of presentational style also has a profound effect on the resulting interpretations. Three options merit brief comment here: rooted dendrograms, unrooted minimum spanning trees (both calculated directly from symmetrical matrices of the Gower similarity values and so summarising all the data input) and principal coordinates ordinations, which abstract from the data those planes that encompass relatively high variation and so effectively employ only a (calculable) proportion of the total variability. Interestingly, experience has taught us that, for morphological data, ordinations are consistently more congruent with prior taxonomy than are either unrooted trees or, especially, rooted trees. A single "errant" maximum similarity value can cause major distortions to relationships inferred among plants placed distally to the suspect maximum similarity value within a tree, whereas such "errors" are in practice more buffered within ordinations, wherein they can at worst only influence the spatial placement of the "errant" individual itself.

In addition, a tree is a singular result that must be either accepted or rejected by the analyst, whereas several principal coordinates can readily be plotted in various combinations, allowing the analyst to search for the most informative plane(s) of variation. Admittedly, as demonstrated here, valuable information is rarely obtained from coordinates of lower order than the third. Superimposing a minimum spanning network onto a principal coordinates plot can, in some cases, usefully assist interpretation, though this approach becomes impractical to draft for presentation in ordinations involving large numbers of points; it is most effective when applied to plots ordinating mean values for populations [69,77,82], rather than to plots based on raw data for individual plants, such as those presented herein in Figures 5–16.

4.2. *Strengths and Weaknesses of Our Chosen Morphometric Approach as Applied to Ophrys*

Our standard sampling strategy for morphometric studies focuses on the principle of reciprocal illumination among three contrasting demographic levels: individual plants, local populations, and aggregates of populations that have the potential to be circumscribed as species (or, failing that, as infraspecific taxa). The amount of research time invested in such a study therefore depends on the number of characters measured and relative numbers of putative species, populations per putative species and plants per population that are scored, based on a carefully planned and geographically extensive sampling strategy [29,67,68]. Typical figures for these parameters in our previous morphometric studies have been 2–10 putative species, 5–20 populations per species and 10(–20) plants per population [66,69,82,83]. Unfortunately, it was impractical to bring an equivalent level of sampling rigour to the present study, given that the primary objective was to characterise the broader morphological trends across a genus considered by some specialists to contain approximately 400 (micro)species. The 457 plants sampled for the present study collectively represent 113 microspecies (Table 1), a ratio of individuals per putative taxon that renders the representation of most microspecies perilously close to typological. It therefore became necessary to accept as fixed a priori both the microspecies, previously circumscribed through traditional morphology, and the macrospecies, previously circumscribed through several DNA-based studies culminating in next-generation DNA data [43]. Our primary objective was to explore whether credible circumscriptions can be achieved at the intermediate demographic level of mesospecies [45], beginning with those designated through several iterations by Delforge [38,39].

When designing the present study, we considered the possibility of characterising microscopically the labellum of each measured plant, but time constraints encouraged us to settle for the more typological approach of placing the columns of representative individuals of a few microspecies under the scanning electron microscope (Figure 4). In retrospect, our study would certainly have benefited from the inclusion of an additional phase of data collection conducted under a binocular microscope in order to better detail micromorphological features of the gynostemium, stigmatic surface and proximal labellar 'neck' (such as the paired 'pseudo-eyes' that characterise some groups within *Ophrys*: [12,13,37,43]). Although the present total of 51 scored characters falls well within our previously accepted range of 40–55 characters, it was inevitable that some characters would became inapplicable or invariant in some of the present subordinate analyses that are based on only a single macrospecies; predictably, this phenomenon peaked (at 11 uninformative characters) when analysing the morphologically least complex macrospecies, Insectifera.

In previous morphometric studies of orchid genera such as *Dactylorhiza* [66,77] and *Platanthera* [79,80,83], we have found vegetative characters to be relatively problematic in that they are on average considerably more variable than floral characters [76], collectively reflecting the vigour of the plant in question. The life history of most terrestrial orchids involves the annual replacement of their all-important stem-tuber; the relative success of this process of somatic replacement strongly influences the likelihood that a plant will flower during the following spring/summer and, if so, how vigorously. Obviously, vegetative vigour has a strong underlying genetic influence and so can genuinely reflect taxonomic differences, but it is also strongly influenced by the environment in which the plant is growing—not only in the present growing season but also in the previous growing season. It becomes difficult—arguably impossible—to disentangle genetic from ecophenotypic factors influencing plant size. In contrast, ecophenotypic influences have a much weaker impact on most floral characters because flowers emerge over a much shorter time period and are generated under stronger developmental constraints. It is therefore helpful, at least from this perspective, that vegetative characters have been proven to exert only modest influence on most of the ordinations conducted for the present study. This observation suggests that the ratio between degrees of vegetative variation and floral variation is lower in *Ophrys* than in comparable orchid genera (see Section 4.4), as well as indicating that variation is poorly correlated between these two functional categories of organ.

In the case of *Ophrys*, it is among the flowers rather than the vegetative organs that concerns are raised regarding positive correlation among potentially co-functional characters. Arguably the most extreme example is provided by our decision to quantify the background colours of three different floral organs: the labellum, lateral petals and lateral sepals. Representing colour through the CIE system required us to report each colour as three characters (chroma, hue, reflectivity), thereby causing flower colour to dictate a total of nine of the 51 morphometric characters scored. In fact, little variation was detected in labellum colour, but inevitably, similarities in colour between the lateral petals and lateral sepals were evident in many plants. On the other hand, colour differences between petals and sepals have been judged taxonomically diagnostic among *Ophrys* microspecies within some macrospecies. Although there are no definitively right or wrong answers when addressing such conundra, it is important to understand the potential consequences of such decision-making for the patterns seen in the resulting ordinations.

It is also worth considering the likely morphological transitions that dictated the origin of the genus *Ophrys*. All genera closely related to *Ophrys* [20,84] have labella that are largely two-dimensional, the third dimension being confined to the simple curvature of the lateral margins, except that, proximally, the labella of genera such as *Neotinea*, *Himantoglossum* and *Anacamptis* are consistently invaginated into an elongate cylindrical spur that may or may not secrete nectar [85]. We suspect that a key stage in the emergence and initial diversification of the genus *Ophrys* was the evolutionary loss of that spur, because it liberated the labellum to develop much more complex three-dimensional topographies—topographies that become evident only after the labellum inverted from concave to convex during the opening of the bud [3] (Figures 2 and 3). There is a general trend for labellum topography to be both more three-dimensional and more complex in the later-diverging macrospecies of *Ophrys*, contrasting especially sharply with the comparative simplicity of the early diverging Insectifera [12,43].

Some recent studies of European orchid clades have employed landmark analysis of labella as a time-saving proxy for all morphological variation within the study group [86–88], but this approach is less utilitarian than attempting to comprehensively quantify the entire above-ground portion of the plant. Firstly, it provides little useful information toward improving previous formal taxonomic descriptions of the taxon. Secondly, much controversy still surrounds the relative contributions to successful pollination of labellum topography compared with labellum surface texture, labellum markings, tepal colour, overall flower size and especially the complex biochemistry of volatiles exuded by the flower (features considered in greater detail in Section 4.5). Just which properties qualify labellar topography to be given pride of place in this spectacular phenotypic pantheon?

One obvious question to ask at this point is whether our 51 morphometric characters are collectively as effective as the human eye when seeking morphological differences among *Ophrys* plants? The answer is almost certainly no; the same eye–brain coordination that allows us to discern among innumerable human faces is likely to be almost as adept in distinguishing among a panoply of bee orchid flowers. The advantage of the detailed morphometric approach is that the principal coordinates algorithm is fastidiously objective; it has no prior expectations of groupings, nor is it determinedly seeking minute differences while overlooking numerous similarities.

In summary, despite the panoply of "Faustian bargains" that we found to be necessary in order to bring to eventual fruition this challenging morphometric study, we are confident that the collective ordinations presented here as Figures 5–16 provide a reasonably accurate picture of not only any broad morphological trends that are evident across the genus *Ophrys* as a whole but also any trends that dominate within each of its nine macrospecies. We will first attempt to summarise those trends and then consider their implications for both the taxonomy and evolutionary mechanisms inferred within this highly contentious 'model' genus.

4.3. *Overview of Morphological Variation within Ophrys*

4.3.1. The Search for Discontinuities among Taxa

We hoped, rather than confidently expected, to detect morphological discontinuities among at least the *Ophrys* macrospecies [28,45]. Ideally, they would coincide with consistent discontinuities identified in molecular data. However, among the 11 ordinations performed by us that employed suites of characters that were both complete and unmodified (Figures 5–16), only two—each confined to a single-macrospecies—yielded clustering of taxa that was sufficiently cohesive to separate those taxa by apparent morphological discontinuities. The first such case was the ordination of macrospecies Speculum (Figure 11), which readily distinguished the only two microspecies included in the analysis, thereby implying that *O. speculum* s.s. and *O. regis-ferdinandii* are separated by a morphological discontinuity. However, it is likely that this distinction would have been weakened had we included in the analysis the morphologically intermediate microspecies from Iberia, *O. vernixia* [39].

The second discontinuity was observed within macrospecies Umbilicata (Figure 9), separating the Cypriot microspecies *levantina* and *aphrodite* from the remaining six microspecies analysed here, which have labella that are far more three-dimensional and bear specula that are both more complex and more striking. We suspect that this apparent discontinuity would, uniquely, survive the sampling of all of the microspecies attributed to macrospecies Umbilicata. This is the only credible example that we detected of a modest morphological discontinuity occurring within a single genetically cohesive macrospecies. Otherwise, the clear impression gained at both the macrospecies and mesospecies levels was of a morphological continuum rather than an aggregate of discrete, readily identifiable taxa.

4.3.2. The Search for Trends in Character Correlation

The simultaneous multivariate analysis of the entire genus *Ophrys* (Figure 5) is dominated by some suites of characters that also dominate the majority of single-macrospecies analyses (sepal and lateral petal colour, labellum dimensions), some suites of characters found in only a minority of single-macrospecies analyses (presence and size of a labellar appendix and of the associated apical notch, labellar trichomes) and one character that dominates no other analysis (pale labellar margin). It is challenging to identify any generalised patterns of character variation by comparing the series of analyses of single macrospecies.

Understandably, the colour of sepals and lateral petals contributes most to analyses of macrospecies that contain a mixture of green-sepaled and pink/purple-sepaled plants (Fuciflora, Umbilicata, and Sphegodes, wherein the presence or absence of sepal-stripes also contributes), but flower colour variables also contribute to some macrospecies that maintain contrasting shades of green (e.g., Bombyliflora and Insectifera). Sphegodes and Fuciflora are unusual among macrospecies in that speculum characters contribute comparatively weakly. Plant size plays a greater role within some macrospecies than others, but no overall pattern can be discerned; overall, vegetative characters are subordinate to floral characters. Much to our surprise, sepal dimensions dominate over the dimensions of lateral petals, labella and gynostemia, which contribute substantially to ordinations mainly in macrospecies that are relatively poor in both microspecies and mesospecies—a statement that also applies to lip shape/lobing and trichome distribution, as well as median sepal shape. When seeking generalisations, the main conclusion to be drawn from Figures 5–16 is simply that morphological variation within the genus is exceptionally multidimensional.

4.3.3. Residual Incongruence between Macrospecies and Mesospecies

The four editions of the traditional monograph by Delforge [38,39] have gradually and partially converged on the results of several 21st century molecular studies that together have provided strong evidence for circumscribing nine macrospecies; consequently, each of the 29 mesospecies can now largely be accommodated (often in multiples) within a particular macrospecies. The major exception to this generalisation concerns Delforge's mesospecies *Bornmuelleri*, which substantially transgresses the boundary separating macrospecies Fuciflora from macrospecies Umbilicata (Figure 1). RAD-seq, nrITS and plastid sequences have all shown conclusively that the Levantine microspecies *bornmuelleri* and *levantina* (presumably also the morphologically very similar *aphrodite* and *carduchorum* [39]) belong to macrospecies Umbilicata [21,43], in which they form a monophyletic pairing that diverged after *attica* and *kotschyi* but before *umbilicata* s.s. and its close relatives [G. Sramkó, O. Paun and R. Bateman, unpublished data]. In contrast, other microspecies assigned by Delforge to his mesospecies *Bornmuelleri*, such as *episcopalis, aeoli, candica* and *biancae*, are placed molecularly within macrospecies Fuciflora [21,43,44]. For example, RAD-seq data generated by Sramkó, Paun and Bateman [unpublished data] nest *biancae* within a tight-knit clade of samples from southern Italy that also consists of the microspecies *lacaitae, oxyrrhynchos* and *celiensis*, all three of which were placed by Delforge [39] in his mesospecies *Fuciflora*.

As demonstrated here, the morphologies of microspecies such as *bornmuelleri, levantina* and *aphrodite* deviate considerably from those of other members of macrospecies Umbilicata (Figure 9). Even in the principal coordinates plot encompassing all macrospecies (Figure 5), *levantina* and *aphrodite* did not overlap with the remainder of macrospecies Umbilicata, but neither did they overlap substantially with macrospecies Fuciflora; instead, values of between −0.03 and −0.16 for the second coordinate placed plants of these taxa mid-way between the rest of macrospecies Umbilicata and macrospecies Tenthredinifera on Figure 5. Their molecular phylogenetic placement (Figure 1) suggests strongly that characters causing them to more closely resemble macrospecies Tenthredinifera and some members of macrospecies Fuciflora—their less laterally recurved labellum with a uniformly villose margin, relatively small, simple speculum and relatively large appendix—arose independently. Thus, although their morphological distinctiveness is arguably sufficient to justify Delforge's [38] decision to separate these microspecies as a mesospecies outside the Umbilicata group, mesospecies *Bornmuelleri* as currently conceived by Delforge is not a natural grouping and therefore requires re-circumscription.

The empirical aims of the present study are too modest to encourage us to meddle below the macrospecies level in previous classifications of the family. Our taxonomic principles rely upon the ability to seek congruent discontinuities in multiple datasets, but herein we are reporting on only one (albeit large) new dataset that representing morphometrics. Both species concepts and how they are applied require further clarification if genuine progress is to be made.

4.4. Comparison with Morphological Variation Observed in Other European Orchid Genera

Characterising morphological variation in bee orchids as near-continuous and exceptionally multidimensional encourages us to compare our study of *Ophrys* with those conducted by us previously on other genera of European orchids [45,66,69,76,77,79–83].

In any analysis, the percentage of variance accounted for by the first two principal coordinates provides a useful indication of the relative degrees of dimensionality in the underlying morphometric data. The higher the dimensionality, the more difficult it becomes to partition individuals and populations into well-circumscribed taxa. Here, the analysis that included all nine macrospecies of *Ophrys* (Figure 5) accounted for 34% of the total variance, a figure consistent with most of the remaining analyses involving subsets of the master data. Most plots occupied the range of 30 to 34%, falling to a nadir of 27% in the analysis of macrospecies Fuciflora (Figure 8). Predictably, morphometric dimensionality was lower in three of the four analyses that encompassed only a small number of plants representing just one or two microspecies (macrospecies Speculum, Apifera and Insectifera), in which the total variance increased to 46–50%. A further indication of comparatively high dimensionality in the majority of ordinations was a small difference for separating the first and second coordinates in terms of the amount of variance that each coordinate accounted for—less than 3% in the case of the single-macrospecies analyses conducted on the Sphegodes and Fuciflora groups (Figures 7 and 8).

Comparison with the vital statistics of 22 in situ morphometric surveys previously conducted by us, spanning seven genera of European orchids, showed that the proportion of the total variation encompassed by the first two principal coordinates was weakly negatively correlated with both the number of plants analysed in the study and the number of characters that could usefully be scored for that particular genus. This conclusion is unsurprising, as increasing the number of plants analysed or characters scored inevitably leads to at least modest increase in the spectrum of variation captured during sampling. Nonetheless, the morphological distinctiveness of species in the genus in question was the primary factor that dictated the amount of variation recovered in that first multivariate plane defined by coordinates 1 and 2.

European *Platanthera* species show comparatively little variation in floral shapes and colours, differing mainly in the relative size of various floral organs; consequently, each species forms a discrete cluster [79,80,83]. In common with studies of hybrids between two morphologically distinct parents [89], they yield highly polarised plots of PCo1 versus PCo2 that encompass approximately 70% of the total variation. In addition, much of the variation tends to be encompassed by the first coordinate, creating an unusually large differential between the first coordinate and the second. Studies of genera that contain a mixture of species separated by morphological discontinuities and species possessing overlapping morphologies, such as *Himantoglossum* [6,82] and the anthropomorphic subgenus of *Orchis* [76,81,90], capture 40–50% of the total variance in the first two coordinates. But taxonomic groups in which none of the species are separated by morphological discontinuities yield plots that capture only approximately 30% of the total variation in the first two coordinates. Examples include the *Dactylorhiza majalis* aggregate [66,91], in which species have originated from repeated allopolyploidy (hybridisation combined with genome doubling) events between different ecotypes of the same two parental lineages [92–94], and the *Gymnadenia conopsea* aggregate [69,78], in which three species have diverged substantially in DNA barcode regions, flowering times and preferred habitats but only subtly and unreliably in morphology [69,95].

Morphometric analyses of *Ophrys* [45], notably those presented here, make clear that patterns of morphological variation within this entire genus are comparable with those observed in the *Dactylorhiza majalis* and *Gymnadenia conopsea* aggregates alone. Although there remain vigorous debates regarding the choice of species versus subspecies status for taxa within both the *D. majalis* [36,91] and *G. conopsea* [78,96] aggregates, there is greater consensus regarding the number of such taxa in each: *D. majalis* s.l. encompasses about ten species/subspecies, whereas *G. conopsea* s.l. encompasses only three or four. These figures present a radical contrast with estimates of up to 400 microspecies in the genus *Ophrys* and at least 113 microspecies in just one of the molecularly circumscribed macrospecies, Sphegodes [39,45].

Given (a) the demonstrable morphological overlap of *Ophrys* taxa at all three analytical levels (i.e., macrospecies, mesospecies and microspecies), (b) the multidimensionality of that variation and limited correlation among morphological characters and (c) the genetic overlap that has been repeatedly demonstrated at the mesospecies and microspecies levels [21,43–45], can a single morphological continuum lacking directional evolutionary trends justifiably be claimed to encompass up to 400 species? To address that question, we need to move beyond merely describing variation in order to consider the mechanisms likely to have generated that variation.

4.5. Review of Features Encouraging the Three Phases of Pseudo-Copulatory Pollination

The attraction of naïve male hymenopterans to interact with their flowers, using the sequential olfactory, visual and tactile cues, has long been the cornerstone of bee orchid studies. Herein, we briefly consider the nature of those three cues, addressed in reverse sequence.

4.5.1. Tactile Stimuli

Setting aside very few exceptions, including the comparative SEM images of gynostemia illustrated here (Figure 4), detailed morphological studies of *Ophrys* have focused almost exclusively on describing the labellum. This emphasis is understandable, given its often rugged three-dimensional topology [22,87]; the complexity of its surface textures and markings [97]; and its equally complex internal anatomy [12–14,98]. Even genome size differs among cells within different regions of the labellum, presumably dictating contrasting levels of glandular activity [14]. The labellum undeniably gives every appearance of representing an exceptional aggregate of numerous adaptations, all finely tooled by natural selection to seduce a particular (range of) species of pollinating insect into repeatedly attempting copulation.

However, most accounts of the functional morphology of the labellum remain largely anecdotal. Interpretations have focused on two main issues: (a) the spatial fit of the visiting male insect to the labellum in terms of size and shape, and (b) putative labellar adaptations, especially broadly concentric markings (Figure 2A) and differential expression of surface features such as trichomes (Figures 2A and 3), that encourage the insect to adopt the optimal orientation for collecting pollinaria on its head (most species) or its abdomen (most commonly members of macrospecies Fusca). Variation in the size, density and orientation of trichomes is especially well illustrated in the supposedly abdominally pollinated *O. pallida* featured in Figure 3A–D. SEM imaging was used by Cortis et al. [99] to compare trichomes in the localised labellar regions of two microspecies of macrospecies Fusca on Sardinia.

Rakosy et al. [100] observed the behaviour of bees visiting flowers of *O. leochroma* (macrospecies Tenthredinifera) from which contrasting segments of the labellum had been removed, noting that the removal of those portions used more frequently by the bees as gripping or contact points caused greater reductions in the frequency and effectiveness of pollination, especially in the deposition of pollinaria. These observations led the authors to predict that those regions most important in ensuring mechanical fit between flower and pollinator—in this case, the stigma and "shoulders/horns" of the labellum—would operate under strong stabilising selection within this microspecies, whereas regions such as lateral lobes, appendix and associated apical notch would be less critical and would therefore show greater variation. We applaud such experimental approaches and wonder whether "shaving" critical regions to remove their trichomes might enable a more subtle approach to exploring the relative importance (or otherwise) of tactile stimuli in encouraging successful pollination?

4.5.2. Visual Stimuli: Significance (or Otherwise) of Sepal Colour

Compared with the large number of investigations of *Ophrys* pseudo-pheromones, visual cues such as flower colour have been under-researched. The most notable exception is the series of behavioural experiments conducted by Hannes Paulus and colleagues using the pink-sepaled microspecies *heldreichii* (macrospecies Fuciflora) as their experimental system. Initially, Spaethe et al. [101] recorded the effect on pollinator attraction of removing the sepals and lateral petals. When tested using the eucerid bee *Tetralonia* (*Eucera*) *berlandi*, excision reduced the attractiveness of an actual flower by ca. 50% and that of an imaged flower by ca. 90%, once the insect was within 30 cm distance of the flower [102]. These results were broadly supported by later field testing, which also confirmed low figures for average male (ca. 8%) and female (ca. 4%) reproductive success of intact *heldreichii* flowers, figures typical of the genus [103].

Streinzer et al. [104] then compared microspecies *heldreichii* with the morphologically similar but phylogenetically distant microspecies *dictynnae* (macrospecies Tenthredinifera), which has sepals of a paler pink and "no conspicuous labellar pattern" (better described as a less complex and less extensive speculum; it is actually of roughly equal brightness to that of *heldreichii*). Their results suggested that the prominence of the speculum had little effect on pollinator preference but that having paler pink sepals, closer to medium green hues in green-receptor reflectance, reduced male pollinator preference by ca. 70% [104]. In contrast, Vereecken and Schiestl [105] found no behavioural difference in pollinators presented with a green versus 'white' sepal polymorphism in *O. arachnitiformis* (macrospecies Sphegodes).

In this context, we note that, for humans, white is not a true colour but rather an absence of colour, though this statement does not apply to the contrasting visual spectrum of insects [106]. In any case, once they had been colour-matched by us, supposedly white *Ophrys* sepals prove to be either pale green or pale pink. Moreover, the contrasting colours are located in different tissues within the perianth; pink–purple anthocyanins are typically diffused within the cytoplasm of epidermal cells, whereas green chlorophylls are packaged in chloroplasts that are more strongly concentrated in the underlying mesophyll [107]. Viewed from a genetic perspective, perianth colour characters appear to be not only polymorphic but also developmentally unstable, emphasised by the fact—demonstrated through captive breeding—that a single self-fertilised *Ophrys* flower can be capable of producing both green-flowered and pink-flowered progeny [108].

Our data show that two-thirds of microspecies allocated to macrospecies Sphegodes are predominantly green-sepaled rather than pink-sepaled; a literature search by Spaethe et al. [109], sampling 71 microspecies across several macrospecies, estimated a similar proportion of green-sepalled taxa. It also suggested a strong bias between those species typically pollinated by *Andrena* s.l. bees (91% green-sepaled) and those typically pollinated by the more visually acute *Eucera* s.l. bees (83% pink-sepaled).

The RAD-seq network (Figure 1) implies that there were not one but two phylogenetic origins of anthocyanin in sepals and lateral petals; one origin immediately prior to the divergence of macrospecies Apifera (i.e., permeating through groups F–J) that is a synapomorphic tendency (characterising some but not all microspecies) and one origin during or after the divergence of macrospecies Tenthredinifera (group B) that is a true synapomorphy (characterising all microspecies). Similarly, the evolutionary spread of dark purplish–brown labellar pigment downwards and outwards into green lateral sepals also shows two origins in Figure 1: the pigment forms two stripes (one dorsal and one ventral) in macrospecies Speculum but develops as a single more diffuse zone (ventral only) in some but not all microspecies of mesospecies *Mammosa* and *Reinholdii* (within macrospecies Sphegodes, roughly equating with haplotype clade V in Figure 18). These patterns of phylogenetic transitions do not suggest that flower colour is of especial utility, either evolutionary or taxonomic.

Like flower colour, labellum pattern has also been subjected to experimentation. Labellum patterns are said to be learned by unsatisfied male insects in order to avoid time-wasting re-visitations of females already mated, therefore diverse speculum patterns evolved in *Ophrys* species to slow the learning process [25]. Admittedly, Stejskal et al. [97] were obliged to employ honey bees rather than the preferred pollinator of microspecies *heldreichii* in their otherwise sophisticated behavioural experiments. Evidence that the honey bees eventually learned to avoid particular speculum shapes that had previously proved unrewarding of nectar-substitute is convincing, but the learning process required at least 50 bee–flower interactions to achieve statistical significance and did not plateau until ca. 90 interactions had occurred—a process far too protracted to facilitate *Ophrys* pollination. The authors were therefore required to suggest that sex is a far stronger motivation for male–male competition than food and would encourage far faster learning, and that this process would drive negative frequency-dependent selection, further diversifying the range of specula maintained within bee orchid populations [97].

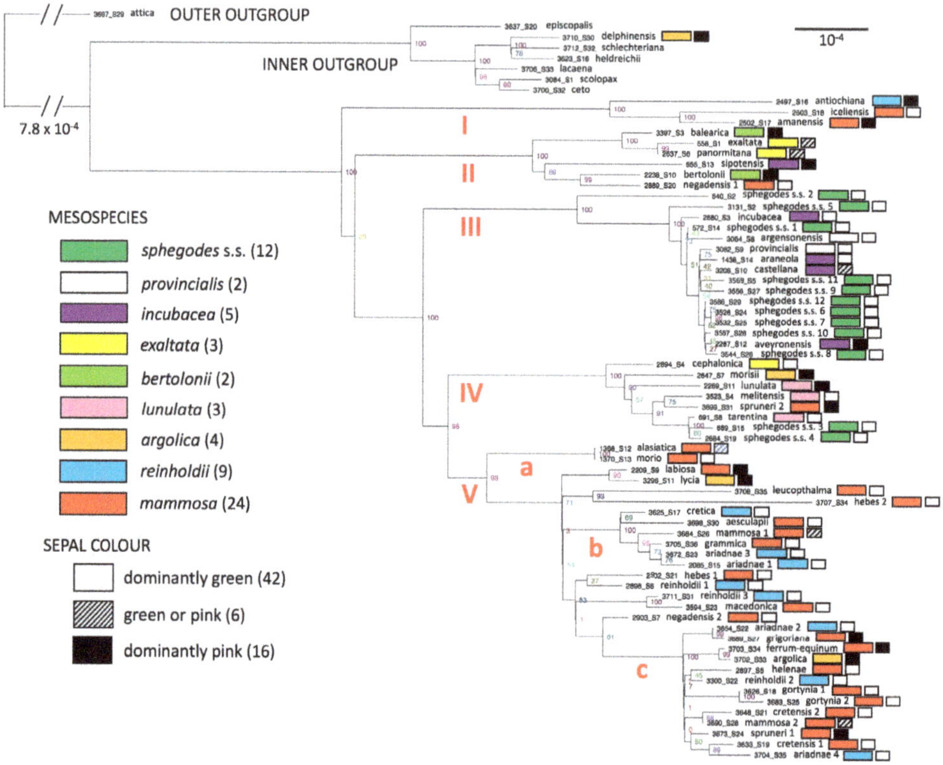

Figure 18. Maximum likelihood phylogeny of whole plastomes obtained via genome skimming from 64 individuals representing 40 microspecies of macrospecies Sphegodes, together with six individuals of macrospecies Fuciflora ('inner outgroup') and one individual of macrospecies Umbilicata (functional 'outer outgroup'). Plants are named according to microspecies, and box-labelled according to both mesospecies attribution and the dominant colour of their sepals. Roman numerals and lower-case letters indicate the main haplotype clusters and subclusters within macrospecies Sphegodes. Figures supporting nodes are bootstrap percentages. Topology from Bateman et al. [45].

In this context, we note that our sample of 24 plants of *heldreichii* from seven populations on Crete collectively occupied a relatively small area of the overall morphospace within macrospecies Fuciflora (Figure 16). Moreover, most of these plants (83%) were scored as possessing the same (most complex) speculum category, 4: single ring with radiating projections or multiple rings (Appendix A). Rather, we detected considerably greater variation in speculum morphology in other microspecies within macrospecies Fuciflora, notably the 30 plants together representing seven populations of microspecies *lacaena* scored by us in the Peloponnese. Does this result mean that potential pollinators of macrospecies Fuciflora occurring in the Peloponnese are more skilled at pattern recognition than their equivalents in Crete? Or are we instead observing random variation of no great evolutionary consequence?

4.5.3. Olfactory Stimuli

Much of the scientific research performed on selected *Ophrys* species has focused increasingly on the complex composition of their pseudo-pheromone cocktails and on determining which groups of insects are excited by which elements present in those cocktails. The relationship between C21–C29 n-alkenes and admixed n-alkanes has attracted particular attention. Some alkenes are more attractive to male than female bees and have consequently been accused of constituting a pre-adaptation that facilitated the diversification of alkenes in *Ophrys*; this in turn is considered to have driven increased reliance on pseudo-copulation for reproductive assurance [110,111]. Experiments suggest that the bee orchid's pseudo-pheromone cocktail must accurately mimic that of the females of the pollinating species, both qualitatively and quantitatively, which raises interesting (and thus far unanswered) questions about how the transition from pre-adaptation to adaptation may have occurred. Research also revealed that "inactive" (perhaps more accurately described as ineffective) compounds in the cocktails show much greater variation among individuals than do "active" compounds and that the flowers are sufficiently rich in these compounds to routinely outcompete for male attention the females of the relevant pollinating species [16,112–114].

Numbers of active compounds per microspecies typically exceed 20, offering ample variation for fine-tooling contrasts among individuals in the amounts of particular components. For example, the cocktail of *O. sphegodes* s.s. is reportedly dominated by 9- and 12-alkenes, whereas that of the closely related microspecies *O. archipelagi* are dominated by 7-alkenes differing in the precise locations of double bonds [18,22,115,116]. Valiant attempts to compare relative degrees of perceived divergence in cocktail compositions, pollinator spectra and genetics have tended to be undermined by inadequate molecular phylogenies [110].

4.5.4. Overview

We conclude that increasing excitement surrounding the novelty and effectiveness of the olfactory cues for pollinators have tended to eclipse research interest in the visual and tactile cues. In practice, morphology has played a largely passive role in attempts to understand the evolution and systematics of bee orchids. Most studies have been confined to just one or two microspecies previously recognised through traditional, authoritarian taxonomy, an approach lacking explicit phases of data collection and subsequent algorithmic analysis. Few populations are sampled, and few if any morphological characters are measured. Morphology has typically been a bystander in a contest for supremacy in species circumscription that has been fought between datasets based on pseudo-pheromones plus pollinator identities versus those based primarily on DNA sequences, now supplemented with the crude overview of morphology presented in this paper. These two highly conflicting species concepts merit reappraisal in the light of this and other recent studies.

4.6. Microspecies versus Macrospecies

4.6.1. Basis of the Ethological Species Concept

The textbook pseudo-copulatory pollination syndrome of *Ophrys* has been subject to innumerable reviews, their increasing sophistication providing a testament to the progressive accumulation of supporting evidence [25,26,31,40,42,117–123]. The most detailed recent overview was provided by Baguette et al. [25], who argued for a mutually beneficial co-evolutionary relationship that is asymmetric—essential for the survival of the orchid but not for that of the insect partner. The scenario requires four predicates:

(1) Male pollinators (most commonly solitary bees) emerge before females, are more likely than females to disperse beyond their foraging range, and are hard-wired to recognise and avoid both kin and non-receptive females.
(2) There exists strong infra-specific competition among *Ophrys* individuals for the attraction of species-specific hymenopteran pollinators. The remarkable learning and memory abilities shown by bees in particular induce strong selection pressure for increased variation, especially in pseudo-pheromone bouquets. The novel orchid bouquets are best regarded as evolutionary darts, thrown randomly at a dartboard of many novel potential pollinators co-occurring with the orchid, in the hope of proving

adaptive by securing a reasonable match with a female insect's bouquet and thus initiating a pollinator shift into a supposed pollinator-free space.

(3) Once the orchid has fortuitously generated an effective bouquet, directional selection then reinforces the attraction of the newly acquired pollinator by adaptively refining the morphology and/or phenology of the flower to better suit the pollinator.

(4) Male members of the pollinating species then increasingly rely on their co-evolving *Ophrys* species to locate, from a distance, suitable habitats within ecologically complex landscapes, thereby increasing the probability that they will encounter a receptive female. Once males have successfully memorised the local diversity of pseudo-pheromone bouquets (and been repulsed by post-pollination alterations in the bouquets), they are more likely to disperse to other suitable habitats, thereby reducing the frequency of self-fertilisation in both the orchid species and its pollinator.

Admittedly, given that (a) bees are vagile whereas orchids are sessile, (b) the decision where to lay her eggs rests with the female bee, and (c) females of any animal species are judged by us to be inherently more intelligent than the equivalent males, one might have predicted from first principles that female solitary bees would instead elect to nest well away from bee orchid populations, in order to avoid unwelcome competition from flowers better-endowed with pseudo-pheromones than their intended paramours [124].

Derived from this widely accepted understanding of the evolutionary process, the ethological species concept relies on pollinator-limited reproduction and consequently reaches its strongest development in cases of the comparatively inefficient pseudo-copulatory mechanisms. At its core is the reportedly intimate relationship between a particular insect species observed to pollinate the flowers and a particular comparable cocktail of pseudo-pheromones in the orchid flower, supported to a lesser degree by an appropriate size and three-dimensional shape and texture of the labellum (and gynostemium). Most studies providing data in support of this species concept were designed primarily to better understand this evolutionary mechanism rather than to circumscribe species (or, more likely, to re-circumscribe species) as an aid to taxonomy. Consequently, such studies were confined to just one or two microspecies, which had previously been recognised through traditional, authoritarian taxonomy. They have typically been based on field observations of insect–flower interactions of few populations and/or laboratory-based experimentation, most commonly emphasising the biochemistry of the pseudo-pheromone cocktails. In cases where DNA data were collected, this was usually done through early fragmentation techniques such as AFLP and microsatellites, the results being graphed with plants labelled simply according to prior microspecies identity rather than more precisely according to source population. In these features, these studies contrast strongly with the integrated analysis of an entire macrospecies Sphegodes performed by Bateman et al. [45] (Figure 18) and with the present broad-brush morphometric analysis of the entire genus.

The ethological model is best viewed as a series of steps that must occur in a particular sequence. It first requires a constant flux of variation in pseudo-pheromone cocktails. Occasionally, a novel cocktail serendipitously attracts a novel pollinator, whose subsequent interactions with the flower gradually improve its ability to repeatedly attract naïve males. Natural selection refines multiple adaptations—optimising the fit to that pollinator of pseudo-pheromone composition, flower size, three-dimensional shape, colour, markings and texture, as well as flowering time. Pre-zygotic isolation has thus been achieved, and at this point, neutral molecular markers can begin to accumulate that will, in time, become fixed within the lineage and thereby provide concrete evidence of that presumed reproductive isolation. The key question is this: At which point in this sequence of events can we be sufficiently confident that speciation has occurred—more precisely, that the presumed speciation process will not subsequently be reversed [29,45]? We will now revisit a selection of pertinent studies, partly in an attempt to answer this question but also to question whether the strength of the conclusions drawn by some previous authors exceeds the strength of the available evidence.

4.6.2. Reassessing the Evidence: Do Prior Assumptions Cloud Objectivity?

We begin this reappraisal with three informative studies reporting gene-flow between *O. lupercalis* (mesospecies *Fusca*, macrospecies Fusca) and other microspecies with which it co-occurs in various regions of the Mediterranean Basin.

Stökl et al. [125] studied Mallorcan populations of *lupercalis* admixed with *O. bilunulata* (also mesospecies *Fusca*) and the later-flowering *O. fabrella* (mesospecies *Obaesa*, macrospecies Fusca), each orchid reputedly pollinated by a different species of *Andrena* bee. Ordinations of volatile compositions show partial overlap between *O. lupercalis* and *O. bilunulata*, with *O. fabrella* intermediate, but AFLP analyses suggest the presence of only two genetic entities, centred on *lupercalis* and *fabrella*. Bizarrely, plants of *bilunulata* were distributed roughly equally between the clouds formed by the other two species, rather than being intermediate. The authors of the study (p. 448) argued that "Most plants of *O. bilunulata* had the genotype of either *O. lupercalis* or *O. fabrella*. This does not mean that *O. bilunulata* does not exist as a species. It has a clearly distinguishable floral morphology and a more important floral odour, which attracts a different pollinator species than the other *Ophrys* species on Majorca. Furthermore, it is widely distributed in the Mediterranean and has the same pollinator throughout its distribution range. We therefore interpret our AFLP data as being the result of ongoing hybridization and backcrossing between the species. The similar sex pheromones mimicked by the flower and the overlapping flowering periods led to a breakdown of reproductive isolation". But both the morphological distinctiveness and contrasting pollinators are assumed rather than rigorously demonstrated, while the absence of intermediate genotypes between the two AFLP clusters suggests the absence of primary hybrids. What evidence do we have that reproductive isolation ever existed among these taxa?

Earlier, Stökl et al. [58] had studied interaction between *O. lupercalis* and *O. iricolor* (also in macrospecies Fusca but in mesospecies *Iricolor*) on Sardinia, which showed poor separation of pseudo-pheromone cocktails and also poor separation on morphometric criteria (17 metric measurements taken from flower images). In contrast, there was stronger genetic separation based on AFLP data, but populations of both the parental species (especially *O. iricolor*) deviated from the genotypes of supposedly conspecific populations located elsewhere in the Mediterranean Basin. Furthermore, one-fifth of the plants analysed proved able to attract naïve males of the preferred pollinators of both microspecies, implying that extensive gene-flow was predictable. The conclusion drawn was that true *O. iricolor* had largely been replaced by hybrids with *O. lupercalis* in Sardinia. This in turn caused some subsequent authors to recognise Sardinian *O. iricolor* as a separate endemic species, *O. eleonorae* [126], and others to argue that the apparently widespread hybridisation between *lupercalis* and *eleonorae* had "contribut[ed] to increase in species numbers" ([25], p. 1643) in the genus. Stökl et al. [49] acknowledged the possibility that *O. iricolor* might form a genetic cline across the Mediterranean and admitted that "*Andrena*-pollinated *Ophrys* species, as [in the case of] the species of the *O. fusca* group, should tend to a high rate of nonlegitimate pollination and consequently to hybridization and introgression." But this realisation of weak divergence in all measured parameters did not deter the authors from stating that "The high similarity of pollinator-attracting scent [among microspecies] could have been a decisive factor for the strong radiation of this group" (p. 478). But by definition, a radiation requires speciation that is concentrated in time, extensive and, most importantly, unequivocal.

Vereecken et al. [127] studied populations in southern France where *O. lupercalis* forms hybrid swarms with *O. arachnitiformis* (synonymised by some authors with *O. exaltata*)—a microspecies that belongs not only to a different mesospecies but also to a different macrospecies, Sphegodes (Figures 1 and 18). Since all authors agree that the two parents represent different bona fide species, it is not surprising that their pseudo-pheromone cocktails are distinct and supposedly attract two different genera of pollinating bees. Although largely intermediate, the cocktails of the resulting hybrids also develop a minority of 23C and 25C alkenes absent from both parents. These novel compounds contribute to cocktails demonstrated to be more attractive to bee species other than those serving the parents, suggesting that new plant–pollinator relationships could arise through such hybridisation. However, the mere existence of frequent hybrids conclusively demonstrates pollinator sharing between parents whose cocktails were far more disparate. Moreover, the authors also described anarchic pollinators that varied in both the position on the labellum in which they attempted to mate with the flowers and the position(s) on the insect where the pollinaria consequently became attached. Given this valuable evidence, we view the author's opinion that *Ophrys'* "reproductive barriers are more permeable than [previously] thought" ([127], p. 5) as seriously understated.

We now broaden our consideration of macrospecies Fusca to contemplate the supposed diversification into eight microspecies of mesospecies *Attaviria* on the island of Crete. As summarised by Baguette et al. ([25], pp. 1652–1653), "All these species are pollinated by different *Andrena* species and their flowering period is remarkably staggered from the beginning of January to the end of May. Only one of these species (*Ophrys cinereophila*) is widely distributed in Crete and in the Eastern part of the Mediterranean basin; the others are all very rare and restricted to mountain massifs in Crete. This [pattern] fits well with an incipient speciation scenario in which species differentiation based on attraction of a new pollinator within a metapopulation is followed by directional selection on flowering period leading to progressive divergence from the parent taxon. Morphological divergence has not yet taken place." Based on our own DNA data, we doubt that molecular divergence has been any greater than morphological divergence. Contrasting pollinators plus a wide range of flowering periods would appear to be regarded as sufficient evidence to recognise several (micro)species, even though considerable altitudinal variation on Crete environmentally exaggerates the phenological spectrum, which must in any case guarantee that the identity of the preferred pollinating insect will change repeatedly through the spring. Is it not more parsimonious to assume that mesospecies *Attaviria* on Crete is represented by a morphological, phenologicial and genetic continuum, serviced by an overlapping sequence of pollinating species—in other words, that it has been artificially divided into microspecies?

But perhaps the more phylogenetically isolated macrospecies are more taxonomically tractable? Arguably the most plesiomorphic *Ophrys* macrospecies, Insectifera (Fly Orchid) was for long universally regarded as a single species, but later it was considered by Delforge [39] and others to contain three microspecies: one widespread and two geographically localised. Each putative species is said to draw its pollinators from a radically different guild of insects: wasps, bees and sawflies, respectively (the sawflies being especially casual in their behaviour; there is no preferred orientation of the fly on the flower, nor is there a preferred location on its body for pollinarium attachment) [11,128]. Sampled Fly Orchid populations demonstrably differed in the very limited sample of three morphological 'traits' that were recorded [59] and also differed subtly in pseudo-pheromone cocktails [129,130], but both plastid and nuclear DNA data constitute "weak but noticeable phylogeographic clustering that correlates only partially with species limits" [25]. These results were considered to "indicate a recent diversification in the three extant Fly Orchid species, which may have been further obscured by active migration and admixture across the European continent" [25,59]. But migration and admixing are exactly the processes that routinely occur *within* species; are such subtle phenotypic distinctions, poorly supported by genetic data, really sufficient to recognise multiple species?

Collectively, these valuable studies demonstrate anarchic pollinator behaviour when interacting with bee orchid flowers, contrasting radically with the adaptive perfectionism inherent in the many admiring descriptions of the extraordinary pollination process. They report supposed species that either lack genetic differentiation or possess modest genetic differentiation that contradicts prior species circumscriptions and overlapping pseudo-pheromone compositions capable of attracting multiple pollinators. Supposed morphological distinctions and pollinator preferences are tested weakly if at all. Only in the case of the hybrid swarms between *O. lupercalis* and *O. arachnitiforms* was hybrid sterility observed, and this resulted from an apparently rare case of not pre-zygotic but rather post-zygotic sterility, due to this particular population of *O. lupercalis* being tetraploid and thus incompatible with the more typically diploid *O. arachnitiformis* [127].

Yet the majority of authors of such studies fail to challenge explicitly their prior assumptions; both the microspecies circumscriptions and preferred/sole pollinator concept fed into the study survive intact in their respective discussions. We acknowledge that opinions differ among ethologists regarding the degree of pollinator specificity enjoyed by microspecies. A recent review of *Ophrys* pollination by Schatz et al. [123] usefully compared the strengths and weaknesses of contrasting methods of directly observing insects visiting *Ophrys* flowers and also offered a relatively measured and opportunistic view of pollinator specificity. Their meta-analysis suggested that the most common situation is to have a primary pollinator species but also multiple (and often taxonomically diverse) secondary pollinators.

Returning to the last of the six examples that we have chosen to highlight, we are told that "The pollinator of *O. speculum* is males of the wasp *Dasyscolia ciliata ciliata* in the western Mediterranean. In the eastern part of its range, the vicariant of *O. speculum* (*Ophrys eos*) is pollinated by the vicariant wasp *Dasyscolia ciliata araratensis*, in which females have a dark-brown body pubescence. Accordingly, the pubescence of the margins of the labellum of the vicariant *Ophrys eos* is conspicuously darker (Paulus, 2006)" ([25], p. 1647). Thus, even a subspecific difference in the identity of preferred pollinators can be cited as justification for further taxonomic division into microspecies of macrospecies Speculum. Moreover, in cases where a named *Ophrys* is shown to possess multiple pollinating species, it is immediately seen as ripe for division into multiple species, as was recently argued by Paulus [42] for *Ophrys bombyliflora* and its seven reported pollinators. From this perspective, the possession of multiple pollinators is regarded as undermining previous (often widely recognised) species circumscriptions, necessitating division into additional microspecies. Each newly minted microspecies is championed by a different pollinating insect, which in effect dictates species recognition among the orchids. Rarely has a taxonomic dog been so determinedly wagged by its own evolutionary tale [27,28,45].

4.6.3. Reproductive Isolation and Lineage Separation Are Only Assumptions

All that is required in order to gain a radically different perspective on the accumulated data is to apply the principles inherent in mainstream neodarwinian microevolution and to view the evolution of *Ophrys* through the same lens as any other genus of vascular plants. First principles predict a constant flux among conspecific populations in genotype and thus in phenotype, in response to a wide range of factors: environmental, neutral and selective. Selection pressures remain capable of inducing either encouraging (directional, disruptive) or discouraging (stabilising) speciation. It is therefore expected that local populations will present at least subtly different values for mean and variance in any measured biological property. Rather, it is the *scale* of the differences among populations under comparison that dictates whether or not they are judged to be conspecific. Moreover, such comparisons should be relative rather than absolute. The expectation is that populations of different species will differ far more than populations attributed to the same species in at least some properties that are considered biologically important. Ideally, reliable discontinuities would be sought, but their emergence requires almost complete reproductive isolation, which is unlikely to be found in groups of species such as *Ophrys* that rely on leaky pre-zygotic mechanisms rooted in pollinator behaviour—a property that leaves no historical footprint and yields observations that are valid only for the present rather than the future. Thus, our demographic species concept, rooted in seeking reliable genetic and at least subtle morphometric boundaries between the same sets of populations [29,68], requires only the severe limitation of gene flow between populations rather than its complete absence.

In their exhaustive review of *Ophrys* microevolution, Baguette et al. ([25] pp. 1657–1658), argued that "Two main species concepts are in conflict: a definition based on DNA sequence homologies, and a definition based on prezygotic isolation by attraction of species-specific pollinators. According to the first definition, there are currently around ten species of bee orchids [actually nine, according to Bateman et al. [43]], whereas adoption of the prezygotic isolation criterion leads to the recognition of several hundred species." It should be clear to any observer that our species concept is just as strongly rooted in isolation as is that employed by the ethologists. The difference is that only gene sequences (and fossils—unknown for *Ophrys*) provide historical documentation of the history of a lineage. In contrast, all of the many remaining categories of information so ably summarised by recent reviewers of *Ophrys* biology implicitly refer to just a single point in time, and most case-studies are also constrained by limited resources to very few points in space. We require the *historical evidence* of longer-term near-isolation that is provided by consistent differences in genotypes among sets of populations that are substantially greater than background levels of disparity. These differences can helpfully be visualised either as comparatively long branches in an appropriate style of tree (Figure 1) or as comparatively large distances across an appropriate style of ordination.

Baguette et al. ([25] p. 1657), then proceeded to "recommend the construction of an exhaustive and reliable molecular phylogeny of the genus *Ophrys*", thus "providing a *definitive response* [our italics] to the endless controversy about species definition in *Ophrys*". If only life were that simple. The repeatedly dichotomous model inherent in any attempt to build a molecular tree can be imposed on any kind of data-matrix, but it does assume that, though time, the species or populations represented by the individuals under analysis are similarly undergoing repeated dichotomies. It is the transition from dominantly reticulate relationships to a level of reproductive isolation sufficient to allow independent evolutionary fates that lies at the crux of speciation. The fundamental problem being that atemporal observations cannot provide information on whether populations (or sets of like individuals within populations) are diverging or converging. In the absence of historical data, divergence is assumed rather than demonstrated.

In the situations such as those observed *within* the nine macrospecies of *Ophrys*, wherein short molecular branches are being compared and phylogenetic resolution is consequently weak, alternative explanations for such patterns are available. For Baguette et al. [25] and other ethologists, those short branches represent a recent and active adaptive radiation, populated by hundreds of species generated on a remarkably short time-scale through strong and persistent pollinator-mediated directional and/or disruptive selection. This process reputedly leads to subtle yet evolutionarily significant changes in phenotype (most notably in the precise composition of pseudo-pheromone cocktails) that are judged (wrongly, in our view) to be sufficient in scale to constitute speciation, but nonetheless are argued to have occurred too recently for genetic signals confirming isolation to have accumulated.

However, any molecular tree based on a character-rich matrix will present such a rake-shaped topology, because all individuals of all sexually reproducing plants and animals show at least modest genetic differences. In our eyes, the phylogenetic rakes/combs evident within the macrospecies represent a range of conspecific populations (or a range of individuals sampled within a single conspecific population—both are demographic contexts in which extensive gene flow is a routine expectation). In such contexts, the sequentially dichotomous framework that is typically imposed on an evolutionary tree is an inappropriate model; an interconnected network and/or multivariate ordination would be more fitting. We conclude with regret that additional molecular phylogenies, even if much better sampled and based on next-generation sequencing, cannot offer "a definitive response to the endless controversy about species definition in *Ophrys*".

As we have repeatedly explained in earlier publications [29,67,68], effective species circumscription requires carefully planned sampling at multiple demographic levels (individual, population and across the range of the hypothesised species) so that reciprocal illumination becomes possible between contrasting levels. The objective is to identify the optimal aggregation of sampled populations at which gene-flow is minimised. It is important to note that, especially for genera such as *Ophrys* that rarely offer any post-zygotic isolation, it is unreasonable to expect gene-flow to be non-existent among the entities recognised as species; what matters is that the genetic disparity demonstrates that gene-flow must have been severely limited, for example to a level where any incoming genes are likely to eventually be lost through drift [28,29,45]. The depression in the degree of gene flow must be sufficiently great and sufficiently prolonged to have resulted in an acceptable percentage of reliable genetic differences. If so, in most cases, those genetic differences will coincide with reliable differences in morphology and other aspects of phenotype, even though they are unlikely to be causally related. As advocated by ethologists, the case for species-level differentiation is, of course, made even stronger if contrasts in flowering time, pollinator preference and/or habitat preference are also evident, but those differences should demonstrably be both substantial and prolonged if they are to seriously impede gene flow. It is essential that prior species circumscriptions initially fed into such studies are not treated as sacrosanct but rather are tested through reciprocal illumination, both among contrasting demographic levels and among contrasting categories of data.

5. Conclusions

Our search for iterative evolutionary trends in *Ophrys* morphology proved at best a qualified success. Despite (or perhaps because of?) our decision to record 51 characters, the genus *Ophrys* appears to approximate a morphological continuum irrespective of the taxonomic scale at which it is analysed as a genus: microspecies, mesospecies and even some macrospecies overlap. We are not the only authors to have reached this conclusion. Pausic et al. [131] found several microspecies of macrospecies Fuciflora in the former Yugoslavia to form a morphometric continuum. Hennecke and Galanos [132] used a contrasting approach to consider morphological and phenological traits as continuous variables within the whole of macrospecies Tenthredinifera, concluding that only a single species could be recognised—the macrospecies itself.

The present results are broadly congruent with our previous broad-brush studies of barcode genetics across the genus [21] and of next-generation whole-plastid sequencing combined with morphometrics across the supposedly hyper-diverse macrospecies Sphegodes [45,133], which failed to resolve either mesospecies or microspecies (Figure 18). Moreover, there are few if any obvious evolutionary trends shared among the nine molecularly self-circumscribing macrospecies, thus giving the impression that many, and perhaps most, phenotypic features of *Ophrys* plants are free to vary independently, rather than operating under strong developmental or selective constraints. This situation is likely to favour microevolution, but it is less clear whether it would facilitate macroevolution (i.e., speciation).

If our search for evolutionary trends in *Ophrys* proved underwhelming, our search for a viable mesospecies concept can only be described as a dismal failure. A biologically valid mesospecies concept capable of circumscribing "natural" species could in theory usefully generate common ground between the two radically different taxonomic views currently available. Our 'lumping' classification is based on nine widespread species that can readily be distinguished using any meaningful category of biological data and that have genetic profiles demonstrating adequate levels of long-term isolation; our work implies that microevolution rarely leads to macroevolution and suggests a speciation rate typical of most other groups of extant flowering plants. Competing 'splitters' classifications recognise several hundred species, most of them geographically localised and many of them hybridogenic, whose circumscription variously relies on subtle differences in morphology and pseudo-pheromones, and/or assumptions of extreme pollinator specificity (either singly or in some combination). The authors of these studies tacitly assume that these properties are stable through evolutionary timescales and that microevolution often leads to macroevolution, thereby implying that the genus is currently undergoing an explosive adaptive radiation.

Does any hope remain for developing a credible mesospecies concept in the future? Relative to the demographic hierarchy, herein we effectively attempted a top-down approach, seeking but failing to find morphological discontinuities within macrospecies, just as we had already failed to find clear molecular discontinuities [45]. In theory, there exists the alternative of a 'bottom-up' approach that, instead of being divisive, attempts to sequentially unite the plethora of microspecies until some kind of biologically meaningful boundary is reached.

The most recent study was devised with the laudable aim of constructing in a more scientific manner mesospecies from microspecies sampled for morphometric analysis (12 metric measurements) and volatiles composition ca. 25 plants each of 12 microspecies within macrospecies Fusca, largely from populations in southern France [134]. Smaller numbers of plants were Sanger-sequenced for nrITS and introns of the (phylogenetically questionable) low-copy nuclear genes *LFY* and *BGP* in order to construct a rooted Bayesian tree. Groupings in the three data-sets were then compared, and any microspecies that could be differentiated in any one of the three datasets was judged to have been adequately circumscribed (in other words, species were allowed to be cryptic in any two of the three data categories). Unfortunately, the design of the study ignored the demographic hierarchy. Genuine circumscription requires constructing species through the concatenation of individuals from multiple sampled populations, whereas only one or at most two populations of each microspecies were sampled. In addition, the ordination method chosen—partial least-squares discriminant analysis (PLS-DA)—is a more complex model than PCA that, crucially, requires the pre-assignment of individuals to taxa and "is prone to overfitting" to prior groupings ([135], p. 1). Thus, in the absence of reciprocal illumination between populations and putative taxa and in the presence of an algorithm that must be pre-programmed with the identity of the plants being analysed and aims to emphasise the distances among those prior groupings, there is no objective circumscription process. Only when two taxa are near-identical in all measured data categories will they be united in such an analysis, so it is unsurprising that Joffard et al. [134] were only able to reduce their initial spectrum of microspecies from 12 to 10.

Future studies employing a bottom-up approach will only be valid if (a) they are conducted within a lineage benefiting from unambiguous molecular circumscription (i.e., a macrospecies); (b) feature a well-conceived sampling strategy encompassing multiple populations spanning the taxon's full geographic extent; (c) employ an analytical arsenal that encompasses extensive morphometric (rather than more superficial, 'trait'-based) morphology, volatiles analyses, next-generation sequencing, pollinator preference and habitat preference; and (d) analyse data using algorithms that take no account of prior groupings. Perhaps the most tractable macrospecies for a such an integrated and intensive study would be Tenthredinifera; it is relatively conspicuous and hence comparatively well-recorded, is confined geographically to the Mediterranean Basin [35] and contains a manageable number of microspecies (so far!) [39]. Admittedly, quantitative studies focusing on this macrospecies have thus far been comparatively few [90,132,136–140].

In our opinion, research on the genus *Ophrys* has so become heavily biased by enthusiasm for what have become widely viewed as the key elements of its predominant evolutionary mechanism, notably the hypothesised selection strengthening the relationship between sexually excited insect and sexually deceptive flower, that some lacunae are left inadequately explored. One area of research that has been surprisingly under-utilised thus far in *Ophrys* studies is functional genomics [18,23,141]. In-depth genomics/transcriptomics studies of the related Eurasian terrestrial orchid genus *Dactylorhiza* have, in contrast, provided valuable insights into the roles played in speciation by both natural selection and non-selective genetic processes such as polyploidy [92–94], as well as elucidating important contributions from epigenetics and ecophenotypy [142,143]. Such data would address our continuing scepticism regarding the supposed fine-tuning of the pseudo-pheromone cocktails of *Ophrys*, as we consider it highly likely that their quantitative compositions are strongly influenced by both epigenetic and ecophenotypic factors that are not amenable to intense, persistent selection. Admittedly, even acquiring such datasets proved insufficient to quell disagreements between 'lumpers' and 'splitters' addressing the taxonomy of *Dactylorhiza* [36,91]. We have long been passionate advocates for superseding the traditional herbarium taxonomy with a more integrated scientific approach that explicitly considers evolutionary mechanisms [29,68], but their inclusion is no panacea; it is clear that acquiring a diversity of datasets can itself prompt a diversity of strongly held opinions.

The suggested number of ca. 400 microspecies is not the only quantitative figure that prompts concern when considering *Ophrys*. In an impressive case of both having your cake and eating it, many ethologists argue that pre-zygotic isolation is adequately effective among microspecies, but where it is demonstrably *not* effective, hybridisation can nonetheless generate a unique phenotype that can itself form a relationship with a new dedicated pollinator and so constitute the basis of yet another new species. When viewed as a symmetrical matrix of prospective parents, the 400 microspecies formally recognised within *Ophrys* represent a theoretical maximum of 79,800 primary hybrid combinations, each a potential further microspecies. Only the extreme geographic endemism of the majority of the microspecies—their localisation often becoming a worryingly self-fulfilling criterion for their initial recognition—currently precludes the erection of an almost infinite number of microspecies. Similarly, if the genus *Ophrys* is judged through molecular phylogenetics to have originated approximately 5 myr ago [24], soon after the Mediterranean Basin flooded, an explanation is needed for why as few as nine species existed after the first 4 myr of evolution but the final ca. 1 myr yielded ca. 400 species in a radiation so rapid that even the term "explosive" becomes an understatement. Lastly, unless the plethora of microspecies originated extremely recently (e.g., post-glacially), sufficient numbers of generations should have passed in isolation to allow DNA barcoding regions to have mutated and at least a very few of those mutations in each *Ophrys* microspecies to have reached fixation through selection or drift [43].

One key question that is ignored by most ethologists because of the emphasis that they implicitly place on the present rather than the past is: What proportion of incipient species predicted by the ethological species model (which, we readily accept, emerge frequently) are rapidly reabsorbed into the ancestral genetic plexus? In other words, how many such events, if viewed through just a relatively short period of time, would be seen to be transient and reticulate rather than long-term and genuinely divergent? At what point does an emerging species cease to be 'incipient' [45,53,57]?

The ongoing debate regarding macrospecies versus microspecies centres on two radically different interpretations of the same body of data accumulated by the *Ophrys* industry. Despite the great volume of research now available on the genus, both sides of the species concept still ultimately rely on belief. In our case, it is the belief than the great majority of the innumerable 'incipient species' of *Ophrys* currently extant are routine products of mainstream microevolution that will not achieve the levels of long-term isolation that should be required for recognition as distinct species; we view the microspecies as comparable with the local variants that can be found within any bona fide species of vascular plant. In the case of the ethologists, they rely on the belief that most of the myriad 'incipient' species currently recognised by them, which remain in the midst of active separation from their ancestral lineages, have nonetheless already become sufficiently reproductively isolated, and have gained enough distinctive features, to be recognised as bona fide species. It is a perspective that requires sceptics such as ourselves to trust that each divergence will become more evident with time as the myriad daughter lineages continue to diverge and dichotomise (perhaps even on a human timescale) into yet more microspecies. If so, a serious challenge awaits writers of field guides to the European flora, as they struggle to summarise innumerable indistinguishable 'species' carved out of morphological continua.

Author Contributions: Conceptualization, R.M.B.; methodology, R.M.B.; sampling and morphometric data collection, R.M.B. and P.J.R.; scanning electron microscopy and anatomy, P.J.R.; data curation and analysis, R.M.B.; writing—original draft preparation, R.M.B.; writing—review and editing, R.M.B. and P.J.R. All authors have read and agreed to the published version of the manuscript.

Funding: This research received no external funding.

Institutional Review Board Statement: Not applicable.

Informed Consent Statement: Not applicable.

Data Availability Statement: The data presented in the current study are not available within the article.

Acknowledgments: We thank many colleagues and bee orchid enthusiasts for their help and encouragement during this lengthy and spasmodic project. R.B. dedicates this manuscript to the late Al Bowley, former Head of the Biology Department at Francis Bacon School—a remarkable man whose view of the future proved clearer than my own.

Conflicts of Interest: The authors declare no conflict of interest.

Appendix A

Table A1. List of 51 morphometric characters measured from 457 individuals collectively spanning the full range of phenotypic variation in the genus *Ophrys*. Numbers of the 15 characters measured in the field are italicised; colours were matched to the RHS colour chart before conversion to CIE coordinates.

(A) Labellum (20 characters)
1. Maximum width
2. Maximum length (excluding appendix)
3. Depth of indentation [if present], from the maximum extent of the lateral portion of the central lobe inward to the base of the notch containing the appendix
4. Maximum length of speculum
5. Maximum width of speculum
6. Speculum position relative to stigma (scale 1–3) (grades into stigma: connected to stigma: not connected to stigma)
7. Pale zone along lower half to entire margin of speculum (scale 0–2) (absent: subdued: prominent)
8. Speculum shape (scale 1–4) (entire + U + W: I I + o o: H: single ring with radiating projections + three rings)
X. Base colour immediately below speculum
9. Colour (x)
10. Colour (y)
11. Colour (Y, %)
12. Width of pale-coloured marginal zone [if present] of labellum
13. Pilosity of central lobe margin of labellum 1 mm inside the margin and immediately above the appendix (scale 0–2) (none/negligible: short: long)
14. Pilosity of central lobe margin of labellum 1 mm inside the margin and at 45° to the vertical (scale 0–2) (none/negligible: short: long)
15. Pilosity of "shoulders"/lateral lobes of labellum 1 mm inside the margin (scale 0–2(none/negligible: short: long)
16. Appendix length [if present]
17. Appendix width [if present]
18. Length of "horns" [if present]
19. Maximum length of lateral lobes [if present] following mounting
20. Degree of curvature of labellum viewed transversely from base (scale 1–3) (±flat: gently convex: lateral lobes strongly recurved)
(B) Lateral petals and sepals (16 characters)
21. Length of lateral petals
22. Maximum width of lateral petals
23. Basal lateral teeth on lateral petals (scale 0–2) (absent: subdued: prominent)
X. Base colour of lateral petals
24. Colour (x)
25. Colour (y)
26. Colour (Y, %)
27. Degree of curvature of lateral petals (scale 1–5) (strongly deflexed: deflexed: ±flat: recurved: strongly recurved)
28. Length of lateral sepals
29. Maximum width of lateral sepals
X. Base colour of upper half of lateral sepals
30. Colour (x)
31. Colour (y)
32. Colour (Y, %)
33. Degree of curvature of lateral sepals (scale 1–5) (strongly deflexed: deflexed: ±flat: recurved: strongly recurved)
34. Degree of curvature of median sepal (scale 1–5) (strongly deflexed: deflexed: ±flat: recurved: strongly recurved)
35. Outline shape of median sepal (scale 1–3) (basally expanded obovate: ovate: apically expanded obovate)
36. Suffusion of dark pigment in lower half of lateral sepal (0: 1 = absent/present)

Table A1. Cont.

(C) Column and ovary (3 characters)
37. Length of ovary
38. Length of column
39. Maximum width of column
(D) Stem and inflorescence (5 characters)
40. Stem height
41. Stem diameter immediately above leaves
42. Inflorescence length
43. Number of flowers/buds
44. Angle subtended by labellum relative to stem (scale 1–3) (0–30° = parallel: 31–60°: 61–90° = perpendicular)
(E) Leaves and bracts (7 characters)
45. Number of basal (spreading) leaves
46. Number of sheathing (± upright) leaves
47. Length of longest basal leaf
48. Maximum width of longest basal leaf
49. Position of maximum with relative to position along length from base (scale 1–2) (<50%: >50%, = ovate-lanceolate: obovate leaf shapes)
50. Length of basal bract
51. Maximum width of basal bract

References

1. Rudall, P.J.; Bateman, R.M. Roles of synorganisation, zygomorphy and heterotopy in floral evolution: The gynostemium and labellum of orchids and other lilioid monocots. *Biol. Rev.* **2002**, *77*, 403–441. [CrossRef] [PubMed]
2. Mondragon-Palomino, M. Perspectives on MADS-box expression during orchid flower evolution and development. *Front. Plant Sci.* **2013**, *4*, 377. [CrossRef] [PubMed]
3. Bateman, R.M.; Rudall, P.J. Hyper-epigyny is the ultimate constraint on orchid floral morphology and an ideal model for testing the Extended Synthesis. In *Perspectives in Evolutionary and Developmental Biology*; Fusco, G., Ed.; Padova University Press: Padova, Italy, 2019; pp. 203–233.
4. Cozzolino, S.; Scopece, G. Specificity in pollination and consequences for postmating reproductive isolation in deceptive Mediterranean orchids. *Phil. Trans. R. Soc. B* **2008**, *363*, 3037–3046. [CrossRef]
5. Claessens, J.; Kleynen, J. *The Flower of the European Orchid: Form and Function*; Jean Claessens & Jacques Kleynen: Voerendaal, The Netherlands, 2011.
6. Bateman, R.M. Systematics and conservation of British and Irish orchids: A "state of the union" assessment to accompany Atlas 2020. *Kew Bull.* **2022**, *77*, 355–402. [CrossRef]
7. Glover, B. *Understanding Flowers and Flowering*, 2nd ed.; Oxford University Press: Oxford, UK, 2014.
8. Johnson, S.D.; Schiestl, F.D. *Floral Mimicry*; Oxford University Press: Oxford, UK, 2016.
9. Kullenberg, B. Studies in *Ophrys* pollination. *Zool. Bidr. Uppsala* **1961**, *34*, 1–340.
10. Agren, L.; Kullenberg, B.; Sensenbaugh, T. Congruences in pilosity between three species of *Ophrys* (Orchidaceae) and their hymenopteran pollinators. *Nova Acta Reg. Soc. Sci. Uppsaliensis V* **1984**, *3*, 5–25.
11. Borg-Karlson, A.-K.; Groth, I.; Agren, L.; Kullenberg, B. Form-specific fragrances from *Ophrys insectifera* L. (Orchidaceae) attract species of different pollinator genera: Evidence of sympatric speciation? *Chemoecology* **1993**, *4*, 39–45. [CrossRef]
12. Bradshaw, E.; Rudall, P.J.; Devey, D.S.; Thomas, M.M.; Glover, B.J.; Bateman, R.M. Comparative labellum micromorphology in the sexually deceptive temperate orchid genus *Ophrys*: Diverse epidermal cell types and multiple origins of structural colour. *Bot. J. Linn. Soc.* **2010**, *162*, 502–540. [CrossRef]
13. Francisco, A.; Porto, M.; Ascensao, L. Morphological phylogenetic analysis of *Ophrys* (Orchidaceae): Insights from morpho-anatomical floral traits into the interspecific relationships in an unresolved clade. *Bot. J. Linn. Soc.* **2015**, *179*, 454–476. [CrossRef]
14. Bateman, R.M.; Guy, J.J.; Rudall, P.J.; Leitch, I.J.; Pellicer, J.; Leitch, A.R. Evolutionary and functional potential of ploidy increase occurring within individual plants: Somatic ploidy mapping of the complex labellum of sexually deceptive bee orchids. *Ann. Bot.* **2018**, *122*, 133–150. [CrossRef]
15. Ayasse, M.; Schiestl, F.P.; Paulus, H.F.; Löfstedt, C.; Hansson, B.; Ibarra, F.; Francke, W. Evolution of reproductive strategies in the sexually deceptive orchid *Ophrys sphegodes*: How does flower-specific variation of odor signals influence reproductive success? *Evolution* **2000**, *54*, 1999–2006.
16. Ayasse, M.; Schiestl, F.P.; Paulus, H.F.; Ibarra, F.; Francke, W. Pollinator attraction in a sexually deceptve orchid by means on unconventional chemicals. *Proc. R. Soc. Lond. B* **2003**, *270*, 517–522. [CrossRef] [PubMed]
17. Ayasse, M.; Stökl, J.; Francke, W. Chemical ecology and pollinator-driven speciation in sexually deceptive orchids. *Phytochemistry* **2011**, *72*, 1667–1677. [CrossRef]

18. Sedeek, K.E.M.; Whittle, E.; Guthörl, D.; Grossniklaus, U.; Shanklin, J.; Schlüter, P.M. Amino acid change in an orchid desaturase enables mimicry of the pollinator's sex pheromone. *Curr. Biol.* **2016**, *28*, 1505–1511. [CrossRef] [PubMed]
19. Soliva, M.; Kocyan, A.; Widmer, A. Molecular phylogenetics of the sexually deceptive orchid genus *Ophrys* (Orchidaceae) based on nuclear and chloroplast DNA sequences. *Mol. Phylog. Evol.* **2001**, *20*, 78–88. [CrossRef] [PubMed]
20. Bateman, R.M.; Hollingsworth, P.M.; Preston, J.; Luo, Y.-B.; Pridgeon, A.M.; Chase, M.W. Molecular phylogenetics and evolution of Orchidinae and selected Habenariinae (Orchidaceae). *Bot. J. Linn. Soc.* **2003**, *142*, 1–40. [CrossRef]
21. Devey, D.S.; Bateman, R.M.; Fay, M.F.; Hawkins, J.A. Friends or relatives? Phylogenetics and species delimitation in the controversial European orchid genus Ophrys. *Ann. Bot.* **2008**, *101*, 385–402.
22. Sedeek, K.E.M.; Scopece, G.; Staedler, Y.M.; Schönenberger, J.; Cozzolino, S.; Schiestl, F.P.; Schlüter, P.M. Genic rather than genomewide differences between sexually deceptive *Ophrys* orchids with different pollinators. *Mol. Ecol.* **2014**, *23*, 6192–6205. [CrossRef]
23. Pineiro-Fernandez, L.; Byers, K.J.R.P.; Cai, J.; Sedeek, K.E.M.; Kellenberger, R.T.; Russo, A.; Qi, W.; Aquino-Fournier, C.; Schlüter, P.M. Phylogenomic analysis of the floral transcriptomes of sexually deceptive and rewarding European orchids, *Ophrys* and *Gymnadenia*. *Front. Plant Sci.* **2019**, *10*, 1553. [CrossRef]
24. Breitkopf, H.; Onstein, R.E.; Cafasso, D.; Schlüter, P.M.; Cozzolino, S. Multiple shifts to different pollinators fuelled rapid diversification in sexually deceptive *Ophrys* orchids. *New Phytol.* **2015**, *207*, 377–386. [CrossRef]
25. Baguette, M.; Betrand, J.; Stevens, V.M.; Schatz, B. Why are there so many bee-orchid species? Adaptive radiation by intraspecific competition for mnemonic pollinators. *Biol. Rev.* **2020**, *95*, 1630–1663. [CrossRef]
26. Bertrand, J.; Baguette, M.N.; Schatz, B. Challenges inherent in the systematics and taxonomy of genera that have recently experienced explosive radiation: The case of orchids of the genus *Ophrys*. In *Systematics and the Exploration of Life*; Grandcolas, P., Maurel, M.-C., Eds.; Wiley/Iste: New York, NY, USA, 2021; pp. 113–134.
27. Bateman, R.M.; Bradshaw, E.; Devey, D.S.; Glover, B.J.; Malmgren, S.; Sramkó, G.; Thomas, M.M.; Rudall, P.J. Species arguments: Clarifying concepts of species delimitation in the pseudo-copulatory orchid genus *Ophrys*. *Bot. J. Linn. Soc.* **2011**, *165*, 336–347. [CrossRef]
28. Bateman, R.M. Two bees or not two bees? An overview of *Ophrys* systematics. *Ber. Arbeitskr. Heim. Orchideen* **2018**, *35*, 5–46.
29. Bateman, R.M. Species circumscription in 'cryptic' clades: A nihilist's view. In *Cryptic Species: Morphological Stasis, Circumscription and Hidden Diversity*; Monro, A.K., Mayo, S.J., Eds.; Cambridge University Press: Cambridge, UK, 2022; pp. 36–77.
30. Vereecken, N.J.; Streinzer, M.; Ayasse, M.; Spaethe, J.; Paulus, H.F.; Stökl, J.; Cortis, P.; Schiestl, F.P. Integrating past and present studies on *Ophrys* pollination—A comment on Bradshaw et al. *Bot. J. Linn. Soc.* **2011**, *165*, 329–335. [CrossRef]
31. Cuypers, V.; Reydon, T.A.C.; Artois, T. Deceiving insects, deceiving taxonomists? *Making theoretical sense of taxonomic disagreement in the European orchid genus Ophrys. Persp. Plant Ecol. Evol. Syst.* **2022**, *56*, 125686. [CrossRef]
32. Nelson, E. *Monographie und Ikonographie der Gattung Ophrys*; E. Nelson: Chernex-Montreux, Switzerland, 1962.
33. De Soó, R. Dactylorhiza Necker ex Nevski. In *Flora Europaea 5*; Tutin, T.G., Heywood, V.H., Burgess, N.A., Moore, D.M., Valentine, D.H., Walters, S.M., Webb, D.A., Eds.; Cambridge University Press: Cambridge, UK, 1980; pp. 333–337.
34. Sundermann, H. *Europäische und Mediterrane Orchideen—Ein Bestimmungsflora*, 3rd ed.; Schmersow: Hildesheim, Germany, 1980.
35. Pedersen, H.; Faurholdt, N. *Ophrys: The Bee Orchids of Europe*; Kew Publishing: Kew, UK, 2007.
36. Kühn, R.; Pedersen, H.A.; Cribb, P. *Field Guide to the Orchids of Europe and the Mediterranean*; Royal Botanic Gardens Kew: Kew, UK, 2019.
37. Devillers, P.; Devillers-Terschuren, J. Essai d'analyse systématique du genre *Ophrys*. *Nat. Belges* **1994**, *75*, 273–400.
38. Delforge, P. *Orchidées d'Europe, d'Afrique du Nord et du Proche-Orient*, 1st ed.; Delachaux et Niéstle: Paris, France, 1994.
39. Delforge, P. *Orchidées d'Europe, d'Afrique du Nord et du Proche-Orient*, 4th ed.; Delachaux et Niéstle: Paris, France, 2016.
40. Paulus, H.F. Deceived males—Pollination biology of the Mediterranean orchid genus *Ophrys* (*Orchidaceae*). *J. Eur. Orch.* **2006**, *38*, 303–353.
41. Paulus, H.F. Bestäuber als Isolationsmechanismen: Freilandbeobachtungen und Experimente zur Spezifität der Bestäuberanlockung in der Gattung *Ophrys* (Orchidaceae und Insecta, Hymenoptera, Apoidea). *Ber. Arbeitskr. Heim. Orchid.* **2015**, *32*, 142–199.
42. Paulus, H.F. Pollinators as isolation mechanisms: Field observations and field experiments regarding specificity of pollinator attraction in the genus *Ophrys* (Orchidaceae und Insecta, Hymenoptera, Apoidea). *Entom. Gener.* **2018**, *37*, 261–316. [CrossRef]
43. Bateman, R.M.; Sramkó, G.; Paun, O. Integrating restriction site-associated DNA sequencing (RAD-seq) with morphological cladistic analysis clarifies evolutionary relationships among major species groups of bee orchids. *Ann. Bot.* **2018**, *121*, 85–105. [CrossRef]
44. Tyteca, D.; Baguette, M. *Ophrys* (Orchidaceae) systematics: When molecular phylogenetics, morphology and biology reconcile. *Ber. Arbeitskr. Heim. Orchideen* **2017**, *34*, 37–103.
45. Bateman, R.M.; Rudall, P.J.; Murphy, A.R.M.; Cowan, R.S.; Devey, D.S.; Perez-Escobar, O.A. Even whole plastomes are not enough: Phylogenomic and morphometric exploration at multiple demographic levels of the bee orchid clade *Ophrys* sect. *Sphegodes*. *J. Exp. Bot.* **2021**, *72*, 654–681. [CrossRef] [PubMed]
46. Kreutz, C.A.J. *Kompendium der Europäischen Orchideen*; Kreutz Publishers: Landgraaf, The Netherlands, 2004.
47. Baumann, H.; Künkele, S.; Lorenz, R. *Orchideen Europas*; Ulmer: Stuttgart, Germany, 2006.
48. Mayr, E. *Systematics and the Origin of Species*; Columbia University Press: New York, NY, USA, 1942.

49. Mayr, E. Speciation and macroevolution. *Evolution* **1982**, *36*, 1119–1132. [CrossRef] [PubMed]
50. Mayden, R.L. A hierarchy of species concepts: The denouement in the saga of the species problem. In *Species: The Units of Biodiversity*; Claridge, M.F., Dawah, H.A., Wilson, M.R., Eds.; Chapman & Hall: London, UK, 1997; pp. 381–421.
51. Wheeler, Q.D.; Meier, R. (Eds.) *Species Concepts and Phylogenetic Theory: A Debate*; Columbia University Press: New York, NY, USA, 2000.
52. Rieseberg, L.H.; Wood, T.E.; Baack, E.J. The nature of plant species. *Nature* **2006**, *440*, 524–527. [CrossRef]
53. de Queiroz, K. Species concepts and species delimitation. *Syst. Biol.* **2007**, *56*, 879–886. [CrossRef]
54. Hart, M.W. The species concept as an emergent property of population biology. *Evolution* **2011**, *65*, 613–616. [CrossRef]
55. Hausdorf, B. Progress toward a general species concept. *Evolution* **2011**, *65*, 923–931. [CrossRef]
56. Mishler, B.D. *What, If Anything, Are Species?* CRC Press: Boca Raton, FL, USA, 2021.
57. Monro, A.K.; Mayo, S.J. (Eds.) *Cryptic Species: Morphological Stasis, Circumscription and Hidden Diversity*; Cambridge University Press: Cambridge, UK, 2022.
58. Stökl, J.; Schlüter, P.M.; Stuessy, T.F.; Paulus, H.F.; Assum, G.; Ayasse, M. Scent variation and hybridization cause the displacement of a sexually deceptive orchid species. *Am. J. Bot.* **2008**, *95*, 472–481. [CrossRef]
59. Triponez, Y.; Arrigo, N.; Pelissier, L.; Schatz, B.; Alvarez, N. Morphological, ecological and genetic aspects associated with endemism in the Fly Orchid group. *Mol. Ecol.* **2013**, *22*, 1431–1446. [CrossRef]
60. Gögler, J.; Stökl, J.; Cortis, P.; Beyrle, H.; Barone Lumaga, M.R.; Cozzolino, S.; Ayasse, M. Increased divergence in floral morphology strongly reduces gene flow in sympatric sexually deceptive orchids with the same pollinator. *Evol. Ecol.* **2015**, *29*, 703–717. [CrossRef]
61. Gögler, J.; Zitari, A.; Paulus, H.; Cozzolino, S.; Ayasse, M. Species boundaries in the *Ophrys iricolor* group in Tunisia: Do local endemics always matter? *Plant Syst. Evol.* **2016**, *302*, 481–489. [CrossRef]
62. Véla, E.; Rebbas, K.; Martin, R.; Premorel, G.; de Tison, J.-M. Waiting for integrative taxonomy: Morphospecies as an operational proxy for the radiative and reticulate genus *Ophrys* L. (Orchidaceae)? *Eur. J. Environ. Sci.* **2015**, *5*, 153–157. [CrossRef]
63. Lussu, M.; De Agostini, A.; Marignani, M.; Cogoni, A.; Cortis, P. *Ophrys annae* and *Ophrys chestermanniae*: An impossible love between two orchid sister species. *Nordic J. Bot.* **2018**, *36*, e01798. [CrossRef]
64. Cozzolino, S.; Scopece, G.; Cortis, P.; Barone Lumaga, M.R.; Cafasso, D.; Ayasse, M. Reproductive character displacement allows two sexually deceptive orchids to coexist and attract the same specific pollinator. *Evol. Ecol.* **2022**, *36*, 217–232. [CrossRef]
65. Turco, A.; Medagli, P.; Wagensommer, R.P.; D'Emerico, S.; Gennaio, R.; Albano, A. A morphometric study on *Ophrys* sect. *Pseudophrys in Apulia (Italy) and discovery of Ophrys japigiae. Plant Biosyst.* **2022**, *156*, 560–571.
66. Bateman, R.M.; Denholm, I. A reappraisal of the British and Irish dactylorchids, 1. The tetraploid marsh-orchids. *Watsonia* **1983**, *14*, 347–376.
67. Bateman, R.M. Evolution and classification of European orchids: Insights from molecular and morphological characters. *J. Eur. Orch.* **2001**, *33*, 33–119.
68. Bateman, R.M. The perils of addressing long-term challenges in a short-term world: Making descriptive taxonomy predictive. In *Climate Change, Ecology and Systematics*; Hodkinson, T.R., Jones, M.B., Waldren, S., Parnell, J.A.N., Eds.; Cambridge University Press: Cambridge, UK, 2011; pp. 67–95.
69. Bateman, R.M.; Rudall, P.J.; Denholm, I. In situ morphometric survey elucidates the evolutionary systematics of the orchid genus *Gymnadenia* in the British Isles. *Syst. Biodiv.* **2021**, *19*, 571–600. [CrossRef]
70. Payne, R.W.; Harding, S.A.; Murray, D.A.; Souter, D.M.; Baird, D.B.; Glaser, A.I.; Welham, S.J.; Gilmour, A.R.; Thompson, R.; Webster, R. (Eds.) *Genstat v14*; VSN International: Hemel Hempstead, UK, 2011.
71. Gower, J.C. A general coefficient of similarity and some of its properties. *Biometrics* **1971**, *27*, 857–872. [CrossRef]
72. Gower, J.C.; Ross, G.J.S. Minimum spanning trees and single linkage cluster analysis. *J. R. Stat. Soc. C* **1969**, *18*, 54–64. [CrossRef]
73. Gower, J.C. Some distance properties of latent root and vector methods used in multivariate analysis. *Biometrika* **1966**, *52*, 325–338. [CrossRef]
74. Gower, J.C. Measures of similarity, dissimilarity and distance. In *Encyclopedia of Statistical Sciences 5*; Wiley: New York, NY, USA, 1985; pp. 397–405.
75. Devey, D.S.; Bateman, R.M.; Fay, M.F.; Hawkins, J.A. Genetic structure and systematic relationships within the *Ophrys fuciflora* aggregate (Orchidinae: Orchidaceae): High diversity in Kent and a wind-induced discontinuity bisecting the Adriatic. *Ann. Bot.* **2009**, *104*, 483–495. [CrossRef] [PubMed]
76. Bateman, R.M.; Rudall, P.J. The life and death of a mythical British endemic, *Orchis militaris* L. var. tenuifrons P.D. Sell: Why infraspecific taxonomy requires a field-based morphometric approach. *N. J. Bot.* **2011**, *1*, 98–110.
77. Bateman, R.M.; Denholm, I.; McLeod, L.; Craig, W.; Ennos, R.A. Systematic reappraisal of Marsh-orchids native to Scotland. *Kew Bull.* **2023**, *79*, 21.
78. Bateman, R.M. Phenotypic versus genotypic disparity in the Eurasian orchid genus *Gymnadenia*: Exploring the limits of phylogeny reconstruction. *Syst. Biodiv.* **2021**, *19*, 400–422. [CrossRef]
79. Bateman, R.M.; James, K.E.; Rudall, P.J. Contrast in morphological versus molecular divergence between two closely related Eurasian species of *Platanthera* (Orchidaceae) suggests recent evolution with a strong allometric component. *New J. Bot.* **2012**, *2*, 110–148. [CrossRef]
80. Bateman, R.M.; Rudall, P.J.; Moura, M. Systematic revision of *Platanthera* in the Azorean archipelago: Not one but three species, including arguably Europe's rarest orchid. *PeerJ* **2013**, *1*, e218. [CrossRef]

81. Bateman, R.M.; Smith, R.J.; Fay, M.F. Morphometric and population-genetic analyses elucidate the origin, evolutionary significance and conservation implications of *Orchis* ×*angusticruris* (*O. purpurea* × *O. simia*), a hybrid orchid new to Britain. *Bot. J. Linn. Soc.* **2008**, *157*, 687–711. [CrossRef]
82. Bateman, R.M.; Molnár, A.V.; Sramkó, G. In situ morphometric survey elucidates the evolutionary systematics of the Eurasian *Himantoglossum* clade (Orchidaceae: Orchidinae). *PeerJ* **2017**, *5*, e2893. [CrossRef]
83. Bateman, R.M.; Rudall, P.J.; Bidartondo, M.I.; Cozzolino, S.; Tranchida-Lombardo, V.; Carine, M.A.; Moura, M. Speciation via floral heterochrony and presumed mycorrhizal host-switching of endemic butterfly orchids on the Azorean archipelago. *Am. J. Bot.* **2014**, *101*, 979–1001. [CrossRef]
84. Jin, W.-T.; Schuiteman, A.; Chase, M.W.; Li, J.-W.; Chung, S.-W.; Hsu, T.-C.; Jin, X.-H. Phylogenetics of subtribe Orchidinae s.l. (Orchidaceae; Orchidoideae) based on seven markers (plastid *matK*, *psaB*, *rbcL*, *trnL-F*, *trnH-psba*, and nuclear nrITS, *Xdh*): Implications for generic delimitation. *BMC Plant Biol.* **2017**, *17*, 214–222. [CrossRef] [PubMed]
85. Bell, A.K.; Roberts, D.L.; Hawkins, J.A.; Rudall, P.J.; Box, M.S.; Bateman, R.M. Comparative morphology of nectariferous and nectarless labellar spurs in selected clades of subtribe Orchidinae (Orchidaceae). *Bot. J. Linn. Soc.* **2009**, *160*, 369–387. [CrossRef]
86. Shipunov, A.B.; Bateman, R.M. Geometric morphometrics as a tool for understanding *Dactylorhiza* (Orchidaceae) diversity in European Russia. *Biol. J. Linn. Soc.* **2005**, *85*, 1–12. [CrossRef]
87. Breitkopf, H.; Schlüter, P.M.; Xu, S.; Schiestl, F.P.; Cozzolino, S.; Scopece, G. Pollinator shifts between *Ophrys sphegodes* populations: Might adaptation to different pollinators drive population divergence? *J. Evol. Biol.* **2013**, *26*, 2197–2208. [CrossRef] [PubMed]
88. Bersweden, L.; Viruel, J.; Schatz, B.; Harland, J.; Gargiulo, R.; Cowan, R.S.; Calevo, J.; Juan, A.; Clarkson, J.J.; Leitch, A.R.; et al. Microsatellites and petal morphology reveal new patterns of admixture in *Orchis* hybrid zones. *Am. J. Bot.* **2021**, *108*, 1388–1404. [CrossRef] [PubMed]
89. Bateman, R.M.; Farrington, O.S. A morphometric study of X*Orchiaceras bergonii* (Nanteuil) Camus and its parents (*Aceras anthropophorum* (L.) Aiton f. and *Orchis simia* Lamarck) in Kent. *Watsonia* **1987**, *16*, 397–407.
90. Bateman, R.M.; Farrington, O.S. Morphometric comparison of populations of *Orchis simia* Lam. (Orchidaceae) from Oxfordshire and Kent. *Bot. J. Linn. Soc.* **1989**, *100*, 205–218. [CrossRef]
91. Bateman, R.M.; Denholm, I. Taxonomic reassessment of the British and Irish tetraploid marsh-orchids. *New J. Bot.* **2012**, *2*, 37–55. [CrossRef]
92. Paun, O.; Bateman, R.M.; Luna, J.A.; Moat, J.; Fay, M.F.; Chase, M.W. Altered gene expression and ecological divergence in sibling allopolyploids of *Dactylorhiza* (Orchidaceae). *BMC Evol. Biol.* **2011**, *11*, 113. [CrossRef]
93. Balao, F.; Trucchi, E.; Wolfe, T.M.; Hao, B.-H.; Lorenzo, M.T.; Baar, J.; Sedman, L.; Kosiol, C.; Amman, F.; Chase, M.W.; et al. Adaptive sequence evolution is driven by biotic stress in a pair of orchid species (*Dactylorhiza*) with distinct ecological optima. *Mol. Ecol.* **2017**, *26*, 3649–3662. [CrossRef]
94. Brandrud, M.K.; Baar, J.; Lorenzo, M.T.; Athanasiadis, A.; Bateman, R.M.; Chase, M.W.; Hedrén, M.; Paun, O. Phylogenomic relationships of diploids and the origins of allotetraploids in *Dactylorhiza* (Orchidaceae): RADseq data track reticulate evolution. *Syst. Biol.* **2020**, *61*, 91–109. [CrossRef] [PubMed]
95. Chapurlat, E.; Le Ronce, I.; Agren, J.; Sletvold, N. Divergent selection on flowering phenology but not on floral morphology between two closely related orchids. *Ecol. Evol.* **2020**, *10*, 5547–5737. [CrossRef] [PubMed]
96. Hedrén, M.; Lorenz, R.; Teppner, H.; Dolinar, B.; Giotta, C.; Griebl, N.; Hansson, S.; Heidtke, U.; Klein, E.; Perazza, G.; et al. Evolution and systematics of polyploid *Nigritella* (Orchidaceae). *Nordic J. Bot.* **2018**, *36*, 01539. [CrossRef]
97. Stejskal, K.; Streinzer, M.; Dyer, A.; Paulus, H.F.; Spaethe, J. Functional significance of labellum pattern variation in a sexually deceptive orchid (*Ophrys heldreichii*): Evidence of individual signature learning effects. *PLoS ONE* **2015**, *10*, e0142971. [CrossRef] [PubMed]
98. Francisco, A.; Ascensao, L. Structure of the osmophore and labellum micromorphology in the sexually deceptive orchids *Ophrys bombyliflora* and *Ophrys tenthredinifera* (Orchidaceae). *Int. J. Plant Sci.* **2013**, *174*, 619–636. [CrossRef]
99. Cortis, P.; Vereecken, N.J.; Schiestl, F.P.; Barone Lumaga, M.R.; Scrugli, A.; Cozzolino, S. Pollinator convergence and the nature of species' boundaries in sympatric Sardinian *Ophrys* (Orchidaceae). *Ann. Bot.* **2009**, *104*, 497–506. [CrossRef] [PubMed]
100. Rakosy, D.; Cuervo, M.; Paulus, H.F.; Ayasse, M. Looks matter: Changes in floral form affect pollination effectiveness in a sexually deceptive orchid. *J. Evol. Biol.* **2017**, *30*, 1978–1993. [CrossRef]
101. Spaethe, J.; Moser, W.H.; Paulus, H.F. Increase of pollinator attraction by means of a visual signal in the sexually deceptive orchid *Ophrys heldreichii* (Orchidaceae). *Plant Syst. Evol.* **2007**, *264*, 31–40. [CrossRef]
102. Streinzer, M.; Paulus, H.F.; Spaethe, J. Floral colour signal increases short-range detectability of a sexually deceptive orchid to its bee pollinator. *J. Exp. Biol.* **2009**, *212*, 1365–1370. [CrossRef]
103. Rakosy, D.; Streinzer, M.; Paulus, H.F.; Spathe, J. Floral visual signal increases reproductive success in a sexually deceptive orchid. *Arthropod-Plant Interac.* **2012**, *6*, 671–681. [CrossRef]
104. Streinzer, M.; Ellis, T.; Paulus, H.F.; Spathe, J. Visual discrimination between two sexually deceptive *Ophrys* species by a bee pollinator. *Arthropod-Plant Interac.* **2010**, *4*, 141–148. [CrossRef] [PubMed]
105. Vereecken, N.J.; Schiestl, F.P. On the roles of colour and scent in a specialized floral mimicry system. *Ann. Bot.* **2009**, *104*, 1077–1084. [CrossRef] [PubMed]
106. Kevan, P.; Giurfa, M.; Chittka, L. Why are there so many and so few white flowers? *Trends Plant Sci.* **1996**, *1*, 280–284. [CrossRef]

107. Vignolini, S.; Davey, M.P.; Bateman, R.M.; Rudall, P.J.; Moyroud, E.; Tratt, J.; Malmgren, S.; Steiner, U.; Glover, B.J. The mirror crack'd: Both structure and pigment contribute to the metallic blue appearance of the Mirror Orchid, *Ophrys speculum*. *New Phyt.* **2012**, *196*, 1038–1047. [CrossRef] [PubMed]
108. Malmgren, S. Are there 25 or 250 *Ophrys* species? *J. Hardy Orchid Soc.* **2008**, *5*, 95–100.
109. Spaethe, J.; Streinzer, M.; Paulus, H.F. Why sexually deceptive orchids have colored flowers. *Commun. Integr. Biol.* **2010**, *3*, 139–141. [CrossRef]
110. Joffard, N.; Arnal, V.; Buatois, B.; Schatz, B.; Montgelard, C. Floral scent evolution in the section *Pseudophrys*: Pollinator-mediated selection or phylogenetic constraints? *Plant Biol.* **2020**, *20*, 881–889. [CrossRef]
111. Schiestl, F.P.; Cozzolino, S. Evolution of sexual mimicry in the orchid subtribe Orchidinae: The role of preadaptations in the attraction of male bees as pollinators. *BMC Evol. Biol.* **2008**, *8*, 27. [CrossRef]
112. Mant, J.; Peakall, R.; Schiestl, F.P. Does selection on floral odor promote differentiation among populations and species of the sexually deceptive orchid genus *Ophrys*? *Evolution* **2005**, *59*, 1449–1463.
113. Vereeckren, N.J.; Mant, J.; Schiestl, F.P. Population differentiation in female sex pheromone and male preferences in a solitary bee. *Behav. Ecol. Sociobiol.* **2007**, *61*, 811–821. [CrossRef]
114. Vereecken, N.J.; Schiestl, F.P. The evolution of imperfect floral mimicry. *Proc. Natl. Acad. Sci. USA* **2008**, *105*, 7484–7488. [CrossRef] [PubMed]
115. Schlüter, P.M.; Ruas, P.M.; Kohl, G.; Ruas, C.F.; Stuessy, T.F.; Paulus, H.F. Evidence for progenitor-driven speciation in sexually deceptive orchids. *Ann. Bot.* **2011**, *108*, 895–906. [CrossRef] [PubMed]
116. Xu, S.; Schlüter, P.M.; Scopece, G.; Breitkopf, H.; Gross, K.; Cozzolino, S.; Schiestl, F.P. Floral isolation is the main reproductive barrier among closely related sexually deceptive orchids. *Evolution* **2011**, *65*, 2606–2620. [CrossRef] [PubMed]
117. Paulus, H.F. Speciation, pattern recognition and the maximization of pollination: General questions and answers given by the reproductive biology of the orchid genus *Ophrys*. *J. Comp. Physiol. A* **2019**, *205*, 285–300. [CrossRef]
118. Vereecken, N.J. Deceptive behaviour in plants. I. Pollination by sexual deception in orchids: A host–parasite perspective. In *Plant–Environment Interactions*; Baluska, F., Ed.; Springer: Berlin, Germany, 2009; pp. 203–222.
119. Ayasse, M.; Gögler, J.; Stökl, J. Pollinator-driven speciation in sexually deceptive orchids of the genus *Ophrys*. In *Evolution in Action*; Glaubrecht, M., Ed.; Springer: Berlin, Germany, 2010; pp. 101–118.
120. Xu, S.; Schlüter, P.M.; Schiestl, F.P. Pollinator-driven speciation in sexually deceptive orchids. *Int. J. Ecol.* **2012**, *2012*, 285081. [CrossRef]
121. Schiestl, F.P.; Johnson, S.D. Pollinator-mediated evolution of floral signals. *Trends Ecol. Evol.* **2013**, *28*, 307–315. [CrossRef]
122. Vereecken, N.J.; Francisco, A. Ophrys pollination: From Darwin to the present day. In *Darwin's Orchids: Then and Now*; Edens-Meier, R., Bernhardt, P., Eds.; University of Chicago Press: Chicago, IL, USA, 2014; pp. 47–69.
123. Schatz, B.; Genoud, D.; Esudié, P.; Geniez, P.; Wünch, K.G.; Joffard, N. Is *Ophrys* pollination more opportunistic than previously thought? Insights from different field methods of pollinator observation. *Bot. Lett.* **2021**, *168*, 333–347. [CrossRef]
124. Wong, B.B.M.; Schiestl, F.P. How an orchid harms its pollinator. *Proc. R. Soc. Lond. B* **2002**, *269*, 1529–1532. [CrossRef]
125. Stökl, J.; Schlüter, P.M.; Steussy, T.F.; Paulus, H.F.; Fraberger, R.; Erdmann, D.; Ayasse, M. Speciation in sexually deceptive orchids: Pollinator-driven selection maintains discrete odour phenotypes in hybridizing species. *Biol. J. Linn. Soc.* **2009**, *98*, 439–451. [CrossRef]
126. Devillers-Terschuren, J.; Devillers, P. Nomenclatural and taxonomic contributions to European orchids (Orchidaceae). *Nat. Belges* **1991**, *72*, 99–101.
127. Vereecken, N.J.; Cozzolino, S.; Schiestl, F.P. Hybrid floral scent novelty drives pollinator shift in sexually deceptive orchids. *BMC Evol. Biol.* **2010**, *10*, 103. [CrossRef] [PubMed]
128. Hermosilla, C.E.; Amardeilh, J.P.; Soca, R. *Sterictiphora furcata* Villers, pollinisateur d'*Ophrys subinsectifera* Hermosilla & Sabando. *L'Orchidophile* **1999**, *139*, 247–254.
129. Joffard, N.; Buatois, B.; Schatz, B. Integrative taxonomy of the fly orchid group: Insights from chemical ecology. *Sci. Nat.* **2016**, *103*, 77. [CrossRef] [PubMed]
130. Gervasi, D.L.; Selosse, M.A.; Sauve, M.; Francke, W.; Vereecken, N.J.; Cozzolino, S.; Schiestl, F.P. Floral scent and species divergence in a pair of sexually deceptive orchids. *Ecol. Evol.* **2017**, *7*, 6023–6034. [CrossRef]
131. Pausic, I.; Lipovsek, M.; Jakely, D.; Pavlec, N.; Ivajnsic, D.; Kaligaric, M. Local climate and latitude affect flower form of *Ophrys fuciflora* (Orchidaceae): Evidence for clinal variation. *Bot. Lett.* **2019**, *166*, 499–512. [CrossRef]
132. Hennecke, M.; Galanos, C. *Ophrys* subgen. Bombyliflorae sect. Tenthrediniferae: A synopsis. *GIROS* **2020**, *63*, 358–419.
133. Cozzolino, S.; Scopece, G.; Roma, L.; Schlüter, P.M. Different filtering strategies of genotyping-by-sequencing data provide complementary resolutions of species boundaries and relationships in a clade of sexually deceptive orchids. *J. Syst. Evol.* **2020**, *58*, 133–144. [CrossRef]
134. Joffard, N.; Buatois, B.; Arnal, V.; Véla, E.; Montgelard, C.; Schatz, B. Delimiting species in the taxonomically challenging orchid section *Pseudophrys*: Bayesian analyses of genetic and phenotypic data. *Front. Ecol. Evol.* **2022**, *10*, 1058550. [CrossRef]
135. Ruiz-Perez, D.; Guan, H.; Madhivanan, P.; Mathee, K.; Narasimhan, G. So you think you can PLS-DA? *BMC Bioinf.* **2020**, *21*, 2. [CrossRef]
136. Gölz, P.; Reinhard, H. Bemerkungen zu "Baumann, H. & S. Künkele (1986): Die Gattung *Ophrys* L.—Eine taxonomische Übersicht". *Mitt. Bl. Arbeitskrs. Heim. Orch. Baden-Württ.* **1987**, *19*, 866–894.

137. Devillers, P.; Devillers-Terschuren, J.; Tyteca, D. Notes on some of the taxa comprising the group of *Ophrys tenthredinifera* Willdenow. *J. Eur. Orch.* **2003**, *35*, 109–161.
138. Gögler, J.; Stökl, J.; Sramkova, A.; Twele, R.; Francke, W.; Cozzolino, S.; Cortis, P.; Scrugli, A.; Ayasse, M. Ménage a trois—Two endemic species of deceptive orchids and one pollinator species. *Evolution* **2009**, *63*, 2222–2234. [CrossRef] [PubMed]
139. Paulus, H.F.; Hirth, M. Bestäubungsbiologie und Systematik der *Ophrys tenthredinifera*-Artengruppe in der Ostägäis (Orchidaceae, Insecta). *J. Eur. Orch.* **2012**, *44*, 625–686.
140. Wucherpfennig, W. Die westmediterranen Arten der *Ophrys tenthredinifera*-Gruppe ein Vergleich. *Ber. Arbeitskrs. Heim. Orch.* **2016**, *33*, 110–150.
141. Sedeek, K.E.M.; Qi, W.; Schauer, M.A.; Gupta, A.K.; Poveda, L.; Xu, S.; Liu, Z.-J.; Grossniklaus, U.; Schiestl, F.P.; Schlüter, P.M. Transcriptome and proteome data reveal candidate genes for pollinator attraction in sexually deceptive orchids. *PLoS ONE* **2013**, *8*, e64621. [CrossRef]
142. Paun, O.; Bateman, R.M.; Fay, M.F.; Hedrén, M.; Civeyrel, L.; Chase, M.W. Stable epigenetic effects impact evolution and adaptation in allopolyploid orchids. *Mol. Biol. Evol.* **2010**, *27*, 2465–2473. [CrossRef] [PubMed]
143. Wolfe, T.M.; Balao, F.; Trucchi, E.; Bachmann, G.; Gu, W.; Baar, J.; Hedrén, M.; Weckwerth, W.; Leitch, A.R.; Paun, O. Recurrent allopolyploidization events diversify eco-physiological traits in marsh orchids. *bioRxiv* **2021**. [CrossRef]

biology

MDPI

Article

Small Leaves, Big Diversity: Citizen Science and Taxonomic Revision Triples Species Number in the Carnivorous *Drosera microphylla* Complex (*D.* Section *Ergaleium*, Droseraceae)

Thilo Krueger [1,*], Alastair Robinson [2], Greg Bourke [3] and Andreas Fleischmann [4,5,*]

1 School of Molecular and Life Sciences, Curtin University, GPO Box U1987, Bentley, WA 6102, Australia
2 National Herbarium of Victoria, Royal Botanic Gardens Victoria, Melbourne, VIC 3004, Australia
3 P.O. Box 3001, Bilpin, NSW 2758, Australia
4 Botanische Staatssammlung München, Menzinger Strasse 67, 80638 Munich, Germany
5 GeoBio-Center LMU, Ludwig Maximilians University, 80638 Munich, Germany
* Correspondence: t.krueger@postgrad.curtin.edu.au (T.K.); fleischmann@bio.lmu.de (A.F.)

Simple Summary: A novel taxonomic treatment is provided for the *Drosera microphylla* complex, which is a group of closely related carnivorous plants endemic to southwest Western Australia. The species that comprise this group are generally rare, micro-endemic, and are potentially threatened by habitat destruction and illegal collection. Resolving the taxonomy and systematics of this complex has been critical to the accurate assessment of its component species under conservation legislation. Following two decades of fieldwork in Western Australia, studies of preserved plant collections, and crucial contributions by citizen scientists and social media, we establish here that the *Drosera microphylla* complex comprises nine distinct species, three times the number previously recognised. Four species are here described and illustrated as new to science. Two previously described varieties are here re-circumscribed as distinct species in light of their rediscoveries via social media posts, allowing them to be studied for the first time since they were described more than 100 years ago. We provide examples from the genus *Drosera* for the impact of social media and citizen science on taxonomic work and biological conservation. This work demonstrates the great potential that citizen science has in supporting rapid advances in taxonomic knowledge in the face of extinction crises worldwide.

Abstract: The carnivorous *Drosera microphylla* complex from southwest Western Australia comprises a group of rare, narrowly endemic species that are potentially threatened by habitat destruction and illegal collection, thus highlighting a need for accurate taxonomic classification to facilitate conservation efforts. Following extensive fieldwork over two decades, detailed studies of both Australian and European herbaria and consideration of both crucial contributions by citizen scientists and social media observations, nine species of the *D. microphylla* complex are here described and illustrated, including four new species: *D. atrata*, *D. hortiorum*, *D. koikyennuruff*, and *D. reflexa*. The identities of the previously described infraspecific taxa *D. calycina* var. *minor* and *D. microphylla* var. *macropetala* are clarified. Both are here lectotypified, reinstated, and elevated to species rank. A replacement name, *D. rubricalyx*, is provided for the former taxon. Key morphological characters distinguishing the species of this complex include the presence or absence of axillary leaves, lamina shape, petal colour, filament shape, and style length. A detailed identification key, comparison figures, and a distribution map are provided. Six of the nine species are recommended for inclusion on the Priority Flora List under the Conservation Codes for Western Australian Flora and Fauna.

Keywords: Australia; carnivorous plants; non-core *Caryophyllales*; Nepenthales; sundews; taxonomy; typification

Citation: Krueger, T.; Robinson, A.; Bourke, G.; Fleischmann, A. Small Leaves, Big Diversity: Citizen Science and Taxonomic Revision Triples Species Number in the Carnivorous *Drosera microphylla* Complex (*D.* Section *Ergaleium*, Droseraceae). *Biology* **2023**, *12*, 141. https://doi.org/10.3390/biology12010141

Academic Editor: Lorenzo Peruzzi

Received: 23 December 2022
Revised: 10 January 2023
Accepted: 11 January 2023
Published: 16 January 2023

1. Introduction

Drosera L. (Droseraceae Salisbury; commonly known as sundews) is a cosmopolitan genus of carnivorous plants comprising ca. 260 herbaceous species, of which ca. 110 are endemic to the Southwest Australian Floristic Region (SWAFR) [1,2]. The SWAFR is recognised as the global centre of diversity for *Drosera* and for carnivorous plants in general [2,3]; this high alpha-diversity is enabled by the region's abundance of nutrient-deficient soils, broad diversity and close geographic proximity of different habitats, and the seasonal Mediterranean climate with long-term climatic stability [4,5].

Within *Drosera*, the geophytes of *D.* section *Ergaleium* (DC.) Planch. ("tuberous sundews"; infrageneric classification following Fleischmann et al. [2]) form the largest and morphologically most diverse evolutionary lineage, comprising 71 currently accepted species [1,2,6]. The members of this monophyletic clade [2,7] can have rosetted, erect, or even climbing habits, various stem branching patterns and leaf arrangements, and many different style shapes [1,8]. However, they all share a swollen, stem-derived subterranean tuber that acts as a storage organ and allows the plants to perennate underground during the very dry summer conditions of the SWAFR [2,8–15] and to survive bushfires [12,15,16].

The complex of morphologically similar (and putatively closely related) species that includes *Drosera microphylla* Endl. consists of freestanding erect, tuberous sundew species that produce long-petiolate, peltate, alternating leaves with long internodes along the stem and terminal inflorescences [1,8,15]. They are further characterised by relatively large sepals that equal or exceed the petals in length and by deeply concave ("cupped") petals [1,8,12,17] and non-ephemeral flowers that open for several days but close every night [1,12,15,18,19]. This combination of features is not paralleled in any other species of *Drosera*.

The *D. microphylla* complex has a complicated taxonomic history, which is summarised in Table 1. The first species of the complex, *Drosera microphylla*, was described by Austrian botanist Stephan Ladislaus Endlicher in 1837 based on a single plant collected by his contemporary, Austrian naturalist Karl ("Charles") von Hügel, from near Albany ("King Georges [George] Sound"; *Hügel s.n.*, W 0009732!) on the south coast of Western Australia [1,20]. Eleven years later, French botanist and *Drosera* monographer Jules Émile Planchon separated *D. calycina* Planch. based on its crescent-like lamina shape [17], in contrast with those of *D. microphylla* that both he and Endlicher [20] described as "orbicular" (circular). In 1864, *D. calycina* var. *minor* Benth. became the first infraspecific taxon of the *D. microphylla* complex, distinguished by smaller leaves and flowers as compared to *D. calycina* [21]. Bentham also synonymised *D. microphylla* with *D. filicaulis* Endl. (a taxon based on *Hügel s.n.* (W 0046809!), which today is considered conspecific with *D. menziesii* R.Br. ex DC., a species that is not part of the *D. microphylla* complex [1], and incorrectly stated that the short diagnosis provided by Endlicher [20] would not allow it to be distinguished from *D. filicaulis* and other species [21].

In his monographic treatment of Droseraceae, Ludwig Diels resurrected *D. microphylla* and synonymised both *D. calycina* and *D. calycina* var. *minor* under the former name [8], resulting in a broad treatment of *D. microphylla* that was generally followed for more than a century (e.g., [12,22,23]). However, Diels did recognise another infraspecific taxon, *D. microphylla* var. *macropetala* Diels, based on its larger petals that are white in dried specimens ("siccata pallida") [8].

Marchant [24] and Marchant et al. [23] subsequently synonymised *D. microphylla* var. *macropetala* with *D. microphylla* and also made the first attempt to lectotypify these names. In common with many plant species described prior to the type method becoming mandatory worldwide in 1935 [25], the names (except for *D. microphylla*) were described without designating a holotype from amongst the numerous specimens (syntypes) comprising the respective type collections. However, Marchant et al. [23] used the term "isotypes" rather than "lectotypes" for *D. calycina*, *D. calycina* var. *minor*, and *D. microphylla* var. *macropetala*, which accordingly did not comprise a valid lectotypification (being the designation of one gathering among the syntypes as the representative, name-bearing type of a species) per

Arts. 7.11 and 9.17 of the International Code of Nomenclature for algae, fungi, and plants (ICN; [26]).

Table 1. Taxonomic history of the *Drosera microphylla* complex. N/A=taxon not yet described by the date of the publication.

Taxon	Endlicher 1837 [20]	Planchon 1848 [17]	Bentham 1864 [21]	Diels 1906 [8]	Marchant et al. 1982 [23]	Lowrie 2014 [1]	Present Work
D. microphylla	Described	Accepted	Under synonymy of *D. filicaulis* [1]	Accepted	Accepted	Accepted	Accepted
D. calycina	N/A	Described	Accepted	Under synonymy of *D. microphylla*	Under synonymy of *D. microphylla*	Accepted	Accepted
D. calycina var. *minor*	N/A	N/A	Described	Under synonymy of *D. microphylla*	Under synonymy of *D. microphylla*	Under synonymy of *D. calycina*	Accepted at species rank [2]
D. microphylla var. *macropetala*	N/A	N/A	N/A	Described	Under synonymy of *D. microphylla*	Under synonymy of *D. calycina*	Accepted at species rank
D. esperensis	N/A	N/A	N/A	N/A	N/A	Described	Accepted

[1] This taxon is today considered conspecific with *D. menziesii*. [2] Under the replacement name *D. rubricalyx*.

In 1987, Australian naturalist Allen Lowrie noted that flower colour can be used to distinguish at least three taxa in the *D. microphylla* complex, suggesting that further studies might separate plants from near Esperance (ca. 400 km east of Albany) based on their white petals [12]. In contrast, plants from near Albany (the area of the type collection of *D. microphylla*) have orange petals, while populations around Perth were described as having dark red petals [12]. This polymorphism was further investigated by Robert Gibson in 2006 [14], who recognised a number of morphological characters (leaf shape, petal colour, sepal and petal length, and plant colour) to distinguish five potentially separate taxa in the complex. He subsequently concluded that "further taxonomic study into this complex appears warranted, and would likely be most rewarding" [14].

The broad circumscription of *D. microphylla* finally changed in 2014, when Lowrie reinstated Planchon's *D. calycina* and described the white-flowered taxon from near Esperance as *D. esperensis* Lowrie [1]. He used some of the same differential characters mentioned earlier by Planchon [17] and Gibson [14] (plant colour, leaf shape, and petal colour) but also noted the stamen and style colour as important characters to distinguish these three taxa [1]. All three species were further described as being geographically well-separated, with *D. microphylla* occurring around Albany, *D. calycina* occurring around and to the north of Perth, and *D. esperensis* occurring east of Esperance [1]. While Lowrie [1] listed "holotypes" for *D. calycina*, *D. calycina* var. *minor*, and *D. microphylla* var. *macropetala*, this did not constitute a valid lectotypification as ICN Art. 7.11 requires the phrase "designated here" for typifications made after 1 January 2001 [26].

Field observations by the authors of the present work from 2002 to 2022, as well as observations on social media and the citizen science website iNaturalist (www.inaturalist.org (accessed on 13 January 2023)), indicated that several distinctive additional taxa in the *D. microphylla* complex should be recognised. The photographic records and increased geographic coverage obtained through observations made by citizen scientists were instrumental in bringing about the formal, scientific documentation of two of the species described in the present work and, crucially, revealed the true identities of *D. calycina* var. *minor* and *D. microphylla* var. *macropetala*. The latter taxon, with reference to its floral display, was described as "The most beautiful of the genus *Drosera*" by its first collector James Drummond [18,19], yet no photographs of it existed until a few were posted on Facebook almost 150 years later. This echoes the discovery of *D. magnifica* Rivadavia &

Gonella, a large and spectacular South American sundew species that was recognised as new from images posted on Facebook [27].

Four new species and two new combinations at species rank are published and illustrated here based on the examination of both herbarium material and living plants in situ, along with the careful re-evaluation of the morphological characters that reliably distinguish the taxa within this complex. We designate lectotypes and provide clarification for the identity of the previously described infraspecific taxa *D. calycina* var. *minor* and *D. microphylla* var. *macropetala*, both of which are reinstated and elevated to species rank. A new name (replacement name) has had to be provided for *D. calycina* var. *minor* at species rank. This raises the total number of species in the *D. microphylla* complex to nine, six of which are rare and potentially threatened. A detailed identification key, comparative figures, and distribution maps are provided.

2. Materials and Methods

Populations of all the taxa from the *Drosera microphylla* complex were studied in situ in Southwest Western Australia from 2002 to 2022 and herbarium specimens (including types) were examined by the authors at B, BM, G, K, M, MEL, PERTH, and W (herbarium acronyms following Index Herbariorum; Thiers, B. M. [updated continuously] https://sweetgum.nybg.org/science/ih/ (accessed on 6 December 2022)). Additional digitised specimen images were obtained and examined from E, FI, KFTA, L, LE, LD, LINN, MPU, NSW, OXF, P, and RSA. New field collections were made for *D. macropetala* (collected by TK under Western Australian flora taking licence FT61000860) and type material was collected for the taxa here described as new to science as *D. reflexa* (collected by GB under scientific licence SW019597), *D. atrata*, and *D. hortiorum* (collected in collaboration with Fred and Jean Hort under Western Australian flora taking licence FT61000255). Scanning electron microscopy (SEM) was carried out to measure seed of *D. hortiorum*, *D. microphylla*, and *D. reflexa* using a TM4000Plus II low-vacuum SEM (Hitachi Co. Ltd., Tokyo, Japan) without sputter coating, using an accelerating voltage of 15 kV, a secondary electron detector, and a 30 Pa vacuum. Macro photographs of seed for all species except *D. koikyennuruff* were taken in situ using 1 mm grid paper for scale. Cultivated material of *D. hortiorum* (originating from the late Allen Lowrie) and *D. reflexa* (originating from the late Phill Mann) was also examined. Measurements and morphological characters were recorded from plants in situ, herbarium material, and cultivated plants. The distribution map was prepared using Google Earth Pro and the DBCA-011 and DBCA-012 datasets published by the Department of Biodiversity, Conservation and Attractions (DBCA).

3. Results

3.1. *Drosera atrata* T.Krueger, A.Fleischm. & G.Bourke *sp. nov.* (Figures 1–5)

Type: AUSTRALIA. Western Australia: Badgingarra [precise locality withheld for conservation purposes], upper hillslope, sandy clay with laterite gravel, 26 June 2022, *F. Hort, J. Hort & T. Krueger FH 4506* (holotype PERTH!).

Diagnosis: *Drosera atrata* differs from all other species of the *D. microphylla* Endl. complex by (contrasting characters in parentheses) (1) its leaf arrangement, with all leaves in groups of 2–5 per node due to the presence of 1–4 slightly shorter axillary leaves in the axils of all cauline leaves (cauline leaves solitary or axillary leaves only found in the upper 1–9 nodes of the stem with the lower 4–15 cauline leaves being solitary); (2) its many-flowered inflorescences typically producing 5–23 flowers per scape (1–8 flowers per scape); and (3) its styles, which mainly branch close to their base into entire or sparsely branched style segments (styles branching near their base and style segments additionally strongly divided). It is further distinguished by its falcate to allantoid seeds (very narrowly obconic, narrowly clavate to acerose seeds, except in *D. calycina* Planch., which has similar but less falcate seed) and its very dark red to blackish-red petals (a similar, but slightly brighter petal colour is also found in *D. hortiorum* T.Krueger & G.Bourke and in *D. koikyennuruff* T.Krueger & A.S.Rob., with the remainder of the species having very different petal colours).

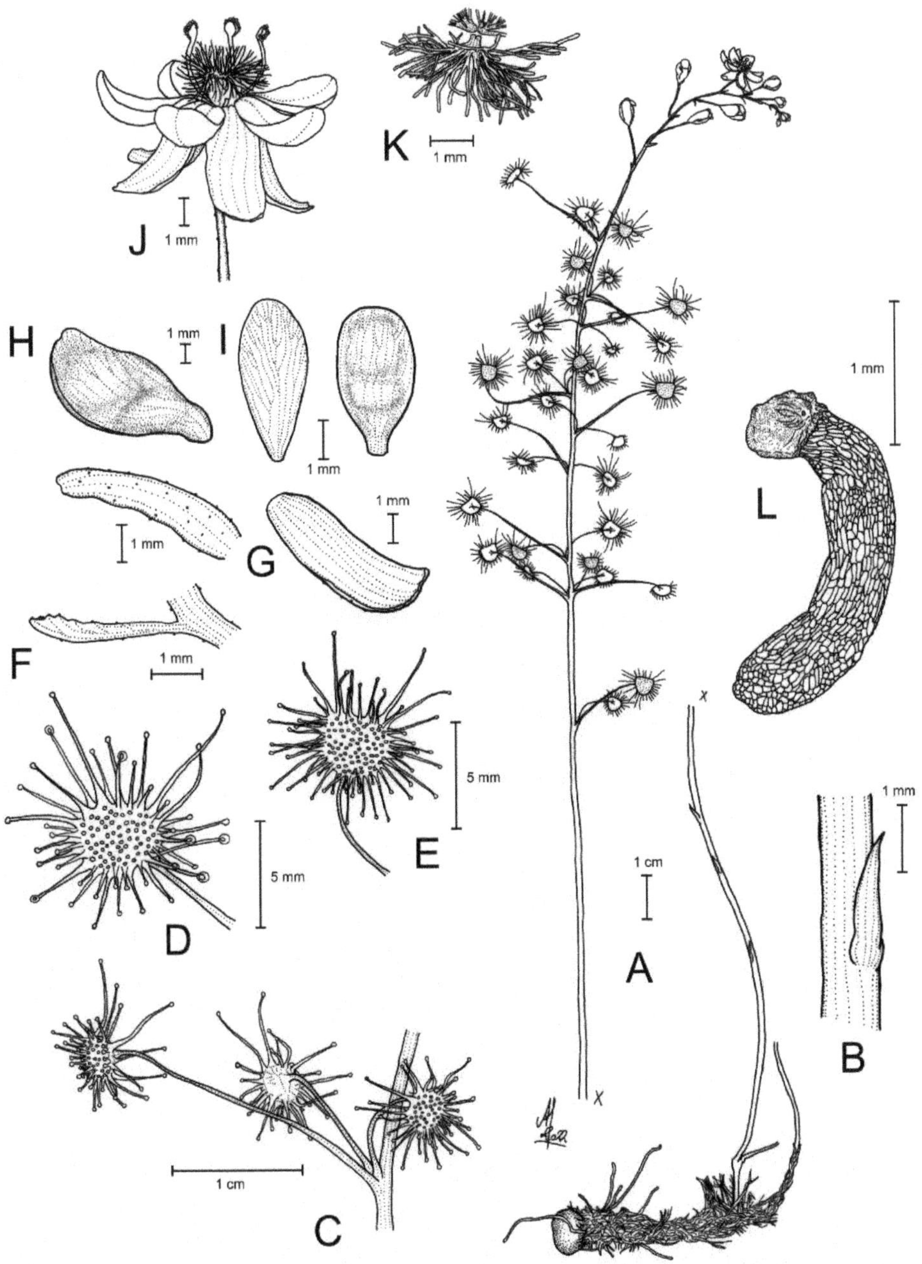

Figure 1. *Drosera atrata* T.Krueger, A.Fleischm. & G.Bourke. (**A**) habit; (**B**) cataphyll from stem base; (**C**) group of leaves from an upper stem internode, comprising one cauline leaf and two axillary leaves; (**D**,**E**) lamina, abaxial surface, (**D**) from cauline leaf, (**E**) from axillary leaf; (**F**) bract; (**G**) sepals, abaxial view (left), adaxial view (right); (**H**,**I**) petals, (**H**) semi-lateral view, (**I**) adaxial view (left example spread, right as in living state); (**J**) flower, lateral view (two stamens removed to reveal the ovary); (**K**) styles, two styles only partially shown; (**L**) seed. (**A**,**B**,**E**,**G**,**I** (left)) from the type (*F. Hort* et al. *FH 4506*), (**C**,**D**,**F**,**H**,**I** (right),**J**–**L**) from in situ photographs. Drawing: A. Fleischmann.

Figure 2. *Drosera atrata* T.Krueger, A.Fleischm. & G.Bourke. (**A**) habit; (**B**) inflorescence; (**C–E**) flowers in diffuse light; (**F**) flowers in bright sunlight, note that thecae of anthers are open to present the yellow pollen on the left flower, while in the younger flower at right they are still closed and orange; (**G**) cataphyll (yellow arrow) with two carnivorous axillary leaves; (**H**) cauline leaf with three smaller axillary leaves; (**I**) flower with closed petals in the late afternoon; (**J**) lamina; (**K**) two stems with cauline leaves, note groups of axillary leaves present throughout on all nodes and the downward-facing laminae; (**L**) flower in bright sunlight, lateral view. (**A**,**C–E**,**I**,**J**) from Coomallo Nature Reserve, Western Australia, 21 July 2019; (**B**) from near Warradarge, Western Australia, 25 June 2021; (**F**,**K**) from east of Warradarge, Western Australia, 22 May 2022; (**G**,**H**,**L**) from Badgingarra, Western Australia, 25 June 2022. Images: T. Krueger.

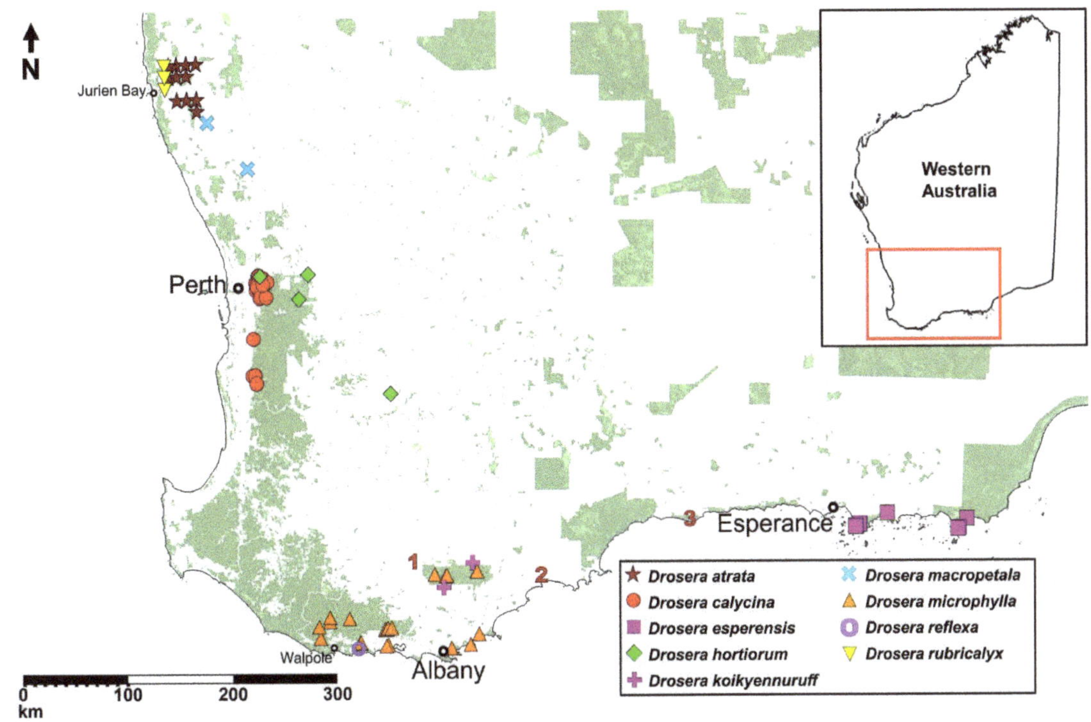

Figure 3. Map showing the known localities of all nine species of the *Drosera microphylla* complex based on herbarium records and field observations by the authors. Locality coordinates of species recommended for inclusion on the Priority Flora List under the Conservation Codes for Western Australian Flora and Fauna have been generalised to the nearest 0.1 degrees. Numbers 1–3 indicate localities of potential undescribed taxa discussed in the Taxonomic notes sections of *D. koikyennuruff* and *D. microphylla*. Background map illustrates protected conservation lands managed by the Department of Biodiversity, Conservation and Attractions (DBCA).

Description: Tuberous perennial herb, (14–)17–44 cm tall above ground including inflorescence. **Tuber** subglobose, ca. 10 mm in diameter, enclosed in black papery sheaths from previous seasons' growth. **Stem** (subterranean part) ca. 5 cm long, 3–6 mm in diameter, enclosed in brown, fibrous tunic formed from previous seasons' stems and roots. **Roots** few, fibrous, emerging laterally from along subterranean part of stem, mostly immediately above tuber. **Stem** (epigeous part) erect, self-supporting, simple, terete, straight, or rarely slightly fractiflex (zig-zag-shaped), glabrous, (10–)12–29 cm tall, 1.2–2.0 mm in diameter near soil surface, 0.7–1.2 mm in diameter at internodes, yellowish green but always reddish orange to red near soil level; sometimes 2–4 stems emerging from the same tuber. **Cataphylls** (often erroneously termed "prophylls" in tuberous *Drosera*) 5–11 on lower part of stem, subulate, 1.7–5.0(–9.0) mm long, ca. 0.5 mm wide, red to orangey yellow or yellowish green with red apex, uppermost 1–4 cataphylls often supporting (1–)2(–3) carnivorous axillary leaves. **Leaves** in groups of 2–5 per node, due to (1–)2(–4) slightly shorter axillary leaves emerging from the axils of all cauline leaves (only rarely lowermost or uppermost 1–2 cauline leaves solitary); internodes (1–)5–20(–24) mm and (1–)3–12 nodes bearing leaves (foliose nodes) present in flowering individuals. **Petioles** terete, semi-erect or horizontal, arcuated abaxially (downwards) with arching usually increasing gradually towards tip, sometimes only arcuated near tip, glabrous, (10–)12–33 mm long, 0.5–1.0 mm wide at base, tapering to 0.2–0.3 mm towards lamina, yellowish green, tip often tinged yellowish pink to orangey yellow. **Lamina** peltate, orbiculate with flattened adaxial lateral margin or reniform, shallowly concave, adaxial surface mostly facing downwards, 2.1–4.0 mm long, 2.3–5.0 mm wide; lamina adaxial surface covered with stalked, carnivorous, secretive capitate glands (tentacles); tentacles 2–6 mm long at lamina margin, decreasing in size towards centre of lamina, with red to greenish yellow stalk; lamina abaxial surface minutely sparsely punctate. **Petioles of axillary leaves** terete, semi-erect, arcuated downwards with arching usually increasing gradually towards tip, glabrous, 5–17(–25) mm long, 0.3–0.5 mm wide at base, tapering to 0.1–0.3 mm towards lamina, yellowish green. **Lamina of axillary leaves** of same shape as the lamina described above, 2.0–3.5 mm long, 2.0–3.8 mm wide. **Inflorescence** a (1–)5–23-flowered scorpioid cyme, terminal, simple or rarely branched, sometimes 1–2 lateral scapes emerging from axils of uppermost leaves, single-sided, (4.5–)5.0–11.0(–15.0) cm long. **Peduncle** terete, 0.4–3.2 cm long, (0.4–)0.6–1.0 mm in diameter, microscopically glandular (appearing glabrous), yellowish green, rarely red. **Pedicels** terete, semi-erect or horizontal in fruit, (3–)4–11(–14) mm long in fruit, 0.3–0.6 mm in diameter, spaced by 2–10 mm along rhachis, microscopically glandular (appearing glabrous), yellowish green. **Bracts** spathulate, narrowly spathulate, narrowly obovate or subulate, often arcuated adaxially (upwards) but not concave, apex entire or irregularly crenulate, 1.5–4.0(–5.0) mm long, 0.3–0.9(–1.7) mm wide, abaxial surface microscopically glandular. **Sepals** 5, narrowly elliptic to narrowly obovate, arcuated adaxially (upwards), slightly concave, often reflexed during anthesis, lateral and/or apical margins sometimes shallowly involute, apex entire, truncate, emarginate or crenulate, 4–6 mm long, 1.5–2.5 mm wide, abaxial surface microscopically glandular, yellowish brown to yellowish green, minute black spots often apparent. **Corolla** 6–9 mm in diameter. **Petals** 5, very dark red to blackish red, minutely punctate with black spots, obovate, broadly obovate, spathulate or broadly spathulate, deeply concave and slightly arcuated adaxially (upwards), apex rounded and entire, 3.0–4.5 mm long, 1.9–2.9 mm wide. **Stamens** 5, 2.0–3.0 mm long. **Filaments** ± linear or only very slightly dilated towards apex, straight or slightly falcate, 0.2–0.3 mm wide, deep red. **Anthers** bithecate, retrorse, 0.5–0.9 mm wide, thecae orange. **Pollen** yellow. **Ovary** obovoid, 3-carpellate, fused, 0.8–1.4 mm in diameter, yellowish green to yellowish brown. **Styles** 3, divided into many filiform segments just above the base, style segments entire or sparsely branched, terete, filiform, extending laterally beyond filaments, 1.4–2.4 mm long, red to dark red. **Stigmas** simple, at tips of style segments, surface appearing smooth, red. **Seeds** falcate to allantoid, flattened, apices obtuse, with funicular base usually present as disc-like appendage, 2.3–2.6 mm long, 0.5–0.6 mm wide, testa black-brown with funicular disc pale brown (sometimes chalazal end also pale brown); testa more or less longitudinally reticulate, with anticlines thin and only shallowly raised.

Etymology: The specific epithet is derived from the Latin *atratus* (=blackened) and refers to the very dark red to blackish red flower colour of this species, which is the darkest petal colour known in the genus *Drosera* (under some lighting conditions appearing almost black).

Taxonomic notes: *Drosera atrata* is arguably the most morphologically distinct species within the *D. microphylla* complex, exhibiting a unique leaf and inflorescence morphology not found in any other species in this group. While the presence of axillary leaves in the axils of all cauline leaves (thus having groups (sometimes incorrectly called "whorls") of 2–5 leaves at each node) is paralleled in other species of *D.* section *Ergaleium* (e.g., in *D. macrantha* Endl. and *D. menziesii* R.Br. ex DC.), its occurrence in *D. atrata* is unique within the *D. microphylla* complex. *Drosera hortiorum*, *D. macropetala*, and *D. rubricalyx* feature similar (but relatively shorter-petioled) axillary leaves in their uppermost nodes, but, in these species, they are never present in all nodes. In *D. atrata*, carnivorous axillary leaves are frequently found even in association with the uppermost cataphylls (Figure 1G).

Figure 4. Comparison of the flowers of all nine species of the *Drosera microphylla* complex. (**A**) *D. atrata*; (**B**) *D. calycina*; (**C**) *D. esperensis*; (**D**) *D. hortiorum*; (**E**) *D. koikyennuruff*; (**F**) *D. macropetala*; (**G**) *D. microphylla*; (**H**) *D. reflexa*; and (**I**) *D. rubricalyx*. Scale bars = 1 mm. Images: T. Krueger.

The number of flowers (5–23) per inflorescence in *D. atrata* is 2–3 times greater than in any other species of the *D. microphylla* complex. The inflorescence sometimes equals or exceeds the foliose part of the stem in length (Figure 1A). In plants with compound inflorescences (lateral flower scapes emerging from the axils of the uppermost leaves), the total number of flowers per plant may exceed 40. Despite its relatively large height of up to 44 cm, *D. atrata* produces the smallest flowers of the *D. microphylla* complex, the corolla measuring just 6–9 mm in diameter. Similarly small flowers are occasionally found in very small individuals of *D. hortiorum*, *D. koikyennuruff*, *D. microphylla*, and *D. reflexa*. In contrast with all other members of the *D. microphylla* complex, the styles of *D. atrata* mainly branch shortly above the base; the style segments themselves are entire or only very sparsely branched.

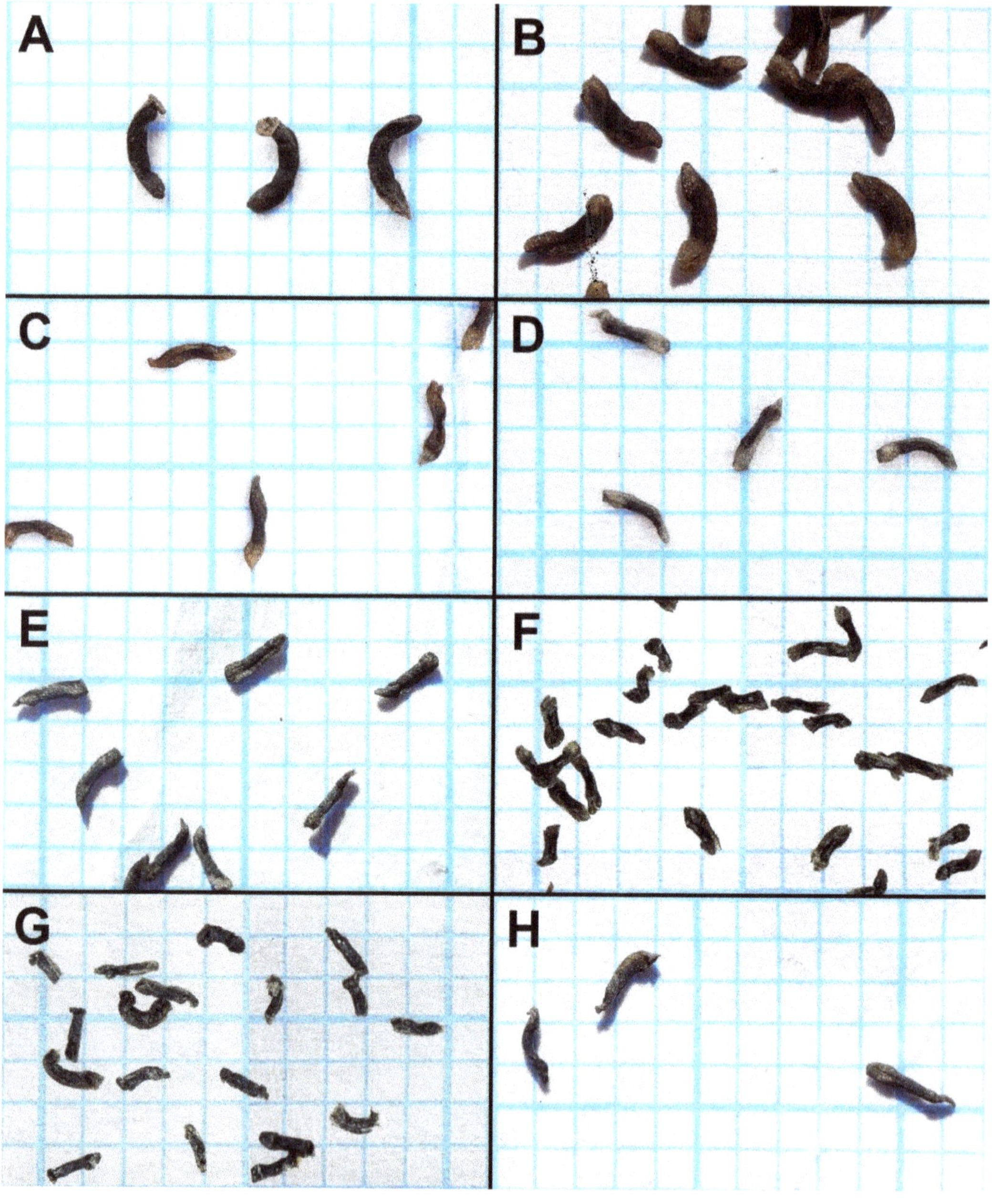

Figure 5. Seed comparison of eight species of the *Drosera microphylla* complex. Seed is placed on 1 mm grid paper. (**A**) *D. atrata*; (**B**) *D. calycina*; (**C**) *D. esperensis*; (**D**) *D. hortiorum*; (**E**) *D. macropetala*; (**F**) *D. microphylla*; (**G**) *D. reflexa*; and (**H**) *D. rubricalyx*. (**A**) from Coomallo Nature Reserve, Western Australia, 28 August 2021; (**B**) from Roleystone, Western Australia, 7 November 2022; (**C**) from Cape Le Grand, Western Australia, 25 November 2021; (**D**) from near York, Western Australia, 14 October 2022; (**E**) from near Dandaragan, Western Australia; (**F**) from near Walpole, Western Australia; (**G**) from near Kentdale, Western Australia; and (**H**) from near Jurien Bay, Western Australia. (**A–E,H**) by T. Krueger; (**F,G**) by G. Bourke, seed digitally superimposed onto grid paper.

The petal colour of *Drosera atrata* is the darkest within the genus with only the tropical rainforest species of north-eastern Australia (*D.* section *Prolifera* C.T.White) producing similarly dark red or dark-pink flowers, in particular certain forms of *D. adelae* F.Muell. While some Australian pygmy sundews (*D.* section *Bryastrum* Planch.) and several members of the South African *D.* section *Ptycnostigma* Planch. produce almost completely black petal bases, these are always paired with a relatively bright colour that comprises the largest part of the petal [28]. In contrast, the dark colour of *D. atrata* is relatively uniform across the entire petal. However, this colour often appears even darker (almost black) when viewed at certain angles. Given the deeply concave petal shape, this often results in the appearance of especially dark areas near the petal margins (Figure 1C,E,F,L) or bases (Figure 1D; here, the strongly reflexed and deeply concave petals result in the bases being viewed at an angle while the margins are viewed ± perpendicular).

The seeds of *D. atrata* differ from all other species of this affinity except *D. calycina* by their falcate to allantoid shape (narrowly obovate, narrowly clavate, or narrowly obtrullate with truncate upper end (=nail- or pin-shaped) and more or less terete in the remainder of species from the *D. microphylla* complex), often with the funiculus still attached to the funicular seed end as a pale brown discus (Figure 5). Only seeds of *D. calycina* are somewhat more similar to those of *D. atrata* in being slightly falcate and flattened. Seed morphology has already been shown to be a reliable taxonomic tool for species delimitation in some other species complexes of tuberous *Drosera* [1].

Distribution and habitat: *Drosera atrata* is known from eleven locations between Warradarge in the north and Badgingarra in the south (Figure 3). It occurs in low kwongan heath on the upper slopes of lateritic hills in poorly drained sandy clay with laterite.

Phenology: Flowering has been recorded from May to August.

Conservation status: Recommended for listing as Priority Three (poorly known species) under Conservation Codes for Western Australian Flora and Fauna (Western Australian Herbarium 1998–, https://florabase.dpaw.wa.gov.au/ (accessed on 6 December 2022)). It is assessed as Vulnerable (VU) under IUCN criterion D1 following IUCN [29]. The populations of *D. atrata* frequently comprise extremely small population sizes of just 1–30 plants. Only a single larger population of ca. 200 plants is known from an unprotected road reserve near Warradarge. Six of the eleven known locations occur on land managed by the Western Australian Department of Biodiversity, Conservation and Attractions (DBCA). Unlicenced collectors and illegal commercial/horticultural trade could pose a threat to *D. atrata* in the future given its extremely small population sizes and the tendency for poachers to target rare carnivorous plant species to supply a demand driven by the horticultural market and carnivorous plant collectors in particular [16]. Further surveys are recommended to gain a better understanding of this taxon's biology, distribution, number and size of populations, and to identify additional potential threats.

Additional specimens examined (paratypes): AUSTRALIA. Western Australia: Coomallo Nature Reserve [precise locality withheld for conservation purposes], breakaway, brown dry ironstone gravel, 24 July 2011, *J.E. Wajon 2435* (PERTH 09050876!); Warradarge [precise locality withheld for conservation purposes], in lateritic sandy soil on lower slope, 16 August 2018, *J. Keeble JK 73* (PERTH 09189807!); Badgingarra [precise locality withheld for conservation purposes], laterite hill top, sand and pebbles, 16 June 2022, *F. Hort & J. Hort FH 4502* (PERTH 09482849!); Badgingarra [precise locality withheld for conservation purposes], white sand with coarse pebbles, 16 June 2022, *F. Hort & J. Hort FH 4499* (PERTH 09482717!); Badgingarra [precise locality withheld for conservation purposes], sand and laterite rise mid slopes, 16 June 2022, *F. Hort & J. Hort FH 4503* (PERTH 09482806!); Badgingarra [precise locality withheld for conservation purposes], small breakaway, weathered stone, clay, sand, gravel, 26 June 2022, *F. Hort & J. Hort FH 4509* (PERTH 09482636!); Badgingarra [precise locality withheld for conservation purposes], upslope from shallow breakaway: sand gravel, 27 June 2022, *F. Hort & J. Hort FH 4514* (PERTH 09482768!); Badgingarra [precise locality withheld for conservation purposes], hill top grey sand with laterite rubble/gravel, 27 June 2022, *F. Hort & J. Hort FH 4513* (PERTH 09482679!).

Additional localities examined: Brand Highway, Badgingarra [precise locality withheld for conservation purposes], mid slope of laterite hill, June 2003, G. Bourke pers. obs.; Lesueur National Park [precise locality withheld for conservation purposes], upper slopes of laterite hill, 11 August 2022, T. Krueger pers. obs.; Cataby [precise locality withheld for conservation purposes], upper slopes of laterite hill, 26 June 2022, T. Krueger pers. obs.

3.2. ***Drosera calycina*** Planch., Ann. Sci. Nat., Bot., sér. 3, 9: 299 (1848). (Figures 3–6)

Lectotype (designated here): [AUSTRALIA. Western Australia:] Swan River, without date [likely part of the Drummond I collection, hence a collection date between 1839 and 1841 is probable [30]], *J. Drummond n. 1* (left individual of K000215039! isolectotype: right individual of K000215039! [both individuals mounted on the same sheet as K000215091 (not a type)]).

Description: Tuberous perennial herb, (8–)14–37(–42) cm tall above ground including inflorescence. **Tuber** subglobose, 6–8 mm in diameter, enclosed in black papery sheaths from previous seasons' growth. **Stem** (subterranean part) 3.5–8.0 cm long, 1.5–3.0 mm in diameter, enclosed in brown, fibrous tunic formed from previous seasons' stems and roots. **Roots** few, fibrous, emerging laterally from along subterranean part of stem, mostly immediately above tuber. **Stem** (epigeous part) erect, self-supporting, simple, terete, slightly to strongly fractiflex, glabrous, (9–)11–35(–39) cm tall, (0.4–)0.7–1.8 mm in diameter near soil surface, 0.5–1.2 mm in diameter at internodes, yellowish green, often irregularly blotched with red, always orange to red near soil level; sometimes 2–6 stems emerging from the same tuber. **Cataphylls** subulate, 4–9(–12) present on lower part of stem, 1–5 mm long, ca. 0.5 mm wide, red to orangey yellow. **Leaves** solitary on each node, alternate, (12–)15–30 present in flowering individuals; internodes (1–)3–17(–26) mm. **Petioles** terete, semi-erect, mostly ± straight or very slightly arcuated abaxially (downwards), strongly arcuated abaxially near tip, glabrous, 9–25(–30) mm long, (0.3–)0.5–0.8 mm wide at base, tapering to 0.1–0.2 mm towards the lamina, yellowish green, often irregularly blotched with red, base yellowish green to red, tip orangey yellow to red. **Lamina** peltate, reniform, or orbiculate with flattened, often truncated adaxial lateral (upper) margin, shallowly concave, adaxial surface facing outwards or slightly downwards, 2.2–3.6 mm long, 2.4–4.1 mm wide; lamina adaxial surface covered with stalked, carnivorous, secretive capitate glands (tentacles); tentacles 2–7 mm long at lamina margin, decreasing in size towards centre of lamina, with red stalk; lamina abaxial surface glabrous. **Inflorescence** a 1–9-flowered scorpioid cyme, terminal, simple or branched, sometimes with an additional scape emerging from axil of uppermost leaf, single-sided, (2.8–)4.0–9.0 cm long. **Peduncle** terete, 1.3–4.0 cm long, 0.4–0.8 mm in diameter, microscopically glandular (appearing glabrous), yellowish green, sometimes blotched with red. **Pedicels** terete, erect in fruit, (6–)9–26(–31) mm long in fruit, 0.3–0.7 mm in diameter, spaced by 3–17 mm along rhachis, microscopically glandular (appearing glabrous), yellowish green to orangey yellow in lower half, reddish orange to red in upper half, uppermost pedicels often completely tinged red. **Bracts** spathulate, narrowly obovate, elliptic, or subulate, often concave and arcuated adaxially (upwards), apex entire or irregularly crenulate, 1.4–3.5 mm long, 0.3–1.2 mm wide, glabrous. **Sepals** 5, narrowly elliptic to narrowly obovate, arcuated adaxially (upwards), slightly concave, often reflexed during anthesis, apex entire or crenulate, 8–13 mm long, 2.8–5.0 mm wide, abaxial surface microscopically glandular, yellowish brown to yellowish green, often with 3–5 red veins, minute black spots sometimes apparent. **Corolla** 14–20 mm in diameter. **Petals** 5, deep red in inner half transitioning to purplish red in outer half, deep red veins often apparent, obovate, deeply concave and slightly arcuated adaxially (upwards), apex rounded and entire, 6.5–9.5 mm long, 4.5–6.0 mm wide. **Stamens** 5, 4.0–6.5 mm long. **Filaments** dilated towards apex, 0.3–0.5 mm wide at base, 0.5–1.1 mm wide near apex, deep red (sometimes purplish red in upper half). **Anthers** bithecate, retrorse, 1.0–1.4 mm wide. **Pollen** yellow. **Ovary** obovoid, 3-carpellate, fused, 1.3–1.7 mm in diameter, deep red. **Styles** 3, divided into a few filiform segments just above the base, style segments again divided into many terete style segments, forming a crowded tuft, not extending laterally beyond filaments,

1.3–1.9 mm long, dark red. **Stigmas** simple, at tips of style segments, papillose, ca. 0.2 mm long, deep red to dark red. **Seeds** slightly falcate to slightly allantoid, flattened, funicular (upper) apex truncate to obtuse, chalazal (lower) end tapering to obtuse apex, 2.4–2.7 mm long, 0.5–0.7 mm wide, testa dark brown with chalazal and funicular ends pale brown); testa longitudinally reticulate, with anticlines thin and only shallowly raised.

Etymology: The specific epithet is derived from the Latin *calycinus* (having a well-developed calyx) and was selected by Planchon [17] to refer to the very large sepals/calyx of this species.

Taxonomic notes: *Drosera calycina* can easily be distinguished from the remainder of the *D. microphylla* complex (especially in herbarium material) by the combination of ± straight petioles (which are only arched near the tip; Figure 6C,E) with the absence of any axillary leaves. It is morphologically similar to *D. microphylla*, *D. hortiorum*, *D. macropetala*, and *D. rubricalyx*. It can be distinguished from *D. microphylla* by (contrasting characters in parentheses) (1) its lamina shape, which is reniform or orbiculate with flattened, often truncated upper margin (lamina orbiculate or sometimes orbiculate with very slightly flattened upper margin); (2) its comparatively large flowers with a corolla diameter of 14–20 mm (corolla diameter 8–15 mm); (3) its petal colour, which is deep red in inner half transitioning to purplish red in outer half (petals reddish orange with deep red bases); (4) its stamen length, which reaches 4.0–6.5 mm (stamens 2.5–3.5 mm long); and (5) its styles, which do not extend laterally beyond the filaments and have deep red stigmas (styles laterally extending beyond the filaments with reddish purple stigmas).

Drosera calycina is further distinguished from *D. hortiorum*, *D. macropetala*, and *D. rubricalyx* by (contrasting characters in parentheses): (1) its solitary leaves (leaves of upper 1–9 nodes in groups of 2–5 due to the presence of usually two shorter axillary leaves); (2) its lamina shape, which is reniform or orbiculate with flattened, often truncated upper margin (lamina orbiculate or orbiculate with slightly flattened upper margin); and (3) its petal colour, which is deep red in inner half transitioning to purplish red in outer half (petals deep red in inner half, dark purplish red in outer half in *D. hortiorum*; white with deep red bases in *D. macropetala*; or deep red in inner half, deep pink in outer half in *D. rubricalyx*; Figure 4). The four species are also ecologically and geographically well separated (Figure 3). While *D. calycina* has been observed growing within a few hundred metres of *D. hortiorum* near Glen Forrest, just east of Perth, they likely do not co-occur due to their different habitat requirements. In this area, *Drosera calycina* is restricted to laterite soils in Jarrah forests while *D. hortiorum* grows in clay loam around granite slopes and boulders.

The distinctive lamina shape of *Drosera calycina* (Figure 6E) has been described as "subtruncate" (Planchon, in annot. K000215039), "suborbiculate-lunate" by Planchon [17], or "crescent-shaped and/or broadly reniform" by Lowrie [1] (the latter likely included both *D. hortiorum* and *D. rubricalyx* in his description of *D. calycina*). In all other species but *D. atrata*, the lamina is usually entirely orbiculate or orbiculate with a slightly flattened upper margin but not truncated. Only *D. atrata* also often produces reniform or truncated laminae but that species is readily distinguished by the presence of axillary leaves in all nodes, its sparsely branching styles extending laterally beyond the filaments, and its very dark red to blackish-red petals (Figure 4). *Drosera atrata* and *D. calycina* additionally share a falcate to allantoid (i.e., slightly depressed and curved) seed shape, which is another taxonomically informative character to distinguish these two species from the remainder of the *D. microphylla* complex, which mostly have straight, pin-shaped, or bone-shaped seeds (Figure 5).

Drosera calycina was previously illustrated by Erickson [22] (p. 40, drawing 2) as "*D. microphylla* var. *macropetala*". While both *D. calycina* and *D. macropetala* indeed are very tall, large-flowered plants, the petals of *D. calycina* are never "drying palish" as stated by Erickson [22] (likely based on Diels' [8] description of *D. microphylla* var. *macropetala*). Indeed, both of Erickson's specimens in the Western Australian Herbarium (PERTH 00666416!,

PERTH 00666874!) are *D. calycina* and it seems unlikely that she observed *D. macropetala* during her studies.

Figure 6. *Drosera calycina* Planch. (**A–C**) habit; (**D**) flowers in diffuse light; (**E**) lamina, abaxial view; (**F**) branched inflorescence; (**G**) flower in diffuse light; (**H**) flower in bright sunlight, lateral view. (**A–D**,**F–H**) from John Forrest National Park, Western Australia, 8 September 2022; (**E**) from Roleystone, Western Australia, 11 July 2021. Images: T. Krueger.

The illustration of *D. calycina* provided by Lowrie in 2014 [1] (p. 355; he previously published a similar illustration as *D. microphylla* in his 1987 book [12] (p. 65)) is evidently based on several different specimens. While most of the illustration matches the cited specimen *A. Lowrie 3043* (PERTH 08988110!, MEL 2443236A!), the presence of axillary leaves on the habit drawing A is puzzling. Crucially, none of the individuals of the *A. Lowrie 3043* collection feature axillary leaves. The specimens also lack tubers, in contrast with the illustration. It is therefore possible that Lowrie's illustration also incorporates specimens of either *D. hortiorum*, *D. macropetala*, or *D. rubricalyx*, all of which have axillary leaves in their upper parts, i.e., the author seems to have used some artistic licence.

Distribution and habitat: Darling Scarp (westernmost part of the Darling Range) between Gidgegannup and Dwellingup (Figure 3). Grows in Jarrah forest mixed with *Banksia sessilis* (Knight) A.R.Mast & K.R.Thiele (Proteaceae), usually on slightly sloping hillsides and hilltops high up on the Darling Scarp. The soils are usually sandy clay with laterite gravel.

Phenology: Flowering has been recorded from August and September.

Conservation status: Not eligible for Conservation Code listing and Least Concern (LC) following Cross [31]. *Drosera calycina* is relatively common in its preferred Jarrah forest habitat and at least twenty localities have been recorded, most of which are on land managed by the Western Australian Department for Biodiversity, Conservation and Attractions (DBCA). While frequently occurring in small population sizes of <50 plants, at least two large populations of >200 plants are known to exist. Unlicensed collection by plant collectors may represent a threatening process but regular monitoring of several populations between 2019 and 2022 indicated there are no current threats to this taxon (T. Krueger pers. obs.).

Notes on the lectotypification: Lectotypification of *D. calycina* is required as Planchon [17] did not select as the type a single specimen out of Drummond's gathering (*J. Drummond n. 1*). He cited two duplicates that he had studied (constituting syntypes), namely "*Drummond* in herb. Hook. et Soc. Linn. Londres" [17] (p. 299), which is the specimen from Herbarium Hookerianum (K000215039) and the one from the Linnean Society of London (Herbarium LINN, some specimens have been transferred to BM [32], and some apparently also to K, see below). However, the second specimen could not be found either at LINN or at BM and it may indeed be lost. However, it is also possible that this second specimen was transferred to K when the "Herbarium Australiense" specimens of the Linnean Society of London herbarium were included in the Kew collections in 1915, including Drummond material ("Herbarium Australiense, presented by the Linnean Society, 1915"; Anonymous in annot. K000843361 photo!). As no other matching specimen could be found at K, this would likely mean that this second specimen was added to the same sheet, which is now K000215039, and that the two plants represented there are indeed the two syntype specimens cited by Planchon, housed at different herbaria at the time. While it might seem counterintuitive to combine specimens this way, the practice was not uncommon at that time and another Drummond collection, *J. Drummond n. 282*, which belongs to the Drummond V collection and represents a different species, *D. microphylla*, was even added to this same sheet at a later date.

These two specimens of *J. Drummond n. 1.* cited by Planchon [17] by definition constitute syntypes, hence lectotypification is required (ICN Arts. 7.11 and 9.17 [26]), even if both are found mounted together on the same herbarium sheet today. K000215039 holds a handwritten personal annotation by Planchon, which represents the sketch of a differential diagnosis noted by the author: "*Drosera calycina* Planch. nov. sp.; Folia nunc subtruncatam. Droserae filicauli Endl. affinis sed petala violacea, et sepala eciliata [leaves now subtruncate. Related to *Drosera filicaulis* Endl. [*D. menziesii*] but with violet petals and hairless sepals]". It is thus clearly evident that K000215039 is original material of *D. calycina*; what cannot be determined is which of the two individuals corresponds to the specimen from Hooker's herbarium, and which may originate from LINN. Accordingly, we have selected the more complete individual as the lectotype, that is, the individual to the left that includes open flowers (the other specimen only has flowers in bud).

Marchant et al. [23] unnecessarily selected an "isotype" at K for *D. calycina*, referring to Planchon's type. In addition, Marchant incorrectly annotated a different specimen at Montpellier Herbarium (MPU1254140) as the "holotype" (Marchant 1985 in sched.) but this was never effectively published (as required by ICN Art 7.10 [26]). Choosing MPU1254140 would have been an incorrect type designation in any case because this specimen is not original material of *Drosera calycina*. It was not cited by Planchon in 1848 [17] and it was not ascribed to the name *D. calycina* by Planchon himself (evident from the label on MPU1254140 in Planchon's hand, which reads, "*Drosera calycina* ? Planch."), and it was annotated by Planchon after he had described *D. calycina* in 1848 (Planchon became assistant professor at Montpellier in 1853 and director of MPU in 1881, while from 1844–1848 he was based at Kew [32]). Thus, MPU1254140 cannot constitute a type for *D. calycina* and, in fact, represents a different species (likely *D. rubricalyx*, see "Notes on Drummond's type collection" under *D. rubricalyx*). Lowrie [1], simply referring to Marchant et al. [23], also incorrectly lists the MPU specimen as the "holotype".

It should also be noted that *J. Drummond n. 1* was likely the only gathering of the *D. microphylla* complex collected by James Drummond that was available to Planchon in 1848 for his revision of Droseraceae. The other specimens *J. Drummond coll. V n. 282*, *J. Drummond coll. VI n. 109*, and *J. Drummond coll. VI n. 110* were collected later (collected in 1847 or 1848 and dispatched to Europe in 1849 for coll. V; and collected in 1850 or 1851 for coll. VI [30]) and, thus, were not considered by Planchon in his 1848 taxonomic treatment of *Drosera* [17].

Notes on Drummond's type collection: Unfortunately, neither the type material nor Drummond's scarce publication records provide any evidence for where exactly in the former Swan River colony the type collection was made. The contemporary botanist Diels [33] (p. 50, literally translated) has already asserted that, "in short, one will never know [exactly] where Drummond's plants were collected; and just in rare cases it can be achieved by the aid of literature to pinpoint at least the approximate habitat". Two of these cases, for which the authors of the present work could trace back the *locus classicus* from Drummond's historic notes [18,19], are *D. macropetala* and *D. rubricalyx* (see "Notes on Drummond's type collection" under the headings of the two respective species).

Additional specimens examined: AUSTRALIA. Western Australia: Bellevue, Darling Range, *C.P. Conigrave s.n.* (E00794030 photo!); Darlington, Darling Range, soil of ironstone gravel, 6 September 1900, *A. Morrison 1238* (PERTH 666920!, E00138677 photo!); Gooseberry Hill, Darling Range, E of Perth, September 1908, *C. Andrews s.n.* (PERTH 666440!); Kalamunda, 1 September 1913, *W.B. Alexander s.n.* (PERTH 666467!); Mundaring Weir, 1 September 1945, *C.D. Hamilton 43* (PERTH 666394!); Pomeroy Road near Welshpool Road, Bickley, laterite soil, 16 August 1965, *N.G. Marchant 6585* (PERTH 666882!); Lesmurdie, near junction of Welshpool and Pomeroy roads, gravelly loam, 16 August 1965, *N.G. Marchant s.n.* (PERTH 666890!); Gooseberry Hill, S of The Knoll near Perth, 28 August 1965, *A.C. Beauglehole ACB 12338* (PERTH 666386!); Gooseberry Hill, Darling Range, E of Perth, 28 August 1965, *R. Erickson s.n.* (PERTH 666874!); Gooseberry Hill, Darling Range, E of Perth, 5 September 1965, *R. Erickson s.n.* (PERTH 666416!); At the intersection of Mundaring Weir Road and Spring Road, Gooseberry Hill, in jarrah forest, 24 August 1974, *S. Carlquist 5398* (RSA0229906 photo!); At the corner of Spring Rd. and Mundaring Weir Rd. in Kalamunda. On the Darling Scarp, Growing in laterite soil with some sand mixture in Eucalyptus forest, 14 September 1974, *L. Debuhr 3606* (RSA0229907 photo!); On Gooseberry Hill, Darling Scarp, 15 September 1974, *S. Carlquist 5631* (RSA0229905 photo!); Junction of Canning Mills Road and Canning Road, Kalamunda, 25 km E of Perth, lateritic sand, 3 September 1984, *G.J. Keighery 7370* (PERTH 5863031!, CANB 363020.1); NE side of minor track in Park Forest Block,W of Stawell Road—Waroona Road intersection, Quadrat P7/1, on black gravel soil, 19 September 1994, *K. McDougall 414* (PERTH 6141110!); Site 46, ca 4 km W of Teesdale Hill, bearing NE, upland, very disturbed, soil surface: littered, gravelly, soil colour: dark brown, soil texture: sandy loam, 4 September 1997, *A. Gundry 1309* (PERTH 4828135!); Between Kalamunda and Mundaring Weir on Mundaring Weir Road, hillside, brown lateritic loam, dense litter cover, 5 September 2001, *K. Macey 380* (PERTH 5910048!); Bodhinyana Monastery, 216 Kingsbury Drive, Serpentine, topography: plain and ridge, soil colour: brown, soil: ironstone gravel, 7 September 2002, *B. Nyanatusita 140* (PERTH 681050!); Pinjarra—Dwellingup Road, grows in laterite-loam soils, 10 September 2004, *A. Lowrie 3043* (PERTH 8988110!, MEL 2443236A!); Beelu National Park, off of Moola road, ironstone gravels, Jarrah woodland with open shrub and sedge/grass understory, 28 August 2019, *D.E. Murfet & A. Lowrie 9406* (MEL 2477153A); West. Australia, without date, *C.A. Gardner 9584* (L.1858657 photo!); without locality, without date, without collector (PERTH 666424!); without locality, without date, without collector (PERTH 666432!).

3.3. ***Drosera esperensis*** Lowrie, Carniv. Pl. Austral. Magnum Opus 3: 1270 (2014). (Figures 3–5 and 7)

Type: AUSTRALIA. Western Australia: Cape Le Grand, E of Esperance, 31 August 2000, *A. Lowrie 2566* (holotype PERTH 08988307 photo!; isotype MEL 2457584!).

Description: Tuberous perennial herb, often forming dense colonies, 7–20(–33) cm tall above ground including inflorescence. **Tubers** not seen. **Stem** (epigeous part) erect, self-supporting, simple, terete, strongly fractiflex, glabrous, 4–17(–29) cm tall, 0.8–1.3(–1.8) mm in diameter near soil surface, 0.5–0.9 mm in diameter at internodes, red or rarely yellowish green. **Cataphylls** subulate, 2–7 present on lower part of stem, (1.0–)1.7–6.5(–9.0) mm long, ca. 0.5 mm wide, red. **Leaves** solitary on each node, rarely uppermost 1–5 nodes with 2 shorter axillary leaves, alternate, 8–22 present in flowering individuals; internodes 2–18 mm. **Petioles** terete, semi-erect, arcuated abaxially (downwards) along whole length or rarely straight, glabrous, 8–22 mm long, 0.3–0.9 mm wide at base, tapering to 0.1–0.3 mm towards the lamina, red. **Lamina** peltate, orbiculate or sometimes orbiculate with slightly flattened adaxial lateral (upper) margin, shallowly to deeply concave, adaxial surface facing downwards or sometimes outwards, 2.0–4.4 mm long, 2.1–4.5 mm wide; lamina adaxial surface covered with stalked, carnivorous, secretive capitate glands (tentacles); tentacles 2.0–5.5 mm long at lamina margin, decreasing in size towards centre of lamina, with red stalk; lamina abaxial surface glabrous. **Inflorescence** a 1–5(–7)-flowered scorpioid cyme, terminal, simple, single-sided, 1.5–4.2 cm long. **Peduncle** terete, (0.2–)0.5–2.7 cm long, 0.5–0.9 mm in diameter, glabrous, red. **Pedicels** terete, erect or semi-erect in fruit, 10–20 mm long in fruit, 0.5–0.8 mm in diameter, spaced by (1–)2–5 mm along rhachis, glabrous, red. **Bracts** spathulate, narrowly spathulate or subulate, often slightly concave and arcuated adaxially (upwards), apex entire or irregularly crenulate, sometimes truncate, 1.8–4.4 mm long, 0.3–1.4 mm wide, glabrous. **Sepals** 5, narrowly obovate to narrowly elliptic, arcuated adaxially (upwards), slightly concave, often reflexed during anthesis, apex entire or crenulate, 5–10 mm long, 2.6–4.1 mm wide, abaxial surface microscopically glandular (appearing glabrous), red or rarely yellowish green, minute black spots often apparent. **Corolla** 10–14 mm in diameter. **Petals** 5, white with pale purplish red base, obovate to broadly obovate, deeply concave and slightly arcuated adaxially (upwards), apex rounded and entire, 5.2–6.5 mm long, 3.9–5.3 mm wide. **Stamens** 5, 2.8–3.4 mm long. **Filaments** ± linear, 0.3–0.6 mm wide, white (often pale purplish red at base). **Anthers** bithecate, retrorse, 0.6–0.9 mm wide, thecae pale yellow. **Pollen** yellow to orangey yellow. **Ovary** obovoid, 3-carpellate, fused, 1.0–1.5 mm in diameter, deep red. **Styles** 3, divided into a few filiform segments just above the base, style segments again divided into many terete style segments, forming a crowded tuft, extending laterally to reach or slightly exceed the filaments, 1.2–2.0 mm long, red at base, gradually transitioning to white near stigma. **Stigmas** shortly branched or simple, at tips of style segments, papillose, 0.2–0.5 mm long, white. **Seeds** narrowly obtrullate to narrowly obovate, outline sinuate, rarely straight, with slight ellipsoid swelling in the proximal and distal half ("bone-shaped seed"), funicular (upper) end truncate (rarely acute), lower (chalazal) end pointed with obtuse tip, 1.9–2.2 mm long,

0.3–0.5 mm wide, testa pale brown, only the median with a blackish-brown rectangular part; testa longitudinally reticulate, with anticlines thin and only shallowly raised.

Figure 7. *Drosera esperensis* Lowrie. (**A–D**) habit; (**E**) flower in diffuse light; (**F**) flower with observed pollinator (a pollinivorous beetle of the family Dermestidae); (**G**) lamina; (**H**) stamens and styles (one stamen is missing). (**A**,**C**,**E–H**) from Cape Le Grand National Park, Western Australia, 19 September 2022; (**B**,**D**) from Cape Le Grand National Park, Western Australia, 16 September 2014. Images: T. Krueger.

Etymology: The specific epithet refers to the Esperance region of southern Western Australia where this species is endemic.

Taxonomic notes: *Drosera esperensis* is morphologically similar to *D. koikyennuruff*, *D. microphylla*, and *D. reflexa*. It is distinguished from these three species by (contrasting characters in parentheses): (1) its tendency to form dense, clonal, mat-like colonies (plants not colony forming or only forming relatively sparse [not mat-like] colonies); (2) its ± linear filament shape (filaments increasing in width towards apex); (3) its petal colour, which is white with a pale purplish red base (dark red in *D. koikyennuruff*, reddish orange with deep red base in *D. microphylla*, or purplish pink with deep red base in *D. reflexa*; Figure 4); and (4) its style and filament colour, which is white with red or purplish red base (styles and filaments red, deep red, purplish red, or reddish purple). The distinctive white petal colour of *D. esperensis* is paralleled in *D. macropetala*, from which it can be distinguished by (contrasting characters in parentheses): (1) its mostly solitary leaves (leaves of upper 1–9 nodes in groups of 3 due to the presence of two shorter axillary leaves); (2) its tendency to form dense, clonal, mat-like colonies (plants not colony forming); (3) its ± linear filament shape (filaments dilated towards apex); and (4) its style colour, which is white with red or purplish red base (styles very dark red).

Drosera esperensis is geographically the most isolated species of the *D. microphylla* complex, occurring ca. 350 km east of the nearest confirmed population of *D. microphylla* (Figure 3; for discussion of the more proximate collection from Hopetoun, see Taxonomic notes under *D. microphylla*).

Plants from the Cape Arid area have been observed to frequently produce axillary leaves. Further studies of these populations are recommended to determine whether they represent a taxon distinct from *D. esperensis* (the type of which was collected from the Cape Le Grand area, where this species almost never produces axillary leaves).

Drosera esperensis was previously illustrated by Gibson [14] (p. 41) and Lowrie [1] (p. 435).

Distribution and habitat: Only known to occur within the Cape Le Grand and Cape Arid National Parks, east of Esperance (Figure 3). Grows in wet, mossy areas on and near granite hills in sandy clay or peat.

Phenology: Flowering has been recorded from August to October. In exceptionally wet habitats or seasons, flowering has been observed to continue until at least December (T. Krueger pers. obs.).

Conservation status: Not eligible for Western Australia Flora and Fauna Conservation Code listing and Least Concern (LC) under IUCN classification, following Cross [34]. *Drosera esperensis* frequently forms very large populations on the large, coastal granite hills east of Esperance. At least nine populations have been recorded, all of which are located on land managed by the Western Australian Department for Biodiversity, Conservation and Attractions (DBCA).

Additional specimens examined: AUSTRALIA. Western Australia: Cape Le Grande [Grand], 6 October 1966, *T.B. Muir 4246* (MEL 0097050A!); Frenchman Peak, in granitic sand on granite outcrops, 20 September 1991, *I. Solomon 512* (PERTH 01675931!); Cape Arid National Park, Mt Arid, SW from Thomas Fisheries, Hillside aspect S, brown loam over granite, 22 August 2014, *M. Hoggart & J. Waters 3/814* (PERTH 08780021!); Cheetup Hill, accessed via track off Saddleback Rd, NE edge of Cape le Grand NP, granite slope aspect SW, mossy brown loam over granite, 26 September 2014, *M. Hoggart 3/914* (PERTH 08780013!); Around the base of Cape Arid, without date, without collector (MEL 0096537A!).

3.4. *Drosera hortiorum* T.Krueger & G.Bourke, *sp. nov.* (Figures 3–5, 8 and 9)

Type: AUSTRALIA. Western Australia: Wandoo National Park [precise locality withheld for conservation purposes], open granitic area, winter damp, semi-shaded, 20 August 2022, *F. Hort, J. Hort & T. Krueger FH 4575* (holotype PERTH!).

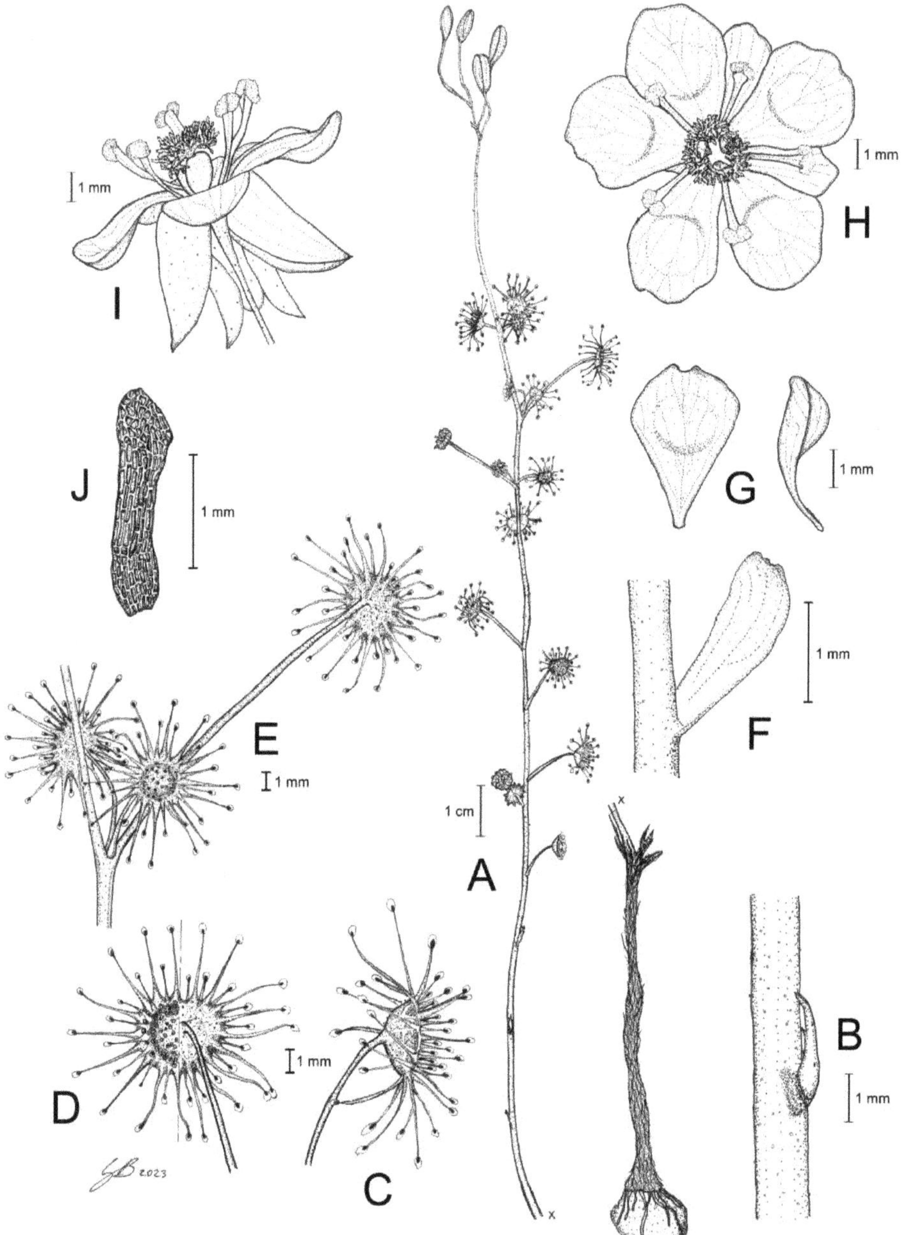

Figure 8. *Drosera hortiorum* T.Krueger & G.Bourke. (**A**) habit; (**B**) stem base with cataphyll; (**C**) lamina, lateral view; (**D**) lamina, left half adaxial view, right half abaxial view; (**E**) group of leaves from upper node of the stem, consisting of one cauline leaf and two axillary leaves; (**F**) bract; (**G**) petals, left adaxial view, right lateral view; (**H**) flower, top view; (**I**) flower, side view; and (**J**) seed. (**A–I**) from photographs of living plants from the type location, Wandoo National Park, Western Australia; (**J**) from near York, Western Australia. Drawing: G. Bourke.

Figure 9. *Drosera hortiorum* T.Krueger & G.Bourke. (**A**) habit; (**B**) leaf on the upper part of the stem exhibiting two smaller axillary leaves emerging from the leaf axil; (**C**) group of flowering plants; (**D**) lamina, this species has an orbiculate lamina shape (sometimes with a sightly flattened upper margin); (**E**) flower in bright sunlight; (**F**) flower in diffuse light; and (**G**) flower, lateral view. (**A**,**B**,**E**) from Wandoo National Park, Western Australia, 20 August 2022. (**C**,**F**,**G**) from near Wickepin, Western Australia, 1 July 2022. (**D**) from near York, Western Australia, 4 September 2022. Images: T. Krueger.

Diagnosis: *Drosera hortiorum* is morphologically most similar to *D. rubricalyx* T.Krueger & A.Fleischm. and *D. macropetala* (Diels) T.Krueger & A.Fleischm. from which it differs by (contrasting characters in parentheses): (1) its small corolla diameter of 8–11 mm (corolla diameter 11–22 mm) and (2) its petal colour, which is deep red in inner half transitioning to dark purplish red in outer half (petals white with deep red base [*D. macropetala*] or petals deep red in inner half transitioning to deep pink in outer half [*D. rubricalyx*]; Figure 4). From *D. macropetala*, it is additionally distinguished by (contrasting characters in parentheses): (1) its filament shape, which are only slightly dilated towards apex, 0.3–0.5 mm wide near the apex (filaments strongly dilated towards apex, 0.5–0.9 mm wide near apex); (2) its tentacle stalk colour, which is greenish yellow (tentacle stalks red in lower half, greenish yellow in upper half, or red throughout); and (3) its filament colour, which is deep red (filaments deep red in lower half, white, or sometimes red in the upper half). *Drosera hortiorum* further shares morphological similarities with *D. calycina* Planch., from which it is distinguished by (contrasting characters in parentheses): (1) the presence of two smaller axillary leaves in the axils of the upper 1–7 cauline leaves (all cauline leaves solitary); (2) its lamina shape, which is orbiculate or orbiculate with a slightly flattened upper margin (lamina reniform or orbiculate with flattened, often truncated upper margin); (3) its filament shape, which only slightly dilated towards apex, 0.3–0.5 mm wide near apex (filaments strongly dilated towards apex, 0.5–1.1 mm wide near apex); and (4) its straight, pin- to bone-shaped seeds (seeds flattened, slightly falcate to slightly allantoid).

Description: Tuberous perennial herb, 14–32(–41) cm tall above ground including inflorescence. **Tuber** subglobose, ca. 10 mm in diameter, enclosed in black papery sheaths from previous seasons' growth. **Stem** (subterranean part) ca. 6 cm long, ca. 2.0 mm in diameter, enclosed in brown, fibrous tunic formed from previous seasons' stems and roots. **Roots** few, fibrous, emerging laterally from along subterranean part of stem, mostly immediately above tuber. **Stem** (epigeous part) erect, self-supporting, simple, terete, slightly fractiflex, glabrous, (10–)14–27(–34) cm tall, 0.7–1.3 mm in diameter near soil surface, 0.4–0.8 mm in diameter at internodes, yellowish green or sometimes red, red near soil level; sometimes 2–5 stems emerging from the same tuber. **Cataphylls** 4–9 on lower part of stem, subulate, 1.4–3.1 mm long, ca. 0.5 mm wide, red to orangey yellow. **Leaves** solitary in lower part of stem but upper (0–)10–50% of leaves in groups of three per node, due to two much shorter axillary leaves emerging from the axils; internodes 4–22 mm and 9–14 nodes bearing leaves (foliose nodes) present in flowering individuals. **Petioles** terete, semi-erect, straight or slightly arcuated abaxially (downwards), strongly arcuated abaxially near tip, glabrous, 8–23(–27) mm long, 0.4–0.7 mm wide at base, tapering to 0.1–0.3 mm towards lamina, yellowish green or sometimes red, tip often tinged orangey yellow. **Lamina** peltate, orbiculate or orbiculate with slightly flattened adaxial lateral margin, shallowly concave, adaxial surface facing outwards or slightly downwards, 2.6–4.0 mm long, 2.7–4.2 mm wide; lamina adaxial surface covered with stalked, carnivorous, secretive capitate glands (tentacles); tentacles 2–5 mm long at lamina margin, decreasing in size towards centre of lamina, with greenish yellow stalk (sometimes red at base); lamina abaxial surface glabrous. **Petioles of axillary leaves** terete, semi-erect, arcuated downwards along whole length, glabrous, 4–6(–8) mm long, 0.2–0.4 mm wide at base, tapering to 0.1–0.2 mm towards lamina, yellowish green or sometimes red. **Lamina of axillary leaves** of same shape as the lamina described above, 2.0–2.9 mm long, 2.0–3.0 mm wide. **Inflorescence** a 2–6-flowered scorpioid cyme, terminal, simple, single-sided, (2.6–)3.2–6.7(–8.8) cm long. **Peduncle** terete, 1.2–4.1 cm long, 0.4–0.6 mm in diameter, microscopically glandular (appearing glabrous), yellowish green, sometimes red. **Pedicels** terete, erect in fruit, 6–21 mm long in fruit, 0.3–0.5 mm in diameter, spaced by 2–8 mm along rhachis, microscopically glandular (appearing glabrous), yellowish green, sometimes red. **Bracts** spathulate, narrowly obovate, elliptic or subulate, arcuated adaxially (upwards), often concave, apex entire or irregularly crenulate, 1.4–3.0 mm long, 0.5–0.9 mm wide, abaxial surface microscopically glandular. **Sepals** 5, narrowly elliptic to narrowly obovate, arcuated adaxially (upwards), slightly concave, often reflexed during anthesis, apex entire or crenulate, 5–9 mm long, 2.5–4.3 mm wide, abaxial surface microscopically glandular, yellowish brown to yellowish green or sometimes red, minute black spots often apparent. **Corolla** 8–11 mm in diameter. **Petals** 5, deep red in inner half transitioning to dark purplish red in outer half, obovate, deeply concave and slightly arcuated adaxially (upwards), apex rounded and entire, 4.1–5.0 mm long, 3.1–4.0 mm wide. **Stamens** 5, 3.0–3.5 mm long. **Filaments** very slightly dilated towards apex, straight or slightly falcate, 0.2–0.4 mm wide at base, 0.3–0.5 mm wide near apex, deep red. **Anthers** bithecate, retrorse, 0.8–1.1 mm wide, thecae reddish orange. **Pollen** yellow. **Ovary** obovoid, 3-carpellate, fused, 1.3–1.6 mm in diameter, deep red or dark olive. **Styles** 3, divided into a few filiform segments just above the base, style segments again divided into many terete style segments, forming a crowded tuft, not extending laterally beyond filaments, 1.0–1.4 mm long, deep red. **Stigmas** simple, at tips of style segments, papillose, ca. 0.2 mm long, deep red. **Seeds** narrowly obtrullate to narrowly obovate, straight or slightly curved, outline rectangular with slight ellipsoid swelling in the proximal and distal half, funicular (upper) end truncate, basal (chalazal) end pointed with obtuse tip, 1.8–2.2 mm long, 0.3–0.5 mm wide, testa pale brown, only the rectangular middle part blackish brown; testa more or less isodiametrically (to slightly longitudinally) reticulate, with anticlines thin and only shallowly raised.

Etymology: The specific epithet honours Fred Hort (1937–) and Jean Hort (1952–), enthusiastic field botanists, nature photographers, and volunteers at the Western Australian Herbarium who found this species at the Wandoo National Park type location in 1987 and brought it to the attention of the authors of the present work. Their prolific collections from the eastern Darling Range have led to the recognition of many new species, several of which have already been named in their honour (e.g., [35–38]).

Taxonomic notes: The presence of axillary leaves in the upper parts of the stem, as well as seed characters (Figure 5), link *D. hortiorum* to the morphologically similar *D. macropetala* and *D. rubricalyx*. However, its corolla is of a much smaller size and its distinctive dark purplish red petal colour easily distinguishes it from these two species (Figure 4). In addition, all three species are geographically well separated, with *D. macropetala* and *D. rubricalyx* occurring well north of Perth while *D. hortiorum* is only known from areas to the east and south-east of Perth (Figure 3).

Despite its usually much smaller size, the corolla shape and colour of *D. hortiorum* closely resembles that of *D. calycina*. Both species further occur in close geographic proximity (Figure 3). However, *D. hortiorum* is easily distinguished from *D. calycina* by the presence of axillary leaves in the upper parts of the stem (D. *calycina* has solitary leaves and always lacks axillary leaves). While *D. hortiorum* has been observed growing within a few hundred metres of *D. calycina* near Glen Forrest, they do not co-occur syntopically due to their different habitat requirements. In that area, *D. calycina* is restricted to laterite soils in Jarrah forests while *D. hortiorum* grows in clay loam around granite slopes and boulders.

A photograph of *D. hortiorum* was published in 1987 by Lowrie [12] (p. 67) who, at the time, treated all taxa of the complex under *D. microphylla*. In his 2014 taxonomic treatment, Lowrie likely included *D. hortiorum* under *D. calycina*, as he described the presence of axillary leaves for this species ("sometimes forming leaves in groups of 2 to 3 in the upper parts" [1] (p. 354)). *Drosera hortiorum* is further illustrated in *Drosera* of the World [15] (p. 228), but with an erroneous location description (Badgingarra). The pictured plant actually represents a specimen cultivated by G. Bourke and originated from the late Allen Lowrie.

Distribution and habitat: Known from Glen Forrest, Wandoo National Park (near York, east of Perth) and two additional sites in the wheatbelt region near York and Wickepin (Figure 3). In the western part of its range, *D. hortiorum* appears to be associated with low granite outcrops and granite slopes where it grows in poorly drained clay loam with *Borya* sp. In the eastern part of its range, *D. hortiorum* has been recorded from within and near shallow drainage channels and moist sandplains in sandy clay.

It is curious to note that *D. hortiorum* has been observed in such a wide range of different habitats, as this is unusual for the complex. Only *D. microphylla* is also known from very different types of habitat.

Phenology: Flowering has been recorded from June to September.

Conservation status: Recommended for listing as Priority Two (poorly known species) under Conservation Codes for Western Australian Flora and Fauna (Western Australian Herbarium 1998–; https://florabase.dpaw.wa.gov.au/ (accessed on 6 December 2022)). Data deficient (DD) following IUCN [29]. Three of the four known locations occur on land managed by the Western Australian Department of Biodiversity, Conservation and Attractions (DBCA). The type population currently comprises ca. 30–40 mature individuals. Additional populations were found by wildflower enthusiasts near Wickepin ("foxydoug" 2022. iNaturalist observation: https://www.inaturalist.org/observations/123515288 (accessed on 9 January 2023)) and near York (photograph posted by Patricia Paull on Facebook). Both populations consist of only ca. 15–30 flowering-sized individuals. The population near Glen Forrest discovered by L. Diels and E. Pritzel in 1901 (*Diels & Pritzel 534/B.59*) was re-located in September 2022 by T. Krueger. At this site, ca. 30 flowering individuals occur in an unprotected area. Given the small number of mature individuals known to occur, *D. hortiorum* could be threatened by unlicensed collection and poaching for the horticultural trade. Further surveys are recommended to gain a better understanding of this taxon's biology, distribution, number and size of populations, and to identify additional threats.

Additional specimens examined (paratypes): AUSTRALIA. Western Australia: Swan Distr.: Smith' Mill [Glen Forrest], Sept. 1901, *Diels & Pritzel 534/B.59* (PERTH 00666904!); Wandoo National Park [precise locality withheld for conservation purposes], open granitic area, winter damp, semi-shaded, 15 August 2022, *F. Hort & J. Hort FH 4574* (PERTH!).

Additional localities examined: Wickepin [precise locality withheld for conservation purposes], poorly drained, seasonally moist drainage channel, 1 July 2022, T. Krueger pers. obs.; York [precise locality withheld for conservation purposes], open Wandoo woodland with low heath, poorly drained seasonally moist sandplain, 4 September 2022, T. Krueger pers. obs.

3.5. ***Drosera koikyennuruff*** T.Krueger & A.S.Rob., ***sp. nov.*** (Figures 3–5, 10 and 11)

Type: AUSTRALIA. Western Australia: Stirling Range National Park [precise locality withheld for conservation purposes], grey clayey sand over sandstone, 23 June 1988, *A. Rose 1029* (holotype PERTH 05812402!).

Diagnosis: *Drosera koikyennuruff* is morphologically most similar to *D. microphylla* Endl., from which it is distinguished by (contrasting characters in parentheses): (1) its much earlier flowering time from June to July (flowering from August to October); (2) its dark red petal colour (petals reddish orange with deep red bases); (3) its deep red stigma colour (stigmas reddish purple), (4) its yellowish green tentacle stalk colour (tentacle stalks red or red in lower half with upper half yellowish green); and (5) its preference for relatively dry sandy habitats in open Mallee woodlands (mossy wet habitat areas on and near granite outcrops, seasonally wet swamps, or rocky mountain slopes). It is further distinguished from the morphologically similar *D. reflexa* G.Bourke & A.S.Rob. by (contrasting characters in parentheses): (1) its sparse populations, which are not colony-forming (plants forming dense populations via adventitious stolons); (2) its petal shape, which is narrowly obovate to broadly spathulate (petals obovate to very broadly obovate); (3) its dark red petal colour (petals purplish pink with deep red base); (4) its yellowish brown to yellowish green sepal colour (sepals red to purplish red); and (5) its preference for relatively dry sandy habitats in open Mallee woodlands in and around the Stirling Range (plants occurring in shallow moss on granite outcrops between Walpole and Denmark).

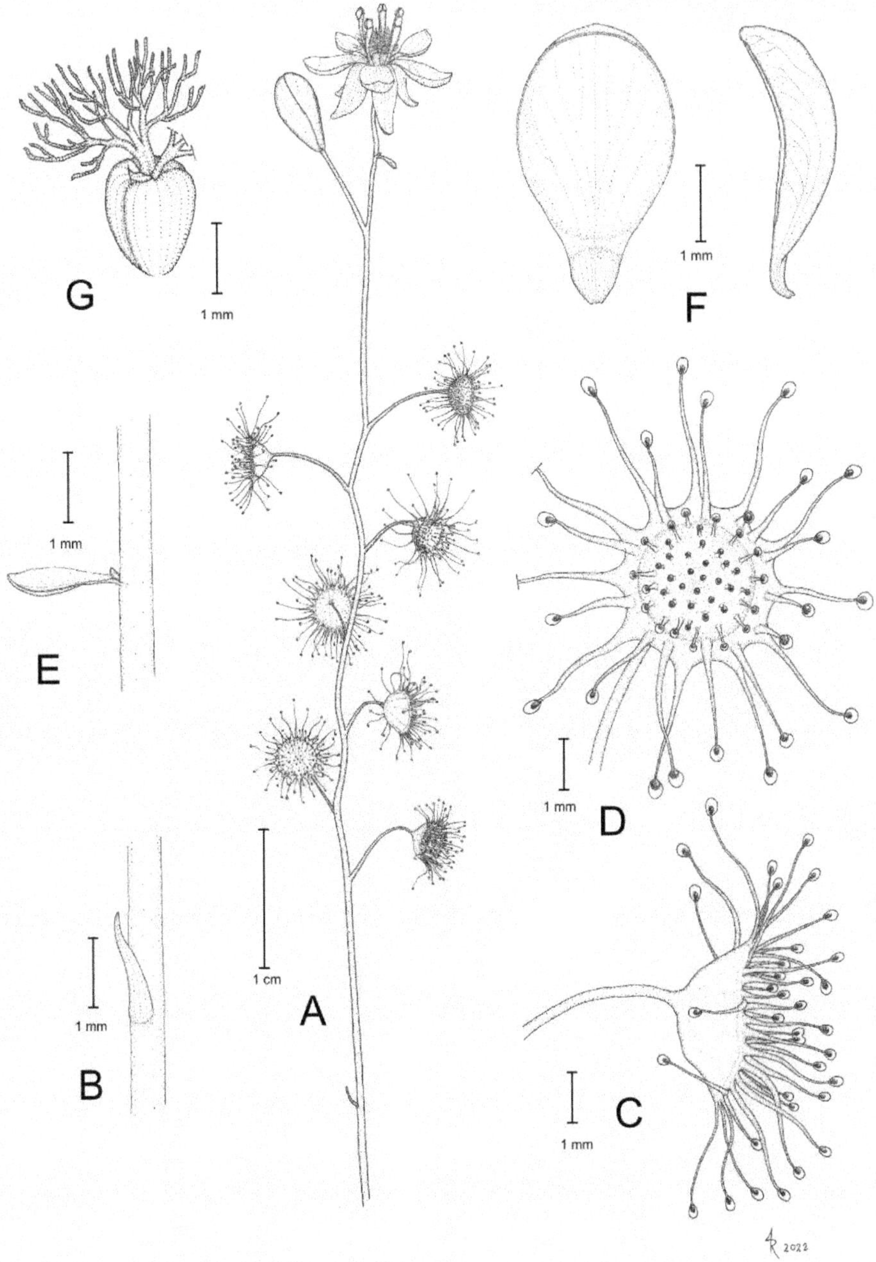

Figure 10. *Drosera koikyennuruff* T.Krueger & A.S.Rob. (**A**) habit; (**B**) cataphyll from stem base; (**C**) lamina, lateral view; (**D**) lamina, adaxial view; (**E**) bract; (**F**) petals, adaxial view (left), lateral view (right); and (**G**) gynoecium, with two styles removed. (**A**,**D**–**G**) from type and photographs of living plants and (**B**) from photographs of living plants only. Drawing: A. Robinson.

Figure 11. *Drosera koikyennuruff* T.Krueger & A.S.Rob. (**A**) habit; (**B**) lamina; (**C**) stem and leaves; (**D**) flower in bright sunlight; (**E**) flower, lateral view; and (**F**,**G**) flowers in diffuse light. (**A–F**) from Stirling Range National Park, Western Australia, 2 July 2022; images by T. Krueger. (**G**) from near Woogenellup, Western Australia, July 2021; image by P. Luscombe.

Description: Tuberous perennial herb, ca. 15 cm tall above ground including inflorescence. **Tuber** not seen. **Stem** (epigeous part) erect, self-supporting, simple, terete, slightly fractiflex, glabrous, 10–12 cm tall, 0.4–0.5 mm in diameter near soil surface, 0.3–0.4 mm in diameter at internodes, yellowish green. **Cataphylls** subulate, few present on lower part of stem, ca. 1.5 mm long, ca. 0.3 mm wide, red to orangey yellow. **Leaves** solitary on each node, alternate, 2–11 present in flowering individuals; internodes 2–11 mm. **Petioles** terete, semi-erect, arcuated abaxially (downwards) along whole length or arching increasing gradually towards the lamina, glabrous, 5–9 mm long, 0.3–0.4 mm wide at base, tapering to 0.1–0.2 mm towards the lamina, yellowish green with orangey yellow or red tip. **Lamina** peltate, orbiculate, shallowly concave, adaxial surface facing outwards or slightly downwards, 2.3–3.4 mm long, 2.3–3.4 mm wide; lamina adaxial surface covered with stalked, carnivorous, secretive capitate glands (tentacles); tentacles 2–4 mm long at lamina margin, decreasing in size towards centre of lamina, with yellowish green stalk; lamina abaxial surface glabrous. **Inflorescence** a 1–2-flowered scorpioid cyme, terminal, simple, single-sided, 2.5–4.2 cm long. **Peduncle** terete, 1.2–1.5 cm long, 0.3–0.4 mm in diameter, glabrous, yellowish green, often blotched with red. **Pedicels** terete, erect in fruit, 5–20 mm long in fruit, 0.2–0.3 mm in diameter, spaced by 5–13 mm along rhachis, glabrous, yellowish green in lower half, orangey yellow to reddish orange in upper half. **Bracts** spathulate, narrowly obovate or subulate, apex entire or irregularly crenulate, 1.5–2.5 mm long, 0.3–0.4 mm wide, glabrous. **Sepals** 5, narrowly obovate, arcuated adaxially (upwards), slightly concave, often reflexed during anthesis, apex entire or sometimes crenulate, 4–6 mm long, 1.8–2.1 mm wide, abaxial surface microscopically glandular (appearing glabrous), yellowish brown to yellowish green, minute black spots often apparent. **Corolla** 8–12 mm in diameter. **Petals** 5, dark red, broadly spathulate to narrowly obovate, deeply concave and slightly arcuated adaxially (upwards), apex rounded and entire, 3.7–5.4 mm long, 2.0–2.2 mm wide. **Stamens** 5, 2.9–3.5 mm long. **Filaments** slightly dilated towards apex, ca. 0.2 mm wide at base, ca. 0.5 mm wide near apex, deep red. **Anthers** bithecate, retrorse, 0.5–0.6 mm wide. **Pollen** yellow. **Ovary** obovoid, 3-carpellate, fused, 1.0–1.6 mm in diameter, deep red. **Styles** 3, divided into a few filiform segments just above the base, style segments again divided into many terete style segments, forming a crowded tuft, extending laterally beyond the filaments, 1.3–2.3 mm long, deep red. **Stigmas** simple or shortly branched, at tips of style segments, ca. 0.2 mm long, deep red. **Seeds** not seen.

Etymology: The specific epithet refers to *koikyennuruff*, the Noongar Aboriginal name for the Stirling Range, where this taxon occurs. The name means "mist over hills" [39].

Taxonomic notes: The overall habit as well as petiole, lamina, and style shape of *D. koikyennuruff* indicate that it is morphologically most similar to *D. microphylla*. Both species grow in close proximity at sites in Stirling Range National Park but favour a different habitat type. While *D. koikyennuruff* grows in low-lying areas with sandy soils in open Mallee woodlands, *D. microphylla* appears (in this area) to be restricted to the middle and upper slopes of the Stirling Range mountains where it typically grows in rocky or lateritic soils. In addition, the two species also differ phenologically and thus are reproductively isolated by non-overlapping flowering times, with *D. koikyennuruff* flowering from June to July while *D. microphylla* flowers from late August to October. *Drosera koikyennuruff* is easily distinguished from *D. microphylla* by its dark red petal colour (*D. microphylla* has reddish orange petals with deep red bases). The flower colour of the type specimen *A. Rose 1029* (PERTH 05812402!) is denoted as "burgandy" (burgundy), which is an apt description for the distinctive dark red petal colour of this species.

Diels 3009 (Plantagenet: westlich des Sucky Peeks [west of "Sucky Peek" =Sukey Hill], B 10 0755996!) represents an intriguing collection from near Cranbrook (marked with "1" in Figure 3). The exceptionally small and consistently single-flowered plants were collected in late May, which is potentially within the flowering time of *D. koikyennuruff*. In addition, the petal colour is described by Diels as "dunkelkarmin" (dark carmine/dark crimson), which might match the dark red flower colour of *D. koikyennuruff*. However, the plants are overall much smaller and appear to have shorter styles. Since this population could not be re-located by the authors prior to submission, it is not currently known whether it represents *D. koikyennuruff* or a closely allied, undescribed species. It is thus not included under *D. koikyennuruff* in the present work.

Distribution and habitat: *Drosera koikyennuruff* is only known from two locations, one in Stirling Range National Park and one from nearby Woogenellup (Figure 3). It grows in low heath amongst Mallee scrub in sandy clay soils.

Phenology: Flowering has been recorded in June and July.

Conservation status: Recommended for listing as Priority Two (poorly known species) under Conservation Codes for Western Australian Flora and Fauna (Western Australian Herbarium 1998–; https://florabase.dpaw.wa.gov.au/ (accessed on 6 December 2022)). Data deficient (DD) following IUCN [29]. The Stirling Range National Park population (recorded in 1988; *A. Rose 1029*) is the only known population on land managed by the Western Australian Department of Biodiversity, Conservation and Attractions (DBCA). This population was surveyed in July 2022 but despite considerable effort, only a single mature individual and four juvenile plants were located (T. Krueger pers. obs.). In July 2021, photos from a second population near Woogenellup were posted on Facebook by local resident Peter Luscombe. The population size of this second population is currently unknown. Given the extremely small number of individuals that are known to exist, any disturbance of the habitat or unlicenced collection could be disastrous to the species' long-term survival. Further surveys are strongly recommended to gain a better understanding of this taxon's distribution, number and size of populations, and to identify potential additional threats.

3.6. *Drosera macropetala* (Diels) T.Krueger & A.Fleischm., *comb. nov. & stat. nov.* (Figures 3–5, 12 and 13)

Basionym: *Drosera microphylla* var. *macropetala* Diels, Das Pflanzenreich Heft 26: 121 (1906).

Lectotype (designated here): [AUSTRALIA]. Westaustralien [Western Australia: "between Moore River and Murchison Rivers"—the collection locality is not provided on the lectotype specimen at B, but on all other syntypes. As evident from Drummond [18,19], the locus classicus is "about 4 miles to the north of Dundaragan", which is ca. 15 km north of today's townsite of Dandaragan, see under "Notes on Drummond's type collection" below], without date [collected 1850 or 1851; [30]], *J. Drummond coll. VI n. 109* (B100755976!; isolectotypes: BM000752962 photo!; E00279841 photo!; FI011168 photo!; G00410322! [one specimen mounted on three sheets, two of them without collector's number and locality, but definitely from the gathering *Drummond 109*]; K000659189!; K000659190!; K000659191!; LD1974467 photo!; LD1971651 photo!; LD1971715 photo! (wrongly labelled as "110"); MEL97059!; NSW146696 photo!; OXF00140703!; P00713916 photo!; P00749106 photo!; W0131702!).

= *Drosera calycina* var. *macropetala* (Diels) N.G.Marchant in annot., nomen nudum.

Description: Tuberous perennial herb, (12–)15–38(–44) cm tall above ground including inflorescence. **Tuber** subglobose, 8–15 mm in diameter, enclosed in black papery sheaths from previous seasons' growth. **Stem** (subterranean part) 4.2–6.8 cm long, 1.5–4.0 mm in diameter, enclosed in brown, fibrous tunic formed from previous seasons' stems and roots. **Roots** few, fibrous, emerging laterally from along subterranean part of stem, mostly immediately above tuber. **Stem** (epigeous part) erect, self-supporting, simple, terete, slightly fractiflex, glabrous, (8–)12–32(–37) cm tall, (0.8–)1.0–1.5 mm in diameter near soil surface, (0.4–)0.6–1.2 mm in diameter at internodes, yellowish green or sometimes red, red near soil level; sometimes 2–5 stems emerging from the same tuber. **Cataphylls** 4–9 on lower part of stem, subulate, 1.5–3.5 mm long, ca. 0.5 mm wide, red to orangey yellow. **Leaves** solitary in lower part of stem but upper (0–)15–50(–65)% of leaves in groups of three per node, due to two much shorter axillary leaves emerging from the axils; internodes 2–21 mm and 10–18(–22) nodes bearing leaves (foliose nodes) present in flowering individuals. **Petioles** terete, semi-erect, slightly arcuated abaxially (downwards), strongly arcuated abaxially near tip, glabrous, 10–30(–45) mm long, 0.5–0.9(–1.1) mm wide at base, tapering to 0.1–0.3 mm towards lamina, yellowish green or sometimes red, tip often tinged orangey yellow or red. **Lamina** peltate, orbiculate or orbiculate with slightly flattened adaxial lateral margin, shallowly concave, adaxial surface facing outwards or slightly downwards, 2.6–3.6(–4.2) mm long, 2.8–4.4 mm wide; lamina adaxial surface covered with stalked, carnivorous, secretive capitate glands (tentacles); tentacles 2–6 mm long at lamina margin, decreasing in size towards centre of lamina, stalk red in lower half, yellowish green in upper half or red throughout; lamina abaxial surface glabrous. **Petioles of axillary leaves** terete, semi-erect, arcuated downwards along whole length, glabrous, 3–7(–9) mm long, 0.3–0.5 mm wide at base, tapering to 0.1–0.3 mm towards lamina, yellowish green or sometimes red. **Lamina of axillary leaves** of same shape as the lamina described above, 1.9–3.0(–3.4) mm long, 2.1–3.1(–3.6) mm wide. **Inflorescence** a 2–8-flowered scorpioid cyme, terminal, simple, single-sided, 3.5–7.5 cm long. **Peduncle** terete, 1.3–4.2 cm long, 0.5–0.9 mm in diameter, microscopically glandular (appearing glabrous), yellowish green or red. **Pedicels** terete, erect in fruit, 6–24 mm long in fruit, 0.4–0.8 mm in diameter, spaced by (1–)2–8 mm along rhachis, microscopically glandular (appearing glabrous), yellowish green or red. **Bracts** spathulate, narrowly obovate or subulate, often arcuated adaxially (upwards) and concave, apex entire or irregularly crenulate, 1.9–4.0 mm long, 0.5–1.3 mm wide, glabrous. **Sepals** 5, obovate, narrowly obovate or narrowly elliptic, arcuated adaxially (upwards), slightly concave, often reflexed during anthesis, apex entire or crenulate, 7–12 mm long, (2.7–)3.0–5.3 mm wide, abaxial surface microscopically glandular, yellowish brown to red, minute black spots often apparent. **Corolla** (11–)13–22 mm in diameter. **Petals** 5, white with deep red base, obovate to very broadly obovate, sometimes broadly spathulate, deeply concave and slightly arcuated adaxially (upwards), apex rounded and entire, (5.5–)6.5–11.0 mm long, (4.0–)4.5–7.5 mm wide. **Stamens** 5, 3.3–4.9 mm long. **Filaments** dilated towards apex, 0.3–0.5 mm wide at base, 0.5–0.9 mm wide near apex, deep red in lower half, white or red in upper half. **Anthers** bithecate, retrorse, 0.8–1.5 mm wide, thecae orange or reddish orange, rarely pale yellow. **Pollen** yellow. **Ovary** obovoid, 3-carpellate, fused, 1.3–1.8 mm in diameter, deep red. **Styles** 3, divided into a few filiform segments just above the base, style segments again divided into many terete style segments, forming a crowded tuft, not extending laterally beyond filaments, 1.2–1.6 mm long, very dark red. **Stigmas** simple or shortly branched, at tips of style segments, papillose, ca. 0.2–0.3 mm long, very dark red. **Seeds** narrowly obtrullate to narrowly obovate, outline narrowly conical with slight ellipsoid swelling in the lower half, funicular (upper) end truncate, basal (chalazal) end obtuse to acute (=pin- or nail-shaped seeds), 1.7–2.0 mm long, 0.3–0.5 mm wide, testa black-brown (often chalazal and funicular ends pale brown); testa more or less isodiametrically reticulate, with anticlines thin and only shallowly raised.

Etymology: The specific epithet, from the Greek *macro-* (=large) and *petalum* (=petal), refers to the comparatively large petals of this species. Indeed, with a length of up to 11 mm and a width of up to 7.5 mm, *D. macropetala* produces the largest petals known in the *D. microphylla* complex. Only *D. calycina* sometimes produces similar-sized (but usually distinctly narrower) petals.

Taxonomic notes: *Drosera macropetala* is morphologically similar to *D. calycina*, *D. hortiorum*, and *D. rubricalyx* from which it can be quickly and reliably differentiated—even in herbarium material—by its petal colour, which is white with a deep red base (petals deep red in inner half transitioning to purplish red in outer half in *D. calycina*; deep red in inner half transitioning to dark purplish red in outer half in *D. hortiorum*; deep red in inner half transitioning to deep pink in outer half in *D. rubricalyx*). It is further distinguished from *D. calycina* by (contrasting characters in parentheses): (1) the presence of two smaller axillary leaves in the axils of the upper 1–9 cauline leaves (all cauline leaves solitary); (2) its lamina shape, which is orbiculate or orbiculate with slightly flattened upper margin (lamina reniform or orbiculate with flattened, often truncated upper margin); and (3) its straight, pin-like seeds (seeds flattened, falcate in *D. calycina*). It is further distinguished from *D. hortiorum* by (contrasting characters in parentheses): (1) its much larger corolla diameter of 13–22 mm (corolla diameter 8–11 mm) and (2) its filament shape, which is markedly dilated towards apex, 0.5–0.9 mm wide near apex (filament width only slightly dilated towards apex, 0.3–0.5 mm wide near apex). *Drosera macropetala* is further distinguished from *D. rubricalyx* by (contrasting characters in parentheses): (1) its broader petals, which are 4.5–7.5 mm wide (petals 3.3–4.8 mm wide); (2) its usually much longer peduncles, which are 1.3–4.2 cm long (peduncles 0.8–2.2 cm long); (3) its filament shape, which is markedly dilated towards apex, 0.5–0.9 mm wide near apex (filaments only slightly dilated towards apex, 0.4–0.6 mm wide near apex); and (4) its yellowish brown or red sepals, which are not strongly contrasting the yellowish green or red stem colour (sepals red, strongly contrasting the yellowish green stem). The distinctive white petal colour of *D. macropetala* is paralleled in *D. esperensis*, from which it can be distinguished by (contrasting characters in parentheses): (1) the presence of two smaller axillary leaves in the axils of the upper 1–9 cauline leaves (all cauline leaves usually solitary); (2) plants not colony forming (plants forming dense, mat-like colonies); (3) its filament shape, which is markedly dilated towards apex, 0.5–0.9 mm wide near apex (filaments ± linear, 0.3–0.6 mm wide near apex); and (4) its style colour, which is very dark red (styles white with red base).

Gibson [14] erroneously lists the petal colour of *D. macropetala* as "purple" even though Diels [8] (p. 121) clearly states "petala [...] siccata pallida (non atropurpurea)" (dried petals pale white [not dark red]) in his description of *D. microphylla* var. *macropetala*. However, the wording of Diels also may indicate that he was not sure of the petal colour in their living state and thus only stated that they are white in the dried condition. The deep red, purplish red, reddish orange, or deep pink petal colours found in other members of the complex are usually well-preserved even in 100+ year-old specimens, provided they have been stored under favourable conditions. Many of the *J. Drummond coll. VI n. 109* specimens of *D. macropetala* still clearly show the reddish inner part of their petals. It is therefore astonishing that the unique flower colour pattern escaped the notice of Diels, the taxon's author, and indeed of later botanists who studied the widely available material; the gathering *J. Drummond coll. VI n. 109* consists of numerous duplicates (the herbarium sheets studied by the authors of the present study [see "Types"] comprise ca. 140 individuals of that taxon) and it was apparently distributed to several major European herbaria by Drummond at the time (syntypes were found in twelve herbaria, see "isolectotypes").

The initial collector, James Drummond, referred to *D. macropetala* as "[t]he most beautiful of the genus *Drosera*" [18,19] and he also clearly described the distinctive flower colour pattern of this taxon as "flowers [...] white with a crimson eye, and they are beautifully variegated with crimson veins" [18,19]. These earlier mentions were however apparently overlooked by Ludwig Diels when he described this distinctive *Drosera* as new to science in 1906 [8].

Distribution and habitat: *Drosera macropetala* is known from the Dandaragan Plateau between Dandaragan and Mogumber, about 100–150 km north of Perth (Figure 3). The species appears to be restricted to the upper slopes of lateritic hills where it grows in low heath in poorly drained, sandy clay with laterite. Possibly also occurs in open *Eucalyptus* ("white gum") forest [18,19].

Phenology: Flowering has only been recorded in August.

Conservation status: Recommended for listing as Priority One (poorly known species) under Conservation Codes for Western Australian Flora and Fauna (Western Australian Herbarium 1998–; https://florabase.dpaw.wa.gov.au/ (accessed on 6 December 2022)). Endangered (EN) under IUCN Red List criteria B1ab(iii,iv,v)+2ab(iii,iv,v) and C2a(i) following IUCN [29]. The extent of occurrence (EOO) and area of occurrence (AOO) of *D. macropetala* is estimated at ca. 200 km^2 and 16 km^2, respectively. These numbers assume that the Mogumber population, last documented in 1904 by Alexander Morrison (see "Specimens examined"), still exists today. Given the extensive vegetation clearing in this area after that year [40], it is possible that this population has been destroyed, in which case both EOO and AOO would be <10 km^2, meeting the Critically Endangered (CR) criteria [29]. *Drosera macropetala* is not known to occur on any land managed by the Western Australian Department of Biodiversity, Conservation and Attractions (DBCA) and is thus potentially threatened by future vegetation clearing.

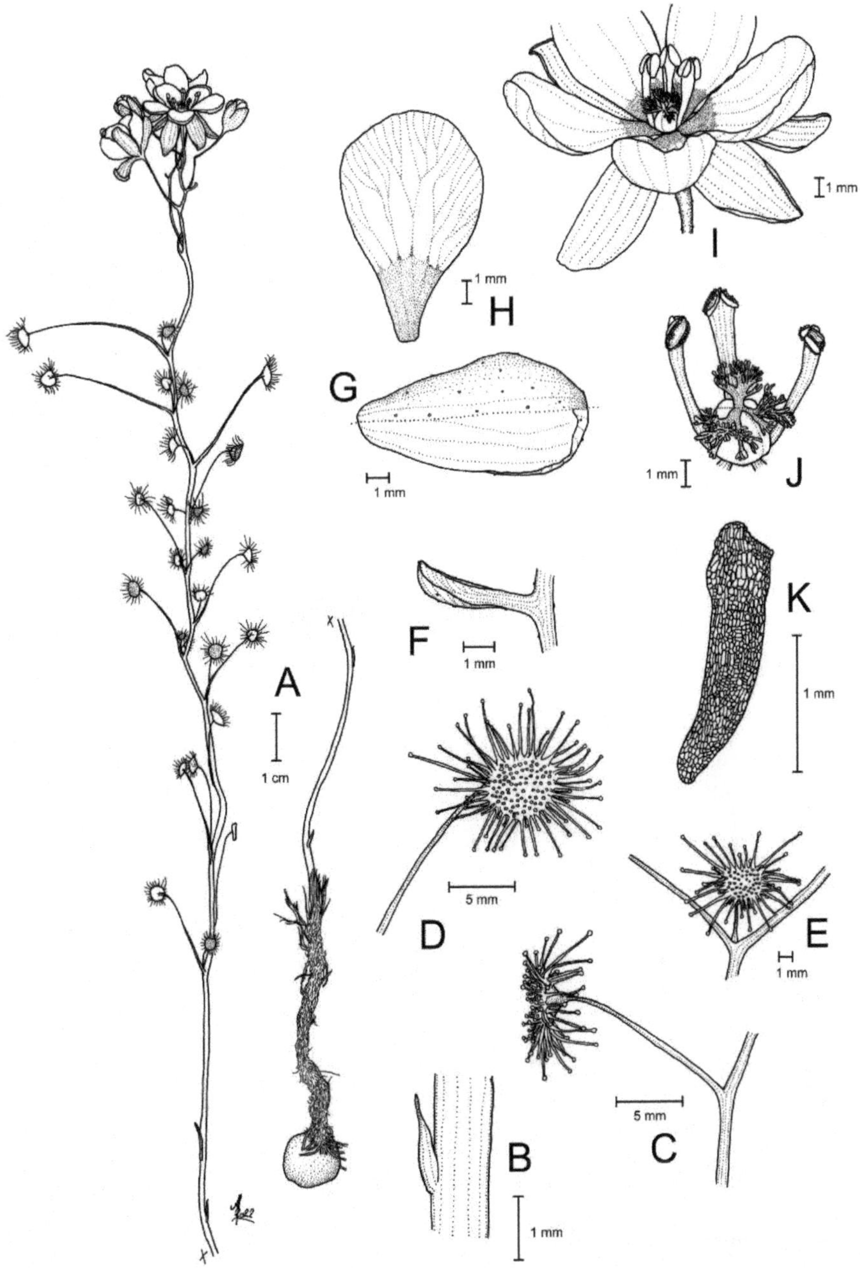

Figure 12. *Drosera macropetala* (Diels) T.Krueger & A.Fleischm. (**A**) habit; (**B**) cataphyll from stem base; (**C**) cauline leaf from lower part of the stem; (**D**) lamina, adaxial view; (**E**) axillary leaf from a stem upper node; (**F**) bract; (**G**) sepal, top half abaxial view, lower half adaxial view; (**H**) petal; (**I**) flower (two stamens removed to reveal the ovary); (**J**) gynoecium, with two stamens removed; and (**K**) seed. (**A**) from the type (*J. Drummond coll. VI n. 109*); (**B**,**D**,**G**,**H**) from *T. Krueger 29*; (**C**,**E**,**F**,**H–K**) from photographs of living plants from near Dandaragan, Western Australia. Drawing: A. Fleischmann.

Figure 13. *Drosera macropetala* (Diels) T.Krueger & A.Fleischm. (**A**) flowers of a single plant in diffuse light, this species often has 3–5 flowers open simultaneously; (**B**) habit of a relatively small individual; (**C**) upper leaf exhibiting two smaller axillary leaves emerging from the leaf axil; (**D**) flower in bright sunlight; (**E**) lamina; (**F**) flower with observed pollinator (a beetle of the family Scarabaeidae); and (**G**) stamens and styles. All from near Dandaragan, Western Australia, 16 August 2021. Images: T. Krueger.

This species was re-located by Declared Rare Flora monitors Gail and Dannielle Reed in August 2020 near Dandaragan, having not been recorded or documented since 1904. Their photographs were uploaded to Facebook and the plants depicted were immediately recognised as an unknown taxon by the authors. Subsequent targeted surveying of this area during 2020, 2021, and 2022 located a total of four (sub-)populations in a single, very narrow strip of unprotected remnant roadside vegetation (T. Krueger pers. obs.). These narrow, linear vegetation corridors, which transect completely cleared agricultural and urban areas, are highly susceptible to road maintenance and construction, altered hydrology, and weed infestation [16]. Population sizes of these four (sub-)populations vary from 2 to ca. 200 mature individuals and the total population in this area is estimated to consist of ca. 500 mature individuals. The historically reported population(s) from near Mogumber (which is ca. 50 km south-east of Dandaragan) have not yet been re-located by the authors as of this publication and their population size and persistence is currently unknown. Unlicenced collectors and illegal commercial/horticultural trade could pose an additional threat to *D. macropetala* in the future given its extremely small population sizes and the unfortunate tendency for poachers to target rare carnivorous plant species [16]. Further surveys are strongly recommended to gain a better understanding of this taxon's distribution, number and size of populations, and to identify further potential threats.

Notes on the lectotypification: Lectotypification of *D. macropetala* is required as Diels [8] did not select a type specimen from the duplicates of *J. Drummond coll. VI n. 109*, which all constitute syntypes. Marchant et al. [23] designated an "isotype" (which is not an inadvertent lectotypification following ICN Arts. 7.11 and 9.10 [26]) and, while K000659191 was labelled as the "holotype" by Marchant in 1985, this was not effectively published and also does not constitute a lectotypification (ICN Arts. 7.10, 7.11, and 9.10 [26]). Even if it had been validly published, the K specimen was incorrectly selected by Marchant as it cannot be the holotype (i.e., the material consulted by the taxon author for the description); Diels visited K to annotate the specimen after the publication of his *D. microphylla* var. *macropetala* in 1906 (evident from the fact that his annotation slip on the K specimen—in contrast with that on the B material—does not bear the label head "bearbeitet für das "Pflanzenreich"" [seen for Diels's taxonomic revision of *Drosera*, i.e., [8]]), which Diels made strict use of for all specimens he examined for his *Drosera* monograph [8]. Lowrie [1] (p. 354) erroneously assumed that both Bentham's *D. calycina* var. *minor* and Diels' *D. microphylla* var. *macropetala* are based on the same sheet (K000215038), which is *J. Drummond coll. VI n. 110*, even though Diels [8] clearly states in his description of *D. microphylla* var. *macropetala* that it is based on *J. Drummond coll. VI n. 109*.

Specimen B 100755976 features an identification slip in the hand of the taxon's author Ludwig Diels and, also given that he worked at B, represents the obvious choice for a lectotype.

At KFTA herbarium, a Drummond specimen has been indicated as "type material" of *D. macropetala* (KFTA0003370 photo!; identified as an "isotype" of "*Drosera macrosepala*" [*sic.!*] in 2013), however, this specimen is not *J. Drummond coll. VI n. 109*, nor does it agree with the locus classicus for the taxon. Rather, it corresponds to *Drummond s.n.*, a collection of *D. menziesii* (the original label reads "Swan River Drummond", to which a pencil-written "n. 109" has been added in error later).

Notes on Drummond's type collection: The syntypes of *J. Drummond coll. VI n. 109* only provide the rough locality information "Western Australia, between Moore River and Murchison Rivers" (locality not indicated on the lectotype specimen in B). However, more precise information on the locus classicus comes from Drummond's newspaper contributions "The Botany of the North-western Districts of Western Australia" [18], republished by Hooker [19]. There, he describes a *Drosera* species with white flowers with crimson centres, large glabrous sepals exceeding the petals in size, and flowers that close at night or during rainy weather, a description that exactly matches *D. macropetala*. Drummond [18,19] mentions that this species "[...] grows abundantly in a White Gum forest about four miles to the north of Dundaragan [Dandaragan]", a locality very close to where it still can be found today (T. Krueger pers. obs.). As *J. Drummond coll. VI n. 109* is the only collection of *D. macropetala* provided by Drummond, it is safe to conclude that this is the collection locality. Additional support for this comes from Barker [41], who evidenced that Drummond's newspaper contribution [18] is referring to Drummond's VI collection series, i.e., the series containing the type collection of *D. macropetala*. This means that the year of collection (not given on any of the syntype specimens) is 1850 or 1851, as for all specimens comprising the VI collection [30].

Additional specimens examined: AUSTRALIA. Western Australia: Between Gillingara [*Gillingarra*] + Mogumber, Moore River, 18 August 1904, *A. Morrison s.n.* (E00794029 photo!); Mogumber, Moore River, 18 August 1904, *A. Morrison s.n.* (E00794031 photo!); Dandaragan [precise locality withheld for conservation purposes], upper hillslope, low kwongan heath, sandy clay with laterite gravel, 22 August 2021, *T. Krueger 29* (PERTH!).

3.7. *Drosera microphylla* Endl., Enum. Pl. (Endlicher): 6 (1837). (Figures 3–5 and 14)

Holotype: AUSTRALIA. Western Australia: King Georges [George] Sound, without date [likely collected in 1833 or 1834], *C.A.A. von Hügel s.n.* (W 0009732!).

≡ *Sondera microphylla* (Endl.) Chrtek & Slavíková, Novit. Bot. Univ. Carol. 13: 44 (2000).

Description: Tuberous perennial herb, (7–)11–32(–51) cm tall above ground including inflorescence. **Tuber** subglobose, ca. 5 mm in diameter, enclosed in black papery sheaths from previous seasons' growth. **Stem** (subterranean part) 2.2–8.0 cm long, 1.5–2.2 mm in diameter, enclosed in brown, fibrous tunic formed from previous seasons' stems and roots. **Roots** few, fibrous, emerging laterally from along subterranean part of stem, mostly immediately above tuber. **Stem** (epigeous part) erect, self-supporting, simple, terete, straight or slightly to strongly fractiflex, glabrous, (6–)10–24(–43) cm tall, 0.4–0.9 mm in diameter near soil surface, 0.3–0.9 mm in diameter at internodes, red to yellowish green, always red near soil surface. **Cataphylls** 3–7 on lower part of stem, subulate, 1.2–2.5 mm long, ca. 0.5 mm wide, red. **Leaves** solitary on each node, alternate, (0–)4–23 present in flowering individuals, caducous, leaves frequently detach randomly from stem before and during anthesis (i.e., leaving large gaps along stem); internodes 2–23 mm. **Petioles** terete, semi-erect, arcuated abaxially (downwards) along whole length or arching increasing gradually towards the lamina, glabrous, 4–12 mm long, 0.3–0.7 mm wide at base, tapering to 0.1–0.3 mm towards the lamina, red or sometimes yellowish green. **Lamina** peltate, orbiculate or sometimes orbiculate with very slightly flattened adaxial lateral margin, shallowly concave, adaxial surface facing outwards or slightly downwards, (1.5–)2.0–3.5 mm long, (1.5–)2.0–3.8 mm wide; lamina adaxial surface covered with stalked, carnivorous, secretive capitate glands (tentacles); tentacles 1.5–4.5 mm long at lamina margin, decreasing in size towards centre of lamina, stalk red or sometimes red in lower half with yellowish green upper half; lamina abaxial surface glabrous but microscopically punctate. **Inflorescence** a 1–7-flowered scorpioid cyme, terminal, simple, single-sided, (1.0–)2.5–6.0(–7.5) cm long. **Peduncle** terete, (0.5–)1.2–4.5 cm long, (0.2–)0.3–0.7 mm in diameter, glabrous, red, reddish orange, or yellowish green. **Pedicels** terete, erect or semi-erect in fruit, (5–)7–12(–15) mm long in fruit, 0.2–0.7 mm in diameter, spaced by 3–8 mm along rhachis, glabrous, red, reddish orange, or yellowish green. **Bracts** spathulate, narrowly obovate, elliptic, lanceolate, subulate or oblong, often slightly concave and arcuated adaxially (upwards), apex entire or rarely irregularly crenulate, 1.5–2.8 mm long, 0.4–1.1 mm wide, glabrous. **Sepals** 5, narrowly elliptic to narrowly obovate, arcuated adaxially (upwards), slightly concave, often reflexed during anthesis, apex entire or crenulate, 5–9 mm long, 2.2–3.1(–3.7) mm wide, abaxial surface microscopically glandular (appearing glabrous), red to yellowish brown, often with 3–5 red veins. **Corolla** 8–15 mm in diameter. **Petals** 5, reddish orange with deep red base, narrowly obovate to spathulate, deeply concave and slightly arcuated adaxially (upwards), apex rounded and entire, 4.5–7.3 mm long, 2.4–4.1 mm wide. **Stamens** 5, 2.5–3.5 mm long. **Filaments** slightly dilated towards apex, 0.2–0.3 mm wide at base, 0.3–0.6 mm wide near apex, deep red in lower half, reddish purple in upper half. **Anthers** bithecate, retrorse, 0.6–1.0 mm wide, thecae yellow to orangey yellow. **Pollen** yellow to orangey yellow. **Ovary**

obovoid, 3-carpellate, fused, 1.2–1.7 mm in diameter, deep red. **Styles** 3, divided into a few filiform segments just above the base, style segments again divided into many terete style segments, forming a crowded tuft, extending laterally beyond the filaments, 1.6–2.5 mm long, deep red, reddish purple near stigmas. **Stigmas** simple or shortly branched, at tips of style segments, 0.3–0.7 mm long, reddish purple. **Seeds** narrowly obtrullate to narrowly obovate, straight or slightly curved, outline rectangular with slight ellipsoid swelling in the proximal and distal half, funicular (upper) end truncate, basal (chalazal) end pointed with obtuse tip, 1.6–1.8(–2.0) mm long, 0.40–0.65 mm wide, testa pale brown, only the middle part blackish brown; testa more or less isodiametrically reticulate, with anticlines thin and only shallowly raised.

Etymology: The specific epithet is derived from the Greek *micros* (=small) and *phyllon* (=leaf), referring to the small leaves of this species.

Taxonomic notes: *Drosera microphylla* is morphologically similar to *D. koikyennuruff* and *D. reflexa*, from which it is distinguished by (contrasting characters in parentheses): (1) its reddish orange petal colour (petals dark red in *D. koikyennuruff* or purplish pink with deep red base in *D. reflexa*); (2) its tendency to detach leaves during and prior to anthesis, often leaving large leaf-free gaps along stem, except in populations from the Stirling Range and Mt. Lindesay (leaves do not detach before or even after anthesis); (3) its late flowering time from August to October (from June to early September); and (4) its reddish purple stigma colour (stigmas deep red or dark red). It is further distinguished from *D. koikyennuruff* by: (1) its tentacle stalk colour, which is red or red in lower half with upper half yellowish green (tentacle stalks yellowish green), and (2) its preference for mossy wet areas on and near granite outcrops, seasonally wet swamps, or rocky mountain slopes (relatively dry sandy habitats in open Mallee woodlands).

Drosera microphylla has historically often been confused with *D. calycina*, from which it can be distinguished by (contrasting characters in parentheses): (1) its lamina shape, which is orbiculate or sometimes orbiculate with very slightly flattened upper margin (lamina reniform or orbiculate with flattened, often truncated upper margin); (2) its comparatively small flowers with a corolla diameter of 8–15 mm (corolla diameter 14–20 mm); (3) its petal colour, which is reddish orange with deep red base (petals deep red in inner half transitioning to purplish red in outer half); (4) its stamen length, which reaches only 2.5–3.5 mm (stamens 4.0–6.5 mm long); and (5) its styles, which extend laterally beyond the filaments and have reddish purple stigmas (styles not laterally extending beyond the filaments, with deep red stigmas).

The leaves of *D. microphylla* frequently detach from near the petiole bases, mostly shortly before and during anthesis, resulting in large, leaf-free "gaps" along the stem that are distinctive for this species (Figure 14B,F). In some cases, mature flowering individuals have been observed with only two or three widely separated leaves still attached to the stem, the remainder having been shed. This is also very apparent in the majority of herbarium specimens of *D. microphylla*. For example, *Cranfield & Ward 25110* (PERTH 08507929!) has at least ten leaf nodes but only three with leaves still attached to them. However, in populations from Stirling Range National Park and Mt. Lindesay, the leaves do not appear to detach at all (T. Krueger pers. obs.).

Leaf detachment in most *D. microphylla* populations appears to serve a role in clonal propagation. In at least four populations near Denmark and Walpole, a red, prostrate, adventitious dropper shoot was observed to emerge from near the centre of the adaxial (tentacle-bearing) lamina surface, directly opposite the point of petiole attachment on the abaxial side (so-called epiphyllous budding; Figure 14D). This is congruent with observations made of naturally occurring asexual regrowth from both basal rosette and stem leaves in the tuberous *Drosera auriculata* Backh. ex Planch. and *D. peltata* Thunb., both likewise from *D.* section *Ergaleium* (depicted and described in detail by Vickery [42]) and from artificially detached leaf cuttings reported for other erect tuberous *Drosera* in cultivation [43] (A. Fleischmann and G. Bourke pers. obs.). Since the detached leaves of *D. microphylla* fall on soils that are still wet at that time of year (from July to September), these adventitious droppers can form a new tuber from their tips once they penetrate the soil, prior to the onset of the dry summer conditions. Leaf detachment thus appears to be a strategy for *D. microphylla* to quickly colonise the bare mossy or sandy soils of its preferred open habitats and may have arisen in connection with the usually very wet seepage soils this species grows in.

The ability to propagate clonally by epiphyllous budding from the leaves mirrors that of perennial *Drosera* in South Africa and Latin America, where only those species growing in rather wet habitats have the capacity to readily multiply asexually via leaf cuttings, something closely related species from drier habitats are incapable of [44]. Gibson [43] observed that, in cultivation, some erect tuberous *Drosera* do multiply through adventitious epiphyllous budding from artificially detached leaves, while others do not; again, there seems to be some connection with whether the species' natural habitats are wet or not. Although clonal propagation by budding from leaves still attached to the basal rosette and stem has been reported to occur naturally in *D. auriculata* and *D. peltata* [42], *D. microphylla* thus far is the only tuberous *Drosera* known to employ its stem leaves for vegetative propagation after shedding them from the mother plant and, generally, it seems to be the only tuberous *Drosera* species with caducous leaves.

Drosera microphylla is a very variable species and at least two collections exist that potentially represent additional undescribed taxa, the precise status of which could not be determined during this work. *Coffey 103 A* (PERTH 08468338!) comprises a collection from near Hopetoun, about 200 km east of the confirmed distribution area of *D. microphylla* (this collection is marked with "3" in Figure 3). While the specimen appears to match *D. microphylla* in morphological detail and could indeed represent an outlying population of this species, additional in situ observations are required to confirm its identity. Despite multiple attempts between 2019 and 2022, the authors of this work failed to re-locate this population.

Newbey 4204 (PERTH 00666459!; marked with "2" in Figure 3) represents a collection from near Boxwood Hill, collected in June 1974. While the flowering time would match that of *D. koikyennuruff*, the petal colour appears to be orange, while the overall habit more closely resembles that of *D. microphylla*. Further in situ observations of this population are required to determine whether it represents *D. microphylla*, *D. koikyennuruff*, or an undescribed taxon.

In addition, red-flowered plants with a different seed shape are known to co-occur with *D. microphylla* at sites along the south coast [15] (G. Bourke pers. obs.). While these have been included within *D. microphylla* in the present work, additional studies are recommended to evaluate whether these represent intraspecific variation of *D. microphylla* or a distinct taxon.

Previously published illustrations labelled as "*D. microphylla*" depict *D. calycina* [22] (p. 40, illustration 2) and a mixture of *D. calycina* with either *D. hortiorum*, *D. macropetala*, or *D. rubricalyx* [12] (p. 65) (see Taxonomic notes under *D. calycina*). Lowrie [1] (p. 609) correctly depicts *D. microphylla*, although the large leaf-free gaps evident on the cited specimen *A. Lowrie 3044* (PERTH 08692637!) are not illustrated. The illustration by Diels [8] (p. 120, illustration E, F) appears to be based on *Diels 3009* (B 100755996!), the precise identity of which could not be determined during this work (see Taxonomic notes under *D. koikyennuruff*).

Distribution and habitat: *Drosera microphylla* occurs from Walpole to Cheynes Beach and throughout the Stirling Range (Figure 3). It grows in mossy wet areas on and near granite outcrops, seasonally wet seepage slopes near large swamp systems, rocky mountain ridges, and slopes. Soil usually sandy clay or peat, sometimes with laterite gravel or rocks.

Phenology: Flowering has been recorded from August to early October.

Conservation status: Not eligible for Conservation Code listing and Least Concern (LC) according to the IUCN Red List Criteria, following Cross [45]. Not threatened, locally common, and widespread, occurring across many reserves managed by the Western Australian Department of Biodiversity, Conservation and Attractions (DBCA).

Figure 14. *Drosera microphylla* Endl. (**A**) group of plants; (**B**) detached leaves; (**C**) lamina, abaxial view; (**D**) detached leaf with adventitious dropper shoot emerging from central adaxial side of the lamina; (**E**) habit of a group of flowering plants; (**F**) habit of a single plant with developing flower buds, note the large leaf-free gaps along the stem due to detached leaves; (**G**,**H**) flowers in bright sunlight; (**I**) flower in diffuse light; and (**J**) stamens and styles. (**A**,**E**,**H**) from near Kentdale, Western Australia, 26 August 2022. (**B**,**D**) from near Denmark, Western Australia, 15 September 2021. (**C**,**I**,**J**) from Mount Frankland North National Park, Western Australia, 13 September 2021. (**F**) from near Walpole, Western Australia, 25 August 2022. (**G**) from near Denmark, Western Australia, 4 September 2014. Images: T. Krueger.

Notes on the type collection: The type collection of *D. microphylla*, *Hügel s.n.* (W 0009732!), comprises a single plant collected from near Albany (King George Sound). The relatively long, straight (i.e., not fractiflex) stem and narrow petals indicate that it does indeed belong to the orange-flowered plants from the area and not to those newly described here as *D. koikyennuruff* and *D. reflexa*. While the type specimen appears to feature leaf-free gaps along the stem, it is unknown whether this is a result of the distinctive leaf detachment habit or whether they detached later in the brittle herbarium material. The large, separated leaf lying next to the lower part of the stem is a leaf of *D. macrantha*, which often co-occurs with *D. microphylla* in its preferred granite outcrop habitat around Albany where this collection was made.

Additional specimens examined: AUSTRALIA. Western Australia: King George's Sound, August 1898, *Goadby s.n.* (PERTH 666939!); 3 miles W of Denmark on Nornalup Road, large sloping granite outcrop, 19 July 1965, *N.G Marchant 6570* (PERTH 666408!); 3 miles W of Denmark on Nornalup Road, large sloping granite outcrop, 19 July 1965, *N.G Marchant 6570* (PERTH 666408!); Tick Flat, halfway between Mount Gardner and the Reserve Office on the lower slopes of Mount Gardner, Two Peoples Bay Nature Reserve, 29 August 1973, *G.T. Smith & L.A. Moore s.n.* (PERTH 5294738!); on granitic dome ca. 5 miles W of Denmark along rd. to Walpole, 17 October 1974, *L. Debuhr 4145* (RSA0229908 photo!); Lower slopes of Mount Manypeaks, on granite rocks, 25 August 1980, *D. Davidson 30 A* (PERTH 04546628 photo!); W of Waychinicup, moss over granite, 27 August 1980, *D. Davidson s.n.* (PERTH 04546601photo!); corner Narrikup Road and Albany Highway, grey sandy loam soil, in association with a Eucalyptus sp. and Banksia sp. grove, 27 August 1984, *E.J. Croxford 3425* (PERTH 04546636 photo!); 50 m up walk-track from Denmark River, Denmark, Mount Lindesay, granitic sand, on and around granitic slabs on steep slope, 3 September 1990, *B.G. Hammersley* (PERTH 1188100!); Thompson Road, open flat, grey peaty sands, 5 August 1994, *R.W. Hearn ARA 4384* (PERTH 4127528!); Foot of Bluff Knoll (mountain), 500 m from the 'peak walk' carpark, Stirling Range, plants growing on ridge with water running shallow soils, brown humous sandy loam, 25 September 1994, *W. Bopp 119* (PERTH 4284968!); Mount Lindesay summit (207), Denmark, grey sand on granite, 11 October 1994, *S. Barrett 176* (PERTH 4213327!); Mount Lindesay, Q207, hillside—summit, grey clayey sand over granite, 31 August 1995, *S. Barrett 616* (PERTH 4273699!); Centre Road from South Western Highway, hillside, lateritic pebbles, granite rocks, brown dry soils, 27 September 2003, *S.C. Coffey 19* (PERTH 7998236!); Near junction South Coast Highway and Lapko Road, Shadforth, near Denmark, grows in skeletal, gritty, black silt soils covered with moss on the aprons of granite outcrops, 3 September 2004, *A. Lowrie 3044* (PERTH 8692637!); Stirling Range National Park, below Bluff Knoll, valley and rocky, brown with green layer (possibly algae) with clayey loam soil, 9 September 2009, *S.C. Coffey 101* (PERTH 8468362!); Ca. 330 m W along Little Lindesay walk trail from Stan Road to first granite outcrop, N of trail, granite outcrop, reserve, dry yellow sand/loam, 1 September 2010, *J. Liddelow JAL 141* (PERTH 8951691!); Surprise forest block, 1.8 km ENE of Western Road along Mountain Road, hill slope, bare to littered dry yellow to brown clayey sand soil over granite, 22 September 2010, *R.J. Cranfield & B.G. Ward 25110* (PERTH 8507929!). King George's Sound, without date, *A. Collie s.n.* (BM014605114 photo!); De Swan River au Cap[e] Riche [most duplicates only mention as locality "Sw. R."=Swan River; only the specimen from de Candolle's herbarium at G provides the full information on the collection locality, the one from Boissier's herbarium does not], without date [collected in 1847 or 1848 [30]], *J. Drummond* [*coll. V*] *n. 282* (BM014605112 photo!, G-6988-420! [inventory number, not a barcode], G s.n.!, P04963085 photo!, K000215091!, W0131692!).

*3.8. **Drosera reflexa*** G.Bourke & A.S.Rob., ***sp. nov.*** (Figures 3–5, 15 and 16)

Type: AUSTRALIA. Western Australia: Kentdale [precise locality withheld for conservation purposes], shallow peaty soil on and near the margins of a granite outcrop, 31 August 2018, *G.J. Bourke 458* (holotype PERTH!; isotypes MEL2500514A!; MEL 2500512A! [spirit collection]).

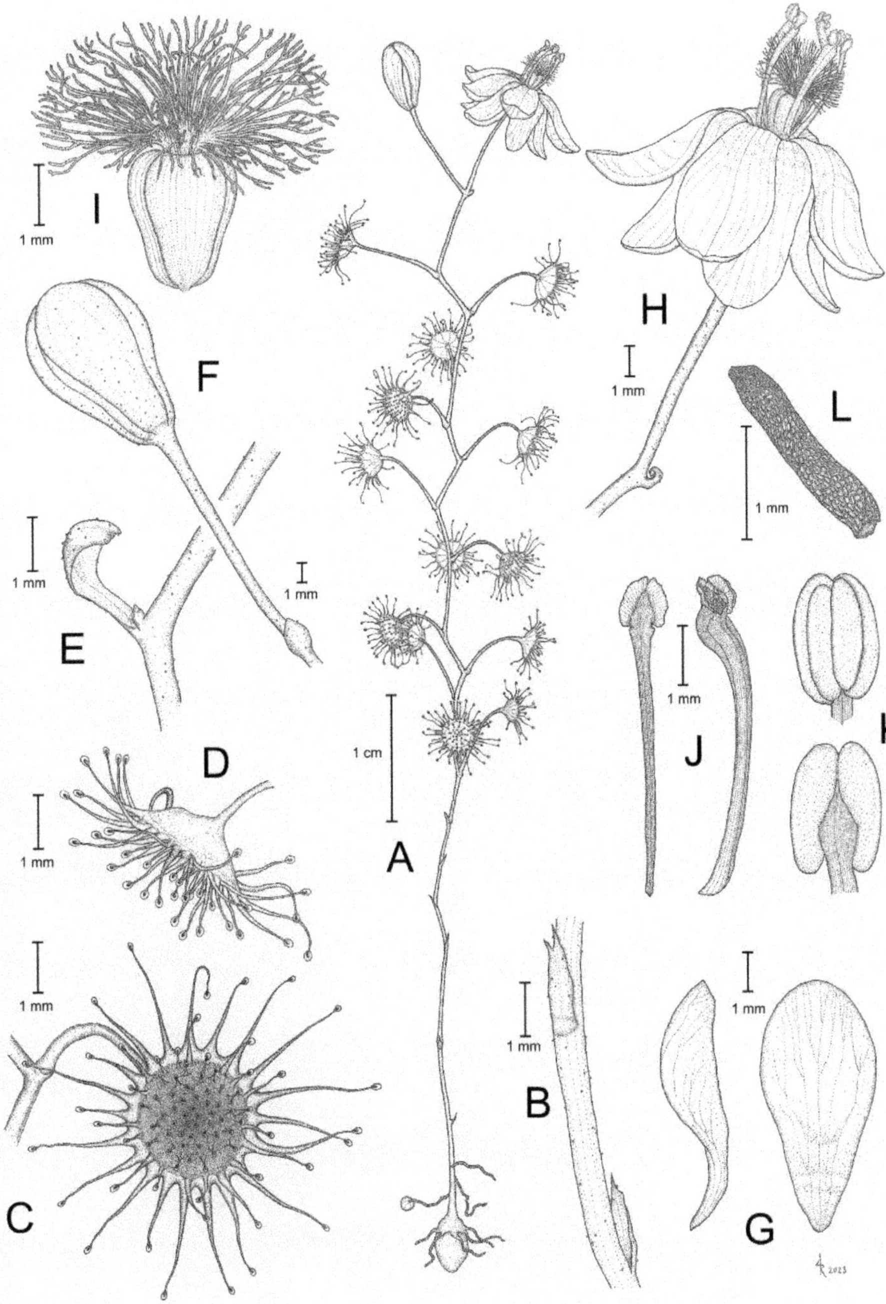

Figure 15. *Drosera reflexa* G.Bourke & A.S.Rob. (**A**) habit (tunic not depicted); (**B**) stem base with two cataphylls; (**C**) lamina, adaxial view; (**D**) lamina, lateral view; (**E**) peduncle and bract; (**F**) pedicel and flower in bud; (**G**) petals, left side view, right adaxial view; (**H**) flower, lateral view; (**I**) gynoecium; (**J**) stamens, left dorsal view, right lateral view; (**K**) thecae, top ventral view, bottom dorsal view; and (**L**) seed. (**A–K**) from the type (*G.J. Bourke 458*, spirit material), (**L**) from SEM images. Drawing: A. Robinson.

Figure 16. *Drosera reflexa* G.Bourke & A.S.Rob. (**A**) group of plants; (**B**) flower in bright sunlight; (**C**) flower in diffuse light; (**D**) habit; (**E**) lamina, abaxial view; (**F**) stamens and styles; (**G**) petiole and lamina, lateral view; (**H**) juvenile lamina, adaxial view; and (**I**) flower, lateral view. (**A–E**,**G**,**I**) from near Kentdale, Western Australia, 26 August 2022; images by T. Krueger. (**F**,**H**) from cultivated material originating from near Kentdale, Western Australia; images by G. Bourke.

Diagnosis: *Drosera reflexa* is morphologically most similar to *D. esperensis* Lowrie, *D. microphylla* Endl. and *D. koikyennuruff* T.Krueger & A.S.Rob. It differs from *D. esperensis* by (contrasting characters in parentheses): (1) its filament shape, which is strongly dilated towards apex, 0.5–0.9 mm wide near apex (filaments ± linear, 0.3–0.6 mm wide near apex); (2) its petal colour, which is purplish pink with a deep red base (petals white with pale purplish red base); and (3) its flowering time from June to early September (flowering late August to October, sometimes until December). It differs from *D. microphylla* by (contrasting characters in parentheses): (1) its leaves, which remain attached to the stem even post-anthesis (leaves often detaching during and prior to anthesis, leaving large leaf-free gaps along stem, except in populations from the Stirling Range and Mt. Lindesay); (2) its stamen length of 3.8–4.4 mm (stamens 2.5–3.5 mm long); (3) its petal colour, which is purplish pink with deep red base (petals reddish orange with deep red base); (4) its stigma colour, which is dark red (stigmas reddish purple); and (5) its flowering time from June to early September (flowering from August to October). It differs from *D. koikyennuruff* by (contrasting characters in parentheses): (1) its tendency to form dense populations via adventitious stolons (plants in sparse populations, not colony-forming); (2) its petal shape, which is obovate to very broadly obovate (petals narrowly obovate to broadly spathulate); (3) its petal colour, which is purplish pink with deep red base (petals dark red); (4) its red to purplish red sepals (sepals yellowish brown to greenish yellow); and (5) its habitat preference of shallow moss on granite outcrops between Walpole and Denmark (plants occurring in sandy soils in open Mallee woodlands in and around the Stirling Range).

Description: Tuberous perennial herb, 5–15(–25) cm tall above ground including inflorescence. **Tuber** subglobose, 2–4 mm in diameter, enclosed in black papery sheaths from previous seasons' growth, pink to red. **Stem** (subterranean) 1.8–5.0 cm long, 0.8–1.2 mm in diameter, enclosed in brown, fibrous tunic formed from previous seasons' stems and roots. **Roots** few, fibrous, emerging laterally from along subterranean vertical stem. **Stolons** few, laterally produced on subterranean stem producing tubers as plants enter dormancy. **Stem** (epigeous part) erect, self-supporting, simple, or occasionally branching from the base, slightly to strongly fractiflex, glabrous throughout, 4–15 cm tall, 0.3–0.8 mm in diameter, yellowish green fading to red towards the end of the season, often red near soil surface. **Cataphylls** 2–8 on lower part of stem, subulate, 1.2–2.3 mm long, ca. 0.5 mm wide, red. **Leaves** solitary on each node, irregularly alternate, 7–18 present in flowering individuals; internodes 2–8 mm. **Petioles** terete, semi-erect, arcuated abaxially (downwards) along whole length, occasionally arching increasing gradually towards the lamina, glabrous to microscopically punctate, 3–8 mm long, 0.3–0.5 mm wide above slightly thickened base, tapering to 0.1–0.2 mm in diameter towards the lamina, yellowish green to red, often darker near lamina. **Lamina** peltate, orbiculate, occasionally with very slightly flattened adaxial lateral margin, shallowly concave, adaxial surface facing outwards or slightly downwards, 1.5–3.5 mm long, 1.8–3.5 mm wide; lamina adaxial surface covered with stalked, carnivorous, secretive capitate glands (tentacles); tentacles 1.7–4.0 mm long at margin, decreasing in size towards centre of lamina, stalk red throughout or red in lower half with greenish yellow upper half; lamina abaxial surface glabrous but microscopically, sparsely punctate. **Inflorescence** a 1–3(–5)-flowered scorpioid cyme, terminal, simple, single-sided, 1.0–3.2(–4.9) cm long. **Peduncles** terete, 0.3–1.9(–4.0) cm long, 0.2–0.5 mm in diameter, microscopically glandular (appearing glabrous), reddish orange to red or rarely yellowish green. **Pedicels** terete, semi-erect or erect in fruit, 6–18 mm long in fruit, 0.2–0.4 mm in diameter, spaced by 2–9 mm along rhachis, microscopically glandular (appearing glabrous), usually more reddish than peduncle. **Bracts** spathulate or narrowly obovate, concave, arcuated adaxially (upwards), apex entire or crenulate, 1.7–3.0 mm long, 0.6–1.0 mm wide, abaxial surface minutely glandular (appearing glabrous). **Sepals** 5, narrowly obovate to narrowly elliptic, arcuated adaxially (upwards), slightly concave, often reflexed during anthesis, apex entire or crenulate, 5.2–8.3 mm long, 2.2–3.3 mm wide, abaxial surface minutely glandular (appearing glabrous), red to purplish red. **Corolla** 9–15 mm in diameter. **Petals** 5, purplish pink with large, deep red blotch towards the base, obovate to very broadly obovate, rarely broadly spathulate, deeply concave and slightly arcuated adaxially (upwards), margins entire, 4.2–7.0 mm long, 2.5–4.3 mm wide. **Stamens** 5, 3.8–4.4 mm long. **Filaments** dilated towards apex, 0.2–0.3 mm wide at base, 0.5–0.9 mm wide near apex, deep red. **Anthers** bithecate, retrorse, 0.9–1.1 mm wide, thecae pale yellow to orangey yellow. **Pollen** yellow to orangey yellow. **Ovary** obovoid, 3-carpellate, fused, 1.5–1.9 mm in diameter, deep red. **Styles** 3, divided into a few filiform segments just above the base, style segments again divided into many terete to distally flattened segments, forming a crowded tuft, extending laterally just beyond filaments, 1.5–1.8 mm long, dark red. **Stigmas** simple or shortly branched, at tips of style segments, papillose, ca. 0.2 mm long, dark red. **Seeds** narrowly elliptical to narrowly obovate, outline more or less narrowly rectangular, rarely curved, funicular (upper) end truncate or with shallow funicular disc, basal (chalazal) end acute to obtuse, 1.3–1.8 mm long, 0.3–0.4 mm wide, dark brown, appearing black, tips pale brown; testa more or less longitudinally reticulate, with anticlines thin and only shallowly raised.

Etymology: The specific epithet is derived from the Latin *reflexus* (=turned back or away) and refers to the often strongly reflexed (by up to ca. 140–170° with respect to the floral axis) sepals and petals.

Taxonomic notes: The overall habit as well as petiole, lamina, and style shape of *D. reflexa* indicate that it is morphologically most similar to *D. esperensis*, *D. koikyennuruff*, and *D. microphylla*. In living specimens, the petal colour can be easily used to differentiate the species within this group (purplish pink with deep red base in *D. reflexa*, white with pale purplish red base in *D. esperensis*, dark red in *D. koikyennuruff*, and reddish orange with red base in *D. microphylla*). *Drosera reflexa* is found in close proximity to *D. microphylla* (ca. 200 m) but the two taxa do not co-occur (no syntopic occurrence) despite their very similar habitat preferences. No hybrids or intermediates between the two species have been observed. This taxon was first mentioned by Lowrie et al. [15] (p. 266) under *D. microphylla* as "a diminutive form with bi-coloured red and pink flowers".

Distribution and habitat: Kentdale (between Walpole and Denmark near the south coast of Western Australia; Figure 3). Occurs in shallow decomposed granitic soils over granite lenses in mosses.

Phenology: Flowering has been recorded from June to early September.

Conservation status: Listed as Priority Two (poorly known species) under Conservation Codes for Western Australian Flora and Fauna (Western Australian Herbarium 1998–; https://florabase.dpaw.wa.gov.au/ (accessed on 6 December 2022)), under the phrase-name "*Drosera* sp. Kentdale (G.J. Bourke 458)". It is Critically Endangered (CR) under IUCN Red List criteria B1ab(iii,v)+2ab(iii,v) following IUCN [29]. *Drosera reflexa* is only known from a single population that is partially located on land managed by the Western Australian Department of Biodiversity, Conservation and Attractions (DBCA). Targeted surveys in the region in 2002, 2017, 2019, 2020, 2021, and 2022 were unable to identify any additional populations despite considerable areas of apparently suitable habitat being surveyed (G. Bourke and T. Krueger pers. obs.). Suitable habitat on nearby private land was not surveyed and may yield additional remnant populations. The total population size is estimated at ca. 1000 mature individuals. Damage to the habitat by recreational vehicles has been observed (G. Bourke pers. obs.) and significant invasive weed infestation is apparent in parts of the habitat (T. Krueger pers. obs.). Further surveys are recommended to gain a better understanding of this taxon's distribution, number and size of populations, and to identify additional potential threats.

*3.9. **Drosera rubricalyx** T.Krueger & A.Fleischm., **nom. nov.** (Figures 3–5, 17 and 18)*

Type: AUSTRALIA. W. [Western] Australia. Between Moore & Murchison Rivers, without date ["1853" written on the herbarium label is the accession date at K, not the actual collection date], the year of collection is either 1850 or 1851 [30]], *J. Drummond* [*coll. VI n.*] *110* (holotype K000215038!; isotypes: BM014605113 photo!; FI011165 photo!; G00410323!; LD1971779 photo! (mixed collection, two individuals belong to *J. Drummond coll. VI n. 110*, the remainder being *J. Drummond coll. VI n. 109, D. macropetala*); MEL 97061!; OXF00140704!; P00713914 photo!; P00749101 photo!).

≡ *Drosera calycina* var. *minor* Benth., Fl. Austral. 2: 469 (1864).

Lectotype (designated here): AUSTRALIA. W. [Western] Australia. Between Moore & Murchison Rivers, without date [actual year of collection is either 1850 or 1851 [30]], *J. Drummond* [*coll. VI n.*] *110* (K000215038!; isolectotypes: BM014605113 photo!; FI011165 photo!; G00410323!; LD1971779 photo! (mixed collection, two individuals belong to *J. Drummond coll. VI n. 110*, the remainder being *J. Drummond coll. VI n. 109, D. macropetala*); MEL 97061!; OXF00140704!; P00713914 photo!; P00749101 photo!).

Diagnosis: *Drosera rubricalyx* is morphologically most similar to *D. hortiorum* T.Krueger & G.Bourke and *D. macropetala* (Diels) T.Krueger & A.Fleischm. from which it differs by (contrasting characters in parentheses): (1) its petal colour, which is deep red in inner half transitioning to deep pink in outer half (petals deep red in inner half transitioning to dark purplish red in outer half in *D. hortiorum* or white with deep red base in *D. macropetala*); (2) its short peduncles, which are 0.8–2.2 cm long (peduncles 1.2–4.2 mm long); and (3) its red sepals, which contrast strongly with the yellowish green stem (sepals yellowish green, yellowish brown, or red, not contrasting strongly with the yellowish green or red stem). It is further distinguished from *D. macropetala* by (contrasting characters in parentheses): (1) its narrower petals, which are 3.3–4.8 mm wide (petals 4.5–7.5 mm wide); (2) its filament shape, which is only slightly dilated towards apex, 0.4–0.6 mm wide near apex (filaments strongly dilated towards apex, 0.5–0.9 mm wide near apex); and (3) its yellowish green tentacle stalk colour (tentacle stalks red in lower half, yellowish green in upper half or red throughout). *Drosera rubricalyx* further shares morphological similarities with *D. calycina* Planch. (of which it was initially described as an infrataxon in 1864 by Bentham), from which it is distinguished by (contrasting characters in parentheses): (1) the presence of 2 smaller axillary leaves in the axils of the upper 2–10 cauline leaves (all cauline leaves solitary); (2) its lamina shape, which is orbiculate or orbiculate with slightly flattened upper margin (lamina reniform or orbiculate with flattened, often truncated upper margin); and (3) its filament shape, which is only slightly dilated towards apex, 0.4–0.6 mm wide near apex (filaments strongly dilated towards apex, 0.5–1.1 mm wide near apex).

Description: Tuberous perennial herb, 14–35(–45) cm tall above ground including inflorescence. **Tuber** subglobose, ca. 12–14 mm in diameter, enclosed in black papery sheaths from previous seasons' growth. **Stem** (subterranean part) 3.5–6.5 cm long, 1.6–4.0 mm in diameter, enclosed in brown, fibrous tunic formed from previous seasons' stems and roots. **Roots** few, fibrous, emerging laterally from along subterranean part of stem, mostly immediately above tuber. **Stem** (epigeous part) erect, self-supporting, simple, terete, slightly fractiflex, glabrous, (11–)14–30(–41) cm tall, 0.8–1.2 mm in diameter near soil surface, 0.5–1.0 mm in diameter at internodes, yellowish green, red near soil level; sometimes 2–4 stems emerging from the same tuber. **Cataphylls** 5–9 on lower part of stem, subulate, 1.5–3.5 mm long, ca. 0.5 mm wide, red to orangey yellow. **Leaves** solitary in lower part of stem but upper (20–)30–60% of leaves in groups of three per node, due to two much shorter axillary leaves emerging from the axils; internodes 3–20 mm and 12–20 nodes bearing leaves (foliose nodes) present in flowering individuals. **Petioles** terete, semi-erect, straight or slightly arcuated abaxially (downwards), strongly arcuated abaxially near tip, glabrous, 9–25(–29) mm long, 0.5–0.9 mm wide at base, tapering to 0.1–0.3 mm towards the lamina, yellowish green, sometimes blotched with red. **Lamina** peltate, orbiculate or orbiculate with slightly flattened adaxial lateral margin, shallowly concave, adaxial surface facing outwards or slightly downwards, 2.2–3.5 mm long, 2.4–3.9 mm wide; lamina adaxial surface covered with stalked, carnivorous, secretive capitate glands (tentacles); tentacles 2–6 mm long at lamina margin, decreasing in size towards centre of lamina, with greenish yellow stalk (sometimes slightly at base); lamina abaxial surface glabrous. **Petioles of axillary leaves** terete, semi-erect, arcuated downwards along whole length, glabrous, 3–7 mm long, 0.3–0.4 mm wide at base, tapering to 0.1–0.2 mm towards lamina, yellowish green. **Lamina of axillary leaves** of same shape as the lamina described above, (1.7–)1.9–2.7 mm long,

(1.8–)2.0–3.0 mm wide. **Inflorescence** a 1–8-flowered scorpioid cyme, terminal, simple, single-sided, 2.0–5.5 cm long. **Peduncle** terete, 0.8–2.2 cm long, 0.5–0.8 mm in diameter, microscopically glandular (appearing glabrous), yellowish green, sometimes blotched with red. **Pedicels** terete, erect in fruit, 5–17 mm long in fruit, 0.4–0.7 mm in diameter, spaced by 2–7 mm along rhachis, microscopically glandular (appearing glabrous), yellowish green, usually transitioning to red in upper half. **Bracts** spathulate, narrowly obovate, or subulate, arcuated adaxially (upwards), often concave, apex entire or irregularly crenulate, 1.9–4.0(–5.0) mm long, 0.6–1.1 mm wide, glabrous. **Sepals** 5, narrowly obovate to narrowly elliptic, arcuated adaxially (upwards), slightly concave, often reflexed during anthesis, apex entire or crenulate, 6–12 mm long, (2.0–)2.5–4.2 mm wide, abaxial surface microscopically glandular, red, sometimes with yellowish brown in upper half, minute black spots sometimes apparent. **Corolla** (10–)11–16(–18) mm in diameter. **Petals** 5, deep red in inner half transitioning to deep pink in outer half, obovate or narrowly obovate, deeply concave and slightly arcuated adaxially (upwards), apex rounded and entire, (5.0–)5.5–8.0 mm long, 3.3–4.8 mm wide. **Stamens** 5, 3.4–5.0 mm long. **Filaments** very slightly dilated towards apex, 0.4–0.5 mm wide at base, 0.4–0.6 mm wide near apex, deep red. **Anthers** bithecate, retrorse, 0.8–1.3 mm wide. **Pollen** yellow. **Ovary** obovoid, 3-carpellate, fused, 1.3–1.9 mm in diameter, deep red. **Styles** 3, divided into a few filiform segments just above the base, style segments again divided into many terete style segments, forming a crowded tuft, not extending laterally beyond filaments, 1.2–1.7 mm long, very dark red. **Stigmas** simple or shortly branched, at tips of style segments, papillose, ca. 0.2 mm long, very dark red. **Seeds** narrowly obtrullate to narrowly obovate, outline narrowly conical with slight ellipsoid swelling in the upper (funicular) half, funicular (upper) end truncate (=pin- or nail-shaped seeds), basal (chalazal) end pointed with short conical to fusiform and often slightly curved appendage, 1.8–2.4 mm long, 0.3–0.5 mm wide, testa black-brown, chalazal and funicular ends pale brown; testa longitudinally reticulate, with anticlines thin and only shallowly raised.

Etymology: The specific epithet is derived from the Latin *ruber* (=red) and *calyx* in reference to the red sepal colour of this species that provides a distinct colour contrast to the yellowish green stem and peduncle.

Taxonomic notes: The distinctive characters distinguishing *D. rubricalyx* from the morphologically most similar taxa (*D. hortiorum* and *D. macropetala*) are detailed under those respective subheadings. The unique petal colour combination of deep red and deep pink usually quickly allows identification of *D. rubricalyx* in the field (Figure 4). Only *D. calycina* may occasionally produce a similar petal colour, although that species always has solitary cauline leaves, while *D. rubricalyx* consistently produces two smaller axillary leaves in the axils of the upper 2–10 cauline leaves. Additionally, *D. calycina* has a different lamina shape, which is reniform to orbiculate with flattened, often truncated, upper margin vs. lamina orbiculate or orbiculate with slightly flattened upper margin in *D. rubricalyx*.

Bentham [21] distinguished *D. calycina* var. *minor* (=*D. rubricalyx*) from *D. calycina* based on its smaller leaves and flowers. Indeed, the petals of *D. rubricalyx* are usually much smaller, especially in width, compared with those of *D. calycina* and *D. macropetala* (the latter species was included in Bentham's description of *D. calycina* as he cites *J. Drummond coll. VI n. 109*, the type of *D. macropetala*). Additionally, Bentham [21] already noted that *D. rubricalyx* has "rather less dilated" filaments when compared to these two species, which is indeed a reliable distinguishing floral feature. However, it is notable that Bentham was able to detect this feature in the dried herbarium specimens he studied, as the filaments in dried specimens of *D. microphylla* complex species often considerably shrink in width (T. Krueger pers. obs.).

The infrataxon epithet of *D. calycina* var. *minor*, published by Bentham in 1864 [21], cannot be elevated to species rank as an older homonym with nomenclatural priority exists (*Drosera minor* Schumach. & Thonn. in Schumach., published in 1827 and generally treated as a synonym of *D. indica* L.). Therefore, a new name, *D. rubricalyx*, had to be coined for *D. calycina* var. *minor* at the rank of species.

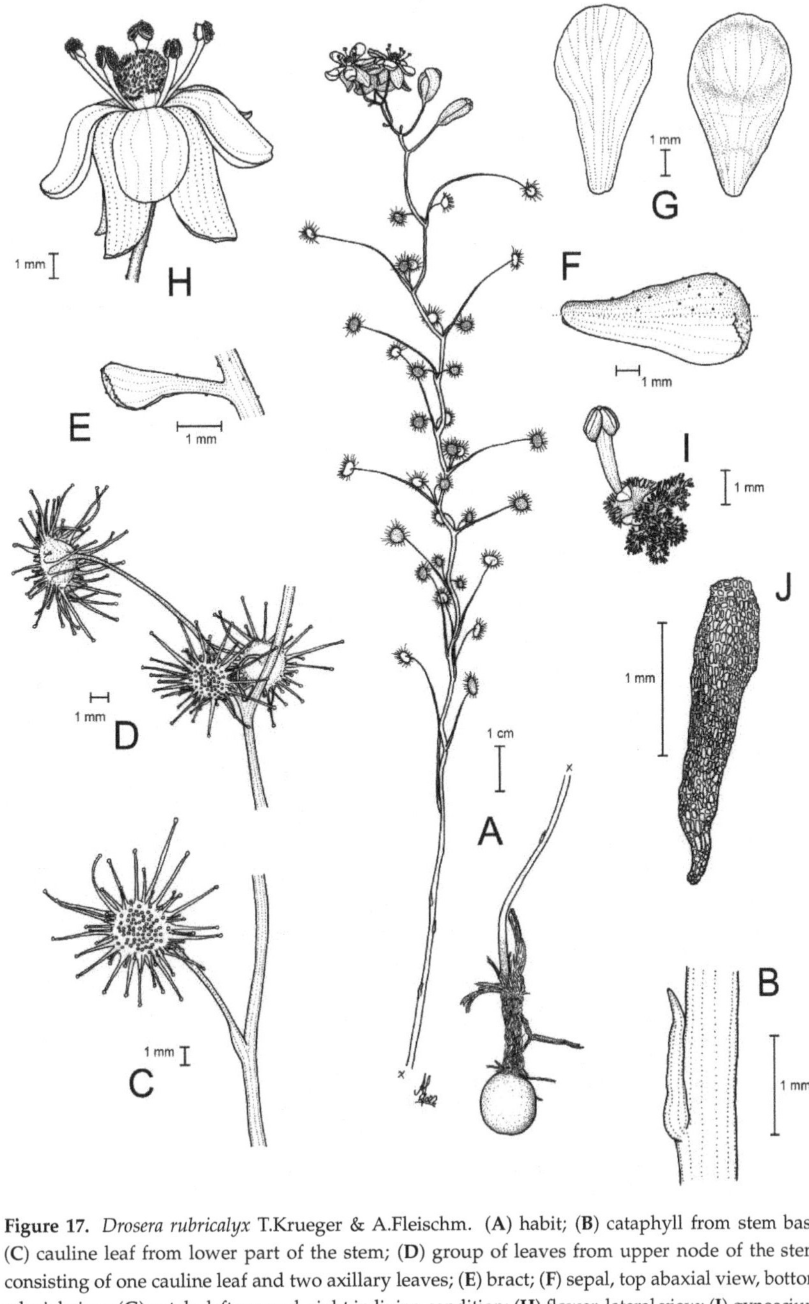

Figure 17. *Drosera rubricalyx* T.Krueger & A.Fleischm. (**A**) habit; (**B**) cataphyll from stem base; (**C**) cauline leaf from lower part of the stem; (**D**) group of leaves from upper node of the stem, consisting of one cauline leaf and two axillary leaves; (**E**) bract; (**F**) sepal, top abaxial view, bottom adaxial view; (**G**) petals, left pressed, right in living condition; (**H**) flower, lateral view; (**I**) gynoecium, top view, with two styles only partially drawn, one stamen from the androecium additionally depicted; and (**J**) seed. (**A**,**B**,**G**-left) from the type (*J. Drummond coll. VI n. 110*), (**C**–**G**-right,**H**–**J**) from photographs taken near Jurien Bay, Western Australia. Drawing: A. Fleischmann.

Figure 18. *Drosera rubricalyx* T.Krueger & A.Fleischm. (**A**) habit; (**B**) sepals showing distinctive red colouration; (**C**) leaf in upper part of the stem with a pair of smaller axillary leaves emerging from the axil; (**D**) flower in bright sunlight, lateral view; (**E**) lamina, most leaves are either orbiculate or orbiculate with flattened upper margin (as shown here); (**F**) flower in bright sunlight; and (**G**) flower in diffuse light. Images: T. Krueger from near Jurien Bay, Western Australia, 28 August 2021.

Drosera rubricalyx was possibly illustrated by Erickson [22] (p. 40, illustration 1, as "*D. microphylla*"), who had examined specimens of this species from Mt. Lesueur (*C.A. Gardner 9350*).

Distribution and habitat: *Drosera rubricalyx* is known only from Lesueur National Park and a population in the same general vicinity (Figure 3). Both sites are located near the coastal town of Jurien Bay, ca. 200 km north of Perth. It occurs in low heath in poorly drained, seasonally moist flats, depressions, and hillslopes with *Calothamnus quadrifidus* R.Br. (Myrtaceae) and *Drosera gigantea* Lindl.

Phenology: Flowering has been recorded in August and early September.

Conservation status: Listed as Priority Two (poorly known species) under Conservation Codes for Western Australian Flora and Fauna (Western Australian Herbarium 1998–; https://florabase.dpaw.wa.gov.au/ (accessed on 6 December 2022)), under the phrase-name "*Drosera* sp. Lesueur National Park (C.A. Gardner 9350)". Vulnerable (VU) under the IUCN Red List criteria D1+2 following IUCN [29]. *Drosera rubricalyx* has only recently been observed from a single roadside population in a reserve managed by the Western Australian Department of Biodiversity, Conservation and Attractions (DBCA) near Jurien Bay where it was photographed and shared online by local wildflower photographer Daniel Anderson. The images of these plants were posted on Facebook where the authors identified them as a possible new species. A survey of this population was subsequently conducted on 28 August 2021 and ca. 250 plants were counted growing scattered in an area of ca. 3000 m^2 (T. Krueger pers. obs.). During this survey (and also during two subsequent surveys in 2022), grasshoppers were observed eating large numbers of flower buds. It is not known whether these grasshoppers are native or whether they pose a significant long-term threat.

The other populations historically reported from Lesueur National Park (*C.A. Gardner 9350* and *E.A. Griffin 4207*) could not be re-located during recent surveys and it is not known if any other populations exist. *Drosera rubricalyx* is potentially vulnerable to unlicenced collection. Further surveys are recommended to gain a better understanding of this taxon's distribution, number and size of populations, and to identify additional potential threats.

Notes on the lectotypification: Very little information is provided by Bentham's description of *D. calycina* var. *minor* and none of the known syntypes of *J. Drummond coll. VI n. 110* appear to be annotated by him. However, since he was based at Kew, selecting K000215038 as the lectotype is a reasonable course of action. This specimen was already mentioned as the "holotype" for *D. calycina* var. *minor* by Lowrie [1], but this does not represent a valid lectotypification as it lacked the phrase "designated here" in accordance with ICN Art. 7.11. [26]. Further support for the choice of the K material as the lectotype comes from Moore [46] (pp. 29–30), who explains that Bentham only consulted the Kew collections for his Flora Australiensis [21] but not any other herbaria and specifically not the (often more accurately labelled) duplicates of the Drummond collections at BM [46].

Notes on Drummond's type collection: The syntypes of *J. Drummond coll. VI n. 110* only provide the rough locality information of "Western Australia, between Moore River and Murchison Rivers". However, more precise information on the locus classicus comes from the collector Drummond himself [18,19]. There, he describes a species that he refers to as "Another new *Drosera*, with bright scarlet flowers and glabrous sepals larger than the petals", a description which matches *D. rubricalyx*. Drummond [18,19] mentions that this species "[...] is found near the Yandyait Spring, to the east of the Hill river". This locality could not be precisely pinpointed, but it is likely within 15–20 km of the known population near Jurien Bay. As *J. Drummond coll. VI n. 110* is the only collection of *D. rubricalyx* by Drummond, it is safe to conclude that this is the collection locality. Additional support for this comes from Barker [41], who evidenced that Drummond [18] is referring to Drummond's VI collection series, i.e., the series comprising the type collection of Bentham's *D. calycina* var. *minor* and hence also *D. rubricalyx*. This means that the year of collection (not provided on any of the syntype specimens) is 1850 or 1851, as for all specimens comprising the VI collection [30].

While the Drummond specimen at MPU (MPU1254140 photo!) has been erroneously annotated as the "holotype" of *D. calycina* by Marchant 1985 (in sched.) (which did not constitute a valid lectotypification, see Notes on the lectotypification under that species), the plants clearly exhibit axillary leaves in the upper parts and a much more orbiculate lamina shape than that typically found in *D. calycina*. Together with the apparently dark red or purplish red petal colour and the overall habit, this indicates that MPU1254140 is most likely *D. rubricalyx* and thus another syntype of *J. Drummond coll. VI n. 110*, although the MPU specimen lacks a collector's number (therefore, it is not included as a syntype here).

It should be noted that Drummond did not number his collections sequentially [33,46], i.e., the type collection of *D. rubricalyx* (*J. Drummond coll. VI n. 110*) was likely not made immediately after that of *D. macropetala* (*J. Drummond coll. VI n. 109*). Generally, it is often difficult to georeference and trace back Drummond's collections, as complained about by Diels [33] (pp. 49–50, literally translated), who pointed out that Drummond's "enormous collections are not labelled. Their numbering is unreliable, and the individual sets do not always correspond to each other regarding their numerics". Diels [33] (p. 50) further wrote: "During sorting and distributing of the exsiccates, various mistakes and confusions arose [...]", which Diels blames on the long transport times and difficult communication between "Western Australia and the outside world". This might explain why some of the collections bear false collection numbers and/or comprise mixed collections from two different gatherings made by Drummond (e.g., LD1971779, labelled "110", consists of two individuals of *J. Drummond coll. VI n. 110, D. rubricalyx*, and four individuals of *J. Drummond coll. VI n. 109, D. macropetala*)—however, these mistakes could also have been made at the respective herbaria later during mounting or (re)labelling of the specimens (which historically was often done inaccurately, e.g., at K, according to Moore [46]).

Additional specimens examined (paratypes): AUSTRALIA. Western Australia: Mount Lesueur, 20 August 1949, *C.A. Gardner 9350* (PERTH 00666955!; PERTH 00661805!; PERTH 00661791!); Proposed Lesueur National Park [precise locality withheld for conservation purposes], upper slope, poorly drained, sandy clay, 5 September 1985, *E.A. Griffin 4207* (PERTH 01613472!).

Additional localities examined: Jurien Bay [precise locality withheld for conservation purposes], poorly drained, seasonally moist flat, 28 August 2021, T. Krueger pers. obs.

3.10. Identification Key to the Species of the Drosera microphylla Complex (See Table 2 for Multiple Access to the Morphological Characters)

1. Axillary leaves absent in adult flowering plants, all cauline leaves solitary.................**2**
    - Axillary leaves present at least on uppermost 1–9 nodes (cauline leaves in groups of 3 [2–5] per node with one larger main leaf and usually two smaller axillary leaves) (Note: white-flowered plants from Cape Arid with axillary leaves belong to unusual *D. esperensis*, see under that species which normally has solitary cauline leaves)...**6**
2. Lamina reniform or orbiculate with strongly flattened, often truncated upper margin; petioles straight or only slightly arched; corolla diameter 14–20 mm; petals deep red in inner half transitioning to purplish red in outer half; stamens 4.0–6.5 mm long; plants occurring in Jarrah forest in the western Darling Range.....................***D. calycina***
    - Lamina orbiculate or orbiculate with slightly flattened (but not truncated) upper margin; petioles arched along whole length or arching gradually increasing towards tip; corolla diameter 8–15 mm; petals reddish orange with deep red base, white with pale purplish red base, dark red throughout, or purplish pink with deep red base; stamens 2.5–4.4 mm long; plants occurring in moss or low heath on granite outcrops, swamps, Mallee sandplains, or mountain slopes near the WA south coast or around the Stirling Range...**3**
3. Petals white with pale purplish red base; plants frequently forming dense, mat-like colonies; stem, petiole, lamina and inflorescence red; filaments $\pm$ linear (0.3–0.6 mm wide throughout); filaments and styles white with red or purplish red base; stigmas white; plants occurring on coastal granite outcrops east of Esperance.....***D. esperensis***

- Petals reddish orange with deep red base, dark red throughout, or purplish pink with deep red base; plants not colony-forming or only forming relatively sparse (not mat-like) colonies; stem, petiole, lamina and inflorescence yellowish green or red; filaments dilated towards apex (from 0.2–0.3 mm near base to 0.3–0.9 mm near apex); filaments and styles deep red or purplish red; stigmas reddish purple, deep red, or dark red; plants occurring on granite outcrops, swamps, Mallee sandplains, or mountain slopes along the south coast between Walpole and Cheynes Beach or around the Stirling Range...**4**

4. Petals reddish orange with deep red bases; leaves frequently detach before anthesis, leaving large leaf-free gaps along stem (except in populations from the Stirling Range and Mt. Lindesay); stigmas reddish purple; flowering from August to October..***D. microphylla***
   - Petals dark red throughout or purplish pink with deep red base; leaves do not detach before or after anthesis; stigmas deep red or dark red; flowering from June to early September..**5**

5. Petals dark red; plants not colony-forming, occurring in very sparse populations; stem slightly fractiflex; sepals yellowish brown to greenish yellow; petal shape narrowly obovate to broadly spathulate; plants occurring in sandy soils in open Mallee woodlands in and around the Stirling Range..***D. koikyennuruff***
   - Petals purplish pink with deep red base; plants colony-forming via adventitious stolons, occurring in dense populations; stem usually strongly fractiflex; sepals red to purplish red; petal shape obovate to very broadly obovate; plants occurring in shallow moss on granite outcrops between Walpole and Denmark..***D. reflexa***

6. Axillary leaves present on all nodes, often even on upper cataphylls, all leaves in groups of 2–5 per node; inflorescence with 5–23 flowers; corolla diameter 6–9 mm; petals very dark red to blackish red; styles branched mostly above base, style segments then only sparsely branched; style segments extending laterally beyond the filaments; seeds falcate to allantoid and 2.3–2.6 mm long...***D. atrata***
   - Axillary leaves present only on uppermost 1–9 nodes, never on cataphylls, lower 4–15 leaves solitary; inflorescence with 1–8 flowers; corolla diameter 8–22 mm; petals deep red, dark purplish red, white, or deep pink; styles branched above base and style segments themselves then further branched, forming a crowded tuft; style segments not laterally extending beyond filaments; seeds very narrowly obconic, narrowly clavate to acerose, 1.7–2.4 mm long...**7**

7. Petals deep red in inner half transitioning to dark purplish red in outer half; corolla diameter 8–11 mm; plants occurring east and south-east of Perth...........***D. hortiorum***
   - Petals white with deep red base or deep red in inner half transitioning to deep pink in outer half; corolla diameter 11–22 mm; plants occurring on and just west of the Dandaragan Plateau north of Perth...**8**

8. Petals white with deep red base; plants yellowish green with yellowish brown sepals or red with reddish brown sepals; tentacle stalk colour red or red in lower half with greenish yellow upper half; peduncle length 1.3–4.2 cm; petal width 4.5–7.5 mm; filaments strongly dilated towards apex (from 0.3–0.5 mm to 0.5–0.9 mm); plants occur in laterite gravel near tops of hills on the Dandaragan Plateau, between Mogumber and Dandaragan..***D. macropetala***
   - Petals deep red in inner half transitioning to deep pink in outer half; plants yellowish green with contrasting red sepals; tentacle stalk colour greenish yellow (only rarely with reddish base); peduncle length 0.8–2.2 cm; petal width 3.3–4.8 mm; filaments only slightly dilated towards apex (from 0.4–0.5 mm to 0.4–0.6 mm), plants occur in sandy clay in seasonally moist flats, depressions, and hillslopes between Badgingarra and Lesueur National Park.............***D. rubricalyx***

Table 2. Morphological comparison of all nine species of the Drosera microphylla complex. Abbreviations: *D. calyc.*=*D. calycina*; *D. esper.*=*D. esperensis*; *D. hort.*=*D. hortiorum*; *D. koiky.*=*D. koikyennuruff*; *D. macro.*=*D. macropetala*; *D. micro.*=*D. microphylla*; and *D. rubri.*=*D. rubricalyx*.

Character	*D. atrata*	*D. calyc.*	*D. esper.*	*D. hort.*	*D. koiky.*	*D. macro.*	*D. micro.*	*D. reflexa*	*D. rubri.*
Colony forming, clonal growth	no	no	yes	no	no	no	yes (sometimes no)	yes	no
Stem shape	± straight	slightly to strongly fractiflex	strongly fractiflex	slightly fractiflex	slightly fractiflex	slightly fractiflex	slightly to strongly fractiflex	strongly fractiflex	slightly fractiflex
Axillary leaves	present (all nodes)	absent	absent (except at Cape Arid)	present (upper 1–7 nodes)	absent	present (upper 1–9 nodes)	absent	absent	present (upper 2–10 nodes)
Petiole shape	arching increasing towards tip	±straight	arched along whole length	±straight or slightly arched	arched along whole length	slightly arched along whole length	arched along whole length	arched along whole length	±straight or slightly arched
Lamina shape	orbiculate with flattened margin	reniform/orbiculate with flattened margin	orbiculate	orbiculate/ orbiculate with slightly flattened margin	orbiculate	orbiculate/ orbiculate with slightly flattened margin	orbiculate	orbiculate	orbiculate/ orbiculate with slightly flattened margin
Flowers per inflorescence	5–23	1–9	1–5	2–6	1–2	2–8	1–7	1–3	1–8
Sepal colour	yellowish brown	yellowish brown	red	yellowish brown	yellowish brown	yellowish brown to red	red to yellowish brown	red to purplish red	red
Corolla diameter (mm)	6–9	14–20	10–14	8–11	8–12	13–22	8–15	9–15	11–16
Petal colour (simplified)	blackish red	deep red	white	dark purplish red	dark red	white + deep red	reddish orange	purplish pink + deep red	deep pink + deep red
Filament shape	not dilated	dilated	not dilated	very slightly dilated	dilated	dilated	dilated	dilated	very slightly dilated
Stamen length (mm)	2.0–3.0	4.0–6.5	2.8–3.4	3.0–3.5	2.9–3.5	3.3–4.9	2.5–3.5	3.8–4.4	3.4–5.0
Relative style length	extending laterally beyond filaments	not extending laterally beyond filaments	extending laterally ± to filaments	not extending laterally beyond filaments	extending laterally beyond filaments	not extending laterally beyond filaments	extending laterally beyond filaments	extending laterally just beyond filaments	not extending laterally beyond filaments
Stigma colour	dark red	dark red	white	deep red	deep red	blackish red	reddish purple	dark red	blackish red
Habitat	lateritic hills, kwongan heath	lateritic hills, Jarrah forest	granite hills, mossy seepages	granite outcrops, moist sandplains	sandplains, Mallee scrub	lateritic hills, kwongan heath	granite outcrops, swamps, mountain slopes	granite outcrops, moss	moist flats or hillslopes, kwongan heath
Flowering time	May to August	August to September	August to October	June to September	June to July	August	August to October	June to September	August to September

4. Discussion

4.1. Endemism and Species Conservation in the Drosera microphylla Complex

The species of the *Drosera microphylla* complex are among the most narrowly endemic and most threatened species of *D.* section *Ergaleium* (tuberous sundews). Six of the nine species of the complex have known distribution areas with a maximum diameter of less than 100 km (*D. atrata*, *D. calycina*, *D. koikyennuruff*, *D. macropetala*, *D. reflexa*, and *D. rubricalyx*; Figure 3). By contrast, only 3 of the 68 remaining species of *D.* section *Ergaleium* occur across such small areas of distribution (these are *D. graniticola* N.G.Marchant, *D. orbiculata* N.G.Marchant & Lowrie, and *D. prostratoscaposa* Lowrie & Carlquist; [1]; https://florabase.dpaw.wa.gov.au/ (accessed on 22 December 2022)). Additionally, most members of the *D. microphylla* complex are uncommon even within their distribution areas, being highly localised and often present in very small populations of fewer than 100 individuals (or even fewer than ten individual plants, as is often the case for *D. atrata*). These very small distribution areas and population sizes, in combination with the threats of habitat loss (including presumed reductions in gene flow as a result of habitat fragmentation) and illegal collection—identified as threats for seven of the nine species—indicate a strong necessity for targeted conservation efforts to ensure their long-term survival in nature.

Drosera macropetala and *D. reflexa* are here assessed as the most threatened members of the *D. microphylla* complex, with a recommended Western Australian Conservation Code status of Priority One and Priority Two, respectively (Western Australian Herbarium 1998–; https://florabase.dpaw.wa.gov.au/ (accessed on 22 December 2022)); and an IUCN category of Endangered (EN) and Critically Endangered (CR), respectively [29]. Despite considerable survey efforts, both species are currently each only known from a single roadside location (with *D. macropetala* being historically also recorded from a second location further south; Figure 3). Such roadside habitats are particularly vulnerable to threats from road maintenance and construction, altered hydrology, and weed infestation [16].

Drosera atrata and *D. rubricalyx* are both assessed as Vulnerable (VU) [29] given the available data, with a Western Australian Conservation Code status of Priority Three and Priority Two, respectively. While both are known from multiple locations, their distribution areas are very small (ca. 20–50 km; Figure 3).

Drosera hortiorum is unusual within the *D. microphylla* complex as it has a relatively large distribution area spanning at least 160 km but is only known from four locations (Figure 3), each with a population size of fewer than 50 individuals. It is recommended for a Western Australian Conservation Code status of Priority Two, but insufficient survey efforts for this species means it cannot yet be assessed under IUCN criteria (=Data Deficient, DD) [29]. Similarly, *D. koikyennuruff* (which is only known from two locations near the Stirling Range; Figure 3) could not be assessed under IUCN criteria given the lack of available survey data, although it is assessed as Priority Two under the Western Australian Conservation Code.

The only species of the *D. microphylla* complex that are not currently assessed as potentially threatened are *D. calycina*, *D. esperensis*, and *D. microphylla*. All three species are known from numerous sites and are generally quite common within their preferred habitat (i.e., they tend to form relatively large populations). In addition, *D. esperensis* and *D. microphylla* have relatively large distribution areas extending over >100 km (Figure 3).

Knowledge of the distributions and range extensions of the species of the *D. microphylla* complex for this study were not only gained through field studies and herbarium research, but also from photographs from citizen science and social media (see Section 4.3). This highlights the importance of citizen scientist contributions for nature conservation (see Section 4.3 and [47–49]).

4.2. Flower Biology of the Drosera microphylla Complex

The members of the *D. microphylla* complex rank among the comparatively few sundew species that have non-ephemeral flowers, i.e., flowers that last for longer than one day. Within that group, they represent a smaller group still of species whose flowers close every night until they finally fade after about 3–5 days (A. Fleischmann pers. obs.). The daily opening and closing of the flowers is achieved by one-sided petal growth (common in many other plants with flowers that cyclically open and close), as evidenced by the fact that the petals of all members of the *D. microphylla* complex increase in size during anthesis as they slightly enlarge with each new opening event. The non-ephemeral nature of the flowers of species from this affinity, as well as their nocturnal closure, was first reported by James Drummond [18,19]. The initial opening of the flower from bud takes some time, as both sepals and petals must spread open on the first occasion. Once a flower has fully opened for the first time, the concave petals close during the late afternoon (see Figure 1I), the sepals also subsequently close at night or during unfavourable weather such as on cold and/or rainy days, covering the reproductive organs. The re-opening of an individual flower is light-dependant, but also strongly temperature-dependant, and covering individual, closed flowers with plastic bags on cold sunny days will often induce them to open within just a few minutes (F. Hort, J. Hort, and T. Krueger pers. obs. for *D. hortiorum*). The flowers of members of the *D. microphylla* complex are non-fragrant (A. Fleischmann pers. obs.; Gibson [14] for *D. esperensis*), which is an exception among members of *D.* section *Ergaleium*, which usually have strongly fragrant flowers [50] (A. Fleischmann pers. obs.). The combination of non-fragrant, non-ephemeral flowers that close each night is unique among Australian *Drosera* and restricted to the *D. microphylla* complex—it represents an ecological apomorphy of that complex. The strongly concave petals, which are shorter than the large sepals, are likewise an apomorphy for this complex. Within the genus, both characters are only paralleled in the very distantly related neotropical *Drosera biflora* Willd. from *D.* section *Drosera* [51].

It is interesting to note that the three species with exceptionally dark flower colours (*D. atrata*, *D. hortiorum*, and *D. koikyennuruff*) have unusually early flowering times, much earlier than most other species of the complex (only *D. reflexa* has also been observed to flower as early as June, G. Bourke pers. obs.) and also much earlier than most other erect tuberous *Drosera* species. This might be an adaptation to certain pollinators that are active during this period, both as temporal reproductive isolation from sympatric taxa and possibly a favoured or easy-to-spot colour among pollinators active at this time of year.

4.3. The Role of Citizen Science and Social Media in Taxonomy and Species Discovery—An Example from Drosera

The present work serves as another excellent example of citizen science, social media, and online photo platforms facilitating or even driving improvements in taxonomic and ecological knowledge. These networks have in several cases alerted botanists to new discoveries, resulted in the relocation of 'lost' or poorly known taxa, or simply extended the ranges of known species beyond those distributions established through herbarium and museum records alone [48,49,52–55]. *Drosera macropetala*, *D. rubricalyx*, *D. hortiorum*, and *D. koikyennuruff* were first (re-)discovered via online photographs or significant contributions to their range and distribution gained from images shared on the iNaturalist website and app and other social networking platforms. Carnivorous plants such as *Drosera* are usually well-represented on naturalist photographic databases and platforms as these peculiar plants are frequently photographed [56]. For example, iNaturalist (Research-Grade observations only) hosts georeferenced photographs of 236 (89.7%) out of the ca. 260 *Drosera* species known to science to date, representing 62,344 individual observations (most of which consist of several photographs) made by 17,477 observers (https://www.inaturalist.org/observations?place_id=any&taxon_id=51935 (accessed 21 December 2022)). As of December 2022, the Global Biodiversity Information Facility (GBIF; www.gbif.org (accessed on 14 January 2023)) provides 453,239 occurrence records of 263 species of *Drosera* (=100% taxon coverage); 79.4% of these records are based on citizen science observations ("human observation"), as opposed to 18.5% that come from georeferenced, databased herbarium specimens ("preserved specimen") (GBIF.org (accessed on 9 January 2023) GBIF Occurrence Download https://doi.org/10.15468/dl.h4fwxq).

This coverage is not restricted only to widespread or common species (the most commonly observed carnivorous plant on GBIF is *Drosera rotundifolia* L. [56], which is as of 9 January 2023 represented in that database by 200,188 occurrences, 89.8% of which are citizen science observations (GBIF.org (accessed on 9 January 2023) GBIF Occurrence Download https://doi.org/10.15468/dl.9w2ded). For some formerly rarely encountered *Drosera* species, more "photo vouchers" exist as geographic records than are available as herbarium specimens. For example, the South African *Drosera xerophila* was known only from three herbarium specimens and seven records on iNaturalist at the time of its description (all of them cited in Fleischmann [44]); the number of herbarium specimens has not changed since the publication of the species in April 2018, but the number of observations for that species on iNaturalist had risen to 307 by January 2023, made by 131 different observers (iNaturalist community. Observations of *Drosera xerophila*. Exported from https://www.inaturalist.org (accessed on 9 January 2023)).

This illustrates very well the scientific value of these image repositories for taxonomic and biogeographic work and for nature conservation [47,54,57]. In other organismic groups that are frequently photographed and uploaded to citizen science social networks, such as iNaturalist, by enthusiasts, these records represent by far the largest contribution to our knowledge of the distribution of these taxa. This is particularly true for "charismatic animals", such as mammals and birds, for which 70% (mammals) and 87% (birds) of the total records on GBIF in 2016 comprised online citizen science records [48]. As the global citizen science naturalist networks continuously and quickly increase their amount of data (e.g., in 2019 alone, the total number of observations on iNaturalist doubled from 25 million to 50 million [55]), the rich dataset of occurrence records provided by citizen science by far outweighs those gained annually from herbarium specimens and literature revisions, though of course the latter usually provide persistent, high-quality taxonomic data along with physical reference specimens that are additionally suitable for DNA extraction and genetic analyses, microscopic examinations, and digitised associated metadata.

Regarding species discoveries and range extensions through social media and citizen science, the present taxonomic revision provides an excellent example. Four out of the six species newly described or newly classified as species here were initially (re-)discovered on social media, with only *D. atrata* and *D. reflexa* (co-)discovered in situ by the authors of the present work. Citizen science and social media networks have also provided the first known photographs of *D. koikyennuruff*, *D. macropetala*, and *D. rubricalyx*, which were previously only known from decades-old herbarium collections. Another relatively well-publicised example of a *Drosera* species discovered on social media is *D. magnifica*, which was first spotted and identified as a species new to science from photographs posted on Facebook in 2014 [27]. An example for significant range extensions in *Drosera* provided by citizen science is *D. biflora*, the first record of which from Colombia was made by photographs posted on iNaturalist; the rediscovery of that species in Brazil was also facilitated by photographs posted online (republished in Gonella et al. [51]). These, at the same time, also represented the rediscovery and first known photographs of this species, which was previously only known from 200-year-old herbarium collections made in Venezuela. The citizen scientist photographs of living *D. biflora* specimens also provided additional unique biological and morphological details for this species (curiously, including the unique character of patent to reflexed sepals, a character this species shares with the unrelated species from the *D. microphylla* complex treated here), which could not be discerned from the historic herbarium material and helped to increase the knowledge about the taxonomy and relationships of this species [51].

Even organismic interactions can be revealed or documented for the first time through social media photographs and citizen science [58], such as plant-pollinator relationships. There are numerous examples where floral visitors and pollinators of *Drosera* species have been first documented by citizen scientists via photographs shared online. Carnivorous plant-prey relationships can also be documented by citizen science, because the different interacting organisms can often be identified from photographs (*Drosera* prey, for example, can often be identified from photographs [59]). These platforms effectively connect taxonomic experts from different fields, such as entomologists and botanists. An example involving *Drosera* is an iNaturalist photograph that was used as a voucher in a citizen science approach for mosquito species monitoring in Australia (published as Figure 4B in Braz Sousa et al. [60]). It shows the mosquito *Aedes camptorhynchus* captured on a leaf of *Drosera planchonii* Hook.f. ex Planch. in South Australia (based on an iNaturalist photograph by observer "frank_prinz": https://www.inaturalist.org/observations/54139335 (accessed on 21 December 2022)). While the *Drosera* expert or trained botanist would have been able to correctly identify the targeted plant taxon in the field and the mosquito expert will name the insect, the advantage of the social media platforms is that both experts, and additionally the observer, are linked via a photograph showing different target taxa of interest. Many of the citizen scientist photographs do not just represent single-species observations but in fact are (often unnoticed) documents of species interactions [58].

An increasing number of applications are being developed to use citizen science and social media data for biodiversity exploration and flora monitoring, linking them with taxonomy (e.g., [56,61]). Machine learning approaches are frequently improved to automate species recognition and thus enhance data mining for images suitable for taxonomy (e.g., [62,63]).

Potential negative effects of providing locality data, especially of rare flora, online in social networks arise in plant groups that are of horticultural interest, such as cacti, orchids, or carnivorous plants [64,65]. For some carnivorous plants, pitcher plants in particular, populations in the wild are mainly threatened by overcollection and poaching for the illegal trade to meet the horticultural demand [16]. However, *Drosera* species, in particular tuberous *Drosera* from Western Australia, have also been and continue to be heavily poached (including from protected areas such as nature reserves) to be sold illegally on social media or internet marketplaces to carnivorous plant enthusiasts and growers worldwide [16,65]. This is a major potential threat, particularly to rare, micro-endemic taxa such as the majority of species from the *D. microphylla* complex treated in this paper. iNaturalist automatically obscures the geographic information of observations of threatened taxa, which greatly helps with mitigating this threat (https://www.inaturalist.org/pages/help#geoprivacy (accessed on 23 December 2022)). For this reason, exact localities have been withheld for conservation purposes for the species of the *D. microphylla* complex recommended to be listed as Priority under the Western Australian Conservation Codes and the authors of the present work generally do not share locality information for threatened flora.

5. Conclusions

This study highlights the importance of both citizen science and careful herbarium examination for taxonomic research and conservation efforts. Of the six species newly recognised here, four were (re-)discovered on social media and all but *D. reflexa* had already been represented in herbaria for many decades. Crucially, these six species are rare, narrowly endemic, and potentially threatened, thus the accurate taxonomic classification provided here is expected to contribute to their conservation.

Author Contributions: Conceptualisation, A.F., A.R., G.B. and T.K.; methodology, A.F., A.R., G.B. and T.K.; formal analysis, A.F., G.B. and T.K.; literature review, research about collection history, and lectotypifications, A.F. and T.K.; investigation: herbarium studies, A.F., A.R., G.B. and T.K.; field work, A.F., A.R., G.B. and T.K.; writing—original draft preparation, T.K. and A.F. (identification key); writing—review and editing, A.F., A.R., G.B. and T.K.; visualisation, figures, T.K., drawings, A.F., A.R. and G.B.; funding acquisition, A.F. and T.K. All authors have read and agreed to the published version of the manuscript.

Funding: T.K. was supported by a Postgraduate Research Stipend Scholarship from Curtin University, Western Australia. Fieldwork by A.F. and T.K. in 2022 was supported by a 2019 research grant from the German Carnivorous Plant Society (G.F.P.).

Institutional Review Board Statement: Not applicable.

Informed Consent Statement: Not applicable.

Data Availability Statement: Not applicable.

Acknowledgments: Jean and Fred Hort are thanked for assisting T.K. with collections. Skye Coffey and Shelley James (Western Australian Herbarium) are thanked for providing access to the specimens housed at PERTH and for assisting with herbarium work. Aaron McArdle, Alison Vaughan, Catherine Gallagher, Eugenia Pacitti, Helen Barnes, and Nimal Karunajeewa (National Herbarium of Victoria) are thanked for assistance with accessioning material and for providing access to the specimens housed at MEL. Royal Botanic Gardens Victoria for in-kind support, in particular access to their SEM facilities. Gail and Dannielle Reed, Peter Luscombe, Daniel Anderson, and Patricia Paull are thanked for sharing observations of the *Drosera microphylla* complex through Facebook. Adam Cross is thanked for fruitful discussions on the conservation of carnivorous plants. The staff of the visited herbaria are thanked for providing study access to the specimens housed at their collections. Christian Bräuchler (W) is thanked for sending on loan specimens from W, Norbert Holstein is thanked for providing images of specimens from BM, Juan Carlos Zamora Señoret and Fred Stauffer for images and information on Drummond collections at G, Serena Marner for sending images from OXF, Arne Thell and Patrik Frödén for images from LD, Leonie Paterson for providing scans of personal correspondence kept at E of Alexander Morrison sent to his supervisor Baley Balfour, and Robert Vogt, Norbert Kilian, and Juraj Paule (B) for information on Ludwig Diels' correspondence and field notes at B.

Conflicts of Interest: The authors declare no conflict of interest.

References

1. Lowrie, A. *Carnivorous Plants of Australia: Magnum Opus*; Redfern Natural History Productions: Dorset, UK, 2014; Volume 1–3.
2. Fleischmann, A.; Cross, A.T.; Gibson, R.; Gonella, P.M.; Dixon, K.W. Systematics and evolution of Droseraceae. In *Carnivorous Plants: Physiology, Ecology, and Evolution*; Ellison, A.M., Adamec, L., Eds.; Oxford University Press: Oxford, UK, 2018; pp. 45–57. [CrossRef]
3. Clarke, C.; Cross, A.; Rice, B. Conservation of carnivorous plants. In *Carnivorous Plants: Physiology, Ecology, and Evolution*; Ellison, A.M., Adamec, L., Eds.; Oxford University Press: Oxford, UK, 2018; pp. 375–388. [CrossRef]
4. Hopper, S.D.; Gioia, P. The Southwest Australian Floristic Region: Evolution and Conservation of a Global Hot Spot of Biodiversity. *Annu. Rev. Ecol. Evol. Syst.* **2004**, *35*, 623–650. [CrossRef]
5. Brundrett, M.C. One biodiversity hotspot to rule them all: Southwestern Australia—An extraordinary evolutionary centre for plant functional and taxonomic diversity. *J. R. Soc. West. Aust.* **2021**, *104*, 91–122.
6. De Salas, M.F. *Drosera gunniana* comb. et stat. nov., a species in the *Drosera peltata* (Droseraceae) complex. *Muelleria* **2018**, *36*, 97–106. [CrossRef]
7. Rivadavia, F.; Kondo, K.; Kato, M.; Hasebe, M. Phylogeny of the sundews, *Drosera* (Droseraceae), based on chloroplast *rbcL* and nuclear 18S ribosomal DNA Sequences. *Am. J. Bot.* **2003**, *90*, 123–130. [CrossRef] [PubMed]
8. Diels, L. Droseraceae. In *Das Pflanzenreich IV.112, Heft 26*; Engler, A., Ed.; Engelmann: Leipzig, Germany, 1906; pp. 1–136.
9. Morrison, A. Note on the formation of the bulb in Western Australian species of *Drosera*. In *Transactions and Proceedings of The Botanical Society of Edinburgh*; Taylor & Francis Group: Oxfordshire, UK, 1905; Volume 22, pp. 419–422.
10. Diels, L. Blattrhizoiden bei *Drosera*. *Ber. Dtsch. Bot. Ges.* **1906**, *24*, 189–191.
11. Dixon, K.; Pate, J. Phenology, morphology and reproductive biology of the tuberous sundew, *Drosera erythrorhiza* Lindl. *Aust. J. Bot.* **1978**, *26*, 441–454. [CrossRef]
12. Lowrie, A. *Carnivorous Plants of Australia*; University of Western Australia Press: Perth, Australia, 1987; Volume 1.
13. Conran, J.G. Aestivation organ structure in *Drosera* subgen. Ergaleium (Droseraceae): Corms or tubers; roots or shoots? *Aust. J. Bot.* **2008**, *56*, 144–152. [CrossRef]
14. Gibson, R. White-petalled *Drosera microphylla* Endl. from near Esperance, Western Australia. *Carniv. Plant Newsl.* **2006**, *35*, 39–42. [CrossRef]
15. Lowrie, A.; Nunn, R.; Robinson, A.S.; Bourke, G.; McPherson, S.R.; Fleischmann, A. *Drosera of the World 1: Oceania*; Redfern Natural History Productions: Dorset, UK, 2017.
16. Cross, A.T.; Krueger, T.A.; Gonella, P.M.; Robinson, A.S.; Fleischmann, A.S. Conservation of carnivorous plants in the age of extinction. *Glob. Ecol. Conserv.* **2020**, *24*, e01272. [CrossRef]
17. Planchon, J.E. Sur la famille des Droséracées. *Ann. Des Sci. Nat. Bot. Sér. 3* **1848**, *9*, 285–309.
18. Drummond, J. The Botany of The North-western Districts of Western Australia. *The Perth Gazette and Independent Journal of Politics and News*, 30 April 1852, p. 5. Available online: https://trove.nla.gov.au/newspaper/article/3173380 (accessed on 6 December 2022).
19. Drummond, J. On the Botany of the North-western Districts of Western Australia. *Hooker's J. Bot. Kew Gard. Misc.* **1853**, *5*, 139–145.

20. Endlicher, S.L. Droseraceae. In *Enumeratio Plantarum quas in Novae Hollandiae ora Austro-Occidentali ad Fluvium Cygnorum et in Sinu Regis Georgii Collegit Carolus Liber Baro de Hügel*; Endlicher, S.L., Fenzl, E., Bentham, G., Schott, H.W., Eds.; Apud F. Beck Universitatis: Vienna, Austria, 1837; pp. 5–7.
21. Bentham, G. *Flora Australiensis: A Description of The Plants of The Australian Territory. Leguminosae to Combretaceae*; Reeve: London, UK, 1864; Volume 2.
22. Erickson, R. *Plants of Prey in Australia*; Lamb Publications: Osborne Park, Australia, 1968.
23. Marchant, N.G.; Aston, H.I.; George, A.S. Droseraceae. In *Flora of Australia Volume 8*; George, A.S., Ed.; Australian Government Publishing Service: Canberra, Australia, 1982; pp. 9–66.
24. Marchant, N.G. Droseraceae. In *Flora of Australia Volume 8*; George, A.S., Ed.; Australian Government Publishing Service: Canberra, Australia, 1982; pp. 383–385.
25. Briquet, J. International Rules of Botanical Nomenclature Adopted by the Internat. Botanical Congresses of Vienna, 1905, and Brussels, 1910. In *International Botanical Congress. 1930: Cambridge, UK*; G. Fischer: Albershausen, Germany, 1935.
26. Turland, N.J.; Wiersema, J.H.; Barrie, F.R.; Greuter, W.; Hawksworth, D.L.; Herendeen, P.S.; Knapp, S.; Kusber, W.-H.; Li, D.-Z.; Marhold, K.; et al. *International Code of Nomenclature for Algae, Fungi, and Plants (Shenzhen Code) Adopted by The Nineteenth International Botanical Congress Shenzhen, China, July 2017*; Koeltz Botanical Books: Glashütten, Germany, 2018; Volume 159. [CrossRef]
27. Gonella, P.M.; Rivadavia, F.; Fleischmann, A. *Drosera magnifica* (Droseraceae): The largest New World sundew, discovered on Facebook. *Phytotaxa* **2015**, *220*, 257–267. [CrossRef]
28. Robinson, A.S.; Cross, A.T.; Meisterl, M.E.; Fleischmann, A. A new pygmy sundew, *Drosera albonotata* (Droseraceae), from the western Wheatbelt and an updated diagnostic key to the orange-flowered pygmy *Drosera* of Western Australia. *Phytotaxa* **2018**, *346*, 221–236. [CrossRef]
29. IUCN. *Red List Categories and Criteria: Version 3.1*, 2nd ed.; IUCN: Gland, Switzerland; Cambridge, UK, 2012.
30. Erickson, R. Drummond, James (1787–1863). *Aust. Dict. Biogr.* **1966**, *1*. Available online: https://adb.anu.edu.au/biography/drummond-james-1995/text2433 (accessed on 10 January 2023).
31. Cross, A. Drosera calycina (errata version published in 2019). In *The IUCN Red List of Threatened Species 2018*; e.T100217225A143999940; IUCN: Gland, Switzerland, 2018. [CrossRef]
32. Stafleu, F.; Cowan, R.S. *Taxonomic Literature*, 2nd ed.; Regnum Veg. 98; Bohn, Scheltema & Holkema: Utrecht, The Netherlands, 1979.
33. Diels, L. Geschichte und Literatur der botanischen Erforschung des extratropischen West-Australiens. In *Vegetation der Erde VII*; Engler, A., Drude, O., Eds.; Engelmann: Leipzig, Germany, 1906; pp. 41–69.
34. Cross, A. Drosera esperensis. In *The IUCN Red List of Threatened Species 2018*; e.T97492495A97718522; IUCN: Gland, Switzerland, 2020. [CrossRef]
35. Hopper, S.D.; Brown, A.P. Australia's wasp-pollinated flying duck orchids revised (*Paracaleana*: Orchidaceae). *Aust. Syst. Bot.* **2006**, *19*, 211–244. [CrossRef]
36. Thiele, K.R. *Darwinia hortiorum* (Myrtaceae: Chamelaucieae), a new species from the Darling Range, Western Australia. *Nuytsia* **2010**, *20*, 277–280.
37. Hislop, M. A taxonomic update of *Conostephium* (Ericaceae: Styphelioideae: Styphelieae). *Nuytsia* **2013**, *23*, 313–335.
38. Olde, P.M. *Grevillea hortiorum* Olde (Proteaceae: Grevilleoideae: Hakeinae), an uncommon species from winter-damp woodlands in the Avon Wheatbelt, south-west Western Australia. *Telopea* **2021**, *24*, 1–6. [CrossRef]
39. McInnes, B. *Noongar Place Names Albany and Plantagenet Shires*; Arts Practicum Research Project; University of Western Australia: Crawley, WA, Australia, 2016.
40. Bradshaw, C.J.A. Little left to lose: Deforestation and forest degradation in Australia since European colonization. *J. Plant Ecol.* **2012**, *5*, 109–120. [CrossRef]
41. Barker, R.M. James Drummond's newspaper accounts of his collecting activities, in particular his 4th Collection and *Hakea victoria* (Proteaceae). *Nuytsia* **1996**, *11*, 1–9.
42. Vickery, J.W. Vegetative reproduction in *Drosera peltata* and *D. auriculata*. *Proc. Linn. Soc. NSW* **1933**, *58*, 245–269.
43. Gibson, R. Propagation of some tuberous *Drosera* by leaf cutting. *Carniv. Plant Newsl.* **2018**, *47*, 28–31. [CrossRef]
44. Fleischmann, A. *Drosera xerophila* (Droseraceae), a new species from Overberg District, South Africa, and an overview of the rosetted hemicryptophyte sundew species from Western Cape Province. *Willdenowia* **2018**, *48*, 93–107. [CrossRef]
45. Cross, A. Drosera microphylla (errata version published in 2019). In *The IUCN Red List of Threatened Species 2018*; e.T66442862A143979181; IUCN: Gland, Switzerland, 2018. [CrossRef]
46. Moore, S.M. Some new species from Australia. *J. Bot. Br. Foreign* **1902**, *40*, 25–30.
47. Maes, D.; Isaac, N.J.B.; Harrower, C.A.; Collen, B.; van Strien, A.J.; Roy, D.B. The use of opportunistic data for IUCN Red List assessments. *Biol. J. Linn. Soc.* **2015**, *115*, 690–706. [CrossRef]
48. Chandler, M.; See, L.; Copas, K.; Bonde, A.M.; López, B.C.; Danielsen, F.; Legind, J.K.; Masinde, S.; Miller-Rushing, A.J.; Newman, G.; et al. Contribution of citizen science towards international biodiversity monitoring. *Biol. Conserv.* **2017**, *213*, 280–294. [CrossRef]

49. Chamberlain, J. Using Social Media for Biomonitoring: How Facebook, Twitter, Flickr and Other Social Networking Platforms Can Provide Large-Scale Biodiversity Data. In *Advances in Ecological Research*; Academic Press: Cambridge, MA, USA, 2018; pp. 133–168. [CrossRef]
50. Gibson, R. Variation in floral fragrance of tuberous *Drosera*. *Carniv. Plant Newsl.* **2013**, *42*, 117–121. [CrossRef]
51. Gonella, P.M.; Sano, P.T.; Rivadavia, F.; Fleischmann, A. A synopsis of the genus *Drosera* (Droseraceae) in Brazil. *Phytotaxa* **2022**, *553*, 1–76. [CrossRef]
52. Marcenò, C.; Cubino, J.P.; Chytrý, M.; Genduso, E.; Gristina, A.S.; La Rosa, A.; Salemi, D.; Landucci, F.; Pasta, S.; Guarino, R. Plant hunting: Exploring the behaviour of amateur botanists in the field. *Biodivers. Conserv.* **2021**, *30*, 3265–3278. [CrossRef]
53. Marcenò, C.; Cubino, J.P.; Chytrý, M.; Genduso, E.; Salemi, D.; La Rosa, A.; Gristina, A.S.; Agrillo, E.; Bonari, G.; del Galdo, G.G.; et al. Facebook groups as citizen science tools for plant species monitoring. *J. Appl. Ecol.* **2021**, *58*, 2018–2028. [CrossRef]
54. Feldman, M.J.; Imbeau, L.; Marchand, P.; Mazerolle, M.J.; Darveau, M.; Fenton, N.J. Trends and gaps in the use of citizen science derived data as input for species distribution models: A quantitative review. *PLoS ONE* **2021**, *16*, e0234587. [CrossRef]
55. Jaume-Schinkel, S.; Mengual, X. New geographic distribution of *Chrysomya megacephala*, the Oriental latrine blow fly (Diptera: Calliphoridae), in Mexico using citizen science and social media. *Rev. Mex. Biodivers.* **2022**, *93*, e934166. [CrossRef]
56. Blanco, R.; Salazar, Z.; Isenberg, T. Exploring carnivorous plant habitats based on images from social media. In Proceedings of the IEEE VIS 2019: IEEE Conference on Visualization, Vancouver, BC, Canada, 21–25 October 2019; Available online: https://hal.inria.fr/hal-02196764 (accessed on 10 January 2023).
57. Wilson, J.S.; Pan, A.D.; General, D.E.M.; Koch, J.B. More eyes on the prize: An observation of a very rare, threatened species of Philippine Bumble bee, *Bombus irisanensis*, on iNaturalist and the importance of citizen science in conservation biology. *J. Insect Conserv.* **2020**, *24*, 727–729. [CrossRef]
58. Groom, Q.; Pernat, N.; Adriaens, T.; de Groot, M.; Jelaska, S.D.; Marčiulynienė, D.; Martinou, A.F.; Skuhrovec, J.; Tricarico, E.; Wit, E.C.; et al. Species interactions: Next-level citizen science. *Ecography* **2021**, *44*, 1781–1789. [CrossRef]
59. Krueger, T.; Cross, A.T.; Fleischmann, A. Size matters: Trap size primarily determines prey spectra differences among sympatric species of carnivorous sundews. *Ecosphere* **2020**, *11*, e03179. [CrossRef]
60. Sousa, L.B.; Fricker, S.; Webb, C.E.; Baldock, K.L.; Williams, C.R. Citizen Science Mosquito Surveillance by Ad Hoc Observation Using the iNaturalist Platform. *Int. J. Environ. Res. Public Health* **2022**, *19*, 6337. [CrossRef]
61. Boho, D.; Rzanny, M.; Wäldchen, J.; Nitsche, F.; Deggelmann, A.; Wittich, H.C.; Seeland, M.; Mäder, P. Flora Capture: A citizen science application for collecting structured plant observations. *BMC Bioinform.* **2020**, *21*, 576. [CrossRef]
62. Rzanny, M.; Mäder, P.; Deggelmann, A.; Chen, M.; Wäldchen, J. Flowers, leaves or both? How to obtain suitable images for automated plant identification. *Plant Methods* **2019**, *15*, 77. [CrossRef]
63. Picek, L.; Šulc, M.; Patel, Y.; Matas, J. Plant recognition by AI: Deep neural nets, transformers, and kNN in deep embeddings. *Front. Plant Sci.* **2022**, *13*, 787527. [CrossRef]
64. Rice, B. Stop sharing information, Dammit! You're not helping! *Carniv. Plant Newsl.* **2019**, *48*, 174–177. [CrossRef]
65. Fleischmann, A. Karnivoren und Naturschutz-die Rolle von Karnivorenliebhabern. *Das Taublatt* **2021**, *89*, 6–35.

MDPI
St. Alban-Anlage 66
4052 Basel
Switzerland
Tel. +41 61 683 77 34
Fax +41 61 302 89 18
www.mdpi.com

Biology Editorial Office
E-mail: biology@mdpi.com
www.mdpi.com/journal/biology

www.ingramcontent.com/pod-product-compliance
Lightning Source LLC
Chambersburg PA
CBHW061547220926
51850CB00005B/275
9783036575582